Molecularly Imprinted Polymers: Path to Artificial Antibodies

Santanu Patra • Sudheesh K. Shukla •
Mika Sillanpää
Editors

Molecularly Imprinted Polymers: Path to Artificial Antibodies

Springer

Editors
Santanu Patra
IDUN, Department of Health Technology
Technical University of Denmark
Kongens Lyngby, Denmark

Mika Sillanpää
Department of Chemical Engineering Technology
University of Johannesburg
Johannesburg, South Africa

Sudheesh K. Shukla
Centre for Nanoscience and Nanotechnology, School of Chemical Engineering and Physical Sciences
Lovely Professional University
Phagwara, Punjab, India

Department of Applied Chemistry
University of Johannesburg
Johannesburg, South Africa

ISBN 978-981-97-4378-0 ISBN 978-981-97-4379-7 (eBook)
https://doi.org/10.1007/978-981-97-4379-7

This Springer imprint is published by the registered company Springer Nature Singapore Pte Ltd.
The registered company address is: 152 Beach Road, #21-01/04 Gateway East, Singapore 189721, Singapore

Preface

Artificial antibodies like molecularly imprinted polymers (MIPs) have gained tremendous research interest as synthetic receptors due to their advantages over natural receptors. They possess the same selectivity but with more stability and cost-effectiveness. The use of MIPs in healthcare monitoring allows for more accurate and sensitive disease detection, monitoring, and personalized medicine. They can be used to target-specific molecules, biomarkers, cells, or tissues, which makes them useful for future healthcare monitoring devices. This book discusses the importance of MIPs as synthetic or engineered antibodies in various medical applications such as disease detection, monitoring, and personalized medicine.

The objective of this book is to provide a comprehensive overview of the current state-of-the-art and future directions of MIPs in healthcare monitoring. The book explores the design and development of MIPs, their use in diagnostics and therapeutics, as well as their applications as synthetic or engineered antibodies in disease detection, monitoring, and drug delivery. The book has been divided into different chapters with each chapter written by experts in their respective fields. The target audience for this book are researchers, students, professionals, and practitioners in bioengineering, biotechnology, materials scientist, and healthcare. The first two chapters discuss the overview of MIPs as artificial antibodies, synthesis, and novel characterization methods. Chapters 3–8 provide an overview of MIP-based electrochemical sensors and their diverse applications. Chaps. 9 and 10, delve into the importance of MIP-based optical sensors, while the significance of chip sensors is covered in Chap. 11. The usefulness of MIPs in therapeutic applications is covered in Chaps. 12–16 with a focus on their role in cancer detection and monitoring, tissue engineering, and drug delivery.

We hope the book will be an important resource for anyone interested in the emerging field of MIPs as artificial antibodies in healthcare monitoring and its potential to revolutionize medical diagnosis and treatment.

Kongens Lyngby, Denmark
Phagwara, Punjab, India
Johannesburg, South Africa
February 2024

Santanu Patra
Sudheesh K. Shukla
Mika Sillanpää

Contents

Editors and Contributors

About the Editors

Santanu Patra is working in the field of materials chemistry, specializing in the integration of functional materials with macro-, micro-, and nano-systems and their multifaceted applications. He received his PhD in chemistry from the Indian Institute of Technology (ISM) in Dhanbad, India, where he gained expertise in cutting-edge technologies related to molecularly imprinted nanotechnology and electrochemistry. He then worked as a group leader/researcher in two R&D companies and at Khon Kaen University, Thailand. Currently, Dr. Patra is associated with the Technical University of Denmark, where he is continuing his research on multifaced nanotechnological approaches toward biomedical applications. Through his research, Dr. Patra aims to develop advanced and innovative technologies that can address some of the most pressing challenges for sustainable development. He has authored or co-authored more than 60 research and review articles, as well as book chapters, and holds two patents. Dr. Patra serves as an editor and editorial board member of several international peer-reviewed journals and has edited several important books.

Sudheesh K. Shukla works in the area of translational research and development of bioelectronics devices for disease detection and disorder management, with a major focus on interfacing chemistry (material science) and engineering for better healthcare (biology) and environmental applications. Currently, Dr. Shukla's research involves real-sample analysis of biochemical markers for personalized healthcare monitoring. In particular, Dr. Shukla is interested in integrating biomaterials with micro- and nano-systems for sensing and actuation technologies. Dr. Shukla is the author of numerous articles in peer-reviewed journals and a prolific author and editor of several scientific monographs in healthcare and environmental disciplines, published in several prominent publications. Dr. Shukla serves as an editor or editorial board member of several international peer-reviewed journals, including Chemosensor, Sensor International, Coronavirus Reports, Reports in Electrochemistry, Chemistry Africa, Topics in Catalysis, and Chemical Physics Impact, among others.

Mika Sillanpää received his M.Sc. (Eng.) and D. Sc. (Eng.) degrees from Aalto University, where he also completed an MBA degree in 2013. Since 2000, he has been a full professor/adjunct professor at the University of Oulu, the University of Eastern Finland, LUT University, and the University of Johannesburg. He has supervised over 60 PhDs and has served as a reviewer for more than 250 academic journals. Mika Sillanpää has published more than 1000 articles in peer-reviewed international journals. He has served on the editorial boards of several scholarly publications. With an h-index of 129, his publications have been cited over 80000 times, according to Google Scholar. Mika Sillanpää has received numerous awards for research and innovation. He also received a literature award from the Water Association of Finland in 2018. In 2019, 2020, and 2021, he was recognized as a highly cited researcher by Thomson Reuters in two different disciplines, joining an elite group of approximately 200 top researchers across all fields of science. He was listed as one of the world's top 2% Scientists by Stanford University. In 2022, he was awarded the title of Provincial Innovative Talent of Zhejiang Province, China.

Contributors

Kamran Banan Department of Pharmaceutics and Pharmaceutical Nanotechnology, School of Pharmacy, Shahid Beheshti University of Medical Sciences, Tehran, Iran

Parveen Bansal University Centre of Excellence in Research, Baba Farid University of Health Sciences, Faridkot, Punjab, India

Kiarash Bastani Nano-Biopolymers Research Laboratory, School of Chemical Engineering, College of Engineering, University of Tehran, Tehran, Iran

Sara Arjomand Bigdeli Halal Research Center of IRI, Food and Drug Administration, Ministry of Health and Medical Education, Tehran, Iran

Jannu Ricardo Casanova-Moreno Centro de Investigación y Desarrollo Tecnológico en Electroquímica S. C., Queretaro, Mexico

Michał Cegłowski Faculty of Chemistry, Adam Mickiewicz University, Poznań, Poland

Abraham Ulises Chávez-Ramírez Centro de Investigación y Desarrollo Tecnológico en Electroquímica S. C., Queretaro, Mexico

Agampreet Singh Dadial Department of Biochemistry, School of Chemical and Life Sciences, Jamia Hamdard, New Delhi, India

Mohammad Mahdi Feizollahi Department of Pharmaceutics and Pharmaceutical Nanotechnology, School of Pharmacy, Shahid Beheshti University of Medical Sciences, Tehran, Iran

Soane Fernández-Puig Centro de Investigación y Desarrollo Tecnológico en Electroquímica S. C., Queretaro, Mexico

Juan de Dios Galindo-de-la-Rosa Universidad Autónoma de Querétaro, Queretaro, Mexico

Fatemeh Ghorbani-Bidkorpeh Department of Pharmaceutics and Pharmaceutical Nanotechnology, School of Pharmacy, Shahid Beheshti University of Medical Sciences, Tehran, Iran

Mikko Haavisto Department of Teacher Education, University of Turku, Rauma, Finland

Parmida Harirchi Nano-Biopolymers Research Laboratory, School of Chemical Engineering, College of Engineering, University of Tehran, Tehran, Iran

Nor Munira Hashim Department of Toxicology, Advanced Medical and Dental Institute, Universiti Sains Malaysia, Kepala Batas, Pulau Pinang, Malaysia

Sepehr Izadi Department of Pharmaceutics and Pharmaceutical Nanotechnology, School of Pharmacy, Shahid Beheshti University of Medical Sciences, Tehran, Iran

Nik Nur Syazni Nik Mohamed Kamal Department of Toxicology, Advanced Medical and Dental Institute, Universiti Sains Malaysia, Kepala Batas, Pulau Pinang, Malaysia

Kuldeep Kaur Department of Chemistry, Mata Gujri College, Fatehgarh Sahib, Punjab, India

Simrat Kaur Department of Chemistry, Mata Gujri College, Fatehgarh Sahib, Punjab, India

Ipek Kucuk Department of Analytical Chemistry, Faculty of Pharmacy, Ankara University, Ankara, Turkey

The Graduate School of Health Sciences, Ankara University, Ankara, Turkey

Cansu İlke Kuru Department of Biotechnology, Graduate School of Natural and Applied Sciences, Ege University, Izmir, Turkey

Buca Municipality Buca Science and Art Center, Buca, Izmir, Turkey

Abdulaziz Umar Kurya Department of Biotechnology, School of Life and Allied Health Sciences, Glocal University, Saharanpur, Uttar Pradesh, India

Tomi Laurila Department of Electrical Engineering and Automation, School of Electrical Engineering, Aalto University, Aalto, Finland

Department of Chemistry and Materials Science, School of Chemical Engineering, Aalto University, Aalto, Finland

Tamara Lazarević-Pašti VINČA Institute of Nuclear Sciences—National Institute of the Republic of Serbia, University of Belgrade, Belgrade, Serbia

Aleksandra Lusina Faculty of Chemistry, Adam Mickiewicz University, Poznań, Poland

Seema Maheshwari Department of Chemistry, Mata Gujri College, Fatehgarh Sahib, Punjab, India

Ashok Kumar Malik Department of Chemistry, Punjabi University, Patiala, Punjab, India

Ashish Mathur Centre for Interdisciplinary Research and Innovation (CIDRI), University of Petroleum and Energy Studies, Dehradun, Uttarakhand, India

Vedran Milanković VINČA Institute of Nuclear Sciences—National Institute of the Republic of Serbia, University of Belgrade, Belgrade, Serbia

Annu Mishra Centre for Interdisciplinary Research and Innovation (CIDRI), University of Petroleum and Energy Studies, Dehradun, Uttarakhand, India

Bahar Mostafiz Department of Mechanical and Materials Engineering, University of Turku, Turku, Finland

Tomasz Nazim Faculty of Chemistry, Adam Mickiewicz University, Poznań, Poland

Khadijeh Nekoueian Department of Electrical Engineering and Automation, School of Electrical Engineering, Aalto University, Aalto, Finland

Goldie Oza Centro de Investigación y Desarrollo Tecnológico en Electroquímica S. C., Queretaro, Mexico

Igor A. Pašti Faculty of Physical Chemistry, University of Belgrade, Belgrade, Serbia

Hooman Pourbala Department of Pharmaceutics and Pharmaceutical Nanotechnology, School of Pharmacy, Shahid Beheshti University of Medical Sciences, Tehran, Iran

Muggundha Raoov Ramachandran Department of Chemistry, Faculty of Science, Universiti Malaya, Kuala Lumpur, Malaysia

Melkamu Biyana Regasa Chemistry Department, College of Natural and Computational Sciences, Wollega University, Nekemte, Ethiopia

Apurva Rohokale Centro de Investigación y Desarrollo Tecnológico en Electroquímica S. C., Queretaro, Mexico

Selenay Sadak Department of Analytical Chemistry, Faculty of Pharmacy, Ankara University, Ankara, Turkey

The Graduate School of Health Sciences, Ankara University, Ankara, Turkey

Behnoosh Safaei Department of Pharmaceutics and Pharmaceutical Nanotechnology, School of Pharmacy, Shahid Beheshti University of Medical Sciences, Tehran, Iran

Department of Toxicology, School of Pharmacy, Shahid Beheshti University of Medical, Tehran, Iran

Kiana Sakhaee Nano-Biopolymers Research Laboratory, School of Chemical Engineering, College of Engineering, University of Tehran, Tehran, Iran

Kavirajaa Pandian Sambasevam Advanced Materials for Environmental Remediation (AMER) Research Group, Faculty of Applied Sciences, Universiti Teknologi MARA, Kuala Pilah, Negeri Sembilan, Malaysia

Mohammad-Ali Shahbazi Department of Biomedical Engineering, University Medical Center Groningen, University of Groningen, Groningen, The Netherlands

W.J. Kolff Institute for Biomedical Engineering and Materials Science, University of Groningen, Groningen, The Netherlands

Sudheesh K. Shukla Centre for Nanoscience and Nanotechnology, School of Chemical Engineering and Physical Sciences, Lovely Professional University, Phagwara, Punjab, India

Department of Applied Chemistry, University of Johannesburg, Johannesburg, South Africa

Tamara Tasić VINČA Institute of Nuclear Sciences—National Institute of the Republic of Serbia, University of Belgrade, Belgrade, Serbia

Abdulhakim Umar Toro Department of Biomedical Engineering, Shobhit Institute of Engineering and Technology (Deemed to-be-University), Modipuram, Meerut, Uttar Pradesh, India

Fulden Ulucan-Karnak Medical Biochemistry Department, Institute of Health Sciences, Ege University, Bornova, Izmir, Türkiye

Bengi Uslu Department of Analytical Chemistry, Faculty of Pharmacy, Ankara University, Ankara, Turkey

Arístides Camilo Valdés-González Instituto Nacional de Higiene, Epidemiología y Microbiología, Ministerio de Salud Pública de Cuba, Havana, Cuba

Vanessa Vallejo-Becerra Universidad Autónoma de Querétaro, Queretaro, Mexico

Noorfatimah Yahaya Department of Toxicology, Advanced Medical and Dental Institute, Universiti Sains Malaysia, Kepala Batas, Pulau Pinang, Malaysia

Payam Zahedi Nano-Biopolymers Research Laboratory, School of Chemical Engineering, College of Engineering, University of Tehran, Tehran, Iran

Nur Nadhirah Mohamad Zain Department of Toxicology, Advanced Medical and Dental Institute, Universiti Sains Malaysia, Kepala Batas, Pulau Pinang, Malaysia

An Overview of MIPs as Artificial Antibodies: Fundamentals and Various Applications

1

Annu Mishra, Ashish Mathur, and Agampreet Singh Dadial

Abstract

Chemical antibody mimics known as molecularly imprinted polymers (MIPs) are created by nano-moulding a synthetic polymer with the desired target's three-dimensional (3D) form and chemical properties. As a result, they have exceptional molecular recognition cavities that allow them to engage the target molecule, frequently with an affinity and a specificity comparable to that of antigen–antibody interactions. The molecular imprinting technique is a potent method for developing synthetic materials with antibody-like molecular recognition. Traditional MIPs are primarily single-function materials produced utilizing techniques created between the 1970s and 2000s, which significantly limits their potential in comparison to naturally occurring, versatile materials. The multifunctionality of proteins that are produced through biosynthesis, in contrast, is the result of intricate production procedures, such as post-translational modifications. Post-imprinting modifications (PIMs) make it possible to site-specifically introduce new functionalities into molecularly imprinted cavities. This method involves chemically altering the monomer residues within the cavities and producing MIPs with characteristics, such as on/off binding activity switching, fluorescence signalling, photo-responsiveness, and precisely controlled binding properties. Molecular recognition—a property shared by numerous molecules, including antibodies, aptamers, and MIPs—has a wide range of potential uses in medical and pharmaceutical research, diagnostics, theranostics, therapy, and drug delivery. As a result, numerous research studies have been conducted to

A. Mishra (✉) · A. Mathur
Centre for Interdisciplinary Research and Innovation (CIDRI), University of Petroleum and Energy Studies, Dehradun, Uttarakhand, India

A. S. Dadial
Department of Biochemistry, School of Chemical and Life Sciences, Jamia Hamdard, New Delhi, India

S. Patra et al. (eds.), *Molecularly Imprinted Polymers: Path to Artificial Antibodies*, https://doi.org/10.1007/978-981-97-4379-7_1

examine the usefulness of MIPs in related applications. When compared to traditional antibodies, MIPs offer a number of benefits, including increased adaptability and affordability. With a focus on novel polymer materials, this chapter offers a clear overview of the main ideas behind molecular imprinting technology (MIT), including initiators, cross-linking agents, functional monomers, template molecules, and characterization methods.

Keywords

Molecularly imprinted polymer (MIP) · Post-imprinting modification (PIM) · Molecularly imprinted SPE (MISPE)

1.1 Introduction

Molecular imprinting is a promising approach to accomplish the crucial goal of generating molecularly imprinted polymers (MIPs) with both affinity and specificity since it provides a multitude of synthetic options for artificial recognition systems. Molecularly imprinted polymers (MIPs) are a type of artificial materials that are intended to replicate the enzymes and antibodies that act as natural recognition elements in terms of both specificity and selectivity.

MIPs have the ability to encapsulate and release particular medications in response to particular stimuli. This could ensure accurate distribution to targeted tissues or cells by imitating the regulated medication release mechanisms present in the human body. MIPs with tailored properties might be utilized to replicate the operations of particular tissues or organs. For instance, MIPs may be able to carry out some detoxifying or filtering tasks if they are designed with characteristics analogous to those of the liver or kidneys.

They are produced using a procedure known as molecular imprinting, in which polymer structures containing molecular recognition sites that are specifically designed to bond with target molecules are formed [1]. The synthesis of synthetic receptors that mimic the recognition characteristics of natural antibodies or enzymes is the fundamental process behind MIPs. The target molecule, also known as the template, and functional monomers—small molecules with the ability to build polymer chains—create a complex at the start of this process. These functional monomers encircle the template and join forces with one another throughout the polymerization process to create a three-dimensional (3D) matrix or network [2].

MIPs are functional porous materials with high-affinity binding sites that may be tailored in terms of size, shape, and functionality to enable analyte-based targeting [2, 3]. Polyakov [4] created molecular imprinting for the first time in the 1930s. Wulff and Sarhan [5] made more progress in the 1970s, but Mosbach and colleagues [6] created the non-covalent imprinting method in the 1990s, which increased the use of MIPs. They proposed MIPs based on the non-covalent interactions between the target and host, whereas Wulff and Sarhan used covalent bonds to generate MIPs. The distinction between the two is attributed to the templated molecules being extracted from the MIPs [7]. Non-covalent molecular imprinting technology

(MIT) was successfully used by Vlatakis et al. to create MIPs [8], and early research has indicated that MIT has the capacity to rapidly expand and is thus considered the most promising method for producing synthetic molecular recognition compounds for various analytes, such as drugs, toxins, biomolecules, and pollutants [9, 10]. The three broad groups of synthetic polymer-based imprinting strategies are covalent, non-covalent, and semi-covalent imprinting procedures, which are determined by the type of contact that occurs between the functional monomer(s) (T/M) and the template [11] (Fig. 1.1). It is predicated on a templating method that enables the production of certain molecular recognition sites in networks of artificial porous polymers [12].

Biological receptors that can recognize and bind to target molecules selectively are the source of two main characteristics of MIPs [13]. MIPs differ from biological receptors in terms of size, rigidity, and solubility; in contrast, natural counterparts are typically smaller, more flexible, and highly soluble in aqueous conditions. Depending on their size, MIPs can bind to several binding sites, which allows them to be recognized by biological receptors with a single or small number of reactive centres [14]. Furthermore, the effect of equilibrium that governs the formation of the template–monomer complex and the dynamics of the growing polymer chain before its co-polymerization result in a heterogeneous amount of MIP binding sites, particularly those imprinted by non-covalent template–monomer interactions [15].

1.2 Principle Behind using MIPs as Artificial Antibodies

Molecular imprinting is generally effective in creating synthetic receptors that offer the necessary geometrical and chemical fit for binding substrate molecules with great selectivity and thus catalysing chemical processes [16]. It is possible to modify the host cavities' form and function in imprinted polymers to get the best possible interactions with the target molecules. Molecularly imprinted polymers, or MIPs [2], have found widespread use in the last 10 years [17] in a variety of

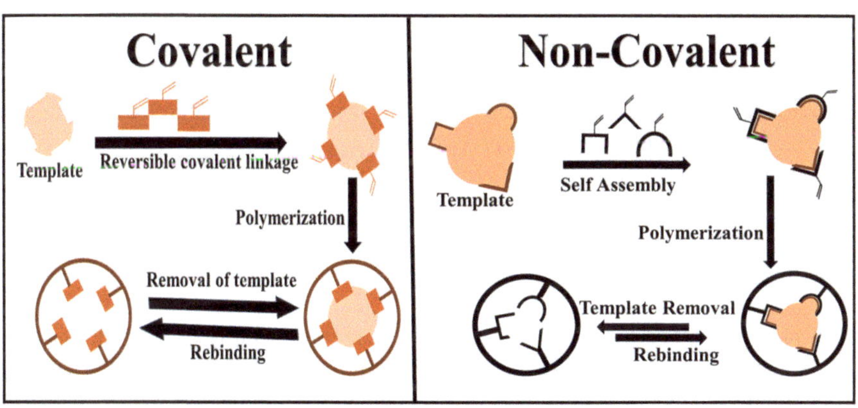

Fig. 1.1 MIP techniques include covalent and non-covalent imprinting

applications such as solid-phase extractions [18, 19], immunoassays [20], chemical sensor coatings [21, 22], sophisticated drug delivery systems (DDSs) [23, 24], and, crucially, enzymatic catalysis [5]. It is believed that molecularly imprinted polymers are synthetic antibodies [25] that do not lose their sensitivity and selectivity even after several cycles of analysis. Additionally, MIPs have exceptional storage stability since they do not require specific humidity, pH, or temperature levels when stored for several months at room temperature. Considering these features, they are far less expensive than are natural antibodies.

For instance, the cost of commercially available MIPs for a given target is approximately USD $0.1–0.5 per mg, whereas that of conventional natural antibodies is between USD $100 and 1000 per mg [26]. Depending on the host–guest species structures, which can be suitably adjusted, molecular imprinting provides a variety of bonding contact opportunities [27, 28].

The basic principle of MIT is based on the formation of a complex between a functional monomer and an analyte (template). A considerable excess of a cross-linking agent causes the formation of a three-dimensional polymer network [16]. Following the polymerization process, the template is separated from the polymer, leaving behind certain recognition sites that are complementary to the template molecule in terms of size, shape, and chemical functionality (Fig. 1.2). Molecular recognition phenomena are usually driven by intermolecular interactions between functional groups found in the polymer matrix and the template molecule, such as hydrogen bonds, dipole–dipole interactions, and ionic interactions. This means that only the template molecules are recognized and selectively bound by the resulting polymer.

1.3 Methodologies for Synthesizing Molecularly Imprinted Polymers

The ultimate aim of molecular imprinting is to produce MIPs with biological receptor-like functionality, which may enable them to take up the role of these naturally occurring substances in some situations. The molecular imprinting process requires the employment of three crucial reagents in order to produce MIPs with a high degree of specificity and affinity to a specific target molecule: cross-linkers, functional monomers, and template molecules (Fig. 1.3). The degree of affinity and specificity attained in the finished product is greatly influenced by the choice of suitable reagents.

1.3.1 Templates

The MIP structure plays a pivotal role in molecular imprinting by serving as a template that guides the organization of the functional group tethered to the targeted functional monomers. Functional groups, including amide, carboxyl, hydroxyl, ester, and amino, are essential for the functionality of molecularly imprinted

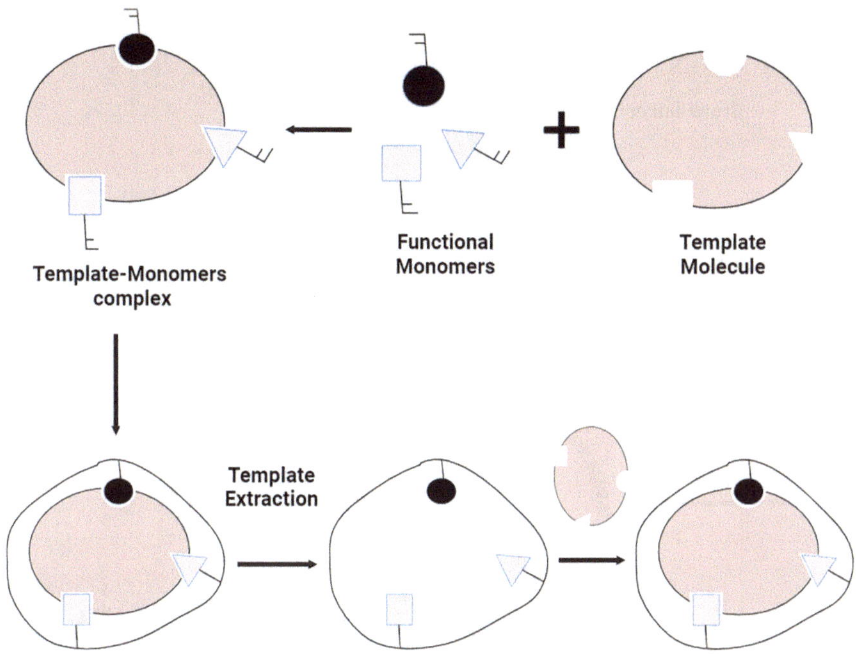

Fig. 1.2 Scheme of molecular imprinting

polymers and should be present in the template [29, 30]. In general, the perfect template must meet three criteria. Prior to anything else, the template molecule needs to show outstanding chemical stability during the polymerization process. Second, it needs to have abundant functional groups that can unite with other functional monomers to form template–monomer complexes. Ultimately, it is crucial to guarantee that the binding between the monomer and template does not cause instability or interfere with the polymerization process. The complexity of these molecules, the non-specific recognition sites on their surfaces, the restricted polymerization methods available for native macromolecules, and the possibility of smaller non-target molecules binding to the relatively larger imprinted sites, thus leading to decreased sensitivity, are all some of the major obstacles facing this imprinting strategy [31–33]. The use of synthetic receptors, such as glycan, monosaccharide, oligosaccharide, epitope, or aptamer, as templates for the molecular imprinting of target proteins has recently been developed as a solution to these problems.

1.3.2 Functional Monomers

MIPs have a wide range of applications, the features of the template (charge, size, and chemical identity) are taken into consideration when selecting functional monomers and cross-linking agents to be used in MIP synthesis, as shown in Fig. 1.4a

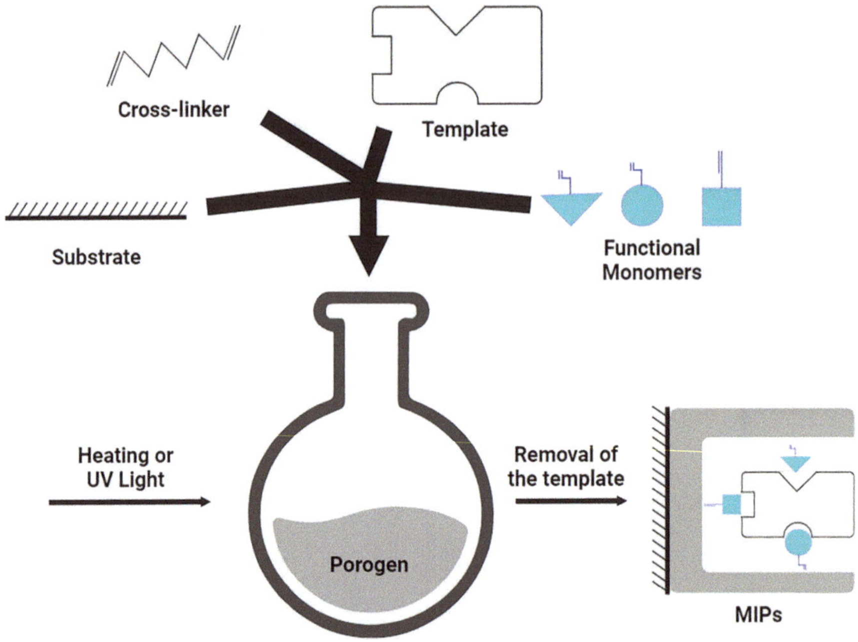

Fig. 1.3 Synthesis of molecularly imprinted polymers

[34]. Functional monomers are tiny compounds that have the ability to interact with the template molecule in covalent or non-covalent approaches. They are chosen according to their capacity to form particular binding sites that are complementary to the dimensions, composition, and molecular functions of the template molecule. They proceed through polymerization to become the MIP's structural component. The two independent types of units that are usually present in monomers are recognition units and polymerization units. Functional groups with the ability to interact with the template are a crucial prerequisite for the selection of monomers. To create a template–monomer complex, these groups can engage with the template through covalent or non-covalent means, such as hydrogen bonding, dipole, van der Waals forces, or π–π contact. Molecular recognition is induced by the interaction of the template molecule with the functional group found in the polymer matrix [35, 36]. This procedure, which takes place in the pre-polymerization reaction, establishes the number and quality of a MIP's recognition unit [37]. When functional monomers react with template molecules enough, complexes are formed. Cross-linking agents are then added to the imprinted molecules to immobilize the functional groups of the functional monomers. This process creates molecularly imprinted polymers. A three-dimensional polymer network is subsequently developed only for the target template. For the binding site to stay intact even after the template is removed, the polymer needs to be accurately cross-linked [38].

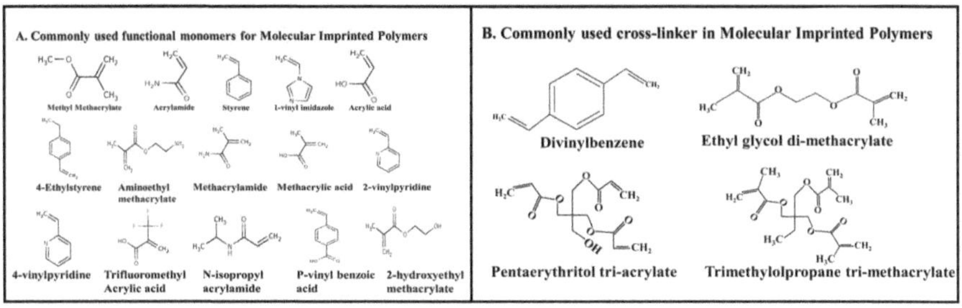

Fig. 1.4 Functional and Crosslinking monomers that are frequently utilized in the molecular printing technique

1.3.3 Cross-Linking Agents

Cross-linking agents are important components of molecularly imprinted polymer (MIP) development. They are essential for sustaining the stability and integrity of the imprinted sites as well as the formation of the polymer matrix's three-dimensional network structure (Fig. 1.4b).

These chemicals aid in the creation of a strong matrix that preserves the molecular recognition sites that are imprinted during the polymerization process. The amount of cross-linkers used during the polymerization process affects both the mechanical stability and the number of recognition sites in MIPs; excessive cross-linkers can result in unstable mechanical properties, whereas an insufficient amount of cross-linkers can reduce the number of recognition sites per unit mass of MIPs [29, 39]. For optimal polymerization outcomes, it is therefore essential to optimize the basic cross-linker ratio in advance. The yield and excellence of MIPs following polymerization are also influenced by the kind of cross-linking agent used. A polymer's mechanical stability and porosity are determined by the proper arrangement and orientation of its monomers, which also improves its affinity and selectivity for the target template. For the synthesis of MIPs, a variety of cross-linking agents can be employed [40], as mentioned in Table 1.1.

1.3.4 Solvents

The fabrication of molecularly imprinted polymers (MIPs) relies heavily on solvents since they make it easier for different ingredients to dissolve and mix together before creating the polymer matrix. The polymerization process and the resulting MIP structure and characteristics can be greatly influenced by the solvent(s) used in the process. A number of solvents are frequently used for the synthesis of MIPs [37], as shown below in Table 1.2:

The nature of the template molecule and functional monomers, their solubility, the preferred polymerization method (bulk, precipitation, or suspension polymerization, for example), and the intended use of the MIP all influence the solvent selection.

Table 1.1 Various kinds of cross-linking agents that can be employed

S. no.	Cross-linking agents	Examples	Functions
1.	Divinyl compounds	Divinylbenzene (DVB) and ethylene glycol dimethacrylate (EGDMA)	These compounds possess two reactive vinyl groups that participate in the polymerization reaction, forming cross-links between the functional monomers. They are commonly used due to their ability to create a stable and highly cross-linked polymer network
2.	Di- or polyfunctional acrylates/ methacrylates	Trimethylolpropane trimethacrylate (TRIM) or trimethylolpropane triacrylate (TMPTA)	These contain multiple acrylate or methacrylate groups. They serve as multifunctional cross-linkers, aiding in the formation of a densely cross-linked polymer structure
3.	Bifunctional compounds	Diacrylates, dimethacrylates, or disacrylamides	These possess two functional groups that can link to functional monomers, thus promoting cross-linking and polymerization
4.	Polyfunctional agents	Pentaerythritol tetraacrylate (PETRA) and glycerol dimethacrylate (GDMA)	Higher-functionality cross-linking agents with more than two reactive sites can be utilized to create a highly cross-linked network

Table 1.2 Numerous types of solvents for the fabrication of MIPs

S. no.	Solvents	Examples	Functions
1.	Protic solvents	Water, methanol, ethanol, and acetic acid	Accessibility, low cost, and an environmentally friendly nature
2.	Aprotic solvents	Chloroform, dichloromethane (DCM), tetrahydrofuran (THF), acetonitrile, and dimethyl sulphoxide (DMSO)	These solvents are preferred when the template molecule and functional monomers are hydrophobic or have limited solubility in water. Aprotic solvents often provide better compatibility with hydrophobic monomers
3.	Mixed solvents	Mixture of water and an organic solvent like DCM	Combinations of solvents are sometimes employed to optimize the solubility of components, control the polymerization process, or enhance the morphology of the resulting MIP

1.3.5 Initiators/Catalysts/ Iniferter Based Polymerization in MIP Synthesis

Initiators or catalysts are frequently employed in the manufacture of molecularly imprinted polymers (MIPs) in order to start and accelerate the polymerization process. These substances cause the functional monomers to form polymer chains more quickly, thus resulting in the development of the MIP's three-dimensional network structure. It is important to consider the compatibility of the chosen monomers and

Table 1.3 Common catalysts and initiators for MIP synthesis

S. no.	Initiators/ catalysts	Examples	Functions
1.	Thermal initiators	Azobisisobutyronitrile (AIBN), benzoyl peroxide (BPO), and thermal decomposition of persulphates	These initiators induce polymerization through heat. Thermal initiators require elevated temperatures to initiate the reaction and are often used in bulk or solution polymerizations
2.	Photochemical initiators	Benzoin methyl ether (BME), benzophenone, and various types of photoinitiators	These initiators induce polymerization upon exposure to light of a specific wavelength. Photochemical initiators are widely employed in processes like photopolymerization or photo-induced polymerization because they provide benefits like precise control over the initiation process and spatial/temporal control of polymerization
3.	Redox initiators	A combination of potassium persulphate and *N,N,N′,N′*-tetramethylethylenediamine (TEMED) or ammonium persulphate and sodium metabisulphite	With this kind of initiator, two chemicals are used, which together start the polymerization process. An oxidizing agent and a reducing agent are the respective roles played by the two reagents

solvent system with the sort of polymerization technique being used when choosing initiators or catalysts. Some of the common catalysts and initiators for MIP synthesis [41] are as follows (Table 1.3):

1.3.5.1 Iniferter-based Polymerization in MIP Synthesis

An "initiator–transfer agent–terminator", or an "iniferter", is a type of polymerization process that is employed in the production of polymers. Molecularly imprinted polymers (MIPs) can be manufactured in a varied and controlled manner using iniferter-based polymerization processes. This allows for the optimization of a target molecule's recognition by the polymer. High control over the final polymer's molecular weight is possible using iniferter-based polymerization. This parameter is useful for optimizing the size and structure of the imprinted cavities in the context of MIPs. A constant molecular weight helps improve repeatability in the properties of MIP identification.

Polymers having different structures, including block co-polymers or graft co-polymers, can be synthesized using initiator-based methods. The ability to construct unique polymer topologies to optimize the accessibility and binding affinity of imprinted sites makes this adaptability useful in the development of molecularly imprinted polymers (MIPs). Different functional groups can be incorporated into the polymer structure through the use of initiator-based polymerization. This can be particularly important in MIP synthesis because it allows for the development of unique chemical functionalities that improve the polymer–target molecule interaction by imitating the recognition sites present in biological systems. Surface changes are made possible by the iniferter process, which makes it easier to introduce particular end groups on the polymer chains. When it comes to MIP applications, surface modification is essential since interactions with the environment like biological fluid are important. It enables more customized interactions with the environment and enhanced biocompatibility. A potent method for modifying the structure and characteristics of MIPs is iniferter-based polymerization, which aids in the creation of extremely effective and selective molecular recognition systems for a range of uses, such as pharmaceutical administration, separation, and sensing.

1.3.6 Porous Supports or Matrices

While developing molecularly imprinted polymers (MIPs), a porous support or matrix is used as a scaffold or substrate to help specify the binding sites inside the polymer matrix onto which the MIP is synthesized. The characteristics and functionality of the resultant MIP can be affected by the support or matrix selection. The following (Table 1.4) types of porous supports or matrices can be used in the manufacturing of MIPs [42]:

Table 1.4 Types of porous supports or matrices

S. no.	Porous support or matrix	Examples	Functions
1.	Silica-based supports	Silica granules or gels	Due to its high porosity and surface area, silica makes it easier for imprinting sites to develop and functional monomers to get immobilized. It is possible to functionalize silica beads or nanoparticles in order to anchor the MIP components
2.	Polymeric supports	Microspheres, beads, or porous membranes	Materials like polystyrene, polyacrylates, or other polymers can be used to build these supports. In addition to providing an organized surface for imprinting, they can be functionalized to improve interoperability with MIP components
3.	Magnetic supports	Iron oxides (Fe_3O_4) like magnetite and magnetic nanoparticles	These magnetic supports are especially helpful in applications involving separation or extraction and help with the recovery and reusability of the MIP
4.	Other inorganic supports	Metal–organic frameworks (MOFs), zeolites, titania, and alumina	Specific surface functions, tunable porosity, and a large surface area are among the distinctive qualities offered by these materials that can affect the MIP's performance

1.4 Evaluating MIPs Against Natural Antibodies

MIPs are more stable, less expensive, and easier to produce than are natural antibodies, yet natural antibodies are much more specific and work better with biological systems. Combining the benefits of natural antibodies and MIPs may improve applications in a number of domains, including environmental monitoring, medication delivery, and diagnostics. A new generation of MIPs, known as polymer-based artificial antibodies with functions beyond those of natural antibodies, may arise owing to the potential dependability, affordability, and convenience of the MIP–post-imprinting modification (PIM) method. Comparing MIPs to natural antibodies involves evaluating the various aspects presented in Table 1.5.

The term "smart imprinting" describes the process of generating molecularly imprinted polymers (MIPs) with extra features that let them react to particular stimuli or changes in their surroundings. These "smart" or stimuli-responsive MIPs are extremely appealing for a variety of applications because they can modify their structure or other characteristics in a reversible manner. Compared to bulk MIPs, nano-sized MIPs (nano-MIPs) have a number of benefits, most notably increased performance and adaptability. The surface-to-volume ratio of nano-MIPs is larger than that of bulk MIPs because of their smaller size. Improved sensitivity and binding efficiency result from this increased surface area's ability to interact with target molecules. Mass transport and binding kinetics can occur more rapidly with nano-MIPs due to their smaller size. Nano-MIPs become more sensitive and effective in applications like sensors or assays as a result of the faster detection and binding of target molecules. The smaller size of nano-MIPs makes them more appropriate for applications involving biological systems. This is especially crucial in industries like medicine since nano-MIPs may be made to interact with biological molecules without posing a serious risk of disruption or barriers. Numerous methods, such as solid-phase synthesis, emulsion polymerization, and precipitation polymerization, can be used to synthesize nano-MIPs. These techniques provide precise control over the distribution and size of the particles, thus enabling the formation of distinct nanostructures with customized characteristics. Improved regeneration and reusability are usually the result of nano-MIPs' smaller size and larger surface area. Reusing the imprinted material multiple times is crucial for cost-effectiveness in applications such as separations, where this is advantageous. Multifunctional applications are possible with nano-MIPs since they may be built to have different capabilities, including fluorescent or magnetic labelling. Their adaptability renders them appropriate for use in targeted medication delivery systems, imaging, and advanced sensing. It is feasible to synthesize smart nano-MIPs that react to outside stimuli such as variations in pH, temperature, or certain ions. Target molecule binding and release may be dynamically controlled owing to this responsiveness, which broadens their range of uses in medication delivery and smart materials.

Table 1.5 Comparison between MIPs and natural antibodies

Features	Molecularly imprinted polymers (MIPs)	Natural antibodies
1. Selectivity and specificity	MIPs are selective and specific because during synthesis, they imprint binding sites that are particular to a given template molecule. The binding sites identify the functional groups, size, and shape of the template	Natural antibodies are highly specific to their target antigens due to their variable regions, allowing them to selectively bind to a particular molecular structure
2. Synthesis and production	Using chemical synthesis techniques is a reasonably simple way to create MIPs in the lab. They provide an affordable and expandable production method	Living things produce natural antibodies through intricate biological mechanisms Antibody production can be more expensive and time-consuming and may need the use of animal or cell culture techniques
3. Consistency and durability	MIPs are resilient and stable in a range of environmental circumstances, such as variations in pH and temperature. They can resist extreme circumstances and are less prone to degeneration	Changes in pH and temperature can have an impact on antibodies. They might become more brittle and break down more quickly
4. Regeneration and adaptability	The binding capacity of MIPs remains relatively intact even after repeated regeneration and reuse. Over several cycles, they provide good stability and reusability	Reusable antibodies may become less active with repeated usage or regeneration
5. Customization and adaptability	MIPs are adaptable to a broad range of target molecules, such as pollutants, biomolecules, and tiny compounds, thus enabling their customization for a variety of applications	Natural antibodies are extremely versatile, but they can only be altered or customized through biological processes, such as the engineering of monoclonal antibody synthesis
6. Biocompatibility	The direct application of some MIPs in biological systems may be limited by their lack of biocompatibility	Natural antibodies can be used in medicine since they are biologically compatible and biocompatible by nature

1.5 Applications of MIPs in the Biomedical Field

Molecular imprinting is currently a recognized technique used in analytical and medicinal chemistry. MIPs have been applied in chromatography for enantiomeric separations [43], solid-phase extraction [44], catalysis, sensor design [45], protein separation [46], and as receptors [47], antibodies [48], and enzymes [49]. Most recently, mimics [50] and drug delivery systems (DDSs) [51] have emerged.

1.5.1 Medication Delivery Systems Based on MIPs

One intriguing and novel field of research is the role of molecular imprinting technology in producing effective polymeric DDSs with stimuli responsiveness and a high specificity for recognizing target biomolecules [52–54]. Despite the lack of clinical trials for MIP-based DDSs, the molecular imprinting technology holds considerable promise for developing optimal dosage forms. A useful indicator is the steadily rising number of articles [55–57] devoted to the application of MIPs in the design of innovative DDSs and other devices.

Additionally, MIPs have demonstrated intriguing promise in medication delivery due to their great stability and durability in challenging circumstances. MIPs can, in fact, be kept for several years at room temperature in a dry form without losing their recognition qualities [58]. Moreover, MIPs are attractive drug delivery vehicles due to their great stability and robustness, as the imprinted cavities can shield drug molecules from harsh human body conditions like problems affecting the gastrointestinal tract [59].

It is crucial to take into consideration the safety and biocompatibility of MIP-based DDSs. Due of their biocompatibility, acrylic and methacrylic monomers have found extensive usage in a variety of MIP applications, including the development of DDSs [60]. Further research is necessary to determine the long-term toxicity of these compounds when ingested by humans. Without a doubt, the usage of naturally biodegradable polymers might be taken into consideration in the development of MIP-based DDSs in order to reduce the potential of long-term toxicity. Furthermore, it is crucial to take the MIP's water compatibility into account, as drug delivery vehicles often need to be able to function in an aqueous milieu in order to guarantee effective drug release in the body.

In 1998, Norell et al. [61] published the first paper suggesting that MIPs may be employed as long-term drug delivery systems to release theophylline. They revealed that compared to the similar non-imprinted polymers (NIPs), synthesized MIPs display a longer sustained release and a higher binding affinity to theophylline than to caffeine.

High levels of cross-linking are necessary for MIP matrices to maintain complementary holes that serve as drug repositories [62], although this process frequently restricts the mobility of polymer chains [63]. As a result, MIPs with less cross-linking are typically more beneficial for drug delivery applications. Additionally, the addition of external components like magnetite [64], thermosensitive polymers [65], and particular moieties like disulphide bonds [66] enables MIPs to respond appropriately to external stimuli like changes in temperature, pressure, pH, ionic strength, electric fields, chemicals, and light. The following table (Table 1.6) lists some molecularly imprinted polymer (MIP)-based drug delivery methods in conjunction with particular medications or pharmacological classes:

Table 1.6 Medication delivery systems based on MIPs

S. no.	MIP-based drug delivery system	Drug	Applications	References
1.	Oral formulation based on MIPs	Theophylline	Sustained release, chronic obstructive pulmonary disease (COPD), and asthma treatment (bronchodilator)	[67]
2.	Magnetic MIP nanoparticles (Fe_3O_4 NPs)	Doxorubicin hydrochloride (DOX)	Targeted drug delivery, cancer therapy	[68]
3.	Implantable MIP devices	Insulin	Controlled release, diabetes management	[69]
4.	MIP hydrogels	Ibuprofen	Localized delivery, pain management	[70]
5.	MIP-based nanocarriers (MoS_2@MIP core–shell nanocomposites)	Antiviral drugs (e.g. favipiravir (FAV))	Targeted delivery, antiviral therapy	[71]
6.	Stents coated with MIPs	Sirolimus	Drug-eluting stents, cardiology	[72]
7.	MIP-based ocular inserts	Timolol	Ocular drug delivery, glaucoma treatment	[73]
8.	Polymer (MIP)-based wound dressings	Growth factors (e.g. transforming growth factor-β (TGF-β), fibroblast growth factor (FGF), epidermal growth factor (EGF), basic fibroblast growth factor (bFGF), keratinocyte growth factor (KGF), vascular endothelial growth factor (VEGF), and platelet-derived growth factor (PDGF)	Controlled release, wound healing	[74]
9.	MIP-based patches	Lidocaine	Transdermal delivery, local anaesthesia	[75]
10.	Pharmacological MIPs	Biomarkers (e.g. peptides, proteins)	Targeted therapy, diagnostics, personalized medicine	[76]

1.5.2 MIPs for Biomolecule Sensing and Identification

Synthetic materials known as MIPs (molecularly imprinted polymers) are engineered to identify and bind to particular target molecules in a selective manner. They are made by first imprinting a template molecule onto a polymer matrix and then by removing it to reveal cavities that are complementary in shape and function to the target molecule. Because of their specificity, stability, and affordability, MIPs have attracted a lot of attention in a variety of domains, including biomolecule sensing and detection. An example list of MIPs designed for particular biomolecules and their uses is provided below (Table 1.7):

1.6 Applications of MIPs in Plant Analysis

The method utilizing MIPs is quick, inexpensive, and has several uses in the field of biological compound identification. In this sense, use of plant metabolite-derived MIPs is another helpful technique for locating and isolating plant components. The following is a description of a few instances of MIP application in plant material analysis.

Identifying and measuring gallic acid, a phenolic secondary metabolite, is known to be problematic in dietary items like orange juice and medicinal plant extracts. The reliability of the enlarged pipette-tip, molecularly imprinted, silica-based monolithic high-performance liquid chromatography (HPLC) methodology was demonstrated through practical application in the trace analysis of gallic acid in orange juice samples. The extraction performance of spiking samples at three concentrations (0.05, 0.50, and 2.00 mg/L) with a standard gallic acid solution was used to assess the accuracy of this approach. Gallic acid concentrations in the tampered orange juice samples ranged from 95.6–100.5% to 92.0–97.0%. The findings demonstrated how effective and dependable the standard procedure was in tracking gallic acid levels in orange juice samples. The surface of the MIP monolith did not exhibit any interposition with other orange juice ingredients [87].

Molecularly imprinted solid-phase extraction (MISPE) has shown, through the study of solutions containing plant materials, that the constituents of the Traditional Chinese Medicine (TCM) extracts are almost entirely adsorbed onto the MIPs [88]. This approach yielded an extraction performance of 80.9% for the terpenoid kirenol and a recovery rate of 91.5% ± 3.2 ($n = 3$) for the spiked solution. These results suggest that the MISPE approach works well for both the selective cleaning of plant extracts and the concentration and quantification of kirenol in TCM. An imprinted membrane was used to extract up to 48% of paclitaxel from the yew tree extract in an experiment by Ghasemi and Nematollahzadeh [89]. This method can be used to identify paclitaxel instead of more complex and expensive approaches because of its simplicity and cost-effectiveness [90]. Additionally, MIPs have been effectively employed to extract glucosamine from chicory roots with encouraging results and to detect polyphenols in plant extracts [90, 91].

The technique known as magnetic molecularly imprinted polymers (MMIPs) for coumarin, or MMIP-coumarin, has been used to detect and extract coumarin from plant and food samples [92]. Ginkgo leaves include flavonoids such as quercetin and kaempferol, which can be captured from the Gingko leaf extract using MIPs specifically made for these compounds [93]. MIPs intended for the efficient extraction of specific protocatechuic acid (PA) from the extracts of *Homalomena occulta* and *Cynomorium songaricum* were employed [94].

Table 1.7 MIPs for biomolecule sensing with its applications

S. no.	Biomolecules	Examples	Applications	References
1.1.	Proteins	Trypsin	Proteomics and enzyme detection	[77]
		Bovine serum albumin (BSA)	Protein tests, identification of biomarkers	[78]
1.2.	DNA/RNA	μRNA	Identification of disease biomarkers and cancer diagnosis	[79]
		DNA oligonucleotides	DNA sequencing and genetic analysis	[80]
1.3.	Pathogens	Influenza virus	Food safety and bacterial detection	[81]
		Escherichia coli	Diagnostics for infectious diseases and viral detection	
1.4.	Drugs and small molecules	Cocaine	Forensic examination and drug testing	[82]
		Antibiotics (e.g. penicillin)	Pharmaceutical analysis, drug monitoring	[83]
5.	MIP-based nanocarriers (MoS_2@MIP core–shell nanocomposites)	Antiviral drugs (e.g. favipiravir (FAV))	Targeted delivery, antiviral therapy	[71]
6.	MIP-based ocular inserts	Timolol	Ocular drug delivery, glaucoma treatment	[73]
7.	Polymer (MIP)-based wound dressings	Growth factors (e.g. transforming growth factor-β (TGF-β), fibroblast growth factor (FGF), epidermal growth factor (EGF), basic fibroblast growth factor (bFGF), keratinocyte growth factor (KGF), vascular endothelial growth factor (VEGF), and platelet-derived growth factor (PDGF)	Controlled release, wound healing	[84]

(continued)

Table 1.7 (continued)

S. no.	Biomolecules	Examples	Applications	References
8.	MIP-based patches	Lidocaine	Transdermal delivery, local anaesthesia	[85]
9.	Pharmacological MIPs	Biomarkers (e.g. peptides, proteins)	Targeted therapy, diagnostics, personalized medicine	[86]

Chicoric acid was also effectively extracted from *Cichorium intybus* using MIPs. When applied against the structural analogues of the template, such as CA, caftaric acid, and chlorogenic acid, this MIP demonstrated a high degree of selectivity [95]. Additionally, tea extract's epigallocatechin gallate was separated using MIPs. This approach produced epigallocatechin gallate with a high degree of purity; the compound's mean recovery rate was 87.42%. Therefore, the aforementioned plant chemical may be acceptable for extraction and purification using the MIP method (T. [96]). Using MIPs, emodin is extracted with a 95% purity from the alcoholic extract of *Rheum palmatum* L. [97]. In *Chamaecyparis obtusa* extract, MIPs can act as adsorbents to extract quercitrin, myricetin, and amentoflavone [98]. Under optimal conditions, 3 g of an adsorbent produced 0.45 mg/g of quercitrin, 0.18 mg/g of myricetin, and 0.12 mg/g of amentoflavone. The study's findings indicate that the strategy for separating the metabolites is highly selective, repeatable, and has a minimal deviation error. In one investigation, chemicals in *Melissa officinalis* were identified using PA as a model for the creation of MIPs. Gentisic acid was not detectable by the MIP, and the syringic acid's recovery power was just 16.5%. On the other hand, more than 56% of the extracted phenolic acids, including vanillic acid, gallic acid, p-hydroxybenzoic acid, and protocatechuic acid, were extracted [99]. A study was conducted to see whether MIPs could be used to extract theophylline and caffeine from green tea [99]. A combination of caffeine and theophylline was used to create one type of MIP, and a mixture of pentoxifylline and theophylline was used in the other. The findings showed that caffeine and theophylline may be extracted from green tea solutions using MIPs. A study was conducted to see whether MIPs could be used to extract theophylline and caffeine from green tea [100]. A blend of caffeine and theophylline and a mixture of pentoxifylline and theophylline were the two distinct patterns used in the construction of MIPs. The findings showed that theophylline and caffeine could be extracted from green tea solutions using MIPs.

Sinomenium acutum's sinomenine can be separated using MIPs. As a result, MIPs produce nearly pure sinomenine. Additionally, sinomenine in biological materials can be identified using MIPs specifically made for sinomenine. Furthermore, findings showed that the MIP technique has broad applicability in the domains of biopharmaceutical analysis and plant metabolite extraction [101].

1.7 Applications of MIPs in Wastewater Treatment

The application of MIPs in wastewater treatment has been the subject of various studies in recent years. Four categories can be used to categorize MIP applications in contaminated water: (1) as selective tools to determine the pollutants present in wastewater; (2) as a kind of efficient sorbent to purify pollutants; (3) as a way to increase the catalytic efficiency for organic purposes; and (4) as a means of recognizing and directly quantifying target analytes using MIP-based fluorescence sensors. MIPs are becoming an effective technique for the selective identification, separation, quantification, and purification of contaminants in wastewater because to their strong affinity and excellent selectivity.

Conversely, because of their favourable molecular recognition capabilities and stability, MIPs have already been effectively used as selective tools in numerous analytical domains, including as sensors, in solid-phase extraction (SPE), and in chromatography. Specifically, SPE represents the pinnacle of MIP applications, offering a straightforward and efficient pre-treatment technique for complex materials [2, 101, 102]. According to Turiel and Martín-Esteban [103], a novel avenue for the selective extraction and identification of target molecules from complex matrices has been made possible by molecularly imprinted SPE (MISPE).

A number of studies published recently have reported the success of MISPE and determination of a wide range of compounds from water samples, such as carbaryl [104], bisphenol A [105], water-soluble acid dyes [106], catechol [107], carbamazepine [108], benzimidazole compounds [109], β-estradiol [110], herbicides [111], mercury [112], and cyanide [113].

Pollutants can be successfully separated from water samples using MISPE in conjunction with a high-performance liquid chromatograph. To separate and determine the amount of trace estrone in the environment, for instance, Xu et al. [114] conducted research using an MISPE connected to a high-performance liquid chromatograph. MIPs exhibit excellent selectivity for estrone because of the unique recognition sites from the molecularly imprinted process [114].

The relative standard deviations for nine repeat extractions of 0.5 $\mu g\ L^{-1}$ estrone were <10.0%, and the results also revealed that the limit of detection was 5.7 $ng\ L^{-1}$ (S/N = 3). This recently developed technique can be used to track the amount of estrone residue currently present in water.

Table 1.8 provides an overview of the many types of molecularly imprinted polymers (MIPs) and explains how they are used in wastewater treatment.

1.8 Difficulties and Prospects for the Future of MIP Development

Molecularly imprinted polymers (MIPs) have shown great potential in a number of disciplines for development and application. Notwithstanding, a number of obstacles continue to exist, and prospective developments in MIP technology entail tackling these obstacles while investigating novel approaches to augment their

Table 1.8 Molecularly imprinted polymers (MIPs) and their uses in wastewater treatment

S. no.	MIP classification	Examples	Applications	References
1.	MIPs for heavy metals	Imprinted for cadmium, lead, and mercury	Removing heavy metals from wastewater	[115]
2.	MIPs for organic pollutants	Imprinted for pesticides, dyes, and bisphenol A (BPA)	Removing organic waste from wastewater	[116]
3.	MIPs for pharmaceuticals and antibiotics	Imprinted for pharmaceutical compounds	Removing drugs from wastewater	[83]
4.	MIPs for endocrine disruptors	Imprinted for phthalates and bisphenol A	Eliminating endocrine-disrupting substances from wastewater	[117]
5.	MIPs for microorganisms	Imprinted for viruses and *E. coli*	Removing microorganisms from sewage	[118]
6.	MIPs for hydrocarbons and oils	Imprinted for hydrocarbons and oils	Removing oils and hydrocarbons from industrial wastewater	[119]
7.	MIPs for nutrients	Imprinted for nitrogen and phosphorus	Removing nutrients to avoid eutrophication	[120]
8.	MIPs for micropollutants	Designed for emerging contaminants and microplastics	Removing microplastics and emerging pollutants	[121]
9.	MIPs for water	Designed to purify water	Enhancing the quality of water by reusing and recycling wastewater	[122]
10.	MIPs for recovery	Imprinted for metal extraction	Recovering and extracting metallic substances from wastewater	[123]

capabilities and uses. It is still difficult to improve the MIP's binding kinetics and affinity for target molecules. The main goals are to enhance the imprinting procedure to produce high-affinity binding sites and to maximize the interactions between functional monomers and template molecules. It is crucial to achieve increased selectivity and specificity, particularly in intricate biological contexts. The goal is to create MIPs that can distinguish between compounds that are structurally similar while maintaining a high level of selectivity. It is imperative to develop MIPs that are both biocompatible and biodegradable in order to employ them in biomedical applications. Furthermore, there are numerous opportunities and viewpoints for future MIP development, as shown in Fig. 1.5. The goal of research is to develop MIPs with the least amount of possible toxicity and compatibility for biological systems. One of the ongoing challenges is scaling up MIP synthesis without sacrificing repeatability and consistency in performance. For large-scale production and common applications, it is essential to optimize synthesis processes and guarantee batch-to-batch reproducibility. A controlled release and targeted delivery system

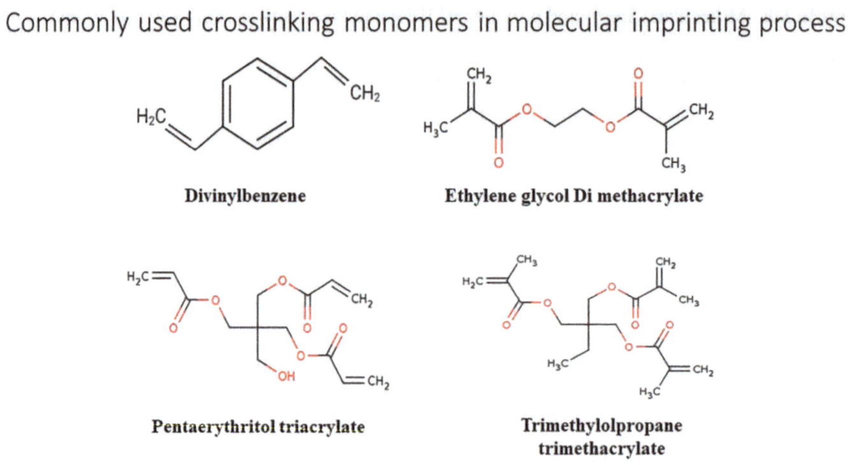

Fig. 1.5 Future prospects of MIP development

with on-demand functions may be made possible by designing MIPs with stimuli-responsive properties, such as pH, temperature, or light responsiveness; by investigating sustainable and ecologically friendly ways to synthesize MIPs, minimizing the usage of dangerous chemicals, and enhancing the process's environmental friendliness; and by incorporating MIPs into biosensors and diagnostic tools to enable quick and precise identification of infections, analytes, or disease indicators. Developing MIP-based nanoparticles or nanomaterials with special qualities, such as enhanced targeting, multifunctionality for a range of applications, and a larger surface area, is the main goal of nanotechnology advancements. In order to guarantee the MIP's dependability, safety, and acceptance across a range of sectors and therapeutic situations, it is imperative to establish standardized protocols, characterization techniques, and regulatory frameworks. The MIP's usefulness in a variety of applications could be increased by investigating the development of multifunctional MIPs, such as those that enable simultaneous drug administration and detection.

1.9 Conclusions: MIPs as Artificial Antibodies—Advantages and Drawbacks

The concept of "greenification" of molecularly imprinted polymers, or MIPs, refers to the design and use of sustainable and ecologically friendly procedures across the full MIP longevity, from synthesis to application. The use of natural polymers derived from biomass wastes, such as chitosan, cellulose, or cyclodextrins, in place of the accepted extremely hazardous chemicals often needed in MIPs is thoroughly outlined. The potential to alter these natural resources to create novel green substrates for the synthesis of MIPs, which can serve as creative functional monomers, initiators, cross-linking agents, or supporting materials, is specifically emphasized. The molecules that can recognize particular targets are the subject of this chapter.

This characteristic of antibodies comes from nature and has led to a wide range of their uses in areas such as drug delivery, biological and pharmacological research, immunotherapy, immunoassays, and biosensing for diagnostics.

In recent years, we have seen a constant reporting of significant advancements in the synthesis and use of MIPs in the literature. MIPs are the most promising artificial materials for molecular recognition in a variety of scientific domains due to their low cost, outstanding stability, and consistently bettering performance. The development of MIPs in many application domains should be supported by new processes for bulk polymerization that can enhance the particle characteristics. The study of alternative methods, such as aptamers and, most importantly, molecularly imprinted polymers, has become necessary due to certain limitations of these types of molecules, such as limited stability, high cost, and immunogenicity. The final ones have demonstrated numerous potential advantages over their biological counterparts, including low cost, adaptability, and stability, even though they have not had much commercial success recently.

A larger number of publications on MISPE have been published in recent years. As an alternative to conventional solid-phase extraction, MISPE offers lower preparation costs and greater stability for imprinted materials. For the selective extraction of several compounds, MISPE cartridges are already on sale in the market [124, 125].

The development of MIPs as selective materials for sensors and biosensors has received a lot of interest in recent years. MIP-based biomimetic sensors still fall short of biosensors in terms of sensitivity, though, as they need more MIP and transducer optimization. Therefore, the commercial development of molecular imprinting sensors is still in its early stages, despite the exciting ongoing work in this field.

Applications such as medication administration and catalysis have not yet been adequately studied. Applications in these fields are still in their beginning stage, but there are a number of promising directions that need to be investigated in order to make progress in the next years.

References

1. Bahrani S, Aslani R, Hashemi SA, Mousavi SM, Ghaedi M (2021) Introduction to molecularly imprinted polymer. In: Interface science and technology, vol 33. Elsevier, pp 511–556. https://www.sciencedirect.com/science/article/pii/B9780128188057000060
2. Haupt K, Mosbach K (2000) Molecularly imprinted polymers and their use in biomimetic sensors. Chem Rev 100(7):2495–2504. https://doi.org/10.1021/cr990099w
3. Spivak DA (2005) Optimization, evaluation, and characterization of molecularly imprinted polymers. Adv Drug Deliv Rev 57(12):1779–1794
4. Polyakov MV (1931) Adsorption properties and structure of silica gel. Zhur Fiz Khim 2:799–805
5. Wulff G (2002) Enzyme-like catalysis by molecularly imprinted polymers. Chem Rev 102(1):1–28. https://doi.org/10.1021/cr980039a
6. Vlatakis G, Andersson LI, Müller R, Mosbach K (1993) Drug assay using antibody mimics made by molecular imprinting. Nature 361(6413):645–647
7. Rao TP, Kala R, Daniel S (2006) Metal ion-imprinted polymers—novel materials for selective recognition of inorganics. Anal Chim Acta 578(2):105–116

8. Branger C, Meouche W, Margaillan A (2013) Recent advances on ion-imprinted polymers. React Funct Polym 73(6):859–875
9. Ahmad OS, Bedwell TS, Esen C, Garcia-Cruz A, Piletsky SA (2019) Molecularly imprinted polymers in electrochemical and optical sensors. Trends Biotechnol 37(3):294–309
10. Gu Y, Wang Y, Wu X, Pan M, Hu N, Wang J, Wang S (2019) Quartz crystal microbalance sensor based on covalent organic framework composite and molecularly imprinted polymer of poly (o-aminothiophenol) with gold nanoparticles for the determination of aflatoxin B1. Sensors Actuators B Chem 291:293–297
11. Caro E, Masqué N, Marcé RM, Borrull F, Cormack PA, Sherrington DC (2002) Non-covalent and semi-covalent molecularly imprinted polymers for selective on-line solid-phase extraction of 4-nitrophenol from water samples. J Chromatogr A 963(1–2):169–178
12. Wulff G (1972) The use of polymers with enzyme-analogous structures for the resolution of racemates. Angew Chem Int Ed 11:341
13. Turiel E, Esteban AM (2020) Molecularly imprinted polymers. In: Solid-phase extraction. Elsevier, pp 215–233. https://www.sciencedirect.com/science/article/pii/B978012816906300008X
14. Fan D, Jia L, Xiang H, Peng M, Li H, Shi S (2017) Synthesis and characterization of hollow porous molecular imprinted polymers for the selective extraction and determination of caffeic acid in fruit samples. Food Chem 224:32–36
15. Paruli EI, Griesser T, Merlier F, Gonzato C, Haupt K (2019) Molecularly imprinted polymers by thiol–yne chemistry: making imprinting even easier. Polym Chem 10(34):4732–4739
16. Ramström O, Mosbach K (1999) Synthesis and catalysis by molecularly imprinted materials. Curr Opin Chem Biol 3(6):759–764
17. Alexander C, Andersson HS, Andersson LI, Ansell RJ, Kirsch N, Nicholls IA, O'Mahony J, Whitcombe MJ (2006) Molecular imprinting science and technology: a survey of the literature for the years up to and including 2003. J Mol Recognit 19(2):106–180. https://doi.org/10.1002/jmr.760
18. Andersson LI (2000) Molecular imprinting for drug bioanalysis: a review on the application of imprinted polymers to solid-phase extraction and binding assay. J Chromatogr B Biomed Sci Appl 739(1):163–173
19. Stevenson D (1999) Molecular imprinted polymers for solid-phase extraction. TrAC Trends Anal Chem 18(3):154–158
20. Ansell RJ (2004) Molecularly imprinted polymers in pseudoimmunoassay. J Chromatogr B 804(1):151–165
21. Mujahid A, Iqbal N, Afzal A (2013) Bioimprinting strategies: from soft lithography to biomimetic sensors and beyond. Biotechnol Adv 31(8):1435–1447
22. Mujahid A, Lieberzeit PA, Dickert FL (2010) Chemical sensors based on molecularly imprinted sol-gel materials. Materials 3(4):2196–2217
23. Alvarez-Lorenzo C, Concheiro A (2004) Molecularly imprinted polymers for drug delivery. J Chromatogr B 804(1):231–245
24. Sellergren B, Allender CJ (2005) Molecularly imprinted polymers: a bridge to advanced drug delivery. Adv Drug Deliv Rev 57(12):1733–1741
25. Haupt K, Mosbach K (1998) Plastic antibodies: developments and applications. Trends Biotechnol 16(11):468–475
26. Whitcombe MJ, Chianella I, Larcombe L, Piletsky SA, Noble J, Porter R, Horgan A (2011) The rational development of molecularly imprinted polymer-based sensors for protein detection. Chem Soc Rev 40(3):1547–1571
27. Dickert FL, Bauer PA (1991) The detection of halogenated hydrocarbons via host–guest chemistry—a mass-sensitive sensor study with QMB- and SAW-devices. Adv Mater 3(9):436–438. https://doi.org/10.1002/adma.19910030906
28. Dickert FL, Haunschild A (1993) Sensor materials for solvent vapor detection—donor–acceptor and host–guest interactions. Adv Mater 5(12):887–895. https://doi.org/10.1002/adma.19930051203

29. Hasanah AN, Safitri N, Zulfa A, Neli N, Rahayu D (2021) Factors affecting preparation of molecularly imprinted polymer and methods on finding template-monomer interaction as the key of selective properties of the materials. Molecules 26(18):5612
30. Yan H, Row KH (2006) Characteristic and synthetic approach of molecularly imprinted polymer. Int J Mol Sci 7(5):155–178
31. Arabi M, Ostovan A, Li J, Wang X, Zhang Z, Choo J, Chen L (2021) Molecular imprinting: green perspectives and strategies. Adv Mater 33(30):2100543. https://doi.org/10.1002/adma.202100543
32. Ertürk G, Mattiasson B (2017) Molecular imprinting techniques used for the preparation of biosensors. Sensors 17(2):288
33. Saylan Y, Yilmaz F, Özgür E, Derazshamshir A, Yavuz H, Denizli A (2017) Molecular imprinting of macromolecules for sensor applications. Sensors 17(4):898
34. Reville EK, Sylvester EH, Benware SJ, Negi SS, Berda EB (2022) Customizable molecular recognition: advancements in design, synthesis, and application of molecularly imprinted polymers. Polym Chem 13(23):3387–3411
35. Schirhagl R (2014) Bioapplications for molecularly imprinted polymers. Anal Chem 86(1):250–261. https://doi.org/10.1021/ac401251j
36. Takeuchi T, Haginaka J (1999) Separation and sensing based on molecular recognition using molecularly imprinted polymers. J Chromatogr B Biomed Sci Appl 728(1):1–20
37. Vasapollo G, Sole RD, Mergola L, Lazzoi MR, Scardino A, Scorrano S, Mele G (2011) Molecularly imprinted polymers: present and future prospective. Int J Mol Sci 12(9):5908–5945. https://doi.org/10.3390/ijms12095908
38. Parlak O, Keene ST, Marais A, Curto VF, Salleo A (2018) Molecularly selective nanoporous membrane-based wearable organic electrochemical device for noninvasive cortisol sensing. Sci Adv 4(7):eaar2904. https://doi.org/10.1126/sciadv.aar2904
39. Sajini T, Mathew B (2021) A brief overview of molecularly imprinted polymers: highlighting computational design, nano and photo-responsive imprinting. Talanta Open 4:100072. https://doi.org/10.1016/j.talo.2021.100072
40. Włoch M, Datta J (2019) Chapter 2—Synthesis and polymerisation techniques of molecularly imprinted polymers. In: Marć M (ed) Comprehensive analytical chemistry, vol 86. Elsevier, pp 17–40. https://doi.org/10.1016/bs.coac.2019.05.011
41. Adumitrăchioaie A, Tertiş M, Cernat A, Săndulescu R, Cristea C (2018) Electrochemical methods based on molecularly imprinted polymers for drug detection. A review. Int J Electrochem Sci 13(3):2556–2576. https://doi.org/10.20964/2018.03.75
42. Fresco-Cala B, Batista AD, Cárdenas S (2020) Molecularly imprinted polymer micro- and nano-particles. A review. Molecules 25(20):4740. https://doi.org/10.3390/molecules25204740
43. Brüggemann O, Visnjevski A, Burch R, Patel P (2004) Selective extraction of antioxidants with molecularly imprinted polymers. Anal Chim Acta 504(1):81–88. https://doi.org/10.1016/j.aca.2003.08.033
44. Haupt K, Mayes AG, Mosbach K (1998) Herbicide assay using an imprinted polymer-based system analogous to competitive fluoroimmunoassays. Anal Chem 70(18):3936–3939. https://doi.org/10.1021/ac980175f
45. Mosbach K (2001) Towards the development of artificial recognition elements in sensor technology with emphasis on molecular imprinting. Electrochemistry 69(12):919–919. https://doi.org/10.5796/electrochemistry.69.919
46. Hansen DE (2007) Recent developments in the molecular imprinting of proteins. Biomaterials 28(29):4178–4191. https://doi.org/10.1016/j.biomaterials.2007.06.017
47. Haupt K (2003) Imprinted polymers—tailor-made mimics of antibodies and receptors. Chem Commun 2:171–178. https://doi.org/10.1039/b207596b
48. Ye L, Mosbach K (2008) Molecular imprinting: synthetic materials as substitutes for biological antibodies and receptors. Chem Mater 20(3):859–868. https://doi.org/10.1021/cm703190w

49. Turiel E, Martin-Esteban A (2004) Molecularly imprinted polymers: towards highly selective stationary phases in liquid chromatography and capillary electrophoresis. Anal Bioanal Chem 378(8):1876–1886. https://doi.org/10.1007/s00216-003-2331-1
50. Yu Y, Ye L, Haupt K, Mosbach K (2002) Formation of a class of enzyme inhibitors (drugs), including a chiral compound, by using imprinted polymers or biomolecules as molecular-scale reaction vessels. Angew Chem Int Ed 41(23):4459–4463. https://doi.org/10.1002/1521-3773(20021202)41:23<4459::AID-ANIE4459>3.0.CO;2-2
51. Alvarez-Lorenzo C, Concheiro A (2008) Intelligent drug delivery systems: polymeric micelles and hydrogels. Mini-Rev Med Chem 8(11):1065–1074. https://doi.org/10.2174/138955708785909952
52. Ayari MG, Kadhirvel P, Favetta P, Plano B, Dejous C, Carbonnier B, Agrofoglio LA (2019) Synthesis of imprinted hydrogel microbeads by inverse Pickering emulsion to controlled release of adenosine 5′-monophosphate. Mater Sci Eng C Mater Biol Appl 101:254–263. https://doi.org/10.1016/j.msec.2019.03.102
53. Sedghi R, Yassari M, Heidari B (2018) Thermo-responsive molecularly imprinted polymer containing magnetic nanoparticles: synthesis, characterization and adsorption properties for curcumin. Colloids Surf B Biointerfaces 162:154–162. https://doi.org/10.1016/j.colsurfb.2017.11.053
54. da Silva MS, Casimiro T, da Silva MS, Casimiro T (2012) High affinity polymers by molecular imprinting for drug delivery. In: Polymerization. IntechOpen. https://doi.org/10.5772/50162
55. He S, Zhang L, Bai S, Yang H, Cui Z, Zhang X, Li Y (2021) Advances of molecularly imprinted polymers (MIP) and the application in drug delivery. Eur Polym J 143:110179. https://doi.org/10.1016/j.eurpolymj.2020.110179
56. Liu Z, Huang Y, Yang Y (eds) (2021) Molecularly imprinted polymers as advanced drug delivery systems: synthesis, character and application. Springer Singapore. https://doi.org/10.1007/978-981-16-0227-6
57. Puoci F, Cirillo G, Curcio M, Parisi OI, Iemma F, Picci N (2011) Molecularly imprinted polymers in drug delivery: state of art and future perspectives. Expert Opin Drug Deliv 8(10):1379–1393. https://doi.org/10.1517/17425247.2011.609166
58. Zaidi SA (2016) Molecular imprinted polymers as drug delivery vehicles. Drug Deliv 23(7):2262–2271. https://doi.org/10.3109/10717544.2014.970297
59. Paul PK, Treetong A, Suedee R (2017) Biomimetic insulin-imprinted polymer nanoparticles as a potential oral drug delivery system. Acta Pharm 67(2):149–168. https://doi.org/10.1515/acph-2017-0020
60. Bodoki AE, Iacob B-C, Bodoki E (2019) Perspectives of molecularly imprinted polymer-based drug delivery systems in cancer therapy. Polymers 11(12):2085. https://doi.org/10.3390/polym11122085
61. Norell MC, Andersson HS, Nicholls IA (1998) Theophylline molecularly imprinted polymer dissociation kinetics: a novel sustained release drug dosage mechanism. J Mol Recogn 11(1–6):98–102. https://doi.org/10.1002/(SICI)1099-1352(199812)11:1/6<98::AID-JMR399>3.0.CO;2-Y
62. Tuwahatu CA, Yeung CC, Lam YW, Roy VAL (2018) The molecularly imprinted polymer essentials: curation of anticancer, ophthalmic, and projected gene therapy drug delivery systems. J Control Release 287:24–34. https://doi.org/10.1016/j.jconrel.2018.08.023
63. Gomes ADS (2012) Polymerization. BoD—Books on Demand
64. Cazares-Cortes E, Nerantzaki M, Fresnais J, Wilhelm C, Griffete N, Ménager C (2018) Magnetic nanoparticles create hot spots in polymer matrix for controlled drug release. Nanomaterials 8(10):850. https://doi.org/10.3390/nano8100850
65. Li L, Chen L, Zhang H, Yang Y, Liu X, Chen Y (2016) Temperature and magnetism bi-responsive molecularly imprinted polymers: preparation, adsorption mechanism and properties as drug delivery system for sustained release of 5-fluorouracil. Mater Sci Eng C Mater Biol Appl 61:158–168. https://doi.org/10.1016/j.msec.2015.12.027

66. Zhu Y, Liu R, Huang H, Zhu Q (2019) Vinblastine-loaded nanoparticles with enhanced tumor-targeting efficiency and decreasing toxicity: developed by one-step molecular imprinting process. Mol Pharm 16(6):2675–2689. https://doi.org/10.1021/acs.molpharmaceut.9b00243
67. Crapnell RD, Banks CE (2021) Electroanalytical overview: the electroanalytical detection of theophylline. Talanta Open 3:100037. https://doi.org/10.1016/j.talo.2021.100037
68. Saboury A, Mohammadi R, Javanbakht S, Ghorbani M (2023) Doxorubicin imprinted magnetic polymethacrylamide as a pH-sensitive anticancer nanocarrier. J Drug Deliv Sci Technol 79:103998. https://doi.org/10.1016/j.jddst.2022.103998
69. Psoma SD, Kanthou C (2023) Wearable insulin biosensors for diabetes management: advances and challenges. Biosensors 13(7):719. https://doi.org/10.3390/bios13070719
70. Lusina A, Cegłowski M (2022) Molecularly imprinted polymers as state-of-the-art drug carriers in hydrogel transdermal drug delivery applications. Polymers 14(3):640. https://doi.org/10.3390/polym14030640
71. Wang S, Wang C, Xin Y, Li Q, Liu W (2022) Core-shell nanocomposite of flower-like molybdenum disulfide nanospheres and molecularly imprinted polymers for electrochemical detection of anti COVID-19 drug favipiravir in biological samples. Mikrochim Acta 189(3):125. https://doi.org/10.1007/s00604-022-05213-9
72. Lim T-H (2015) Practical textbook of cardiac CT and MRI. Springer
73. Hiratani H, Fujiwara A, Tamiya Y, Mizutani Y, Alvarez-Lorenzo C (2005) Ocular release of timolol from molecularly imprinted soft contact lenses. Biomaterials 26(11):1293–1298. https://doi.org/10.1016/j.biomaterials.2004.04.030
74. Alven S, Peter S, Mbese Z, Aderibigbe BA (2022) Polymer-based wound dressing materials loaded with bioactive agents: potential materials for the treatment of diabetic wounds. Polymers 14(4):724. https://doi.org/10.3390/polym14040724
75. Yang G, Zhao F (2014) A novel electrochemical sensor for the determination of lidocaine based on surface-imprinting on porous three-dimensional film. J Mater Chem C 2(47):10201–10208. https://doi.org/10.1039/C4TC02039A
76. Kang MS, Cho E, Choi HE, Amri C, Lee J-H, Kim KS (2023) Molecularly imprinted polymers (MIPs): emerging biomaterials for cancer theragnostic applications. Biomater Res 27(1):45. https://doi.org/10.1186/s40824-023-00388-5
77. Ertürk G, Hedström M, Mattiasson B (2016) A sensitive and real-time assay of trypsin by using molecular imprinting-based capacitive biosensor. Biosens Bioelectron 86:557–565. https://doi.org/10.1016/j.bios.2016.07.046
78. Arcadio F, Zeni L, Perri C, D'Agostino G, Chiaretti G, Porto G, Minardo A, Cennamo N (2021) Bovine serum albumin protein detection by a removable SPR Chip combined with a specific MIP receptor. Chemosensors 9(8):218. https://doi.org/10.3390/chemosensors9080218
79. Sfragano PS, Pillozzi S, Palchetti I (2021) Electrochemical and PEC platforms for miRNA and other epigenetic markers of cancer diseases: recent updates. Electrochem Commun 124:106929. https://doi.org/10.1016/j.elecom.2021.106929
80. Zhang Z, Liu J (2016) Molecularly imprinted polymers with DNA aptamer fragments as macromonomers. ACS Appl Mater Interfaces 8(10):6371–6378. https://doi.org/10.1021/acsami.6b00461
81. Yadav AK, Verma D, Dalal N, Kumar A, Solanki PR (2022) Molecularly imprinted polymer-based nanodiagnostics for clinically pertinent bacteria and virus detection for future pandemics. Biosens Bioelectron 12:100257. https://doi.org/10.1016/j.biosx.2022.100257
82. D'Aurelio R, Chianella I, Goode JA, Tothill IE (2020) Molecularly imprinted nanoparticles based sensor for cocaine detection. Biosensors 10(3):22. https://doi.org/10.3390/bios10030022
83. Zhao G, Zhang Y, Sun D, Yan S, Wen Y, Wang Y, Li G, Liu H, Li J, Song Z (2023) Recent advances in molecularly imprinted polymers for antibiotic analysis. Molecules 28(1):335. https://doi.org/10.3390/molecules28010335
84. Zahedi L, Ghourchi Beigi P, Shafiee M, Zare F, Mahdikia H, Abdouss M, Abdollahifar M-A, Shokri B (2021) Development of plasma functionalized polypropylene wound dressing for

betaine hydrochloride controlled drug delivery on diabetic wounds. Sci Rep 11(1):9641. https://doi.org/10.1038/s41598-021-89105-7
85. Huang M, Huang Y, Liu H, Tang Z, Chen Y, Huang Z, Xu S, Du J, Jia B (2022) Hydrogels for the treatment of oral and maxillofacial diseases: current research, challenges, and future directions. Biomater Sci 10(22):6413–6446. https://doi.org/10.1039/d2bm01036d
86. Mostafa AM, Barton SJ, Wren SP, Barker J (2021) Review on molecularly imprinted polymers with a focus on their application to the analysis of protein biomarkers. TrAC Trends Anal Chem 144:116431. https://doi.org/10.1016/j.trac.2021.116431
87. Arabi M, Ghaedi M, Ostovan A (2017) Synthesis and application of in-situ molecularly imprinted silica monolithic in pipette-tip solid-phase microextraction for the separation and determination of gallic acid in orange juice samples. J Chromatogr B Analyt Technol Biomed Life Sci 1048:102–110. https://doi.org/10.1016/j.jchromb.2017.02.016
88. Chen F-F, Wang R, Shi Y-P (2012) Molecularly imprinted polymer for the specific solid-phase extraction of kirenol from Siegesbeckia pubescens herbal extract. Talanta 89:505–512. https://doi.org/10.1016/j.talanta.2011.12.080
89. Ghasemi S, Nematollahzadeh A (2018) Molecularly imprinted ultrafiltration polysulfone membrane with specific nano-cavities for selective separation and enrichment of paclitaxel from plant extract. React Funct Polym 126:9–19. https://doi.org/10.1016/j.reactfunctpolym.2018.02.012
90. Gomes C, Sadoyan G, Dias R, Costa M (2017) Development of molecularly imprinted polymers to target polyphenols present in plant extracts. Processes 5(4):72. https://doi.org/10.3390/pr5040072
91. Henry N, Favetta P, Delépée R, Seigneuret J-M, Agrofoglio LA (2015) Synthesis of a molecularly imprinted polymer to isolate glucosamine from plant extracts by an ionic–non-covalent dual approach. Int J Cosmet Sci 37(2):196–206. https://doi.org/10.1111/ics.12182
92. Machyňáková A, Hroboňová K (2017) Preparation and application of magnetic molecularly imprinted polymers for the selective extraction of coumarins from food and plant samples. Anal Methods 9(14):2168–2176. https://doi.org/10.1039/C7AY00502D
93. Xie J, Zhu L, Luo H, Zhou L, Li C, Xu X (2001) Direct extraction of specific pharmacophoric flavonoids from gingko leaves using a molecularly imprinted polymer for quercetin. J Chromatogr A 934(1–2):1–11. https://doi.org/10.1016/S0021-9673(01)01294-8
94. Saad EM, Madbouly A, Ayoub N, Mohamed El Nashar R (2015) Preparation and application of molecularly imprinted polymer for isolation of chicoric acid from Chicorium intybus L. medicinal plant. Anal Chim Acta 877:80–89. https://doi.org/10.1016/j.aca.2015.03.047
95. Sun X, Zhang C, Huang Y-P, Liu Z-S (2015) Separation of epigallocatechin gallate from natural plant extracts using crowding agents—assisted imprinted polymers. Chromatographia 78(15–16):995–1003. https://doi.org/10.1007/s10337-015-2914-y
96. Chen T, Gu J, Wang H, Yuan G, Chen L, Xu X, Xiao W (2014) Semi-preparative scale separation of Emodin from plant extract by using molecularly imprinted polymer as stationary phase. Chromatographia 77(13–14):893–899. https://doi.org/10.1007/s10337-014-2691-z
97. Tian M, Bi W, Row KH (2012) Simultaneous extraction and separation of flavonols and flavones from *Chamaecyparis obtusa* by multi-phase extraction using an ionic liquid-modified microsphere polymer. Phytochem Anal 23(6):576–581. https://doi.org/10.1002/pca.2357
98. Karasová G, Lehotay J, Sádecká J, Skačáni I, Lachová M (2005) Selective extraction of derivates of p-hydroxy-benzoic acid from plant material by using a molecularly imprinted polymer. J Sep Sci 28(18):2468–2476. https://doi.org/10.1002/jssc.200500190
99. Wang D, Hong SP, Row KH (2004) Solid extraction of caffeine and theophylline from green tea by molecular imprinted polymers. Korean J Chem Eng 21(4):853–857. https://doi.org/10.1007/BF02705530
100. Lin L, Zhang J, Fu Q, He L, Li Y (2006) Concentration and extraction of sinomenine from herb and plasma using a molecularly imprinted polymer as the stationary phase. Anal Chim Acta 561(1–2):178–182
101. Sellergren B (2000) Molecularly imprinted polymers: man-made mimics of antibodies and their application in analytical chemistry. Elsevier. https://books.google.com/books?hl=en&

lr=&id=tKztDDfX1SMC&oi=fnd&pg=PP1&dq=108.%09Sellergren+B+(2000)+Molecularly+imprinted+polymers:+man-made+mimics+of+antibodies+and+their+application+in+analytical+chemistry.+Elsevier,+Amsterdam&ots=B6DDdDlv2S&sig=TFPRbirXZiGIR8NEguPyay0oVWs
102. Whitcombe MJ, Vulfson EN (2001) Imprinted polymers. Adv Mater 13(7):467–478. https://doi.org/10.1002/1521-4095(200104)13:7<467::AID-ADMA467>3.0.CO;2-T
103. Turiel E, Martín-Esteban A (2010) Molecularly imprinted polymers for sample preparation: a review. Anal Chim Acta 668(2):87–99
104. Sánchez-Barragán I, Karim K, Costa-Fernández JM, Piletsky SA, Sanz-Medel A (2007) A molecularly imprinted polymer for carbaryl determination in water. Sensors Actuators B Chem 123(2):798–804
105. Ou J, Hu L, Hu L, Li X, Zou H (2006) Determination of phenolic compounds in river water with on-line coupling bisphenol A imprinted monolithic precolumn with high performance liquid chromatography. Talanta 69(4):1001–1006
106. Luo X, Zhan Y, Tu X, Huang Y, Luo S, Yan L (2011) Novel molecularly imprinted polymer using 1-(α-methyl acrylate)-3-methylimidazolium bromide as functional monomer for simultaneous extraction and determination of water-soluble acid dyes in wastewater and soft drink by solid phase extraction and high performance liquid chromatography. J Chromatogr A 1218(8):1115–1121
107. Tarley CRT, Kubota LT (2005) Molecularly-imprinted solid phase extraction of catechol from aqueous effluents for its selective determination by differential pulse voltammetry. Anal Chim Acta 548(1):11–19. https://doi.org/10.1016/j.aca.2005.05.055
108. Beltran A, Caro E, Marcé RM, Cormack PAG, Sherrington DC, Borrull F (2007) Synthesis and application of a carbamazepine-imprinted polymer for solid-phase extraction from urine and wastewater. Anal Chim Acta 597(1):6–11. https://doi.org/10.1016/j.aca.2007.06.040
109. Cacho C, Turiel E, Pérez-Conde C (2009) Molecularly imprinted polymers: an analytical tool for the determination of benzimidazole compounds in water samples. Talanta 78(3):1029–1035. https://doi.org/10.1016/j.talanta.2009.01.007
110. Watabe Y, Kubo T, Nishikawa T, Fujita T, Kaya K, Hosoya K (2006) Fully automated liquid chromatography-mass spectrometry determination of 17beta-estradiol in river water. J Chromatogr A 1120(1–2):252–259. https://doi.org/10.1016/j.chroma.2006.01.057
111. Sambe H, Hoshina K, Haginaka J (2007) Molecularly imprinted polymers for triazine herbicides prepared by multi-step swelling and polymerization method: their application to the determination of methylthiotriazine herbicides in river water. J Chromatogr A 1152(1–2):130–137
112. Liu Z, Huan S, Jiang J, Shen G, Yu R (2006) Molecularly imprinted TiO2 thin film using stable ground-state complex as template as applied to selective electrochemical determination of mercury. Talanta 68(4):1120–1125
113. Say R, Ersöz A, Türk H, Denizli A (2004) Selective separation and preconcentration of cyanide by a column packed with cyanide-imprinted polymeric microbeads. Sep Purif Technol 40(1):9–14
114. Xu Z, Chen S, Huang W, Fang G, Pingzhu H, Wang S (2009) Study on an on-line molecularly imprinted solid-phase extraction coupled to high-performance liquid chromatography for separation and determination of trace estrone in environment. Anal Bioanal Chem 393(4):1273–1279. https://doi.org/10.1007/s00216-008-2544-4
115. Tchekwagep PMS, Crapnell RD, Banks CE, Betlem K, Rinner U, Canfarotta F, Lowdon JW, Eersels K, van Grinsven B, Peeters M (2022) A critical review on the use of molecular imprinting for trace heavy metal and micropollutant detection. Chemosensors 10(8):296
116. Bhogal S, Kaur K, Mohiuddin I, Kumar S, Lee J, Brown RJC, Kim K-H, Malik AK (2021) Hollow porous molecularly imprinted polymers as emerging adsorbents. Environ Pollut 288:117775. https://doi.org/10.1016/j.envpol.2021.117775
117. Wang D, Liu Y, Xu Z, Zhao D, Liu Y, Liu Z (2020) Multitemplate molecularly imprinted polymeric solid-phase microextraction fiber coupled with HPLC for endocrine disruptor analysis in water samples. Microchem J 155:104802. https://doi.org/10.1016/j.microc.2020.104802

118. Roy E, Patra S, Tiwari A, Madhuri R, Sharma PK (2017) Retracted: single cell imprinting on the surface of Ag–ZnO bimetallic nanoparticle modified graphene oxide sheets for targeted detection, removal and photothermal killing of E. coli. Biosens Bioelectron 89:620–626. https://doi.org/10.1016/j.bios.2015.12.085
119. Krupadam RJ, Nesterov EE, Spivak DA (2014) Highly selective detection of oil spill polycyclic aromatic hydrocarbons using molecularly imprinted polymers for marine ecosystems. J Hazard Mater 274:1–7. https://doi.org/10.1016/j.jhazmat.2014.03.050
120. Storer C, Coldrick Z, Tate D, Donoghue J, Grieve B (2018) Towards phosphate detection in hydroponics using molecularly imprinted polymer sensors. Sensors (Basel) 18(2):531. https://doi.org/10.3390/s18020531
121. Zarejousheghani M, Rahimi P, Borsdorf H, Zimmermann S, Joseph Y (2021) Molecularly imprinted polymer-based sensors for priority pollutants. Sensors 21(7):2406. https://doi.org/10.3390/s21072406
122. Huang D-L, Wang R-Z, Liu Y-G, Zeng G-M, Lai C, Xu P, Lu B-A, Xu J-J, Wang C, Huang C (2015) Application of molecularly imprinted polymers in wastewater treatment: a review. Environ Sci Pollut Res 22(2):963–977. https://doi.org/10.1007/s11356-014-3599-8
123. Kyzas GZ, Bikiaris DN (2014) Molecular imprinting for high-added value metals: an overview of recent environmental applications. Adv Mater Sci Eng 2014:1–8. https://doi.org/10.1155/2014/932637
124. Bjarnason B, Chimuka L, Ramström O (1999) On-line solid-phase extraction of triazine herbicides using a molecularly imprinted polymer for selective sample enrichment. Anal Chem 71(11):2152–2156. https://doi.org/10.1021/ac9810314
125. Crescenzi C, Bayoudh S, Cormack PA, Klein T, Ensing K (2001) Determination of clenbuterol in bovine liver by combining matrix solid-phase dispersion and molecular imprinted solid-phase extraction followed by liquid chromatography/electrospray ion trap multiple-stage mass spectrometry. Anal Chem 73(10):2171–2177. https://doi.org/10.1021/ac0014360

Synthesis and Characterization of MIPs

2

Aleksandra Lusina, Tomasz Nazim, and Michał Cegłowski

Abstract

Molecularly imprinted polymers (MIPs) represent a unique class of polymeric matrices, notable for their specific recognition capabilities toward chosen template molecules via dedicated chemical functionalities. These polymers have garnered significant interest in developing advanced sensor systems due to their exceptional ability to selectively bind target molecules. A key attribute of MIPs lies in their stability, which plays a pivotal role in enhancing the longevity and efficacy of sensor applications. This chapter provides a comprehensive overview of the synthesis, fabrication, and characterization techniques of MIPs. Emphasis is placed on detailed analysis and understanding of MIPs as specialized materials, highlighting their synthesis methodologies and the nuances of their physicochemical properties. This exploration aims to deepen the understanding of MIPs' unique characteristics and their potential applications in selective sensing technologies.

Keywords

Molecularly imprinted polymers · Synthetic techniques · Surface imprinting Physicochemical characterization · Adsorption properties

2.1 Introduction

Molecularly imprinted polymers, often abbreviated as MIPs, are polymeric materials with the ability to selectively recognize specific templates at the molecular level. This recognition property arises directly from the synthesis method of these

A. Lusina · T. Nazim · M. Cegłowski (✉)
Faculty of Chemistry, Adam Mickiewicz University, Poznań, Poland
e-mail: michal.ceglowski@amu.edu.pl

S. Patra et al. (eds.), *Molecularly Imprinted Polymers: Path to Artificial Antibodies*, https://doi.org/10.1007/978-981-97-4379-7_2

materials, in which the template molecule plays a central role in shaping molecular interactions, resulting in the formation of molecular cavities with recognition capabilities. In general, MIPs are synthesized by mixing a monomer, a template, and a cross-linking agent in an appropriate solvent (Fig. 2.1).

During the polymerization process, interactions between the template molecule and polymer chains occur, giving rise to the transitional complex. Following polymerization, the template can be easily removed. The voids formed in this manner, after removing the template molecule, have the ability to rebind it through identical interactions and are referred to as molecular cavities [2]. Leveraging their valuable attributes, including remarkable selectivity, efficient recovery, extended shelf stability, and reliable repeatability, MIPs are extensively employed in real-time analytical processes [3]. Furthermore, due to their ability to analyze complex samples, MIPs can be utilized in processing biological samples, making them distinctive compared to other established recognition systems [4]. Furthermore, MIPs find applications across several domains. The most common applications include sensing [5, 6], purification [7, 8], separation [9, 10], catalysis [11], and degradation processes [12]. The pervasive adoption of MIPs stems from their advantageous attributes, including robust physical stability under demanding chemical and physical conditions, facile preparation, exceptional resilience, extensive reusability, and cost-effective synthesis [13–15].

Although MIPs offer numerous advantages, it is important to acknowledge certain drawbacks. One of the most significant drawbacks is the time-consuming and cost-intensive nature of designing a new MIP-based system tailored to a specific template molecule [16]. Conversely, designing MIPs demands a thorough understanding of chemical equilibrium states and the fundamentals of molecular recognition, thermodynamics, and polymer chemistry. This knowledge is essential for creating a polymer with the desired properties [17]. Additionally, the synthesis of MIPs is heavily reliant on the selected template molecule, with some being amenable to imprinting, whereas others present challenges [18]. Among the other prevalent challenges are the presence of heterogeneous binding sites, template bleeding, protein imprinting, or even imprinting of hydrophilic templates and subsequent application in aqueous media [19–21].

The primary aim of this chapter is to provide a detailed and comprehensive analysis of MIP materials. In the ensuing sections, we will methodically explore the synthesis and fabrication processes integral to the development of MIPs. This will

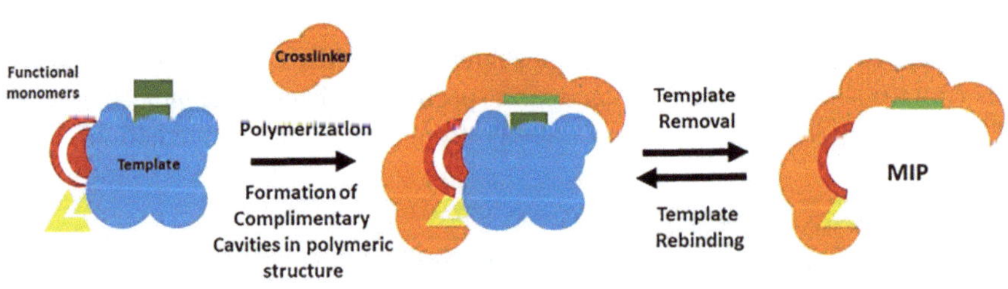

Fig. 2.1 A general overview of MIP synthesis. (Reprinted from Lowdon et al. [1]. Copyright (2023), with permission from Elsevier (or applicable society copyright owner))

be followed by a thorough characterization of these materials, elucidating their unique properties and capabilities. Our goal is to offer an in-depth understanding of MIPs, from their initial creation to the final assessment of their functional attributes, and highlighting their potential applications in various fields.

2.2 The Fundamentals of Molecularly Imprinted Polymers

2.2.1 Methods of Molecular Imprinting

2.2.1.1 Covalent

The primary distinction between classes of molecular imprinting lies in the interactions occurring within the transitional complex. In the first reported method introduced by Wulff [22], within the covalent imprinting class, the foundation of recognition ability is the establishment of a covalent bond between the polymer chains and the template molecules. After the polymerization process, the formed covalent bonds can be cleaved, allowing for the easy removal of the template molecules from the complex. This process leads to the formation of molecular cavities capable of rebinding template molecules by reestablishing the same covalent linkages. A notable limitation of this class is the restricted range of template molecules that can be effectively imprinted in succession. This limitation arises from the necessity for the establishment of quickly reversible covalent interactions between the molecular components. Furthermore, an additional constraint stems from the inherent nature of covalent interactions. Although considered some of the most durable interactions, they also hinder the attainment of thermodynamic equilibrium within the mixture due to their slow dissociation and binding kinetics [21].

2.2.1.2 Non-covalent

Alternatively, non-covalent interactions can occur between the polymer chains and template molecules, shaping the material's recognition property. This method was first proposed by Mosbach [23] and is considered more favorable and, as a result, is more commonly utilized due to its simplicity [24]. Moreover, this broadens the spectrum of applicable compounds that can be imprinted through non-covalent imprinting, thus making it the prevailing and widely adopted synthesis strategy for molecular imprinting technology (MIT). In the course of this process, a template–monomer complex is forged through a series of diverse interactions, encompassing hydrogen bonds, π–π interactions, ionic attractions, van der Waals forces, and more. Upon removing the template from the resulting transitional complex, the established interactions can be readily reconstructed using the same non-covalent interactions as were employed prior to polymerization [21].

2.2.1.3 Semi-covalent

When seeking to reestablish a non-covalent bond after removing a covalently bound template, the semi-covalent approach is employed. In general, the semi-covalent method, initially introduced by Whitcombe et al. [25], presents an intermediate solution in which the template is covalently attached to a functional monomer, whereas the template rebinding primarily relies on non-covalent interactions. Semi-covalent bonds exhibit a notable combination of the strong affinity found in covalent binding and the gentle operational conditions characteristic of non-covalent rebinding [13].

2.2.2 Essential Elements of Molecular Imprinting

2.2.2.1 Monomer

The choice of monomer is a critical factor in determining the interactions within the transition complex, which, in turn, influences the ultimate recognition properties. Generally, the choice of monomer is tied to the templates employed, as both the type and strength of the formed interactions have to be considered. In the context of covalent imprinting, the use of esters or amides derived from acrylic or methacrylic acids proves advantageous due to their potential for forming robust covalent bonds between the template molecule and the vinyl moieties present in the monomer structure. Alternatively, the same vinyl monomers can be employed when non-covalent imprinting is the chosen approach. This is because non-covalent interactions necessitate the presence of functional groups capable of engaging with the template molecule, such as through the formation of hydrogen bonds, among others [13]. As a result, the most frequently used monomers are methacrylic acid (MAA) [26, 27], methyl methacrylate [28], acrylic acid (AA) [29, 30], acrylamide (AM) [31, 32], vinyl pyridines (VP) [33], and styrene and its derivatives [34].

2.2.2.2 Template

One of the most crucial components of the MIP structure is the template, which plays a pivotal role in arranging functional groups derived from the functional monomers during the molecular imprinting process. However, profound knowledge of the template's chemical structure and properties is required. Certainly, the template must contain accessible functional groups that facilitate the occurrence of strong interactions with the polymer chain. Furthermore, it is essential that the selected template molecule does not interfere with the polymerization reaction. Another essential requirement is the stability of the template. The template molecule must remain unchanged throughout the polymerization process. This implies that the selected templates should undergo preliminary testing, maintaining the same reaction conditions, including temperature or pH, prior to the polymerization process [14].

2.2.2.3 Cross-Linker

The primary function of a cross-linker is to create a densely interconnected polymer structure. The ultimate objective is to achieve a robust cross-linked polymer, even after the templates have been removed. Usually, both the quantity and type of cross-linker used are critical factors that determine the selectivity and binding capacity of MIPs. As a result, the quantity of cross-linkers employed in the polymerization process also holds significant sway over the properties of MIPs. An insufficient amount of cross-linker can lead to unstable mechanical properties, whereas an excessive amount can diminish the number of recognition sites per unit mass of MIPs. The most frequently used cross-linkers are ethylene glycol dimethacrylate (EGDMA) [35, 36], divinylbenzene (DVB) [37], diethylenetriamine (DETA) [15, 38], and *N,N′*-methylenebisacrylamide (MBA) [39, 40].

2.2.2.4 Solvent

During polymerization, the solvent also plays a crucial role, serving two essential functions: it acts as a dispersing medium and as a pore-forming agent in the resulting MIPs. Initially, solvent molecules get trapped within the polymer chains, and they can be easily removed through drying, leading to pore formation. These pores greatly enhance the template's ability to access the cavities. Hence, the chosen solvent should encourage the establishment of bonds between the template and monomer. In the case of non-covalent methods, it is recommended to employ nonpolar aprotic solvents such as chloroform, tetrahydrofuran, toluene, and acetonitrile [41]. In general, the selected solvent should have specific attributes. First, it should facilitate excellent solubility for all the components employed. Moreover, the used solvent should facilitate the creation of sizable pores, ensuring optimal flow-through characteristics for the resulting polymer. Finally, it should exhibit low polarity to minimize any disruptions in the intricate process of complex formation. This careful selection of solvent properties plays a pivotal role in achieving the desired high selectivity in MIPs [24].

2.2.2.5 Initiator

Given that the most effective methods for preparing MIPs typically involve free-radical polymerization (FRP), photopolymerization, and electropolymerization, the utilization of initiators becomes essential. Generally, the quantity of initiator employed in the polymerization mixture is relatively low. Unlike monomers, initiators are typically used at levels such as 1 wt.% or 1 mol.% relative to the total number of moles of polymerizable bonds [24]. The most frequently employed azo initiators in free-radical polymerization are 2,2′-azobis(isobutyronitrile) (AIBN) and 2,2′-azobis(2,4-dimethylvaleronitrile) (ADVN). Typically, the initiation of the reaction occurs through thermal means within a temperature range spanning from 50 to 70 °C. However, it is important to note that this temperature range can pose potential hazards to the stability of the monomer/template adduct or to the template itself, especially in cases involving sensitive compounds like biological samples. Hence, in such instances, the utilization of photochemical initiation through ultraviolet (UV) light becomes a viable alternative [41]. In summary, the rate control and mode of decomposition of the initiator can vary significantly based on its chemical composition. This control can be achieved through various mechanisms, including thermal activation, exposure to light, and chemical or electrochemical processes [24].

In Table 2.1, an overview of the methodologies employed in molecular imprinting is provided. This includes examples of the monomers, cross-linkers, imitators, solvents, and templates that can be utilized in the process.

Table 2.1 A summary of the fundamentals of molecularly imprinted polymers

Methods of molecular imprinting		
Covalent	**Non-covalent**	**Semi-covalent**
This relies on establishing covalent bonds between polymer chains and template molecules and reestablishing the same covalent linkages after cavity formation	This relies on establishing non-covalent bonds between polymer chains and template molecules and reestablishing the same covalent linkages after cavity formation	This relies on establishing covalent bonds between polymer chains and template molecules and reestablishing non-covalent linkages after cavity formation
Essential elements of molecular imprinting		
Monomer	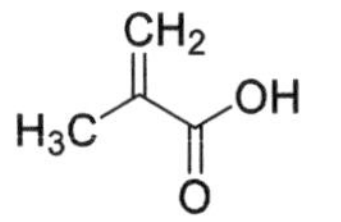	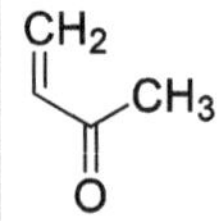
	Methacrylic acid	Acrylic acid
	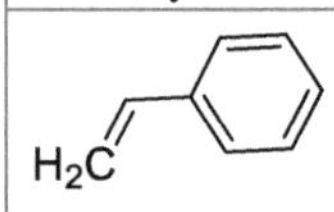	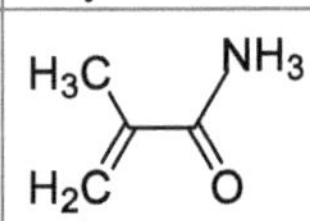
	4-Vinylpyridine	Methacrylamide
	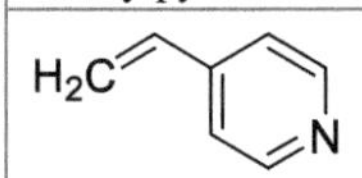	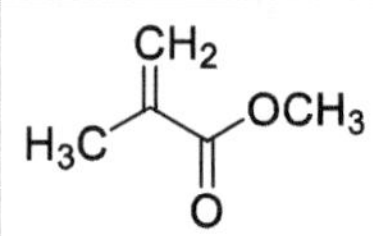
	2-Vinylpyridine	Methyl methacrylate
Template	Herbicides [38, 42]; pharmaceuticals [43]; phenol derivatives [15, 44]; explosive materials [45]; antibiotics [46, 47]; drugs [48, 49]	
Cross-linker	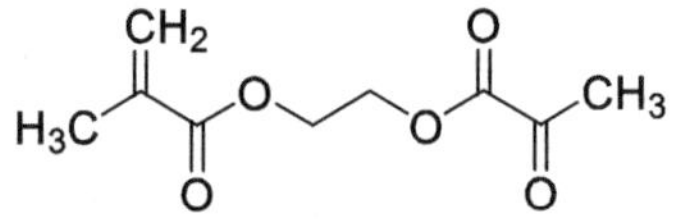	
	Ethylene glycol dimethacrylate	
	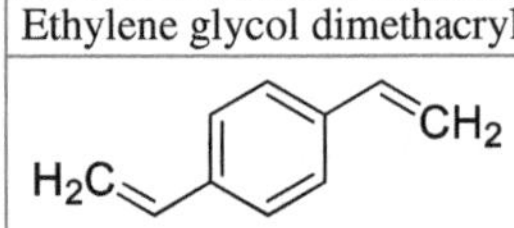	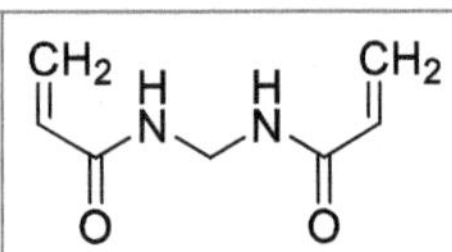
	Divinylbenzene	*N,N′*-Methylenediacrylamide
Solvent	Methanol, acetonitrile; tetrahydrofuran; *N,N′*-dimethylformamide; toluene; chloroform; dichloromethane	
Initiator	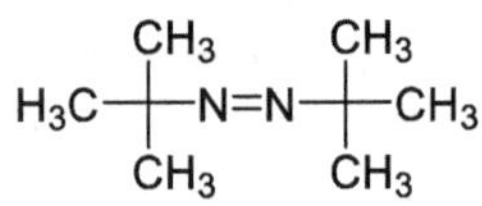	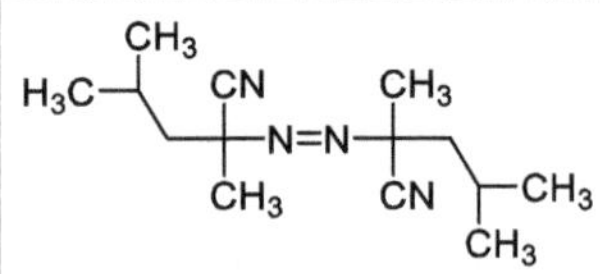
	Azobisisobutyronitrile (AIBN)	2,2′-Azobis(2,4-dimethylvaleronitrile) (ADVN)

2.3 Preparation Procedures: Methods of Molecular Imprinting

The field of synthetic opportunities for MIPs has seen remarkable progress since the introduction of polymerization methods. Following the initial report on bulk polymerization, various alternative pathways have arisen, addressing the limitations of this technique and providing novel approaches to create valuable MIPs. The carefully selected methods are thoroughly elaborated in the subsequent sections. These chosen instances illustrate the core characteristics of the particles, demonstrating the MIPs' ability to recognize specific target templates. Moreover, they display considerable variety, thus underscoring the versatility of MIP formation.

2.3.1 One-Pot Polymerization Methods

This paragraph will delve into the one-pot polymerization methods that have garnered significant attention in both the literature and commercial applications. Our primary focus of the section will be to describe various techniques, including *bulk polymerization*, *precipitation polymerization*, *emulsion polymerization*, *suspension polymerization*, and *multistep swelling polymerization*. Subsequently, the paragraph will explore each of these techniques in separate subsections to provide a comprehensive understanding of their principles and applications.

2.3.1.1 Bulk Polymerization

Bulk polymerization is a straightforward and fast method of molecular imprinting, well-suited for detecting small molecules at the nanoscale [50]. In this process, a mixture of monomers, cross-linking agents, initiators, and template molecules is combined in a suitable solvent, followed by free-radical polymerization. During polymerization, template molecules are imprinted within the emerging cavities of the three-dimensional polymer matrix, creating highly specific binding sites for the target molecules [51]. Bulk polymerization is preferred for creating easily accessible and template-specific interaction sites within chemoselective materials. These materials can take the form of a complete polymer matrix [52] or inorganic nanoparticles [53]. After polymerization, the template can be extracted from the polymer matrix, and the resulting MIPs can be ground to achieve the desired particle size. Bulk-imprinted synthetic materials typically exhibit rapid and fully reversible recognition and release characteristics, making them suitable for multiple analyses in liquid or gaseous states. This property is especially valuable for sensing molecular and ionic analytes [54, 55]. However, it is important to note that bulk-imprinting techniques have limitations when it comes to larger, micro-sized templates, which can result in diffusion and regeneration issues [56]. Larger templates may lead to prolonged release or recognition times and can be time-consuming. Additionally, larger imprinting cavities may attract smaller, structurally similar analytes, leading to cross-reactivity and reduced overall selectivity [57]. Grinding can result in irregular particle size distribution, thus reducing the practicality of the technique. Furthermore, it may lead to the mechanical disintegration of particles, creating heterogeneous binding sites and significantly reducing sensitivity [58].

2.3.1.2 Precipitation Polymerization

Precipitation polymerization was first developed by Goh and Stöver [59], and its application in MIPs for the synthesis of sub-micrometer MIP beads was first proposed by Ye et al. [60]. Precipitation polymerization is based on the utilization of a template mixture with a high concentration, a substantial volume of the solvent, and vigorous stirring at elevated speeds. This results in polymerization occurring in a diluted solution, with growth initiating from nanogel seed particles and subsequent entrapment of larger oligomers. Another approach involves initially achieving the desired particle size and then commencing the precipitation from the solution [61]. This is a straightforward polymerization method that yields microspheres characterized by pristine and smooth surfaces, along with particle sizes that are well-suited for various applications [16, 62]. Therefore, several critical factors must be carefully considered when designing a precipitation polymerization process to control particle size effectively. These factors include optimal stirring speed, solvent polarity, reaction temperature, and monomer concentration, with each playing a pivotal role in achieving the desired particle size [19]. Additionally, it is worth noticing that the application of high stirring speeds during the precipitation process, in conjunction with dilute solutions, enables the production of smaller particle sizes compared to those achieved by other polymerization techniques [61]. The most important advantages of this polymerization technique encompass its straightforward preparation process, negating the necessity for stabilizers or any other additives. It also exhibits compatibility with high degrees of cross-linking agents and the use of polar aprotic solvents [63]. These advantages have contributed to the extensive utilization of precipitation polymerization across a spectrum of domains, enabling the fabrication of polymer particles characterized by captivating internal structures and also the possibility of imprinting both small- [64] and larger-sized templates [65]. Additionally, it finds application in creating electrophoretic displays and environmentally responsive polymers, further enhancing its appeal and versatility [66, 67]. Nevertheless, this technique is not without its drawbacks. It necessitates the use of substantial quantities of solvents, which can be environmentally detrimental. Additionally, the requirement for high-concentration template mixtures may not be cost-effective, particularly in cases involving expensive or rare template molecules [19, 61].

2.3.1.3 Emulsion Polymerization

The emulsion polymerization technique relies on the use of surfactants or cyclodextrins to create a polymerization mixture that alternates between water-in-oil and oil-in-water phases. This method involves the careful combination of two solutions, which are vigorously mixed to produce polymer particles. In the initial step, a homogeneous mixture is prepared, containing monomers, cross-linking agents, and initiators. In parallel, the template molecule is combined with surfactants or cyclodextrins in a suitable solvent, creating a phase opposite to the first mixture. These two distinct mixtures are then brought together through intense agitation, resulting in very small, typically in the tens to hundreds of nanometers range [57]. It is important to note that the molecular weight of the polymer obtained through emulsion polymerization is significantly higher, typically an order of magnitude higher, compared to a homogeneous free-radical polymerization process. This difference arises from polymer chains undergoing decay into smaller particles and the generation of radicals within the aqueous phase, contributing to an increase in the radical's lifespan. In homogeneous free-radical polymerization, elevating the initiator concentration often accelerates termination reactions, leading to lower molecular weights in the final polymer product. In contrast, emulsion polymerization allows for the control of propagating polymer chain decay, enabling high polymerization rates and the production of high-molecular-weight polymers, which is a significant advantage compared to other free-radical polymerization methods [68]. Emulsion polymerization can be conducted in three different process types: *batch emulsion polymerization*, *semi-batch emulsion polymerization*, and *continuous emulsion polymerization*, each with its specific characteristics.

Batch Emulsion Polymerization

The most straightforward approach in the emulsion polymerization technique entails adding all the reactants to the reactor right from the start. Typically, the process begins with the preparation of the monomer emulsion, which is then heated to the desired reaction temperature. Subsequently, a solution of initiator in monomer is introduced to initiate the polymerization reaction.

Semi-batch Emulsion Polymerization

In a semi-batch process, the main difference lies in the timing of reactant additions. Typically, some of the reactants are initially introduced into the reactor, whereas the remaining components are fed in a controlled manner at a later stage. This semi-batch approach is known for its exceptional versatility, making it the most commonly employed technique among the three.

Continuous Emulsion Polymerization

In a continuous process, the constituent components of the formulation are continuously supplied into a reactor system, from which the product is continuously extracted. These processes are primarily well-suited for the large-scale production of a single polymer, although they have also found utility in academic investigations of polymerization mechanisms and kinetics [68].

This technique does have some clear drawbacks worth considering. One notable limitation is its reliance on both water and surfactants. The use of these components introduces the potential for disruption of the bonds that form between the template molecules and the polymer matrix. Furthermore, the presence of surfactants might lead to the isolation of template molecules within their core, thus making it difficult to establish the necessary interactions for successful imprinting. Consequently, this may hinder the overall effectiveness of the imprinting process [69].

2.3.1.4 Suspension Polymerization

This type of polymerization is widely used in commercial applications, as it is used for manufacturing polymers, including poly(vinyl chloride), poly(methyl methacrylate), polystyrene, styrene acrylonitrile copolymers, and others [70]. In typical suspension polymerization, a water-insoluble monomer containing an initiator is dispersed in the aqueous phase through strong stirring and the use of small amounts of stabilizers. Stabilizers play a crucial role in preventing the coalescence of monomer droplets and in stabilizing polymer beads that tend to agglomerate during polymerization when they become sticky [71]. Additionally, ultrasonic waves are often applied to disperse the mixture into fine droplets within the dispersion medium. Although water is commonly used as a dispersion medium, it may disrupt hydrogen bonds in the template–monomer adduct. Therefore, mineral oils or liquid surfactants are the preferred alternatives [58]. Suspension polymerization is particularly advantageous when small beads are the desired final polymer form, typically yielding spherical polymer beads in the micrometer to millimeter range. However, potential product contamination can occur if the stabilizers are not effectively removed. It is essential to note that suspension polymerization typically requires larger reactor volumes compared to bulk processes, making it more practical for large-scale MIP synthesis [70]. One of the primary drawbacks of suspension polymerization is its potential negative impact on template recognition [58]. Other disadvantages include wastewater issues and lower productivity for the same reactor capacity compared to bulk polymerization. On the other hand, the key advantages include easy heat removal and temperature control, low dispersion viscosity, and low impurity levels and separation costs compared to emulsion procedures [71]. Presently, suspension polymerization plays a significant role in various commercial processes for synthesizing drugs and chemicals and addressing environmental pollutants [72].

2.3.1.5 Multistep Swelling Polymerization

Multistep swelling polymerization is a combinational procedure that results in uniform and spherical MIP particles. Polymerization occurs in incremental stages, starting with small polymeric seeds like polystyrene. Swelling progressively increases at each step until uniform MIP particles of the desired size are achieved. It is worth noting that the multistep polymerization technique has seen relatively limited use due to its intricate procedural demands and the need for meticulous optimization at every stage [73]. However, it can be found in some recent research reports [74–76].

2.3.2 Surface Imprinting Methods

In fact, bulk polymerization, the most common variant of one-pot polymerization methods, has notable limitations, particularly when dealing with larger analytes like living cells and microorganisms. This is due to the slow diffusion rates of bioanalytes, resulting in significantly extended response times. The resulting materials may exhibit diminished recognition capabilities due to this slow diffusion [77]. Consequently, in the 2000s, a new imprinting technique called surface imprinting was implemented [19]. Surface imprinting revolves around creating selective sensing layers where recognition cavities are directly formed on the surface of a polymeric or sol–gel material. This approach results in both chemical and geometrical matching between the targeted analyte and the created cavities, ultimately leading to highly specific recognition ability [78]. However, surface imprinting may exhibit lower sensitivity due to a reduced number of available binding sites per unit area, and it is often time-consuming because of its multiple-step synthesis [19]. On the flip side, the surface imprinting technique offers several advantages compared to other methodologies, including easily accessible binding sites, swift and straightforward binding or release processes, and the potential for imprinting larger molecules [79, 80]. The following subsections will focus on the most reported surface imprinting techniques, including grafting polymerization, soft lithography, and surface imprinting after template immobilization.

2.3.2.1 Soft Lithography

Soft lithography, also known as the stamping technique, involves creating MIP films on solid substrates using a soft polymeric stamp. It is important to ensure that the template units are closely packed to achieve a high imprinting density. Typically, a template suspension of the appropriate concentration is allowed to self-organize on the stamp's surface, thus forming a continuous layer. Including the stamped patterned network on an MIP film enhances the exposed surface area, consequently increasing its binding capacity [81]. In general, a pre-polymerization mixture is applied to the surface of a solid substrate to generate surface patterns with various feature sizes [82]. After placing the reaction mixture on a solid substrate, the solution can be compressed between a glass slide and a soft poly(dimethylsiloxane) mold. Applying the right pressure along with UV radiation initiates the formation of patterned polymeric films. After polymerization, the template molecule can be extracted from the polymer matrix, resulting in thin-layer MIP films with highly accessible and complementary cavities [83].

In 2015, Yang et al. [84] reported a pioneering MIP pattern matrix for the detection of atrazine, which was synthesized through the utilization of soft lithography and UV-initiated polymerization. In the course of this investigation, the successful patterns comprising poly(MAA-EGDMA) were obtained. The obtained MIP films not only exhibited a noteworthy sensitivity value of −9.98 Hz/μM, as determined by gravimetric quartz crystal microbalance (QCM) measurements, but also displayed remarkable selectivity over other herbicides. Consequently, a conclusion was drawn that the regulation of binding sites through the proposed soft lithographic technique could play a pivotal role in advancing the development of effective MIP-sensing platforms. A general overview of the proposed MIP synthesis using the soft lithography technique is presented in Fig. 2.2.

2.3.2.2 Surface Imprinting After Template Immobilization

Another imprinting technique involves immobilizing a template molecule on a solid support via chemical linkages. The template-fixed stamp is then exposed to functional monomers to initiate the polymerization process. Afterward, both the stamp and template units are extracted from the polymeric matrix, creating surface-imprinted cavities capable of selectively recognizing the target analyte [85]. This technique, introduced by Shi et al. [86], has proven useful for fixed protein stamping due to several crucial factors. These include the proximity of the imprinted cavities to the surface, mitigating issues related to steric hindrance, and preserving the conformation of the template molecule throughout the imprinting process. This allows protein molecules to readily approach and interact with the

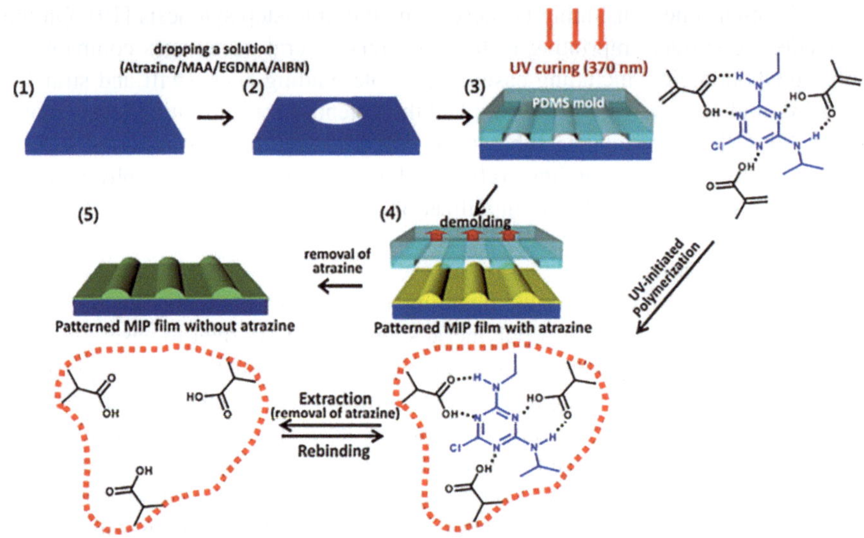

Fig. 2.2 The fabrication process of MIP film using UV-initiated soft lithography polymerization. (Reprinted from Yang et al. [84]. Copyright (2023), with permission from Elsevier (or applicable society copyright owner))

surface effectively. As a result, surface imprinting after template immobilization offers several significant advantages, including applicability for insoluble templates, facilitating complete and straightforward template removal after polymerization without the need for excessive washing cycles, and ensuring proper orientation of the binding sites for template rebinding [55]. In the evolution of surface imprinting after the template immobilization technique, new variants have emerged, including *micro-contact imprinting*, *semi-covalent imprinting*, and *solid-phase imprinting*, which will be discussed in the following section. It is worth noting that the majority of surface imprinting after template immobilization variants focus on protein imprinting.

Micro-contact Imprinting

Micro-contact imprinting is a rapid and efficient technique for simultaneous polymerization using multiple templates. This method was introduced in 2005 as a simple and effective way for template immobilization imprinting [87]. It relies on immobilizing templates using covalent linkages onto a glass cover slip. The procedure begins with exposing the glass cover slip to the functional monomer to create the pre-polymerization mixture. During this step, the monomer's functional groups orient themselves toward the template molecules, facilitating specific inorganic interactions. Separately, a mixture of a cross-linker and an initiator is prepared in an appropriate solvent, and a small drop of this mixture is applied to the glass slide. The imprinting process is initiated by bringing the cover slip in contact with the glass slide and exposing it to UV radiation. After polymerization, the cover slip can be removed, and any unreacted monomers or template molecules can be easily washed away.

Semi-covalent Imprinting

This technique actually combines three independent methods, including (1) inverse opal structuring, (2) surface imprinting, and (3) semi-covalent imprinting. This innovative approach to MIP synthesis is relatively new, but it shows promise due to its high selectivity. One of its main disadvantages is the need for a significant amount of template, typically a protein. The ensuing discussion covers the foundational concepts of this approach, commencing with the utilization of silica beads as a substrate for generating an opal-like structure. Functionalized template molecules, derivatized with functional monomers, are immobilized on this structure. Subsequently, the functionalized template molecules are cross-linked with an excess of a cross-linking monomer. The removal of the silica base by dissolution results in a porous material containing cavities that are selective for the targeted template [88, 89].

Solid-Phase Imprinting

The solid-phase imprinting technique involves immobilizing the template molecule using covalent bonds onto the surface of functionalized activated glass beads. These prepared glass beads are then brought into contact with the pre-polymerization mixture, initiating the polymerization reaction. In this type of imprinting technique, the result is not thin MIP films but rather the formation of nanoparticles. The formation of nanoparticles is favored by high dilution and short polymerization times, among other factors. After polymerization, the glass beads continue to play a significant role. Unreacted species can be eluted from the system by washing the glass beads under specific conditions. The obtained nanoparticles, which remain attached to the glass beads, can be considered as affinity mediums. Subsequently, the resulting nanoparticles can be eluted from the glass beads using different conditions, such as a higher temperature or a different solvent. One important drawback of this surface imprinting technique is its relatively small yield. However, it offers a much higher binding affinity compared to MIPs produced using other imprinting methods. To address the yield issue, researchers have explored the possibility of improving it by cycling the same technique multiple times [90, 91].

2.3.2.3 Grafting Polymerization

Due to the limitations of crushing and grinding materials obtained via the conventional bulk polymerization technique, which often result in the loss of the final material's targeted affinity, there was a need to develop a new method to overcome this significant drawback. The proposed grafting polymerization technique provides a solution to these challenges. Similar to the methods described in this section, grafting polymerization is carried out in the presence of a template molecule on solid substrates, which can be beads or particles. The polymerization reaction in this method leads to the formation of a thin MIP film on their surface [92]. One may wonder about the differences between template immobilization and grafting polymerization procedures. The key to understanding this procedure lies in its general approach. The grafting polymerization technique relies on the adsorption phenomenon of a template molecule, which is based on inorganic interactions with the functional groups in the polymer side chain. Another significant aspect of this approach is the sequence of the steps involved: first, grafting the polymer onto the surface and subsequently forming interactions, rather than an immediate covalent binding with the template molecules [55]. In contrast to the soft lithography technique, which forms MIPs on flat solid surfaces, the grafting polymerization method generates MIPs on the surfaces of numerous small particles. This leads to a substantial increase in the exposed surface area. Additionally, using solid particles offers an additional advantage: enabling precise control over the size, morphology, and surface area of the resulting MIPs [80]. Grafting is a versatile technique that can be successfully applied to both small and larger template molecules. Moreover, it provides strict control over the final polymer shape and morphology, addressing the issues associated with bulk polymers. As a result, this technique offers improved affinity interactions with templates, and the polymer matrix exhibits high analyte mobility, leading to faster mass transfer rates. This is particularly advantageous when imprinting larger templates, such as proteins, as it enhances their binding efficiency and reversibility. The enhanced mobility is a crucial benefit in the process [55]. However, the grafting technique does have some drawbacks. One important concern is the possibility of forming thick layers, leading to an uncontrolled increase in the chances of local bulk polymerization. To avoid this scenario, initiators should first be grafted onto silica beads, and, then, the obtained particles can be exposed to a mixture containing reactive monomers, cross-linkers, and template molecules [93].

The general procedure for preparing surface-grafted beads was initially documented by Tan and Tong [94]. Their report marked the beginning of investigations into the structure of the particles used in grafting polymerization. Although the reported beads were based on silica, it is important to note that these types of nanoparticles remain highly popular and extensively utilized in various applications. Silica's favorable properties, such as inertness, biocompatibility, and mechanical and chemical stability, make it an ideal substrate for biological applications. Additionally, the chemical structure of silica, which contains numerous hydroxyl functionalities, allows for easy polymerization on its surface [95–97]. Although silica remains one of the most common substrates for grafting MIPs, other particles have also been explored. Examples of alternative surfaces reported in the literature include polystyrene beads [98], carbon nanotube surfaces [99], polysaccharide beads [100], and even grapheme oxide surface [101].

2.3.3 Other Methods Using Different Polymerization Mechanisms

This section will shine a spotlight on other noteworthy polymerization techniques that have a significant influence on the advancement of MIPs. These techniques have proven their worth through their utility and documented commercial applications. The subsequent subsections will delve into two techniques: *electropolymerization* and *controlled/living radical polymerization (CRP)*. This exploration aims to provide a deeper insight into their principles and their crucial roles in MIP synthesis.

2.3.3.1 Electropolymerization

In the process of electropolymerization, a solution of functional monomers in a suitable solvent is directly deposited onto the surface of a conductive electrode, such as gold, platinum, or glassy carbon, in the presence of a template molecule [102]. Subsequently, this reaction mixture is exposed to the electrochemical energy of the electrode, leading to molecular imprinting on the electrode's surface [103]. The thickness of the resulting MIP is determined by the quantity of charge transferred during the electropolymerization process. Furthermore, the surface morphology can be controlled by selecting an appropriate solvent and supporting electrolyte [102]. After the formation of MIP films, the template can be removed, creating cavities complementary to the targeted template molecule. Moreover, the removal of the template generates a signal at the electrode, which can be measured using a suitable electrochemical analytical device [19]. A general overview of the electropolymerization technique is presented in Fig. 2.3.

The electropolymerization process offers significant advantages, such as shorter reaction times and improved reproducibility when compared to bulk polymerization. This technique can be conducted directly on the working electrode's surface, allowing the entire process to be observed in real time on a computer screen. Additionally, by managing the potential, the film's thickness and accuracy can be controlled, thus enhancing the reproducibility of the synthesized MIPs. However, it is essential to acknowledge certain drawbacks, including template leaching, potential issues with selectivity, and reduced effectiveness of MIPs toward specific targeted template molecules [102].

According to the current literature reports, specific monomers are more efficient in the electropolymerization process compared to others. The guiding principle is that the chosen monomer should readily polymerize at a low potential and demonstrate the ability to engage in non-covalent interactions with various template molecules. Pyrrole, phenol, scopoletin, and *o*-phenylenediamine are among the most commonly used monomers in the electropolymerization technique. The choice of monomer typically depends on the specific electrochemical analytical method being used. For instance, pyrrole is predominantly employed in direct voltammetric assays, including potentiometry, square wave voltammetry (SWV), cyclic voltammetry (CV), and others. Conversely, less conductive monomers like *o*-phenylenediamine are more frequently used in electrochemical impedance spectroscopy (EIS) assays [19, 103].

2.3.3.2 Controlled/Living Radical Polymerization (CRP)

Radical polymerization offers numerous advantages that can be highly significant for the commercialization process. This method can effectively prepare high-molecular-weight polymers using a wide range of vinyl monomers under mild reaction conditions and within a broad temperature range, spanning from −20 to 200 °C [104]. Moreover, many monomers can readily undergo radical copolymerization, leading to a variety of final copolymers whose properties depend on the ratios of the co-monomers. The important disadvantage is the lack of control. Free-radical polymerization relies on the presence of a high concentration of generated radicals, resulting in rapid chain propagation. Consequently, controlling premature chain termination becomes challenging, ultimately leading to the formation of polymer chains with varying lengths [19]. Furthermore, the lack of control over product formation results not only in uncontrolled degrees of polymerization but also in unregulated polydispersity, end functionalities, or even final chain compositions.

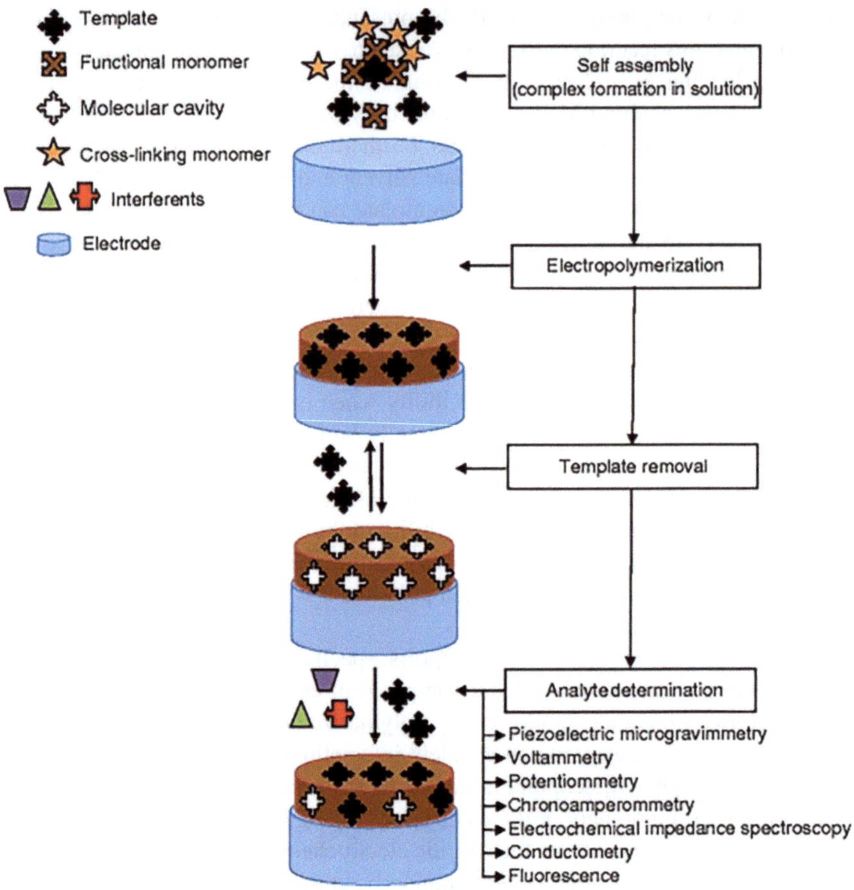

Fig. 2.3 A broad overview of the process of molecular imprinting using an electroactive functional monomer and the common signal transduction methods used in detection. (Reprinted from Sharma et al. [102]. Copyright (2023), with permission from Elsevier (or applicable society copyright owner))

The aim is to use radical polymerization to prepare novel, precisely defined polymers that have not been attainable through ionic living polymerization methods [105]. Consequently, controlled/living radical polymerization (CRP) can be characterized as a method that facilitates the creation of well-defined polymers from a broad spectrum of monomers under simpler reaction conditions compared to those required by ionic processes [106]. For these reasons, CRP techniques have been widely introduced into the realm of molecular imprinting. Moreover, the remarkable versatility of CRPs has positioned CRP-based "grafting from" techniques as the most widely adopted methods for creating advanced functional polymers [107]. The fundamentals of CRP have paved the way for precise macromolecular synthesis, incorporating the principles of living polymerization. In this process, the impact of chain-breaking reactions is minimized, allowing for nearly instantaneous initiation and the simultaneous growth of all chains [108]. The CRP mechanism relies on the use of temporary capping agents known as the dormant species. Their purpose is to protect the radicals from premature termination steps. With these dormant species present in the reaction mixture, polymer chains are allowed to propagate at a consistent rate until their scheduled termination. Consequently, this process enables the production of polymers with uniform sizes and an extremely narrow size distribution range. An additional advantage of this technique is that it allows for the introduction of specific functional groups to the ends of polymer chains, which can subsequently be utilized for further modifications of MIP structures [19].

Currently, the most extensively researched CRPs encompass infester-induced "living" radical polymerization [109], nitroxide-mediated polymerization (NMP) [110, 111], atom transfer radical polymerization (ATRP) [112, 113], and reversible addition–fragmentation chain transfer (RAFT) polymerization [114–116].

Table 2.2 provides a summary of the molecular imprinting methods outlined within this chapter.

Table 2.2 A summary of the preparation procedures: methods of molecular imprinting

Preparation procedures			
Method	Technique	Sub-technique	Description
One-pot polymerization	Bulk polymerization	In this technique, a mixture of monomers, cross-linking agents, initiators, templates, and a solvent undergoes free-radical polymerization [51]	
	Precipitation polymerization	This technique involves initially combining all reactants, starting with preparing a heated monomer emulsion, followed by introducing an initiator solution to initiate polymerization [60]	
	Emulsion polymerization	Batch emulsion polymerization	This sub-technique involves combining all reactants initially in a heated monomer emulsion, followed by introducing an initiator solution [68]
		Semi-batch emulsion polymerization	This is a sub-technique that involves initially introducing reactants into the reactor, followed by adding the remaining components in a controlled manner at a later stage [68]
		Continuous emulsion polymerization	This sub-technique relies on continuously supplying the constituent components of the formulation, from which the product is continuously extracted [68]
	Suspension polymerization	This is a technique based on vigorous stirring that involves dispersing a water-insoluble monomer containing an initiator in the aqueous phase [70]	
	Multistep swelling polymerization	Polymerization proceeds incrementally, with swelling increasing at each stage until uniform MIP particles of the desired size are attained [73]	

(continued)

Table 2.2 (continued)

Preparation procedures			
Method	Technique	Sub-technique	Description
Surface imprinting	Soft lithography	This is a stamping technique that involves the fabrication of MIP films on solid substrates using a soft polymeric stamp [81]	
	Surface imprinting after template immobilization	Micro-contact imprinting	This is a sub-technique that involves the immobilization of templates by forming covalent linkages onto a glass cover slip [87]
		Semi-covalent Imprinting	This sub-technique integrates three distinct methods: 1. Inverse opal structuring 2. Surface imprinting 3. Semi-covalent imprinting [88]
		Solid-phase imprinting	This sub-technique involves immobilizing the template molecule using covalent bonds onto the surface of functionalized, activated glass beads [90]
	Grafting polymerization	This technique relies on the adsorption phenomenon of a template molecule, which is based on inorganic interactions with the functional groups in the polymer side chain [92]	
Others	Electropolymerization	This technique involves the direct deposition of a solution containing functional monomers in a suitable solvent onto the surface of a conductive electrode in the presence of a template molecule [102]	
	Controlled/living radical polymerization	This technique is defined by its dependence on a high concentration of generated radicals, leading to swift chain propagation in the polymerization process [104]	

2.4 Physical Analysis of Molecularly Imprinted Polymers

2.4.1 Morphology

The physical properties of MIPs, such as size, shape, and distribution, play a crucial role in determining their performance in various applications. Scanning electron microscopy (SEM) serves as a powerful tool for investigating the morphology of MIPs. With its high-resolution imaging capabilities, SEM provides detailed insights into the particle size, shape, and surface roughness of MIPs. These morphological features observed using SEM directly influence the accessibility of the imprinted cavities and the overall surface area of the MIPs. For example, smaller particles generally exhibit a higher surface area per unit volume, which can result in higher binding capacities. Recently, there has been a notable trend toward reducing the size of MIP particles, as this is associated with a greater number of binding sites per gram of material [117–120].

In a similar vein, the shape of MIP particles can influence their packing density in chromatographic columns, which, in turn, impacts their performance in separation processes [121]. A study conducted by González et al. [122] highlights several factors that affect the final morphology of MIPs and non-imprinted polymers (NIPs). Scanning electron microscopy (SEM) imaging revealed that MIPs exhibited a more intricate, porous architecture, whereas NIPs presented a smoother, more homogeneous appearance. The research also emphasized morphological differences among polymers synthesized using various functional monomers and solvents (porogens) during the synthesis process. Another crucial and somewhat evident factor affecting the morphology of the produced materials is the type of polymerization employed, its conditions, and the choice of cross-linking agent and solvent (porogen) used [123–125].

On the other hand, transmission electron microscopy (TEM) provides highly detailed images of the internal structure of materials at a much higher resolution than does SEM. One crucial capability of TEM is its ability to reveal the core–shell structure of the synthesized MIPs [126–128].

Atomic force microscopy (AFM) is a valuable technique that provides nanoscale resolution of the surface of MIPs. AFM works by scanning a sharp probe over the polymer surface, creating a detailed topographical map that reveals surface features, irregularities, and roughness. This information is crucial because even subtle variations at the nanoscale level can significantly impact the binding properties and interactions between target molecules and MIPs.

When MIPs are prepared in the form of a thin layer, membrane, or film, AFM becomes a feasible method for imaging. This technique offers a relatively quick way to discern topography and pore structures. Analyzing these images can provide an in-depth insight into the surface's porous nature, enabling a quantitative assessment of pore dimensions. In a specific study [129], AFM cantilevers with covalently bonded bovine cytochrome *c* molecules were employed to directly assess the interaction force between the protein and the synthetic receptor sites created through molecular imprinting. Scanning the polymer surfaces with these particular AFM cantilevers revealed specific binding interactions, with force magnitudes ranging from 85 to 95 pN. In another study [130], MIPs were constructed on graphene oxide (GO) nanosheets for desulfurization purposes, using dibenzothiophene as the template. The AFM technique was used to investigate the thickness of both the GO sheets and the MIPs/GO composite. Cross-sectional analysis revealed that the GO sheet thickness was approximately 3 nm, indicating double-layered nanosheets. Meanwhile, AFM imaging of the MIPs/GO showed that the polymers grafted onto the GO surface had an average thickness of about 22 nm, consistent with the data obtained from TEM imaging. Furthermore, numerous studies have utilized AFM to characterize the topography of electrochemical sensors based on MIPs [131–134] and surface plasmon resonance (SPR) sensors [135–138].

2.4.2 Porosity

The porous structure of molecularly imprinted polymers (MIPs) is a defining feature that directly impacts their functional performance. Higher porosity typically translates to an increased surface area, providing more binding sites for the target molecule. Various techniques are employed to examine porosity in detail.

The BET (Brunauer–Emmett–Teller) method is rooted in the physical principle of gas molecules adsorbing onto a solid in multiple layers. One of the primary outputs of the BET method is the determination of the specific surface area of a material. This quantifies the total surface area available for adsorption or chemical reactions per unit mass or volume of the material. In contrast, the BJH (Barrett, Joyner, and Halenda) method is based on the physical concept of capillary condensation occurring in mesopores. It focuses on determining the size distribution of these mesopores using gas adsorption and desorption data, in conjunction with the Kelvin equation to relate the pore size to the condensation pressure. Understanding both specific surface area and pore size distribution is vital for assessing the properties of MIPs. The latter, in particular, can indicate the accessibility and number of imprinted sites, along with the diffusion kinetics of the target molecules within the polymer matrix.

Mercury intrusion porosimetry is another important method for characterizing the porosity and pore structure of materials. By utilizing the non-wetting nature of mercury under pressure, MIPs allow for the determination of pore size distributions, total porosity, pore volume, and pore area. It can assess a broad range of pore sizes, from meso- to macropores. It is important to note that while mercury intrusion porosimetry provides valuable information about the porous structure of a material, it does have limitations. For instance, it may not be suitable for materials with tiny nanopores, and the invasive nature of mercury can alter some delicate structures. Additionally, it does not provide information about pore connectivity or differentiate between open and closed porosity [139].

Standard surface area measurements for imprinted polymers typically fall within the range of 100–400 m^2/g. The key factors influencing the surface area and distribution of pores include the proportion of the cross-linking monomer, the type and quantity of the porogen used, and the synthesis temperature [140].

An illustrative example highlighting the differences between MIPs and NIPs can be found in the study authored by Chrzanowska et al. [141]. In her work, the influence of various functional monomers on the total surface area of the resulting MIPs was examined, among other parameters. Nitrogen sorption techniques demonstrated that all produced polymers possessed intricate pore networks. Predominantly, the surface of these materials was characterized by mesopores, accounting for 73–77% of the overall pore structure. Mesopores generally offer optimal recognition capabilities and the necessary surface area for effective interactions with the template molecule. In the conducted tests, regardless of the monomer used, MIPs always exhibited a greater surface area than did their NIP counterparts. The reduced surface area of the NIPs likely signifies the absence of readily available selective binding locations within these polymers. In the study, the measured BET surface areas for the polymers ranged from 239.0 to 385.6 m^2/g.

Interesting results were obtained in the research work of Holland et al. [142]. As indicated by the conducted nitrogen sorption studies, those polymers with a higher cross-linker concentration (40-mmol EGDMA) during polymerization exhibited a higher surface area compared to those with a lesser amount of the cross-linking reagent (10-mmol EGDMA). Similarly, when comparing average pore volumes, the trend was consistent: polymers with a higher cross-linking degree had larger pore volumes than did their less cross-linked counterparts. The BET surface area was higher for the 40-mmol EGDMA polymers (ranging from 293.9 to 347.2 m^2/g) than for the 10-mmol EGDMA polymers (ranging from 67.1 to 93.1 m^2/g). The cumulative surface area of pores for the 40-mmol EGDMA polymers varied between 260.9 and 289.6 m^2/g, whereas for the 10-mmol EGDMA polymers, it ranged from 64.6 to 92.4 m^2/g.

More recently, alongside these traditional methods, ^{129}Xe (xenon) nuclear magnetic resonance (NMR) has emerged as a powerful tool for studying the porosity of MIPs. This method leverages the heightened sensitivity of xenon gas to its surrounding environment [143]. Using variable-temperature ^{129}Xe NMR, the research establishes how morphology varies based on the degree of cross-linking. ^{1}H time domain (TD)-NMR, on the other hand, helps determine polymer chain mobility. As the research indicated, templated pores in MIPs in their dry state are too minute to house xenon atoms. However, these atoms can probe deeper levels in the porous structure, thus enabling an in-depth study. Both the degree of cross-linking and the presence of a template influence polymer morphology. Variations in polymer composition lead to distinct differences in polymer microstructure and rigidity. This is established through TD-NMR and then correlated with polymer surface area and pore size distribution. The spectral region with higher confinement showcased broad signals with a temperature-dependent chemical shift, suggesting a robust interaction where xenon is possibly dissolved in a surface layer distinct from the main polymer matrix but still retains its structural rigidity. Xe NMR showcases its capability to analyze ordered systems, even in soft matter. This signifies that not only MIPs but also a range of systems such as polymeric foams and core–shell particles can be examined using a similar methodology.

2.5 Chemical Characterization

2.5.1 Nuclear Magnetic Resonance (NMR) Spectroscopy

Since MIPs are in solid form, they cannot be analyzed using standard polymer characterization techniques that require solutions, such as solution-oriented NMR methods. However, there are references that describe the use of solid-state magic angle spinning (MAS) NMR for characterizing cross-linked polymers [144–146]. Understanding how MIPs interact with target molecules is crucial for fully harnessing the capabilities of these materials. In particular, solid-state MAS NMR spectroscopy can provide insights into the structure of solid samples. However, due to challenges like low sensitivity and broad signals, it is not always effective in discerning the less abundant binding sites in imprinted polymers. To date, it has been successful in cases where templates are bound to the MIP through strong non-covalent or reversible covalent ties. In such situations, the concentration of binding sites is high enough to produce distinguishable signals. Still, solid-state NMR has not yet provided detailed structural information about the binding sites [145].

The traditional ^{1}H NMR technique can be employed before the polymerization process to determine the mechanism of molecular imprinting and to identify the interactions between the template and the monomer. An example of this can be found in the following study [147]. ^{1}H NMR spectroscopy was used to examine the extent of complexation between the functional monomer and the template in non-covalent MIP pre-polymerization mixtures. The primary purpose of using the NMR technique was to observe the stabilities of these complexes and correlate them with the recognition capacity of the resulting polymer toward the template. NMR titration studies were conducted using analogues for the functional monomers, specifically acetic acid for MAA and pyridine for 4-vinylpyridine (4-VP), to determine the apparent dissociation constant for atrazine complexation in chloroform. The observations from the NMR studies provided insights into the binding affinities of the functional monomers, guiding the subsequent synthesis and evaluation of the MIPs. The results from the NMR study were consistent with the recognition characteristics of the synthesized polymers, showing that MAA MIPs had better recognition and binding capacities compared to those prepared with 4-VP. In another example [148], ^{1}H NMR analysis was conducted to investigate the interactions between the template molecule and MAA under conditions mimicking those before polymerization. By monitoring changes in the line widths for the exchangeable proton of the amino group, the study aimed to determine the formation of complexes at different concentrations of MAA. The observed peak changes in the NMR spectra provided insights into the existence of two predominant complexes. The first peak was attributed to the formation of a 1:1 complex, whereas the second indicated a 1:2 complex.

2.5.2 Ultraviolet–Visible (UV–Vis) Spectroscopy

UV–vis spectroscopy provides valuable insights into electronic transitions within molecules. In the context of MIPs, UV–vis spectroscopy can be utilized before polymerization to understand the interactions between the template and functional monomers. Shifts in absorbance peaks or changes in their intensities can indicate complex formations or alterations in the electronic environment of the molecules. This is particularly useful for verifying the selectivity of the functional monomers toward the template, ensuring the optimal recognition sites in the final polymer. Complexation mechanisms of interactions between the template and functional monomers frequently lead to a hypsochromic shift, where the peak of maximum optical absorbance moves to shorter wavelengths. To observe this complexation, various ratios of the template and the monomer(s) are diluted in a chosen solvent. Subsequently, the UV absorbance of the mixture is measured and compared against the spectra of the uncombined template and monomers.

In addition to the applications mentioned, UV–vis spectroscopy is valuable when we need to confirm that the entire template has been removed during the purification of MIPs. The UV–vis method can detect the presence of the template molecule by measuring its characteristic absorption peaks [118]. An interesting application of the UV–vis method was proposed in the study by Lanza and Sellergren [149]. Synthesized polymers were processed directly in situ and subsequently assessed for their rebinding capabilities via high-performance liquid chromatography (HPLC) or UV absorbance techniques. A set of six monomers was tested against the terbuthylazine template, combined with ethylene glycol dimethacrylate and azobisisobutyronitrile in a dichloromethane solution. During the testing phase, any polymers that did not release the template were set aside, leaving only the effective ones for further rebinding studies. In another study conducted by Jie et al. [150], an investigation into how acrylamide-based MIPs could differentiate between amino acid hydantoins' enantiomers in aqueous solutions was highlighted. They utilized (5*R*)-5-benzylhydantoin as a guiding template. Their results demonstrated that these MIPs not only had the ability to recognize peptides and proteins but also exhibited enantiomeric selectivity. In a similar approach introduced in this study [151], researchers employed MAA and 2-diethylaminoethyl methacrylate (DEAEMA) as the key monomers to produce MIPs. Through UV–vis titrations, they observed a stronger bond between shikonin and the basic functional monomers compared to the acidic ones.

2.5.3 Fourier Transform Infrared (FT-IR) Spectroscopy

FT-IR spectroscopy is a valuable tool for assessing the functional groups present in a sample. Before polymerization, FT-IR can be employed to confirm the presence of desired monomers and the absence of contaminants that could interfere with the imprinting process. After polymerization, FT-IR spectra can offer insights into the successful incorporation of the monomers into the polymer matrix. Additionally, comparing the pre- and post-polymerization spectra can provide hints about any chemical changes or reactions that occurred during the process. Specifically, the disappearance or appearance of certain characteristic peaks can confirm the formation of new bonds or the breaking of existing ones. FT-IR is commonly used to confirm molecular imprinting, involving the comparison of the spectra of MIPs before removing the template molecules with the corresponding spectra of NIPs [152–157]. Another crucial post-polymerization application is verifying the successful removal of the template from the polymer matrix by looking for the absence of its characteristic IR peaks.

Duffy et al. [158] examined the relationship between thymine and diacryloyl-2,6-diaminopyridine moieties present within synthesized imprinted polymer. They measured the frequencies of OH and NH stretching vibrations to track template–monomer interactions, as these often shift to lower frequencies due to hydrogen bonding. The group compared NH stretching vibration frequencies before and after polymerization, demonstrating that the complex stabilities remained consistent throughout the process. FT-IR studies [159] were also employed to understand the behavior of theophylline molecularly imprinted methacrylic acid–ethylene glycol dimethacrylate co-polymers at elevated temperatures. These polymers demonstrated an ability to maintain their affinity for the template up to 150 °C. Beyond this temperature, a decline in template affinity was observed, but notable affinity persisted up to 200 °C when compared to a reference non-imprinted polymer. The primary reduction in activity is believed to be due to the polymer undergoing structural changes, leading to temperature-induced decarboxylation. This hypothesis is supported not only by FT-IR studies but also by observed weight variations and elemental assessments.

2.6 Thermal Stability

Thermal stability is a critical parameter when assessing the robustness and potential applications of MIPs. MIPs find utility across diverse fields, often exposed to varying environmental conditions, with temperature playing a significant role. Therefore, comprehending how MIPs react to or degrade under elevated temperatures is essential to ensure their efficacy and longevity in real-world applications. Thermogravimetric analysis (TGA) is one of the primary techniques used to assess the thermal stability of polymers, including MIPs. In TGA, the weight loss of the sample is monitored as a function of increasing temperature or time. Information from TGA studies also aids in determining whether residues of the solvent used for synthesis, monomers, or cross-linking agents remain in the material. TGA can be performed under different gases, allowing the investigation of the effect of an oxidizing atmosphere on material stability. Differential scanning calorimetry (DSC) measures the heat flow associated with phase transitions in the polymer, such as melting or crystallization. For MIPs, DSC can provide insights into the glass transition temperature (T_g), melting temperature (T_m), and any exothermic or endothermic processes that may occur due to the presence of the template or other additives.

For example, consider the following research [160]. The authors synthesized MIPs using 2-vinylpyridine (2-VP) as the functional monomer, EGDMA as the cross-linker, 1,1′-azobis(cyclohexanecarbonitrile) as the initiator, a porogenic mixture of toluene/acetonitrile (9:1, v/v), and ketoprofen as the template. At 40 °C, these polymers experienced a roughly 4% decrease in mass, possibly due to residual methanol. These polymers began to thermally degrade around 290 °C, indicating the degradation of their primary structure. Interestingly, while the NIPs showed complete thermal decomposition at 425 °C, the MIPs exhibited only a 90% decrease. This deviation could result from structural modifications during the template extraction process.

2.7 Adsorption Property Characterization

2.7.1 Binding Parameters

To assess the effectiveness of MIPs, corresponding NIPs are typically prepared as references. This process helps determine the impact of imprinting and verify the specificity of the cavities formed in the MIP structure for the intended analyte. Both MIP and NIP particles are immersed in an analyte solution until equilibrium is reached and then isolated. The residual analyte concentration is assessed using techniques like UV–vis or HPLC. This enables the calculation of equilibrium binding capacity (affinity) (q_{eq}) or the distribution coefficient (K_d) using the following formulas:

$$q_{eq} = \frac{C_i - C_{eq}}{m} V$$

$$K_d = \frac{C_i - C_{eq}}{m C_{eq}} V$$

where C_i is the initial concentration, C_{eq} is concentration at equilibrium, m is mass of the polymer, and V is volume of the solution.

The imprinting factor (IF) provides a quantitative measure of the MIP's recognition ability, calculated as the ratio of the amount of analyte bound to the MIP to the amount of the same analyte bound to the corresponding NIP:

$$\text{IF} = \frac{q_{\text{MIP}}}{q_{\text{NIP}}}$$

where q_{MIP} is amount of the analyte bound to the MIP and q_{NIP} is amount of the analyte bound to the NIP.

Another crucial parameter to consider is selectivity, which measures the MIP's preference for the template molecule over other structurally similar compounds. Competitive binding assays are typically employed, where the MIP is exposed to a solution containing both the template and a competitor molecule. The amounts of each molecule bound to the MIP are then determined. A higher selectivity is indicated if the MIP predominantly binds to the template over the competitor. Selectivity (α) can be expressed either as the ratio of the binding capacity of the analyte to the competitor in the MIP or as the ratio of the imprinting factors of the analyte to the imprinting factor of the competitor:

$$\alpha = \frac{q_{\text{ANALYTE}}}{q_{\text{COMPETITOR}}}$$

$$\alpha = \frac{\text{IF}_{\text{ANALYTE}}}{\text{IF}_{\text{COMPETITOR}}}$$

where q_{ANALYTE} is the amount of the analyte bound to the MIP, $q_{\text{COMPETITOR}}$ is the amount of the competitor bound to the MIP, $\text{IF}_{\text{ANALYTE}}$ is the imprinting factor of the analyte, and $\text{IF}_{\text{COMPETITOR}}$ is the imprinting factor of the competitor.

2.7.2 Adsorption Capacity

Adsorption capacity refers to the maximum amount of template that the MIPs can bind per unit mass once all available binding sites are occupied. To determine the adsorption capacity, MIPs are exposed to increasing concentrations of the template molecule until no further increase in bound template per mass unit is observed. This capacity is crucial as it helps determine the amount of adsorbent required to almost completely extract the analyte from a specific solution. The relationship between a sorbent's adsorption capacity and the remaining template concentration in a specific model is represented by an adsorption isotherm. Assuming uniform affinity across all adsorption sites and no influence between neighboring sites, the Langmuir isotherm model outlines monolayer adsorption. The linear form of the Langmuir isotherm can be expressed as:

$$\frac{C_{\text{eq}}}{q_{\text{eq}}} = \frac{C_{\text{eq}}}{q_{\text{m}}} + \frac{1}{Kq_{\text{m}}}$$

where K is the binding equilibrium constant, C_{eq} is the adsorbate concentration at equilibrium, q_{m} is the maximum adsorption capacity, and q_{eq} is the adsorbate amount adsorbed at a particular C_{eq} concentration.

Another commonly used model in adsorption studies is the Freundlich isotherm. This model assumes a heterogeneous surface and posits that adsorption intensity varies with the energy of the adsorption sites. Unlike the Langmuir model, which describes monolayer adsorption, the Freundlich isotherm does not assume uniform affinity across all adsorption sites. The linear form of the isotherm can be represented by the equation:

$$\ln q_{\text{eq}} = \ln K_{\text{f}} + \frac{1}{n} \ln C_{\text{eq}}$$

where K_{f} is the Freundlich isotherm constant related to sorption capacity, C_{eq} is the adsorbate concentration at equilibrium, q_{eq} is the adsorbate amount adsorbed at a particular C_{eq} concentration, and $\frac{1}{n}$ is the heterogeneity factor.

Another model used to explain adsorption phenomena is the Sips isotherm. The Sips isotherm, also known as the Sips or Langmuir–Freundlich isotherm, is a hybrid model that combines features of both the Langmuir and Freundlich isotherms. It is particularly useful for describing adsorption processes on heterogeneous surfaces. The Sips isotherm bridges these two concepts and is characterized by its ability to describe both monolayer and multilayer adsorption on heterogeneous surfaces. It has an adjustable parameter that allows it to transition from Langmuir-type behavior at low concentrations to Freundlich-type behavior at higher concentrations. Mathematically, the Sips isotherm can be represented as:

$$q_{\text{eq}} = \frac{q_{\text{m}} K_{\text{s}} C_{\text{eq}}^{1/n}}{1 + K_{\text{s}} C_{\text{eq}}^{1/n}}$$

where K_s is the Sips constant related to the energy of adsorption, C_{eq} is the adsorbate concentration at equilibrium, q_m is the maximum adsorption capacity, q_{eq} is the adsorbate amount adsorbed at a particular C_{eq} concentration, and $\frac{1}{n}$ is the Sips heterogeneity parameter.

After fitting the experimental data to the isotherm models, the R^2 coefficient indicates the data's alignment with the applied model. Consequently, it aids in identifying the nature of the adsorption phenomenon.

In a study conducted by Sadeghi and Jahani [156], MIPs were fabricated using 4-VP with florfenicol as the template. The resultant MIP exhibited a superior adsorption capacity compared to its non-imprinted counterpart. Specifically, the peak static adsorption capacities for florfenicol on the MIP and NIP were noted to be 4.32 and 2.88 mg/g, respectively. The experimental q_m value closely aligned with the theoretical value derived from the Langmuir adsorption model. This suggests that the Langmuir model aptly characterizes the adsorption pattern of florfenicol on the imprinted polymer.

2.7.3 The Thermodynamics of Adsorption

Understanding the thermodynamics of adsorption is crucial as it provides insights into the nature of interactions and the spontaneity of the binding process. Thermodynamic parameters, including changes in enthalpy (ΔH), entropy (ΔS), and Gibbs free energy (ΔG), offer a comprehensive view of the forces driving the adsorption process. ΔH represents the heat evolved or absorbed during the adsorption process. It indicates whether the process is exothermic (negative ΔH) or endothermic (positive ΔH). The following equation can be utilized to determine the values of ΔH and ΔS:

$$\ln K_d = \frac{\Delta S}{R} - \frac{\Delta H}{RT}$$

where K_d is the distribution coefficient, T is the absolute temperature of the adsorption process, R is the universal gas constant, ΔH is the change in enthalpy, and ΔS is the change in entropy.

The following equation allows for the computation of ΔG:

$$\Delta G = -RT \ln K_d$$

where K_d is the distribution coefficient, T is the absolute temperature of the adsorption process, R is the universal gas constant, ΔH is the change in enthalpy, ΔS is the change in entropy, and ΔG is the Gibbs free energy.

2.7.4 Adsorption Kinetics

Adsorption kinetics delves into the dynamics of interactions between a target molecule and its corresponding binding site within the MIP matrix. Understanding these kinetics is vital as it governs the pace of molecular recognition, which, in turn, dictates the efficiency and responsiveness of MIPs in practical applications. To conduct adsorption kinetics measurements, samples are prepared and collected at specific time intervals. Subsequently, the template concentrations in each sample are determined using appropriate analytical methods. This approach allows for the monitoring of the progression of the adsorption process over time, thus providing insights into the rate and efficiency of the adsorbate binding to the adsorbent.

To comprehend the kinetics of the adsorption process, the observed data should be analyzed using various kinetic models. The most commonly used models for characterizing adsorption kinetics are the pseudo-first-order kinetic model and the pseudo-second-order kinetic model. The pseudo-first-order model, proposed by Sten Lagergren [161], can be represented by the following equation:

$$\ln\left(q_{\text{eq}} - q_t\right) = \ln q_{\text{eq}} - \frac{k_1}{2.303} t$$

where q_{eq} is the amount of adsorbate adsorbed at equilibrium, q_t is the amount of template bound to the polymer at time t, and k_1 is the pseudo-first-order rate constant.

This model assumes that the rate of adsorption is directly proportional to the number of unoccupied sites available on the adsorbent. It is worth noting that while this model can fit experimental data for many adsorption systems, it is often more applicable to the initial stages of the adsorption process when the concentration of unoccupied sites is the highest.

The pseudo-second-order kinetic model, often associated with the chemisorption processes, is represented by the following equation:

$$\frac{t}{q_t} = \frac{1}{k_2 q_{\text{eq}}^2} + \frac{1}{q_{\text{eq}}} t$$

where q_{eq} is the amount of adsorbate adsorbed at equilibrium, q_t is the amount of template bound to the polymer at time t, and k_2 is the pseudo-second-order rate constant.

This model assumes that the rate of occupation of adsorption sites is proportional to the square of the number of unoccupied sites and has been successfully applied to many adsorption systems to describe the kinetics.

The study conducted by Yang et al. [162] aimed to monitor the adsorption of 2,4-dichlorophenoxyacetic acid (2,4-D) on the MIP coating using in situ fiber optic sensing. The effects of contact time, temperature (298–318 K), and initial concentration (20–150 mg/L) of 2,4-D on the adsorption properties of the coating were investigated. The changes in thermodynamic parameters, ΔH, ΔS, and ΔG, were calculated to understand the nature of the adsorption process. The results indicated that the adsorption of 2,4-D at a low concentration by the coating is an exothermic and spontaneous process. The study found that the pseudo-second-order model was most suitable for describing the adsorption kinetics, suggesting that the adsorption process was likely chemisorption.

Table 2.3 outlines the characterization techniques that are applicable for investigating the structure and properties of MIPs.

Table 2.3 A summary of the characterization methods of MIPs

	Physical parameters		
	Morphology	**Porosity**	
Analysis techniques	SEM, TEM, AFM	BJH model, BET isotherm, mercury intrusion, ^{129}Xe NMR	
Application	Size, shape, and distribution of the MIP particles [122, 126, 129]	Determination of the specific surface area of MIP particles Determination of the pore size and their distribution [139–143]	
	Chemical characterization		
Spectroscopic techniques	Nuclear magnetic resonance (NMR) spectroscopy	Ultraviolet–visible (UV–vis) spectroscopy	Fourier transform infrared (FT-IR) spectroscopy
Application	Chemical composition (MAS-NMR for solid samples, limited applicability) [144–146] (^{1}H NMR for pre-polymerization mixtures) [147, 148]	Interactions between the template and functional monomers in pre-polymerization mixtures Detection of the template molecule during the purification stage of MIP synthesis [149–151]	Confirmation of the presence of desired monomers and the absence of contaminants in pre-polymerization mixtures Analysis of chemical changes or reactions that occurred during the process of polymerization Confirmation of the molecular imprinting by comparison of the spectra of MIPs before removing the template molecules with the corresponding spectra of NIPs [152–159]
	Thermal stability		
Analysis techniques	TGA	DSC	
Application	Analysis of degradation under elevated temperatures (efficacy and longevity in real-world applications) Determination of the residues of the solvent, monomers, or cross-linking agents used for synthesis [160]	Determining exothermic or endothermic processes that may occur due to the presence of the template or other additives	
	Adsorption property characterization		

(continued)

Table 2.3 (continued)

Parameters	Binding parameters: • Equilibrium binding capacity (affinity) • Distribution coefficient • Imprinting factor • Selectivity	Adsorption capacity: The maximum amount of template that the MIPs can bind per unit mass The most popular models for calculating adsorption capacity include: • Langmuir isotherm • Freundlich isotherm • Sips isotherm	Adsorption thermodynamics: Changes in: • Enthalpy (ΔH) • Entropy (ΔS) • Gibbs free energy (ΔG)	Adsorption kinetics: Investigation of the dynamics of the interactions between a target molecule and its corresponding binding site within the MIP matrix The most commonly used models for characterizing adsorption kinetics are as follows: • Pseudo-first-order model • Pseud-second-order model

2.8 Conclusions and Future Perspectives

The synthesis and application of MIPs present a dynamic and challenging field, one that is crucial for the advancement of analytical chemistry and related scientific domains. This chapter has thoroughly explored the various methodologies and materials involved in MIP synthesis, emphasizing the importance of careful selection and optimization of synthesis parameters. Key challenges and the emerging trend of "greenification" in MIP synthesis are highlighted below.

The selection of synthesis methods significantly impacts the morphology and functional properties of MIPs. Traditional methods like mechanical crushing to achieve a specific surface area are limited and can be inefficient. Modern approaches focus on direct synthesis of micro- or nanoparticles, thus eliminating the need for size reduction post-synthesis. The choice of monomers is pivotal, as it determines the bonding efficiency with templates and influences the overall properties of MIPs. Uniformity in binding sites, especially for biomolecules like proteins, and ensuring water compatibility remain significant challenges. Customizing MIPs for specific applications, such as targeted drug delivery, requires optimizing not only the synthesis components but also surface modifications to achieve the desired in vivo performance.

In line with global sustainability efforts, green chemistry principles are increasingly being integrated into MIP synthesis. This shift involves the use of benign chemicals, biodegradable or natural polymers, and energy-efficient synthesis methods. Techniques such as ultrasound-assisted synthesis and supercritical CO_2 processes are being explored for their environmental benefits and efficiency. These green approaches aim to minimize the ecological footprint of MIP production while maintaining their effectiveness in various applications. Various review papers provide valuable insights into these emerging trends [163–165].

The field of MIPs is rapidly evolving, with ongoing research focusing on enhancing selectivity, stability, and applicability in diverse fields such as biosensing, environmental monitoring, and drug delivery. The integration of nanotechnology and smart materials into MIP synthesis offers exciting possibilities for developing highly sensitive and selective sensors. Furthermore, the adoption of green synthesis methods is expected to grow, aligning MIP production with environmental sustainability goals.

In summary, while the synthesis of MIPs involves complex decision-making regarding functional monomers, cross-linkers, and solvents, the continuous evolution of these materials promises vast potential. The comprehensive characterization of MIPs, including their physical, thermal, chemical, and functional properties, is essential for ensuring their efficacy in the intended applications. As we advance, the fusion of innovative synthesis techniques, green chemistry, and detailed characterization will continue to drive the field of MIPs toward new horizons in analytical chemistry and beyond.

References

1. Lowdon JW et al (2020) MIPs for commercial application in low-cost sensors and assays—an overview of the current status quo. Sens Actuators B Chem 325:128973
2. Haupt K, Mosbach K (2000) Molecularly imprinted polymers and their use in biomimetic sensors. Chem Rev 100(7):2495–2504
3. Basak S, Venkatram R, Singhal RS (2022) Recent advances in the application of molecularly imprinted polymers (MIPs) in food analysis. Food Control 139:109074
4. Selvolini G, Marrazza G (2017) MIP-based sensors: promising new tools for cancer biomarker determination. Sensors 17(4):718
5. Li G et al (2022) Ultrasensitive, label-free voltammetric determination of norfloxacin based on molecularly imprinted polymers and au nanoparticle-functionalized black phosphorus nanosheet nanocomposite. J Hazard Mater 436:129107
6. Arabi M et al (2022) Chiral molecular imprinting-based SERS detection strategy for absolute enantiomeric discrimination. Nat Commun 13(1):5757
7. Su Y et al (2022) Surface ion-imprinted brewer's spent grain with low template loading for selective uranyl ions adsorption from simulated wastewater. J Hazard Mater 440:129682
8. Caro E et al (2006) Application of molecularly imprinted polymers to solid-phase extraction of compounds from environmental and biological samples. TrAC Trends Anal Chem 25(2):143–154
9. Ramström O, Ye L, Gustavsson P-E (1998) Chiral recognition by molecularly imprinted polymers in aqueous media. Chromatographia 48:197–202
10. Sun Y, Gu Y, Zhang P (2022) Adsorption properties and recognition mechanisms of a novel surface imprinted polymer for selective removal of Cu(II)-citrate complexes. J Hazard Mater 424:127735
11. Martín-Esteban A (2013) Molecularly-imprinted polymers as a versatile, highly selective tool in sample preparation. TrAC Trends Anal Chem 45:169–181
12. Lai C et al (2016) Synthesis of surface molecular imprinted TiO2/graphene photocatalyst and its highly efficient photocatalytic degradation of target pollutant under visible light irradiation. Appl Surf Sci 390:368–376
13. Lusina A, Cegłowski M (2022) Molecularly imprinted polymers as state-of-the-art drug carriers in hydrogel transdermal drug delivery applications. Polymers 14(3):640
14. Chen L et al (2016) Molecular imprinting: perspectives and applications. Chem Soc Rev 45(8):2137–2211
15. Bogdanowicz N et al (2023) Rapid quantification of 2, 4-dichlorophenol in river water samples using molecularly imprinted polymers coupled to ambient plasma mass spectrometry. J Hazard Mater 450:131068
16. Vasapollo G et al (2011) Molecularly imprinted polymers: present and future prospective. Int J Mol Sci 12(9):5908–5945
17. Lok C, Son R (2009) Application of molecularly imprinted polymers in food sample analysis—a perspective. Int Food Res J 16(2):127–140
18. Wulff G, Knorr K (2001) Stoichiometric noncovalent interaction in molecular imprinting. Bioseparation 10:257–276
19. Mostafa AM et al (2021) Review on molecularly imprinted polymers with a focus on their application to the analysis of protein biomarkers. TrAC Trends Anal Chem 144:116431
20. Advincula RC (2011) Engineering molecularly imprinted polymer (MIP) materials: developments and challenges for sensing and separation technologies. Korean J Chem Eng 28:1313–1321
21. Chen L, Xu S, Li J (2011) Recent advances in molecular imprinting technology: current status, challenges and highlighted applications. Chem Soc Rev 40(5):2922–2942
22. Wulff G (1995) Molecular imprinting in cross-linked materials with the aid of molecular templates—a way towards artificial antibodies. Angew Chem Int Ed Engl 34(17):1812–1832
23. Mosbach K (1994) Molecular imprinting. Trends Biochem Sci 19(1):9–14

24. Yan H, Row KH (2006) Characteristic and synthetic approach of molecularly imprinted polymer. Int J Mol Sci 7(5):155–178
25. Whitcombe MJ et al (1995) A new method for the introduction of recognition site functionality into polymers prepared by molecular imprinting: synthesis and characterization of polymeric receptors for cholesterol. J Am Chem Soc 117(27):7105–7111
26. Liang Y et al (2017) High sensitive and selective graphene oxide/molecularly imprinted polymer electrochemical sensor for 2, 4-dichlorophenol in water. Sens Actuators B Chem 240:1330–1335
27. Peng S et al (2022) Technology for rapid detection of cyromazine residues in fruits and vegetables: molecularly imprinted electrochemical sensors. Biosensors 12(6):414
28. Marć M et al (2018) Preparation and characterization of dummy-template molecularly imprinted polymers as potential sorbents for the recognition of selected polybrominated diphenyl ethers. Anal Chim Acta 1030:77–95
29. Huang Y et al (2017) Rapid and sensitive detection of trace malachite green and its metabolite in aquatic products using molecularly imprinted polymer-coated wooden-tip electrospray ionization mass spectrometry. RSC Adv 7(82):52091–52100
30. Guć M, Schroeder G (2019) Application of molecularly imprinted polymers (MIP) and magnetic molecularly imprinted polymers (mag-MIP) to selective analysis of quercetin in flowing atmospheric-pressure afterglow mass spectrometry (FAPA-MS) and in electrospray ionization mass spectrometry (ESI-MS). Molecules 24(13):2364
31. Motia S et al (2019) Electrochemical sensor based on molecularly imprinted polymer for sensitive triclosan detection in wastewater and mineral water. Sci Total Environ 664:647–658
32. Sawetwong P et al (2021) Enhanced selectivity and sensitivity for colorimetric determination of glyphosate using Mn–ZnS quantum dot embedded molecularly imprinted polymers combined with a 3D-microfluidic paper-based analytical device. Talanta 225:122077
33. Pawlaczyk M, Guć M, Schroeder G (2021) Adsorption and selectivity studies of direct and magnetite-cored molecularly imprinted polymers (MIPs and magMIPs) towards chosen chalcones investigated with various analytical methods. RSC Adv 11(41):25334–25347
34. Xu J et al (2021) Simultaneous self-assembly of molecularly imprinted magnetic nanoparticles to construct a magnetically responsive photonic crystals sensor for bisphenol A. Sens Actuators B Chem 338:129858
35. da Silva ATM et al (2021) Terephthalic acid-based magnetic molecularly imprinted polymer for enantioselective capillary electrophoresis determination of atenolol in human plasma. Sep Purif Technol 261:118257
36. Karim K, Lamaoui A, Amine A (2022) Acetazolamide smartphone-based detection via its competition with sulfamethoxazole on molecularly imprinted polymer: a proof-of-concept. J Pharm Biomed Anal 219:114954
37. Pesavento M et al (2021) SPR-optical fiber-molecularly imprinted polymer sensor for the detection of furfural in wine. Biosensors 11(3):72
38. Cegłowski M et al (2021) Molecularly imprinted polymers with enhanced selectivity based on 4-(aminomethyl) pyridine-functionalized poly (2-oxazoline) s for detecting hazardous herbicide contaminants. Chem Mater 34(1):84–96
39. Rizvi AS et al (2020) Development of molecularly imprinted 2D photonic crystal hydrogel sensor for detection of L-kynurenine in human serum. Talanta 208:120403
40. Wang H et al (2020) Oriented boronate affinity–imprinted inverse opal hydrogel for glycoprotein assay via colorimetry. Microchim Acta 187:1–9
41. Takeuchi T, Mukawa T, Asanuma H (2003) Molecular imprinting. From fundamentals to applications. Wiley-VCH, Weinheim
42. Lusina A, Cegłowski M (2023) Determination of MCPA in environmental samples using an analytical on-line system based on imprinted poly [2-(3-butenyl)-2-oxazoline] and ambient ionization mass spectrometry. Sens Actuat B Chem 403:135140
43. Cegłowski M, Jerca VV, Jerca FA, Hoogenboom R (2020) Reduction-responsive molecularly imprinted poly (2-isopropenyl-2-oxazoline) for controlled release of anticancer agents. Pharmaceutics 12:506

44. Cegłowski M, Schroeder G, Hoogenboom R (2020) Porous poly (2-oxazoline)-based polymers for removal and quantification of phenolic compounds. Chem Mater 32:6425–6436
45. Cegłowski M, Otłowski T, Gierczyk B, Smeets S, Lusina A, Hoogenboom R (2023) Explosives removal and quantification using porous adsorbents based on poly (2-oxazoline) s with various degree of functionalization. Chemosphere 340:139807
46. Liu B, Zhu H, Feng R, Wang M, Hu P, Pan J, Niu X (2022) Facile molecular imprinting on magnetic nanozyme surface for highly selective colorimetric detection of tetracycline. Sens Actuators B Chem 370:132451
47. Wu Z, He D, Cui B, Jin Z (2019) Ultrasensitive detection of microcystin-LR with gold immunochromatographic assay assisted by a molecular imprinting technique. Food Chem 283:517–521
48. Tavares LS, Carvalho TC, Romão W, Vaz BG, Chaves AR (2017) Paper spray tandem mass spectrometry based on molecularly imprinted polymer substrate for cocaine analysis in oral fluid. J Am Soc Mass Spectrom 29:566–572
49. Díaz-Liñán M, García-Valverde M, Lucena R, Cárdenas S, López-Lorente A (2021) Dual-template molecularly imprinted paper for the determination of drugs of abuse in saliva samples by direct infusion mass spectrometry. Microchem J 160:105686
50. Lieberzeit PA et al (2005) Imprinting as a versatile platform for sensitive materials–nanopatterning of the polymer bulk and surfaces. Sens Actuators B Chem 111:259–263
51. Włoch M, Datta J (2019) Synthesis and polymerisation techniques of molecularly imprinted polymers. In: Comprehensive analytical chemistry. Elsevier, pp 17–40
52. Lieberzeit PA et al (2008) Polymers imprinted with PAH mixtures—comparing fluorescence and QCM sensors. Anal Bioanal Chem 392:1405–1410
53. Lieberzeit PA et al (2007) Molecularly imprinted sol–gel nanoparticles for mass-sensitive engine oil degradation sensing. Anal Bioanal Chem 389:441–446
54. Rao TP, Kala R, Daniel S (2006) Metal ion-imprinted polymers—novel materials for selective recognition of inorganics. Anal Chim Acta 578(2):105–116
55. Mujahid A, Iqbal N, Afzal A (2013) Bioimprinting strategies: from soft lithography to biomimetic sensors and beyond. Biotechnol Adv 31(8):1435–1447
56. Yoshida M et al (2000) Surface imprinted polymers recognizing amino acid chirality. J Appl Polym Sci 78(4):695–703
57. Quinn T (2017) Molecularly imprinted polymers (MIPs). Nova Science Publishers, Incorporated
58. Ertürk G, Mattiasson B (2017) Molecular imprinting techniques used for the preparation of biosensors. Sensors 17(2):288
59. Goh EC, Stöver HD (2002) Cross-linked poly (methacrylic acid-co-poly (ethylene oxide) methyl ether methacrylate) microspheres and microgels prepared by precipitation polymerization: a morphology study. Macromolecules 35(27):9983–9989
60. Ye L, Cormack PA, Mosbach K (2001) Molecular imprinting on microgel spheres. Anal Chim Acta 435(1):187–196
61. Crapnell RD et al (2019) Recent advances in electrosynthesized molecularly imprinted polymer sensing platforms for bioanalyte detection. Sensors 19(5):1204
62. Lai J-P et al (2007) Molecularly imprinted microspheres and nanospheres for di (2-ethylhexyl) phthalate prepared by precipitation polymerization. Anal Bioanal Chem 389:405–412
63. Wackerlig J, Lieberzeit PA (2015) Molecularly imprinted polymer nanoparticles in chemical sensing–synthesis, characterisation and application. Sens Actuators B Chem 207:144–157
64. Dai C-M et al (2011) Synthesis by precipitation polymerization of molecularly imprinted polymer for the selective extraction of diclofenac from water samples. J Hazard Mater 198:175–181
65. Zhai YY, Yun YB, Li CL (2020) Preparation of bovine serum albumin molecularly imprinted polymer by precipitation polymerization. J Dispers Sci Technol 41(9):1371–1380
66. Jiang H et al (2011) Preparation of monodisperse poly (St-co-PETEA) microspheres by precipitation polymerization using ethanol as solvent. Polym Adv Technol 22(12):1555–1562

67. Pardeshi S, Singh SK (2016) Precipitation polymerization: a versatile tool for preparing molecularly imprinted polymer beads for chromatography applications. RSC Adv 6(28):23525–23536
68. Lovell PA, Schork FJ (2020) Fundamentals of emulsion polymerization. Biomacromolecules 21(11):4396–4441
69. Chern C (2006) Emulsion polymerization mechanisms and kinetics. Prog Polym Sci 31(5):443–486
70. Brooks B (2010) Suspension polymerization processes. Chem Eng Technol 33(11):1737–1744
71. Yuan H, Kalfas G, Ray W (1991) Suspension polymerization. J Macromol Sci Polym Rev 31(2–3):215–299
72. Liu X et al (2019) Synthesis of molecularly imprinted polymer by suspension polymerization for selective extraction of p-hydroxybenzoic acid from water. J Appl Polym Sci 136(3):46984
73. Sambe H, Hoshina K, Haginaka J (2007) Molecularly imprinted polymers for triazine herbicides prepared by multi-step swelling and polymerization method: their application to the determination of methylthiotriazine herbicides in river water. J Chromatogr A 1152(1–2):130–137
74. Nakamura Y et al (2017) Preparation of molecularly imprinted polymers for warfarin and coumachlor by multi-step swelling and polymerization method and their imprinting effects. J Chromatogr A 1516:71–78
75. Nishimura K et al (2019) Evaluation of molecularly imprinted polymers for chlorpromazine and bromopromazine prepared by multi-step swelling and polymerization method—the application for the determination of chlorpromazine and its metabolites in rat plasma by column-switching LC. J Pharm Biomed Anal 174:248–255
76. Nishimura K, Haginaka J (2019) Preparation and evaluation of molecularly imprinted polymers for promazine and chlorpromazine by multi-step swelling and polymerization: the application for the determination of promazine in rat serum by column-switching LC. Anal Sci 35(6):659–664
77. Hayden O, Dickert FL (2001) Selective microorganism detection with cell surface imprinted polymers. Adv Mater 13(19):1480–1483
78. Jenik M et al (2009) Sensors for bioanalytes by imprinting—polymers mimicking both biological receptors and the corresponding bioparticles. Biosens Bioelectron 25(1):9–14
79. Eersels K, Lieberzeit P, Wagner P (2016) A review on synthetic receptors for bioparticle detection created by surface-imprinting techniques: from principles to applications. ACS Sensors 1(10):1171–1187
80. Ding X, Heiden PA (2014) Recent developments in molecularly imprinted nanoparticles by surface imprinting techniques. Macromol Mater Eng 299(3):268–282
81. Dickert FL et al (2004) Bioimprinted QCM sensors for virus detection—screening of plant sap. Anal Bioanal Chem 378:1929–1934
82. Whitesides GM et al (2001) Soft lithography in biology and biochemistry. Annu Rev Biomed Eng 3(1):335–373
83. Lalo H et al (2010) Nanopatterning molecularly imprinted polymers by soft lithography: a hierarchical approach. Lab Chip 10(10):1316–1318
84. Yang JC et al (2015) Lithographically patterned molecularly imprinted polymer for gravimetric detection of trace atrazine. Sens Actuators B Chem 216:476–481
85. Yilmaz E, Haupt K, Mosbach K (2000) The use of immobilized templates—a new approach in molecular imprinting. Angew Chem Int Ed 39(12):2115–2118
86. Shi H et al (1999) Template-imprinted nanostructured surfaces for protein recognition. Nature 398(6728):593–597
87. Chou P-C, Rick J, Chou T-C (2005) C-reactive protein thin-film molecularly imprinted polymers formed using a micro-contact approach. Anal Chim Acta 542(1):20–25
88. Dabrowski M et al (2017) Hierarchical templating in deposition of semi-covalently imprinted inverse opal polythiophene film for femtomolar determination of human serum albumin. Biosens Bioelectron 94:155–161

89. Kalecki J et al (2020) Oriented immobilization of protein templates: a new trend in surface imprinting. ACS Sensors 5(12):3710–3720
90. Poma A et al (2013) Solid-phase synthesis of molecularly imprinted polymer nanoparticles with a reusable template–"plastic antibodies". Adv Funct Mater 23(22):2821–2827
91. Medina Rangel PX et al (2019) Solid-phase synthesis of molecularly imprinted polymer nanolabels: affinity tools for cellular bioimaging of glycans. Sci Rep 9(1):3923
92. Gao B et al (2009) Preparation and recognition performance of cholic acid-imprinted material prepared with novel surface-imprinting technique. Polymer 50(14):3275–3284
93. Sulitzky C et al (2002) Grafting of molecularly imprinted polymer films on silica supports containing surface-bound free radical initiators. Macromolecules 35(1):79–91
94. Tan CJ, Tong YW (2007) Molecularly imprinted beads by surface imprinting. Anal Bioanal Chem 389:369–376
95. Roshan S et al (2019) Molecularly imprinted polymer-silica hybrid particles for biomimetic recognition of target drugs. Adv Polym Technol 2019:1
96. Yin Y-M et al (2012) Dummy molecularly imprinted polymers on silica particles for selective solid-phase extraction of tetrabromobisphenol A from water samples. J Chromatogr A 1220:7–13
97. Shirani MP et al (2021) Development of an eco-friendly fluorescence nanosensor based on molecularly imprinted polymer on silica-carbon quantum dot for the rapid indoxacarb detection. Food Chem 339:127920
98. Bossi A et al (2001) Surface-grafted molecularly imprinted polymers for protein recognition. Anal Chem 73(21):5281–5286
99. Lee H-Y, Kim B (2009) Grafting of molecularly imprinted polymers on iniferter-modified carbon nanotube. Biosens Bioelectron 25(3):587–591
100. Li F, Li J, Zhang S (2008) Molecularly imprinted polymer grafted on polysaccharide microsphere surface by the sol–gel process for protein recognition. Talanta 74(5):1247–1255
101. Li Y et al (2010) A graphene oxide-based molecularly imprinted polymer platform for detecting endocrine disrupting chemicals. Carbon 48(12):3427–3433
102. Sharma PS et al (2012) Electrochemically synthesized polymers in molecular imprinting for chemical sensing. Anal Bioanal Chem 402:3177–3204
103. Goncalves LM (2021) Electropolymerized molecularly imprinted polymers: perceptions based on recent literature for soon-to-be world-class scientists. Curr Opin Electrochem 25:100640
104. Moad G, Solomon DH (1995) The chemistry of free radical polymerization. Pergamon
105. Webster OW (1991) Living polymerization methods. Science 251(4996):887–893
106. Matyjaszewski K (1998) Overview: fundamentals of controlled/living radical polymerization
107. Zhang H (2013) Controlled/"living" radical precipitation polymerization: a versatile polymerization technique for advanced functional polymers. Eur Polym J 49(3):579–600
108. Braunecker WA, Matyjaszewski K (2007) Controlled/living radical polymerization: features, developments, and perspectives. Prog Polym Sci 32(1):93–146
109. Otsu T (2000) Iniferter concept and living radical polymerization. J Polym Sci A Polym Chem 38(12):2121–2136
110. Hawker CJ, Bosman AW, Harth E (2001) New polymer synthesis by nitroxide mediated living radical polymerizations. Chem Rev 101(12):3661–3688
111. Boonpangrak S et al (2006) Preparation of molecularly imprinted polymers using nitroxide-mediated living radical polymerization. Biosens Bioelectron 22(3):349–354
112. Ramakers G et al (2019) Laser-grafted molecularly imprinted polymers for the detection of histamine from organocatalyzed atom transfer radical polymerization. Macromolecules 52(6):2304–2313
113. Kamigaito M, Ando T, Sawamoto M (2001) Metal-catalyzed living radical polymerization. Chem Rev 101(12):3689–3746
114. Gao S-P et al (2017) Molecularly imprinted polymer prepared with polyhedral oligomeric silsesquioxane through reversible addition–fragmentation chain transfer polymerization. Anal Bioanal Chem 409:3741–3748

115. Steinhauer W et al (2013) Block and gradient copolymers of 2-hydroxyethyl acrylate and 2-methoxyethyl acrylate via RAFT: polymerization kinetics, thermoresponsive properties, and micellization. Macromolecules 46(4):1447–1460
116. Tang T et al (2006) Thermo-responsive poly (methyl methacrylate)-block-poly (N-isopropylacrylamide) block copolymers synthesized by RAFT polymerization: micellization and gelation. Macromol Chem Phys 207(19):1718–1726
117. Dai J et al (2012) Preparation of molecularly imprinted nanoparticles with superparamagnetic susceptibility through atom transfer radical emulsion polymerization for the selective recognition of tetracycline from aqueous medium. J Hazard Mater 205-206:179–188
118. Yoshimatsu K et al (2012) Influence of template/functional monomer/cross-linking monomer ratio on particle size and binding properties of molecularly imprinted nanoparticles. J Appl Polym Sci 124(2):1249–1255
119. Sajini T, Mathew B (2021) A brief overview of molecularly imprinted polymers: highlighting computational design, nano and photo-responsive imprinting. Talanta Open 4:100072
120. Shahar T, Tal N, Mandler D (2016) Molecularly imprinted polymer particles: formation, characterization and application. Colloids Surf A Physicochem Eng Asp 495:11–19
121. Haginaka J (2008) Monodispersed, molecularly imprinted polymers as affinity-based chromatography media. J Chromatogr B 866(1):3–13
122. González GP, Hernando PF, Alegría JSD (2006) A morphological study of molecularly imprinted polymers using the scanning electron microscope. Anal Chim Acta 557(1):179–183
123. Yin J, Yang G, Chen Y (2005) Rapid and efficient chiral separation of nateglinide and its l-enantiomer on monolithic molecularly imprinted polymers. J Chromatogr A 1090(1):68–75
124. Ndunda EN (2020) Molecularly imprinted polymers—a closer look at the control polymer used in determining the imprinting effect: a mini review. J Mol Recognit 33(11):e2855
125. Song X et al (2009) Quercetin molecularly imprinted polymers: preparation, recognition characteristics and properties as sorbent for solid-phase extraction. Talanta 80(2):694–702
126. Tang H et al (2012) Selective photocatalysis mediated by magnetic molecularly imprinted polymers. Sep Purif Technol 95:165–171
127. Carter SR, Rimmer S (2004) Surface molecularly imprinted polymer core–shell particles. Adv Funct Mater 14(6):553–561
128. Gao R et al (2011) Preparation and characterization of uniformly sized molecularly imprinted polymers functionalized with core–shell magnetic nanoparticles for the recognition and enrichment of protein. J Mater Chem 21(44):17863–17871
129. El Kirat K, Bartkowski M, Haupt K (2009) Probing the recognition specificity of a protein molecularly imprinted polymer using force spectroscopy. Biosens Bioelectron 24(8):2618–2624
130. Duan F et al (2014) Efficient adsorptive removal of dibenzothiophene by graphene oxide-based surface molecularly imprinted polymer. RSC Adv 4(3):1469–1475
131. Kor K, Zarei K (2016) Development and characterization of an electrochemical sensor for furosemide detection based on electropolymerized molecularly imprinted polymer. Talanta 146:181–187
132. Sehit E et al (2020) Ultrasensitive nonenzymatic electrochemical glucose sensor based on gold nanoparticles and molecularly imprinted polymers. Biosens Bioelectron 165:112432
133. Gupta VK et al (2013) Molecular imprinted polypyrrole modified glassy carbon electrode for the determination of tobramycin. Electrochim Acta 112:37–43
134. Rosy HC, Goyal RN (2014) Molecularly imprinted sensor based on o-aminophenol for the selective determination of norepinephrine in pharmaceutical and biological samples. Talanta 125:167–173
135. Yola ML, Atar N, Eren T (2014) Determination of amikacin in human plasma by molecular imprinted SPR nanosensor. Sens Actuators B Chem 198:70–76
136. Atar N, Eren T, Yola ML (2015) A molecular imprinted SPR biosensor for sensitive determination of citrinin in red yeast rice. Food Chem 184:7–11
137. Atar N et al (2015) A sensitive molecular imprinted surface plasmon resonance nanosensor for selective determination of trace triclosan in wastewater. Sens Actuators B Chem 216:638–644

138. Kara M et al (2013) Combining molecular imprinted nanoparticles with surface plasmon resonance nanosensor for chloramphenicol detection in honey. J Appl Polym Sci 129(4):2273–2279
139. Zhong D-D et al (2013) Rapid preparation of molecularly imprinted polymer by frontal polymerization. Anal Bioanal Chem 405(10):3205–3214
140. Spivak DA (2005) Optimization, evaluation, and characterization of molecularly imprinted polymers. Adv Drug Deliv Rev 57(12):1779–1794
141. Chrzanowska AM, Poliwoda A, Wieczorek PP (2015) Characterization of particle morphology of biochanin A molecularly imprinted polymers and their properties as a potential sorbent for solid-phase extraction. Mater Sci Eng C 49:793–798
142. Holland N et al (2010) The influence of polymer morphology on the performance of molecularly imprinted polymers. Polymer 51(7):1578–1584
143. Boventi M et al (2022) Porosity of molecularly imprinted polymers investigated by 129Xe NMR spectroscopy. ACS Appl Polym Mater 4(12):8740–8749
144. Courtois J et al (2006) Molecularly imprinted polymers grafted to flow through poly(trimethylolpropane trimethacrylate) monoliths for capillary-based solid-phase extraction. J Chromatogr A 1109(1):92–99
145. Skogsberg U et al (2007) A solid-state and suspended-state magic angle spinning nuclear magnetic resonance spectroscopic investigation of a 9-ethyladenine molecularly imprinted polymer. Polymer 48(1):229–238
146. Pereira LA, Rath S (2009) Molecularly imprinted solid-phase extraction for the determination of fenitrothion in tomatoes. Anal Bioanal Chem 393(3):1063–1072
147. Svenson J et al (2005) The role of functional monomer–template complexation on the performance of atrazine molecularly imprinted polymers. Anal Lett 38(1):57–69
148. Lu Y et al (2003) Study on the mechanism of chiral recognition with molecularly imprinted polymers. Anal Chim Acta 489(1):33–43
149. Lanza F, Sellergren B (1999) Method for synthesis and screening of large groups of molecularly imprinted polymers. Anal Chem 71(11):2092–2096
150. Jie Z, Xi-Wen H, Hong-Sheng G (2000) A molecularly imprinted polymer receptor for the enantiomeric recognition of amino acid hydantoins mimicking cooperative hydrogen bonds between nucleotide bases. Chin J Chem 18(4):482–488
151. Tsermentseli SK et al (2013) Molecularly imprinted polymers for the isolation of bioactive naphthoquinones from plant extracts. J Chromatogr A 1315:15–20
152. Bakhtiar S, Bhawani SA, Shafqat SR (2019) Synthesis and characterization of molecular imprinting polymer for the removal of 2-phenylphenol from spiked blood serum and river water. Chem Biol Technol Agric 6(1):15
153. Cegłowski M et al (2022) EGDMA- and TRIM-based microparticles imprinted with 5-fluorouracil for prolonged drug delivery. Polymers (Basel) 14(5):1027
154. Shaikh H et al (2012) Preparation and characterization of molecularly imprinted polymer for di(2-ethylhexyl) phthalate: application to sample clean-up prior to gas chromatographic determination. J Chromatogr A 1247:125–133
155. Li L et al (2015) Molecularly imprinted polymers for extraction of malachite green from fish samples prior to its determination by HPLC. Microchim Acta 182(9):1791–1796
156. Sadeghi S, Jahani M (2013) Selective solid-phase extraction using molecular imprinted polymer sorbent for the analysis of Florfenicol in food samples. Food Chem 141(2):1242–1251
157. Simões M et al (2014) Tailor-made molecularly imprinted polymers for dimethoate and deltamethrin recognition: synthesis, characterization and chromatographic evaluation. J Polym Res 21(3):368
158. Duffy DJ et al (2002) Binding efficiency and transport properties of molecularly imprinted polymer thin films. J Am Chem Soc 124(28):8290–8296
159. Svenson J, Nicholls IA (2001) On the thermal and chemical stability of molecularly imprinted polymers. Anal Chim Acta 435(1):19–24
160. Zunngu SS et al (2017) Synthesis and application of a molecularly imprinted polymer in the solid-phase extraction of ketoprofen from wastewater. C R Chim 20(5):585–591

161. Lagergren S (1898) Zur theorie der sogenannten adsorption geloster stoffe, Kungliga svenska vetenskapsakademiens. Handlingar 24:1–39
162. Yang X et al (2020) In-situ kinetic and thermodynamic study of 2,4-dichlorophenoxyacetic acid adsorption on molecularly imprinted polymer based solid-phase microextraction coatings. Sensors Actuators A Phys 313:112190
163. Ostovan A, Arabi M, Wang Y, Li J, Li B, Wang X, Chen L (2022) Greenificated molecularly imprinted materials for advanced applications. Adv Mater 34:2203154
164. Madikizela LM, Tavengwa NT, Tutu H, Chimuka L (2018) Green aspects in molecular imprinting technology: from design to environmental applications. Trends Environ Anal Chem 17:14–22
165. Del Sole R, Mele G, Bloise E, Mergola L (2021) Green aspects in molecularly imprinted polymers by biomass waste utilization. Polymers 13:13

MIP-Based Electrochemical Sensors

3

Ipek Kucuk, Selenay Sadak, and Bengi Uslu

Abstract

The latest technique for creating receptors with excellent selectivity against particular compounds is molecular printing of polymers. Functional monomers and the target analyte interact during molecularly imprinted polymer (MIP) synthesis. The development of a polymer network all around the analyte covers it. Following the removal of the analyte, binding cavities in the polymer develop that show the analyte's size, shape, and function. With basic electrochemical instrumentation, electrochemical MIP sensors integrate the potentials of synthetic binders. An electrochemical sensor is a device that converts an analytical signal into a meaningful analyte–receptor interaction on the surface of an electrode. The electrochemical signal depends on the mass transfer rate of the target analytes to the MIP-coated electrode, in other words, the coupling of the analyte with the MIP. Due to their simplicity, ease of manufacture, and affordability of equipment and devices, the use of nanoparticles or nanostructured coatings is a current approach to the development of electrochemical MIP sensors. The use of nanoparticles provides MIP-based sensors with good sensitivity and selectivity and fast analytical responses. MIP-based electrochemical sensors have received increasing attention in recent decades. The unique properties of MIPs have made them highly popular for different areas of application, including electrochemical sensors, biosensors, catalysis, and drug delivery. This chapter discusses in detail the development of MIP over the past decade and the advantages of electrochemical sensing based on MIPs.

I. Kucuk · S. Sadak
Department of Analytical Chemistry, Faculty of Pharmacy, Ankara University, Ankara, Turkey

The Graduate School of Health Sciences, Ankara University, Ankara, Turkey

B. Uslu (✉)
Department of Analytical Chemistry, Faculty of Pharmacy, Ankara University, Ankara, Turkey
e-mail: buslu@pharmacy.ankara.edu.tr

S. Patra et al. (eds.), *Molecularly Imprinted Polymers: Path to Artificial Antibodies*, https://doi.org/10.1007/978-981-97-4379-7_3

Keywords

Sensor · Electrochemical · Nanoparticles · Molecularly imprinted polymer Functional monomers

3.1 Introduction

The electrical signal produced by the reaction of the analyte and the recognition layer in electrochemical sensors is used to derive analytical information. Electrochemical sensors are advanced devices that demonstrate a high degree of specificity toward the analyte and offer either qualitative or quantitative analytical data using an electrochemical transducer. Electrochemical sensors enable the sensitive and rapid detection of analytes with high sensitivity and selectivity. Additionally, electrochemical sensors have successfully resolved the difficulties associated with operating in small spatial dimensions at the nanometer or even micro scale [1]. The results obtained using these small detection devices are superior to other conventional analytical methods in terms of fast analysis time, remarkable sensitivity over a wide concentration range, and the ability to selectively detect specific metallic ions with two different oxidation steps [2].

Analytes oxidize and reduce during electrochemical processes, which is the basis for the electrochemical sensor's method of operation. Following the reduction-amplification reactions, various electrical signals are produced, including voltammetric/amperometric, impedance, conductometric, and potentiometric signals. These signals are directly proportional to the concentration of the analyte. Given that the electrochemical processes take place specifically on the surface of the working electrode, this electrode provides an essential function in the sensor system. It possesses distinctive benefits, including specificity, cost-effectiveness, practicality, adaptability for detecting and analyzing targets in intricate systems, and low concentration samples.

These sensors have notable sensitivity, a low detection limit, a wide linear range, and excellent selectivity, allowing for the potential of detecting multiple substances simultaneously [3]. Their approach to integrating polymeric materials into these sensors to enhance their speed, sensitivity, and stability has been particularly intriguing. Conducting polymers are used to create electrochemical sensors made of polymers due to their remarkable mechanical qualities, favorable electrical conductivities, environmental durability, affordability, low operating temperature, lightweight nature, and large surface areas [2].

Molecular Imprinting Technology (MIT) is a synthetic technique that produces certain molecular recognition materials that can mimic natural recognition components like antibodies, antigens, enzymes, substrates, hormones, and biological receptors [4, 5]. Molecularly imprinted materials are uniquely designed polymers that offer molecular recognition sites for a particular or closely related target molecule [6]. In comparison to biological materials, molecular printed polymers are simple to produce and more stable, which has made their usage in electroanalytical applications highly popular in recent years [7].

MIT, which date back to the 1920s, are of great interest to researchers with the discovery of the advantageous and promising properties of molecularly imprinted polymers (MIPs). MIPs are a type of antigen–antibody mimic that can selectively bind to an intrinsic template molecule that is specific to a target. Utilizing MIPs in electro-chemical sensing strategies is possible due to their inexpensive cost and ease of manufacturing processes, which make them alternative synthetic receptors for recognition.

MIP consists of three basic structures that include a monomer, a template, a cross-linker, a polymerization initiator with a solvent. Initially, an interaction occurs between the target analyte, that is the template, and the functional monomer. In molecular imprinting, functional monomers constitute the main component of the 3D structure formed after polymerization. The template molecule is alternatively covalently or non-covalently linked to the functional monomer in the synthesis process with the correct linker groups and then polymerized with an excess of cross-linker. Polymerization is carried out in a solvent in the presence of a crosslinker and initiator [4, 8]. The template then forms recognition cavities whose morphology and functionality are unique to the template molecule through extraction or chemical cleavage with a suitable solvent. The template can be added to the mixture during polymerization or modified on the electrode surface prior to polymerization. The template and monomer interact chemically through non-covalent connections such hydrogen bonding, electrostatic interactions, van der Waals force, hydrophobic contact, and π–π interaction, among others. Importantly, for a stable network to be produced following polymerization, the functional monomer and the crosslinker should have a good chemical connection. If you want your pre-polymerization complex to rebind and form active memory sites on your MIPs, you need a functional monomer that increases the likelihood of its intermolecular interaction with the template molecule. A monomer with this functionality is ideal [9]. An illustration of the imprinting process is shown in Fig. 3.1. When looking for a bioanalogue of antibodies or enzymes, MIP NPs are a strong contender.

3.1.1 Electrochemical Techniques and Sensors

Modern analytical chemistry is becoming more and more interested in chemicals and biosensors because of the range of methods and strategies needed to meet evolving standards and criteria. Interest areas include environmental analysis, food analysis, and clinical diagnostics. A translating physicochemical transducer and a recognition element, or "receptor," are the two fundamental functional components of a sensor. The transducer, which can be mechanical, optical, or electrochemical, is in charge of turning the chemical or biological recognition into a signal that can be measured. The receptor is in charge of selectively recognizing the target analyte.

MIP-based sensors can be used in different techniques, including electrochemical, optical, and piezoelectric sensing. In this chapter, we will focus on the use of MIP-based sensors in electrochemical techniques. To date, MIPs are well integrated into electrochemical sensors as selective recognition layers [11].

The use of electrochemical methods enables rapid analyte detection. Parameters such as detection limit, recovery, and standard deviation indicate their applicability. The interaction of the analyte with a receptor on the electrode surface is the basis of electrochemical sensors. Different electroanalytical techniques are used with these

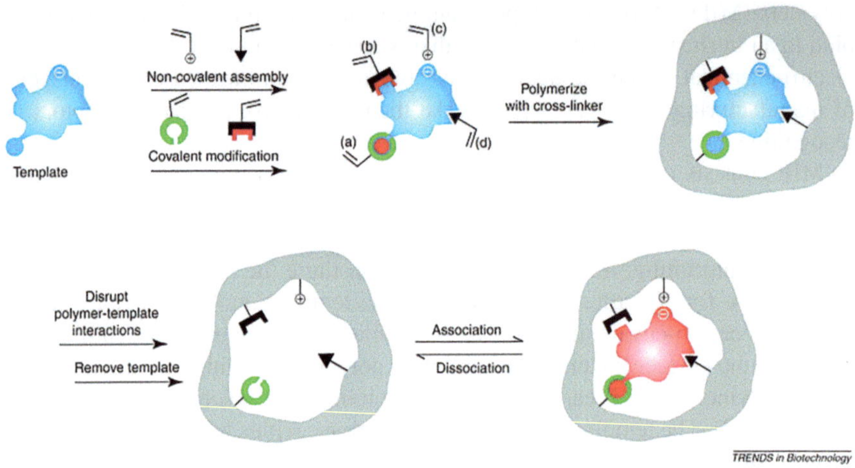

Fig. 3.1 The molecular imprinting process. (Adapted with permission from Poma et al. [10]. Reprinted with permission)

devices. Electrochemical methods consist of two main parts, a receiver and a transducer. The electrochemical transducer converts the chemical recognition event into a measurable analytical signal, that is current (*A*), potential (*V*), resistance (*U*), or conductance (*S*).

Electrochemical techniques can be grouped into two primary types when used for analysis: interfacial methods and noninterfacial approaches. The interfacial methods can be divided into two primary groups: static methods and dynamic methods. When there is no disturbance and there is zero electric current ($i = 0$), the method is said to be static. When there is an electron transfer between the electrode and the analyte, it is said to be dynamic. Potentiometry is a static interface method. Potentiometric MIP sensors, which are typically used in situations with zero current, assess the potential difference between the working and reference electrodes [12]. Dynamic interfacial approaches involve applying an electric stimulus with a nonzero current ($i \neq 0$) to the surface of the working electrode, indicator electrode, or sensor. Of these, amperometry and voltammetry stand out.

The expectations for the development of smaller devices for the monitoring of several drug sample groups and the requirement for the detection of lower detection analyte concentrations have been achieved by electrochemical sensors. The ease of integration with handheld and portable devices provided by the newly developed analytical methods allows for on-site sample analysis, increasing the speed of the method.

Voltammetric sensors record faradayic current as a function of the potential. They measure the faradaic current of a redox reaction under a suitable potential. The linear measurement of the faradaic current is contingent upon the concentration of the analyte involved in the electrochemical reaction. Cyclic (C), differential pulse (DP), square wave (SW), linear sweep (LS), and stripping are a few electrochemical sensors that are based on the voltammetry approach. The differential pulse technique, in particular, provides a higher sensitivity because it provides a high signal-to-noise ratio. Differential pulse voltammetry (DPV) is a method that includes the application of amplitude potential pulses on a linear ramp potential. Pulse techniques are used for quantitative determinations. Potential pulse methods, such as pulse voltammetry and square wave voltammetry (SWV), were developed to enhance the sensitivity of voltammetric methods by minimizing the contribution of the capacitive current. One of the most sensitive and quick pulse voltammetry methods is SWV.

LSV and CV are the most commonly used potential sweep techniques. In potential sweep methods, also called voltammetric methods, a working electrode is subjected to a potential that changes over time. This causes electroactive species in the solution to undergo oxidation or reduction reactions. LSV and CV are widely utilized techniques for identifying oxidation and reduction processes of specific molecular species and obtaining insights into chemical events initiated by electron transport. CV not only allows for the synthesis and characterization of electroactive species by revealing the relationship between their current-potential characteristic, but it also explains the oxidation or reduction reaction of various electroactive species [13].

Electrochemical impedance spectroscopy (EIS) is a technique used in the analysis of electrochemical processes that occur in the electrode/electrolyte solution interface. This approach identifies and determines parameters from a model created using the electrochemical system under study's frequency response.

Another popular electrochemical method that is frequently used in electrochemical sensors is amperometry. The amperometric sensors record the current as a function of time, with the current flowing as a result of the working electrode receiving a constant voltage. The basis of amperometric and chronoamperometric sensors is the measurement of the change in current over time due to diffusion from the solution onto the MIP film-coated electrode. The amount of electroactive species in the solution directly correlates with the current's amplitude. Amperometric sensors offer the benefit of selectivity due to the variation in oxidation or reduction potentials used for analyzing different species of the analyte [3]. Several researchers have investigated the amperometry technique with the MIP sensor they developed [14, 15].

The basic idea behind MIP-based electrochemical sensors is to conduct different electrochemical measurements on the surface of the transducer using immobilized or electropolymerized molecularly imprinted polymers. Two typical approaches are put forth by MIP electrochemical sensors to identify target analytes. For electroactive analytes in the potential range of their electroactivity, that is, when direct electron transfer takes place between the analyte and the transducer surface, direct analytical analysis is carried out. When it comes to determining electroactive targets quantitatively, voltammetry and amperometry are superior. Potentiometry is typically used to determine charged analytes. Furthermore, it is possible to monitor the adsorption of electroinactive analytes onto the electrode surface by measuring changes in the electro-chemical double layer capacity. For electroactive analytes in the potential range of their electroactivity, that is, when direct electron transfer takes place between the analyte and the transducer surface, direct analytical analysis is carried out. An external redox probe, such as ferrocene, hexaaminruthenium(III) chloride, etc., is introduced to the test solution in order to quantify the indirect analytical signal and identify the electro-inactive analytes. The redox probe's faradaic current serves as the output signal, and the measurement is carried out in the potential range where it is electroactive [16].

3.1.2 Advantages of Polymers

The production methods of polymers are quite simple and relatively low cost. Modification of their physical and chemical properties (in order to make them more reactive, biocompatible, degradation-resistant, and flexible) provides high affinity for the detection of target analytes.

To achieve sensors with better performance, polymer based technologies have recently become very popular. Polymeric materials used in sensors include MIP, conducting polymers, and their composites.

Polymer molecules

- Improve recognition of the target molecule.
- Serves as support for immobilization of metal nanoparticles.

3.1.2.1 The Use of Nanoparticles Together with Polymers

Increasing the sensitivity of the developed method can be achieved by optimizing the electrode surface. In addition to MIP, the use of nanoparticles such as metal, magnetic, carbon nanotube, graphene oxide, etc. increases the surface selectivity. By increasing the surface-to-volume ratio, nanoparticles enhance the detection limit of sensors, the adsorption capacity of separation, and the catalytic activity. These substances enhance the sensing element's electrical characteristics, making it easier to integrate them into transducers and other electrical components. In particular, the use of molecular imprinting in combination with metal ions is widely used by researchers to selectively detect a specific pattern. The population of active chemical recognition sites and the electron transfer rate are both considerably increased by the synergistic effect of the used nanomaterials. This enhances the signal of the sensor, increasing selectivity, reproducibility, and stability. Metal ions enable the formation of ionic bonds instead of weak hydrogen bonds to facilitate the interaction between monomer and template in pre-polymerization [17].

3.1.3 Synthetic Antibodies-MIPs

The use of highly selective molecules such as antibodies and enzymes in pharmaceutical diagnostics is important, but these natural receptors are expensive and difficult to produce. There are also limitations in the lifetime and applicability of these receptors. Molecular imprinting methods offer an alternative to traditional biomanufacturing methods such as antibodies and immunosensor preparation to overcome these challenges. The intrinsic stability of MIPs across a wide range of media and pH levels sets them apart from biological systems. The polymer's identification skills are preserved for many decades at room temperature, demonstrating its long-life period storage. One well-known method for creating artificial materials with "molecular memory" is molecular imprinting. In the synthesis step, the key–lock mechanism of enzyme substrate systems is mimicked for selective recognition of the target molecule. MIPs provide a number of benefits, including simple production, chemical and physical stability, compatibility with organic media, reusability, and tailored architectures that allow for the identification of a broad variety of target analytes or chemicals [18]. In addition to being inexpensive to produce, MIPs exhibit a strong affinity for the stamping process's target molecules, and their extended degradation time permits the recognition sites to remain at room temperature for an extended period. At the same time, MIPs are inexpensive relative to the cost of natural antibodies and are relatively cost-effective materials given their long-term storage properties and reusability. Systems known as MIP-based sensors work in concert to combine the inherent benefits of sensors with the benefits of MIPs, particularly their selectivity. The user-friendly sensors provide simplicity, sensitivity, portability, and quick analysis [19].

The selective and sensitive response of a sensor through a precisely designed biological recognition element can be achieved through MIPs. Their high stability and resistance to degradation, low cost, and simple fabrication techniques have brought MIPs into the field of research. Recently, affinity-based sensing device developers have started to base their development strategies on MIPs due to their advantages over natural bioreceptors, such as low cost, good reproducibility, and good stability. MIPs are one of the several supramolecular biomimetic recognition schemes that have demonstrated their potential as synthetic receptors in many application analyses and sensor technologies [20].

3.1.4 MIP Structure

There are various types of interactions between template and monomer, such as covalent, semi-covalent, non-covalent, and metal-coordinated bonds. Metal-coordinated bonds between template and monomer are the most preferred and provide versatile bonding.

The type of interaction between the template and the monomer changes the amount of energy required to remove the template. At the same time, the separation of the template from the monomer creates defining cavities with different shapes, structures, and functional groups, called specific binding sites [21].

3.1.4.1 Monomers

The monomer structure's functional groups are crucial to the polymerization process. Due to their tendency to oxidize, amino and hydroxyl groups in particular help in polymerization by altering the bond between the monomer and template. The monomers can be split into two sections: the portion that polymerizes and interacts with the template, and the portion that does not. Aniline, pyrrole, methacrylic acid (MAA), acrylamide, and methyl methacrylate (MMA) are a few examples of functional monomers. In order to successfully substitute biological receptors in practical applications and implementation, it is crucial to select target templates that meet the specificity and affinity requirements of the molecular imprinting method [22].

3.1.4.2 Cross-Linkers

Crosslinkers are commonly defined as organics that are introduced to the polymerization mixture in order to fix the monomers around the template. The use of crosslinkers is crucial for the stability of polymers. A large percentage of crosslinkers is used to create polymers with notable stiffness and strength. The crosslinker fixes recognition sites, establishes hydrophobicity, and gives the polymer enough porosity and mechanical stability. During polymerization, a porogenic solvent forms pores in the structure and homogenizes all of the constituents. These materials' primary characteristics include their resistance to a wide range of solvents, mechanical stress, high temperatures and pressures, and chemical and physical conditions. They are also repeatable and resistant to processes involving metal ions, bases, and acids [23].

The reversible binding of the template functional monomer to the proper binding groups initiates the synthesis process. With crosslinkers, polymerization then proceeds. Prior to polymerization, there is a contact between the monomer and crosslinker, which results in the condensation of the template molecule on the electrode surface. The molecule in the MIP film becomes even more immobile due to the strength of this interaction. After that, polymerization is initiated by heating, redox, or UV radiation in the presence of an initiator and a crosslinker in a porogenic solvent. Subsequently, the template can be extracted to return the morphology and functionality of the molecule to those of the template molecule through chemical digestion or extraction using a suitable solvent. Ethylene glycol dimethacrylate (EGDMA) and azobis (isobutyronitrile) (AIBN) are the most commonly used crosslinkers in polymerization [20].

3.1.4.3 Initiators

The activation of monomer species usually depends on the presence of an initiator in the polymerization mixture. They can be classified as thermal, redox, and photoinitiators. Benzoyl peroxide, azo-bis-isobutyronitrile, ammonium persulfite, ethyl 2-chloro propionate, and 2,2′-azo-bis-(2,4-dimethyl valeronitrile). MIP can selectively recognize the target analyte in template-generated regions. Other analytes do not match the template-imprinted cavities. In MIP sensors, at low analyte concentration, the polymers form high-affinity cavities, while in concentrated analyte solutions, we can observe lower affinity cavities [24].

3.1.5 Characterization of MIPs

To determine printing efficiency when a template molecule is not available, it is necessary to create a control polymer, a non-imprinted polymer (NIP) that is matched to a MIP. Physical, chemical, and functional evaluation of MIP and NIP materials thus frequently employ a variety of analytical techniques. Various physical characterization techniques are employed to determine the surface and morphological properties of MIPs. These techniques include electrochemiluminescence, atomic force microscopy (AFM), differential scanning calorimetry (DSC), laser diffraction, specific surface area and pore size calculation (Brunauer-Emmett-Teller, or BET) analysis, Raman spectroscopy, swelling tests, scanning electron microscopy (SEM), thermogravimetric analysis (TGA), transmission electron microscopy (TEM), X-ray photoelectron spectroscopy, X-ray powder diffraction, and zeta potential analysis. But chemical characterization methods include nuclear magnetic resonance (NMR) spectroscopy, elemental analysis, and infrared (IR) spectroscopy. Furthermore,

NMR, IR, and UV-Vis spectroscopy reveal functional interactions between monomers and slab molecules inside complexes prior to polymerization [25].

3.1.6 Polymerization Methods

In MIP synthesis, several fundamental polymerization methods are employed. The most popular polymerization method, free radical polymerization, gives you very little control over how much MIP is produced and how it looks. However, the amount of MIP deposited on the electrode surface needs to be regulated in order to ensure excellent repeatability of the MIP. Consequently, the more favored technique for depositing MIP coatings on electrode surfaces is electropolymerization. For electropolymerization to take place, the transducer surface must come into touch with a template, functional monomer, solvent (porogen), and, occasionally, a crosslinker monomer. Furthermore, a notable benefit of this method is the absence of any requirement for an initiator, UV light, or heat in the electropolymerization process of MIPs. The polymerization area becomes limited by this approach to the electrode surface alone. The resulting MIP film thickness may be accurately controlled by varying the amount of charge that passes through the electrode. In addition, the properties of a MIP film, including its conductivity or porosity, can be controlled by choosing the right electropolymerization settings. CV is an electrochemical method that is commonly used to prepare MIPs. Changing the scan rate and number of scans allows one to control the synthesized polymer layer's thickness and compactness. Although the film compactness cannot be adjusted during potentiostatic deposition, the charge can be used to accurately regulate the film thickness. In electropolymerization, the analyte passes through the polymer matrix and interacts with the monomer units to form specific recognition sites on the electrode surface. The amount of monomer added to the polymerization mixture and the ratio between the crosslinker affect the sensitivity of the sensor [19, 26].

3.1.7 Latest Developments About MIP

Numerous applications, including food control, environmental monitoring, and clinical diagnostics, have demonstrated the effectiveness and affordability of MIPs as materials that allow selective detection. Molecular imprinting is a multidisciplinary technology that is continually evolving because of advancements in allied domains such as environmental science, analytical chemistry, biotechnology, polymer science, and nanotechnology. MIP sensors can save time and money while revolutionizing drug testing in the pharmaceutical industry. Even in areas where there is a lot of cost pressure, attractive prices can be attained with the right application [27].

The results from numerous investigations on the sensitive and focused detection of analytes such as hormones, pesticides, and pharmaceutical ingredients in pharmaceuticals utilizing MIP-based electrochemical sensors are compiled. It has proven possible to identify target analytes from genuine samples, including commercial pharmaceuticals, serum, urine, and a wide range of foods, by investigating different monomers and nanomaterials (Tables 3.1 and 3.2).

Table 3.1 MIP-based electrochemical sensors-DPV technique

Analyte	Polymer	Surface layer content	Method	Linear range	Limit of detection	Real sample	Reference
β-Lactoglobulin	Pyrrole	Au NPs-rGO-MIP/GCE	DPV	4–100 ng/mL	3.58 ng/mL	Camel milk	[28]
Entacapone	4-Aminophenol	4-AP@MIP/GCE	DPV	1.0–10.0 pM	0.24 pM	Tablet dosage form	[29]
Dibutyl phthalate	Chitosan	MIP-DBP-CTS/F-CC_3/GCE	DPV	0–1.8 μM	0.0026 μM	Rice wine	[30]
5-Hydroxyindole-3-acetic acid	Pyrrole	MIP/GCE	DPV	5×10^{-11}–5×10^{-5} M	15×10^{-12} M	Human biological fluid	[31]
Indoxyl sulphate	Aniline	PANI-G-NiS_2-MIP/ITO	DPV	1.0 pM–6.0 mM	0.286 pg/mL	Urine samples	[32]
Amoxicillin and clavulanic acid	*o*-Phenylenediamine	MIP/Bi_2Se_3 NPs@ rGO/GCE	DPV	1.42–1100 μM 0.2–1100 μM	0.43 μM 0.06 μM	Pharmaceutical tablets and human serum	[33]
Salidroside	Acrylamide	MWCNT-COOH/MIP/GCE	DPV	1 nM–1 μM	0.51 nM	–	[34]
Codeine	*N*-methacryloyl-L-tryptophan	MA-Trp@MIP/GCE	DPV	0.1–1.0 pM	0.0150 pM	Tablet dosage form, spiked serum, and urine	[35]
Sulfadiazine	Pyrrole	MIP@$CuInS_2$/ZnS/GCE	DPV	0.05–25 μM	2.1 nM	Bovine milk, beef, and fish	[36]
Acesulfame-K	*o*-Phenylenediamine	MIP/GCE	DPV	0.1–17.0 μM	0.35 μM	Cola drink, tabletop, sweetener, and candy	[37]

Ezetimibe	Benzoyl peroxide	Fe_3O_4@MIP/SPGE	DPV	1.0 nM–10.0 μM	0.7 nM	Human serum	[38]
HRP	*tert*-Butyl acrylate	Co, Mo_2C-CNF/MIPs	DPV	1×10^{-14}–1×10^{-7} g/mL	7.4 fg/mL	Human serum	[39]
Octreotide	Aniline	P (ANI)@MIP/GCE	DPV	10–80 fM	0.801 fM	Blood samples	[40]
Bisphenol A	*para*-Aminobenzoic acid	Au@MIP/Au/GCE	DPV	0.5–100 μM	52 nM	Milk and orange juice	[41]
Ciprofloxacin	Pyrrole	MIP/S-CoFe-MOFs/AuNPs/GCE	DPV	0.05–500 ng/mL	0.0033 ng/mL	Cow's milk	[42]
Endosulfan	3-Aminopropyl triethoxysilane	MIP/g-C_3N_4/GCE	DPV	0–96 μM	20 nM	Celery	[43]
Citalopram	Pyrrole	MIP/hNiNS/AMWCNT@GONRs/GCE	DPV	0.5–10 and 10–190 μM	0.042 nM	Human urine, human serum, and pharmaceutical samples	[44]
Sulfadimethoxine	Aniline	GQDs/ZIF-8/MIP/GCE	DPV	0.01–275 nM	0.0063 nM	Milk and river water	[45]
Tenofovir	2-Aminophenol	MIP/Pt@g-C_3N_4/F--MWCNTs/SPE	DPV	0.005–0.69 μM	0.0030 μM	Biological and water samples	[46]
Ethanethiol	Polypyrrole	AuNPs/MIP/GCE	DPV	0.3–3.1 mg/L	0.4 mg/L	Wine samples	[47]
Methyl-parathion	L-Arginine	MIP/ZnOHS/GCE	DPV	5×10^{-9}–0.1×10^{-4} M	$0.5 x 10^{-9}$ M	Strawberry, tomato, cabbage and green beans	[48]

(continued)

Table 3.1 (continued)

Analyte	Polymer	Surface layer content	Method	Linear range	Limit of detection	Real sample	Reference
Chloramphenicol	*o*-Phenylenediamine	MIPs/CS-MWCNTs/GCE	DPV	0.1–1000 ng/mL	0.033 ng/mL	Milk samples	[49]
Luteolin	Methacrylic acid	Fe_3O_4@MIP/rGO/GCE	DPV	2.5 pM–0.1 μM	1 pM	Lotus leaves extract	[50]
Deoxynivalenol	Arginine	P-Arg-MIP/COOH-MWCNTs	DPV	0.1–70 μM	0.07 μM	Wheat flour samples	[51]
Dopamine	4-Vinylphenylboric acid	CQDs@MIPs/SPCE	DPV	5×10^{-9}–5×10^{-4} M	5×10^{-9} M	Goat blood serum	[52]
Tetraobisphenol-S	Methacrylic acid and 4-aminothiophenol	MIP(DiM)@AuNPs/CPE	DPV	0.1–10 nM	0.029 nM	Water samples	[53]
Dimetridazole	Poly-arginine	P-Arg@MIP/GCE	DPV	0.1×10^{-9}–10×10^{-6} M	0.1×10^{-9} M	Egg, milk, honey samples	[54]
Azithromycin	Polyaniline	MIP/GNU/GO/GCE	DPV	0.3–920.0 nM	0.2 nM	Blood serum samples	[55]
Bisphenol A	Ferrocenylmethyl methacrylate	MIP/SPE	DPV	4.7–8 nM	3.2 nM	–	[56]
Dopamine	Polypyrrole	S-$MoSe_2$/NSG/Au/MIPs/GCE	DPV	0.05–1000 μM	0.02 μM	Human blood sample	[57]
Sulfadimidine	3,4-Ethylenedioxythiophene and methacrylic acid	MIP/GQDs-PtNPs/GCE	DPV	0.1 μM–0.1 nM	0.023 nM	Milk and pork samples	[58]
Adenosine	Poly(bithiophene) and poly(3-indolacetic acid)	MIP	DPV	0.37–37.4 μM	0.21 μM	Human urine samples	[59]
Glycerol	Acrylamide/bisacrylamide	Au-SPE/MIP	DPV	20.0–227.81 μM	0.001 μM	Domestic wastewater samples	[60]

α-Synuclein	Pyrrole	MIP/CMP/GS/GCE	DPV	1×10^{-4}–8 nM	3.5×10^{-5} nM	Human serum samples	[61]
Paracetamol	pdhes	MIP-MWCNTs	DPV	0.1–2500 μM	0.02 μM	Human urine samples	[62]
Levofloxacin	Poly(L-cysteine)	rGO/AuNPs/GCE	DPV	0.1–100 μM	0.000003 μM	Human serum samples	[63]
Colchicine	Methacrylic acid	MIM/G@AuNPs/GCE	DPV	1.2×10^{-8}–1.0×10^{-4} M	4.8×10^{-9} M	Tablet and human serum samples	[64]
Pefloxacin	*o*-Phenylenediamine	MIP/AuNPs/rGO/SWCNTs/GCE	DPV	5.0×10^{-7}–2.0×10^{-5} M	1.6×10^{-8} M	Milk samples	[65]
2,4,6-Trinitrotoluene	Poloxamer	Ag_2S/MWCNTs/GCE	DPV	0.22–15.4 μM	3.23 nM	River water samples	[66]
Rutin	Acrylamide	rGO/Fe_3O_4/Ag-MIP/SPE	DPV	1.0×10^{-2}–10.0×10^{-6} M	4.2×10^{-3} M	Tablet samples	[67]
Oxaliplatin	Methacrylic acid	CQDs@HBNNS/UiO-66-NH_2/MIP/GCE	DPV	1.0–250.0 nM	0.37 nM	Human serum and urine samples	[68]
Thrombin	Polythionine	MIP/AuNPs/rGO/GCE	DPV	0.27 pM–0.27 nM	27 fM	Bovine serum samples	[69]
Zileuton	Methacrylic acid Ethylene glycol dimethacrylate	MIP/CPE	DPV	0.1–25 μM	0.189 μM	Human plasma samples	[70]
Aflatoxin B2	Polypyrrole	ZnO-NPs/CS/Au-SPE	DPV	1×10^{-7}–1×10^{-3} M	0.6×10^{-7} M	Dairy milk samples	[71]
Rifampicin and isoniazid	Polypyrrole	Cu-MOF/MC/MIP/GCE	DPV	0.08–85 μM	0.28 nM	Human blood serum and urine samples	[72]

(continued)

Table 3.1 (continued)

Analyte	Polymer	Surface layer content	Method	Linear range	Limit of detection	Real sample	Reference
Sodium lauryl sulfate	2-ATP	MIP/au-SPE	DPV	0.1–1000 pg/mL	0.18 pg/mL	Waste water, river water, shampoo, shower gel	[73]
4-Ethylphenol	Polypyrrole	MIP/GCE	DPV	0.2–34.8 μM	0.1 μM	Wine samples	[47]
Malathion	Acrylamide	Au-SPE/MIP	DPV	0.1–1000 pg/mL	0.06 pg/mL	Olive fruits and oils samples	[74]
Luteolin	β-Cyclodextrin	MIP/ITO	DPV	5.0×10^{-8}–3.0×10^{-5} M	2.4×10^{-8} M	Capsule samples	[50]
p-Nonylphenol	Pyrrole	TiO_2/MIP/Nafion/GCE	DPV	1.0×10^{-8}–1.0×10^{-6} M	3.91×10^{-9} M	Nestle milk powder sample	[75]
Dopamine and chlorpromazine	Nicotinamide	Au/N-GOQDs/NiS_2/BC/MIP/GCE	DPV	0.005–2 μM	0.25 nM	Tablet, human serum and urine samples	[76]
L-Tryptophan	4-Vinylbenzoic acid	Nafion-MIP-MWCNTs@IL/GCE	DPV	0.008–26 μM	0.006 μM	Human serum samples and amino acid oral liquid samples	[77]

Table 3.2 MIP-based electrochemical sensors—other techniques

Cyanazine	*o*-Phenylenediamine	MIP/ZnO/CPE	AMP	0–120 μM	5 nM	Lettuce	[78]
Dibutyl phthalate	Methacrylic acid	SiO_2-COOH@MIP/AuNPs/MWCNTs/GCE	CV	10^{-7}–10^{-2} g/L	5.09×10^{-9} g/L	Tap water and Chinese baijiu samples	[79]
Cannabidiol	Methacrylic acid	mag-MIP/Gr-UiO-66/SPE	CV	5–00 μM	0.05 μM	*Cannabis sativa* L.	[80]
Semicarbazide	*o*-Phenylenediamine	MIP/SWNTs-COOH/CS	CV	0.04–7.6 nM	0.025 nM	Honey, pork, shrimp, sheep casings samples	[81]
Azithromycin	3-Thienyl boronic acid	MIP/GCE	CV	13.33 nM–66.67 μM	0.85 nM	Urine, plasma, tears samples	[82]
Rutin	Polypyrrole	rGO/ZIF-8/MIP/GCE	CV	0.05–100 and 0.0005–0.05 μM	0.0001 μM	Orange juice samples	[83]
Erythromycin	*m*-Phenylenediamine	MIP/SPE	CV	2–16 nM	0.12 nM	Water samples	[84]
Dopamine	Resorcinol	MIP/hAgNS/GE	CV	8.0×10^{-11}–6.0×10^{-8} M	3.2×10^{-11} M	Rat serum and cell samples	[85]
2-Isopropoxyphenol	Poly-pyrrole	MIP/GCE	CV	0.09–45 μM	0.09 μM		[86]
Cloxacillin	Methacrylic acid	MIP/GO/AuNPs/SPCE	CV	110–750 nM		Milk samples	[87]
Phenol	Polyaniline	MIP/CPE	CV	1.0×10^{-7}–1.0×10^{-1} M	2.5×10^{-7} M	–	[88]
Florfenicol	Indole-5-carboxylic acid	MIP/Pd-Pt NPs-P-N-r-GO/GCE	CV	5.0×10^{-8}–8.0×10^{-6} M	1.0×10^{-9} M	Feedstuff, milk, and porcine muscle sample	[89]
Mitoxantrone	β-Cyclodextrin	MIP/GCE	CV	6×10^{-8}–1×10^{-5} M	3×10^{-8} M	Tablet and human urine samples	[90]
Genistein	Carbazole	MIP/cMWCNTs/GCE	CV	0.02–7.0 μM	0.006 μM	Tablet and human serum samples	[91]
Amoxicillin	*meta*-Phenylenediamine	MIP QCM	CV	2–40 nM	0.2 nM	–	[92]

(continued)

Table 3.2 (continued)

Tetracycline	Acrylic acid	Fe_3O_4@SiO_2–C=C-MIP/CPE	SWV	5.0×10^{-7}–4.0×10^{-5} M	1.5×10^{-7} M	Market and natural milk	[93]
Catechol	Pyrrole	MIP/pTHi/MXene/Fe@Ti-MOF-NH_2/GCE	SWV	1.0–300.0 and 300.0–4000.0 μM	0.54 μM	Tap water and human urine	[94]
Chlorpromazine	Nicotinamide	MIP/pTHi/Ni-MOF/Fe-MOF-5/AuNPs/GCE	SWV	0.001–40 and 40–900 μM	0.025 μM	Human serum and urine samples	[95]
Insulin	Chitosan	MIP cryogel/f-MWCNTs/AuE	SWV	0.050–1.40 pM	33 fM	Blood serum samples	[96]
Cortisol	*o*-Phenylenediamine	Au@MIP/GCE	SWV	1 pM–500 nM	200 fM	Artificial saliva samples	[97]
Folic acid	Methacrylic acid	Fe_3O_4@MIP-GO/CPE	SWV	2.5–48 μM	0.65 μM	River water	[98]
Zearalenone	*o*-Phenylenediamine	MIP/poly(MB)/BP-GO/GCE	SWV	0.05–13 μM	12.7 nM	Human serum samples	[99]
Glucose	*o*-Phenylenediamine	AuNP-MIP	SWV	1.25 nM–2.56 μM	1.25 nM	Human blood serum sample	[100]
Capecitabine	Capecitabine	Fe_3O_4@GO@MIP	SWV	1.0–100.0 nM	0.324 nM	Health human plasma sample and capecitabine tablet	[101]
Nevirapine	Methacrylic acid	MIP/CPE	SWSV	0.05–300 μM	10 nM	Human blood serum and tablet sample	[102]
3,4-Methylenedioxy-methamphet-amine	Poly(*o*-phenylenediamine)	MIP/SPCE	SWV	Up to 0.2 mM	0.79 μM	Human blood and urine samples	[103]

Diltiazem	3-Amino-4-hydroxybenzoic acid	MIP/GCE	SWV	1–5 mM	0.12 μM	Cocaine samples	[104]
Ethinylestradiol	Methacrylic acid	(mag@MIP)-GQDs-FG-NF/SPE	SWV	10×10^{-3}–2.5 μM	2.6 nM	Human urine, serum samples and river water samples	[105]
Dopamine	Carbazol-aniline	MIP/Au electrode	SWV	10^{-4}–10^{-5} M	2.0×10^{-6} M	Bovine serum samples	[106]
β-Caryophyllene	Acrylic acid	MIP/CPE	SWV	1.5–7.5×10^{-7} M	–	–	[107]
4-Chlorophenol	Poly(*o*-phenylenediamine)	GCE/ZnO/GNPs/MIP	SWV	0.2–170 μM	0.04 μM	Real water samples	[108]
Imiquimod	1-Allyl-3-vinylimidazolium bromide	Au/GO/MIP/GCE	SWV	0.02–20 μM	0.006 μM	Cream and tablet samples	[101]
Quetiapine	2,2-Azobisisobutyronitrile	MIP-CPE	SWV	1.6×10^{-8}–2.5×10^{-6} M	5.04×10^{-9} M	Tablet and human urine samples	[109]
Sertraline	Methacrylic acid	MIP/Gr-SPCE	SWV	5.0×10^{-9}–7.5×10^{-7} M	1.99×10^{-9} M	Tablet and human serum samples	[110]
Melamine	Poly(aniline-*co*-itaconic acid)	MIPANI-PIA/GCE	SWV	0.25–100 nM	127 pM	Infant formula and raw milk samples	[111]
Asulam	Dopamine	ePDA-MIP/g-C_3N_4/GCE	EIS	0.5–20 pM	0.17 pM	–	[112]
Ketamine	Methacrylic acid	PANI/MIP/C-SPE	EIS	10^{-16}–10^{-2} M	9.29×10^{-7} M	Plasma, urine and beverages	[113]
Paracetamol	Methylphenenic acid	MIP/GCE	EIS	1 pM–0.1 mM	3.6 pM	Paracetamol tablet	[114]
FSH	Methacrylic acid	$NiCo_2O_4$/rGO/MIP/ITO	EIS	0.1 pM–1 μM	0.1 pM	Human serum samples	[115]

(continued)

Table 3.2 (continued)

Aflatoxin B1	Dopamine	MIP-Apt/Cu_2O NCs/GCE	EIS	50.0 pg/L–40.0 ng/L	12 pg/L	Milk	[116]
Vancomycin	Dopamine	MIPDA/peptide/GNPs/GCE	EIS	0.17–2.0 μM	10 pM	Fetal calf serum, probiotic drink and honey	[117]
Cholestanol	Chitosan	MIP/AuPd NPs/PDA/MWCNTs-CS-IL/GCE	EIS	0.1–60 pM	0.05 pM	Human serum samples	[118]
Atrazine	Methacrylic acid	MIP/GCE	EIS	5.0×10^{-9}–1.4×10^{-7} M	1×10^{-9} M	Water samples	[119]
Serotonin		Fe_3O_4@Au@SiO_2-MIP/GPE/PAD	LSV	0.01–1000 μM	0.002 μM	Human urine samples and diet capsules	[120]
L-Tryptophan		AuNPs@PVP@SiO_2MIP/Gr electrode	LSV	1–350 μM	0.3 μM	Pharmaceutical preparations	[121]
Perfloxacin	Poly(pyrrole)	MIP/BPNS-AuNP/GCE	LSV	0.005–10 μM	0.80 nM	Milk and orange juice	[122]

Abbreviations: *Ag_2S* silver sulfide, *AMWCNT* activated multi walled carbon nanotube, *Au* gold, *AuE* gold electrode, *AuNPs* gold nanoparticles, *BC* biomass, *BP-GO* black phosphorus graphene oxide, *BPNS* black phosphorus nanocomposite, *CMP* conjugated microporous polymer, *CNF* carbon nanofibers, *Co* cobalt, *COOH-MWCNT* carboxylic acid functionalized multiwalled carbon nanotube, *CPE* carbon paste electrode, *CQDs* carbon quantum dots, *CS* chitosan, *CS-MWCNTs* chitosan-multiwalled carbon nanotube, *C-SPE* carbon screen printed electrode, *Cu_2O NCs* copper(I) oxide nanocubes, *$CuInS_2$* copper indium oxide, *DA* dopamine, *DiM* functional monomers, *Fe@Ti-MOF-NH_2* amino-functionalization bimetallic organic framework materials, *Fe_3O_4* iron(II, III) oxide, *f-MWCNT* carboxylated multiwalled carbon nanotube, *FSH* Follicle-Stimulating Hormone, *G* graphene, *g-C_3N_4* graphitic carbon nitride, *GCE* glassy carbon electrode, *GNPs* gold nanoparticles, *GO* graphene oxide, *GONRs* graphene oxide nanoribbons, *GQDs* graphene quantum dots, *Gr electrode* graphite rod electrode, *Gr-UiO-66* graphene-UiO-66 composites, *GS* graphene nanosheets, *HBNNS* hexagonal boron nitride nanosheets, *hNiNS* hollow nickel nanospheres, *HRP* horseradish peroxidas, *IL* ionic liquid, *ITO* indium-tin oxide, *Mag-MIP* magnetic molecularly imprinted polymer, *MA-Trp* *N*-methacryloyl-L-tryptophan, *MC* mesoporous carbon, *MIM* molecularly imprinted membrane, *MIP* molecularly imprinted polymer, *Mo_2C* molybdenum carbide, *MOF* metal organic framework, *MWCNTs* multi-walled carbon nanotubes, *MXene* titanium carbide, *NC* nanocomposite, *N-GOQDs* nitrogen-doped graphene oxide quantum dots, *Ni* nickel, *$NiCo_2O_4$* nickel cobaltite, *NiS_2* nickel disulfide, *P (ANI)@MIP* molecularly imprinted polymer film, *PANI* polyaniline, *P-Arg* poly arginine, *Poly(MB)* poly methylene blue, *Pt* platinum, *pTHi* polythionine, *PtNPs* Pt nanoparticles, *PVP* polyvinylpyrrolidone, *rGO* reduced graphene oxide, *Si_2O* silicone dioxide, *SPE* screen printed electrode, *SPE* screen-printed carbon electrodes, *SPGE* screen printed gold electrode, *SWNTs-COOH* carboxylated single-walled carbon nanotube, *TiO_2* titanium dioxide, *ZIF-8* zeolitic imidazolate framework-8, *ZnOHS* zinc oxide hollow spheres, *ZnO-NPs* zinc oxide nanoparticles, *ZnS* zinc sulfide

3.2 Recent Applications with MIP

The remarkable optical and electrical capabilities, along with their exceptional chemical stability, make carbon quantum dots an extremely interesting class of nanomaterials. Graphene quantum dots, a subclass of carbon quantum dots, are hybrid nanomaterials possessing characteristics of both graphene and carbon dots. These nanomaterials have been found to have structurally unique features, which have contributed to their increased utilization in recent years. Graphene quantum dots find widespread uses in various fields, including fuel cells, photoelectronics, supercapacitors, electronics, biomedicine, and sensors in the electrochemistry domain [123–125]. Indeed, graphene is a material possessing semiconductor characteristics. This constraint places restrictions on graphene's application in electrochemistry. However, this impact is removed and remarkable stability along with exceptional optical and distinctive electrical properties are seen due to the confinement of graphene and edge effects in graphene quantum dots. Furthermore, it may be claimed that these materials are very biocompatible and nearly nontoxic. They can be used in a variety of application areas thanks to their excellent solubility [126, 127]. Because of their high conductivity, platinum nanoparticles find application in the chemical industry, the manufacture of biomedical equipment, the automotive sector, and the field of electrochemistry. Owing to its catalytic properties, platinum is widely used in industries including gas sensors, bioimaging, and nanomedicine. However, due to its large drop in global reserves, platinum is extremely hard to come by. Platinum optimization is crucial as a result. For this reason, platinum's size and shape are tailored to its intended applications. Because of their excellent conductivity, platinum nanoparticles are employed in electrochemical research [128–130]. Platinum nanoparticles (PNPs) and graphene quantum dots (GQDs) were used to modify a glassy carbon electrode in a study by Zhang et al. that was carried out in 2020 to determine sulfadimidine. Methacrylic acid and 3,4-ethylenedioxythiophene were used to polymerize the electrode surface as soon as it was modified, as shown in Fig. 3.2. Utilizing TEM and SEM techniques, the electrode's surface was characterized. On the electrode, the MIP/GQDs-PtNPs layer produced a rough surface, based on the results. For determining the analyte, this rough surface offered a sizable and analyte-specific surface. The analyte was detected using DPV, and the results have shown the sensor has a linear concentration range as 0.1 nM–0.1 mM and a detection limit of 0.023 nM. The sensor was successfully applied to detect sulfadimidine milk and pork samples [58].

Graphene quantum dots were also the preferred materials in another study. In 2021, Santos et al. were studied a molecule-imprinted polymer-based screen printed electrode coated with functionalized graphene, graphene quantum dots, and magnetic nanoparticles. Magnetic nanoparticles, which have a wide range of uses due to their combination of high magnetization and paramagnetic properties, are one of the popular nanomaterials of recent years. These nanomaterials, which have a large surface area due to their small particle size, also have highly advanced physicochemical properties. The most commonly used magnetic nanoparticles are Fe_3O_4 and γ-Fe_2O_3. It has many different application areas such as radiotherapy, drug

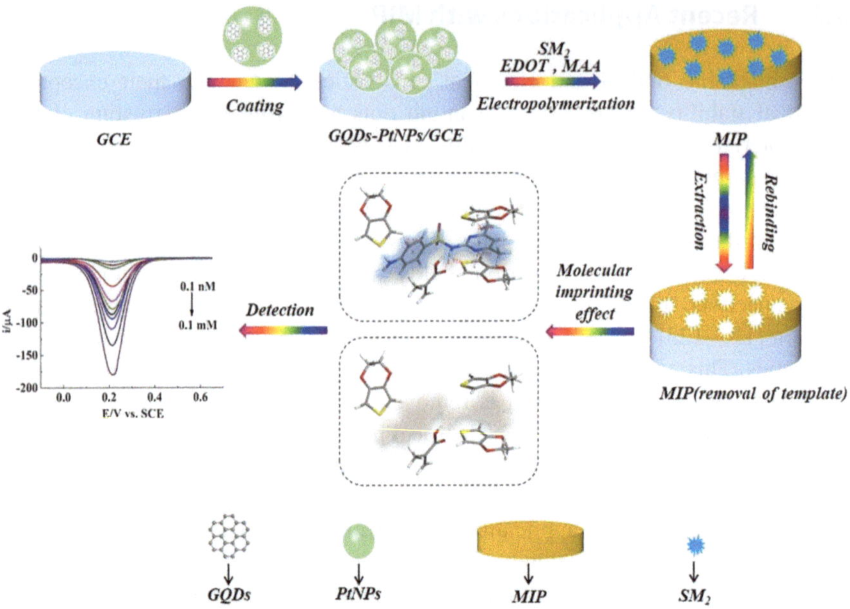

Fig. 3.2 Schematic diagram of the MIP sensor to detect SM_2. (Reprinted with permission)

formulations, chemotherapy, and drug encapsulation [131, 132]. Fe_3O_4 was synthesized for use as magnetic nanoparticles in the investigation. Using ethinylestradiol as the template molecule and methacrylic acid as the functional monomer, photopolymerization was utilized to form the MIP layer on the modified screen printed electrode surface. Photopolymerization was initiated using a 125 W high-pressure mercury lamp. Surface characterization was carried out with EDX spectrum and SEM. Nevertheless, current analytical techniques are unable to clearly visualize the holes of ethinylestradiol on the MIP surface since they occur at the molecular level. Because of this, it was determined using BET analysis that, in contrast to the mag@NIP alteration, the mag@MIP modification produced a greater surface area on the electrode surface. When ethinylestradiol was determined by SWV, linear concentration range was found as 10 nM–2.5 μM with a detection limit of 2.6 nM. The developed method was utilized for detection of the analyte in urine, serum, and water samples. According to the results, the recovery rates were calculated as 96–105%. When it was compared with the other methods developed for the detection of ethynilestradiol, it was seen that the newly prepared sensor had a wider linear range and a lower detection limit. It could be said a sensitive and selective sensor was developed in this study [105].

The size and electrical characteristics of nanoparticles make them a prominent material in today's world. Advanced electrical, magnetic, and optical properties are possessed by nickel disulfide nanoparticles. With a strong electrocatalytic effect, this material finds extensive application in the fields of electrochemistry and supercapacitors [133, 134]. As nanotechnology has advanced, one type of nanoparticle that has seen widespread use is gold nanoparticles. Nanoscale gold materials in the field of electrochemistry have excellent optical properties, chemical stability, advantageous size, low toxicity, and high biocompatibility. These properties enable these materials to be used in many areas, including therapeutic delivery systems, air and water purification, immunotherapy-like treatment methods, and electrochemical nanosensors. It offers chances in many fields. Gold nanoparticles enhance conductivity and accelerate electron transmission, particularly in electrochemical investigations. It consequently results in a rise in electrochemical signals [135, 136]. A molecularly printed electrochemical sensor was created in a 2020 study by Lu et al. In the study, dopamine and chlorpromazine were detected. As shown in Fig. 3.3, gold nanoparticles (AuNPs), NiS_2 nanoparticles (NiS_2/BC), and nitrogen-doped graphene oxide quantum dots (N-GQDs) have been attached to a glassy carbon electrode (GCE). Then, using the CV technique to electrodeposit nicotinamide (NA) monomer, a MIP layer was applied to the modified GCE surface. SEM and TEM were used to characterize the electrode surface. The MIP layer-covered surface was seen to be smoother and more durable based on the characterization results. Dopamine and chlorpromazine were analyzed by DPV and linear concentration range was calculated as 0.05–8 and 8–40 μM for dopamine with a detection limit of 2.8 nM and found as 0.005–2 μM for chlorpromazine with a detection limit of 0.25 nM. According to the results, it was seen that a stable, reproducible, selective, and repeatable sensor was developed by this method. The sensor was successfully utilized to determine analytes in human serum and urine. The developed method has a potential for use in biomedical utilizations with the recovery between 93.9% and 106.15% and RSD value between 1.5% and 6.6% [76].

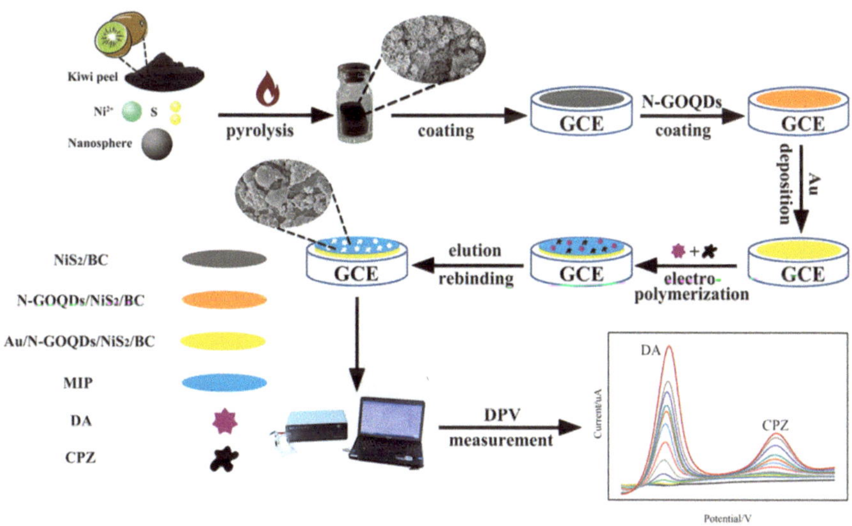

Fig. 3.3 Schematic of the preparation process and electrochemical determination of the Au/N--GOQDs/NiS_2/BC/MIP/GCE composite. (Reprinted with permission)

In a different investigation employing the MIP-based sensor, Jafari et al. coated the screen-printed carbon electrode surface with a polymer layer in 2019 by utilizing a gold nanocomposite and graphene oxide. For many years, the structure and unique features of graphene have been the focus of several studies. A substance used in semiconductors is graphene. The layered structure of graphene oxide is identical to that of graphene. Graphene oxide exhibits superior optical qualities, a huge surface area, and good electrical and thermal conductivity [137–139]. Methacrylic acid was employed as a monomer in the study to form the MIP layer following electrode surface modification. Figure 3.4 has shown the process of MIP fabrication and detection of the target molecule cloxacillin. Through analysis of variables like pH, contact time, and MIP layer composition, the analyte removal from the electrode surface was optimized. SEM was used to analyze the sensor's surface morphology. Upon comparing SEM images of MIP and NIP (MIP layer without cloxacillin) samples, it was revealed that while the NIP-coated surface appeared smoother, the MIP-coated surface had a rougher surface because of analyte-specific cavities. Nonetheless, the MIP surface's pores yielded a more useful surface area for the analyte's detection. By utilizing CV, cloxacillin was examined. The voltammograms that were obtained demonstrated that this method could detect the analyte at a low detection limit of 36 nM and in a wide concentration range of 110–750 nM. The constructed sensor also made it possible to analyze the antibiotic in milk samples, and it was seen that the sensor had benefits including high sensitivity and the ability to produce an electrocatalytic effect against cloxacillin [87].

With semiconductor qualities similar to those of graphene oxide, zinc oxide finds numerous uses in antibacterial agents, photoelectron devices, sensors, and catalysts, among other fields. Superior chemical and physical qualities are possessed by zinc oxide. These characteristics include their electrical, photoelectronic, catalytic, and

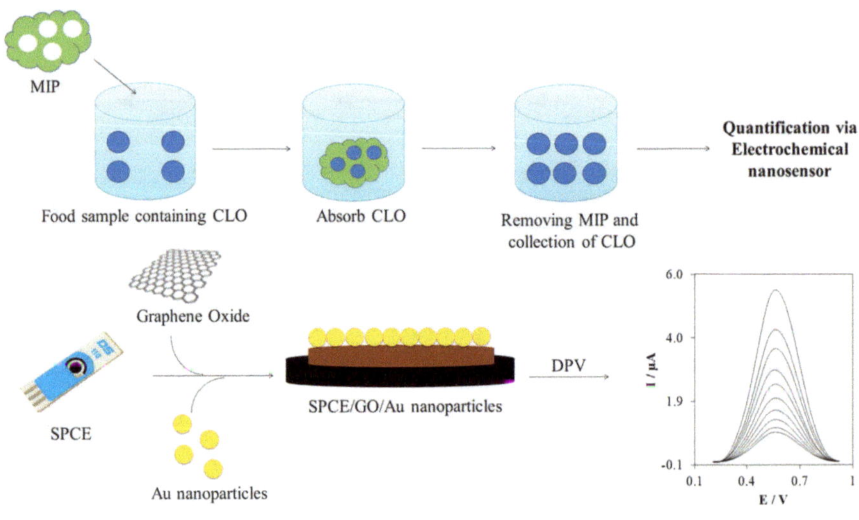

Fig. 3.4 Representation of the MIP fabrication, pre-concentration and detection of the CLO using the electrochemical nanosensor. (Reprinted with permission)

chemical stability. Especially because of their high surface area and catalytic activities, they speed up a lot of reactions [140, 141]. You et al. created a carbon paste electrode based on MIP that had been modified with ZnO nanoparticles in 2022. This sensor was employed to measure cyanazine. MIP was applied to the electrode using 5 mM *o*-phenylenediamine as the monomer, and electrochemical polymerization was carried out using CV. Following XRD and FESEM results, the surface was characterized, and it was found that the MIP-based surface had a more porous structure. NaOH was selected to be the removal solution. The linear concentration range was discovered to be 0–120 μM and the detection limit to be 5 nM as a consequence of the amperometry determination. The sensor has also been utilized to detect cyanazine in lettuce samples. The technique for determining cyanazine in food and agricultural wastewater is valid and accurate, as demonstrated by the creation of a sensor with a sensitivity of 0.5034 μA/μM [78].

3.3 Conclusion

In order to achieve better sensitivity and selectivity, researchers are developing electrochemical sensors with a wide variety of nanomaterials. Recently, manufacturers of affinity-based sensing devices have found that they are low compared to natural bioreceptors. They began to base their development strategies on MIPs because of their advantages such as cost, good repeatability, and good stability. MIP-based electrochemical sensors, with their relatively low cost, robustness, and miniaturization properties have increased their use, especially in drug analysis. These sensors are of great importance today, as traditional analytical techniques such as chromatographic and spectroscopic require complex instrumentation and trained personnel, and their use in resource-limited areas is limited. Electrochemical sensors contain numerous advantages compared to conventional analytical methods, including selectivity, sensitivity, user-friendliness, quick measurement, and transportation. MIP-based sensors are very common in the detection of pesticides, hormones, and similar analytes as well as drug analysis. In this section, the current state of use of MIPs in electrochemical detection is discussed.

Acknowledgments Ipek Kucuk thanks to the financial support from TUBITAK under the BIDEB/2211-A doctoral scholarship program. Selenay Sadak thanks to the financial support from TUBITAK under the BIDEB/2210-A master scholarship program.

References

1. Dumitrescu E, Andreescu S (2017) Bioapplications of electrochemical sensors and biosensors, 1st edn. Elsevier Inc.
2. Ghanam A, Mohammadi H, Amine A et al (2022) Chemical sensors: voltammetric and amperometric electrochemical sensors. In: Encyclopedia of sensors and biosensors, vol 1–4, 1st edn. Elsevier, pp 161–177. https://doi.org/10.1016/B978-0-12-822548-6.00032-7
3. Zhu X, Shi L (2018) Electrochemistry. Elsevier Inc.

4. Mostafa AM, Barton SJ, Wren SP, Barker J (2021) Review on molecularly imprinted polymers with a focus on their application to the analysis of protein biomarkers. TrAC Trends Anal Chem 144:116431. https://doi.org/10.1016/j.trac.2021.116431
5. Roy E, Patra S, Tiwari A et al (2017) Introduction of selectivity and specificity to graphene using an inimitable combination of molecular imprinting and nanotechnology. Biosens Bioelectron 89:234–248. https://doi.org/10.1016/j.bios.2016.02.056
6. BelBruno JJ (2019) Molecularly imprinted polymers. Chem Rev 119:94–119. https://doi.org/10.1021/acs.chemrev.8b00171
7. Pan J, Chen W, Ma Y, Pan G (2018) Molecularly imprinted polymers as receptor mimics for selective cell recognition. Chem Soc Rev 47:5574–5587. https://doi.org/10.1039/C7CS00854F
8. Elugoke SE, Adekunle AS, Fayemi OE et al (2021) Molecularly imprinted polymers (MIPs) based electrochemical sensors for the determination of catecholamine neurotransmitters—review. Electrochem Sci Adv 1:1–43. https://doi.org/10.1002/elsa.202000026
9. Thakur N, Das TR, Gupta S et al (2023a) Material selection for the fabrication of MIP-based devices: an overview. Elsevier Inc.
10. Poma A, Turner APF, Piletsky SA (2010) Advances in the manufacture of MIP nanoparticles. Trends Biotechnol 28:629–637. https://doi.org/10.1016/j.tibtech.2010.08.006
11. Farré M, Barceló D (2007) Sensor, biosensors and MIP based sensors. Woodhead Publishing Limited
12. Ayerdurai V, Cieplak M, Kutner W (2023) Molecularly imprinted polymer-based electrochemical sensors for food contaminants determination. TrAC Trends Anal Chem 158:116830. https://doi.org/10.1016/j.trac.2022.116830
13. Simões FR, Xavier MG (2017) Electrochemical sensors. Elsevier Inc.
14. Bao J, Zhang Q, Huang W (2022) Determination of metandienone using molecularly imprinted based electrochemical sensor in human urine. Int J Electrochem Sci 17:22129. https://doi.org/10.20964/2022.12.12
15. Nur Ashakirin S, Zaid MHM, Haniff MASM et al (2023) Sensitive electrochemical detection of creatinine based on electrodeposited molecular imprinting polymer modified screen printed carbon electrode. Measurement 210:112502. https://doi.org/10.1016/j.measurement.2023.112502
16. Lahcen AA, Amine A (2019) Recent advances in electrochemical sensors based on molecularly imprinted polymers and nanomaterials. Electroanalysis 31:188–201. https://doi.org/10.1002/elan.201800623
17. Alberti G, Zanoni C, Losi V et al (2021) Current trends in polymer based sensors. Chemosensors 9. https://doi.org/10.3390/chemosensors9050108
18. Akgönüllü S, Kılıç S, Esen C, Denizli A (2023) Molecularly imprinted polymer-based sensors for protein detection. Polymers (Basel) 15. https://doi.org/10.3390/polym15030629
19. Kadhem AJ, Gentile GJ, de Cortalezzi MMF (2021) Molecularly imprinted polymers (MIPs) in sensors for environmental and biomedical applications: a review. Molecules 26:6233
20. Ahmad OS, Bedwell TS, Esen C et al (2019) Molecularly imprinted polymers in electrochemical and optical sensors. Trends Biotechnol 37:294–309. https://doi.org/10.1016/j.tibtech.2018.08.009
21. Adumitrăchioaie A, Tertiş M, Cernat A et al (2018) Electrochemical methods based on molecularly imprinted polymers for drug detection. A review. Int J Electrochem Sci 13:2556–2576. https://doi.org/10.20964/2018.03.75
22. Thakur N, Das TR, Gupta S et al (2023b) Chapter 5—Material selection for the fabrication of MIP-based devices: an overview. In: Singh M (ed) Molecularly imprinted polymers (MIPs). Elsevier, pp 101–114
23. Yarman A, Kurbanoglu S, Erkmen C et al (2021) Chapter 4—Quantum dot-based electrochemical molecularly imprinted polymer sensors: potentials and challenges. In: Uslu B (ed) Electroanalytical applications of quantum dot-based biosensors, Micro and nano technologies. Elsevier, pp 121–153

24. Rahman S, Bozal-Palabiyik B, Unal DN et al (2022) Molecularly imprinted polymers (MIPs) combined with nanomaterials as electrochemical sensing applications for environmental pollutants. Trends Environ Anal Chem 36:e00176. https://doi.org/10.1016/j.teac.2022.e00176
25. He S, Zhang L, Bai S et al (2021) Advances of molecularly imprinted polymers (MIP) and the application in drug delivery. Eur Polym J 143:110179. https://doi.org/10.1016/j.eurpolymj.2020.110179
26. Vasapollo G, Del Sole R, Mergola L et al (2011) Molecularly imprinted polymers: present and future prospective. Int J Mol Sci 12:5908–5945. https://doi.org/10.3390/ijms12095908
27. Wang L, Zhang W (2023) Molecularly imprinted polymer (MIP) based electrochemical sensors and their recent advances in health applications. Sens Actuators Rep 5:100153. https://doi.org/10.1016/j.snr.2023.100153
28. Meng F, Duan M, Wu W et al (2024) Enzymatic construction Au NPs-rGO based MIP electrochemical sensor for adulteration detection of bovine-derived allergen in camel milk. Food Chem 436:137638. https://doi.org/10.1016/j.foodchem.2023.137638
29. Budak F, Cetinkaya A, Kaya SI, Ozkan SA (2023) MIP-based electrochemical sensor for highly selective and sensitive determination of entacapone from the triple mixture in tablet dosage form. J Pharm Biomed Anal Open 1:100005. https://doi.org/10.1016/j.jpbao.2023.100005
30. Zhou Q, Guo M, Wu S et al (2021) Electrochemical sensor based on corncob biochar layer supported chitosan-MIPs for determination of dibutyl phthalate (DBP). J Electroanal Chem 897:115549. https://doi.org/10.1016/j.jelechem.2021.115549
31. Moncer F, Adhoum N, Catak D, Monser L (2021) Electrochemical sensor based on MIP for highly sensitive detection of 5-hydroxyindole-3-acetic acid carcinoid cancer biomarker in human biological fluids. Anal Chim Acta 1181:338925. https://doi.org/10.1016/j.aca.2021.338925
32. Dalal N, Dhiman TK, Lakshmi GBVS et al (2022) MIP-based sensor for the detection of gut microbiota-derived indoxyl sulphate using PANI-graphene-NiS2. Mater Today Chem 26:101157. https://doi.org/10.1016/j.mtchem.2022.101157
33. Ali R, El-Wekil MM (2023) Construction of MIP/Bi2S3 nanoparticles/rGO nanoprobe for simultaneous electrochemical determination of amoxicillin and clavulanic acid. J Alloys Compd 962:2–9. https://doi.org/10.1016/j.jallcom.2023.171180
34. Zheng H, Yan W, Zuo Z, Liu X (2023) Fabrication of an electrochemical sensor based on molecular imprinting strategy for detecting salidroside. Int J Electrochem Sci 18:100367. https://doi.org/10.1016/J.IJOES.2023.100367
35. Yence M, Cetinkaya A, Çorman ME et al (2023) Fabrication of molecularly imprinted electrochemical sensors for sensitive codeine detection. Microchem J 193:109060. https://doi.org/10.1016/j.microc.2023.109060
36. Yu X, Yang Y, Shen Q et al (2024) A novel differential ratiometric molecularly imprinted electrochemical sensor for determination of sulfadiazine in food samples. Food Chem 434:1–9. https://doi.org/10.1016/j.foodchem.2023.137461
37. Singh R, Singh M (2023) Molecularly imprinted electrochemical sensor for highly selective and sensitive determination of artificial sweetener acesulfame-K. Talanta Open 7:100194. https://doi.org/10.1016/j.talo.2023.100194
38. Hashkavayi AB, Alizadeh A, Azimi R et al (2023) Development of core-shell magnetic molecularly imprinted polymer-based electrochemical sensor for sensitive and selective detection of ezetimibe. Heliyon 9:e17169. https://doi.org/10.1016/j.heliyon.2023.e17169
39. Chen F, Lv C, Xing Y et al (2023) Electrospinning carbon fibers based molecularly imprinted polymer self-supporting electrochemical sensor for sensitive detection of glycoprotein. Sens Actuators B Chem 396:134552. https://doi.org/10.1016/j.snb.2023.134552
40. Ozkan E, Ozcelikay G, Gök Topak ED et al (2023) Molecularly imprinted electrochemical sensor for the selective and sensitive determination of octreotide in cancer patient plasma sample. Talanta 263:124679. https://doi.org/10.1016/j.talanta.2023.124679
41. Han E, Pan Y, Li L, Cai J (2023) Bisphenol A detection based on nano gold-doped molecular imprinting electrochemical sensor with enhanced sensitivity. Food Chem 426:136608. https://doi.org/10.1016/j.foodchem.2023.136608

42. Xiong Y, Zhang D, Ye C et al (2023) Ultra-sensitive detection of ciprofloxacin hydrochloride in milk by molecularly imprinted electrochemical sensor based on S-CoFe-MOFs/AuNPs. J Food Compos Anal 122:105439. https://doi.org/10.1016/j.jfca.2023.105439
43. Ma C, Wu D, Xu E, Fan Y (2022) Molecularly imprinted electrochemical sensor based on g-C3N4 for determination of endosulfan insecticide in agricultural food. Int J Electrochem Sci 17. https://doi.org/10.20964/2022.12.34
44. Aminikhah M, Babaei A, Taheri A (2022) A novel electrochemical sensor based on molecularly imprinted polymer nanocomposite platform for sensitive and ultra-selective determination of citalopram. J Electroanal Chem 918:116493. https://doi.org/10.1016/j.jelechem.2022.116493
45. Hatamluyi B, Sadeghzadeh S, Balouch Zehi Z et al (2023) A rapid and recyclable analysis method for sulfadimethoxine detection based on molecularly imprinted electrochemical sensor reinforced by GQDs/ZIF-8 nanocomposite. Microchem J 185:1–8. https://doi.org/10.1016/j.microc.2022.108266
46. Erk N, Bouali W, Mehmandoust M, Soylak M (2022) An electrochemical sensor for molnupiravir based on a metal-organic framework composited with poly(3,4-ethylene dioxythiophene): poly(styrene sulfonate). ChemistrySelect 7:e202203325. https://doi.org/10.1002/slct.202203325
47. Domínguez-Renedo O, Navarro-Cuñado AM, Arnáiz-Lozano V, Alonso-Lomillo MA (2020) Molecularly imprinted polypyrrole based electrochemical sensor for selective determination of 4-ethylphenol. Talanta 207:120351. https://doi.org/10.1016/j.talanta.2019.120351
48. Daizy M, Ali MR, Bacchu MS et al (2021) ZnO hollow spheres arrayed molecularly-printed-polymer based selective electrochemical sensor for methyl-parathion pesticide detection. Environ Technol Innov 24:101847. https://doi.org/10.1016/j.eti.2021.101847
49. Geng L, Huang J, Zhai H et al (2022) Molecularly imprinted electrochemical sensor based on multi-walled carbon nanotubes for specific recognition and determination of chloramphenicol in milk. Microchem J 182:107887. https://doi.org/10.1016/j.microc.2022.107887
50. Wei M, Geng X, Liu Y et al (2019) A novel electrochemical sensor based on electropolymerized molecularly imprinted polymer for determination of luteolin. J Electroanal Chem 842:184–192. https://doi.org/10.1016/j.jelechem.2019.04.074
51. Li W, Diao K, Qiu D et al (2021) A highly-sensitive and selective antibody-like sensor based on molecularly imprinted poly(L-arginine) on COOH-MWCNTs for electrochemical recognition and detection of deoxynivalenol. Food Chem 350:129229. https://doi.org/10.1016/j.foodchem.2021.129229
52. Zhang W, Zhang L, Sun S et al (2023) An effective and simple strategy for highly selective and anti-interference detection of dopamine based on the carbon quantum dots-molecularly imprinted polymers modified electrode. Microchem J 195:109340. https://doi.org/10.1016/j.microc.2023.109340
53. Sarpong KA, Zhang K, Luan Y et al (2020) Development and application of a novel electrochemical sensor based on AuNPS and difunctional monomer-MIPs for the selective determination of tetrabromobisphenol-S in water samples. Microchem J 154:104526. https://doi.org/10.1016/j.microc.2019.104526
54. Ali MR, Bacchu MS, Daizy M et al (2020) A highly sensitive poly-arginine based MIP as an electrochemical sensor for selective detection of dimetridazole. Anal Chim Acta 1121:11–16. https://doi.org/10.1016/j.aca.2020.05.004
55. Jafari S, Dehghani M, Nasirizadeh N, Azimzadeh M (2018) An azithromycin electrochemical sensor based on an aniline MIP film electropolymerized on a gold nano urchins/graphene oxide modified glassy carbon electrode. J Electroanal Chem 829:27–34. https://doi.org/10.1016/j.jelechem.2018.09.053
56. Rebocho S, Cordas CM, Viveiros R, Casimiro T (2018) Development of a ferrocenyl-based MIP in supercritical carbon dioxide: towards an electrochemical sensor for bisphenol A. J Supercrit Fluids 135:98–104. https://doi.org/10.1016/j.supflu.2018.01.006
57. Zang Y-J, Nie J, He B et al (2020) Fabrication of S-MoSe2/NSG/Au/MIPs imprinted composites for electrochemical detection of dopamine based on synergistic effect. Microchem J 156:104845. https://doi.org/10.1016/j.microc.2020.104845

58. Zhang L, He L, Wang Q et al (2020a) Theoretical and experimental studies of a novel electrochemical sensor based on molecularly imprinted polymer and GQDs-PtNPs nanocomposite. Microchem J 158:105196. https://doi.org/10.1016/j.microc.2020.105196
59. Gliga LE, Iacob BC, Cheşcheş B et al (2020) Electrochemical platform for the detection of adenosine using a sandwich-structured molecularly imprinted polymer-based sensor. Electrochim Acta 354:136656. https://doi.org/10.1016/j.electacta.2020.136656
60. Motia S, Bouchikhi B, Llobet E, El Bari N (2020) Synthesis and characterization of a highly sensitive and selective electrochemical sensor based on molecularly imprinted polymer with gold nanoparticles modified screen-printed electrode for glycerol determination in wastewater. Talanta 216:120953. https://doi.org/10.1016/j.talanta.2020.120953
61. Ma Y, Hu Q, Liu C, Wang L (2020) A nanospherical conjugated microporous polymer-graphene nanosheets modified molecularly imprinted electrochemical sensor for high sensitivity detection of α-Synuclein. J Electroanal Chem 862:113994. https://doi.org/10.1016/j.jelechem.2020.113994
62. Tan K, Ma Q, Luo J et al (2018) Water-dispersible molecularly imprinted nanohybrids via co-assembly of carbon nanotubes with amphiphilic copolymer and photocrosslinking for highly sensitive and selective paracetamol detection. Biosens Bioelectron 117:713–719. https://doi.org/10.1016/j.bios.2018.07.009
63. Ghanbari MH, Khoshroo A, Sobati H et al (2019) An electrochemical sensor based on poly (L-cysteine)@AuNPs @ reduced graphene oxide nanocomposite for determination of levofloxacin. Microchem J 147:198–206. https://doi.org/10.1016/j.microc.2019.03.016
64. Bai H, Wang C, Chen J et al (2018) Graphene@AuNPs modified molecularly imprinted electrochemical sensor for the determination of colchicine in pharmaceuticals and serum. J Electroanal Chem 816:7–13. https://doi.org/10.1016/j.jelechem.2018.02.061
65. Shi X, Zuo Y, Jia X et al (2020) A novel molecularly imprinted sensor based on gold nanoparticles/reduced graphene oxide/single-walled carbon nanotubes nanocomposite for the detection of pefloxacin. Int J Electrochem Sci 15:9683–9697. https://doi.org/10.20964/2020.10.38
66. Li X, Ma Y, Zhong XL, Wang LS (2020) Silver sulfide nanoparticles on MWCNTs stabilized by poloxamer: an enhanced electrochemical sensor for high sensitivity detection of 2,4,6-trinitrotoluene. Microchem J 159:105488. https://doi.org/10.1016/j.microc.2020.105488
67. Yang S, Yang X, Tang R et al (2018b) A novel rutin electrochemical sensor using reduced graphene oxide/magnetite/silver nanoparticle-molecularly imprinted polymer composite modified electrode. Int J Electrochem Sci 13:2483–2497. https://doi.org/10.20964/2018.03.55
68. Hatamluyi B, Hashemzadeh A, Darroudi M (2020) A novel molecularly imprinted polymer decorated by CQDs@HBNNS nanocomposite and UiO-66-NH2 for ultra-selective electrochemical sensing of Oxaliplatin in biological samples. Sens Actuators B Chem 307:127614. https://doi.org/10.1016/j.snb.2019.127614
69. Yang S, Yang J, Cao Q et al (2018a) A molecularly imprinted polythionine-modified electrode based on a graphene-gold nanoparticle composite (MIP/AuNPs/rGO/GCE) for the determination of thrombin. Int J Electrochem Sci 13:9333–9345. https://doi.org/10.20964/2018.10.53
70. Baezzat MR, Bagheri M, Abdollahi E (2019) Molecularly imprinted polymer based sensor for measuring of zileuton: evaluation as a modifier for carbon paste electrode in electrochemically recognition. Mater Today Commun 19:23–31. https://doi.org/10.1016/j.mtcomm.2018.12.013
71. El Alami El Hassani N, Bouchikhi B, El Bari N (2020) Recent development of an electrochemical imprinted sensor for the detection of trace-level of unmetabolized aflatoxin B2 in dairy milk. J Electroanal Chem 865:114123. https://doi.org/10.1016/j.jelechem.2020.114123
72. Rawool CR, Srivastava AK (2019) A dual template imprinted polymer modified electrochemical sensor based on Cu metal organic framework/mesoporous carbon for highly sensitive and selective recognition of rifampicin and isoniazid. Sens Actuators B Chem 288:493–506. https://doi.org/10.1016/j.snb.2019.03.032
73. Motia S, Tudor IA, Madalina Popescu L et al (2018) Development of a novel electrochemical sensor based on electropolymerized molecularly imprinted polymer for selective detection of sodium lauryl sulfate in environmental waters and cosmetic products. J Electroanal Chem 823:553–562. https://doi.org/10.1016/j.jelechem.2018.03.069

74. Aghoutane Y, Diouf A, Österlund L et al (2020) Development of a molecularly imprinted polymer electrochemical sensor and its application for sensitive detection and determination of malathion in olive fruits and oils. Bioelectrochemistry 132:107404. https://doi.org/10.1016/j.bioelechem.2019.107404
75. Yu M, Wu L, Miao J et al (2019) Titanium dioxide and polypyrrole molecularly imprinted polymer nanocomposites based electrochemical sensor for highly selective detection of p-nonylphenol. Anal Chim Acta 1080:84–94. https://doi.org/10.1016/j.aca.2019.06.053
76. Lu Z, Li Y, Liu T et al (2020) A dual-template imprinted polymer electrochemical sensor based on AuNPs and nitrogen-doped graphene oxide quantum dots coated on NiS2/biomass carbon for simultaneous determination of dopamine and chlorpromazine. Chem Eng J 389:124417. https://doi.org/10.1016/j.cej.2020.124417
77. Xia Y, Zhao F, Zeng B (2020) A molecularly imprinted copolymer based electrochemical sensor for the highly sensitive detection of L-tryptophan. Talanta 206:120245. https://doi.org/10.1016/j.talanta.2019.120245
78. You L, You L, Qi W et al (2022) Fabrication of electrochemical sensor based on molecularly imprinted polymer/zinc oxide nanoparticle-nanocomposite (MIP/ZnO/CPE) for determination of cyanazine in food samples. Int J Electrochem Sci 17:1–12. https://doi.org/10.20964/2022.12.33
79. Wang S, Pan M, Liu K et al (2022) A SiO2@MIP electrochemical sensor based on MWCNTs and AuNPs for highly sensitive and selective recognition and detection of dibutyl phthalate. Food Chem 381:132225. https://doi.org/10.1016/j.foodchem.2022.132225
80. Tang X, Gu Y, Tang P, Liu L (2022) Electrochemical sensor based on magnetic molecularly imprinted polymer and graphene-UiO-66 composite modified screen-printed electrode for cannabidiol detection. Int J Electrochem Sci 17:1–15. https://doi.org/10.20964/2022.05.64
81. Zhang L, He L, Wang Q et al (2020b) Preparation of a carboxylated single-walled carbon-nanotube-chitosan functional layer and its application to a molecularly imprinted electrochemical sensor to quantify semicarbazide. Food Chem 333:127524. https://doi.org/10.1016/j.foodchem.2020.127524
82. Stoian IA, Iacob BC, Dudaş CL et al (2020) Biomimetic electrochemical sensor for the highly selective detection of azithromycin in biological samples. Biosens Bioelectron 155:112098. https://doi.org/10.1016/j.bios.2020.112098
83. El Jaouhari A, Yan L, Zhu J et al (2020) Enhanced molecular imprinted electrochemical sensor based on zeolitic imidazolate framework/reduced graphene oxide for highly recognition of rutin. Anal Chim Acta 1106:103–114. https://doi.org/10.1016/j.aca.2020.01.039
84. Ayankojo AG, Reut J, Ciocan V et al (2020) Molecularly imprinted polymer-based sensor for electrochemical detection of erythromycin. Talanta 209:120502. https://doi.org/10.1016/j.talanta.2019.120502
85. Chen C, Wang Y, Ding S et al (2019) A novel sensitive and selective electrochemical sensor based on integration of molecularly imprinted with hollow silver nanospheres for determination of carbamazepine. Microchem J 147:191–197. https://doi.org/10.1016/j.microc.2019.03.024
86. Qader B, Baron M, Hussain I, Gonzalez-Rodriguez J (2018) Electrochemical determination of 2-isopropoxyphenol in glassy carbon and molecularly imprinted poly-pyrrole electrodes. J Electroanal Chem 821:16–21. https://doi.org/10.1016/j.jelechem.2017.11.027
87. Jafari S, Dehghani M, Nasirizadeh N et al (2019) Label-free electrochemical detection of cloxacillin antibiotic in milk samples based on molecularly imprinted polymer and graphene oxide-gold nanocomposite. Measurement 145:22–29. https://doi.org/10.1016/j.measurement.2019.05.068
88. Setiyanto H, Rahmadhani S, Sukandar S et al (2020) The performance of molecularly imprinted polymers (MIPs)-modified carbon paste electrode and its application in detecting phenol. Int J Electrochem Sci 15:5477–5486. https://doi.org/10.20964/2020.06.66
89. Fan A, Yang G, Yang H, Zhao F (2020) Synthesis and application of dendritic Pt-Pd bimetallic nanoparticles in imprinted electrochemical sensor for the determination of florfenicol. Mater Today Commun 25:101448. https://doi.org/10.1016/j.mtcomm.2020.101448

90. Liu Y, Wei M, Hu Y et al (2018) An electrochemical sensor based on a molecularly imprinted polymer for determination of anticancer drug Mitoxantrone. Sens Actuators B Chem 255:544–551. https://doi.org/10.1016/j.snb.2017.08.023
91. Niu LT, Li GG, Li HF et al (2019) A novel genistein electrochemical sensor based on molecularly imprinted polycarbazole/carboxylated multiwalled carbon nanotubes nanocomposite. Chin J Anal Chem 47:e19095–e19103. https://doi.org/10.1016/S1872-2040(19)61176-X
92. Ayankojo AG, Reut J, Boroznjak R et al (2018) Molecularly imprinted poly(meta-phenylenediamine) based QCM sensor for detecting amoxicillin. Sens Actuators B Chem 258:766–774. https://doi.org/10.1016/j.snb.2017.11.194
93. Zeb S, Wong A, Khan S et al (2021) Using magnetic nanoparticles/MIP-based electrochemical sensor for quantification of tetracycline in milk samples. J Electroanal Chem 900:115713. https://doi.org/10.1016/j.jelechem.2021.115713
94. Lu Z, Wei K, Ma H et al (2023a) Bimetallic MOF synergy molecularly imprinted ratiometric electrochemical sensor based on MXene decorated with polythionine for ultra-sensitive sensing of catechol. Anal Chim Acta 1251:340983. https://doi.org/10.1016/j.aca.2023.340983
95. Lu Z, Wei K, Ma H et al (2023b) Nanoarchitectonics of on–off ratiometric signal amplified electrochemical sensor for chlorpromazine with molecularly imprinted polymer based on Ni-MOF/Fe-MOF-5 hybrid Au nanoparticles. Sep Purif Technol 327:124858. https://doi.org/10.1016/j.seppur.2023.124858
96. Wardani NI, Kangkamano T, Wannapob R et al (2023) Electrochemical sensor based on molecularly imprinted polymer cryogel and multiwalled carbon nanotubes for direct insulin detection. Talanta 254:124137. https://doi.org/10.1016/j.talanta.2022.124137
97. Yeasmin S, Wu B, Liu Y et al (2022) Nano gold-doped molecularly imprinted electrochemical sensor for rapid and ultrasensitive cortisol detection. Biosens Bioelectron 206:114142. https://doi.org/10.1016/j.bios.2022.114142
98. Garcia SM, Wong A, Khan S, Sotomayor MDPT (2021) A simple, sensitive and efficient electrochemical platform based on carbon paste electrode modified with Fe3O4@MIP and graphene oxide for folic acid determination in different matrices. Talanta 229:122258. https://doi.org/10.1016/j.talanta.2021.122258
99. Hu X, Tang Y, Xia Y et al (2022) Antifouling ionic liquid doped molecularly imprinted polymer-based ratiometric electrochemical sensor for highly stable and selective detection of zearalenone. Anal Chim Acta 1210:339884. https://doi.org/10.1016/j.aca.2022.339884
100. Sehit E, Drzazgowska J, Buchenau D et al (2020) Ultrasensitive nonenzymatic electrochemical glucose sensor based on gold nanoparticles and molecularly imprinted polymers. Biosens Bioelectron 165:112432. https://doi.org/10.1016/j.bios.2020.112432
101. Afzali M, Mostafavi A, Shamspur T (2020) A novel electrochemical sensor based on magnetic core@shell molecularly imprinted nanocomposite (Fe3O4@graphene oxide@MIP) for sensitive and selective determination of anticancer drug capecitabine. Arab J Chem 13:6626–6638. https://doi.org/10.1016/j.arabjc.2020.06.018
102. Massumi S, Ahmadi E, Akbari A, Gholivand MB (2020) Highly sensitive and selective sensor based on molecularly imprinted polymer for voltammetric determination of Nevirapine in biological samples. J Electroanal Chem 876:114508. https://doi.org/10.1016/j.jelechem.2020.114508
103. Couto RAS, Costa SS, Mounssef B et al (2019) Electrochemical sensing of ecstasy with electropolymerized molecularly imprinted poly(o-phenylenediamine) polymer on the surface of disposable screen-printed carbon electrodes. Sens Actuators B Chem 290:378–386. https://doi.org/10.1016/j.snb.2019.03.138
104. Sposito HGM, Lobato A, Tasić N et al (2022) Swift electrochemical sensing of diltiazem employing highly-selective molecularly-imprinted 3-amino-4-hydroxybenzoic acid. J Electroanal Chem 911:1–7. https://doi.org/10.1016/j.jelechem.2022.116207
105. Santos AM, Wong A, Prado TM et al (2021) Voltammetric determination of ethinylestradiol using screen-printed electrode modified with functionalized graphene, graphene quantum dots and magnetic nanoparticles coated with molecularly imprinted polymers. Talanta 224:1–9

106. Sağlam Ş, Arman A, Üzer A et al (2020) Selective electrochemical determination of dopamine with molecularly imprinted poly(carbazole-co-aniline) electrode decorated with gold nanoparticles. Electroanalysis 32:964–970. https://doi.org/10.1002/elan.201900646
107. de Assis IM, de Moraes MOS, da Conceição RC et al (2019) Novel electrochemical sensor based on molecularly imprinted polymer for selective recognition of sesquiterpene β-caryophyllene. Spectrochim Acta A Mol Biomol Spectrosc 217:271–277. https://doi.org/10.1016/j.saa.2019.03.097
108. Al-Ammari RH, Ganash AA, Salam MA (2019) Electrochemical molecularly imprinted polymer based on zinc oxide/graphene/poly(o-phenylenediamine) for 4-chlorophenol detection. Synth Met 254:141–152. https://doi.org/10.1016/j.synthmet.2019.06.015
109. Motaharian A, Naseri K, Mehrpour O, Shoeibi S (2020) Electrochemical determination of atypical antipsychotic drug quetiapine using nano-molecularly imprinted polymer modified carbon paste electrode. Anal Chim Acta 1097:214–221. https://doi.org/10.1016/j.aca.2019.11.020
110. Khosrokhavar R, Motaharian A, Milani Hosseini MR, Mohammadsadegh S (2020) Screen-printed carbon electrode (SPCE) modified by molecularly imprinted polymer (MIP) nanoparticles and graphene nanosheets for determination of sertraline antidepressant drug. Microchem J 159:105348. https://doi.org/10.1016/j.microc.2020.105348
111. Regasa MB, Soreta TR, Femi OE et al (2020) Novel multifunctional molecular recognition elements based on molecularly imprinted poly (aniline-co-itaconic acid) composite thin film for melamine electrochemical detection. Sens Biosensing Res 27:100318. https://doi.org/10.1016/j.sbsr.2019.100318
112. Roushani M, Zalpour N (2021) Selective detection of Asulam with in-situ dopamine electropolymerization based electrochemical MIP sensor. React Funct Polym 169:105069. https://doi.org/10.1016/j.reactfunctpolym.2021.105069
113. Soliman SS, Mahmoud AM, Elghobashy MR et al (2023) Point-of-care electrochemical sensor for selective determination of date rape drug "ketamine" based on core-shell molecularly imprinted polymer. Talanta 254:124151. https://doi.org/10.1016/j.talanta.2022.124151
114. Zheng B, He X, Zhang Q, Duan M (2022) A novel molecularly imprinted membrane for highly sensitive electrochemical detection of paracetamol. Int J Electrochem Sci 17:1–11. https://doi.org/10.20964/2022.07.37
115. Pareek S, Jain U, Balayan S, Chauhan N (2022) Ultra-sensitive nano-molecular imprinting polymer-based electrochemical sensor for follicle-stimulating hormone (FSH) detection. Biochem Eng J 180:1–10. https://doi.org/10.1016/j.bej.2021.108329
116. Roushani M, Farokhi S, Rahmati Z (2022) Development of a dual-recognition strategy for the aflatoxin B1 detection based on a hybrid of aptamer-MIP using a Cu2O NCs/GCE. Microchem J 178:1–7. https://doi.org/10.1016/j.microc.2022.107328
117. Tan F, Zhai M, Meng X et al (2021) Hybrid peptide-molecularly imprinted polymer interface for electrochemical detection of vancomycin in complex matrices. Biosens Bioelectron 184:113220. https://doi.org/10.1016/j.bios.2021.113220
118. Jalalvand AR, Zangeneh MM, Jalili F et al (2020) An elegant technology for ultrasensitive impedimetric and voltammetric determination of cholestanol based on a novel molecularly imprinted electrochemical sensor. Chem Phys Lipids 229:104895. https://doi.org/10.1016/j.chemphyslip.2020.104895
119. Che Lah NF, Ahmad AL, Low SC, Shoparwe NF (2019) The role of porogen-polymer complexation in atrazine imprinted polymer to work as an electrochemical sensor in water. J Environ Chem Eng 7:103500. https://doi.org/10.1016/j.jece.2019.103500
120. Amatatongchai M, Sitanurak J, Sroysee W et al (2019) Highly sensitive and selective electrochemical paper-based device using a graphite screen-printed electrode modified with molecularly imprinted polymers coated Fe3O4@Au@SiO2 for serotonin determination. Anal Chim Acta 1077:255–265. https://doi.org/10.1016/j.aca.2019.05.047
121. Rezaei F, Ashraf N, Zohuri GH (2023) A smart electrochemical sensor based upon hydrophilic core–shell molecularly imprinted polymer for determination of L-tryptophan. Microchem J 185:108260. https://doi.org/10.1016/j.microc.2022.108260

122. Li G, Qi X, Wu J et al (2024) Highly stable electrochemical sensing platform for the selective determination of pefloxacin in food samples based on a molecularly imprinted-polymer-coated gold nanoparticle/black phosphorus nanocomposite. Food Chem 436:137753. https://doi.org/10.1016/j.foodchem.2023.137753
123. Li M, Chen T, Gooding JJ, Liu J (2019) Review of carbon and graphene quantum dots for sensing. ACS Sensors 4:1732–1748. https://doi.org/10.1021/acssensors.9b00514
124. Sun H, Wu L, Wei W, Qu X (2013) Recent advances in graphene quantum dots for sensing. Mater Today 16:433–442. https://doi.org/10.1016/j.mattod.2013.10.020
125. Tian P, Tang L, Teng KS, Lau SP (2018) Graphene quantum dots from chemistry to applications. Mater Today Chem 10:221–258. https://doi.org/10.1016/j.mtchem.2018.09.007
126. Feng J, Dong H, Pang B et al (2018) Theoretical study on the optical and electronic properties of graphene quantum dots doped with heteroatoms. Phys Chem Chem Phys 20:15244–15252. https://doi.org/10.1039/c8cp01403e
127. Feng J, Dong H, Yu L, Dong L (2017) The optical and electronic properties of graphene quantum dots with oxygen-containing groups: a density functional theory study. J Mater Chem C 5:5984–5993. https://doi.org/10.1039/c7tc00631d
128. Mostafa S, Behafarid F, Croy JR et al (2010) Shape-dependent catalytic properties of Pt nanoparticles. J Am Chem Soc 132:15714–15719. https://doi.org/10.1021/ja106679z
129. Pedone D, Moglianetti M, De Luca E et al (2017) Platinum nanoparticles in nanobiomedicine. Chem Soc Rev 46:4951–4975. https://doi.org/10.1039/c7cs00152e
130. Stepanov AL, Golubev AN, Nikitin SI, Osin YN (2014) A review on the fabrication and properties of platinum nanoparticles. Rev Adv Mater Sci 38:160–175
131. Issa B, Obaidat IM, Albiss BA, Haik Y (2013) Magnetic nanoparticles: surface effects and properties related to biomedicine applications. Int J Mol Sci 14:21266–21305. https://doi.org/10.3390/ijms141121266
132. Mohammed L, Gomaa HG, Ragab D, Zhu J (2017) Magnetic nanoparticles for environmental and biomedical applications: a review. Particuology 30:1–14
133. Batool SS, Sher F (2016) Determination of density of states, conduction mechanisms and dielectric properties of nickel disulfide nanoparticles. AIP Adv 6:055120
134. El-wekil MM, Hayallah AM, Abdelgawad MA et al (2022) Nanocomposite of gold nanoparticles@nickel disulfide-plant derived carbon for molecularly imprinted electrochemical determination of favipiravir. J Electroanal Chem 922:116745
135. Hammami I, Alabdallah NM, Al Jomaa A (2021) Gold nanoparticles: synthesis properties and applications. J King Saud Univ Sci 33:101560
136. Sardar R, Funston AM, Mulvaney P, Murray RW (2009) Gold nanoparticles: past, present, and future. Langmuir 25:13840–13851. https://doi.org/10.1021/la9019475
137. Farjadian F, Abbaspour S, Abdolahi A (2020) Recent developments in graphene and graphene oxide: properties, synthesis, and modifications: a review. ChemistrySelect 5:10200–10219. https://doi.org/10.1002/slct.202002501
138. Pei S, Cheng H (2011) The reduction of graphene oxide. Carbon 50. https://doi.org/10.1016/j.carbon.2011.11.010
139. Zhu BY, Murali S, Cai W et al (2010) Graphene and graphene oxide: synthesis, properties, and applications. Adv Mater 22:3906–3924. https://doi.org/10.1002/adma.201001068
140. Parihar V, Raja M, Paulose R (2018) A brief review of structural, electrical and electrochemical properties of zinc oxide nanoparticles. Rev Adv Mater Sci 53:119–130. https://doi.org/10.1515/rams-2018-0009
141. Sharmila PP, Tharayil NJ (2014) DNA assisted synthesis, characterization and optical properties of zinc oxide nanoparticles. Int J Mater Sci Eng:1–6. https://doi.org/10.12720/ijmse.2.2.147-151

122. Li G, Qi X, Wu J et al (2024) Highly stable electrochemical sensing platform for the selective determination of pefloxacin in food samples based on a [illegible] coated gold nanoparticle/black phosphorus nanocomposite [illegible]

MIP-Based Electrochemical Sensors in the Detection of Opioids, Benzodiazepines and Psychoactive Compounds

4

Khadijeh Nekoueian and Tomi Laurila

Abstract

The introduction of modern molecular imprinting technology into electrochemical sensors has resulted in a new era of smart diagnostic sensors. Molecularly imprinted polymers (MIPs) are a specialized group of polymers engineered to contain precise "artificial receptors" or "artificial antibodies" capable of highly selective recognition of target analytes. Adding this unique molecular recognition property to the advantages of electrochemical sensors has paved the way for developing advanced sensing platforms to selectively interact with a diverse range of target analytes, spanning from small biomolecules to organic macromolecules, even within complicated environments. These electrochemical (bio) sensors have extraordinary sensitivity, allowing them to detect trace concentrations of target analytes, and establishing them as key players in advancing health assessment and ecological surveillance, among other applications. Furthermore, their potential for miniaturization paves the way for the design of lab-on-a-chip, biomedical assays, and wearable sensors.

As the advancements in this field continue to progress, the potential for these smart diagnostic assays to revolutionize various industrial sectors becomes increasingly evident. The MIP-based electrochemical sensors hold immense promise for the future, offering unprecedented opportunities for precise, rapid,

K. Nekoueian (✉)
Department of Electrical Engineering and Automation, School of Electrical Engineering, Aalto University, Aalto, Finland
e-mail: Khadijeh.Nekoueian@Aalto.Fi

T. Laurila
Department of Electrical Engineering and Automation, School of Electrical Engineering, Aalto University, Aalto, Finland

Department of Chemistry and Materials Science, School of Chemical Engineering, Aalto University, Aalto, Finland

S. Patra et al. (eds.), *Molecularly Imprinted Polymers: Path to Artificial Antibodies*, https://doi.org/10.1007/978-981-97-4379-7_4

and accessible diagnostics in diverse scientific domains. This chapter discusses the latest developments in the use of MIP-based electrochemical (bio)sensors for detecting and monitoring opioids and psychoactive compounds, with a focus on biofluids including saliva, urine, sweat and blood as well as synthetic samples and beyond.

Keywords

Molecularly imprinted polymers · Electrochemical sensors · Opioids and opioid analgesics · Antipsychotics · Antidepressants

4.1 Introduction

The demand for the development of accurate and efficient biomedical techniques to analyze and monitor various opioids, antidepressant medications, and psychoactive compounds is paramount in today's medical landscape, spurring extensive global research efforts. Real-time monitoring of these compounds in real sample matrix is increasingly regarded as essential for medication diagnosis and tracking.

Various highly sensitive analysis methods have emerged to meet this demand, including the high-performance liquid chromatography (HPLC) method [1], multi-analyte liquid chromatographic tandem mass spectrometric technique (LC-MS/MS) [2], and gas chromatography-tandem mass spectrometer (GC–MS/MS) [3]. However, while these techniques offer precision, they often entail high costs and require trained personnel to execute time-consuming sample preparation and cleaning procedures, limiting their suitability for real-time monitoring applications [4].

In recent decades, electrochemical techniques and electrochemical (bio)sensors have developed as promising alternatives, addressing the limitations of traditional spectrometric and chromatographic methods [5]. These electrochemical approaches offer swift, easily transportable, on-location monitoring capabilities, and presenting a paradigm shift in biomedical analysis [6]. Moreover, their potential for miniaturization makes them applicable in both laboratory-based (in vitro) and living organism-based (in vivo) environments, effectively integrating electronics and biology [5, 6]. Voltammetric methods such as cyclic voltammetry (CV), pulse voltammetry (PV), and stripping voltammetry (SV) have been instrumental in elucidating the redox processes of analytes via tracking the response current in relation to the applied potential [7]. Meanwhile, electrochemical impedance spectroscopy (EIS), measuring electrical impedance across frequencies, provides valuable insights into charge transfer kinetics on electrode surfaces [8]. EIS's simplicity, cost-effectiveness, and label-free nature have facilitated the development of biosensors suitable for point-of-use applications [9], enhancing its utility in portable opioid detection by offering real-time information on biochemical processes at the electrode interface [9]. Amperometry, another electrochemical technique, evaluates current signals at constant potentials, proving invaluable for real-time monitoring and biomolecule detection [7, 10]. Electrochemiluminescence (ECL) combines electrochemistry and luminescence, offering high sensitivity in trace analyte detection with stability, low background noise, and cost-effectiveness [11, 12]. Its simplicity and compatibility with electrochemical assays enable simultaneous analysis of multiple analytes, making it a preferred choice in sensitive bioassays and immunosensing applications [11, 12]. These adaptable electrochemical technologies have applications in environmental monitoring, clinical diagnostics, and materials research [13]. The benefits of electrochemical methods are found in their outstanding precision and rapid signal processing, which are vital for identifying minute quantities of opioids in biofluids [14, 15]. Furthermore, their miniaturization and portability enhance their versatility, making them ideal for diverse environments such as medical facilities and wearable devices, while remaining cost-effective [16–18].

In recent decades, the preference for electrochemical sensors over other techniques has grown due to their compatibility with biological samples, making them well suited for detecting pharmaceutical compounds in physiological fluids or tissues [17–19]. Their minimal interference from complex matrices ensures accurate detection in clinical settings, while their seamless integration with advanced signal processing techniques enhances accuracy and reliability. This lays the foundation for the creation of intelligent sensing systems for continuous real-time monitoring [17].

The effectiveness of electrochemical methods depends largely on the choice of electrode material. There is a need to find efficient electrode materials to overcome the limitations of mercury-based and traditional noble metal electrodes like platinum, iridium, and gold. Carbon-based materials emerged as a solution because of their distinctive characteristics: they have a wide electrochemical potential range, produce minimal background signals, have low electrical resistance, and are highly conductive [4, 20]. Their electrochemical performance can also be easily customized with various modifiers, have a customizable electrochemically active surface area, and are more cost-effective and sustainable. Carbon-based electrodes, particularly carbon paste electrodes, were initially employed as substitutes for mercury-based electrodes in the late 1950s by Professor Ralph Norman Adams [4, 20].

One of the key strengths of electrochemical (bio)sensors is their compatibility with various modified electrode materials, allowing for the customization of sensing platforms tailored to study a wide range of redox (bio)systems and detection targets, spanning from organic compounds to biomolecules [21]. This versatility positions electrochemical sensors as valuable tools for real-time monitoring, both in research and practical applications, offering a promising avenue for future advancements in analytical chemistry and biosensing. In the realm of biosensor development, the strategic incorporation of diverse bio-based receptors onto sensor surfaces, including enzymes like tyrosinase or polyphenol oxidase, as well as DNA in the form of aptamers, has indeed emerged as a promising avenue for augmenting selectivity in analyte detection [22]. Nevertheless, a pivotal challenge encountered in the integration of these biorecognition components within electrode designs is the persistent issue of insufficient sensitivity, which can be attributed to a range of inherent limitations. These constraints encompass factors such as low electrical conductivity, inadequate electrocatalytic activity, and diminished chemical and thermal stability [21, 22].

This collective set of challenges invariably obstructs the efficient identification and quantification of target analytes, thereby confining the broader utility of these biosensors across various analytical and clinical applications. Consequently, a central objective in the ongoing advancement of biosensors remains the pursuit of strategies to surmount these sensitivity limitations while preserving the invaluable selectivity offered by artificial biorecognition elements, ultimately paving the way for the advancement of durable and high-performance MIPs-based diagnostic assays [23]. MIPs have become a reliable method for creating synthetic receptor molecules, achieved through copolymerizing functional and cross-linking monomers alongside a template molecule [24]. MIPs can be synthesized through both bulk chemical polymerization and electrochemical processes [25]. Bulk polymerization of MIPs, a common approach, has various challenges, including partial template elimination, limited availability of template molecules, slow interaction rates, insufficient affinity, and a variety of interaction receptors [25, 26]. To address these challenges, nanotechnology has been integrated with molecular imprinting techniques to create nano-molecularly imprinted polymers (nanoMIPs). NanoMIPs surpass bulk MIPs due to their reduced dimensions and enhanced surface area, resulting in a greater quantity of available recognition receptors for enhanced interactions with target molecules [26]. Consequently, nanoMIPs hold great promise for applications requiring precise molecular recognition and control over release [26]. Additionally, nanoMIPs offer superior rebinding properties, lower mass transfer resistance, and better template accessibility, making them ideal for various applications. Compared to bulk MIPs, nanoMIPs exhibit faster binding kinetics, compatibility with nanotechnology, reduced sample volumes, enhanced stability, and multiple cycle use without significant degradation thus enhancing their cost effectiveness and reliability. Furthermore, nanoMIPs can be precisely tailored during synthesis to optimize properties such as size, shape, and surface chemistry, catering to specific application needs [25, 26].

MIPs have gained significant attention due to their preparation via multiple chemical and electrochemical methods [27, 28]. The electropolymerization method enables the direct creation of a slim film on the sensor's surface, providing a straightforward, reproducible, and environmentally friendly means of depositing MIP films onto support surfaces. Moreover, the thickness, uniformity, and morphology of the deposited layer can be precisely controlled by adjusting polymerization conditions, such as applied potential and cyclic scanning [29, 30]. Furthermore, sol–gel technology has found extensive use in enhancing the performance of electropolymerized MIP films. Sol–gel processing simplifies the generation of a rigid three-dimensional silicate network, efficiently incorporating an organic–inorganic hybrid network [31]. The sol–gel-based MIPs offer numerous advantages, including a straightforward fabrication process, low-temperature reaction conditions, and matrix porosity arising from a highly cross-linked structure. Additionally, the sol–gel procedure substitutes conventional toxic solvents (like chloroform or toluene) with eco-friendly alternatives such as water or ethanol. The non-conductivity of sol–gel materials can be overcome by introducing conducting polymers or metal nanostructures [31].

In recent decades, molecular imprinting techniques have advanced significantly toward the use of smart polymers [32]. The integration of the MIP concept with "smart" polymers, capable of responding to environmental changes, has resulted in MIPs responsive to external triggers such as pH, temperature, ionic strength, UV irradiation, and electric fields [32, 33]. For instance, in drug delivery, researchers design MIPs with temperature-responsive components. Below a certain temperature, these polymers swell, releasing drugs. However, as the temperature rises, they collapse, holding onto the drugs. This controlled release mechanism enables tailored drug delivery, where the release rate can be adjusted by adjusting the temperature. Smart MIPs have potential applications beyond drug delivery, including sensing and catalysis, where precise control over molecular interactions is crucial for performance optimization [34].

The interaction of MIPs with analytes and the subsequent detection methods can adapt to the specific characteristics of the analyte and its electroactivity properties (Fig. 4.1). When MIPs encounter electroactive analytes, their direct interaction typically yields a stronger signal. Conversely, with non-electroactive analytes, detection relies on analyte binding, necessitating the incorporation of external or internal redox probes. In an indirect detection system, the interaction between the analyte and MIPs results in diminished electrical conductivity within the MIP layer, leading to a corresponding decline in the recorded response signal from the redox probe [36].

MIP-based electrochemical sensors and their developed vision as MIP-based point-of-care (POC) diagnosis can be explained as an analytical technique that is performed in patient-based healthcare settings, outside the conventional laboratories. The MIP-based POC methodology allows simple, low-cost, automated analysis, and rapid results.

These sensors are engineered to mimic natural biorecognition receptors and exhibit a high degree of specificity, allowing them to specifically identify and attach to the designated molecules, such as pharmaceutical compounds or disease biomarkers. The key advantage of MIP-based electrochemical sensors lies in their ability to provide rapid, real-time results. For instance, they can continuously monitor

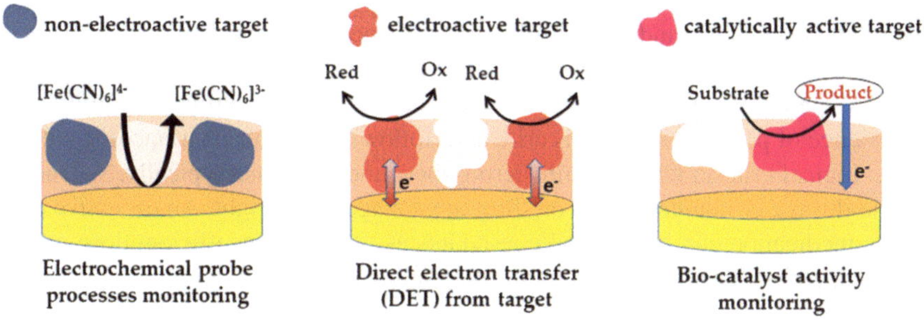

Fig. 4.1 Schemes of various detection methods in MIP-based electrochemical sensors. (Reprinted from Mazzotta et al. [35]. Copyright © 2022, with permission from Elisabetta Mazzotta et al.)

drug concentrations in a patient's bloodstream, facilitating personalized medication management by ensuring patients receive the right dosage at the right time. Moreover, these sensors have far-reaching applications beyond medication monitoring. They can be tailored to detect specific disease biomarkers, offering early diagnosis and continuous disease monitoring. The POC testing capabilities of MIP-based sensors make them invaluable in remote or resource-limited settings, enabling quick and convenient testing for a wide range of health-related factors.

Since the subject of MIP-based electrochemical sensors improves, researchers are seeking ways to boost the electrochemical properties of MIP-based electrodes. The creation of new substances and their integration with microsystem technologies hold great promise for the future of healthcare. These sensors are poised to revolutionize the healthcare industry by providing real-time, personalized data for diagnosis, treatment optimization, and proactive health management. In this section, we will discuss new MIP modification avenues for electrochemical detection and monitoring opioids and opioid analgesics, antidepressant medications, and psychoactive compounds, as well as their application areas, which include street samples, biological fluids (human blood, urine, and saliva), synthetic samples, and so on.

4.2 Opioids and Opioid Analgesic

The global public health crisis related to the misuse and addiction of opioids, including substances like tramadol, codeine, oxycodone, nalbuphine, fentanyl, dextromethorphan, and meperidine, has significant economic and social implications [34]. In 2018, the United States alone incurred an estimated annual "economic burden" of \$78.5 billion due to prescription opioid misuse, encompassing expenses related to healthcare, lost productivity, addiction treatment, and involvement with the criminal justice system. According to the World Health Organization (WHO), in 2017, nearly 115,000 individuals lost their lives to opioid-involved overdoses, and millions of patients worldwide continue to grapple with the lack of effective and safe pain management options [37].

Opioids are a class of compounds utilized to manage the analgesic component of anesthesia by targeting opioid receptors found in the brain and spinal cord. They interact with and activate opioid receptors distributed throughout different areas of the brain, spinal cord, and other bodily organs, especially those linked to feelings of pain and pleasure [38]. Opioids interfere with the release of pain signals from the brain to the body via the interaction with pain receptors, resulting in a diminished experience of pain. However, it is critical to follow acceptable pharmacological and toxicological criteria while administering these medications. Typically, a standard opioid dosage is changed based on the patient's age and weight. An excessive dose of the medication can have lethal consequences, while an insufficient dose proves ineffective [38]. The chemical structures of some opioids and opioid analgesics are shown in Fig. 4.2.

Nalbuphine, a phenanthrene opioid derivative closely related in structure to oxymorphone and naloxone, exhibits an analgesic potency on par with morphine.

Notably, it demonstrates a ceiling effect in both analgesia and respiratory depression, typically observed within the dosage range of 150–300 μg/kg [40]. The plasma therapeutic drug level in humans is reportedly 20 ng/mL [41]. Morphine, a powerful opioid initially obtained from opium by Volkow and McLellan [42], undergoes liver metabolism via conjugation, with its main metabolite being morphine-6-glucuronide, and possesses the same effect as the original medicine [39]. The effective plasma concentration of morphine is 40 ng/mL, and excessive levels can result in respiratory depression and euphoria [40]. Fentanyl, a highly potent opioid, undergoes liver metabolism to form norfentanyl. Its effective plasma concentration has an analgesic range of 1–2 ng/mL, while for anesthesia, it is 10–20 ng/mL [43]. Additionally, synthetic opioids such as methadone, buprenorphine, oxycodone, and codeine find application in surgical settings for various purposes, each having distinct therapeutic ranges of 100–1000 ng/mL [44], 10–40 ng/mL [45], 1–2 ng/mL, and 159 ng/mL, respectively [46]. Codeine, a commonly used opiate, is a versatile pharmacological drug with antitussive, antidiarrheal, and analgesic properties. Along with its therapeutic effects, codeine and its metabolite morphine have analgesic qualities that perform an important part in the management of mild to moderate pain. Nonetheless, the major hazards associated with opioid prescriptions, including codeine, such as abuse, dependency, and overdose deaths, highlight the importance of careful medical administration. In the field of forensic research, many analytical methods for detecting opioids and opioid analgesics in biological matrices have been developed, allowing its detection in samples such as urine and blood, with different nations applying different cut-off values for screening tests from 300 to 2000 ng/mL [47].

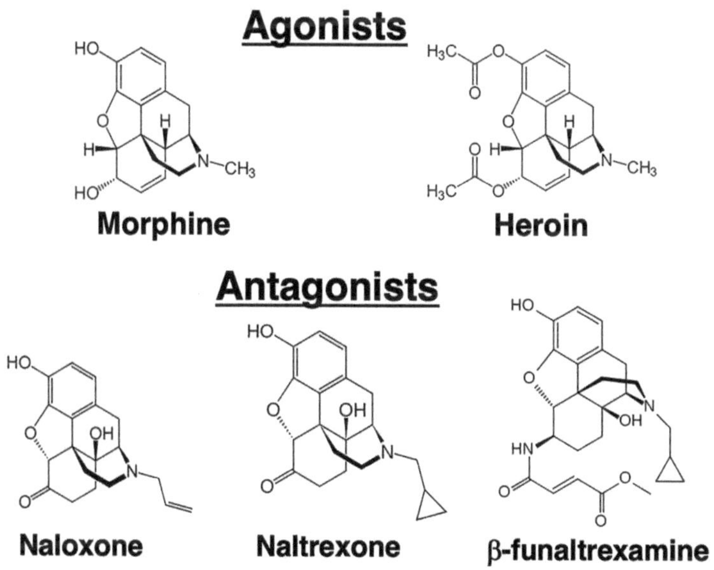

Fig. 4.2 Chemical structures of some opioids and opioid analgesics. (Reprinted from Eisenstein [39] with permission from Frontiers)

4.3 Antipsychotics and Antidepressants

Antidepressant medications and psychoactive compounds like benzodiazepines are designed to impact the central nervous system. They achieve this by either deactivating neuronal transmitters or inhibiting excitatory synaptic transmission. These substances, known for their pronounced relaxing and sedative properties, are commonly prescribed to alleviate conditions such as anxiety, panic disorders, insomnia, and muscle spasms. Ensuring careful monitoring of the prescribed dosage is essential to mitigate the risk of memory impairments, addiction, and drug dependence [48]. In the context of forensic toxicology, benzodiazepines frequently appear in cases of fatal drug intoxication and are also detected in the bloodstream of individuals engaged in fatal crashes on roads. These circumstances underscore the importance of advancing novel analytical techniques to accurately assess their presence and concentration in real time [49]. The chemical structures of some benzodiazepines are shown in Fig. 4.3.

Chlorpromazine hydrochloride (CPZ) is an antipsychotic drug used to treat psychotic disorders. It effectively regulates symptoms of agitation, excitement, and other psychomotor disturbances in individuals with schizophrenia, as well as mitigates the manic phase of manic-depressive disorders. Additionally, it is prescribed to control hyperkinetic states, manage aggression, and occasionally address anxiety and tension. Furthermore, CPZ serves as an antiemetic in palliative care settings. The daily dosage of CPZ can vary within the range of 75–300 mg, with instances of higher doses, including up to 1 g or more, prescribed for severe psychotic cases. Approximately 1% of an oral dose of CPZ is excreted unchanged in the urine [51].

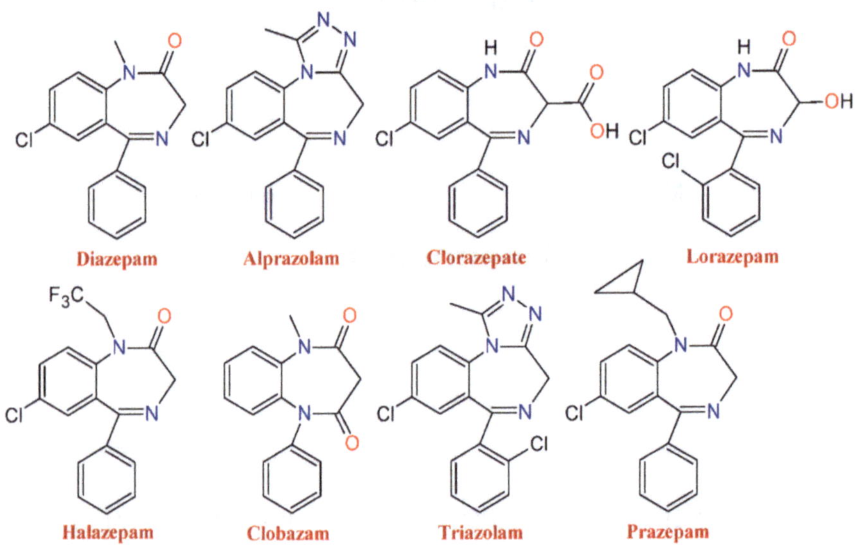

Fig. 4.3 Chemical structures of some benzodiazepines. (Reprinted from Sanabria et al. [50] with permission from MDPI)

The presence of residual CPZ has been associated with various adverse effects, including cholestatic jaundice, leukocytopenia, postural hypotension, extrapyramidal effects, contact dermatitis, and tardive dyskinesia. To promote effective and safe medication therapy, strong quality control must be implemented during the drug manufacturing process [51, 52].

Quetiapine, known by its brand name Seroquel, is categorized as an atypical antipsychotic medication and has been a top-performing drug in its class since its introduction in 1997 [53]. It is used to treat various mental health conditions, such as bipolar disorder, insomnia, psychotic disorders, schizophrenia, obsessive–compulsive disorder, depression, post-traumatic stress disorder, anxiety, schizoaffective disorder, and more [53]. Normally, the recommended daily dose of quetiapine ranges from 300 to 450 mg, split into two doses, with a maximum daily limit of 750 mg. However, elderly patients typically start with a lower initial dose of 25 mg. The therapeutic dosage of quetiapine is recognized at 0.4 mg/L, while concentrations exceeding 7 mg/L are considered potentially lethal [53, 54]. Sertraline (STR) is an antidepressant medicine often recommended for depression due to its low side effects. However, recent research has revealed that selective serotonin reuptake inhibitors (SSRIs) like sertraline have the ability to penetrate cell membranes, significantly elevating the risk of cancer development. Consequently, precise tests for sertraline and its metabolites in biological samples such as blood, plasma, and urine are necessary [55].

4.4 MIP-Based Electrochemical (Bio)sensors in the Detection of Opioids, Opioid Analgesics, Benzodiazepines, and Psychoactive Compounds

Personalized medication administration is a scientific approach to addressing the opioid and psychoactive compounds crisis effectively. Ensuring the precise dosage of pain relievers for patients suffering from chronic or acute pain in real time is of utmost importance. This not only facilitates efficient pain management but also prevents the administration of toxic and potentially fatal doses of pain relievers thereby ensuring safe patient care. The utilization of cost-effective and user-friendly electrochemical techniques in biological fluids encounter various challenges, primarily centered around selectivity, electrode durability, and issues related to electrode passivation and biofouling. Achieving adequate sensitivity and selectivity remains a persistent challenge, primarily due to the existence of coexisting biomolecules like ascorbic acid (AA), proteins, and other small molecules in blood samples. The main hurdle arises from the proximity of the oxidation potentials of these compounds to one another, leading to overlapping voltammetric responses and making it challenging to distinguish their signals. To combat this interference, tailored carbon allotropes, such as carbon nanotubes, can be employed to shift the oxidation potentials of AA away from the target opioid or benzodiazepine.

Another obstacle in measuring low concentrations of target analyte in complex biological fluids like blood is biofouling, which occurs when matrix species, including proteins or lipids, passively adsorb to the electrode surface, forming an insulating layer that diminishes sensitivity toward the target detection. This layer significantly hampers measurements in complex biological fluids (such as serum, urine, blood, plasma, and saliva) and compromises the electrode's sensitivity to the target of interest [56]. The adverse effects of electrode biofouling and passivation can be mitigated by depositing various polymers on the surfaces of carbon-based sensing platforms. For instance, coating single-walled carbon nanotubes with nafion has demonstrated effective antifouling properties in the detection of morphine in blood tests [57].

Successful modification of electrode materials and the utilization of various carbon structures, including carbon nanotubes, graphene, and nanodiamonds, present encouraging avenues for augmenting the sensitivity of the engineered MIP-based sensor. These efforts involve harnessing a diverse range of carbon-based nanostructured electrode materials, known for their outstanding attributes such as high electrical conductivity, mechanical stability, and thermal resilience. These materials provide a substantial effective surface area, impressive adsorption capabilities, and the adaptability to enhance or tailor their electrocatalytic and photocatalytic activities through functionalization or by incorporating various substances, including functional agents and biomolecules. Notably, carbon-based nanostructures like tetrahedral amorphous carbon (ta-C), carbon nanotubes (CNTs), and carbon nanofibers (CNFs) exhibit exceptional characteristics, including high electrical conductivity, biocompatibility, inertness, and exceptional mechanical and thermal stability. These qualities position them as effective charge carriers and reliable substrates for MIPs by facilitating both non-covalent and covalent interactions among diverse substances [6, 20].

The functionalization of multi-walled carbon nanotubes (MWCNTs) with carboxylic acid functional groups develops the possible uses of MWCNTs in fields such as biomedicine and environmental science [58]. Carboxylic acid functional groups (–COOH) are pivotal for enhancing the dispersion and stability of MWCNTs in aqueous media. They achieve this by improving dispersion, preventing agglomeration through electrostatic repulsion, ensuring long-term stability, and providing compatibility with various aqueous solvents and biological fluids. Additionally, these functional groups serve as anchor points for facilitating surface modifications or biomolecule attachment, allowing researchers to customize MWCNT properties for applications as a substrate for MIP [59]. Furthermore, their remarkable electrochemical properties, characterized by a wide water window and minimal background signal, make them highly suitable for the creation of MIP-based sensing architecture. Researchers are actively addressing the challenges inherent in electrochemical techniques by innovating advanced sensing platforms. Their focus lies in mitigating biofouling and harnessing the exceptional properties of carbon-based nanomaterials to enhance sensitivity, selectivity, and overall stability in these systems in complex biological fluids [6]. MIPs offer a cost-effective alternative to natural receptors, exhibiting robust selectivity and durability, effectively serving as "artificial receptors" or "plastic antibodies" for precise interaction and recognition of target opioids. Leveraging advanced molecular imprinting techniques, these mimetic artificial receptors could modify sensing platforms through electrochemical or chemical polymerization processes. This replication of the fundamental functions of analgesic receptors on the sensing platform's surface enables them to bind to target molecules specifically and avidly within their designated binding sites. These three-dimensional printed artificial receptors hold the promise of exceptional selectivity and the ability to conduct real-time monitoring of minute traces of target substances in complex biological fluids [6].

The impressive blend of superior selectivity, sensitivity, and potential stability renders these MIP-based sensors an exceptional option for accurately assessing opioid levels as well as psychoactive compounds. Its practical utility extends to healthcare settings, where it facilitates swift diagnoses of overdose cases among patients [58].

Nalbuphine hydrochloride (NAL), as illustrated in Scheme 4.1, belongs to the phenanthrene derivative class of opioid analgesics. It is employed in the management of various types of pain linked to severe and chronic medical conditions. NAL exerts its therapeutic effects by binding to kappa-opioid receptors, making it a valuable treatment option for ailments like renal colic, cancer-related pain, postsurgical pain, migraines, and vascular headaches [61]. Hassan et al. used a carboxylated MWCNTs/polyaniline (f-MWCNTs/PANI) nanocomposite to create a new screen-printed electrode (SPE) for nalbuphine potentiometry. The synthesized f-MWCNTs/PANI nanocomposite, which served as an ion-to-electron transducer in the building of MIP-based sensors, efficiently inhibited the creation of an unwanted thin water layer across the sensing membrane and the conducting substrate. This avoided membrane delamination and increases potential stability. The nanocomposite was covered with a specific polyvinyl chloride (PVC) membrane containing molecularly imprinted drug–polymer beads for precise detection of NAL. The SPE exhibited an impressive Nernstian response, characterized by a steep slope of 60.3 ± 1.2 mV/decade (R^2 = 0.999), spanning a wide concentration range from 2.4×10^{-7} to 5.0×10^{-2} mol/L. It also boasted an exceptional detection limit of 1.1×10^{-7} mol/L

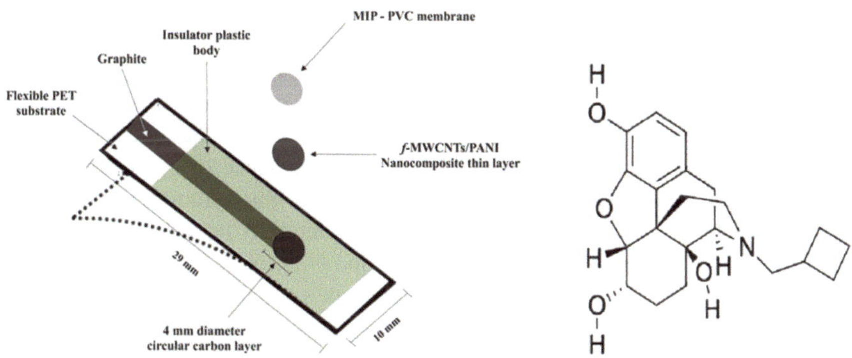

Scheme 4.1 Construction of the screen-printed electrode (NAL-SPE) and NAL chemical structure [60]. (Reprinted from Hassan et al. [60] with permission from Elsevier B.V.)

(equivalent to 0.04 μg/mL), with a rapid response time of less than 20 s (<20 s) facilitated by chronopotentiometry (CP) and EIS [60].

In 2017, Lopez and colleagues introduced a pioneering MIP-based electrochemical sensor designed for the detection of Naloxone (NLX), an opioid antagonist used in the treatment of opioid and heroin overdose. Their innovative approach involved modifying the screen-printed carbon electrode (SPCE) with a layer of MWCNT to enhance sensitivity and establish a more reliable immobilization substrate for molecular imprinting. Subsequently, they electropolymerized the MIP layer on the surface of the MWCNT/SPCE using NLX as a template and 4-aminobenzoic acid (4-ABA) as a functional monomer. This MIP/MWCNT/SPCE sensor was successfully employed to detect NLX in complex biological matrices such as spiked urine and human serum, using the standard addition method. The results were highly encouraging, with recoveries ranging from 92% to 111% and relative standard deviations (RSD) between 3% and 10%, respectively. These findings underscore the method's efficiency, eliminating the need for cumbersome pretreatment steps or methods. The designed MIP/MWCNT/SPCE sensor determined the NLX concentrations within the range of 0.25–10.0 μM, with a limit of detection (LOD) of 0.20 μM and a limit of quantification of 0.67 μM. As a result, this sensor holds significant promise for POC screening of NLX in individuals experiencing opioid overdose, offering a valuable tool for timely intervention [62].

In 2020, Shaabani and their research team advanced the field of MIP-based sensors for NLX detection with an innovative approach. They enhanced the surface of a commercial SPE by employing an in-situ electrochemical method to create a composite of reduced graphene oxide-gold nanoparticles (rGO-AuNPs). This modification served as a substrate to elevate the performance of the MIP. Subsequently, they tailored the rGO-AuNPs/SPE surface by introducing the MIP layer, utilizing NLX as the template and *para*-phenylenediamine (*p*-PDA) as the functional monomer. The presence of $-NH_2$ groups in *p*-PDA facilitated effective interactions with both NLX (via hydrogen bonding and intermolecular forces) and AuNPs (via Au–N bonds). Previous research has shown that PDA monomers were highly suitable for MIP fabrication, yielding dense and durable polymer layers with hydrophilic, hydrophobic, and basic recognition sites. The resultant MIP/rGO-AuNPs/SPE demonstrated outstanding capabilities, detecting NLX with a LOD of 0.16 μM. It exhibited exceptional repeatability from run to run, with relative standard deviations ranging from 5.7% to 9.6% ($n = 4$), and consistent performance reproducibility from batch to batch, with variations of less than 9% ($n = 3$). The proposed sensor demonstrated high selectivity toward NLX in the presence of other opioids and opioid analgesics such as morphine, naltrexone, and noroxymorphone. This outcome underscored the MIP sensor's exceptional selectivity for NLX. Furthermore, the fabricated MIP sensor successfully quantified NLX in synthetic urine samples with a recovery of 92%. This underscores the sensor's practical utility and reliability for NLX detection, suggesting its potential for real-world applications in drug monitoring and healthcare scenarios (Scheme 4.2) [63].

In 2014, Rezaei et al. introduced a sol–gel-based MIP for the detection of lorazepam, as illustrated in Scheme 4.3. This innovative MIP sensor was crafted through electrochemical techniques and demonstrated the capability to detect lorazepam within concentration ranges spanning from 0.2 to 2.0 and 2.0 to 20.0 nM, achieving a remarkable detection limit of 0.09 nM. The sensor's adaptability was confirmed through assessments of lorazepam levels in tablet formulations, plasma, and urine, yielding mean recoveries of 102% [64]. Building upon this pioneering work, Shaabani et al. further advanced sol–gel-based MIP sensors in 2021, with a particular focus on NLX detection. Notably, they broke new ground by developing a sol–gel-based MIP methodology for NLX sensing [65]. This sol–gel MIP solution was created by combining pyrrole as the conducting matrix, NLX as the template molecule, triethoxyphenylsilane as the functional monomer, and tetraethoxysilane as the cross-linker. Subsequently, the surface of the fMWCNT/indium-tin-oxide electrode (fMWCNT/ITO) was modified with sol–gel MIP using the electropolymerization technique. The incorporation of pyrrole into the sol–gel MIP solution not only facilitated adequate conductivity but also established hydrogen bonds with NLX's hydroxyl group. Additionally, interactions occurred between the phenyl group in the TEPS monomer and NLX's aromatic ring, significantly enhancing the MIP-based sensor's effectiveness in NLX detection. The resulting pyrrole@sol-gelMIP/fMWCNT/ITO sensor exhibited outstanding characteristics, including high repeatability, selectivity, and stability against structurally similar substances such as codeine, fentanyl, naltrexone, and noroxymorphone. Under optimal experimental conditions, it displayed a dynamic linear range from 0 to 12 μM, coupled with an exceptional detection limit of 0.02 μM. Moreover, when applied to analyze NLX in synthetic urine samples, this sensor consistently achieved recoveries exceeding 88%. This novel methodology represents a significant advancement in NLX sensing and holds promise for a wide range of applications, particularly in the domains of drug monitoring and healthcare [65].

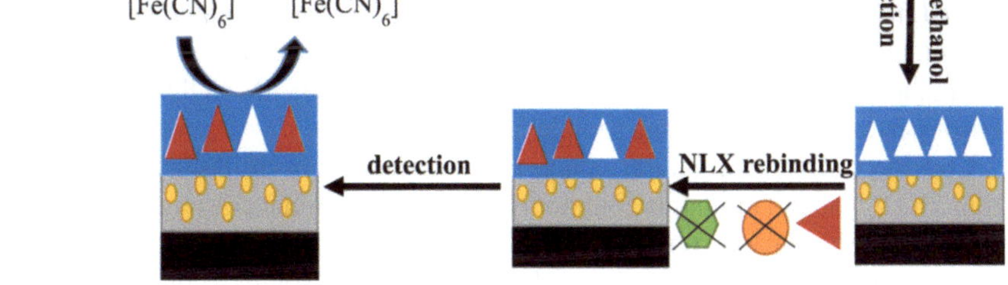

Scheme 4.2 Schematic illustration of the preparation of the MIP/rGO-AuNPs/SPE and the NLX detection. (Reprinted from Shaabani et al. [63] with permission from the Journal of the Electrochemical Society)

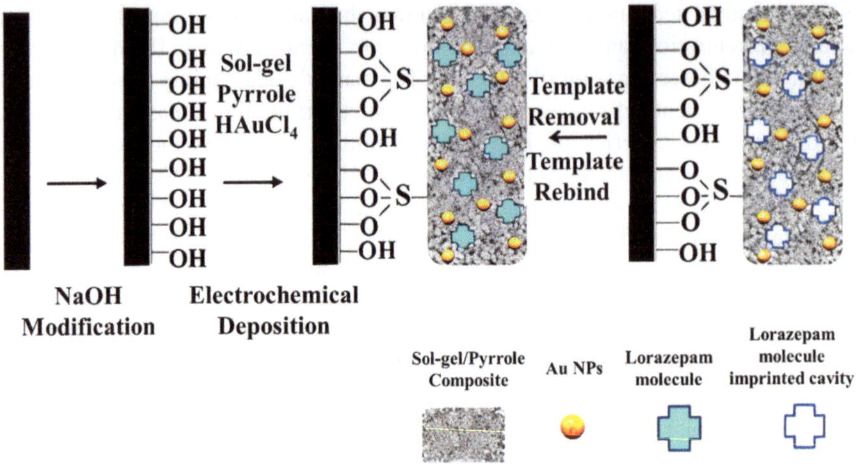

Scheme 4.3 Schematic representation of the proposed sol–gel MIP/PGE sensor for lorazepam. (Reprinted from Rezaei et al. [64] with permission from the Elsevier)

The first MIP suspension for the detection of morphine (MO) was fabricated based on molecular imprinting of L-phenylalanine anilide using photolytic hemolysis [66, 67]. Rezaei and his colleagues introduced an effective MIP-based sensor to detect minute quantities of MO. They utilized f-MWCNTs as a substrate for the electropolymerization of MO-imprinted polymers onto the surface of a pencil graphite electrode (PGE). This MIP-based sensor exhibited the capability to detect MO within two linear ranges, spanning from 0.008 to 5 μmol/L, with an impressive detection limit of 2.9 nM. Additionally, this electrochemical sensor technique demonstrated exceptional performance in the detection of MO in actual samples of undiluted human urine and plasma, all without requiring any preliminary extraction steps [68]. Salajegheh et al. designed another MIP electrochemical sensor for MO detection by electropolymerizing L-lysine on a glassy carbon electrode (GCE) surface modified with activated carbon and sodium alginate. The sensor featured a molecularly imprinted composite film, designed computationally for specific MO detection. The electrode modification involved electropolymerization of L-lysine in the presence of MO on a sodium alginate and activated carbon (SA–AC) composite applied to the GCE. The SA–AC's unique surface area enhances sensitivity. Differential pulse voltammetry (DPV) showed good sensitivity, detecting MO within the 0.1–1000.0 M concentration range with a low 48 nM LOD (S/N = 3). The sensor performed well in urine sample analysis [69].

Codeine (COD) is a natural opiate alkaloid derived from hashish or produced from MO by methylation. It is commonly used as a pain reliever, cough suppressant, and anti-diarrheal agent. Yence et al. designed an efficient sensor for COD detection in tablets and biological fluids. The MIP polymerization solution consisted of COD as a template molecule for the MIP, along with an amino acid *N*-methacryloyl-L-tryptophan-based functional monomer (MA-Trp), a basic monomer 2-hydroxyethyl methacrylate (HEMA), a cross-linker ethylene glycol dimethacrylate (EGDMA), and a pore-forming material (ZnO). Its electrochemical performance was assessed using DPV in a dynamic linear range of 0.1–1.0 pM, and a detection limit of 0.0150 pM was obtained under ideal conditions. It also had a high recovery rate in human urine (98.5%), synthetic serum samples (101.4%), and tablets (101.5%). Furthermore, for COD electrochemical measurement, the MA-Trp@MIP/GCE sensor demonstrated high reproducibility, repeatability, stability, and quick reaction times, making it a viable instrument for COD detection in biological fluids [70].

Ansari et al. expedited the advancement of molecular imprinting for the separation and analysis of oxycodone. This MIP was designed using magnetic graphene oxide (MGO) and carbon dot nanoparticles (CDs), employing ultrasonic-assisted dispersive solid-phase microextraction in conjunction with HPLC. The adsorption behavior of MIP@MGO/CDs adhered to the Langmuir model, demonstrating an impressive maximum adsorption capacity of 99 mg/g, along with an imprinting factor of 3.39. By employing a central composite design within the response surface technique, the optimal conditions were achieved, leading to a LOD calculated as 0.80 ng/mL within the linear range of 1–2000 ng/mL. In the analysis of human urine samples, the average oxycodone recoveries exhibited a remarkable range from 92.50% to 103.20%, with a commendably low RSD not exceeding 3.65% [71]. The proposed MIP@MGO/CDs composite has a high potential for electrode surface modification and the design of novel magnetic MIP-based sensors [72].

Diouf and collaborators engineered an electrochemical sensor designed for tramadol (TRA) detection. Their approach involved the creation of a self-assembled poly-aniline layer coated with silver nanoparticles (PANI-AgNPs) on a screen-printed gold electrode (Au-SPE). This groundbreaking sensor featured the self-polymerization of a TRA-imprinted poly(2-aminothiophenol) (PATH) layer on the PANI-AgNPs/Au-SPE platform, resulting in a sensitive MIP sensor for TRA detection. Furthermore, they developed a voltammetric electronic tongue (VE-Tongue) combined with advanced chemometric techniques to detect TRA within a concentration range spanning from 0.01 to 100 μg/mL. Impressively, the sensor demonstrated detection and quantification limits of 9.42 and 28.55 μg/mL, respectively, while maintaining exceptional selectivity, even in the presence of potential interferences such as paracetamol and citric acid. The sensor's efficacy was rigorously evaluated using spiked saliva and urine samples, consistently achieving substantial recovery rates exceeding 90% and remarkably low standard deviations below 4.5%. The designed VE-Tongue performed well in urine samples as a complex matrix for drugs [73].

To effectively utilize MIPs in sensors, enhancing binding kinetics, reducing analysis duration, and ensuring full template removal is essential. The potential for designing nano-sized MIPs (nanoMIPs) has proven to be a crucial factor in addressing these challenges, primarily by increasing the surface-to-volume ratio. This improvement makes binding sites more readily accessible to analytes, resulting in quicker and more efficient sensor performance [74]. Among various MIP nanostructures, nanoMIPs have gained significant attention due to their preparation via multiple chemical and electrochemical methods. Among the various methods proposed for synthesizing nanoMIPs, solid-phase synthesis stands out as highly advantageous. This method enables the creation of nanoMIPs with pseudo-monoclonal binding characteristics, consistent binding sites, and a strong affinity for the template molecule. An advantage of this imprinting strategy lies in its flexibility, allowing for easy adjustment of the polymerization mixture's composition by introducing specific monomers. This customization can impart distinct properties to the resulting nanoparticles, free from the solubility limitations associated with precipitation and mini-emulsion polymerization [27, 28].

Smolinska-Kempisty et al. developed a nanoMIP-based potentiometric sensor for detecting cocaine for the first time in 2017. These nanoMIPs were synthesized using the solid-phase imprinting technique, involving chemical polymerization in water and UV-initiated polymerization in an organic solvent. Following this, the prepared nanoMIPs were incorporated into a conductive membrane designed to be placed within an ion-selective electrode (ISE) for interacting with cocaine. The sensor exhibited the capability to quantify cocaine within a concentration range of 0.6–5.3 nM. Furthermore, it was successfully employed to measure cocaine levels in blood serum samples, covering concentrations ranging from 1 nM to 1 mM [75]. In 2020, D'Aurelio and colleagues introduced a remarkably selective nanoMIP-based sensor designed for the precise detection of cocaine. This advanced sensor was crafted by modifying the surfaces of gold electrodes. Initially, the gold electrode surfaces underwent treatment with 11-mercapto decanoic acid (MUDA) to functionalize them with carboxylic groups, forming a self-assembled monolayer (SAM). Subsequently, the carboxylic groups of MUDA were activated through a treatment involving a mixture of *N*-hydroxysuccinimide (NHS) and 1-ethyl-3-(3-dimethylaminopropyl) carbodiimide (EDC). This activation process enabled these groups to interact effectively with nanoMIPs using amine coupling chemistry (Scheme 4.4). This developed nano MIP-based sensor exhibited significant potential for the selective detection of cocaine compared to morphine and levamisole within a concentration range spanning from 100 pg/mL to 50 ng/mL, demonstrating its applicability in both buffered solutions and water environments [76].

The ability to easily modify the composition of nanoMIPs while maintaining their binding properties was harnessed in the development of redox-label-doped nanoMIPs. This was accomplished by introducing redox-active probes, including ferrocene electroactive groups or hexacyanoferrate (HCF), into the polymerization mixture [77]. The inclusion of a redox probe within the MIPs structure eliminated the need for an external redox probe for detection of non-electroactive analyte, simplifying the sensor's POC application. This strategy has been used in the design of several MIP-based sensors.

Alanazi et al. introduced a disposable electrochemical sensor of remarkable sensitivity for paracetamol, capitalizing on redox-label-doped nanoMIPs. These nanoMIPs were synthesized through solid-phase synthesis, comprising itaconic acid as a dedicated functional monomer and ferrocene as a redox label thereby rendering the nanoparticles electroactive. Sensor fabrication involved the covalent attachment of nanoMIPs onto SPCE, subsequently facilitating electrochemical paracetamol determination via DPV (Scheme 4.5). The sensor's performance was successfully

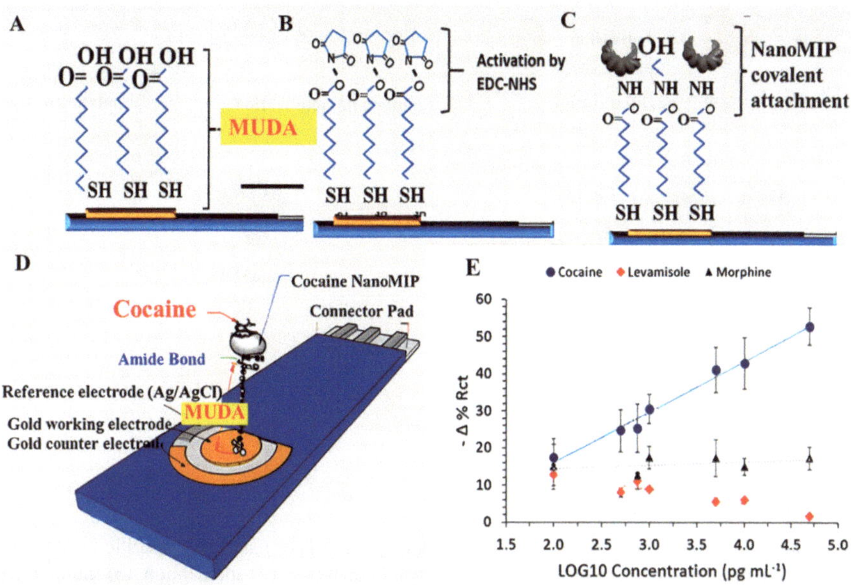

Scheme 4.4 The process of affixing the nanoMIPs to the gold working electrode surface involves several key steps: (**a**) formation of a self-assembled monolayer (SAM), (**b**) activation of carboxylic groups using EDC-NHS, (**c**) covalent attachment of nanoMIPs through amine coupling. Additionally, (**d**) a 3D representation illustrates the final nanoMIPs sensor for cocaine detection. Finally, (**e**) a selectivity study is presented, showcasing a comparison of linear calibration curves for cocaine, levamisole, and morphine, with data obtained from five repetitions ($n = 5$). (Reprinted from D'Aurelio et al. [76] with permission from MDPI)

assessed in spiked human plasma, boasting recoveries ranging from 94% to 108%, and showcasing a sensitivity of 6.18 ± 0.22 μA/mM. Notably, the sensor demonstrated a LOD of 50 μM and a limit of quantification of 167 μM, operating within a linear concentration range spanning 0.1–1 mM. Exceptional selectivity with no interference detected even in the presence of compounds such as caffeine, procainamide, or ethyl 4-aminobenzoate. Furthermore, the sensors exhibited notable reproducibility (RSD, 4.8%), swift response times (~8 s), and acceptable shelf life (90 days), affirming their suitability for POC applications [78].

Truta et al. engineered redox-label-doped nanoMIPs through computational methods to precisely target amphetamine and then synthesized via solid-phase synthesis. A composite film consisting of chitosan, nanoMIPs, and graphene oxide was strategically employed to immobilize nanoMIPs onto SPCE. The incorporation of ferrocenyl methyl methacrylate into the nanoMIPs resulted in the formation of redox-labeled nanoMIPs, enabling electrochemical detection. The electrochemical signal linked to ferrocene oxidation was notably influenced by the interaction of amphetamine with nanoMIPs. Thorough examination using street samples demonstrated outstanding sensitivity and satisfactory recoveries, ranging from 100.9% to 107.6%. These findings were corroborated through UPL-MS/MS analysis. This

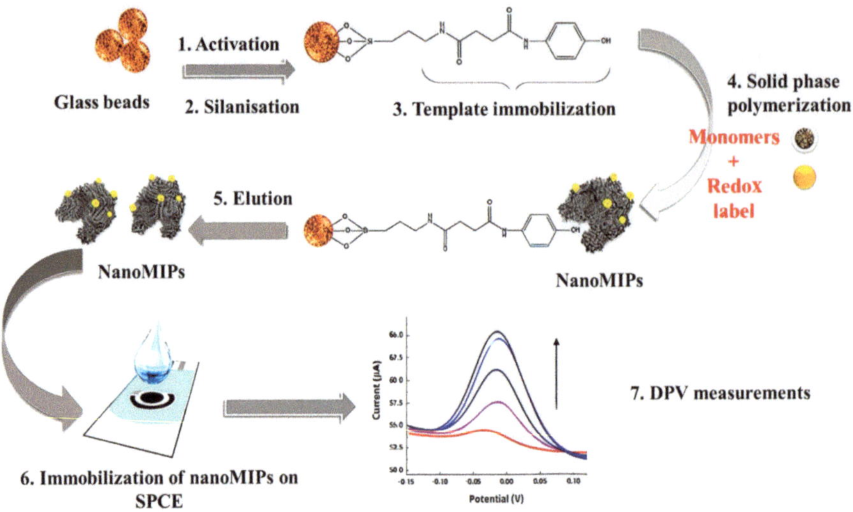

Scheme 4.5 Experimental protocol for solid-phase synthesis of nanoMIPs specific for amphetamine. The process includes (1) activation of glass microspheres, (2) silanization, (3) immobilization of amphetamine on the solid phase, (4) polymerization, (5) elution of the nanoMIPs (6) immobilization of nanoMIPs on SPCE, and (7) DPV evaluation. (Reprinted from Alanazi et al. [78] with permission from Elsevier)

pioneering method shows significant potential for forensic applications, especially in the selective detection of amphetamine in street samples [79].

Continuing in this line of research, Almabadi et al. introduced a graphene-assisted sensor incorporating redox-label-doped nanoMIPs for amphetamine detection. These nanoparticles were made electroactive by incorporating ferrocene, making them specific actuators in electrochemical sensors. The inclusion of a ferrocene redox probe within the nanoMIPs structure facilitated the detection of non-electroactive amphetamine. Almabadi et al. explored two immobilization methods (Scheme 4.5): one involving the covalent immobilization of nanoparticles onto electrochemical sensors through drop-casting using silanes, while the other utilized 3D printing alongside a graphene ink composite. The electrochemical performance of both approaches was thoroughly assessed. Remarkably, the 3D printed nanoMIPs/graphene sensors exhibited superior selectivity in spiked human plasma, with a sensitivity of 73 nA/nM and a LOD of 68 nM (RSD 2.4%), surpassing the silane drop-cast electrodes. The primary advantage of the optimized 3D printing technology lies in its ability to quantitatively detect amphetamine, a non-electroactive drug challenging to detect using conventional electrochemical sensors. Moreover, the cost-effectiveness of 3D printing facilitates easy sensor manufacturing, resulting in robust, highly selective, and sensitive sensors. As proof of concept, these sensors were assessed using street specimens and clinically relevant samples, with their performance successfully validated using UPLC-MS [80].

The utilization of redox-label MIPs in combination with a ratiometric detection approach has proven to be an effective method for the simultaneous dual detection of both non-electroactive and electroactive substances. A ratiometric detection approach incorporates an internal reference or a redox probe, typically an electroactive substance with a distinct redox potential compared to the analyte or probe. The signal from this internal reference serves to correct for the impact of both internal and external variations. Depending on how internal reference signals relate to analytes, ratiometric electrochemical sensors can be categorized into two types: signal on/off and signal on–off. In the signal on/off type, the reference signal remains independent of the analytes and remains relatively constant [81]. Conversely, in the signal on–off type, both electrochemical signals are contingent on the analytes, typically exhibiting opposite responses, where one signal increases while the other decreases [82].

In 2022, Lio et al. introduced a groundbreaking ratiometric MIP-based electrochemical sensor designed for the detection of CPZ. This innovative sensor was fabricated by electrodeposition of a dual-monomer molecularly imprinted polymer (DMMIP) film onto the surface of a Pt/Co_3O_4 nanoparticle-modified GCE. The template molecule employed was CPZ, while methylene blue and catechol served as functional monomers. The copolymerization of these two monomers expanded the range of functional groups available for binding template molecules, enhancing overall stability. The concentration of CPZ was quantified through DPV, employing the peak current of poly(methylene blue) as a reference signal and the peak current of CPZ as the indicator signal. The outcomes demonstrated that the developed DMMIP sensor exhibited exceptional selectivity and sensitivity and displayed robust resistance to interference. Under optimal conditions, the sensor yielded a linear detection range spanning from 0.005 to 9 μmol/L ($R^2 = 0.9962$) and achieved a low LOD of 2.6 nmol/L. Furthermore, the sensor exhibited excellent reproducibility and stability when applied to CPZ detection. It was successfully employed for the detection of CPZ in serum and pharmaceutical samples, consistently delivering satisfactory recoveries within the range of 95.3–108.0% [83]. Continuing in this line of research, Lu et al. developed a resilient on–off ratiometric MIP-based electrochemical sensing for the detection of CPZ (Scheme 4.6). The procedure commences with the electroplating of AuNPs onto a pristine GCE surface. Subsequently, a mixture of Ni-MOF and Fe-MOF-5 is drop-casted onto the electrode's surface, followed by the electropolymerization of polythionine (pTHi), serving as an internal reference to ensure a ratiometric on–off response. A notable innovation in this study is the method for acquiring the pTHi reference signal, which involves boiling water-bath-assisted in-situ electropolymerization. The Ni-MOF/Fe-MOF-5 hybrid AuNPs (Ni-MOF/Fe-MOF-5/AuNPs) serve a dual purpose, enhancing the output signal and providing an expanded surface area for the immobilization of pThi and MIPs. Furthermore, they act as catalytically active sites for the oxidation of CPZ. The MIPs membrane is then fabricated through in-situ electropolymerization, using CPZ as the template molecule and nicotinamide as the functional monomer. Under optimized conditions, the ratiometric signals (expressed as the peak current ratio ICPZ/IpTHi) exhibit linearity within CPZ concentration ranges of 0.001–40 and 40–900 μM, with an outstandingly low detection limit of 0.025 μM (S/N = 3). Moreover, when applied to biological samples, this sensor demonstrates exceptional

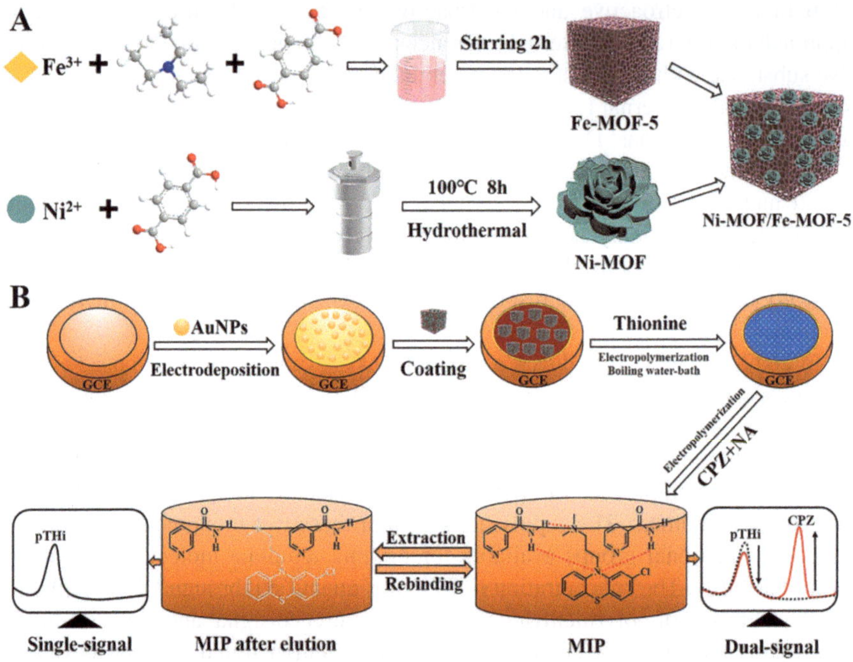

Scheme 4.6 Schematic illustration of the preparation of the Fe-MOF-5 and Ni-MOF (**a**) and MIP/pTHi/Ni-MOF/Fe-MOF-5/AuNPs on–off electrochemical sensor (**b**). (Reprinted from Lu et al. [84] with permission from © 2023 Elsevier B.V.)

interference resistance, repeatability, stability, and remarkable performance, yielding favorable recovery values ranging from 98.11% to 100.81%. In conclusion, this on–off ratiometric MIP sensing platform holds great promise for the rapid and accurate identification of antipsychotic medications and is marked by high selectivity and stability [84].

Vahidifar and Eshaghi developed a dual-template MIP sensor to simultaneously measure oxazepam and diazepam in buffer solutions, pharmaceutical items, and biological samples. This MIP-based sensor with a pencil graphite electrode was fabricated uniquely by using a composite of MWCNTs decorated with magnetite iron nanoparticles (Fe_3O_4) as the MIP substrate. This dual-template MIP-based sensor could detect low concentrations of oxazepam and diazepam within linear ranges of 0.01–200 and 0.05–150 μM, respectively. It provided great detection limits for oxazepam at 59 nM and diazepam at 21 nM. The findings highlight the sensor's positive characteristics, such as stability, repeatability, and reproducibility [85].

Tetrahydrocannabinol (THC) is the primary chemical component responsible for marijuana's notable psychological effects and has the potential to cause central nervous system depression. Zhao et al. developed an electrochemical biosensor employing MIP/MWCNTs/SPCE for the exact detection of THC, a forbidden drug discovered in athletes' biological fluid samples, in their investigation. The electrodeposition of MWCNT nanostructures onto an SPCE was followed by the electropolymerization of MIP onto the MWCNTs/SPCE surface. Electrochemical studies demonstrated a considerable improvement in the electrocatalytic activity and selectivity of MIP/MWCNTs nanostructured electrodes, which was due to the synergistic interaction between MWCNT nanostructures and MIP molecules. The MIP/MWCNTs/SPCE THC sensor has a sensitivity of 0.00155 A/ng/mL and a minimum detection limit for THC of 0.37 ng/mL (S/N = 3). The sensor demonstrated linearity throughout a concentration range of 0–3150 ng/mL, with standard curve correlation values above 0.99. The new approach was used to test THC levels in a cyclist's blood plasma sample to confirm its practical applicability, displaying high agreement with electrochemical and ELISA assays. Furthermore, it demonstrated acceptable relative standard deviation and relative recovery values, confirming the method's efficacy in identifying THC levels in clinical samples [86].

Motaharian et al. utilized MIP materials to develop an innovative electrochemical sensor for detecting the drug quetiapine (QTP). NanoMIPs were meticulously fabricated using a precipitation polymerization technique and applied in the design of modified CPE with nanoMIPs. This designed sensing electrode could detect QTP in linear concentration ranges of 1.6×10^{-8}–2.5×10^{-6} M ($R^2 = 0.9964$), with an outstandingly low detection limit of 5.04×10^{-9} M and quantification limit of 1.68×10^{-8} M, respectively. Additionally, assessments of the repeatability and reproducibility of the proposed sensors yielded %RSD values between 2.19% and 3.02%, respectively. The method was effectively employed to determine QTP in pharmaceutical formulations and human urine samples [87].

Khosrokhavar and colleagues created a unique MIP-based sensing device designed to evaluate small concentrations of STR's antidepressant drug. The synthesis of MIP nanoparticles employed the precipitation polymerization method, using STR as the template molecule. The sensor was crafted by incorporating an exceptionally thin MIP/graphene layer onto custom-made SPCEs. To characterize the sensor, the HCF redox probe was utilized by employing voltammetric techniques. The affinity tests of STR revealed that MIP/graphene-modified SPCEs exhibited superior adsorption capabilities compared to their non-imprinted polymer (NIP)/graphene-modified counterparts. Several factors influencing sensor response, such as the composition of the MIP/graphene suspension, extraction pH, preconcentration time, and stirring rate of the sample solution, were meticulously optimized. Under these ideal conditions, the sensor demonstrated commendable sensitivity (177.25 μA L/μmol) for sertraline, spanning a linear range from 5.0×10^{-9} to 7.5×10^{-7} M ($R^2 = 0.9971$), and achieving a detection limit of 1.99×10^{-9} M. The proposed MIP-based SPCE detected STR in tablet and human serum samples, consistently yielding recovery values within the range of 97.98–101.33% [88]. A comparison of recent MIP-based electrochemical sensors for the analysis of opioids and opioid analgesics is presented in Table 4.1, while Table 4.2 provides a compilation of data related to the detection of antipsychotics and antidepressants using similar sensors.

4.5 Conclusions and Future Prospects

In this chapter, we explored the versatile applications of recently designed MIP-based electrochemical sensors for the detection and monitoring of opioids and psychoactive compounds in diverse biological fluids such as human blood, urine, and saliva, as well as synthetic samples and more. We discussed electrochemical sensors modified with MIP-thin layers, magnetic MIP, nanoMIPs, ratiometric MIPs, and the latest innovations involving the integration of MIPs with carbon materials. We provided concise evaluations of the advantages, disadvantages, and capabilities inherent in each contemporary MIP-based approach. These insights underscore how these customized materials can significantly improve extraction processes, enhancing their selectivity, environmental friendliness, and sensitivity to a broad spectrum of opioids and psychoactive compounds found in biological fluids. Tables 4.1 and 4.2 serve as valuable references, presenting a convenient summary of recent studies employing MIP-based electrochemical sensors for the analysis of opioids and psychoactive compounds. The passion for the design of high-performance MIP-based electrochemical sensors will continue to gain momentum in the future. Subsequent investigations and advancements in the field of MIPs are poised to focus on the following key areas:

- Innovative imprinting techniques: The development of novel imprinting techniques to synthesize MIP-based architectures with elevated selectivity, capacity, and binding affinity uniformity thus broadening the range of applications for these materials [101].
- Functional monomer optimization: Acquiring optimal functional monomers through advanced molecular modelling techniques for MIPs, ensuring efficient design for effective imprinting of active recognition sites, and high-performance temple removal procedures [102].
- Scale-up production: Exploration of large-scale synthesis methods to facilitate the industrialization of MIP-based sensing assays, making them more accessible for widespread use.
- Macro(bio)molecule targets: Tackling the analysis and detection of large proteins, bacteria, hormones, toxins, and other macromolecule targets in the real matrices of target samples represents a pivotal research direction.
- Advanced polymerization: Introducing new polymerization techniques and harnessing advanced monomers with responsive functionalities to craft analytical techniques suited for diverse drug analytes.
- Enhanced performance: Continuous enhancements in the performance of introduced techniques to streamline extraction and separation processes within analytical laboratories.

The greenification of MIPs is significantly shaping the development of advanced MIPs for the future. Integrating greenification principles into MIP research enhances the sustainability of chemical sensing technologies. Versatile applications of MIP-based electrochemical sensors for detecting opioids and psychoactive compounds underscore the importance of advancing greenification efforts [103]. Exploring innovative green imprinting techniques, optimizing biowaste-derived functional

Table 4.1 Comparison of recent MIP-based electrochemical sensor for analysis opioids and opioid analgesics

Sample	Sample matrix	Electrode	DLR (M)	LOD (M)	Method	Reference
Amphetamine	Street sample	NanoMIPs/CS[a]-based SPEs	1×10^{-9}–250×10^{-9}	0.3×10^{-9}	DPV	[79]
Amphetamine	Street sample	NanoMIPs/chitosan/GPHOx[b]-based SPEs	1×10^{-9}–200×10^{-9}	0.3×10^{-9}	DPV	[79]
Amphetamine	Buffer Plasma Street sample	NanoMIP/G[c] NanoMIP/G NanoMIP/G	75×10^{-9}–220×10^{-9} 75×10^{-9}–220×10^{-9} 25×10^{-9}–220×10^{-9}	62×10^{-9} 68×10^{-9} 37.5×10^{-9}	DPV DPV DPV	[80]
Paracetamol	Buffer	NanoMIP/G	0.1×10^{-3}–1×10^{-3}	56×10^{-6}	DPV	[78]
Tramadol	Buffer	MIP-MWCNT	9.0×10^{-6}–$30. \times 10^{-6}$	1.4×10^{-6}	DPV	[89]
Buprenorphine	Buffer	NanoMIP-MWCNT	1.0×10^{-9}–5.0×10^{-5}	6.0×10^{-10}	DPV	[90]
Methamphetamine	Buffer	MIP/CNT-CPE	1.0×10^{-8}–1.0×10^{-6}	8.3×10^{-10}	FFT-SWV[d]	
Paracetamol	Buffer	MIP/CPE	0.01×10^{-6}–0.1×10^{-6}	4.63×10^{-6}	DPV	[91]
Tapentadol	Buffer	GCPE	17.0–136.3 ng/mL	5.48 ng/mL	DPV	[92]

[a] CS: Chitosan
[b] *GPHOx*: Graphene oxide
[c] G: Graphene
[d] FFT-SWV: Fast Fourier transform square wave voltammetric

Table 4.2 Comparison of recent MIP-based electrochemical sensor for analysis of antipsychotics and antidepressants

Sample	Sample matrix	Electrode	DLR (M)	LOD (M)	Method	Reference
Promethazine	Buffer	NanoMIP-MWCNT	4×10^{-9}–4×10^{-7} 4×10^{-7}–7×10^{-6}	1.4×10^{-9}	DPV	[93]
Chlordiazepoxide	Buffer	NanoMIPs/GCPE	6×10^{-10}–7.5×10^{-8}	2.6×10^{-10}	DPV	[94]
Midazolam	Buffer Ampoule urine	NanoMIP/GCPE	5×10^{-10}–1×10^{-7} 5×10^{-10}_ 1×10^{-7}	62×10^{-9} 1.77×10^{-10}	SWV	[95]
CPZ	Buffer	TVGCPE[a]	3.30×10^{-8}–8.5×10^{-4}	5.3×10^{-9}	DPV	[96]
CPZ	Buffer	[b]Au/N-GOQDs/NiS_2/BC/MIP/GCE	0.005×10^{-6}–2×10^{-6}	0.25×10^{-9}	DPV	[97]
CPZ	Buffer	MIP/Au/Cu/ANE[c]	0.1×10^{-6}–100×10^{-6}	0.07×10^{-6}	DPV	[98]
Trimipramine	Buffer	NanoMIP/MWCNTs	1.0×10^{-10}–2.5×10^{-8}	4.5×10^{-11}	SWV	[99]
Trimipramine	Buffer	[d]CeO_2NP–RGO/Ru(bpy)$_3^{2+}$–MIP–CS	0.2×10^{-12}–100×10^{-12}	0.025×10^{-12}	ECL[e]	[100]

[a] TVGCPE: The carbon paste electrode was modified with vanadium pentoxide (V_2O_5) and titanium dioxide (TiO_2) nanoparticles and graphene

[b] Au/N-GOQDs/NiS_2/BC: Gold nanoparticles nitrogen-doped graphene oxide quantum dots: N-GOQDs, nickel sulfide nanoparticles: NiS_2, biomass carbon

[c] ANE: Acupuncture needle electrode

[d] CeO_2NP–RGO/Ru(bpy)$_3^{2+}$: Ceria nanoparticles/tris(2,2′-bipyridyl)ruthenium(II)

[e] ECL: Electrochemiluminescence sensor

monomers, and scaling up production methods broaden applications while minimizing environmental impact, resulting in the development of more effective and environmentally sustainable analytical methodologies. Strategies such as utilizing biomass waste and green synthesis methods contribute to sustainability. Despite recent advancements in biomass-derived MIPs, further development is needed, particularly in large-scale production and adopting green principles in MIP synthesis [103]. Overall, the integration of greenification into MIP technology drives innovation and contributes to the ongoing refinement of green analytical methodologies, promising a greener and more sustainable future.

This transformative approach focuses on integrating sustainable practices into MIP synthesis and application, leading to several promising avenues for innovation in the following important subjects:

- Green synthesis methods: One key aspect of greenifying MIP research involves utilizing eco-friendly synthesis methods, such as electrochemical approaches and microwave-based reactors. Electrochemical methods offer several advantages, including reduced use of hazardous chemicals, energy efficiency, and precise control over polymerization conditions. By harnessing electrochemical techniques, researchers can synthesize MIPs with improved environmental friendliness and tailored properties. Moreover, these methods enable the development of MIPs with enhanced selectivity and sensitivity, paving the way for more efficient molecular recognition and sensing applications [104, 105].
- Utilization of biomass-based materials: Another strategy for greenifying MIP synthesis involves leveraging biomass-based materials as starting components. Biomass waste, such as chitosan, cellulose, or cyclodextrins, offers a sustainable alternative to traditional monomers, reducing reliance on fossil-based resources and minimizing waste generation. By incorporating biomass-derived materials into MIP synthesis, researchers can create environmentally friendly polymers while simultaneously managing waste and reducing resource consumption. This approach not only contributes to sustainability but also diversifies the range of available materials for MIP fabrication, potentially leading to novel properties and applications [103].
- Application of machine learning for smart modelling of MIPs: Machine learning techniques have gained high power in MIP research for modeling molecular interactions and optimizing MIP design. By harnessing computational tools, researchers can predict optimal monomer compositions, template–monomer interactions, and polymerization conditions, thereby streamlining the MIP development process. Machine learning models enable rapid screening of large chemical libraries, accelerating the identification of promising MIP formulations with desired properties [102]. Additionally, these models facilitate the design of MIPs tailored for specific analytes or applications, enhancing their performance and efficiency. By integrating machine learning into MIP research, scientists can expedite the development of advanced MIPs while minimizing experimental costs and resource consumption [104].

In conclusion, the greenification of MIPs represents a significant opportunity for the advancement of eco-friendly MIP-based sensors. By harnessing biomass-derived materials and employing machine learning techniques, researchers can craft sustainable MIPs with heightened functionality and efficacy [102–104]. These breakthroughs not only support environmental preservation but also foster innovation in molecular recognition and sensing applications.

References

1. Midha KK, Hubbard JW, Cooper JK, Gurnsey T, Hawes EM, McKay G, Chakraborty BS, Yeung PKF (1987) Therapeutic monitoring of chlorpromazine IV: comparison of a new high-performance liquid chromatographic method with radioimmunoassays for parent drug and some of its major metabolites. Ther Drug Monit 9:358
2. Kim K-Y, Kang J-H, Kim C-W, Nam M (2011) Application of liquid chromatography tandem mass spectrometry for the simultaneous quantification of multiple non-opioid drugs in human plasma. Mol Cell Toxicol 7:185–189
3. Caramelo D, Rosado T, Oliveira V, Rodilla JM, Rocha PMM, Barroso M, Gallardo E (2019) Determination of antipsychotic drugs in oral fluid using dried saliva spots by gas chromatography-tandem mass spectrometry. Anal Bioanal Chem 411:6141–6153
4. Wang J (2006) Analytical electrochemistry. Wiley
5. Wang J (1999) Towards genoelectronics: electrochemical biosensing of DNA hybridization. Chem Eur J 5:1681–1685
6. Brett C, Brett AMO (1993) Electrochemistry: principles, methods, and applications
7. Guan J-G, Miao Y-Q, Zhang Q-J (2004) Impedimetric biosensors. J Biosci Bioeng 97:219–226
8. Paul A, Banga IK, Muthukumar S, Prasad S (2024) Naloxone-AuNPs@ZIF-8-based impedimetric sensor platform for ultrasensitive detection of fentanyl and fabrication of fen-track prototype for real-field analysis. ACS Appl Mater Interfaces 16:190–200
9. Laurila T, Sainio S, Caro MA (2017) Hybrid carbon based nanomaterials for electrochemical detection of biomolecules. Prog Mater Sci 88:499–594
10. Wu J, Wu Y, Bian H, Peng Z, Liu Y, Yin Y, Du J, Lu X (2023) Fabrication of a ratiometric electrochemiluminescence biosensor using single self-enhanced nanoluminophores for the detection of spermine. Talanta 253:123880
11. Zanut A, Palomba F, Scota MR, Rebeccani S, Marcaccio M, Genovese D, Rampazzo E, Valenti G, Paolucci F, Prodi L (2020) Dye-doped silica nanoparticles for enhanced ECL-based immunoassay analytical performance. Angew Chem Int Ed 59:21858–21863
12. Meng C, Knežević S, Fangxin D, Guan Y, Kanoufi F, Sojic N, Guobao X (2022) Recent advances in electrochemiluminescence imaging analysis. eScience 2:591–605
13. Mohammadi S, Taher MA, Beitollahi H (2020) A hierarchical 3D camellia-like molybdenum tungsten disulfide architectures for the determination of morphine and tramadol. Microchim Acta 187:312
14. Verrinder E, Wester N, Leppänen E, Lilius T, Kalso E, Mikladal B, Varjos I, Koskinen J, Laurila T (2021) Electrochemical detection of morphine in untreated human capillary whole blood. ACS Omega 6:11563–11569
15. Caldevilla R, Morais SL, Cruz A, Delerue-Matos C, Moreira F, Pacheco JG, Santos M, Barroso MF (2023) Electrochemical chemically based sensors and emerging enzymatic biosensors for antidepressant drug detection: a review. Int J Mol Sci 24:8480
16. Zhang W, Wang R, Luo F, Wang P, Lin Z (2020) Miniaturized electrochemical sensors and their point-of-care applications. Chin Chem Lett 31:589–600
17. Tiwari N, Chatterjee S, Kaswan K, Chung J-H, Fan K-P, Lin Z-H (2022) Recent advancements in sampling, power management strategies and development in applications for non-invasive wearable electrochemical sensors. J Electroanal Chem 907:116064

18. Imanzadeh H, Khataee A, Hazraty L, Amiri M (2024) Broken hollow carbon spheres decorated by gold nanodendrites as the advanced electrochemical sensing platform for sensitive tracing of morphine in human serum and saliva. Sens Actuators B Chem 398:134738
19. Yang Y, Yang X, Yang Y, Yuan Q (2018) Aptamer-functionalized carbon nanomaterials electrochemical sensors for detecting cancer relevant biomolecules. Carbon 129:380–395
20. Liu X, Liu J (2021) Biosensors and sensors for dopamine detection. VIEW 2:20200102
21. Martins GV, Khan MAR, Piloto AML, Guerreiro JRL, Castaño-Guerrero Y, Moreira FTC, Sales MGF (2023) Chapter 8—Molecularly imprinted polymers in electrochemical sensing—a step closer to achieving technology readiness. In: Singh M (ed) Molecularly imprinted polymers (MIPs). Elsevier
22. Roy E, Patra S, Tiwari A, Madhuri R, Sharma PK (2017) Introduction of selectivity and specificity to graphene using an inimitable combination of molecular imprinting and nanotechnology. Biosens Bioelectron 89:234–248
23. Thakur N, Das TR, Gupta S, Shukla SK, Patra S, Kumar D (2023) Chapter 5—Material selection for the fabrication of MIP-based devices: an overview. In: Singh M (ed) Molecularly imprinted polymers (MIPs). Elsevier
24. Wackerlig J, Lieberzeit PA (2015) Molecularly imprinted polymer nanoparticles in chemical sensing—synthesis, characterisation and application. Sens Actuators B Chem 207:144–157
25. Poma A, Guerreiro A, Whitcombe MJ, Piletska EV, Turner APF, Piletsky SA (2013) Solid-phase synthesis of molecularly imprinted polymer nanoparticles with a reusable template–"plastic antibodies". Adv Funct Mater 23:2821–2827
26. Haupt K (2003) Molecularly imprinted polymers: the next generation. Anal Chem 75:376A–383A
27. Tarannum N, Kumar D, Khatoon S (2023) Chapter 3—Inefficient removal of templates as a limitation for molecular imprinting of polymers. In: Singh M (ed) Molecularly imprinted polymers (MIPs). Elsevier
28. Yang Y, Fang G, Liu G, Pan M, Wang X, Kong L, He X, Wang S (2013) Electrochemical sensor based on molecularly imprinted polymer film via sol–gel technology and multi-walled carbon nanotubes-chitosan functional layer for sensitive determination of quinoxaline-2-carboxylic acid. Biosens Bioelectron 47:475–481
29. Ge Y, Butler B, Mirza F, Habib-Ullah S, Fei D (2013) Smart molecularly imprinted polymers: recent developments and applications. Macromol Rapid Commun 34:903–915
30. de las Heras Alarcón C, Pennadam S, Alexander C (2005) Stimuli responsive polymers for biomedical applications. Chem Soc Rev 34:276–285
31. Morsi SMM, El-Aziz MEA, Mohamed HA (2023) Smart polymers as molecular imprinted polymers for recognition of target molecules. Int J Polym Mater Polym Biomater 72:612–635
32. Sharma PS, Garcia-Cruz A, Cieplak M, Noworyta KR, Kutner W (2019) 'Gate effect' in molecularly imprinted polymers: the current state of understanding. Curr Opin Electrochem 16:50–56
33. Mazzotta E, Di Giulio T, Malitesta C (2022) Electrochemical sensing of macromolecules based on molecularly imprinted polymers: challenges, successful strategies, and opportunities. Anal Bioanal Chem 414:5165–5200
34. Warren JA, Carbonell AM, Jones LK, Mcguire A, Hand WR, Cancellaro VA, Ewing JA, Cobb WS (2019) Length of stay and opioid dose requirement with transversus abdominis plane block vs epidural analgesia for ventral hernia repair. J Am Coll Surg 228:680–686
35. Richards G (2021) An evidence-based approach to assess the use of opioids. University of Oxford
36. Paul AK, Smith CM, Rahmatullah M, Nissapatorn V, Wilairatana P, Spetea M, Gueven N, Dietis N (2021) Opioid analgesia and opioid-induced adverse effects: a review. Pharmaceuticals 14:1091
37. Koyyalagunta D, Waldman SD (2011) Chapter 122—Opioid analgesics. In: Waldman SD (ed) Pain management, 2nd edn. W.B. Saunders, Philadelphia

38. Kick BL, Shu P, Wen B, Sun D, Taylor DK (2017) Pharmacokinetic profiles of nalbuphine after intraperitoneal and subcutaneous administration to C57BL/6 mice. J Am Assoc Lab Anim Sci 56:534–538
39. Rosenblum LA, Marsch HJ, Portenoy RK (2008) Opioids and the treatment of chronic pain: controversies, current status, and future directions. Exp Clin Psychopharmacol 16(5):405–416
40. Graves DA, Arrigo JM, Foster TS, Baumann TJ, Batenhorst RL (1985) Relationship between plasma morphine concentrations and pharmacologic effects in postoperative patients using patient-controlled analgesia. Clin Pharm 4(1):41–47
41. Miller RS, Peterson GM, Abbott F, Maddocks I, Parker D, McLean S (1995) Plasma concentrations of fentanyl with subcutaneous infusion in palliative care patients. Br J Clin Pharmacol 40:553–556
42. Volkow ND, McLellan AT (2016) Opioid abuse in chronic pain — misconceptions and mitigation strategies. N Engl J Med 374(13):1253–1263
43. Bell J, Seres V, Bowron P, Lewis J, Batey R (1988) The use of serum methadone levels in patients receiving methadone maintenance. Clin Pharmacol Ther 43:623–629
44. Furo H, Schwartz DG, Sullivan RW, Elkin PL (2021) Buprenorphine dosage and urine quantitative buprenorphine, norbuprenorphine, and creatinine levels in an office-based opioid treatment program. Subst Abuse 15:11782218211061749
45. Webster LR, Dove B (2007) Avoiding opioid abuse while managing pain. Sunrise River Press
46. Kirchheiner J, Schmidt H, Tzvetkov M, Keulen JTHA, Lötsch J, Roots I, Brockmöller J (2007) Pharmacokinetics of codeine and its metabolite morphine in ultra-rapid metabolizers due to CYP2D6 duplication. Pharmacogenomics J 7:257–265
47. Eisenstein TK (2019) The role of opioid receptors in immune system function. Front Immunol 10:2904
48. Morin CM, Bastien C, Guay B, Radouco-Thomas M, Leblanc J, Vallières A (2004) Randomized clinical trial of supervised tapering and cognitive behavior therapy to facilitate benzodiazepine discontinuation in older adults with chronic insomnia. Am J Psychiatry 161:332–342
49. Lozano-Chaves ME, Palacios-Santander JM, Cubillana-Aguilera LM, Naranjo-Rodríguez I, Hidalgo-Hidalgo-de-Cisneros JL (2006) Modified carbon-paste electrodes as sensors for the determination of 1,4-benzodiazepines: application to the determination of diazepam and oxazepam in biological fluids. Sens Actuators B Chem 115:575–583
50. Sanabria E, Cuenca RE, Esteso MÁ, Maldonado M (2021) Benzodiazepines: their use either as essential medicines or as toxics substances. Toxics 9:25
51. Moffat AC, David Osselton M, Widdop B, Watts J (2011) Clarke's analysis of drugs and poisons. Pharmaceutical Press, London
52. Kokulnathan T, Kumar JV, Chen S-M, Karthik R, Elangovan A, Muthuraj V (2018) One-step sonochemical synthesis of 1D β-stannous tungstate nanorods: an efficient and excellent electrocatalyst for the selective electrochemical detection of antipsychotic drug chlorpromazine. Ultrason Sonochem 44:231–239
53. Daly EJ, Trivedi MH (2007) A review of quetiapine in combination with antidepressant therapy in patients with depression. Neuropsychiatr Dis Treat 3:855–867
54. Ignácio ZM, Calixto AV, da Silva RH, Quevedo J, Réus GZ (2018) The use of quetiapine in the treatment of major depressive disorder: evidence from clinical and experimental studies. Neurosci Biobehav Rev 86:36–50
55. Nouws HPA, Delerue-Matos C, Barros AA, Rodrigues JA (2005) Electroanalytical study of the antidepressant sertraline. J Pharm Biomed Anal 39:290–293
56. Russo MJ, Han M, Desroches PE, Manasa CS, Dennaoui J, Quigley AF, Kapsa RMI, Moulton SE, Guijt RM, Greene GW, Silva SM (2021) Antifouling strategies for electrochemical biosensing: mechanisms and performance toward point of care based diagnostic applications. ACS Sensors 6:1482–1507
57. Wester N, Mikladal BF, Varjos I, Peltonen A, Kalso E, Lilius T, Laurila T, Koskinen J (2020) Disposable nafion-coated single-walled carbon nanotube test strip for electrochemical quan-

titative determination of acetaminophen in a finger-prick whole blood sample. Anal Chem 92:13017–13024
58. Nekoueian K, Amiri M, Sillanpää M, Marken F, Boukherroub R, Szunerits S (2019) Carbon-based quantum particles: an electroanalytical and biomedical perspective. Chem Soc Rev 48:4281–4316
59. Nekoueian K, Akhoundian M, Wester N, Laurila T (2023) An ultra-sensitive dopamine measurement platform based on molecularly imprinted polymer-carbon hybrid nanomaterials for in vitro use. Electrochim Acta 445:142029
60. Kujala J, Wester N, Lohela TJ, Kurkela M, Backman JT, Mikladal B, Laurila T, Koskinen J, Lilius TO, Kalso EA (2023) Introduction of an electrochemical point-of-care assay for quantitative determination of paracetamol in finger-prick capillary whole blood samples. Br J Clin Pharmacol 89:2933–2938
61. Zhang Z, Hu Y, Zhang H, Yao S (2010) Novel layer-by-layer assembly molecularly imprinted sol–gel sensor for selective recognition of clindamycin based on Au electrode decorated by multi-wall carbon nanotube. J Colloid Interface Sci 344:158–164
62. Narver HL (2015) Nalbuphine, a non-controlled opioid analgesic, and its potential use in research mice. Lab Anim 44:106–110
63. Hassan SSM, Kamel AH, Fathy MA (2022) A novel screen-printed potentiometric electrode with carbon nanotubes/polyaniline transducer and molecularly imprinted polymer for the determination of nalbuphine in pharmaceuticals and biological fluids. Anal Chim Acta 1227:340239
64. Lopes F, Pacheco JG, Rebelo P, Delerue-Matos C (2017) Molecularly imprinted electrochemical sensor prepared on a screen printed carbon electrode for naloxone detection. Sens Actuators B Chem 243:745–752
65. Shaabani N, Chan NWC, Lee WE, Jemere AB (2020) Electrochemical determination of naloxone using molecularly imprinted poly(para-phenylenediamine) sensor. J Electrochem Soc 167:137508
66. Rezaei B, Boroujeni MK, Ensafi AA (2014) A novel electrochemical nanocomposite imprinted sensor for the determination of lorazepam based on modified polypyrrole@sol-gel@gold nanoparticles/pencil graphite electrode. Electrochim Acta 123:332–339
67. Shaabani N, Chan NWC, Jemere AB (2021) A molecularly imprinted sol-gel electrochemical sensor for naloxone determination. Nanomaterials 11:631
68. Kriz D, Mosbach K (1995) Competitive amperometric morphine sensor based on an agarose immobilised molecularly imprinted polymer. Anal Chim Acta 300:71–75
69. O'Shannessy DJ, Ekberg B, Mosbach K (1989) Molecular imprinting of amino acid derivatives at low temperature (0 °C) using photolytic homolysis of azobisnitriles. Anal Biochem 177:144–149
70. Rezaei B, Foroughi-Dehnavi S, Ensafi AA (2015) Fabrication of electrochemical sensor based on molecularly imprinted polymer and nanoparticles for determination trace amounts of morphine. Ionics 21:2969–2980
71. Salajegheh M, Kazemipour M, Foroghi MM, Ansari M (2019) Morphine sensing by a green modified molecularly imprinted poly L-lysine/sodium alginate-activated carbon/glassy carbon electrode based on computational design. Electroanalysis 31:468–476
72. Yence M, Cetinkaya A, Çorman ME, Uzun L, Caglayan MG, Ozkan SA (2023) Fabrication of molecularly imprinted electrochemical sensors for sensitive codeine detection. Microchem J 193:109060
73. Ansari S, Masoum S (2021) A hybrid imprinted polymer based on magnetic graphene oxide and carbon dots for ultrasonic assisted dispersive solid-phase microextraction of oxycodone. Microchem J 164:105988
74. Yáñez-Sedeño P, Campuzano S, Pingarrón JM (2017) Electrochemical sensors based on magnetic molecularly imprinted polymers: a review. Anal Chim Acta 960:1–17
75. Diouf A, Aghoutane Y, Burhan H, Sen F, Bouchikhi B, El Bari N (2021) Tramadol sensing in non-invasive biological fluids using a voltammetric electronic tongue and an electrochemical sensor based on biomimetic recognition. Int J Pharm 593:120114

76. Malitesta C, Mazzotta E, Picca RA, Poma A, Chianella I, Piletsky SA (2012) MIP sensors—the electrochemical approach. Anal Bioanal Chem 402:1827–1846
77. Mazzotta E, Turco A, Chianella I, Guerreiro A, Piletsky SA, Malitesta C (2016) Solid-phase synthesis of electroactive nanoparticles of molecularly imprinted polymers. A novel platform for indirect electrochemical sensing applications. Sensors and Actuators B: Chemical 229:174–180
78. Smolinska-Kempisty K, Sheej Ahmad O, Guerreiro A, Karim K, Piletska E, Piletsky S (2017) New potentiometric sensor based on molecularly imprinted nanoparticles for cocaine detection. Biosens Bioelectron 96:49–54
79. D'Aurelio R, Chianella I, Goode JA, Tothill IE (2020) Molecularly imprinted nanoparticles based sensor for cocaine detection. Biosensors 10:22
80. Alanazi K, Cruz AG, Di Masi S, Voorhaar A, Ahmad OS, Cowen T, Piletska E, Langford N, Coats TJ, Sims MR, Piletsky SA (2021) Disposable paracetamol sensor based on electroactive molecularly imprinted polymer nanoparticles for plasma monitoring. Sens Actuators B Chem 329:129128
81. Truta F, Cruz AG, Tertis M, Zaleski C, Adamu G, Allcock NS, Suciu M, Ştefan M-G, Kiss B, Piletska E, De Wael K, Piletsky SA, Cristea C (2023) NanoMIPs-based electrochemical sensors for selective detection of amphetamine. Microchem J 191:108821
82. Almabadi MH, Truta FM, Adamu G, Cowen T, Tertis M, Drăgan A-M, Alanazi KDM, Ştefan M-G, Piletska E, Kiss B, Cristea C, De Wael K, Piletsky SA, Cruz AG (2023) Integration of smart nanomaterials for highly selective disposable sensors and their forensic applications in amphetamine determination. Electrochim Acta 446:142009
83. Zhu T, Ren H, Liang W, Li Y, Yuwen X, Dai M, Ye B-C (2020) Ratiometric electrochemical sensing based on Mo2C for detection of acetaminophen. Analyst 145:7609–7615
84. Yang J, Hu Y, Li Y (2019) Molecularly imprinted polymer-decorated signal on-off ratiometric electrochemical sensor for selective and robust dopamine detection. Biosens Bioelectron 135:224–230
85. Liu Y, Hu X, Xia Y, Zhao F, Zeng B (2022) A novel ratiometric electrochemical sensor based on dual-monomer molecularly imprinted polymer and Pt/Co3O4 for sensitive detection of chlorpromazine hydrochloride. Anal Chim Acta 1190:339245
86. Lu Z, Wei K, Ma H, Xiong Q, Li Y, Sun M, Wang X, Wang Y, Wu C, Su G, Bai Y, Deng R, Ye J, Zhou C, Rao H (2023) Nanoarchitectonics of on–off ratiometric signal amplified electrochemical sensor for chlorpromazine with molecularly imprinted polymer based on Ni-MOF/Fe-MOF-5 hybrid Au nanoparticles. Sep Purif Technol 327:124858
87. Vahidifar M, Es'haghi Z (2022) Magnetic nanoparticle-reinforced dual-template molecularly imprinted polymer for the simultaneous determination of oxazepam and diazepam using an electrochemical approach. J Anal Chem 77:625–639
88. Zhao Y, Moon Y, Savari R (2022) Molecularly imprinted electrochemical sensor for determination of tetrahydrocannabinol in human blood plasma. Int J Electrochem Sci 17:221185
89. Motaharian A, Naseri K, Mehrpour O, Shoeibi S (2020) Electrochemical determination of atypical antipsychotic drug quetiapine using nano-molecularly imprinted polymer modified carbon paste electrode. Anal Chim Acta 1097:214–221
90. Khosrokhavar R, Motaharian A, Hosseini MRM, Mohammadsadegh S (2020) Screen-printed carbon electrode (SPCE) modified by molecularly imprinted polymer (MIP) nanoparticles and graphene nanosheets for determination of sertraline antidepressant drug. Microchem J 159:105348
91. Tarley CRT, de Cássia Mendonça J, da Rocha LR, Capelari TB, Prete MC, Fonseca MC, de Oliveira FM, Pereira AC, Scheel GL, Borges KB, Segatelli MG (2020) Development of a molecularly imprinted poly(acrylic acid)-MWCNT nanocomposite electrochemical sensor for tramadol determination in pharmaceutical samples. Electroanalysis 32:1130–1137
92. Alizadeh T, Atashi F, Akhoundian M, Ganjali MR (2019) Highly selective extraction and voltammetric determination of the opioid drug buprenorphine via a carbon paste electrode impregnated with nano-sized molecularly imprinted polymer. Microchim Acta 186:654

93. Aryasa IWT, Noviandri I (2022) Voltammetric determination of paracetamol with carbon paste electrode modified with molecularly imprinted electropolymer. Molekul 17:30
94. El-Shal MA, Hendawy HAMH, Eldin GMG, El-Sherif ZA (2019) Application of nano graphene-modified electrode as an electrochemical sensor for determination of tapentadol in the presence of paracetamol. J Iran Chem Soc 16:1123–1130
95. Akhoundian M, Alizadeh T (2020) In situ voltammetric determination of promethazine on carbon paste electrode modified with nano-sized molecularly imprinted polymer. Anal Bioanal Electrochem 12:1014–1024
96. Motaharian A, Hosseini MRM (2016) Electrochemical sensor based on a carbon paste electrode modified by graphene nanosheets and molecularly imprinted polymer nanoparticles for determination of a chlordiazepoxide drug. Anal Methods 8:6305–6312
97. Panahi Y, Motaharian A, Hosseini MRM, Mehrpour O (2018) High sensitive and selective nano-molecularly imprinted polymer based electrochemical sensor for midazolam drug detection in pharmaceutical formulation and human urine samples. Sens Actuators B Chem 273:1579–1586
98. Tavakkoli N, Soltani N, Salavati H, Talakoub M (2018) New carbon paste electrode modified with graphene/TiO2/V2O5 for electrochemical measurement of chlorpromazine hydrochloride. J Taiwan Inst Chem Eng 83:50–58
99. Lu Z, Li Y, Liu T, Wang G, Sun M, Jiang Y, He H, Wang Y, Zou P, Wang X, Zhao Q, Rao H (2020) A dual-template imprinted polymer electrochemical sensor based on AuNPs and nitrogen-doped graphene oxide quantum dots coated on NiS2/biomass carbon for simultaneous determination of dopamine and chlorpromazine. Chem Eng J 389:124417
100. Chen J, Liu H, Wang C, Fan K, Li L, Zhang Y, Fang L, Yin Z-Z, Lü Z (2023) An electrochemical chlorpromazine sensor based on a gold–copper bimetallic synergetic molecularly imprinted interface on an acupuncture needle electrode. Analyst 148:2214–2224
101. Akhoundian M, Alizadeh T, Ganjali MR, Rafiei F (2018) A new carbon paste electrode modified with MWCNTs and nano-structured molecularly imprinted polymer for ultratrace determination of trimipramine: the crucial effect of electrode components mixing on its performance. Biosens Bioelectron 111:27–33
102. Khonsari YN, Sun S (2021) A novel MIP-ECL sensor based on RGO–CeO2NP/Ru(bpy)32+-chitosan for ultratrace determination of trimipramine. J Mater Chem B 9:471–478
103. Das TR, Bhattacherjee P, Patra S, Govender PP, Mangala B, Shukla SK (2021) Metal–organic framework-based artificial antibodies for sensing applications. In: Metal–organic frameworks for environmental sensing. American Chemical Society
104. Del Sole R, Mele G, Bloise E, Mergola L (2021) Green aspects in molecularly imprinted polymers by biomass waste utilization. Polymers 13:2430
105. Madikizela LM, Tavengwa NT, Tutu H, Chimuka L (2018) Green aspects in molecular imprinting technology: from design to environmental applications. Trends Environ Anal Chem 17:14–22

Molecularly Imprinted Polymers-Based Electrochemical Sensors: Development, Integration, and Analytical Applications

5

Melkamu Biyana Regasa

Abstract

Molecularly imprinted polymer (MIP) is an artificial receptor that has molecular memory widely in use for the fabrication of various chemical sensors. Among these sensors, MIP-based electrochemical sensors attracted a tremendous attention from the researchers and industries due to their vigorous features including being user-friendly, fast response, sensitivity, and selectivity. These types of sensors combine the advantages of electroanalytical methods with that of MIPs such as cost-effectiveness, freedom for preparing it for different target analytes, selectivity, and robust nature. Electrochemical sensors based on MIPs as receptor are widely applicable in different fields. The broad applications of such sensors for the detection or determination of various analytes are owing to their selectivity and sensitivity as well as suitability for real-time analysis. In this sensor technology, the key step is the proper integration of MIP films with suitable electrodes for better performance of the device platforms. Thus, this chapter will present the development of MIPs, its advantages, and integration into electrochemical sensors as a key element in the electrochemical sensors to create good interface and high signal-to-noise ratio, and its applications in various fields such as environmental inspection, food safety control, clinical diagnosis, precision agriculture, and security system are discussed.

Keywords

MIP · Electrochemical sensor · Template · Voltammetry · Natural receptors · Synthetic receptor

M. B. Regasa (✉)
Chemistry Department, College of Natural and Computational Sciences, Wollega University, Nekemte, Ethiopia

S. Patra et al. (eds.), *Molecularly Imprinted Polymers: Path to Artificial Antibodies*, https://doi.org/10.1007/978-981-97-4379-7_5

5.1 Introduction

Electrochemical sensors are devices that convert the interaction between the electrode and analyte into meaningful qualitative and quantitative analytical signals. These sensors in particular are very important in the modern society due to their high selectivity and sensitivity, fast response, low cost, user-friendly and possible to miniaturization [1, 2]. There are different modes of measurement in electrochemical sensor system (e.g. conductometry, potentiometry, amperometry and voltammetry) that allow the development of novel sensor devices using both natural and artificial molecular recognition elements. The preparation of electrochemical sensor involves the surface modification of electrode (transducer) using suitable receptors based on the nature of analyte to be detected [3, 4]. Thus, recognition elements or sensing elements are considered as the heart of chemical sensors because they perform the recognition of analytes of interest. The natural receptors have been widely used but challenged by the poor stability and high price as well as its unavailability for some analytes apart from their extremely selectivity in electrochemical sensors. Following this, there have been many efforts made to replace them with synthetic specific recognition elements [5]. The discovery of new recognition elements greatly improved the selectivity, sensitivity, limit of detection, and signal-to-noise ratio of chemical sensors with the help of nanotechnology [6]. For example, the development and integration of artificial receptors like molecularly imprinted polymer (MIP) in to electrochemical system revealed the increase in sensitivity and lower detection limit with enhanced signal-to-noise ratio and good stability than those use natural receptors [7]. MIP has advantages over conventional biological receptors that were practically tested. Contrary to this, MIPs are characterized by their high physical and chemical stability, easy to prepare tailor-made functionality, cheap, and have analyte binding affinity comparable to biological receptors. Thus, electrodes modified with MIPs are meaningfully eye-catching sensing devices replacing biosensors [8]. Three issues are critically important for successful design of MIP sensors such as development or selection of suitable electrode, synthesis of MIPs, and its integration with the electrode [9]. The development of MIP involves the polymerization of functional monomer in the presence of imprint molecule (template) using suitable solvents, initiator, and cross linker at specified experimental conditions. The template removal from the polymer matrix formed can be done using chemical method or electrochemical methods to create cavity that has complementarity in size, shape and functionality [10]. For the preparation of electrochemical sensors using MIPs as recognition elements with good selectivity and sensitivity for analyte one can use chemical, electrochemical and enzymatic methods [5, 11]. The electrode surfaces can be coated with MIPs using chemical, electrochemical and enzymatic methods [5, 12]. These kinds of sensors are used commonly for the determination of organic molecules, metal ions, evolving pollutants and biomolecules with sensitivity and selectivity [13]. MIPs-based electrochemical sensors have wider applications including food quality monitoring [12, 14], clinical diagnosis [11, 15], health [16], environmental monitoring [17], precision agriculture [1, 18] and security [19]. Therefore, this chapter presents the insight in the advancements in the combination of MIP receptors and electrochemical transducers for the construction of electrochemical sensors for various requests. Particularly, it focuses on the general introduction, development of MIPs and its application in electrochemical sensor development, and the various application areas.

5.2 Development of MIPs

The concept of MIT was first coined by Polyakov [20] and repeated by Dickey. In their work, silica matrix assembles together in the presence of organic dyes and then showed unusual affinity towards these molecules in a similar fashion of antibody–antigen affinity. Later on, a great achievement in the MIT was reported by Wulff and Sarhan [21] through the introduction of covalent imprinting method to prepare MIPs. Interestingly, another approach based on the non-covalent bonding was presented by Mosbach and Arshady [22]. Synthetic receptors like MIPs are more attractive than their counter biological receptors because they can be prepared for every guest or analyte to be bound [23, 24]. MIPs are synthetic molecular recognition elements with binding sites created in polymer matrices for specific recognition and catalysis [25]. The synthesis of MIPs is performed from a mixture consisting functional monomers and template (an analyte) as the main components and other components such as templates/analytes, functional monomer, porogen, initiator and cross linker. The essential components required to prepare MIPs for different applications are template, functional monomers (FM), porogen/solvent, initiator and cross linker mixed in a defined ratio. The synthesis involves different steps including the formation of complexes between templates and FM first and then polymerization to form three-dimensional imprinted polymer or plastic anti-bodies (Fig. 5.1). After this, the template is washed away using solvent extraction or electrochemical methods. Here it is important to note that the performances of MIPs-based electrochemical sensors mainly depends on the optimized ratios of the components mentioned [26]. As a result, it is important to get familiar with important parameters affecting the properties and performances of the final MIPs matrices. These are optimal template, functional monomer, crosslinker or the porogen to exploit the recognition performances of the MIPs. Here, the selection of proper functional monomer plays a great role in the synthesis of MIPs for specific purposes. Also, it is important to know the possible effects of porogen on the final performances of the desired MIPs [27]. The polymerization of functional monomers during MIP preparation occurs through the generation of free radicals. These free radicals form polymers using various fabrication methods including bulk, precipitation, emulsion, suspension techniques, precipitation, multi-step swelling, surface imprinting and electrochemical polymerization methods [28]. The natural phenomenon such as enzyme–substrate and antigen–antibody interaction has been simulated to prepare MIPs with good selectivity for specific target molecules through the development of MIT [29]. In MIT, the template molecule interacts with the functional monomer through covalent, non-covalent and semi-covalent bonding to form pre-polymerization complex. It is worthy to select functional monomer that

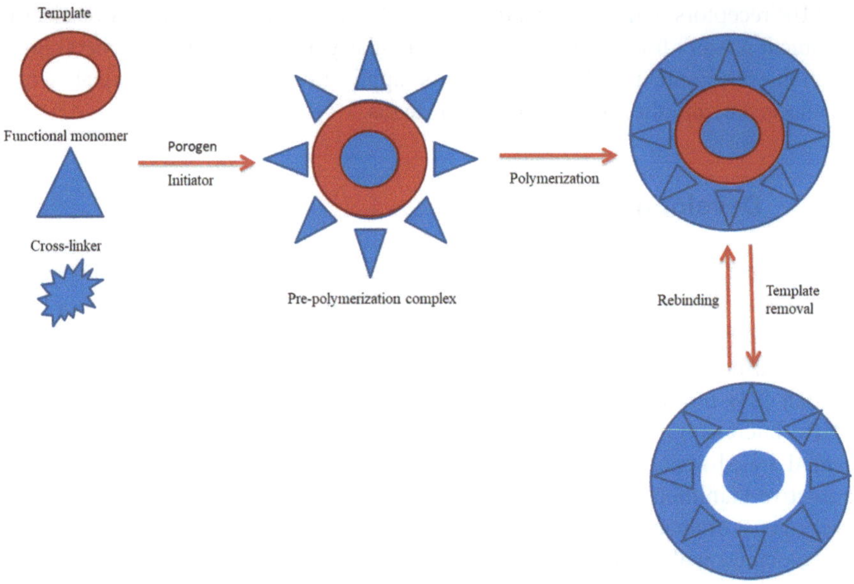

Fig. 5.1 Steps in molecular imprinting during the preparation of MIPs

will have sufficient interaction with the template to obtain MIP with available recognition sites [30]. There are four important consecutive steps in the molecular imprinting of MIPs. These are

(a) Formation of complexes through the interaction between the template and functional monomer before polymerization starts. The pre-polymerization complex is formed through covalent bonding, non-covalent bonding, π–π stacking, electrostatic attractions and van der Waals forces. The self-organization of functional monomer through functional group–template and functional group–monomer interactions.
(b) Polymerization of the complex formed in the presence of initiator and cross linker. The polymer particles grow around the template through encapsulating. Finally, the three dimensional polymer with rigid structure is formed.
(c) Template extraction or removal through the application of solvent extraction process or electrochemical methods. This process creates empty cavities in/or on the polymer matrices which have harmonizing in extent, figure and functional units to memorize the template upon re-interaction again.
(d) Recognition of the template through the immersion of MIPs in the solution containing the target or its analogue. During this step, MIP will recognize the template through the interaction in different mechanisms. The selectivity of the MIP can be evaluated through the addition of some interferents into the solution comprising the target.

The effectiveness of the imprinting strategy can be confirmed through the characterization of the MIPs before and after template elimination as well as after the rebinding with the template.

Characterization techniques such as spectroscopic techniques (ultraviolet-visible, infrared, nuclear magnetic resonance, energy dispersive X-ray) for functional units' analyses, microscopic techniques (scanning electron microscopy, atomic force microscopy) for morphological analyses, surface area analysers, particle size analyser and pore size determination are commonly used [28, 31].

5.3 Advantages of MIPs as Recognition Elements

Natural receptors (e.g. antibodies, enzymes) and synthetic aptamers were in use as precise receptors to analyte in a different testing and sensing technology. Though, there are numerous problems for their application due to the deficiency in strength and reusability. Furthermore, it is expensive to prepare biological receptors. Due to the major challenging accompanying natural recognition elements, researchers have struggled to develop artificial-specific receptors as a substitute for natural receptors. From different synthetic molecular recognition elements, MIPs are the promising functional materials owing to their exceptional properties, high physicochemical stability, low cost, availability and cheap production processes [32, 33]. The integration of artificial receptor materials into biochemical detection methods has become increasingly attractive because of the better engineering possibility. The specific recognition capacity of MIPs to detect chemical and biological compounds in food, clinical, environmental and industrial process control was confirmed [34, 35]. It is evident that the development of new MIPs for the construction of chemical sensors is still the hot research topic today. Biomimetic polymers such as MIPs can be used to recognize larger biological molecules and small organic molecules [36, 37]. Therefore, MIPs are considered promising alternatives to biological sensing elements since they provide cost-effective and better performances accompanied with high stability in modern chemical sensors that are the main requirements for the future commercialization [38]. With this regard, MIPs have many advantages over traditional biological recognition elements in the sensors development [30, 39, 40]. These significant advantages of MIPs include

- Having high discrimination and sympathy for the template molecule used in the imprinting procedure.
- Having higher physical and chemical robustness, strength, resistance to elevated temperature and pressure and inertness towards acids, bases, metal ions and organic solvents when compared to biological recognition elements like enzymes, proteins, nucleic acids, etc.
- They can be produced for several targets of clinical or environmental interest and to many kinds of matrices for which no receptor or enzyme is unavailable or when they are too expensive.

- Easy to prepare and cheap and the long storage life without loss of their recognition capability for numerous years at room temperature.
- The polymerization step is fully compatible with microfabrication used in sensor technology.
- The rebinding of the analyte and features such as rigidity, porosity, molecular weight, resistance in different conditions and responsivity to various stimuli can be modulated.
- The use of steady reagents and the production of low-cost assays free of animal products.
- The availability and high probability of selecting various functional monomers, porogen, cross linkers and initiators allow the freedom and flexibility in predetermining the desired properties of MIPs depending on its application [29].
- MIPs are prepared for both organic (drugs, herbicides, metabolites, hormones and vitamins, toxins) and inorganic (heavy metals such as lead, cadmium, copper, nickel, etc.) templates/analytes. These synthetic functional materials are relatively cheap, reusable, stable and robust though there are rare approved protocols for the preparation. In sensing technology, MIPs have comparable affinity towards analytes like that of biological sensing elements [8, 41, 42].

5.4 Application of MIPs in Electrochemical Sensors

5.4.1 MIP Electrochemical Sensors

The electrochemical sensor consists of recognition element, electrode, signal amplifier, signal processor and readout components. Generally, it is made up of a recognition element and a transducer element [43]. The chemical signal generated after the recognition element specifically binding to the analyte is converted into an output electrical signal by the conversion element, so as to achieve the purpose of highly sensitive and selective analysis. As the core component of electrochemical sensors, electrodes show an important role in the production of electrochemical sensors [44]. The sensing element directly influences the important parameters of electrochemical sensors such as selectivity, sensitivity, stability and response time [38]. MIT is a deep-rooted technique for making synthetic receptors of chemical sensors. The immobilization of MIP as a thin film directly onto the surface of the transduction unit integrates these units to result in an electrochemical sensor [45]. The presentation of MIPs-based recognition elements in the design and fabrication of sensitive and selective electrochemical sensor helps us to address the challenges of using natural recognition elements such as high cost and instability in harsh environmental conditions. Microporous MIP films deposited on the electrode surfaces obtain fast analyte rebinding kinetics when compared to the continuous MIP films [46]. Importantly, the combination of MIP as a recognition element and electrochemical sensors improves both the selectivity and sensitivity. This can be reasoned out as because of the high thermal and chemical stability, simple preparation process, abundant reusability, likely specific recognition and satisfactory selectivity as well as the unique features of electrochemical methods [47]. The interaction and adsorption of the template molecule with the cavity on the MIP films generate electrochemical signals. These are the underlying principles of such kinds of sensors.

5.4.2 Preparation of MIPs-Based Electrochemical Sensors

The recognition elements play a great role in determining the performances of sensors. Such recognition components need to be vigorous, low-cost, penetrating to all targets, proficient for producing a sensor response and appropriate for incorporation into economical and vigorous electrodes. Regrettably, none of the natural recognition elements fulfil these requirements. Recently, MIPs-based sensors become the promising devices to address the mentioned criteria for sensor fabrication [48]. MIP-based electrochemical was prepared for the first time in the early of 1990s by the Mosbach group [49]. After 3 years, MIP thin film-based membrane made up of polyaniline was prepared for L-phenylalanine sensing applications using field-effect capacitor [50]. Basically, there are two main approaches for the preparation of MIP-based electrochemical sensors. The first approach involves the synthesis of MIPs directly on the surface of electrode using electropolymerization and self-assembly approaches. In the second approach, MIPs are prepared separately and deposited on the electrode surface using drop casting, spin coating and dip-coating methods.

5.4.3 Electrode Surface Modification Strategies

Modification of the electrode's surface with electroactive chemicals (chemically modified electrodes, CMEs) was reported in the 1970s to advance the signal of electrochemical sensors [51]. The immobilization of electrochemically active species as thin films controls the physical–chemical nature of the electrode–solution interface by changing the reactivity and selectivity of the electrode's surface [52]. MIP thin films are ideal for assembling and constructing an electrochemical sensor for detecting a range of analytes. These films facilitate analyte mass transfer, which is very important for rapid signal generation and analysis [53, 54]. The integration of MIPs into electrochemical methods involves the preparation of thin films on the surface of different electrodes (Fig. 5.2). Hence, direct deposition of MIP thin film on the electrode surface is regarded as the simple approach for the preparation of electrochemical sensors. The process involves in-situ and ex-situ approaches. For the in-situ methods, the MIPs films are directly synthesized on the active surface of electrode using electropolymerization [55] and self-assembly [56], while in the ex-situ type, the MIP is prepared separately first and then immobilized on the surface of electrode using drop-casting and dip-coating methods [57] and in-situ chemical polymerization followed by spin coating [58]. MIP electropolymerization method has been appealed to be effective because it needs no external initiator to start the reaction and it permits the regulation of the depth of the polymer layer on the surface of electrode (i.e. polymerization). This method lets the creation of MIPs with

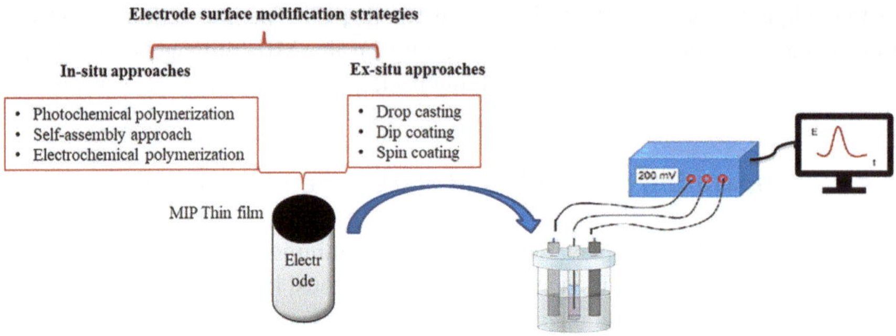

Fig. 5.2 Electrode surface modification strategies for the preparation of MIP-based electrochemical sensor

sufficient accessible binding sites with high selectivity and with fast uptake kinetics in short period, permitting the achievement of nano-molar limits of detection [12].

Surface molecularly imprinted polymers (SMIPs) possess the advantages of high specific surface area, ease of template molecules removal, quick mass transfer, and controllable binding capacity (mainly depending on the physical properties of the substrate) [44]. Condensation and radical polymerization are the common mechanisms used so far [59]. Radical polymerization is the most studied method to prepare MIPs as it can be performed in a wide range of temperatures. This method enables to prepare MIPs in different arrangements, such as bulk form or nano/micro dispersions; there are several functional monomers with different functional groups that can be employed with this strategy. This initiation method opened various approaches such as thermal heating [60], microwave [61], ultrasound [62], photosynthesis [63], etc. The elution of the template is the next step that can be performed using a solvent extraction or electrochemical method with significantly affecting the polymer matrix. The last step is evaluation of the rebinding capacity of the MIP prepared. The MIP prepared away from the transducer will immobilized on the surface of the transducer through different strategies. Lately, the application of magnetic MIPs in the fabrication of electrochemical sensors for the detection of several kinds of target compounds highly enhanced the signal output. This approach has also been used for magnetic MIPs, mainly chemically prepared. Magnetic MIPs pull from the solution containing the analyte to the electrode surface lets selective removal from complex matrices to detect trace level of analyte [64]. Different MIPs are prepared using chemical polymerization methods such as bulk polymerization [65], precipitation polymerization [66], suspension polymerization [67], emulsion polymerization [68], two-step swelling and polymerization [69] and sol–gel methods [70]. Then, the prepared MIPs will be immobilized on the surface of the substrate by drop casting, dip coating and spin coating.

Importantly, the incorporation of magnetic nanomaterials into MIPs had opened a new door for the development of rapid, user-friendly, cheap, reproducible, very sensitive and selective sensors. This achievement was confirmed through the application of external magnetic field. Furthermore, various nanomaterials such as metallic nanoparticles, carbon nanomaterials (single-walled and multi-walled carbon nanotubes, graphene oxides) and ionic liquids have been used to modify the analytical performances of MIPs-based electrochemical sensors. These types of sensors were prepared and employed to detect important analytes in clinical samples, food industry and environmental matrices [71].

5.5 Types of MIP Electrochemical Sensors

All electrochemical sensors encompass two main components, namely, a sensing element and a electrode/transducer. For long time, biological receptors such as enzymes, antibodies, aptamers, receptors, DNA, phages and whole cells have been used in biosensors fabrication. These natural materials have significant limitations and thus replaced by MIPs that serve as selective recognition units [72]. The transducer translates the chemical sensing occasion into a measurable analytical signal [5]. Generally, electrochemical sensor consists of two main elements, namely a recognition elements and electrochemical transducers. The recognition elements interact and recognize the target, while the transducer translates the chemical recognition incident into a computable analytical signal such as current (*A*), charge (*C*), potential (*V*), resistance (Ω) or conductivity (*S*). Therefore, it is possible to classify electrochemical sensors into some groups based on the transduction mechanism and output signals into voltammetric, potentiometric, amperometric, conductometric and impedimetric (Fig. 5.3). Amperometry and voltammetry are the two major types of electrochemical sensors. There are many types of voltammetric techniques including differential pulse voltammetry (DPV), cyclic voltammetry (CV), square

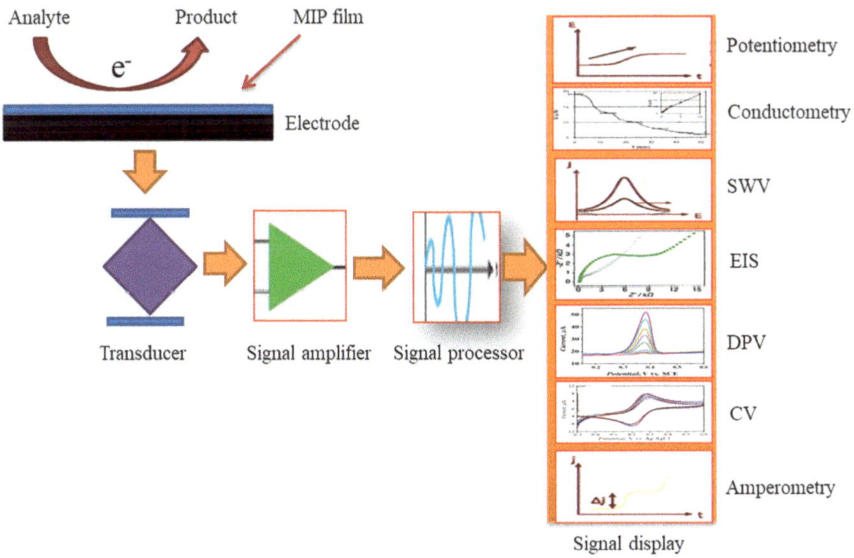

Fig. 5.3 Types of MIP-based electrochemical sensors with their output signals

wave voltammetry (SWV), stripping voltammetry and linear sweep voltammetry (LSV) [73, 74]. MIP-based voltammetric sensors are the best established techniques because they are simple, rapid and sensitive in determining analytes. In these techniques, faradaic current generated from redox reaction is measured as a function of the concentration of the analyte that participates in the reaction [75]. The potential applied during this measurement can be modulated as a linear scanning with time, non-linear modulating with a pulse or sinusoidal excitation, etc. MIP-based voltammetric sensors were developed for the detection of different analytes [55, 76–78]. Also, MIP-based impedimetric sensors [79], conductometric sensors [80], potentiometric sensors [81] and other formats were fabricated to determine targets in different matrices [82].

5.6 Detection Mechanisms

Since Mosbach and his co-workers reported the first MIP-based electrochemical sensor; this technology become popularized. In MIP-based electrochemical sensors, quantitative analyses are performed by measuring the changes in current generated before and after template binding to MIPs. Since they are stable, sensitive and selective, they are widely used [74]. MIP membrane-based sensor has the advantage of easy to operate, computerization, miniaturization, high sensitivity, cost-effective, detection of electroactive and non-electroactive molecules. However, in most cases it needs labelling the analyte to increase the electrochemical reaction at the working electrode. Generally, MIP electrochemical sensors propose two common strategies for determining target analytes (Fig. 5.4). These methods are direct electrochemical readout for electroactive analytes and indirect electrochemical readout used for non-electroactive analytes. The direct electrochemical detection is accomplished for electroactive target molecules in the potential window of the analyte electroactivity, that is, where the electron is directly transferred between the analyte and the

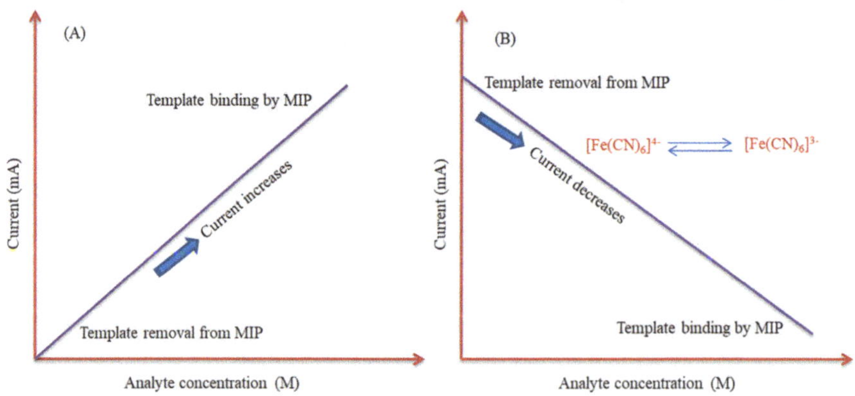

Fig. 5.4 Graphical representation of direct (**a**) and indirect (**b**) electrochemical detection signals for electroactive and non-electroactive analytes

electrode surface occurs. Electroactive targets can be quantified by the magnitude of the current generated in amperometry or voltammetry. Voltammetry [73, 83] and amperometry [84] are appropriate for detecting electroactive analytes, while potentiometry is suitable for the detection of charged analytes [85, 86]. MIP-based electrochemical sensors can detect non-electroactive targets because the rebinding such analytes brings the variation in the conductivity or porosity of the polymer matrix. Such change in property of the MIP can be terraced using CV, conductimetry or EIS using an external redox probes such as hexacyanoferrate, ferrocene and hexaammineruthenium(III) chloride [6, 87]. The indirect method (gate effect) involves the measurement of faradaic current generated due to the presence of probe and the current gradually reduces as more analyte get about to the MIP and block the passage of electron through the interface [88, 89]. For example, Li and his co-workers proposed an electrochemical sensor using indirect detection to recognize thiol-3-indoleacetic acid (IAA) [90].

5.7 Applications of MIP Electrochemical Sensors

The important characteristics such as robust nature, simple preparation processes (easy electrochemical template removal to generate cavities) and fast response attracted a huge interest from different fields. Furthermore, the synergistic advantages of MIPs and electrochemical methods offer user-friendly sensing technology with short response time, high selectivity and sensitivity [16]. These sensors are highly applicable in the fields such as clinical diagnosis, food quality control, environmental monitoring and agriculture as depicted in Fig. 5.5. This implies that the potential application of these sensors for detecting various target analytes [91, 92]. According to Yarman and his co-authors [92], voltammetric sensors showed better sensitivity than others for drug analysis.

Determination of various mycotoxins in agricultural products and food products using MIP electrochemical sensors become the promising approach currently. For instance, nano-sized MIPs were captured in electro-active polymers for voltammetric detection of ephedrine [93], in a polyvinyl chloride matrix for potentiometric determination of cocaine and atrazine [94], in nafion for voltammetric measurements of vancomycin [95] and in cross-linked electrodes using self-assembled monolayers of alkane thiols for voltammetric and impedimetric detection of 4-ethylphenol [96]. Furthermore, the MIP-based electrochemical sensors are applicable in detecting biologically important molecules, pesticide residues and antibiotic, toxins, food additives and adulterants, environmental organic and inorganic pollutants [14, 17, 55, 77].

5.7.1 Food Quality Monitoring

With development in economy and personal safety, food safety is attracting more attention from the food production and distribution companies. Therefore, strong

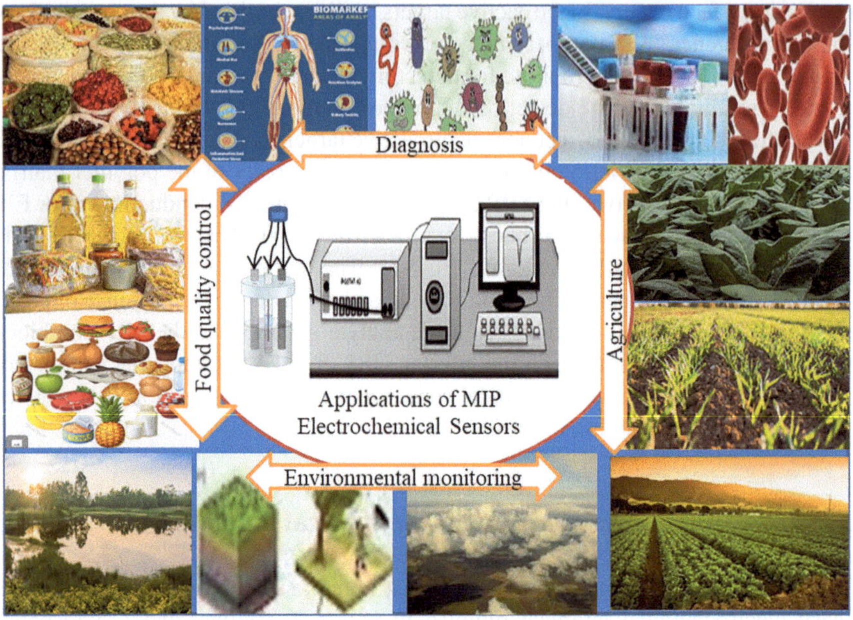

Fig. 5.5 Some application areas of MIPs-based electrochemical sensors

regulatory rules are in place in many countries across the world to safeguard the safety of consumers. These regulatory rules can be implemented only if rapid food analysis techniques are available with affordable prices and good stability. Food contaminants of concern can be chemical, microbial and physical [97]. The selectivity and stability offered by MIP-based electrochemical sensors provide a good way for monitoring target food contaminants and other nutritionally important elements/molecules in complex matrices [12, 98]. In the globalization era, the food safety issue is among the front points raised by consumers to avoid the possible health and economic effects caused by consuming contaminated food stuffs. Thus, MIP-based electrochemical sensors are multipurpose as they can be designed and prepared for various food contaminants without difficulties [5]. Hence, the design and development of MIP sensors ensure food safety because of its numerous advantages.

Milk can be contaminated with both organic and inorganic chemicals from the feed and water the cows consume as well as drugs used to treat them. For example, ciprofloxacin hydrochloride is used to treat animal disease but also affects the quality of milk products. Poly-pyrrole-based MIP was used to develop electrochemical sensor to detect ciprofloxacin hydrochloride in milk. The electron signal was enhanced through the incorporation of gold nanoparticles (AuNPs) to attain the linear range of 0.05–500 ng/mL and detection limit of 0.0033 ng/mL [99]. The same polymer was used to prepare electrochemical sensor to determine Bisphenol A in food in the range of 750–0.5 nM and detection limit of 0.20 nM [100].

Magnetic MIP was introduced into electrochemical techniques to detect scombrotoxin (histamine) in fish food products. In this work, histamine template and 2-vinyl pyridine functional monomer were used. The sensor showed the capacity to sense the target with the detection limit of 1.6 ng/L successfully [73]. In another work, Magnetic MIP, carbon black nanoparticles (CBNPs) and AuNPs were used to prepare an electrochemical sensor for the selective and sensitive measurement of Bisphenol A (BPA) in water with nano-level detection limit [64]. The composite consisting of MIP, multi-walled carbon nanotubes (MWCNTs) and AuNPs was also used to fabricate electrochemical sensor for the selective and sensitive determination of dibutyl phthalate (DBP) in food. The reported sensor showed wide linear range and lower detection limits [74]. Food contaminant such as 3-monochloropropane-1,2-diol (3-MCPD) is produced during food heating. For this contaminant, electrochemical sensor was developed by using nanoporous gold (NPG) capped with MIP as a sensing element to detect it in real soy sauce. At the optimum conditions, the sensor displayed best analytical against the 3-MCPD with a low detection limit of 0.035 fM [101]. Oxytetracycline (OTC) remainder in food products and other environmental components can pose severe health and ecosystem problems. To effectively detect and quantify this pollutant, MIP was combined with AuNPs and carbon nanospheres to fabricate chemical sensor. The surface modifiers were characterized by using different analytical techniques such as Fourier transform infrared absorption spectroscopy, X-ray photoelectron spectroscopy, scanning electron microscopy and transmission electron microscopy. The fabricated sensors showed a wide linear range from 0.02 to 20 μM and detection limit of 5 nM to detect OTC. In addition to this, the sensor exhibited the imprinting factor of 5.50, adsorption capacity of 56.70 mg/g and fast response time of about 15 min [102].

The residues of pesticides in vegetables are among the priority pollutants that need strict monitoring. MIP and AuNPs were used as a recognition element and signal amplification strategies to prepare an electrochemical sensor to selectively detect carbofuran in vegetables. Under optimized conditions, the sensors exhibited a linear range from 50.0 nM to 0.40 mM with a low detection limit of 0.24 nM [103]. Another pesticide called Metribuzin (MTZ) is a hazardous chemical present in the environment. The residues of MTZ in different food samples were detected by the electrochemical sensors based on MIP composites. DPV was used to measure the analyte to obtain the limit of detection of 0.10 pg/mL at the optimized experimental conditions with good recovery range of 97.12–103.41% [104].

The global food production and quality is highly challenged by the occurrence of mycotoxins mainly by aflatoxins. As a result, it is mandatory to regulate the levels of aflatoxins in different food stuffs. To address this issue, MIP-based sensitive electrochemical sensing device was prepared using polyaniline to determine aflatoxin B1 (AFB1) and fumonisin B1 (FuB1). The reported sensor displayed the detection limit of 0.313 and 0.322 pg/mL for targets, good selectivity for the target, fast response and cost-effective preparation when compared to other similar sensors [105]. Different analytes in food matrices were analysed using MIP-based electrochemical sensors. The electrochemical technique used, linear range and limit of detection (LOD) for selected analytes detected in food for quality monitoring applications are shown in Table 5.1.

5.7.2 Environmental Monitoring

Environmental monitoring refers to the monitoring of air, water, noise and biodiversity. The levels of some pollutant chemicals in the environment need to be monitored periodically for proper action to mitigate pollution. Such kinds of action entirely depend on the good data generated by suitable analytical methods like electrochemical sensors. In particular, MIP-based sensing technology has a great impact on the environmental applications due to their capacity to detect molecules at trace levels in complex matrices without the prior need for sample treatment. MIP-based electrochemical sensors have been widely used to detect environmental pollutants released from pharmaceuticals, pesticides, personal care products, detergents, heavy metals and food packaging materials [109, 110]. Hence, these sensors are targeted to provide the users with simple, fast and cheap methodologies for the detection and quantification of a chemical compound in different environmental samples [111]. Particularly, these electrochemical sensors are suitable for field testing and in-situ monitoring of priority pollutants. In the industrialized world, aquatic environments are highly affected by the excessive use of pesticides like endosulfan [112]. This chemical destroys the fish production sources and pollutes drinking water largely. Bow and co-workers [98] developed voltammetric sensor using MIP to recognize the endosulfan pesticide and showed the detection limit of 0.02 mM.

BPA is a toxic chemical widely available in environmental water systems primarily discharged from food storage containers. The combination of MIP with AuNPs offered a good protocol for the preparation of an electrochemical sensor to detect BPA in water medium. The modified electrode was connected to smartphone that

Table 5.1 MIP-based electrochemical sensor for food quality monitoring analytical applications

Analyte	Linear range	LOD	Reference
AFB1	1 pg/mL–500 ng/mL	0.313 pg/mL	[105]
FuB1	1 pg/mL–500 ng/mL	0.322 pg/mL	[105]
MTZ	0.2 ng/mL–21.43 μg/mL	0.10 pg/mL	[104]
BPA	0.07–10 μM	8.8 nM	[64]
3-MCPD	0.10 fM–0.10 μM	0.035 fM	[101]
DBP	0.10 μg/L–10 mg/L	5.09 ng/L	[74]
OTC	0.02–20 μM	5 nM	[102]
Carbofuran	50.0 nM–0.40 mM	0.24 nM	[103]
Ciprofloxacin	0.05–500 ng/mL	3.3 pg/mL	[99]
2,4-Dichlorophenoxyacetic acid	0.1–0.250 nM	0.117 pM	[52]
17β-Estradiol	1 pM–0.1 μM	10 pM	[106]
Sulfadimidine	0.2–1000 ng/mL	0.169 ng/mL	[107]
Streptomycin	0.05–20 ng/mL	10 pg/mL	[108]
Melamine	0.5–200 nM	0.413 nM	[77]

showed high selectivity and sensitivity with linear range from 0.01 to 10 μM and limit of detection of 3.97 nM [113].

Polycyclic aromatic hydrocarbons (PAHs) are persistent organic pollutants that are suspected to be hazardous. Its monitoring is becoming mandatory in many countries recently. In line with this, MIP-based conductometric sensor based on screen printed interdigital gold electrode was reported to measure PAHs (anthracene) in water samples with the detection limit of 1.3 nM [114]. 2,4-Dichlorophenol (2,4-DCP) is another highly toxic chlorophenol compound that has long-term effects on living organism including humans. Liu and his co-authors [115] reported the development of an electrochemical sensors using MIP to determine 2,4-DCP in water sample with high selectivity and sensitivity. These chlorinated hydrocarbons are not electroactive in nature thus why an electroactive ferrocene was used to measure the decrease in the current signal when these compounds are rebound. The electrochemical determination was able to give the detection limit of 0.5 pM [116].

In addition to organic pollutants, there are many inorganic pollutants in the environment. These inorganic pollutants are mainly due to the presence of heavy metals that are lethal and persistent. The exponential growth of industrial activities and urbanization highly increased the accumulation of heavy metals both in soil and water systems. In the last two decades, ion imprinted polymers (IIPs) have widely applied the development of electrochemical sensors for the detection of heavy metals and other anions in the environmental samples. Lead, cadmium, chromium and arsenic are heavy metals that are highly toxic to living things. Thus, their monitoring using IIP composite materials for the fabrication of electrochemical sensors was reported. For example, magnetic ion-imprinted nanoparticles Fe_3O_4@SiO_2@IIP-based electrochemical sensor was developed to measure the level of lead in water and fruit juice to achieve the wide linear range from 480 pM to 0.35 μM and detection limit of 240 pM [117]. Similarly, IIP combined with multi-walled carbon nanotubes (MWCNTs) for the detection of mercury ion in water samples [118]. The concentrations range of mercury from 100 pM to 0.2 nM and detection limit of 200 pM were demonstrated. Other metals such as arsenic(III), cadmium(II) [119] and chromium(III) [120] were detected in various environmental samples using MIP-based electrochemical sensors. Some selected pollutants in the environment were determined by employing various types of MIP-based electrochemical sensors as summarized in Table 5.2.

5.7.3 Clinical Diagnosis

Clinical diagnosis helps to identify and detect the status of illness in patients to increase the success of treatments and to eliminate the possible long-term complications [127]. The proper detection and quantification of biomarkers in body fluids are very important for the early detection, diagnosis and treatment of a variety of diseases [128]. In most patients, the level of biological molecules (biomarkers) in the body fluids that may increase or decrease is considered a good instrument for detecting diseases. The sample matrices such as blood, urine, tears, sweat,

Table 5.2 Analytical applications of MIP-based electrochemical sensor for environmental monitoring

Analyte	Linear range	LOD	Reference
Endosulfan	0.02–0.12 mM	0.02 mM	[98]
BPA	0.01–10 μM	3.97 nM	[113]
Anthracene	–	1.3 nM	[114]
2,4-DCP	3.40–3.9 ppm	0.5 pM	[116]
Lead	480 pM–0.35 μM	240 pM	[117]
Mercury	100 pM–0.2 nM	200 pM	[118]
Cloxacillin	10–750 nM	36 nM	[121]
Toltrazuril	0.43–42.54 μg/L	0.21 μg/L	[122]
Methidathion	40–200 μg/L	5.14 μg/L	
Phosphate	0.66–8 mg/L	0.16 mg/L	[123]
Chlorpyrifos	20–300 μg/L	4.5 μg/L	[124]
Sulfamethoxazole	0.1–100 μM	24.1 nM	[125]
Methyl parathion	0.05–15 μM	0.01 μM	[126]

nasopharyngeal fluids and others are used to detect biomarkers [11]. The detection of important biomarkers in biological fluids, endocrine hormones, proteins, drugs and important chemicals in the body is very important for the treatment and early protection of diseases. With this regard, electrochemical sensors become eye-catching methods.

Different electrochemical sensors based on MIPs and its composites developed to detect dopamine neurotransmitter [129], proparacaine anesthetics drug [130], antipsychotic sulpiride drug [131], cocaine [48], small molecule tetrahydrocannabinol (THC) and a protein (trypsin), paracetamol [96], vancomycin antibiotic [95], trypsin, glucose, paracetamol, C4-homoserine lactone [132], D-arabitol in urine [133] and gram-negative bacterial quorum signalling molecules [134]. Similarly, stress hormone cortisol [135] and sarcosine (SAR) [136, 137] were successfully determined through the proper design, development and application of MIP-based electrochemical sensors.

Glucose electrochemical sensor was developed using MIPs decorated with gold nanoparticles. Such sensor showed the linear concentration range of 1.25 nM–2.56 μM, detection limit of 1.25 nM and good stability with about 95% performance maintained for the storage time of 40 days [138]. The important disease biomarkers such as carcinoembryonic antigen (CEA) and neuron-specific enolase (NSE) were determined at trace level through the enhancement of sensor signals using nanomaterials and MIPs together to modify the surface electrodes [139, 140]. The scenario of the occurrence of COVID-19 pandemic caused by severe acute respiratory syndrome coronavirus-2 (SARS-CoV-2) coronavirus changed the global trends. The testing of this disease using accurate and cheap portable MIP-based electrochemical sensors was reported. The sensor was applied for the electrochemical detection of SARS-CoV-2 nucleoprotein with the linear response of 2.22–111 fM, detection limit of 15 fM and quantification limit of 50 fM [141]. Furthermore, SARS-CoV-2 antigen, SARS-CoV-2 spike protein, and SARS-CoV-2 nucleocapsid protein, C-reactive protein, hepatitis C virus core antigen and E2 protein, and biomarkers of nosocomial infections (*K. pneumoniae*, *P. aeruginosa*, *L. monocytogenes*, and *A. baumanii*) were successfully detected and quantified using MIP-based electrochemical sensors [11]. Generally, MIP-based electrochemical sensor has been widely employed in clinical diagnosis for the determination of disease biomarkers and drugs as indicated in Table 5.3.

5.7.4 Precision Agriculture

Precision agriculture is a smart and sustainable system that maximizes agricultural yields and improves environmental protection by refining resource consumption [143]. The monitoring of important information such as pH and farm soil nutrient level is very important in precision agriculture. For this purpose, MIP-based electrochemical sensors provide vital information demanded for precision agriculture. MIPs-modified electrodes sense particular ion in soil and water by recognizing it. The ion can be hydrogen ion, nitrate, phosphate, metal ions and others. Excessive use of pesticides in agricultural sectors causes food contamination and environmental pollution including water. This might be due to the toxic nature of pesticides to humans and can have both acute and chronic health effects [144]. MIP sensors are promising in detecting pesticides in-situ and real time [145]. Electropolymerization method was used to prepare MIP-based electrochemical sensors to detect paraoxon [146], triazine herbicides [147], methyl parathion [126, 148], chlorpyrifos [124], lindane [149], carbofuran [150], dichlorvos [151], metribuzin herbicide [104], hexazinone herbicide [152] and atrazine [153]. The monitoring of nutrient levels, status and physical condition of soil plays a crucial role in agriculture since it is exposed to various natural and anthropogenic activities [154]. Hence, the knowledge of the status of nutrients in soil plays a key role in agriculture. Thus, nutrients

Table 5.3 Analytical applications of MIP-based electrochemical sensor in clinical diagnosis for the measurement of selected targets

Analyte	Linear range	LOD	Reference
Dopamine	0.395–3.96 nM	0.193 nM	[129]
Proparacaine	0.5–12.50 mg/L	0.11 mg/L	[130]
Cocaine	1.0 nM–1.0 mM	–	[48]
Paracetamol	0.24–5 mg/L	–	[96]
Cortisol	0.1–100 nM	0.036 nM	[135]
SAR	5.0 μM–1.1 mM	0.38 μM	[137]
Glucose	1.25 nM–2.56 μM	1.25 nM	[138]
CEA	0.1–100,000 ng/mL	80 pg/mL	[139]
NSE	25–4000 pg/mL	200 pg/mL	[140]
SARS-CoV-2	2.22–111 fM	15 fM	[141]
Insulin	0.287–11.47 ng/mL	0.149 pg/mL	[132]
Testosterone	0.144–5.76 ng/mL	144 pg/mL	[142]
CEA	1–500 ng/mL	320 pg/mL	[139]

like phosphate [155], nitrate and phosphate in soil [156], and urea [157] were monitored by using MIP-based electrochemical sensors. For accurate and selective determination of different chemicals use in agriculture, different formats of electrochemical sensors using MIPs were reported as shown in Table 5.4.

5.7.5 Security System

More than ever, the homeland security and counterterrorism become the front-line threat to many countries in the world at the present time due to the risks associated with the applications of explosives and explosive-related compounds. To avoid the health, ecological and security risks associated to explosives, it is very important to monitor them in different environmental samples such water, air and soil. Thus, there is a huge demand in the development and applications of electrochemical sensors to detect explosives and chemical warfare agents at trace level in complex environmental samples [164]. It is normal to find research reports on the applications of MIPs in electrochemical sensors for the field-detection and real-time analysis of explosives and warfare agents. This accounts for the fact that MIPs are highly stable, selective, sensitive, cheap, flexible and reproducible if integrated into sensor systems [165]. It shows that the sensing of energy materials and warfare agents is the most important and significant issue in counterterrorism, maintaining national security and environmental protection [19, 166]. 2,4,6-Trinitrotoluene (TNT) is a commonly used explosive compound, and it simply accumulates in soil and water during transportation, use and storage [167]. MIP and carbon paste electrodes were combined to prepare an electrochemical sensor to determine trace level of TNT in the environment. When compared to the non-imprinted polymer, the MIP showed an enhanced recognition capacity towards TNT. The electrochemical measurement performed using differential pulse voltammetry technique showed a dynamic linear range of 5.0 nM–1.0 μM and detection limit of 1.5 nM. The sensor was successfully applied to measure trace level of TNT in soil samples [164]. Electrochemical sensor made up of MIP and AuNPs for the detection of TNT with enhanced sensitivities was reported. The detection mechanism was based on tailoring pi-donor–acceptor interactions between TNT and pi-donor-modified electrodes or pi-donor-cross linked Au nanoparticles linked to the electrode to give the 200 pM sensitivity [168]. Moreover,

Table 5.4 Selected applications of MIP-based electrochemical sensor in precision agriculture

Analyte	Linear range	LOD	Reference
Glycine betaine	1 fM–10 mM	0.707 fM	[1]
L-Tryptophan	10–110 μM	0.037–0.121 μM	[18]
Sunset yellow	2.2 nM–4.64 μM	1.4 nM	[158]
Proline	0.1 fM–10 mM	9.18 aM	[159]
Hypoxanthine	2–50 μg/mL	0.165 μg/mL	[160]
Glyphosate	1 pM–1 nM	1 pM	[161]
Carbofuran	0.2–50 nM	67 pM	[162]
Chlorpyrifos	1.0 fM–0.4 pM	0.35 fM	[163]

electropolymerized polydopamine MIP films were applied in the voltammetric sensor development for the measurement of TNT and 1,3,5-trinitroperhydro-1,3,5-triazine explosives in the range from 0.1 to 10 nM [169]. Furthermore, TNT was determined in tap and natural waters using the MIP microporous-metal-organic framework (MMOF)-based LSV in the linear range of 4.4 fM–44 nM, with a detection limit of 0.04 fM [170]. In another work, MIP composite electrochemical sensor was prepared with enhanced selectivity by using dengue virus particles to assist imprinting effects. This multi-analyte sensor revealed good detection capability towards multiple explosive molecules such as TNT, pentaerythritol tetranitrate (PETN) and 2,4-dinitrotoluene (DNT) in water samples with the detection limit of 0.3 ppt [171]. Dummy MIP was synthesized using dinitrophenol (DNP) as a template to develop an electrochemical sensor for the detection of TNT and DNP. The sensor generated a varied current signal towards DNP and TNT for the concentration range of 1.5–8.0 μM for DNP and 1.3–6.5 μM for TNT and detection limit of 0.59 and 0.29 μM, for DNP and TNT, respectively. This kind of sensor has potential applications in the environmental protection field [172].

Triacetone triperoxide (TATP) is another explosive commonly found in improved explosive devices. Glassy carbon electrode was modified MIP to prepare an electrochemical sensor for the determination of TATP within a concentration range of 82–44,300 μg/L, limit of quantification and limit of detection between 26.9 and 81.6 μg/L, respectively. The sensor was evaluated to have good selectivity for the target in the occurrence of other similar explosives [173].

MIP-based DPV sensors were also constructed to measure peroxide-based explosives such as TATP and hexamethylene triperoxide diamine (HMTD). The electrochemical measurement of the current against various analyte concentrations was linear in the range from 0.1 to 1.0 mg/L. The detection limit of 15 μg/L for both TATP and HMTD was achieved by the reported sensor. The two analytes were measured separately in the existence of nitro-aromatic, nitramine and nitrate ester energetic materials [174]. Similarly, MIP-based EIS sensor was developed to determine 5-chloro-2,4-dinitrotoluene (5-CDNT) in wide dynamic range of 1.0–100 μM with the detection limit of 0.1 μM [175]. Recently, the research and development in MIP-based electrochemical sensors for the selective and sensitive determination of some chemicals of security concern is shown in Table 5.5.

5.8 Conclusion and Future Outlook

The increasing growth in the demand for the fast, selective, user-friendly and affordable analytical methods is observed in the present time. This is because the prevalence of numerous organic and inorganic contaminants in food, water, and environmental segments and need in diagnosis for the efficient patient treatment. To prevent or minimize the health effects of these food and environmental contaminants as well as to enhance the effective patient treatment, the development of electrochemical sensors using MIPs becomes an alternative option nowadays in place of biosensors. The emergence of MIP-based electrochemical sensors combined the

Table 5.5 Development and applications of MIP-based electrochemical sensors in explosive detection

Analyte	Linear range	LOD	Reference
TNT	1.5–8.0 μM	0.29 μM	[172]
TNT	1.3–6.5 μM	0.59 μM	[172]
TNT	4.4 fM–44 nM	0.04 fM	[170]
TNT	4.4 fM–44 nM	0.04 fM	[170]
TNT	5.0 nM–1.0 μM	1.5 nM	[164]
TATP	82–44,300 μg/L	81.6 μg/L	[173]
HMTD	0.1–1.0 mg/L	15 μg/L	[174]
5-CDNT	1.0–100 μM	0.1 μM	[175]
Cocaine	100–500 μM	50 μM	[176]

advantages of electroanalytical methods (e.g. their low cost, high sensitivity, repeatability and accuracy towards the target analyte) and that of MIPs (e.g. high physical and chemical stability, low-cost preparation process, excellent selectivity, sensitivity and fast response towards the target analyte). These kinds of sensors are promising for the selective and sensitive determination of various target analytes in different complex matrices such as food, biological and environmental samples. Therefore, the advancement in the preparation and application of MIP-based electrochemical sensors are presented in this chapter. An electrochemical sensor utilizing MIP as a sensing layer includes various types of formats such as voltammetric, conductometric, aerometric, potentiometric and capacitance or EIS. Voltammetric techniques such as DPV and SWV are particularly useful in the determination of trace amounts of electroactive compounds. These formats of sensors have provided essential tools to detect various analytes in environmental, food, clinical, precision agriculture and security fields due to their high sensitivity, cheapness and miniaturization. However, few MIP-based electrochemical sensors are currently commercialized. Hence, the suitable preparation approach for the fabrication of user-friendly and affordable sensors for field testing and real-time analysis needs to be developed through the selection of appropriate materials in the future.

References

1. Ai G, Zhou Y, Zhang H, Wei Q, Luo B, Xie Y, Wang C, Xue X, Li A (2024) Ultrasensitive molecular imprinted electrochemical sensor for in vivo determination of glycine betaine in plants. Food Chem 435:137554
2. Saputra HA (2023) Electrochemical sensors: basic principles, engineering, and state of the art. Monatshefte Chem Chem Monthly 154:1083–1100
3. Baranwal J, Barse B, Gatto G, Broncova G, Kumar A (2022) Electrochemical sensors and their applications: a review. Chemosensors 10(9):363
4. Cho G, Azzouzi S, Zucchi G, Lebental B (2021) Electrical and electrochemical sensors based on carbon nanotubes for the monitoring of chemicals in water—a review. Sensors 22:218
5. Ayerdurai V, Cieplak M, Kutner W (2023) Molecularly imprinted polymer-based electrochemical sensors for food contaminants determination. TrAC Trends Anal Chem 158:116830

6. Beluomini MA, da Silva J, de Sá AC, Buffon E, Pereira TC, Stradiotto NR (2019) Electrochemical sensors based on molecularly imprinted polymer on nanostructured carbon materials: a review. J Electroanal Chem 840:343–366
7. Justino CIL, Freitas AC, Pereira R, Duarte AC, Rocha TAP, Santos TAP (2015) Recent developments in recognition elements for chemical sensors and biosensors. TrAC Trends Anal Chem 68:2–17
8. Fuchiwaki Y, Kubo I (2010) Electrochemical sensor based on biomimetic recognition utilizing molecularly imprinted polymer receptor. In: Biomimetics learning from nature. InTech. https://doi.org/10.5772/8783
9. Denmark DJ, Mohapatra S, Mohapatra SC (2020) Point-of-care diagnostics: molecularly imprinted polymers and nanomaterials for enhanced biosensor selectivity and transduction. EuroBiotech J 4(4):184–206
10. Dolai J, Ali H, Jana NR (2020) Molecular imprinted poly-cyclodextrin for selective removal of dibutyl phthalate. ACS Appl Polym Mater 2(2):691–698
11. Pilvenyte G, Ratautaite V, Boguzaite R, Ramanavicius S, Chen CF, Viter R, Ramanavicius A (2023) Molecularly imprinted polymer-based electrochemical sensors for the diagnosis of infectious diseases. Biosensors 13:620
12. Elfadil D, Lamaoui A, Della Pelle F, Amine A, Compagnone D (2021) Molecularly imprinted polymers combined with electrochemical sensors for food contaminants analysis. Molecules 26:4607
13. Cui B, Liu P, Liu X, Liu S, Zhang Z (2020) Molecularly imprinted polymers for electrochemical detection and analysis: progress and perspectives. J Mater Res Technol 9(6):12568–12584
14. Shao Y, Duan J, Wang M, Cao J, She Y, Cao Z, Li G, Jin F, Wang J, Abd El-Aty AM (2023) Application of molecularly imprinted electrochemical biomimetic sensors for detecting small molecule food contaminants. Polymers 15:187
15. Mustafa YL, Keirouz A, Leese HS (2022) Molecularly imprinted polymers in diagnostics: accessing analytes in biofluids. J Mater Chem B 10:7418–7449
16. Wang L, Pagett M, Zhang W (2023) Molecularly imprinted polymer (MIP) based electrochemical sensors and their recent advances in health applications. Sens Actuators Rep 5:100153
17. Regasa MB, Nyokong T (2023) Molecularly imprinted polymers for the detection of environmental estrogens. In: Molecularly imprinted polymers for environmental monitoring: fundamentals and applications. Institute of Physics (IOP) Publishing Ltd., pp 10-1–10-23
18. Yang L, Ma Y, Ye J (2021) In vivo detection of L-tryptophan in tomatoes using multi-walled carbon nanotubes and poly (sulfosalicylic acid) film modified graphite rod electrode. Biosens Bioelectron X 9:100086
19. Karadurmus L, Bilge S, Sınağ A, Ozkan SA (2022) Molecularly imprinted polymer (MIP)-based sensing for detection of explosives: current perspectives and future applications. TrAC Trends Anal Chem 155:116694
20. Polyakov M (1931) Adsorption properties and structure of silica gel. Zhur Fiz Khim 2:799–805
21. Wulff G, Sarhan A (1972) The use of polymers with enzyme-analogous structures for the resolution of racemates. Angrew Chem Int Ed 1972(11):341
22. Mosbach K, Arshady R (1980) Synthesis of substrate selective polymers by host-guest. Die Makromol Chem 182(2):687–692
23. Cui L, Wu J, Ju H (2016) Label-free signal-on aptasensor for sensitive electrochemical detection of arsenite. Biosens Bioelectron 79:861–865
24. Evtugyn G, Porfireva A, Stepanova V, Sitdikov R, Stoikov I, Nikolelis D, Hianik T (2014) Electrochemical aptasensor based on polycarboxylic macrocycle modified with neutral red for aflatoxin B_1 detection. Electroanalysis 26:2100–2109
25. Yano K, Karube I (1999) Molecularly imprinted polymers for biosensor applications. TrAC Trends Anal Chem 18:199–204

26. Pratama KF, Manik MER, Rahayu D, Hasanah AN (2020) Effect of the molecularly imprinted polymer component ratio on analytical performance. Chem Pharm Bull (Tokyo) 68(11):1013–1024
27. Gagliardi M (2023) Design and development of molecularly imprinted biodegradable polymers for nanomedicines. Adv Ind Eng Polym Res 6(4):396–406
28. Ramajayam K, Ganesan S, Ramesh P, Beena M, Kokulnathan T, Palaniappan A (2023) Molecularly imprinted polymer-based biomimetic systems for sensing environmental contaminants, biomarkers, and bioimaging applications. Biomimetics 8(2):245
29. Liu H, Jin P, Zhu F, Nie L, Qiu H (2021) A review on the use of ionic liquids in preparation of molecularly imprinted polymers for applications in solid-phase extraction. TrAC Trends Anal Chem 134:116132
30. Parisi OI, Francomano F, Datilo M, Patitucci F, Prete S, Amone F, Puoci F (2022) The evolution of molecular recognition: from antibodies to molecularly imprinted polymers (MIPs) as artificial counterpart. J Funct Biomater 13(1):12
31. Middeleer GD, Dubruel P, Saeger SD (2016) Characterization of MIP and MIP functionalized surfaces: current state-of-the-art. TrAC Trends Anal Chem 76:71–85
32. Cieplak M, Kutner W (2016) Artificial biosensors: how can molecular imprinting mimic biorecognition? Trends Biotechnol 34:922–941
33. Uprety JB, Shah J, Korsen JA, Wayne AE, Galassi TV, Cohen JR, Harvey JD, Jena PV, Ramanathan LV, Jaimes EA, Heller DA (2019) Synthetic molecular recognition nanosensor paint for microalbuminuria. Nat Commun 10:3605
34. Crapnell RD, Hudson A, Foster CW, Eersels K, van Grinsven B, Cleij TJ, Banks CE, Peeters M (2019) Recent advances in electrosynthesized molecularly imprinted polymer sensing platforms for bioanalyte detection. Sensors 19:1204
35. Dickert FL (2014) Biomimetic receptors and sensors. Sensors 14:22525–22531
36. Rahmandhani S, Setiyanto H, Zulfikar MA (2018) Electropolymerized of aniline as a new molecularly imprinted polymer for determination of phenol: a study for phenol sensor. Mater Sci Forum 936:71–76
37. Zhang J, Landry MP, Strano MS (2013) Molecular recognition using corona phase complexes made of synthetic polymers adsorbed on carbon nanotubes. Nat Nanotechnol 8:959–968
38. Moussa FB, Kutner W, Beduk T, Sena-Torralba A, Mostafavi E (2024) Electrochemical bio- and chemosensors for cancer biomarkers: natural (with antibodies) versus biomimicking artificial (with aptamers and molecularly imprinted polymers) recognition. Talanta 267:125259
39. Vasapollo G, Sole RD, Mergola L, Lazzoi MR, Scardino A, Scorrano S, Mele G (2011) Molecularly imprinted polymers: present and future prospective. Int J Mol Sci 12(9):5908–5945
40. Yemis F, Alkan P, Yenigül B, Yenigül M (2013) Molecularly imprinted polymers and their synthesis by different methods. Polym Polym Compos 21(3):145–150
41. Sharma G, Kandasubramanian B (2020) Molecularly imprinted polymers for selective recognition and extraction of heavy metal ions and toxic dyes. J Chem Eng 65(2):396–418
42. Tchekwagep PMS, Crapnell RD, Banks CE, Betlem K, Rinner U, Canfarotta F, Lowdon JW, Eersels K, Van Grinsven B, Peeters M, McClements J (2022) A critical review on the use of molecular imprinting for trace heavy metal and micropollutant detection. Chemosensors 10(8):296
43. Amali RKA, Lim H, Ibrahim I, Ming H, Zainal Z, Ahmad SAA (2021) Significance of nanomaterials in electrochemical sensors for nitrate detection: a review. Trends Environ Anal Chem 31:e00135
44. Dong C, Shi H, Han Y, Yang Y, Wang R, Men J (2021) Molecularly imprinted polymers by the surface imprinting technique. Eur Polym J 145:110231
45. Sharma PS, Dabrowski M, D'Souza F, Kutner W (2013) Surface development of molecularly imprinted polymer films to enhance sensing signals. TrAC Trends Anal Chem 51:146–157
46. Unger C, Lieberzeit PA (2021) Molecularly imprinted thin film surfaces in sensing: chances and challenges. React Funct Polym 161:104855

47. Teng Y, Liu F, Kan X (2017) Voltammetric dopamine sensor based on three-dimensional electrosynthesized molecularly imprinted polymers and polypyrrole nanowires. Microchim Acta 184:2515–2522
48. Mmolinska-Kempisty K, Ahmad OS, Guerreiro A, Karim K, Piletska E, Piletsky S (2017) New potentiometric sensor based on molecularly imprinted nanoparticles for cocaine detection. Biosens Bioelectron 96:49–54
49. Andersson LI, Miyabayashi A, O'Shannessy DJ, Mosbach K (1990) Enantiomeric resolution of amino acid derivatives on molecularly imprinted polymers as monitored by potentiometric measurements. J Chromatogr A 516:323–331
50. Hedborg E, Winquist F, Lundström I, Andersson LI, Mosbach K (1993) Some studies of molecularly-imprinted polymer membranes in combination with field-effect devices. Sens Actuators A Phys 37:796–799
51. Murray RW, Goodenough JB, Albery WJ (1981) Modified electrodes: chemically modified electrodes for electrocatalysis and discussion. Philos Trans R Soc Lond A Math Phys Sci 302(1468):253–265
52. Wang H, Xu Q, Wang J, Du W, Liu F, Hu X (2018) Dendrimer-like amino-functionalized hierarchical porous silica nanoparticle: a host material for 2,4-dichlorophenoxyacetic acid imprinting and sensing. Biosens Bioelectron 100:105–114
53. Feier B, Blidar A, Pusta A, Carciuc P, Cristea C (2019) Electrochemical sensor based on molecularly imprinted polymer for the detection of cefalexin. Biosensors 9:31
54. Gui R, Gou H, Jin H (2019) Preparation and applications of electrochemical chemosensors based on carbon nanomaterials modified molecularly imprinted polymers. Nanoscale Adv 1:3325–3363
55. Regasa MB, Nyokong T (2022) Synergistic recognition and electrochemical sensing of 17β-estradiol using ordered molecularly imprinted polymergraphene oxide-silver nanoparticles composite films. J Electroanal Chem 922:116713
56. Balamurugan S, Spivak DA (2011) Molecular imprinting in monolayer surfaces. J Mol Recognit 24(6):915–929
57. Malitesta C, Mazzotta E, Picca RA, Poma A, Chianella I, Piletsky SA (2012) MIP sensors-the electrochemical approach. Anal Bioanal Chem 402:1827–1846
58. Guan G, Wang S, Zhou H, Zhang K, Liu R, Mei Q, Wang S, Zhang Z (2011) Molecularly imprinted polypyrrole nanonecklaces for detection of herbicide through molecular recognition-amplifying current response. Anal Chim Acta 702:239–246
59. Lahcen AA, Amine A (2019) Recent advances in electrochemical sensors based on molecularly imprinted polymers and nanomaterials. Electroanalysis 31:188–201
60. Zamora-Gálvez A, Lahcen AA, Mercante L, Morales-Narváez E, Amine A, Merkoçi A (2016) Molecularly imprinted polymer-decorated magnetite nanoparticles for selective sulfonamide detection. Anal Chem 88:3578–3584
61. Lamaoui A, Palacios-Santander JM, Amine A, Cubillana-Aguilera L (2021) Fast microwave-assisted synthesis of magnetic molecularly imprinted polymer for sulfamethoxazole. Talanta 232:122430
62. Lamaoui A, Lahcen AA, Guzmán JJG, Palacios-Santander JM, Aguilera LC, Amine A (2019) Study of solvent effect on the synthesis of magnetic molecularly imprinted polymers based on ultrasound probe: application for sulfonamide detection. Ultrason Sonochem 58:104670
63. Breton F, Euzet P, Piletsky SA, Giardi MT, Rouillon R (2006) Integration of photosynthetic biosensor with molecularly imprinted polymer-based solid phase extraction cartridge. Anal Chim Acta 569:50–57
64. Messaoud NB, Lahcen AA, Dridi C, Amine A (2018) Ultrasound assisted magnetic imprinted polymer combined sensor based on carbon black and gold nanoparticles for selective and sensitive electrochemical detection of bisphenol A. Sens Actuators B Chem 276:304–312
65. Bakas I, Salmi Z, Jouini M, Geneste F, Mazerie I, Floner D, Carbonnier B, Yagci Y, Chehimi MM (2015) Picomolar detection of melamine using molecularly imprinted polymer-based electrochemical sensors prepared by UV-graft photopolymerization. Electroanalysis 27:429–439

66. Ye L, Cormack PAG, Mosbach K (2001) Molecular imprinting on microgel spheres. Anal Chim Acta 435:187–196
67. Mayes AG, Mosbach K (1996) Molecularly imprinted polymer beads: suspension polymerization using a liquid perfluorocarbon as the dispersing phase. Anal Chem 68:3769–3774
68. Gao R, Kong X, Wang X, He X, Chen L, Zhang Y (2011) Preparation and characterization of uniformly sized molecularly imprinted polymers functionalized with core-shell magnetic nanoparticles for the recognition and enrichment of protein. J Mater Chem 21:17863–17871
69. Haginaka J, Takekira H, Hosoya K, Tanaka N (1999) Uniform-sized molecularly imprinted polymer for (S)-naproxen selectively modified with hydrophilic external layer. J Chromatogr A 849:331–339
70. Bengamra M, Grayaa-Jaoued N, Khlifi-Riani A, Chehimi MM, Kalfat R (2018) Highly selective molecularly imprinted sol-gel membrane grafted to gold for the detection of melamine. Silicon 11:2267–2274
71. Yáñez-Sedeño P, Campuzano S, Pingarrón JM (2017) Electrochemical sensors based on magnetic molecularly imprinted polymers: a review. Anal Chim Acta 960:1–17
72. Leibl N, Haupt K, Gonzato C, Duma L (2021) Molecularly imprinted polymers for chemical sensing: a tutorial review. Chemosensors 9:123
73. Hassan AHA, Sappia L, Moura SL, Ali FHM, Moselhy WA, Sotomayor MPT, Pividori MI (2019) Biomimetic magnetic sensor for electrochemical determination of scombrotoxin in fish. Talanta 194:997–1004
74. Wang S, Pan M, Liu K, Xie X, Yang J, Hong L, Wang S (2022) A SiO_2@MIP electrochemical sensor based on MWCNTs and AuNPs for highly sensitive and selective recognition and detection of dibutyl phthalate. Food Chem 381:132225
75. Franklin RK, Martin SM, Strong TD, Brown RB (2016) Chemical and biological systems: chemical sensing systems for liquids. Elsevier
76. Blanco-López MC, Lobo-Castañón MJ, Miranda-Ordieres AJ, Tuñón-Blanco P (2003) Voltammetric sensor for vanillylmandelic acid based on molecularly imprinted polymer-modified electrodes. Biosens Bioelectron 18:353–362
77. Regasa MB, Soreta TR, Femi OE, Ramamurthy PC, Subbiahraj S (2020) Development of melamine electrochemical sensor using molecularly imprinted conducting polyanilne-oxalic acid blend as a molecular recognition element. Nano Hybrids Compos J 29:61–73
78. Yoshimi Y, Ohdaira R, Iiyama C, Sakai K (2001) Gate effect of thin layer of molecularly-imprinted poly (methacrylic acid-co-ethyleneglycol dimethacrylate). Sens Actuators B 73:49–53
79. Ratautaite V, Plausinaitis D, Baleviciute I, Mikoliunaite L, Ramanaviciene A, Ramanavicius A (2015) Characterization of caffeineimprinted polypyrrole by a quartz crystal microbalance and electrochemical impedance spectroscopy. Sens Actuator B Chem 212:63–71
80. Suedee R, Intakong W, Dickert FL (2006) Molecularly imprinted polymer-modified electrode for on-line conductometric monitoring of haloacetic acids in chlorinated water. Anal Chim Acta 569:66–75
81. Sadeghi S, Fathi F, Abbasifar J (2007) Potentiometric sensing of levamisole hydrochloride based on molecularly imprinted polymer. Sens Actuators B 122:158–164
82. Ayankojo AG, Boroznjak R, Reut J, Öpik A, Syritski V (2022) Molecularly imprinted polymer based electrochemical sensor for quantitative detection of SARS-CoV-2 spike protein. Sens Actuator B Chem 353:131160
83. Sundhoro M, Agnihotra SR, Amberger B, Augustus K, Khan NG, Barnes A, BelBruno J, Mendecki L (2021) An electrochemical molecularly imprinted polymer sensor for rapid β-lactoglobulin detection. Food Chem 344:128648
84. Limthin D, Leepheng P, Klamchuen A, Phromyothin D (2022) Enhancement of electrochemical detection of gluten with surface modification based on molecularly imprinted polymers combined with superparamagnetic iron oxide nanoparticles. Polymers 14:91
85. Ardalani M, Shamsipur M, Besharati-Seidani A (2020) A new generation of highly sensitive potentiometric sensors based on ion imprinted polymeric nanoparticles/multiwall car-

bon nanotubes/polyaniline/graphite electrode for sub-nanomolar detection of lead(II) ions. J Electroanal Chem 879:114788
86. Zhang H, Yao RQ, Wang N, Liang RN, Qin W (2018) A molecularly imprinted polymer-based potentiometric sensor for determination of bisphenol AF. Anal Chem 90:657
87. Richter EM, Rocha DP, Cardoso RM, Keefe EM, Foster CW, Munoz RAA, Banks CE (2019) Complete additively manufactured (3D-printed) electrochemical sensing platform. Anal Chem 91:12844–12851
88. Ayerdurai Y, Garcia-Cruz A, Piechowska J, Cieplak M, Borowicz P, Noworyta KR, Spolnik G, Danikiewicz W, Lisowski W, Pietrzyk-Le A, D'Souza F, Kutner W, Sharma PS (2021) An advantageous application of molecularly imprinted polymers in food processing and quality control. J Agric Food Chem 69:14689
89. Lach P, Cieplak M, Majewska M, Noworyta KR, Sharma PS, Kutner W (2019) "Gate effect" in p-synephrine electrochemical sensing with a molecularly imprinted polymer and redox probes. Anal Chem 91:7546
90. Li JP, Yin WL, Tan YJ, Pan HC (2014) A sensitive electrochemical molecularly imprinted sensor based on catalytic amplification by silver nanoparticles for 3-indoleacetic acid determination. Sens Actuators B Chem 197:109–115
91. Yarman A, Kurbanoglu S, Zebger I, Scheller FW (2021) Simple and robust: the claims of protein sensing by molecularly imprinted polymers. Sens Actuators B Chem 330:129369
92. Yarman A, Kurbanglu S, Jetzschmann KJJ, Ozkan SA, Wollenberger U, Scheller FW (2018) Electrochemical MIP-sensors for drugs. Curr Med Chem 25:4007–4019
93. Mazzotta E, Picca R, Malitesta C, Piletsky S, Piletska E (2008) Development of a sensor prepared by entrapment of MIP particles in electrosynthesised polymer films for electrochemical detection of ephedrine. Biosens Bioelectron 23:1152–1156
94. Prasad K, Prathish K, Gladis JM, Naidu G, Rao TP (2007) Molecularly imprinted polymer (biomimetic) based potentiometric sensor for atrazine. Sens Actuators B Chem 123:65–70
95. Mazzotta E, Turco A, Chianella I, Guerreiro A, Piletsky SA, Malitesta S (2016) Solid-phase synthesis of electroactive nanoparticles of molecularly imprinted polymers. A novel platform for indirect electrochemical sensing applications. Sens Actuators B Chem 229:174–180
96. Garcia-Mutio D, Gomez-Caballero A, Guerreiro A, Piletsky S, Goicolea MA, Barrio RJ (2016) Solid-phase synthesis of imprinted nanoparticles grafted on gold substrates for voltammetric sensing of 4-ethylphenol. Sens Actuators B Chem 236:839–848
97. Cao Y, Feng T, Xu J, Xue C (2019) Recent advances of molecularly imprinted polymer-based sensors in the detection of food safety hazard factors. Biosens Bioelectron 141:111447
98. Bow Y, Sutriyono E, Nasir S, Iskandar I (2017) Molecularly imprinted polymers (MIP) based electrochemical sensors for detection of endosulfan pesticide. Int J Adv Sci Eng Inf Technol 7(2):662–668
99. Xiong Y, Zhang D, Ye C, Wang Y, Deng X, Deng D, Xiong C, Zhang J, Gu Q, Huang G (2023) Ultra-sensitive detection of ciprofloxacin hydrochloride in milk by molecularly imprinted electrochemical sensor based on S-CoFe-MOFs/AuNPs. J Food Compos Anal 122:105439
100. Karthika P, Shanmuganathan S, Viswanathan S, Delerue-Matos C (2021) Molecularly imprinted polymer-based electrochemical sensor for the determination of endocrine disruptor bisphenol-A in bovine milk. Food Chem 363:130287
101. Cheng W, Zhang Q, Wu D, Yang Y, Zhang Y, Tang X (2022) A facile electrochemical method for rapid determination of 3-chloropropane-1,2-diol in soy sauce based on nanoporous gold capped with molecularly imprinted polymer. Food Control 134:108750
102. Chen Y, Tang Y, Liu Y, Zhao F, Zeng B (2022) Selective and fast removal and sensitive determination of oxytetracycline using surface molecularly imprinted polymer based on ionic liquid and ATRP polymerization. J Hazard Mater 434:128907
103. Qi P, Wang J, Wang X, Wang Z, Xu H, Di S, Wang Q, Wang X (2018) Sensitive and selective detection of the highly toxic pesticide carbofuran in vegetable samples by a molecularly imprinted electrochemical sensor with signal enhancement by AuNPs. RSC Adv 8:25334–25341

104. Fatah MAA, El-Moghny MGA, El-Deab MS, El Nashar RM (2023) Application of molecularly imprinted electrochemical sensor for trace analysis of Metribuzin herbicide in food samples. Food Chem 404(Part B):134708
105. Singh AK, Lakshmi GBVS, Fernandes M, Sarkar T, Gulati P, Singh RP, Solanki PR (2021) A simple detection platform based on molecularly imprinted polymer for AFB1 and FuB1 mycotoxins. Microchem J 171:106730
106. Wen T, Wang M, Luo M, Yu N, Xiong H, Peng H (2019) A nanowell-based molecularly imprinted electrochemical sensor for highly sensitive and selective detection of 17β-estradiol in food samples. Food Chem 297:124968
107. Wei X, Zhang Z, Zhang L, Xu X (2018) Synthesis of molecularly imprinted polymers/NiCo2O4 nanoneedle arrays on 3D graphene electrode for determination of sulfadimidine residue in food. J Mater Sci 54:2066–2078
108. Liu B, Tang D, Zhang B, Que X, Yang H, Chen G (2013) Au(III)-promoted magnetic molecularly imprinted polymer nanospheres for electrochemical determination of streptomycin residues in food. Biosens Bioelectron 41:551–556
109. Rahman S, Bozal-Palabiyik B, Unal DN, Erkmen C, Siddiq M, Shah A, Uslu B (2022) Molecularly imprinted polymers (MIPs) combined with nanomaterials as electrochemical sensing applications for environmental pollutants. Trends Environ Anal Chem 36:e00176
110. Rebelo P, Costa-Rama E, Seguro I, Pacheco JG, Nouws HPA, Cordeiro MNDS, Delerue-Matos CD (2021) Molecularly imprinted polymer-based electrochemical sensors for environmental analysis. Biosens Bioelectron 172:112719
111. Kadhem AJ, Gentile GJ, de Cortalezzi MM (2021) Molecularly imprinted polymers (MIPs) in sensors for environmental and biomedical applications: a review. Molecules 26:6233
112. Wang W, Wang X, Cheng N, Luo Y, Lin Y, Xu W, Du D (2020) Recent advances in nanomaterials-based electrochemical (bio)sensors for pesticides detection. TrAC Trends Anal Chem 132:116041
113. Beduk T, Gomes M, De Oliveira Filho JI, Shetty SS, Khushaim W, Garcia-Ramirez R, Durmus C, Ait Lahcen A, Salama KN (2022) A portable molecularly imprinted sensor for onsite and wireless environmental bisphenol A monitoring. Front Chem 10:833899
114. Latif U, Ping L, Dickert FL (2018) Conductometric sensor for PAH detection with molecularly imprinted polymer as recognition layer. Sensors 18:767
115. Liu Y, Liang Y, Yang R, Li J, Qu L (2019) A highly sensitive and selective electrochemical sensor based on polydopamine functionalized graphene and molecularly imprinted polymer for the 2,4-dichlorophenol recognition and detection. Talanta 195:691–698
116. Zheng X, Li H, Xia F, Tian D, Hua X, Qiao X, Zhou C (2016) An electrochemical sensor for ultrasensitive determination the polychlorinated biphenyls. Electrochim Acta 194:413–421
117. Dahaghin Z, Kilmartin PA, Mousavi HZ (2020) Novel ion imprinted polymer electrochemical sensor for the selective detection of lead(II). Food Chem 303:125374
118. Alizadeh T, Hamidi N, Ganjali MR, Rafiei F (2017) Determination of subnanomolar levels of mercury (II) by using a graphite paste electrode modified with MWCNTs and Hg(II)-imprinted polymer nanoparticles. Microchim Acta 185:16
119. Samandari L, Bahrami A, Shamsipur M, Farzin L, Hashemi B (2019) Electrochemical preconcentration of ultra-trace Cd2+ from environmental and biological samples prior to its determination using carbon paste electrode impregnated with ion imprinted polymer nanoparticles. Int J Environ Anal Chem 99:172–186
120. Alizadeh T, Mirzaee S, Rafiei F (2017) All-solid-state Cr(III)-selective potentiometric sensor based on Cr(III)-imprinted polymer nanomaterial/MWCNTs/carbon nanocomposite electrode. Int J Environ Anal Chem 97:1283–1297
121. Jafari S, Dehghani M, Nasirizadeh N, Baghersad MH, Azimzadeh M (2019) Label-free electrochemical detection of Cloxacillin antibiotic in milk samples based on molecularly imprinted polymer and graphene oxide-gold nanocomposite. Measurement 145:22–29
122. Huang X, Wei S, Yao S, Zhang H, He C, Cao J (2019) Development of molecularly imprinted electrochemical sensor with reduced graphene oxide and titanium dioxide enhanced per-

formance for the detection of toltrazuril in chicken muscle and egg. J Pharm Biomed Anal 164:607–614

123. Warwick C, Guerreiro A, Gomez-Caballero A, Wood E, Kitson J, Robinson J, Soares A (2013) Conductance based sensing and analysis of soluble phosphates in wastewater. Biosens Bioelectron 52:173–179
124. Uygun ZO, Dilgin Y (2013) A novel impedimetric sensor based on molecularly imprinted polypyrrole modified pencil graphite electrode for trace level determination of chlorpyrifos. Sens Actuators B Chem 188:78–84
125. Zhao Y, Yuan F, Quan X, Yu H, Chen S, Zhao H, Liu Z, Hilal N (2015) An electrochemical sensor for selective determination of sulfamethoxazole in surface water using a molecularly imprinted polymer modified BDD electrode. Anal Methods 7:2693–2698
126. Wang FR, Lee GJ, Haridharan N, Wu JJ (2017) Electrochemical sensor using molecular imprinting polymerization modified electrodes to detect methyl parathion in environmental media. Electrocatalysis 9:1–9
127. Knottnerus JA, van Weel C, Muris JW (2002) Evaluation of diagnostic procedures. BMJ 324(7335):477–480
128. Wilkirson EC, Singampalli KL, Li J, Dixit DD, Jiang X, Gonzalez DH, Lillehoj PB (2023) Affinity-based electrochemical sensors for biomolecular detection in whole blood. Anal Bioanal Chem 415:3983–4002
129. Shama NA, Asur S, Ozsoz M, Gokturk I, Turkmen D, Yilmaz F, Denizli A (2022) Gold-modified molecularly imprinted *N*-methacryloyl-(L)-phenylalanine-containing electrodes for electrochemical detection of dopamine. Bioengineering 9(3):87
130. Demir E, Inam O, Inam R (2018) Determintion of Ophthalmic drug proparacaine using multi-walled carbon nanotube paste electrode by square wave stripping voltammetry. Anal Sci 34:771–776
131. Zhang W, She X, Wang L, Fan H, Zhou Q, Huang X, Tang JZ (2017) Preparation, characterization and application of a molecularly imprinted polymer for selective recognition of sulpiride. Materials 10(5):475
132. Cruz AG, Haq I, Cowen T, Di Masi S, Trivedi S, Alanazi K, Piletska E, Mujahid A, Piletsky SA (2020) Design and fabrication of a smart sensor using in silico epitope mapping and electro-responsive imprinted polymer nanoparticles for determination of insulin levels in human plasma. Biosens Bioelectron 169:112536
133. Dabrowski M, Sharma PS, Iskierko Z, Noworyta K, Cieplak M, Lisowski W, Oborska S, Kuhn A, Kutner W (2016) Early diagnosis of fungal infections using piezomicrogravimetric and electric chemosensors based on polymers molecularly imprinted with D-arabitol. Biosens Bioelectron 79:627–635
134. Jiang H, Jiang D, Shao J, Sun X (2016) Magnetic molecularly imprinted polymer nanoparticles based electrochemical sensor for the measurement of Gram-negative bacterial quorum signaling molecules (N-acyl-homoserine-lactones). Biosens Bioelectron 75:411–419
135. Kim M, Park D, Park J, Park J (2023) Bio-inspired molecularly imprinted polymer electrochemical sensor for cortisol detection based on O-phenylenediamine optimization. Biomimetics 8:282
136. Nguy TP, Phi TV, Tram DTN, Eersels K, Wagner P, Truong TN, Lien TTN (2017) Development of an impedimetric sensor for the label-free detection of the amino acid sarcosine with molecularly imprinted polymer receptors. Sens Actuators B Chem 246:461–470
137. Sheydaei O, Khajehsharifi H, Rajabi HR (2020) Rapid and selective diagnose of Sarcosine in urine samples as prostate cancer biomarker by mesoporous imprinted polymeric nanobeads modified electrode. Sens Actuators B Chem 309:127559
138. Sehit E, Drzazgowska J, Buchenau D, Yesildag C, Lensen M, Altintas Z (2020) Ultrasensitive nonenzymatic electrochemical glucose sensor based on gold nanoparticles and molecularly imprinted polymers. Biosens Bioelectron 165:112432
139. Carneiro LPT, Ferreira NS, Tavares APM, Pinto AMFR, Mendes A, Sales MGF (2021) A passive direct methanol fuel cell as transducer of an electrochemical sensor, applied to the detection of carcinoembryonic antigen. Biosens Bioelectron 175:112877

140. Pirzada M, Sehit E, Altintas Z (2020) Cancer biomarker detection in human serum samples using nanoparticle decorated epitope-mediated hybrid MIP. Biosens Bioelectron 166:112464
141. Raziq A, Kidakova A, Boroznjak R, Reut J, Öpik A, Syritski V (2021) Development of a portable MIP-based electrochemical sensor for detection of SARS-CoV-2 antigen. Biosens Bioelectron 178:113029
142. Kellens E, Bové H, Vandenryt T, Lambrichts J, Dekens J, Drijkoningen S, D'Haen J, De Ceuninck W, Thoelen R, Junkers T, Haenen K, Ethirajan A (2018) Micro-patterned molecularly imprinted polymer structures on functionalized diamond-coated substrates for testosterone detection. Biosens Bioelectron 118:58–65
143. Pedersen SM, Lind KM (2017) Precision agriculture: technology and economic perspectives. Springer
144. Nageib AM, Halim AA, Nordin AN, Ali F (2023) Recent applications of molecularly imprinted polymers (MIPs) on screen-printed electrodes for pesticide detection. J Electrochem Sci Technol 14(1):1–14
145. Sadriu I (2021) Development of environmental sensors based on MIP/graphene electrode allowing the detection of pesticides in waters. Material chemistry, PhD thesis. Universitéd'Orléans
146. Li S, Luo Q, Liu Y, Zhang Z, Shen G, Wu H, Chem A, Liu X, Zhang A (2017) Surface molecularly imprinted polymer film with poly(*p*-aminothiophenol) outer layer coated on gold nanoparticles inner layer for highly sensitive and selective sensing paraoxon. Polymers 9(8):359
147. Pogorelova SP, Bourenko T, Kharitonov AB, Willner I (2002) Selective sensing of triazine herbicides in imprinted membranes using ion-sensitive field-effect transistors and microgravimetric quartz crystal microbalance measurements. Analyst 127(11):1484–1491
148. Li Y, Liu J, Zhang Y, Gu M, Wang D, Dang YY, Ye BC, Li Y (2018) A robust electrochemical sensing platform using carbon paste electrode modified with molecularly imprinted microsphere and its application on methyl parathion detection. Biosens Bioelectron 106:71–77
149. Anirudhan TS, Alexander S (2015) Design and fabrication of molecularly imprinted polymer-based potentiometric sensor from the surface modified multiwalled carbon nanotube for the determination of lindane (γ-hexachlorocyclohexane), an organochlorine pesticide. Biosens Bioelectron 64:586–593
150. Tan X, Hu Q, Wu J, Li X, Li P, Yu H, Li X, Lei F (2015) Electrochemical sensor based on molecularly imprinted polymer reduced graphene oxide and gold nanoparticles modified electrode for detection of carbofuran. Sens Actuators B Chem 220:216–221
151. Liu W, Guo Y, Luo J, Kou J, Zheng H, Li B, Zhang Z (2015) A molecularly imprinted polymer based a lab-on-paper chemiluminescence device for the detection of dichlorvos. Spectrochim Acta A Mol Biomol Spectrosc 141:51–57
152. Toro MJU, Marestoni LD, Sotomayor MDPT (2015) A new biomimetic sensor based on molecularly imprinted polymers for highly sensitive and selective determination of hexazinone herbicide. Sens Actuators B Chem 208:299–306
153. Sergeyeva TA, Piletskya SA, Brovkob AA, Slinchenkoa EA, Sergeevab LM, El'Skaya AV (1999) Selective recognition of atrazine by molecularly imprinted polymer membranes: development of conductometric sensor for herbicides detection. Anal Chim Acta 392:105–111
154. Barhoum A, Hamimed S, Slimi H, Othmani A, Abdel-Haleem FM, Bechelany M (2023) Modern designs of electrochemical sensor platforms for environmental analyses: principles, nanofabrication opportunities, and challenges. Trends Environ Anal Chem 38:e00199
155. Storer CS (2017) Molecularly imprinted polymer sensors for the detection of phosphate in agriculture, PhD thesis. University of Manchester
156. Kamat V, Burton L, Venkadesh V, Jayachandran K, Bhansali S (2023) Enabling smart agriculture through sensor-integrated microfluidic chip to monitor nutrient uptake in plants. ECS Sensors Plus 2(4):044201
157. Yarahmadi S, Azadbakht A, Derikvand RM (2019) Hybrid synthetic receptor composed of molecularly imprinted polydopamine and aptamers for impedimetric biosensing of urea. Microchim Acta 186:71

158. Yin ZZ, Cheng SW, Xu LB, Liu HY, Huang K, Li L, Zhai YY, Zeng YB, Liu HQ, Shao Y, Zhang ZL, Lu YX (2018) Highly sensitive and selective sensor for sunset yellow based on molecularly imprinted polydopamine-coated multi-walled carbon nanotubes. Biosens Bioelectron 100:565–570
159. Yan L, Luo B, Wang C, Dong H, Wang X, Hou P, Liu K, Li A (2024) Molecularly imprinted polymer-based electrochemical sensor for in situ detection of free proline in cucumber leaves. ChemElectroChem. https://doi.org/10.1002/celc.202300538
160. Garg D, Verma N, Monika. (2022) Molecularly imprinted polymer-based electrochemical sensor for rapid and selective detection of hypoxanthine. Biosensors 12(12):1157
161. Mazouz Z, Rahali S, Fourati N, Zerrouki C, Aloui N, Seydou M, Yaakoubi N, Chehimi MM, Othmane A, Kalfat R (2017) Highly selective polypyrrole MIP-based gravimetric and electrochemical sensors for picomolar detection of glyphosate. Sensors 17:2586
162. Li S, Li J, Luo J, Xu Z, Ma X (2018) A microfluidic chip containing a molecularly imprinted polymer and a DNA aptamer for voltammetric determination of carbofuran. Microchim Acta 185:295
163. Roushani M, Nezhadali A, Jalilian Z (2018) An electrochemical chlorpyrifos aptasensor based on the use of a glassy carbon electrode modified with an electropolymerized aptamer-imprinted polymer and gold nanorods. Mikrochim Acta 185:551
164. Alizadeh T, Zare M, Ganjali MR, Norouzi P, Tavana B (2010) A new molecularly imprinted polymer (MIP)-based electrochemical sensor for monitoring 2,4,6-trinitrotoluene (TNT) in natural waters and soil samples. Biosens Bioelectron 25(5):1166–1172
165. Wei L, Min X, Zhibin X, Xiao D, Fei X, Fengyan W, Qiuhong W, Zihui M (2015) Molecularly imprinted polymers for the sensing of explosives and chemical warfare agents. Curr Org Chem 19(1):62–71
166. Apak R, Uzer A, Saglam S, Arman A (2023) Selective electrochemical detection of explosives with nanomaterial based electrodes. Electroanalysis 35(1):e202200175
167. Li P, Li X, Chen W (2019) Recent advances in electrochemical sensors for the detection of 2,4,6-trinitrotoluene. Curr Opin Electrochem 17:16–22
168. Riskin M, Ran T, Tatiana B, Eran G, Itamar W (2008) Imprinting of molecular recognition sites through electropolymerization of functionalized Au nanoparticles: development of an electrochemical TNT sensor based on π-donor–acceptor interactions. J Am Chem Soc 130(30):9726–9733
169. Leibl N, Duma L, Gonzato C, Haupt K (2020) Polydopamine-based molecularly imprinted thin films for electro-chemical sensing of nitro-explosives in aqueous solutions. Bioelectrochemistry 135:107541
170. Guo ZZ, Florea A, Cristea C, Bessueille F, Vocanson F, Goutaland F, Zhang A, Săndulescu R, Lagarde F, Jaffrezic-Renault N (2015) 1,3,5-Trinitrotoluene detection by a molecularly imprinted polymer sensor based on electropolymerization of a microporous-metal-organic framework. Sens Actuators B Chem 207(Part B):960–966
171. Tancharoen C, Sukjee W, Yenchitsomanus P, Panya A, Lieberzeit PA, Sangma C (2021) Selectivity enhancement of MIP-composite sensor for explosive detection using DNT-dengue virus template: a co-imprinting approach. Mater Lett 285:129201
172. Herrera-Chacon A, Gonzalez-Calabuig A, del Walle M (2021) Dummy molecularly imprinted polymers using DNP as a template molecule for explosive sensing and nitroaromatic compound discrimination. Chemosensors 9(9):255
173. Mamo SK, Gonzalez-Rodriduez J (2014) Development of a molecularly imprinted polymer-based sensor for the electrochemical determination of triacetone triperoxide (TATP). Sensors 14(12):23269–23282
174. Saglam S, Uzer A, Apak R (2022) Direct determination of peroxide explosives on polycarbazole/gold nanoparticle-modified glassy carbon sensor electrodes imprinted for molecular recognition of TATP and HMTD. Anal Chem 94(50):17662–17669
175. Goud KY, Satyanarayana M, Ready KK, Gobi KV (2016) Development of highly selective electrochemical impedance sensor for detection of sub-micromolar concentrations of 5-chloro-2,4-dinitrotoluene. J Chem Sci 128:763–770
176. Florea A, Cowen T, Piletsky S, De Wael K (2019) Electrochemical sensing of cocaine in real samples based on electrodeposited biomimetic affinity ligands. Analyst 144:4639–4646

158. Yin ZZ, Cheng SW, Xu LB, Liu HY, Huang K, Li L, Zhai YY, Zeng YB, [illegible] Y, Zhang [illegible] (2018) Highly sensitive and selective sensor based on [illegible]

Molecularly Imprinted Polymer-Based Electrochemical (MIP-EC) Sensors for Biomarker Detection and Quantification

6

Kamran Banan, Bahar Mostafiz, Behnoosh Safaei,
Sara Arjomand Bigdeli, Mikko Haavisto,
and Fatemeh Ghorbani-Bidkorpeh

Abstract

Diagnosis and management of diseases have always been challenging in the healthcare system. Infections and illnesses can significantly impact the patient's quality of life, and numerous morbidities and mortalities are due to diagnostic issues. The sooner these conditions could be detected, the better the provided prognosis and management could be. To achieve this, biomarkers (biological molecules that increase in concentration in the case of an illness) show great potential in early diagnosis.

Kamran Banan and Bahar Mostafiz have equal contributions and are shared first authors.

K. Banan · F. Ghorbani-Bidkorpeh (✉)
Department of Pharmaceutics and Pharmaceutical Nanotechnology, School of Pharmacy, Shahid Beheshti University of Medical Sciences, Tehran, Iran
e-mail: f.ghorbani@sbmu.ac.ir

B. Mostafiz
Department of Mechanical and Materials Engineering, University of Turku, Turku, Finland

B. Safaei (✉)
Department of Pharmaceutics and Pharmaceutical Nanotechnology, School of Pharmacy, Shahid Beheshti University of Medical Sciences, Tehran, Iran

Department of Toxicology, School of Pharmacy, Shahid Beheshti University of Medical, Tehran, Iran

S. A. Bigdeli
Halal Research Center of IRI, Food and Drug Administration, Ministry of Health and Medical Education, Tehran, Iran

M. Haavisto
Department of Teacher Education, University of Turku, Rauma, Finland

S. Patra et al. (eds.), *Molecularly Imprinted Polymers: Path to Artificial Antibodies*, https://doi.org/10.1007/978-981-97-4379-7_6

Molecularly imprinted polymers (MIPs) are specifically designed polymers with selective binding sites similar to a target molecule. These polymers could be used as artificial antibodies to detect or quantify biomarkers from biological fluids. Their low-cost fabrication, long shelf-life, and integration with electrochemical sensors could provide the applicant with a precise, selective, cheap, rapid, user-friendly, reliable, and sensitive tool. Additionally, they provide an ethical alternative to harvesting antigens from living cells.

This chapter explores the most recent studies on the application of MIP-based electrochemical sensors in detecting and quantifying various biomarkers. These applications enable the identification and evaluation of diseases, such as cancer, bacterial and viral infections, and other diseases.

Keywords

Molecularly imprinted polymers (MIPs) · Electrochemical sensors · Artificial antibody · Disease biomarkers

6.1 Introduction

The integration of molecularly imprinted polymers (MIPs) and electrochemical (EC) sensors offers abundant possibilities in analytical studies. In this chapter, we explore their applications in biomarker detection. Sensors optimized for detecting various disease biomarkers in recent studies were investigated.

First, the chapter emphasizes cancer biomarkers, specifically the prostate-specific antigen (PSA) and carcinoembryonic antigen (CEA). Cancer is one of the leading causes of mortality worldwide, and thus, the significance of reliable, sensitive, and rapid detection methods cannot be overstated. Early detection of cancer biomarkers allows timely intervention, leading to better treatment outcomes and patient survival rates. Second, recent research on viral biomarkers is reviewed; indeed, due to the emergence of COVID-19, a noticeable number of studies have focused on developing an electrochemical system coupled with synthetic antibodies. Third, this chapter briefly explores other research areas where MIP-based electrochemical sensors have been applied for the detection of other biomarkers. Finally, the chapter investigates the application of these sensors in other fields, such as allergens, Alzheimer's, autism, bacterial pathogens, bone-related disorders, cardiac diseases, inflammation, kidney injuries, and thyroid problems.

6.1.1 Diseases and Biomarkers

Studying biomarkers helps understand the diseases' pathomechanism, diagnosis, and prognosis. It also facilitates the monitoring of the treatment progress and effects. This subchapter first introduces diseases and how they impact people's lives. Then, it explains the role biomarkers play in diseases and how a simple molecule can make a stark difference in diagnosis and therapeutic approaches to treatment [1].

6.1.1.1 Disease Classifications

The International Classification of Diseases (ICD) is a tool for recording and reporting conditions and factors related to disease and disorders. This classification contains health-related conditions, external causes of illness or death, anatomy, and other factors. The ICD makes global data collection (e.g., data on mortality rate) more systematic and straightforward [2]. The primary purpose of the ICD is to provide a standardized and comprehensive way to categorize health-related data for statistical and epidemiological purposes, as well as for billing, research, and clinical management in healthcare systems.

The ICD is maintained and regularly updated by the World Health Organization (WHO). The development of ICD-11 started in 2007 and was released in 2018. The ICD-11 officially came into effect on 1 January 2022.

Alphanumeric ICD codes allow for precise categorization of diseases and conditions. These codes are used by healthcare professionals, medical researchers, insurance companies, and health organizations to ensure consistency and accuracy in describing and documenting health-related data [2, 3].

By analyzing biomarkers, the healthcare system can identify subtypes of diseases, track the progression of conditions, and study the impact of genetic factors on disease development, all of which can be used to specify ICD codes [4], resulting in better classification and management of diseases.

6.1.1.2 Biomarkers and Their Importance

Biomarkers are primarily defined as functional variants or quantitative indices of a biological process that predict or reflect the evolution of or predisposition to a disease or a response to therapy [5]. Another definition of biomarkers is that they are measurable biological indicators or molecules found in tissues, bodily fluids, or other biological samples (Table 6.1). These markers provide information about normal physiological processes, pathogenic conditions, disease states, or responses to therapeutic interventions [1, 4]. They include various molecular, genetic, proteomic, or metabolic entities.

A proper biomarker needs to be accurate, repeatable, rapid, and cost-effective. It must provide information that is not already available from clinical assessment, and finally, the level of the biomarker must aid in medical decision-making [1].

Biomarkers play a crucial role in the emerging field of precision medicine. When specific biomarkers are associated with certain diseases or conditions, they can guide healthcare practitioners to personalize treatments to an individual's genetic or molecular profile. This personalization can lead to better treatment outcomes [1]. These biomarkers can range from small chemical molecules to peptides, proteins, and even whole cells. In the case of small molecules, the analysis usually only requires a sample preparation step to separate the biomarker from the complex environment [6]. However, analyzing the biomarkers will likely become more difficult as the structure becomes more complex. Indeed, when the biomarkers are in the macromolecular range, the molecules have to be extracted from the matrice, and they are also usually susceptible to variations in heat or pH. Therefore, researchers must be cautious to ensure that the integrity of the molecules is preserved through

Table 6.1 Biomarkers and their conventional determination methods

Biomarker	Source	Molecular type	Measurement type	Biomarker measurement
Prostate-specific antigen (PSA)	Serum	Protein	ELISA	Prostate cancer
Carcinoembryonic antigen (CEA)	Serum	Protein	Chemiluminescence ELISA	General cancer evaluation
STEAP1	Serum	Protein	ELISA	Prostate cancer
CA-125	Serum	Protein	ELISA/IRMA Chemiluminescence	Ovarian cancer
CA-15-3	Serum	Protein	IRMA Chemiluminescence	Metastatic breast cancer
Covid 19 genomic sequence Covid 19 antigen	Upper respiratory tract specimen Upper respiratory tract specimen	Nucleic acid Protein	RT-PCR Rapid test	Covid 19
Hepatitis C Hepatitis C viral antibody	Serum Serum	Nucleic acid Protein	RT-PCR RIBA/EIA	Chronic cirrhosis
Dangue genomic sequence Nonstructural protein 1 (NS1)	Serum Serum	Nucleic acid Protein	RT-PCR Immunoassay	Dangue
H. pylori antigen *H. pylori* antibody	Stool Serum	Protein Protein	ELISA/rapid immunochromatographic test EIA/ELISA	Peptic ulcer
E. coli	Stool	Bacteria	Microbial culture	*E. coli* infection
Thyroid-stimulating hormone (TSH) T4 T3	Serum Serum Serum	Protein Chemical compound Chemical compound	IRMA Radioimmuno assay/ ELISA Radioimmuno assay/ ELISA	Thyroid function
C-terminal telopeptide of type I collagen (CTX-I)	Serum	Protein	ELISA	Osteoporosis
β-Amyloids Tau	CSF/blood CSF/blood	Peptide Protein	ELISA/LCMS ELISA	Alzheimer
Cystatin-C (Cys-C) Creatinine	Serum Serum/ urine	Protein Chemical compound	ELISA HPLC	Kidney's function and glomerular filtration rate
Troponin I (cTnI) Myoglobin (Mb)	Serum Serum	Protein Protein	ELISA Immunoassay	Myocardial infection (MI)

sample preparation [7]. Finally, recognizing the whole cell is the most difficult of them all. In this case, which commonly concerns non-histopathologic approaches, the integrity of only the individual macromolecules is not enough, and the integrity of both the cell membrane and the membrane macromolecules as a whole should be preserved through the process of sample preparation and analysis [8].

After diagnosis and treatment, biomarkers can be used to monitor a patient's progress and response to therapy [1, 4]. By assessing biomarkers over time, healthcare providers can adjust treatment plans and assess the effectiveness of interventions, which may result in changes to the assigned ICD codes if the patient's condition changes.

In conclusion, having a deep understanding of biomarkers in healthcare can significantly contribute to decreasing mortality and morbidity rates. Biomarkers provide valuable insights into an individual's health status, disease risk, and treatment response. This information enables healthcare providers to make more accurate diagnoses, personalize treatment plans, and monitor patient progress. Ultimately, the timely and precise use of biomarkers can lead to earlier disease detection, improved interventions, and better outcomes, which, in turn, can result in reduced mortality and morbidity rates in various health conditions.

6.1.1.3 Conventional Methods of Biomarker Detection and Quantification

Numerous analytical methods have been developed for detecting and quantifying biomarkers from bodily matrices. Some of these commonly used methods to quantify biomarkers are enzyme-linked immunosorbent assay (ELISA), enzymatic assays, liquid chromatography–mass spectrometry in multiple reaction mode (MRM-LCMS), protein assay, and real-time polymerase chain reaction (RT-PCR) [9]. These methods are gold standard methods that have real-life clinical applications.

ELISA is considered a low-cost method per sample, but it requires a specific antibody against the biomarker. As a result, it cannot be used for newly discovered biomarkers, and it is time-consuming and expensive to develop a new method for each new biomarker [9].

In preclinical stages with variable potential biomarkers to analyze, MRM-LCMS can be a more suitable method. This method is also known as selected reaction monitoring. By quantifying and measuring protein concentrations, MRM-LCMS is being used to connect new biomarker discovery and their clinical validation. MRM's success relies on the selection of suitable target peptides, their optimal MRM transitions, and the sample not containing amino acids that are readily modified [10].

RT-PCR is a flexible, highly accurate, sensitive, reproducible, and extract-based method for measuring gene expression over a wide dynamic range [11, 12]. RT-PCR can provide quantitative data, but it is applicable in a limited concentration. Without extensive dilution or concentration steps, it may not accurately measure extremely low or high expression levels. Moreover, improper sample handling or storage can result in RNA degradation and unreliable results. RT-PCR can be cost-intensive, especially when analyzing multiple biomarkers in many samples. It also requires a skilled technician to perform and analyze the experiments, making it labor-intensive [13]. Table 6.1 demonstrates the biomarkers discussed in this chapter [14–16].

There are some disadvantages to these conventional methods. Some of these methods could be expensive and require advanced instruments. Therefore, they require experienced personnel. Furthermore, they usually require sample preparation before the analysis, and the overall procedure could be unsuitable for the patient if frequent follow-up monitoring is required.

Point-of-care systems (e.g., sensors and biosensors) usually solve these disadvantages using a precise, easy-to-use, inexpensive, and available method. By integrating MIPs, the lack of selectivity can also be overcome. These polymers can be highly selective and specific. They can target many molecules, including small organic compounds, proteins, and complex biomolecules like DNA.

6.1.2 MIPs and Their Integration with EC Sensors

Molecularly imprinted polymers (MIPs) are becoming increasingly popular as potential replacements for biological antibodies due to their numerous advantages. This subchapter briefly introduces their fabrication steps and examines their unique abilities for selective recognition. Then, it delves deeper into the characteristics of MIPs as artificial antibodies and explores their diverse applications.

The final part of this section is dedicated to familiarizing the reader with electrochemical methods, how EC sensors are modified with MIP particles, and their advantages in synthetic antibodies-related studies.

6.1.2.1 The Advantages of MIPs

MIPs have come a long way since their inception in the 1930s [17]. Wulff and Sarhan laid the groundwork for imprinting small molecules in synthetic matrices [18]. Initially, MIPs were designed to recognize and bind to small molecular targets. However, advancements in polymer chemistry and molecular imprinting techniques in the late twentieth century led researchers to create MIPs that could recognize small molecules and complex macromolecules, mimicking the paratopic specificity of antibodies [19, 20].

The most critical trait associated with MIPs is their specific 3D cavity morphology [21, 22]. Quite a few variables can bring about a highly specific three-dimensional structure. For example, the researcher can alter the MIP building blocks and functional groups and engineer the chemical interactions occurring in the imprinting process and the polymerization methods.

The manipulation of various steps and variables would affect the outcome (Fig. 6.1). MIPs can be optimized for specific applications by iterating through these steps, ensuring maximum specificity and binding efficiency. Below is a summary of MIP fabrication components and routes:

1. Template molecule selection: An appropriate template molecule is essential, considering size, structure, functional groups, stability, solubility, availability, and cost. Sometimes, a similar "dummy" molecule may save costs or reduce toxicity.
2. Functional monomer selection: Functional monomers are chosen based on (a) their interaction with the template molecule and (b) their characteristics, such as

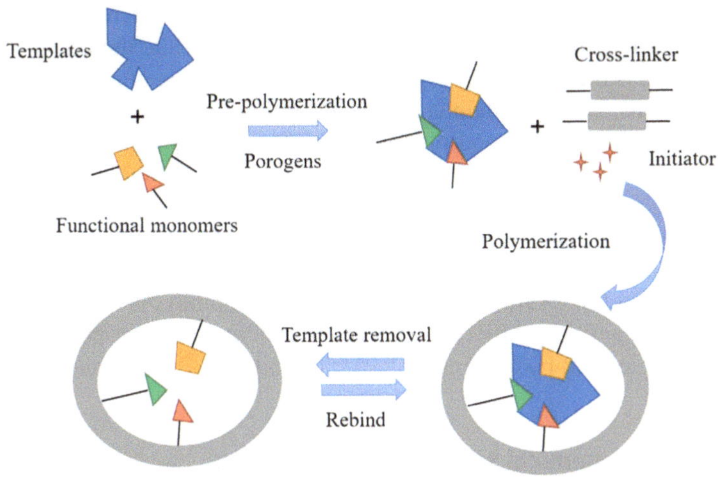

Fig. 6.1 General synthesis route of an MIP. (Reprinted with permission from [31])

conductivity, polarity, reactivity, stability, and biodegradability, and (c) their compatibility with the crosslinker and polymerization conditions [23–25].

3. Crosslinker addition: Crosslinkers add structural stability to the polymer matrix. Choosing crosslinkers compatible with the functional monomer and the polymerization conditions is important. Furthermore, the solvent's effect should be considered [23].
4. Pre-polymerization complex formation: A stable complex between functional monomers and template molecules is formed before polymerization. This step involves careful stoichiometric calculations of solution volume and quantity adjustments and temperature and pressure control [23, 26].
5. Polymerization: Various methods like bulk, emulsion, suspension, precipitation, core–shell, electrochemical, UV-induced, and thermal polymerization are used to solidify the pre-polymerization complex into a polymer matrix, depending on the fabrication requirements [23, 27].
6. Template removal: Removing the template molecule from the final polymer matrix is crucial for the MIP's performance. While there's significant documentation on optimizing polymerization, literature is limited concerning cost-effective template removal methods, such as solvent extraction, Soxhlet extraction, and physically assisted methods [23].
7. Binding and rebinding: The effectiveness of MIPs is measured by their ability to bind and rebind the template molecule selectively. This is tested by exposing the MIPs to the template solution and measuring the binding efficiency. Moreover, the specificity is tested by comparing the binding efficiency with other structurally related molecules [28–30].

MIPs became more comparable to antibodies' functionality, so they were aptly named "artificial antibodies." This term signifies their ability to perform specific molecular recognition without the limitations of biological systems. Artificial antibody MIPs differ from classical MIPs in several ways. While the primary goal of classical MIPs is the selective recognition of smaller targets, often relying on simpler interactions, artificial antibody MIPs prioritize the accurate mimicry of the complex three-dimensional and multifaceted interaction landscapes characteristic of biological antibodies [7, 32, 33]. This involves the integration of multifunctional monomers, cross-linkers, and innovative polymerization techniques to achieve the desired recognition specificity and selectivity [32, 34]. Additionally, synthetic polymeric antibodies can perform consistently even in complex biological matrices, rivaling and sometimes surpassing the recognition capabilities of natural antibodies. Lastly, artificial antibody MIPs are designed to exhibit kinetic properties mirroring those of biological antibodies, ensuring specific and timely recognition and binding [32, 35–37].

6.1.2.2 Electrochemical Sensors and Their Roles

Electrochemical sensors are instrumental devices designed to identify alterations in environmental parameters like temperature fluctuations, external pressures, or compound concentration changes. When diving into electrochemistry, we find that these detectors have been made more and more reactive to analytes, which mediate and modulate various electroanalytical indicators. The genesis of electrochemical sensors traces back to the early 1900s, beginning with the invention of glass lamellas sensitive to pH levels. This innovation laid the groundwork for their application in potentiometric devices. The field took a significant leap in the 1920s with the introduction of polarography, later renamed voltammetry, which became instrumental in analyzing metals such as zinc, cadmium, lead, and copper. The 1960s witnessed a resurgence in the field, marked by new techniques like differential pulse voltammetry (DPV) and the advent of ion-selective membranes for potentiometry. The 1980s saw a shift in focus towards chemically modified electrodes and the use of new materials like glassy carbon and noble metals. Influential developments by Leland Clark, including the first oxygen and glucose sensors, significantly boosted electroanalytical research. Today, we are in a new era of electrochemical sensors, driven by the integration of nanotechnologies like graphene, representing a third wave of innovation in this field since the start of the twenty-first century [38–41].

As mentioned, the methods and techniques used in electrochemical sensing have evolved significantly over time. These methods can be segmented into three main categories. First, there are (a) amperometric and voltammetric techniques, which primarily focus on determining the fluctuations in electric current as a direct consequence of changes in applied voltage, (b) potentiometric approaches that the emphasis is on monitoring the shifts in electric potential, and (c) conductometric systems which are designed to interpret and translate alterations in resistance into discernible outputs. There are various configurations of electrochemical systems. Some might consist of a duo of electrodes—reference and an indicator. These are more common in static methods where the concentration of species remains constant, and only potential shifts are recorded. Others might employ a triad: reference, counter, and working electrodes, typically found in dynamic arrangements (i.e., methods in which there are disturbances in the current due to redox reactions). This exposition pivots towards examining the attributes of working electrodes. Key characteristics of these electrodes encompass a defined chemical interface, established surface metrics, quick response times, and the capability to yield consistent data within specific potential parameters. It is crucial to mention that these electrodes are not monolithic in design. They vary widely in dimensions, architecture, and fabrication routes [42, 43].

To provide a clear comparison between their performances, electrochemical sensors, when compared to optical (including colorimetric sensors) and mass-sensitive sensors, offer several advantages but also come with their disadvantages.

The primary advantage of electrochemical sensors is their exceptional sensitivity and selectivity, which is especially crucial for detecting specific analytes in complex matrices [44]. For instance, these sensors play a pivotal role in glucose monitoring devices for diabetics, where enzyme-based electrochemical sensors selectively respond to glucose in blood despite various interfering substances, using techniques like DPV [45, 46]. These techniques are highly sensitive and selective, enabling the detection of specific analytes even amidst interfering substances. Optical sensors, while useful for broad detections such as pH changes, may not offer the same level of specificity for certain compounds [47–49]. Mass-sensitive sensors are highly sensitive to mass changes but may require additional molecular recognition elements to differentiate between similar molecules [50–52].

Electrochemical sensors are also highly adaptable and can be miniaturized, making them ideal for portable, field-based applications. Portable electrochemical glucose monitors, for example, have transformed diabetes management by allowing patients to perform blood glucose measurements conveniently [53]. In contrast, optical sensors, particularly those requiring sophisticated components like lasers and photodetectors, are often more challenging to miniaturize. Mass-sensitive sensors, though sensitive, typically require controlled environments to mitigate the effects of temperature or pressure fluctuations, limiting their portability [54].

Electrochemical sensors often have lower power requirements and can be more cost-effective to produce and maintain, making them attractive for continuous monitoring applications in health monitoring devices or industrial process control.

However, electrochemical sensors have disadvantages as well. They can be susceptible to environmental factors like temperature, pH, and humidity, which may affect their performance, necessitating complex signal processing or frequent recalibration. Electrode degradation over time is another concern, leading to drift in sensor response and necessitating regular maintenance or replacement. This is particularly problematic in long-term monitoring scenarios. Yet, what is exciting is the pace of innovation; each year, we witness technological advancements that push the boundaries, leading to increasingly efficient performance.

Molecular imprinting is one intriguing method that garners attention in EC sensor enhancement. MIPs paired with electrochemical sensors stand out for their unparalleled precision and selectivity. Also, their adaptability and ability to mirror desired molecular characteristics are noteworthy. However, multiple considerations come into play when incorporating MIPs into sensory devices. These range from the sensor's structural design and rigidity to the methods employed for MIP immobilization. While certain electrode materials resist modification, others are more amenable to changes during their construction phase.

Moreover, each MIP preparation technique carries its own set of challenges and limitations. Effective integration of MIPs into sensing systems can involve chemically growing the MIP using magnetic particles, electropolymerization, and methods like spin-coating. Compatibility between the MIP and the sensor's sensory unit is vital. Any discord can lead to reduced efficiency in mass transfer and binding kinetics [6, 55]. Also critical are the resilience and durability of MIP-integrated sensors [56–58]: Once the measurement process concludes, the MIPs' cavities become occupied with the analytes. Therefore, the sensor surface must be cleaned or renewed for subsequent measurements to ensure repeatable electroanalytical performance. The applications of MIP-ECs have been thoroughly explored in multiple areas, such as (a) the detection of pesticides, heavy metals, mycotoxin, counterfeit drugs, antibiotics, explosives, illicit drugs [59], (b) pharmaceutical quality control, chemical production examination, wastewater treatment [59–61], and (c) the monitoring of herbicides, drug metabolites, and fertilizers [59] However, studies and applications related to the present chapter are done in MIP-EC sensor development and implementation in biomarker detection. In the following sections, we examine the latest developments in the introduction of EC-MIP sensors that can prove effective in detecting cancer, viral, and bacterial biomarkers, as well as the biomarkers of infections and illnesses that cannot be categorized in any of the mentioned groups.

6.2 MIP-EC Sensors

6.2.1 MIP-EC Sensors for Cancer Biomarkers

Almost every cell in the human body can proliferate and spread to other body parts. While normal cells undergo programmed cell death upon reaching senescence or sustaining damage, this regulatory process is sometimes disrupted, leading to uncontrolled cell division and the formation of tumors. These tumors can manifest as either cancerous or non-cancerous entities [62]. Cancer is the first or second leading cause of death before the age of 70 years in more than 60% of countries worldwide [63]. Given this substantial impact on public health, there is a pressing need for a comprehensive approach to global cancer control. Focusing on developing and implementing effective preventive interventions and early diagnostic strategies is paramount in alleviating the burden of cancer within populations [64].

6.2.1.1 Prostate-Specific Antigen (PSA)

Prostate cancer poses a significant threat to men aged 55–80, with a noticeable increase in its occurrence in recent years. This rise has led to the adoption of PSA as a serum biomarker for early detection. While some non-cancerous prostate conditions can cause false positives by elevating serum PSA levels, there is currently no alternative serum biomarker for prostate cancer detection. PSA, despite being produced by other organs like ovaries, kidneys, and lungs, remains the primary choice for early detection [65, 66].

Recent routine analysis methods for detecting PSA in practices are based on immunoassays, such as ELISA [65]. Some commercial PSA detection assays are on the market, such as the Abcam PSA (KLK3) human ELISA kit, Abnova human ELISA kit, and Sigma Aldrich ELISA kits [66]. However, these methods suffer from various disadvantages, and recently, more suitable methods for detecting these antibodies are being considered. One of these novel methods is MIPs, which have shown promising properties in the selective absorption of different targets [65].

A nanobiosensor was created by electrochemically polymerizing pyrrole on a screen-printed gold electrode. Using Differential Pulse Voltammetry (DPV) with $K_3[Fe(CN)_6]/K_6[Fe(CN)_6]$ as an electrochemical marker, the nanobiosensor exhibited high affinity, rapid rebinding, and improved detection limits compared to other techniques. It proved comparable to the gold standard immunoradiometric method [65].

Another study utilized MIPs and nanocomposites with graphene nanoplatelets, gold nanoparticles, and chitosan for an amperometric assay to detect PSA. The MIP, synthesized by electropolymerization of dopamine, showed improved sensitivity, specific binding sites, and good conductivity. Chitosan prevented nanomaterial leakage, but its non-conductive nature posed challenges. The search for suitable substrates and optimized concentrations complicated MIP electrochemical sensor fabrication for PSA [66].

In 2022, a research group successfully developed an MIP electrochemical sensor using *o*-phenylenediamine (*o*-PD) and *o*-aminophenol (*o*-AP) to determine PSA levels. This sensor, deposited on a screen-printed carbon electrode modified with PEDOT: PSS@Fc, demonstrated high selectivity, sensitivity, and reusability for over 12 cycles and long stability for 7 weeks [67].

Another study explored electrochemical sensors based on polypyrrole (PPy) MIP, employing molecular dynamics calculations and experimental procedures. The strong binding observed between the designed MIP and PSA suggested PPy to be a favorable polymeric matrix for electrochemical sensors [68].

A PSA sensor was fabricated using MnNP-MWCNTs (manganese nanoparticle-decorated multi-walled carbon nanotubes). This composite acted as a nano-iniferter for the 3D construction of the MIP for PSA detection. The polymerization method was controlled by radical polymerization. To confirm the analytical behavior of the fabricated sensor, differential pulse stripping voltammetric (DPSV) and squarewave voltammetry (SWV) techniques were used [69].

Also, in a notable work, a dual aptamer-MIP electrochemiluminescence (ECL) sensor was developed. As the sensor was based on the "signal-off" strategy, as a result of PSA binding in imprinted cavities, the sensor exhibited a reduction in ECL response. Furthermore, the sensor demonstrated strong and specific recognition of PSA by incorporating both aptamer and MIP recognition elements. Real samples were analyzed in this experiment, demonstrating results comparable to those obtained using conventional methods. The concurrent use of MIPs and aptamer resulted in higher selectivity and stability for PSA detection [70].

Table 6.2 summarizes the outcomes of the studies above.

6.2.1.2 Carcinoembryonic Antigen (CEA)

One of the most investigated cancer biomarkers is CEA. This biomarker is produced in high concentrations at the embryonic stage, but it declines after birth, and adults have normal concentrations of up to 2.5 ng/mL. This biomarker is mainly produced and secreted to the bloodstream in the case of colorectal cancer, though it is not specific to this disease. Any cancer that could be classified as adenocarcinoma can also produce CEA. These measurements are mostly done to assess the course of treatment or possibilities of metastasis. A low level of CEA with an already diagnosed tumor can be an indicator of a benign one. In contrast, an elevated level of CEA (i.e., greater than 10–20 ng/mL) can be a sign of a large or metastatic tumor. Also, identifying prominent levels of CEA in other organs or bodily fluids, such as cerebrospinal fluid (CSF), indicates tumor metastasis to that site.

There have been numerous studies on the development of CEA-imprinted sensors. A simple electrode based on the surface electropolymerization of dopamine on the glassy carbon electrode surface was investigated. This sensor also utilized gold nanoparticles to increase the electron transfer and enhance the detection signal. This sensor shows good linearity and LOD, which fits the clinically appropriate concentration [71].

Recently an innovative microfluidic paper-based electrochemical device (μPADs) was introduced. This device utilizes a movable valve to transport fluids from one compartment to another. This makes multi-step or long-term reactions feasible on such devices (Fig. 6.2). For example, in this study, the entire process of imprinting and adding other additives happens in situ, making it a low-cost, green, and user-friendly device. One of the advantages of this study is that it compares their device with the gold standard method, chemiluminescent immunoassay (CLIA). It was shown that this method could quantify CEA in real patient's serum samples with RSDs less than 6.5% [72].

One of the central teams working on the subject matter is Professor Goreti Sales' lab. Their latest attempt resulted in an all-in-one sensor in which CEA's adsorption on the polymeric film on the anode surface hinders the methanol access to the anode and its oxidation, resulting in a different polarization curve [73].

Table 6.3 summarizes the outcomes of the studies mentioned above.

6.2.1.3 Other Cancer Biomarkers

Other cancer biomarkers were also investigated. The six-transmembrane epithelial antigen of the prostate 1 (STEAP1) is one of these biomarkers. This protein is overexpressed in the membrane of prostate cancer cells. Compared to PSA, this biomarker has a much lower rate of false positives, and it could be used to detect prostate cancer before entering the adenocarcinoma stage. One of the shortcomings of this biomarker is that it is a membrane protein, and its detection requires a biopsy from the site. An MIP-based biosensor was developed for this biomarker, and it was shown that it could be accurately measured using this system. However, the difficulties, such as cellular extraction and the requirement for biopsy, still stand [74].

Table 6.2 A brief demonstration of studies conducted by MIP-EC sensors on cancer (PSA biomarker)

Organ/disease	Biomarker	MIP composition	Analytical technique	LOD	Linear range	Sample	Recovery	Reference
Cancer	Prostate-specific antigen (PSA)	Polypyrrole	DPV	2 pg/mL	0.01–4 ng/mL	Real patient serum	N/A	[65]
		Polydopamine	DPV	0.15 pg/mL	0.001–100 ng/mL	Spiked human serum	92–103.5%	[66]
		Poly(*o*-phenylenediamine-co-*o*-aminophenol)	DPV	8.3×10^{-8} ng/mL	1.0×10^{-7}–1.0×10^{-4} ng/mL	Spiked human serum	89.4–102.9%	[67]
		Polypyrrole (PPy)	SWD	N/A	3×10^{-8}–300 ng/mL	Real human serum	N/A	[68]
		Polyitaconic acid/EGDMA	DPSV, SWSV	3.04 pg/L	0.01–10 ng/L, 20–62 μg/L	Spiked urine and serum	96.57–100.44%	[69]
		Polydopamine	ECL	3.0 pg/mL	5 pg/mL–50 ng/mL	Human serum	97.30–104.49%	[70]

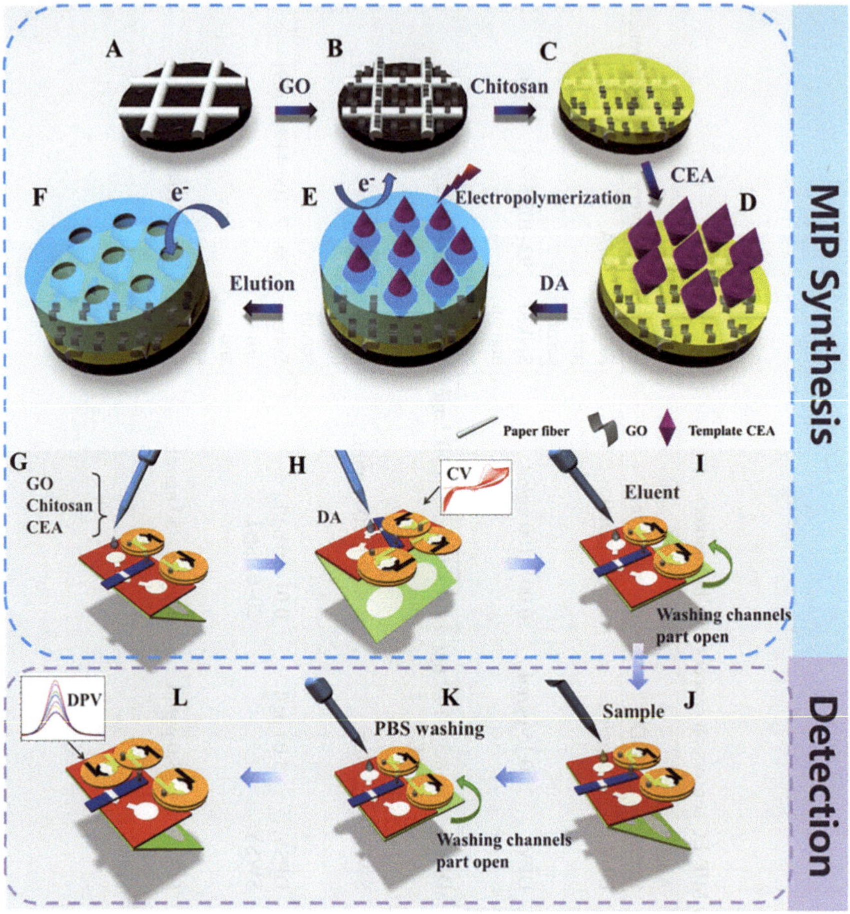

Fig. 6.2 An overview of in-situ MIP synthesis and detection procedure on an electrochemical μPAD device. (Reprinted with permission from [72])

Cancer antigen 125 (CA-125) is a non-selective biomarker, the elevation of which could be an indicator of various cancers such as ovarian, uterus, cervix, pancreas, liver, colon, breast, lung, and digestive system. Other conditions, such as endometriosis and menstrual period, can also increase its concentration in blood. However, the most prominent reason for its increase is usually ovarian cancer, which is why CA-125 is used for screening in people who have a considerable risk of getting diagnosed with the mentioned cancer type.

A study focused on developing an MIP-solid phase extraction (SPE) sensor for CA-125. Two methods were compared: an electrochemical method (i.e., [SWV]) and an optical transduction method (i.e., surface plasmon resonance [SPR]). It was shown that the electrochemical technique had higher sensitivity, lower LOD, and higher linearity range, though the SPR technique has advantages, such as being rapid and a label-free method. It was also shown that although most of these sensors

Table 6.3 A brief demonstration of studies conducted by MIP-EC sensors on cancer (CEA biomarker)

Organ/disease	Biomarker	MIP composition	Analytical technique	LOD	Linear range	Sample	Recovery	Reference
Cancer	CEA	Polydopamine	DPV	0.259 pg/mL	0.001–1000 ng/mL	Spiked human serum	88.7–124.6%	[71]
		Polydopamine/chitosan	CV	0.32 ng/mL	1.0–500.0 ng/mL	Real patient serum	N/A	[72]
		Polypyrol/poly(3,4-ethylenedioxythiophene)	Polarization and power density	0.08 ng/mL	0.1–100 ng/mL	Buffer—Cormay® serum—artificial human serum	N/A	[109]
		Galic acid polymer	EIS	1.0 ng/mL	1–1000 ng/mL	Sample solution	N/A	[110]
		Polyaniline	EIS and SWV	0.1 pg/mL	0.013–1.75 ng/mL	Urine	N/A	[111]
		Polypyrol/poly(3,4-ethylenedioxythiophene)	EIS, polarization and power density	3.52 ng/mL	30–30,000 ng/mL	Cormay® serum	N/A	[73]

are disposable, this one could be used twice with a stable response and an RSD of 5% for the subsequent application [75].

Cancer antigen 15-3 (CA 15-3) is a biomarker mostly used to evaluate the course of treatment in women with advanced breast cancer. An elevated alkaline phosphatase concentration in conjunction with this biomarker indicates early breast cancer recurrence [76]. Two recent studies worked on the point-of-care approaches to determine and quantify this biomarker. Both of these studies used gold SPE as the basis of their probe and coated it with CA 15-3 imprinted polymeric films. The first study created its MIP using electropolymerization of 2-aminophenol, in which the absorption of CA 15-3 onto a polymeric film reduces the electrochemical signal. The selectivity of such a sensor was also investigated using other breast-related biomarkers (i.e., HER2 and Cys-C), confirming the selectivity of this sensor [77]. The second study used a conducting polymer, poly(toluidine blue), as the MIP matrix, greatly enhancing the sensor's performance. It resulted in a sensor with a lower LOD and wider linear range [78]. Both studies concluded that such a tool could be used as a point-of-care device in clinical contexts.

Regenerating protein 1 (REG1), also known as pancreatic stone protein (PSP), is originally a sepsis biomarker that has been shown to be elevated in pancreatic cancer as well. The clinical biomarker for pancreatic cancer screening is CA19-9, and REG1 is not clinically used to screen such cancer. However, the potential application of REG1 for pancreatic cancer has also been investigated [79].

Table 6.4 summarizes the outcomes of the aforementioned studies.

6.2.2 MIP-EC Sensors for Viral Biomarkers

The application of MIPs in the detection of viral biomarkers is an exciting area of research, particularly given the critical need for rapid, accurate, and early detection of viruses in various settings, from clinical diagnostics to border control and outbreak management [80]. This subchapter will mainly discuss the MIP-EC sensor developed for detecting SARS-CoV-2 (COVID-19) because of the massive and relatively recent number of studies. However, there are plenty more viral infections that have been studied to introduce an MIP-EC (e.g., influenza virus, HIV, and hepatitis). In addition to specific virus detection, general viral pathogen surveillance is another area of developing virus-sensitive MIP-EC sensors. Indeed, MIP-based electrochemical sensors can be used in broader viral surveillance, such as screening water sources for enteroviruses or detecting emergent zoonotic viruses at points of the animal–human interface [81]. By targeting specific viral proteins or nucleic acid sequences, these sensors can aid in the early detection of potential public health threats [82].

Nevertheless, while the potential applications of MIP-based electrochemical sensors for viral detection are extensive, several challenges remain. These include the need for high specificity in complex biological fluids, the stability of the MIP under a range of environmental conditions, and the scalability of sensor production. Furthermore, for some viruses, the rapid mutation rates may necessitate frequent redevelopment of MIPs to maintain efficacy.

Table 6.4 A brief demonstration of studies conducted by MIP-EC sensors on cancer (various biomarkers)

Organ/disease	Biomarker	MIP composition	Analytical technique	LOD	Linear range	Sample	Recovery	Reference
Cancer	STEAP1	Poly(pyrrole-2-carboxylic acid)	CV	N/A	130 pg/mL–13 µg/mL	Cellular extract	N/A	[74]
	CA-125	Polypyrol	SWV and SPR	0.01 U/mL	0.01–500 U/mL	Spiked artificial serum	91–105%	[75]
	CA-15-3	Poly(2-aminophenol)	CV and EIS	1.5 U/mL	5–50 U/mL	Spiked human serum	N/A	[77]
		Poly(toluidine blue)	DPV	<0.10 U/mL	0.10–100 U/mL	Spiked human serum	99–102%	[78]
	REG1	Poly(ethylene-co-vinyl alcohol)	CV	N/A	N/A	*E. coli* culture medium	N/A	[79]

6.2.2.1 SARS-CoV-2 (COVID-19)

The SARS-CoV-2 pandemic has highlighted the critical importance of rapid, sensitive, and specific diagnostic tools. MIPs integrated into electrochemical sensors are promising for detecting SARS-CoV-2 and its biomarkers. These sensors can offer several advantages in a pandemic, including rapid diagnostics, portability, and the ability to deploy them at points of care, such as hospitals, airports, and community health centers. For SARS-CoV-2, there are three typical biomarkers of interest.

The first biomarker is the spike (S) protein, which facilitates viral entry into host cells and is the primary target of neutralizing antibodies. MIPs that can selectively bind to the S protein provide an excellent opportunity to detect the virus [83, 84].

A study designed an electrochemical sensor with a molecularly imprinted polymer synthetic receptor (ncovS1-MIP) to detect the SARS-CoV-2 S protein subunit S1 (ncovS1). This sensor achieves high selectivity for ncovS1 by using a covalent imprinting method that relies on the chemical affinity between the diols on the glycosylated S protein and boronic acid groups on APBA, used as a functional monomer. It offers a rapid response, with a rebinding time of 15 min and a measurement time of just 5 min, making it competitive with current antigen tests. The sensor detects the S protein in phosphate-buffered saline and patient nasopharyngeal samples and effectively differentiates against the nucleocapsid protein (ncovNP), another potential biomarker in COVID-19 patient samples [83].

Another study based on EC sensor performance regarding S proteins introduced a novel method to improve the specificity of MIPs for protein recognition. Amino acid-derived monomers are formed into a polymer structure, closely mimicking natural protein ligands like antibodies. The novel monomer, aspartate-umbelliferone (Asp-UMB), enables the creation of MIPs with high affinity for protein epitopes. Using a heptapeptide from the SARS-CoV-2 S protein's receptor binding domain (RBD), MIP microarrays were produced on surface plasmon resonance imaging (SPRi) chips, achieving a highly uniform, electrically insulating polymer film around 2 nm thick. This film was optimal for imprinting and SPRi detection, resulting in MIPs with low picomolar dissociation constants for the target RBD protein. The method allowed for on-chip optimization, producing MIPs without significant nonspecific binding and with excellent selectivity. This indicates the potential for creating a diverse library of protein-selective MIPs [84].

The second biomarker of SARS-CoV-2 is the nucleocapsid (N) protein, which is involved in packaging the viral RNA. Due to its abundance in the virus and its strong immunogenicity, it is another prime target for MIP-based detection [85, 86].

A recent study introduced a sensor that comprises a disposable chip with a thin-film electrode. The sensor is connected to a portable potentiostat (Fig. 6.3) and enhanced with MIPs for high specificity to the SARS-CoV-2 nucleoprotein. This enables it to detect the presence of the virus in nasopharyngeal swabs from infected patients with high selectivity, distinguishing the virus from other proteins. The sensor's portability makes it an excellent candidate for point-of-care testing, significantly improving current antigen tests by providing rapid on-site results [85].

A sensor fabricated using the N protein incorporates gold/graphene nanohybrids on a screen-printed carbon electrode. It utilizes arginine as a functional monomer to

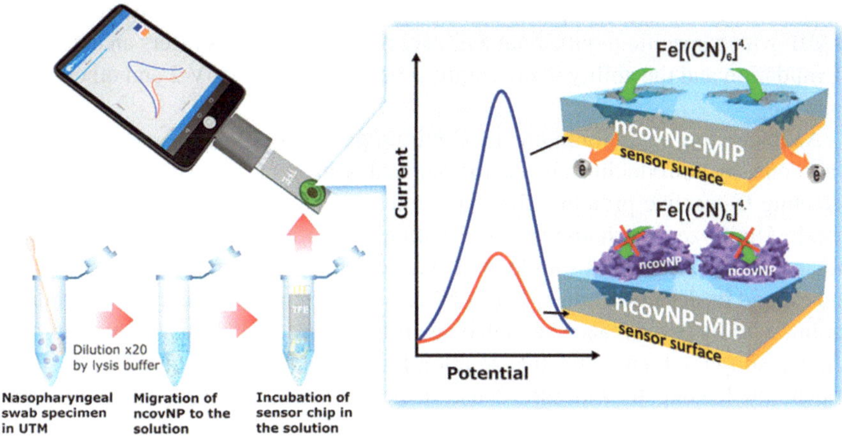

Fig. 6.3 COVID-19 diagnostics principle by nconNP sensor analyzing the samples prepared from patients' nasopharyngeal swab specimens. (Reprinted with permission from [85])

create a simple yet extremely sensitive detection method. The sensor achieves a wide linear range for ncovNP detection by leveraging the specific recognition ability of the MIP as well as the superior electrical properties and large surface area provided by the Au/Gr hybrid. The developed MIP sensor displays notable selectivity, reproducibility, and stability over time, underscoring its potential for SARS-CoV-2 detection in clinical settings [86].

The third biomarker of SARS-CoV-2 is the viral RNA [87], typically detected with amplification techniques like RT-PCR. However, there are several alternatives for detection. First, MIPs could be developed to capture specific RNA sequences directly, although this is more challenging due to the flexibility and structural complexity of RNA molecules. Also, the whole virus can be employed as the template [88]. Finally, as an alternative, studies can focus on a secondary, nonspecific biomarker released in the body as a defensive reaction to COVID-19 viral infection [89].

Researchers have developed an electropolymerized molecularly imprinted polymer (E-MIP) sensor to detect the SARS-CoV-2 virus rapidly. These E-MIPs are crafted to selectively recognize the virus by imprinting with SARS-CoV-2 pseudoparticles and are formed on screen-printed electrodes through reductive electropolymerization. Detecting the COVID-19 virus as the analyte, this method shows promise in distinguishing SARS-CoV-2 from other respiratory viruses owing to its high selectivity and specificity. The sensor can identify the virus in under 10 min using saliva samples, providing a fast alternative to current testing methods like the loop-mediated isothermal amplification (LAMP) technique. Electrochemical impedance spectroscopy (EIS) and principal component analysis (PCA) are employed to differentiate between COVID-19 positive and negative samples. The results exhibit a clear separation, indicating the sensor's potential as a rapid diagnostic tool. This technology streamlines testing as it doesn't require virus manipulation or genetic material amplification, which are steps in conventional testing methods. Additionally, the presence of the virus can be confirmed by treating the E-MIP with a specific monoclonal antibody post-capture. This offers an extra layer of validation and the ability to differentiate between SARS-CoV-2 and other pathogens [88].

An interesting study for indirectly confirming SARS-CoV-2 presence in the body was done by introducing an electrochemical sensor using MIPs developed for detecting C-reactive protein (CRP), a marker of inflammation and cardiovascular health. Optimized for home-care use within lab-on-a-chip devices, the sensor is enhanced with tungsten disulfide (WS2) to increase its electrochemical sensitivity and expand its detection range. This MIP sensor efficiently measures CRP in human serum, with its performance validated against standard ELISA tests. This approach offers a simplified, cost-effective solution for personal health monitoring systems, potentially leading to widespread use in home diagnostics and wearable health devices [89].

MIP-based electrochemical sensors can play a key role in the continued monitoring of COVID-19 and effective testing in future pandemics. The sensors could be integrated into portable devices, enabling rapid and non-invasive point-of-care testing. MIP-based sensors could also be connected to digital data systems, contributing to real-time epidemiological monitoring and aiding public health responses. Continued innovation and research in MIP technology could lead to its routine use in clinical settings, providing a supplementary and perhaps more immediate diagnostic tool alongside traditional methods, such as PCR testing.

6.2.2.2 Other Viral Biomarkers

A novel sensor for hepatitis C virus (HCV) core antigen detection has been developed, leveraging a hybrid approach that combines aptamer-based sensing and molecular imprinting techniques. Using a composite of multi-walled carbon nanotubes and chitosan (MWCNTs-Chit) to modify glassy carbon electrodes enhances the sensor's conductivity, improving its performance characteristics. The aptamer for the HCV core antigen is immobilized on this interface and encapsulated by electropolymerized dopamine (DA), creating a highly sensitive and stable sensing surface. This sensor offers a quick response, is cost-effective, and exhibits good sensitivity, as assessed by various electrochemical techniques, including cyclic voltammetry (CV), DPV, and EIS. The proposed method has been successfully tested on human serum samples, indicating its practical application for real sample analysis and showing promise for clinical use. Its performance outmatches many existing methods regarding sensitivity, ease of use, and cost [90].

Another example of a recent EC sensor developed for detecting a viral disease can be found in a paper on detecting dengue virus (DENV) NS1 protein. It combined molecular imprinting advantages with selected peptide templates' conservational, structural, and immunogenic properties. The sensor employs a peptide from DENV NS1 imprinted onto carbon screen-printed electrodes, with heat denaturation exposing key epitopes for more effective binding. The sensor can distinguish between DENV NS1 and the similar Zika virus (ZIKV) NS1 and selectively detects DENV NS1 in human serum. Although the sensor's detection limit does not surpass all previous studies, it successfully differentiates between DENV and ZIKV without chemical modification of proteins—a challenging task due to the viruses' overlapping symptoms and genetic similarity. The study suggests that enhancing sensor specificity and selectivity could be achieved in the future by using strong, reversible interactions, like reversible covalent bonds, to improve the reliability of antigen detection. This advancement in sensor technology is a step forward in the accurate and label-free diagnosis of flavivirus infections [91].

Table 6.5 summarizes the outcomes of the aforementioned studies.

6.2.3 MIP-EC Sensors for Bacterial Biomarkers

The MIPs-EC sensors developed to recognize bacterial biomarkers are not as commonly researched as the other subcategories mentioned in this chapter. However, several studies have been conducted on gastric bacteria. One of the more investigated fabrication approaches is to select the whole bacterial cell as the template. Imprinting the entire cell forms cavities corresponding to the functional groups attached to the membrane and the overall chemical interaction of compounds in the cell's phospholipid bilayer. A more fruitful method of bacteria-based MIP fabrication is to incorporate bacterial virulence. Virulence is defined as the capacity of an organism to invade and induce pathology in the host organism. Virulence factors refer to the molecular components that facilitate bacterial colonization and interaction at the cellular level within the host. These factors can be categorized based on their localization: secretory, membrane-associated, or cytosolic.

6.2.3.1 *Helicobacter pylori* (*H. pylori*)

H. pylori is a Gram-negative, microaerophilic bacterium that colonizes the human stomach lining. Its spiral shape and multiple flagella enable it to penetrate the stomach's mucous lining and avoid the acidic environment. The bacterium produces urease, an enzyme that breaks down urea into carbon dioxide and ammonia, which helps neutralize stomach acid and provides a protective alkaline environment. *H. pylori* synthesizes cytotoxin-associated gene A (CagA) and vacuolating cytotoxin A (VacA), both associated with gastric disease pathogenesis. Its colonization is associated with the development of chronic gastritis, peptic ulcers, gastric adenocarcinoma, and mucosa-associated lymphoid tissue (MALT) lymphoma. Genomic analysis reveals a high genetic diversity and plasticity level, suggesting its adaptation to different hosts and environments. Eradication therapies for *H. pylori* often combines multiple antibiotics with a proton pump inhibitor to counteract its resistance mechanisms and decrease stomach acidity [92–95].

A cutting-edge sensor for detecting *H. pylori's* VacA toxin was developed using a MIP that leverages silica nanoparticles (SiO_2 NPs) for enhanced sensitivity. When applied to an SPE, this platform serves as a specialized receptor for the VacA protein, a critical factor in *H. pylori* pathogenesis. The sensor's design involves a polymerization process that embeds the VacA antigen template directly onto the SPE. It shows a high sensitivity due to the SiO_2 NPs' ability to create a porous

Table 6.5 A brief demonstration of studies conducted by MIP-EC sensors on viral biomarkers

Organ/disease	Biomarker	MIP composition	Analytical technique	LOD	Linear range	Sample	Recovery	Reference
Covid-19	Spike protein subunit S1 (ncovS1)	Poly(3-aminophenyl-boronic acid (APBA))	CV and SWV	15 fM	0–400 fM	Phosphate buffered saline and spiked patient's nasopharyngeal samples	N/A	[83]
	Spike protein receptor binding domain (RBD)	Poly(asp umbelliferone (asp-UMB))	CV	N/A	N/A	Human serum albumin (HSA)	N/A	[84]
	Nucleocapsid protein (ncovNP)	Poly(*m*-phenylenediamine (PmPD))	CV and DPV	15 fM	2.22–111 fM	ncovNP spiked lysis buffer and nasopharyngeal swab specimens	N/A	[85]
	Nucleocapsid protein (ncovNP)	Poly arginine	DPV and EIS	3.0 fM	10.0–200.0 fM	Saliva and artificial nasal	94.3–102.0%	[86]
	The whole intact virus	Poly(*N*-hydroxmethylacrylamide (NHMA))	CV and DPV and EIS	4.9 log10 pfu/mL	4.1–6.0 log10 pfu/mL	Saliva	N/A	[88]
	C-reactive protein (CRP)	Poly(aniline (AN)-co-*m*-aminobenzenesulfonic acid (MSAN))	CV	N/A	1.0 fg/mL–1.0 ng/mL	Human serum	96%	[89]
Hepatitis C	Core antigen (anti)	Poly dopamine	CV and DPV and EIS	1.67 fg/mL	5.0 fg/mL–1.0 pg/mL	Human serum	99.76–99.89%	[90]
Dengue	Heat-denatured nonstructural protein 1 (NS1)	Polyaminophenol	CV and EIS	29.3 μg/L	50–200 μg/L	Spiked samples of diluted human serum	N/A	[91]

structure for MIP attachment, improving VacA antigen detection. This sensor distinguishes itself from DNA-based sensors by offering a lower cost and high sensitivity. With its potential for early *H. pylori* detection, the sensor, could be a valuable tool for point-of-care diagnostics for gastric cancer. The economic viability of this MIP-based sensor further supports its potential for widespread use in early disease detection [92].

CagA is another common biomarker in *H. pylori* detection. A new biosensor has been engineered to detect the CagA antigen. The biosensor employs a MIP specific to CagA, coated on a gold nanoparticle-enhanced reduced graphene oxide (Au/rGO) screen-printed electrode. The sensor's creation involved molecular dynamics simulations to optimize the MIP's pre-polymerization system, which confirmed the strong affinity of the MIP for the CagA antigen through low binding energy interactions [93].

6.2.3.2 *Escherichia coli* (*E. coli*)

E. coli is a Gram-negative, facultatively anaerobic, rod-shaped bacterium commonly found in warm-blooded organisms' lower intestines. Most strains are harmless and essential to the gut microbiota; however, some pathogenic strains can cause various illnesses, from diarrhea to urinary tract infections. The bacterium's outer membrane contains lipopolysaccharides (LPS) that determine its serotype and can trigger an immune response in the host. Its plasmids often carry antibiotic-resistant genes, which can be transferred horizontally, leading to the proliferation of antibiotic-resistant strains. *E. coli* can metabolize various substrates, and its metabolic capabilities are frequently employed in biotechnology to produce biofuels, pharmaceuticals, and other chemicals. Effective treatment for pathogenic *E. coli* infections often involves fluid replacement and antibiotics. However, antibiotics are used only when necessary due to resistance concerns [96, 97].

A novel chemosensor was developed using MIP technology to rapidly detect a modified strain of *E. coli*, a common bacterium in the human gut. The sensor utilizes 2-aminophenyl boronic acid as the functional monomer to target the *E. coli* strain. The strain has a strong interaction with the boronic acid due to the porous and flexible nature of its cell wall. Scanning electron microscopy (SEM) confirmed the effective entrapment and extraction of the *E. coli* template within the polymeric matrix in a streamlined process. The biosensor identified the K-12 strain of *E. coli*, showing promise in detecting bacterial presence up to a specific concentration. Additionally, interference studies indicated that the MIP could distinguish between different bacteria based on variations in their cell surface, implying a degree of selectivity for the sensor. This technology exemplifies a potential step forward in rapid microbial detection, offering a straightforward and effective method for identifying specific strains of bacteria, which is crucial for managing bacterial infections and contamination [98].

Table 6.6 summarizes the outcomes of the aforementioned studies.

Table 6.6 A brief demonstration of studies conducted by MIP-EC sensors on bacterial biomarkers

Organ/disease	Biomarker	MIP composition	Analytical technique	LOD	Linear range	Sample	Recovery	Reference
H. pylori	Vacuolating cytotoxin A (VacA)	Poly(methyl methacrylate (MMA))	CV and EIS	0.01 ng/mL	0.01–100 ng/mL	Spiked serum samples	97.8–98.6%	[92]
	CagA	Poly(methyl methacrylate (MMA))	CV and DPV and EIS	0.05 ng/mL	0.05–50 ng/mL	Spiked blood samples	96%	[93]
Escherichia coli	*E. coli* K-12 cell	Poly(2-aminophenyl-boronic acid (APBA))	DPV and EIS	0.46 cells/mL	2.9×10^4–3.1×10^7 cells/mL	N/A	N/A	[98]

6.2.4 MIP-EC Sensors for Other Biomarkers

This section presents insights into recent research that developed MIP-based EC sensors for other diseases.

6.2.4.1 Thyroid Diseases

There are some established clinical tests to determine thyroid functionality. The three most prevalent tests for detecting hypothyroidism or hyperthyroidism are T4, T3, and thyroid stimulating hormone (TSH), routinely performed using ELISA or radioimmunoassay. In addition, a physician might order a more in-depth analysis, especially for Graves' disorder TSH receptor antibody (TSI) or imaging using radioactive iodine. These tests usually need to be performed in specified periods to assess the appropriate dose of medication. A rapid validated sensor-based detection could streamline this process. Two magnetic nanoparticles were produced to achieve this goal.

One study focused on the synthesis of core–shell magnetic T4 MIP nanoparticles. A quantum mechanical approach was taken to select the top three functional monomers, and this calculation was experimentally confirmed. It was shown that itaconic acid was the most suitable functional monomer. No specificity was observable in the cases of acrylic acid and acrylamide. Although this research focused on the electrochemical assay of T4, immunoassay, and fluorescent assay were performed using these nanoparticles. The measurement techniques showed that the T4 MIPs suit all three settings. In particular, the electrochemical assay showed that this system could measure T4 in around 30 min in conditions similar to human serum [99].

A subsequent study utilized Janus nanoparticles as an electrochemical signaling probe in a sandwich nanosystem. TSH and T4 were imprinted on the surface of ZnO and CdO nanoparticles, respectively. This enables the system to selectively measure the amount of TSH and T4 at different voltages. Capture probes consisted of TSH and T4 antibodies immobilized on a superparamagnetic nanoparticle's surface. This enables the operator to separate the particles that have captured the target molecule easily. Afterward, washing the system with nitric acid and stripping analysis of Zn^{2+} and Cd^{2+} concentrations could correspond to the concentration of TSH and T4 [100].

6.2.4.2 Osteoporosis

Osteoporosis is one of the most common bone conditions, which has a prevalence of 20% in women and 5% in men over the age of 50. Osteoporosis is often a silent disease; it worsens gradually and remains unnoticed until it leads to a broken bone. The disease is diagnosed by a bone density test, which uses X-ray images to calculate the concentration of calcium and minerals in a bone segment. Recently, the C-terminal telopeptide of type I collagen (CTX-I) has been identified as a potential biomarker for bone turnover. Overall, an increase in this biomarker could be a sign of bone loss. Therefore, an easy-to-use device to measure this biomarker could be a good indicator for further medical interventions that prevent osteoporosis from advancing to higher stages. To achieve this purpose, CTX-I imprinted microbeads were immobilized on a layer of gold electrode using acrylate resin. These sensors were tested in real serum samples using complex nonlinear least squares (CNLS) single frequency measurements, and CNLS yielded higher accuracy [101].

6.2.4.3 Alzheimer's Disease

Alzheimer's disease is a neurodegenerative disorder that is manifested by a gradual decline in memory, resulting in dementia. Genetics plays a significant role in the progression of Alzheimer's disease, and it happens mostly in older people. The prevalent hypothesis is that two main proteins are responsible for this. First, β-amyloids sediment in the neurons and create amyloid plaques, eventually resulting in the death of neurons. Second, tau proteins undergo conformational changes and turn into neurofibrillary tangles that disrupt the neural transport system and cause cell damage. Some kits and tests are available to predict the amyloid plaque formation. PrecivityAD® and Quest AD-Detect™ are tests to determine the ratio of β-amyloids 42/40. This ratio, combined with other factors such as age and apolipoprotein E genotype, determines a score that could be used to estimate the probability of amyloid plaque formation.

Research has been done to measure the concentration of β-amyloid 42 in human blood [102, 103]. β-Amyloid 42 has a considerably low blood concentration and could mostly be found in amyloid plaques in the brain. The overall concentration of this amyloid increases in blood with age. Meanwhile, its decrease in blood could be a sign of its sedimentation. However, its quantification cannot be used as a biomarker due to the lack of a concrete normal concentration range and the fact that its decrease might be an indicator of Alzheimer's progression. Hence, an aptamer-based sensor for detecting both β-amyloids 42 and 40 was proposed. This study demonstrated that these two amyloids have only two amino acid differences, so MIP cavities are not selective enough to differentiate between them [104]. In conclusion, the total amount of amyloid could be measured using this system. However, it could not be a reliable method for assessing Alzheimer's risk.

6.2.4.4 Kidney

Cystatin-C (Cys-C) is a widely known biomarker for monitoring the kidney's functionality. Its blood concentration inversely correlates with the glomerular filtration rate (GFR). The decline in the GFR indicates kidney malfunction and mostly chronic kidney disease. GFR is also important as some medications require dose adjustment based on its decline. Cys-C could also be used to check for acute kidney failures, especially in kidney transplants. This biomarker is more reliable than serum creatinine as it is not codependent on muscle mass or race. It also could detect kidney problems sooner than measuring serum creatinine, as it has been evident that even a 50% decline in GFR could still result in normal serum creatinine. Therefore, a point-of-care device that could rapidly measure this biomarker could be of utmost use in chronic kidney patients and people who have undergone kidney transplants.

A polypyrrole-based MIP for the detection of Cys-C was developed. This design aimed to develop a rapid and inexpensive disposable electrode based on screen-printing technology. Pyrrole was electropolymerized on the surface of carbon screen printed electrodes using ten CV cycles ranging from −0.5 to +0.8 V with a scan rate of 20 mV/s. SWV was performed to analyze the target molecule in the spiked serum and to determine the selectivity of this sensor. The sensor's capability was assessed in the presence of various interfering substances: creatine kinase-MB, ascorbic acid, creatinine, and bovine serum albumin. It was shown that these substances have negligible impact on the sensor's functionality, and the sensor can accurately measure Cys-C in human serum [105].

6.2.4.5 Cardiac

Myocardial infection (MI) has two prominent biomarkers: The cardiac troponin I (cTnI) and myoglobin (Mb). Both biomarkers are released into the bloodstream after damage to the myocardium. Mb has no specificity as it could also be released due to damage to other muscle types. Furthermore, it is rapidly excreted from the body through the kidneys within 24 h. As it has a rapid rise through the onset of cardiac damage, it could be used as an early biomarker for detecting or excluding damage to the heart. However, cTnI is a more sensitive and specific biomarker in evaluating MI and is therefore considered the gold standard method. It is only released from cardiac muscles, and its concentration stays in the blood for 2–3 weeks [106].

The two studies discussed next have investigated biomarkers of MI. The first study focused on the detection and determination of cTnI using a dual method. The probe section consisted of aptamer-conjugated polydopamine structures. These structures had chelated Fe^{3+} inside them. These ions get released in acidic conditions. Through an oxidation process involving $K_3[Fe(CN)_6]$, HCl, and ferrous ions, Prussian blue ($Fe_4[Fe(CN)_6]_3 \cdot xH_2O$) is produced, which is an electroactive compound with a distinct blue color. Therefore, this chemical entity could be measured both by colorimetric and electrochemical measurement techniques (Fig. 6.4). The sensing section of this device consists of magnetic nanoparticles with a cTnI imprinted shell. These particles are fixed using an external magnet, and the probe gets attached to cavities

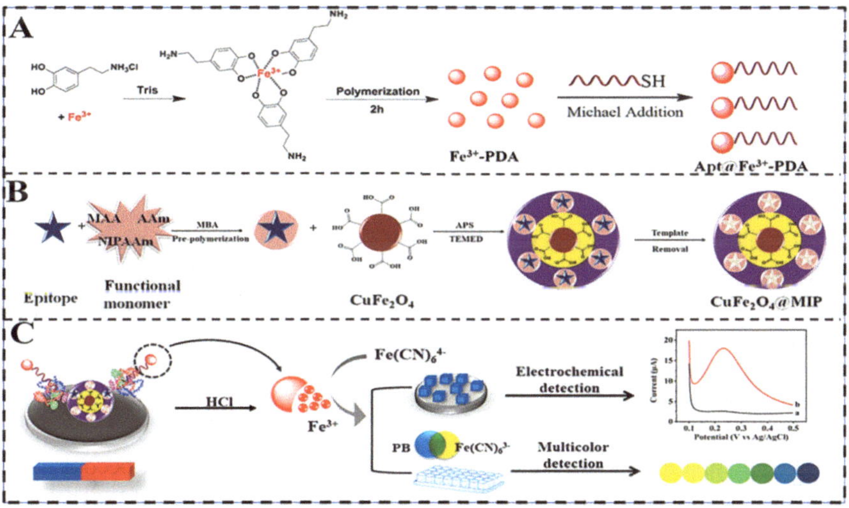

Fig. 6.4 Schematic diagram of dual-mode electrochemical-colorimetric imprinted sensing strategy. (**a**) Preparation of Apt@Fe^{3+}-PDA; (**b**) preparation of $CuFe_2O_4$@MIP; (**c**) dual mode detection of cTnI. (Reprinted with permission from [107])

containing cTnI in a sandwich-like procedure. It was shown that the dual-detecting mode increases selectivity and stability compared to other methods [107].

The second study focused on detecting and quantifying cTnI and Mb as cardiac biomarkers. A dual ECL electrode was prepared by imprinting both biomarkers on a polydopamine film. Two distinct sensing probes were designed for this purpose. First, an aptamer-MoS_2 quantum dots conjugate modified surface was fabricated to detect Mb. Second, an antibody conjugated to $Ru(bpy)_3^{2+}$ entrapped mesoporous carbon nanospheres was employed to detect cTnI. As their intensities are in different potential ranges, a dual ECL signal readout could be performed on a single sample simultaneously [108].

Table 6.7 summarizes the outcomes of the aforementioned studies.

6.3 Future Studies and Conclusion

In conclusion, the application of molecularly imprinted polymers in the development of electrochemical sensors has the potential to revolutionize the detection and quantification of biomarkers. The remarkable selectivity, sensitivity, and stability of these MIPs have enabled the creation of highly specific and efficient sensing platforms for various biomolecules, thereby addressing the critical challenges in clinical diagnostics and biotechnology. Integrating MIPs with electrochemical sensors has paved the way for rapid, cost-effective, and reliable analysis, allowing for the early detection and monitoring of numerous diseases.

To establish these tests as the gold standard, it is imperative to employ rigorous validation processes. This includes conducting comprehensive analytical validation studies to assess sensitivity, specificity, accuracy, precision, and robustness parameters. The following five criteria should be met for these methods and devices to be considered an alternative to the conventional method.

1. It should be clinically valid. The test should have a relevant outcome or be a predictor of one, and its result should remain consistent in different settings.
2. It should be responsive, that is, change in response to therapy, and show improvement or worsening after the treatment.
3. It should have a low LOD and LOQ so that the usually low concentration of a biomarker could be detectable and quantifiable from the background noise.
4. It should be selective and be able to recognize the target molecule in complex matrices where interfering molecules coexist.
5. It should be validated in clinical studies involving large patient cohorts to demonstrate clinical utility and efficacy in real-world settings.

When comparing these methods with conventional techniques, it becomes evident that MIP-based electrochemical sensors can offer distinct advantages,

Table 6.7 A brief demonstration of studies conducted by MIP-EC sensors on other disease biomarkers

Organ/ disease	Biomarker	MIP composition	Analytical technique	LOD	Linear range	Sample	Recovery	Reference
Thyroid	Levothyroxine	Poly(itaconicn acid)/ EGDMA	Magneto-actuated immunoassay and LSV	0.0356 ng/mL	2.40 × 10^{-4}–24.00 μg/mL	Spiked blood serum samples	N/A	[99]
	TSH	Poly(itaconic acid/bis acrylamide)	cc-PSA	0.0002 μU/mL	0.001–100 μU/mL	Spiked and real human serum	96.9–104.1%	[100]
	T4	Poly(itaconic acid/bis acrylamide)	cc-PSA	0.02 ng/dL	0.05–50 ng/dL	Spiked and real human serum	96.2–103.4%	[100]
Bone	CTx-I	MAA/EGDMA	EIS	0.09 ng/mL	0.1–2.5 ng/mL	Real sheep serum	N/A	[101]
Alzheimer	Aβ-42	Polyaniline	SWV	0.4 pg/mL	1.0–66.0 ng/mL	Spiked human serum	88–93%	[102]
		MAA/DAU/DVB/ AEDV	DPV	1.22 pg/mL	5–10 pg/mL	Spiked human serum	93.0–107.7%	[104]
		Polypyrol	CV	1.2 pg/mL	0.003–7 ng/mL	Tris buffer and artificial cerebrospinal fluid	N/A	[103]
Kidney	Cys-C	Poly carboxylated Pyrol	CV and DPV	0.5 ng/mL	0.5–20 ng/mL	Spiked Cormay® HN serum	87.8–98.6%	[105]
Cardiac	cTnI	MAA/AAM/NIPAAm	SWV and EIS	3.2 pg/mL	0.01–1000 ng/mL	Spiked human serum	98.0–101.9%	[107]
		MAA/AAM/NIPAAm	Colorimetric	7.4 pg/mL	0.01–1000 ng/mL	Spiked human serum	97.0–100.1%	[107]
		Polydopamine	ECL	0.0184 ng/mL	0.05–10,000 ng/mL	Spiked human serum	99.8–104.7%	[108]
	Mb	Polydopamine	ECL	0.0492 ng/mL	0.05–10,000 ng/mL	Spiked human serum	93.3–102.3%	[108]

including superior selectivity, rapid response times, and cost-effectiveness. Unlike conventional methods, MIP-based sensors usually do not need complex sample preparation steps, and the biological fluids can be directly applied to them, making them more suitable for point-of-care applications and high-throughput analysis. Additionally, the inherent stability of MIPs allows for prolonged storage in normal conditions. Therefore, they can be used without significant performance loss, ensuring long-term reliability in diagnostic settings. Conversely, conventional methods often suffer from cross-reactivity issues and require elaborate sample handling, increasing analysis time and costs.

Looking ahead, further advancements in this field are anticipated. Current studies have demonstrated the applicability of MIP-EC sensors in this manner. It has been shown that they can quantify biomarkers at such low concentrations, and due to the incorporation of MIPs into the sensory system, they are highly selective toward their respective template molecules. These proposed devices and methods have been primarily utilized on spiked samples, and a lack of comparison to the conventional gold standard methods on real patients is evident. At the next level, large-scale clinical trials should be conducted to ensure their efficacy as point-of-care diagnostic devices.

Overall, integrating MIPs into electrochemical sensors represents a significant leap forward in biomarker detection and quantification, holding immense promise for the advancement of personalized medicine and disease management. As research in this area continues to unfold, the synergy between MIP technology and electrochemical sensing is expected to redefine the landscape of clinical diagnostics, fostering a new era of improved patient outcomes.

References

1. Jain KK (2010) Role of biomarkers in health care. In: Handbook of biomarkers. Humana Press, Totowa, NJ, pp 115–188. https://doi.org/10.1007/978-1-60761-685-6_5
2. WHO (2023) WHO International Classification of Diseases (ICD)
3. Fung KW, Xu J, McConnell-Lamptey S, Pickett D, Bodenreider O (2023) A practical strategy to use the ICD-11 for morbidity coding in the United States without a clinical modification. Journal of the American Medical Informatics Association 30(10):1614–1621. https://doi.org/10.1093/jamia/ocad128
4. Aronson JK, Ferner RE (2017) Biomarkers—a general review. Curr Protoc Pharmacol 2017:9.23.1–9.23.17. https://doi.org/10.1002/cpph.19
5. FitzGerald GA (2016) Measure for measure: biomarker standards and transparency. Sci Transl Med 8. https://doi.org/10.1126/scitranslmed.aaf8590
6. Elugoke SE, Adekunle AS, Fayemi OE, Akpan ED, Mamba BB, Sherif EM, Ebenso EE (2021) Molecularly imprinted polymers (MIPs) based electrochemical sensors for the determination of catecholamine neurotransmitters—review. Electrochem Sci Adv 1. https://doi.org/10.1002/elsa.202000026
7. Mostafa AM, Barton SJ, Wren SP, Barker J (2021) Review on molecularly imprinted polymers with a focus on their application to the analysis of protein biomarkers. TrAC Trends Anal Chem 144:116431. https://doi.org/10.1016/j.trac.2021.116431

8. Pan J, Chen W, Ma Y, Pan G (2018) Molecularly imprinted polymers as receptor mimics for selective cell recognition. Chem Soc Rev 47:5574–5587. https://doi.org/10.1039/C7CS00854F
9. Lee JW, Figeys D, Vasilescu J (2006) Biomarker assay translation from discovery to clinical studies in cancer drug development: quantification of emerging protein biomarkers. Adv Cancer Res 96:269–298. https://doi.org/10.1016/S0065-230X(06)96010-2
10. Chambers AG, Percy AJ, Simon R, Borchers CH (2014) MRM for the verification of cancer biomarker proteins: recent applications to human plasma and serum. Expert Rev Proteomics 11:137–148. https://doi.org/10.1586/14789450.2014.877346
11. Boyle DL, Rosengren S, Bugbee W, Kavanaugh A, Firestein GS (2003) Quantitative biomarker analysis of synovial gene expression by real-time PCR. Arthritis Res Ther 5:R352–R352. https://doi.org/10.1186/ar1004
12. Clark-Langone KM, Wu JY, Sangli C, Chen A, Snable JL, Nguyen A, Hackett JR, Baker J, Yothers G, Kim C, Cronin MT (2007) Biomarker discovery for colon cancer using a 761 gene RT-PCR assay. BMC Genomics 8:279. https://doi.org/10.1186/1471-2164-8-279
13. Hardikar AA, Farr RJ, Joglekar MV (2014) Circulating microRNAs: understanding the limits for quantitative measurement by real-time PCR. J Am Heart Assoc 3. https://doi.org/10.1161/JAHA.113.000792
14. Baron EJ (2015) Specimen collection, transport, and processing. Bacteriology. https://doi.org/10.1128/9781555817381.ch18
15. David DKO, Jacobs S, DeMott WR (2001) Jacobs & DeMott laboratory test handbook. Lexi-Comp
16. Alan W (2006) Tietz clinical guide to laboratory tests. Elsevier Health Sciences. https://books.google.com/books?id=aQqjBQAAQBAJ
17. Alexander C, Andersson HS, Andersson LI, Ansell RJ, Kirsch N, Nicholls IA, O'Mahony J, Whitcombe MJ (2006) Molecular imprinting science and technology: a survey of the literature for the years up to and including 2003. J Mol Recognit 19:106–180. https://doi.org/10.1002/jmr.760
18. Yao J, Li X, Qin W (2008) Computational design and synthesis of molecular imprinted polymers with high selectivity for removal of aniline from contaminated water. Anal Chim Acta 610:282–288. https://doi.org/10.1016/j.aca.2008.01.042
19. Menger M, Yarman A, Erdőssy J, Yildiz H, Gyurcsányi R, Scheller F (2016) MIPs and aptamers for recognition of proteins in biomimetic sensing. Biosensors 6:35. https://doi.org/10.3390/bios6030035
20. Li S, Cao S, Whitcombe MJ, Piletsky SA (2014) Size matters: challenges in imprinting macromolecules. Prog Polym Sci 39:145–163. https://doi.org/10.1016/j.progpolymsci.2013.10.002
21. Byrne ME, Park K, Peppas NA (2002) Biomimetic networks for selective recognition of biomolecules. MRS Proc 724:N9.3. https://doi.org/10.1557/PROC-724-N9.3
22. Zhu Q, Li X, Xiao Y, Xiong Y, Wang S, Xu C, Zhang J, Wu X (2017) Synthesis of molecularly imprinted polymer via visible light activated RAFT polymerization in aqueous media at room temperature for highly selective electrochemical assay of glucose. Macromol Chem Phys 218. https://doi.org/10.1002/macp.201700141
23. Sajini T, Mathew B (2021) A brief overview of molecularly imprinted polymers: highlighting computational design, nano and photo-responsive imprinting. Talanta Open 4:100072. https://doi.org/10.1016/j.talo.2021.100072
24. Boukadida M, Anene A, Jaoued-Grayaa N, Chevalier Y, Hbaieb S (2022) Choice of the functional monomer of molecularly imprinted polymers: does it rely on strong acid-base or hydrogen bonding interactions? Colloid Interface Sci Commun 50:100669. https://doi.org/10.1016/j.colcom.2022.100669
25. Thakur N, Das TR, Gupta S, Shukla SK, Patra S, Kumar D (2023) Material selection for the fabrication of MIP-based devices: an overview. In: Molecularly imprinted polymer. Elsevier, pp 101–114. https://doi.org/10.1016/B978-0-323-91925-8.00006-5

26. Sanglar C, Jansen T, Silaghi MC, Mernier J, Mignon P, Chermette H (2012) Study of Prepolymerization complex formation in the synthesis of steroid-based molecularly imprinted polymers. Anal Chem 84:4481–4488. https://doi.org/10.1021/ac3003159
27. Orowitz TE, Sombo PPAAA, Rahayu D, Hasanah AN (2020) Microsphere polymers in molecular imprinting: current and future perspectives. Molecules 25. https://doi.org/10.3390/molecules25143256
28. Ali MM, Zhu S, Amin FR, Hussain D, Du Z, Hu L (2022) Molecular imprinting of glycoproteins: from preparation to cancer theranostics. Theranostics 12:2406–2426. https://doi.org/10.7150/thno.69189
29. Chen H, Guo J, Wang Y, Dong W, Zhao Y, Sun L (2022) Bio-inspired imprinting materials for biomedical applications. Adv Sci 9. https://doi.org/10.1002/advs.202202038
30. Zhao G, Zhang Y, Sun D, Yan S, Wen Y, Wang Y, Li G, Liu H, Li J, Song Z (2023) Recent advances in molecularly imprinted polymers for antibiotic analysis. Molecules 28:335. https://doi.org/10.3390/molecules28010335
31. Liu Y, Lian Z, Li F, Majid A, Wang J (2021) Review on molecular imprinting technology and its application in pre-treatment and detection of marine organic pollutants. Mar Pollut Bull 169:112541. https://doi.org/10.1016/j.marpolbul.2021.112541
32. Dixit CK, Bhakta S, Reza KK, Kaushik A (2022) Exploring molecularly imprinted polymers as artificial antibodies for efficient diagnostics and commercialization: a critical overview. Hybrid Adv 1:100001. https://doi.org/10.1016/j.hybadv.2022.100001
33. Parisi OI, Francomano F, Dattilo M, Patitucci F, Prete S, Amone F, Puoci F (2022) The evolution of molecular recognition: from antibodies to molecularly imprinted polymers (MIPs) as artificial counterpart. J Funct Biomater 13:12. https://doi.org/10.3390/jfb13010012
34. Takeuchi T, Sunayama H (2018) Beyond natural antibodies—a new generation of synthetic antibodies created by post-imprinting modification of molecularly imprinted polymers. Chem Commun 54:6243–6251. https://doi.org/10.1039/C8CC02923G
35. Xu J, Miao H, Wang J, Pan G (2020) Molecularly imprinted synthetic antibodies: from chemical design to biomedical applications. Small 16:e1906644. https://doi.org/10.1002/smll.201906644
36. Vaneckova T, Bezdekova J, Han G, Adam V, Vaculovicova M (2020) Application of molecularly imprinted polymers as artificial receptors for imaging. Acta Biomater 101:444–458. https://doi.org/10.1016/j.actbio.2019.11.007
37. BelBruno JJ (2019) Molecularly imprinted polymers. Chem Rev 119:94–119. https://doi.org/10.1021/acs.chemrev.8b00171
38. Power AC, Morri A (2013) Electroanalytical sensor technology. In: Electrochemistry. InTech. https://doi.org/10.5772/51480
39. Kalcher K (2017) Renaissances and current trends with electrochemical sensors and biosensors. Int J Biosens Bioelectron 3. https://doi.org/10.15406/ijbsbe.2017.03.00083
40. Murthy HCA, Gebretsadik A, Haftu M, Nemera T, Bhattacharya T, Ravikumar CR (2023) Polymer composites for electrochemical sensor applications. In: Polymeric nanocomposite materials for sensor applications. Elsevier, pp 405–436. https://doi.org/10.1016/B978-0-323-98830-8.00016-3
41. Roy E, Patra S, Tiwari A, Madhuri R, Sharma PK (2017) Introduction of selectivity and specificity to graphene using an inimitable combination of molecular imprinting and nanotechnology. Biosens Bioelectron 89:234–248. https://doi.org/10.1016/j.bios.2016.02.056
42. Ivaska A (2008) Electrochemical detection. In: Comprehensive analytical chemistry. Elsevier, pp 441–459. https://doi.org/10.1016/S0166-526X(08)00616-8
43. Majeed S, Naqvi STR, ul Haq MN, Ashiq MN (2022) Electroanalytical techniques in biosciences: conductometry, coulometry, voltammetry, and electrochemical sensors. In: Analytical techniques in biosciences. Elsevier, pp 157–178. https://doi.org/10.1016/B978-0-12-822654-4.00004-X
44. Li B, Xie X, Meng T, Guo X, Li Q, Yang Y, Jin H, Jin C, Meng X, Pang H (2024) Recent advance of nanomaterials modified electrochemical sensors in the detection of heavy metal ions in food and water. Food Chem 440:138213. https://doi.org/10.1016/j.foodchem.2023.138213

45. Govindaraj M, Srivastava A, Muthukumaran MK, Tsai P-C, Lin Y-C, Raja BK, Rajendran J, Ponnusamy VK, Arockia Selvi J (2023) Current advancements and prospects of enzymatic and non-enzymatic electrochemical glucose sensors. Int J Biol Macromol 253:126680. https://doi.org/10.1016/j.ijbiomac.2023.126680
46. Jarnda KV, Wang D, Qurrat-Ul-Ain R, Anaman VE, Johnson GP, Roberts PS, Johnson BW, Jallawide T, Kai PD (2023) Recent advances in electrochemical non-enzymatic glucose sensor for the detection of glucose in tears and saliva: a review. Sens Actuators A Phys 363:114778. https://doi.org/10.1016/j.sna.2023.114778
47. Herrera-Domínguez M, Morales-Luna G, Mahlknecht J, Cheng Q, Aguilar-Hernández I, Ornelas-Soto N (2023) Optical biosensors and their applications for the detection of water pollutants. Biosensors 13:370. https://doi.org/10.3390/bios13030370
48. Ebralidze II, Laschuk NO, Poisson J, Zenkina OV (2019) Colorimetric sensors and sensor arrays. In: Nanomaterials design for sensing applications. Elsevier, pp 1–39. https://doi.org/10.1016/B978-0-12-814505-0.00001-1
49. Liu B, Zhuang J, Wei G (2020) Recent advances in the design of colorimetric sensors for environmental monitoring. Environ Sci Nano 7:2195–2213. https://doi.org/10.1039/D0EN00449A
50. Mujahid A, Dickert FL (2012) Molecularly imprinted polymers for sensors. In: Molecularly imprinted sensors. Elsevier, pp 125–159. https://doi.org/10.1016/B978-0-444-56331-6.00006-2
51. Irshad M, Iqbal N, Mujahid A, Afzal A, Hussain T, Sharif A, Ahmad E, Athar MM (2013) Molecularly imprinted nanomaterials for sensor applications. Nanomaterials 3:615–637. https://doi.org/10.3390/nano3040615
52. Dickert FL, Hayden O (2001) Non-covalent molecularly imprinted sensors for vapours, polyaromatic hydrocarbons and complex mixtures. In: Techniques and instrumentation in analytical chemistry. Elsevier, pp 503–525. https://doi.org/10.1016/S0167-9244(01)80024-X
53. Mohan JM, Amreen K, Javed A, Dubey SK, Goel S (2022) Emerging trends in miniaturized and microfluidic electrochemical sensing platforms. Curr Opin Electrochem 33:100930. https://doi.org/10.1016/j.coelec.2021.100930
54. Nieuwenhuizen MS, Venema A (1991) Mass-sensitive devices. In: Sensors. Wiley, pp 647–680. https://doi.org/10.1002/9783527620135.ch13
55. Cui B, Liu P, Liu X, Liu S, Zhang Z (2020) Molecularly imprinted polymers for electrochemical detection and analysis: progress and perspectives. J Mater Res Technol 9:12568–12584. https://doi.org/10.1016/j.jmrt.2020.08.052
56. Li Y, Guan C, Liu C, Li Z, Han G (2023) Disease diagnosis and application analysis of molecularly imprinted polymers (MIPs) in saliva detection. Talanta 269:125394. https://doi.org/10.1016/j.talanta.2023.125394
57. Lamaoui A, Lahcen AA, Amine A (2023) Unlocking the potential of molecularly imprinted polydopamine in sensing applications. Polymers (Basel) 15:3712. https://doi.org/10.3390/polym15183712
58. Yang B, Fu C, Li J, Xu G (2018) Frontiers in highly sensitive molecularly imprinted electrochemical sensors: challenges and strategies. TrAC Trends Anal Chem 105:52–67. https://doi.org/10.1016/j.trac.2018.04.011
59. Rebelo P, Costa-Rama E, Seguro I, Pacheco JG, Nouws HPA, Cordeiro MNDS, Delerue-Matos C (2021) Molecularly imprinted polymer-based electrochemical sensors for environmental analysis. Biosens Bioelectron 172:112719. https://doi.org/10.1016/j.bios.2020.112719
60. Radi A-E, Wahdan T, El-Basiony A (2019) Electrochemical sensors based on molecularly imprinted polymers for pharmaceuticals analysis. Curr Anal Chem 15:219–239. https://doi.org/10.2174/1573411014666180501100131
61. Wang L, Pagett M, Zhang W (2023) Molecularly imprinted polymer (MIP) based electrochemical sensors and their recent advances in health applications. Sens Actuators Rep 5:100153. https://doi.org/10.1016/j.snr.2023.100153
62. Helwig NE, Hong S, Hsiao-Wecksler ET (n.d.) What is cancer?. https://www.cancer.gov/about-cancer/understanding/what-is-cancer

63. WHO (2020) Global health estimates 2020: deaths by cause, age, sex, by country and by region, 2000–2019
64. Sung H, Ferlay J, Siegel RL, Laversanne M, Soerjomataram I, Jemal A, Bray F (2021) Global Cancer Statistics 2020: GLOBOCAN estimates of incidence and mortality worldwide for 36 cancers in 185 countries. CA Cancer J Clin 71:209–249. https://doi.org/10.3322/caac.21660
65. Yazdani Z, Yadegari H, Heli H (2019) A molecularly imprinted electrochemical nanobiosensor for prostate specific antigen determination. Anal Biochem 566:116–125. https://doi.org/10.1016/j.ab.2018.11.020
66. Ma Y, Shen XL, Zeng Q, Wang LS (2017) A glassy carbon electrode modified with graphene nanoplatelets, gold nanoparticles and chitosan, and coated with a molecularly imprinted polymer for highly sensitive determination of prostate specific antigen. Microchim Acta 184:4469–4476. https://doi.org/10.1007/s00604-017-2458-y
67. Khumngern S, Thavarungkul P, Kanatharana P, Bejrananda T, Numnuam A (2022) Molecularly imprinted electrochemical sensor based on poly(o-phenylenediamine-co-o-aminophenol) incorporated with poly(styrenesulfonate) doped poly(3,4-ethylenedioxythiophene) ferrocene composite modified screen-printed carbon electrode for highly sensiti. Microchem J 177:107311. https://doi.org/10.1016/j.microc.2022.107311
68. Mazouz Z, Mokni M, Fourati N, Zerrouki C, Barbault F, Seydou M, Kalfat R, Yaakoubi N, Omezzine A, Bouslema A, Othmane A (2020) Computational approach and electrochemical measurements for protein detection with MIP-based sensor. Biosens Bioelectron 151:111978. https://doi.org/10.1016/j.bios.2019.111978
69. Gu Z, Zhao M, Zhang W, Jiang T, Sun M (2017) Preparation of carbon nanotube/MnO2 nanocomposite as an electrode modifier for prostate-specific antigen (PSA) determination. Int J Electrochem Sci 12:10726–10736. https://doi.org/10.20964/2017.11.05
70. Wang Y, Kan X (2021) Sensitive and selective "signal-off" electrochemiluminescence sensing of prostate-specific antigen based on an aptamer and molecularly imprinted polymer. Analyst 146:7693–7701. https://doi.org/10.1039/d1an01645h
71. Lai Y, Deng Y, Yang G, Li S, Zhang C, Liu X (2018) Molecular imprinting polymers electrochemical sensor based on AuNPs/PTh modified GCE for highly sensitive detection of carcinomaembryonic antigen. J Biomed Nanotechnol 14:1688–1694. https://doi.org/10.1166/jbn.2018.2617
72. Qi J, Li B, Zhou N, Wang X, Deng D, Luo L, Chen L (2019) The strategy of antibody-free biomarker analysis by in-situ synthesized molecularly imprinted polymers on movable valve paper-based device. Biosens Bioelectron 142:111533. https://doi.org/10.1016/j.bios.2019.111533
73. Carneiro LPT, Pinto AMFR, Mendes A, Goreti M (2022) An all-in-one approach for self-powered sensing: a methanol fuel cell modified with a molecularly imprinted polymer for cancer biomarker detection. J Electroanal Chem 906:116009. https://doi.org/10.1016/j.jelechem.2022.116009
74. Carvalho M, Gomes RM, Moreira Rocha S, Barroca-Ferreira J, Maia CJ, Guillade L, Correa-Duarte MA, Passarinha LA, Moreira FTC (2023) Development of a novel electrochemical biosensor based on plastic antibodies for detection of STEAP1 biomarker in cancer. Bioelectrochemistry 152:108461. https://doi.org/10.1016/j.bioelechem.2023.108461
75. Rebelo TSCR, Costa R, Brandão ATSC, Silva AF, Sales MGF, Pereira CM (2019) Molecularly imprinted polymer SPE sensor for analysis of CA-125 on serum. Anal Chim Acta 1082:126–135. https://doi.org/10.1016/j.aca.2019.07.050
76. Keshaviah A, Dellapasqua S, Rotmensz N, Lindtner J, Crivellari D, Collins J, Colleoni M, Thürlimann B, Mendiola C, Aebi S, Price KN, Pagani O, Simoncini E, Castiglione Gertsch M, Gelber RD, Coates AS, Goldhirsch A (2007) CA15-3 and alkaline phosphatase as predictors for breast cancer recurrence: a combined analysis of seven International Breast Cancer Study Group trials. Ann Oncol 18:701–708. https://doi.org/10.1093/annonc/mdl492
77. Pacheco JG, Silva MSV, Freitas M, Nouws HPA, Delerue-Matos C (2018) Molecularly imprinted electrochemical sensor for the point-of-care detection of a breast cancer biomarker (CA 15-3). Sens Actuators B Chem 256:905–912. https://doi.org/10.1016/j.snb.2017.10.027

78. Ribeiro JA, Pereira CM, Silva AF, Sales MGF (2018) Disposable electrochemical detection of breast cancer tumour marker CA 15-3 using poly(toluidine blue) as imprinted polymer receptor. Biosens Bioelectron 109:246–254. https://doi.org/10.1016/j.bios.2018.03.011
79. Lee MH, Thomas JL, Liao CL, Jurcevic S, Crnogorac-Jurcevic T, Lin HY (2018) Epitope recognition of peptide-imprinted polymers for Regenerating protein 1 (REG1). Sep Purif Technol 192:213–219. https://doi.org/10.1016/j.seppur.2017.09.071
80. Scheller FW, Zhang X, Yarman A, Wollenberger U, Gyurcsányi RE (2019) Molecularly imprinted polymer-based electrochemical sensors for biopolymers. Curr Opin Electrochem 14:53–59. https://doi.org/10.1016/j.coelec.2018.12.005
81. Bräuer B, Unger C, Werner M, Lieberzeit PA (2021) Biomimetic sensors to detect bioanalytes in real-life samples using molecularly imprinted polymers: a review. Sensors 21:5550. https://doi.org/10.3390/s21165550
82. Hayden O, Lieberzeit PA, Blaas D, Dickert FL (2006) Artificial antibodies for bioanalyte detection—sensing viruses and proteins. Adv Funct Mater 16:1269–1278. https://doi.org/10.1002/adfm.200500626
83. Ayankojo AG, Boroznjak R, Reut J, Öpik A, Syritski V (2022) Molecularly imprinted polymer based electrochemical sensor for quantitative detection of SARS-CoV-2 spike protein. Sens Actuators B Chem 353:131160. https://doi.org/10.1016/j.snb.2021.131160
84. Bognár Z, Kozma J, Kovács N, Gyurcsányi RE (2023) Novel functional monomer for the electrochemical synthesis of highly affine epitope-imprinted polymers. Electroanalysis 35. https://doi.org/10.1002/elan.202300025
85. Raziq A, Kidakova A, Boroznjak R, Reut J, Öpik A, Syritski V (2021) Development of a portable MIP-based electrochemical sensor for detection of SARS-CoV-2 antigen. Biosens Bioelectron 178:113029. https://doi.org/10.1016/j.bios.2021.113029
86. Zhang T, Sun L, Zhang Y (2021) Highly sensitive electrochemical determination of the SARS-COV-2 antigen based on a gold/graphene imprinted poly-arginine sensor. Anal Methods 13:5772–5776. https://doi.org/10.1039/d1ay01478a
87. Kumar N, Shetti NP, Jagannath S, Aminabhavi TM (2022) Electrochemical sensors for the detection of SARS-CoV-2 virus. Chem Eng J 430:132966. https://doi.org/10.1016/j.cej.2021.132966
88. El Sharif HF, Dennison SR, Tully M, Crossley S, Mwangi W, Bailey D, Graham SP, Reddy SM (2022) Evaluation of electropolymerized molecularly imprinted polymers (E-MIPs) on disposable electrodes for detection of SARS-CoV-2 in saliva. Anal Chim Acta 1206:339777. https://doi.org/10.1016/j.aca.2022.339777
89. Liu KH, Lin HY, Thomas JL, Chen CY, Chen YT, Chen CY, Yang CH, Lee MH (2022) Sensing of C-reactive protein using an extended-gate field-effect transistor with a tungsten disulfide-doped peptide-imprinted conductive polymer coating. Biosensors 12. https://doi.org/10.3390/bios12010031
90. Ghanbari K, Roushani M (2018) A nanohybrid probe based on double recognition of an aptamer MIP grafted onto a MWCNTs-Chit nanocomposite for sensing hepatitis C virus core antigen. Sens Actuators B Chem 258:1066–1071. https://doi.org/10.1016/j.snb.2017.11.145
91. Siqueira Silva M, Moreira Tavares AP, Leomil Coelho LF, Morganti Ferreira Dias LE, Chura-Chambi RM, Guimarães da Fonseca F, Ferreira Sales MG, Costa Figueiredo E (2021) Rational selection of hidden epitopes for a molecularly imprinted electrochemical sensor in the recognition of heat-denatured dengue NS1 protein. Biosens Bioelectron 191:113419. https://doi.org/10.1016/j.bios.2021.113419
92. Saxena K, Chauhan N, Malhotra BD, Jain U (2023) A molecularly imprinted polymer-based electrochemical biosensor for detection of VacA virulence factor of H. pylori causing gastric cancer. Process Biochem 130:87–95. https://doi.org/10.1016/j.procbio.2023.03.024
93. Saxena K, Murti BT, Yang PK, Malhotra BD, Chauhan N, Jain U (2022) Fabrication of a molecularly imprinted nano-interface-based electrochemical biosensor for the detection of CagA virulence factors of H. pylori. Biosensors 12. https://doi.org/10.3390/bios12121066

94. Garza-González E (2014) A review of Helicobacter pylori diagnosis, treatment, and methods to detect eradication. World J Gastroenterol 20:1438. https://doi.org/10.3748/wjg.v20.i6.1438
95. Kao C-Y, Sheu B-S, Wu J-J (2016) Helicobacter pylori infection: an overview of bacterial virulence factors and pathogenesis. Biomed J 39:14–23. https://doi.org/10.1016/j.bj.2015.06.002
96. Jang J, Hur HG, Sadowsky MJ, Byappanahalli MN, Yan T, Ishii S (2017) Environmental Escherichia coli: ecology and public health implications—a review. J Appl Microbiol 123:570–581. https://doi.org/10.1111/jam.13468
97. Allocati N, Masulli M, Alexeyev M, Di Ilio C (2013) Escherichia coli in Europe: an overview. Int J Environ Res Public Health 10:6235–6254. https://doi.org/10.3390/ijerph10126235
98. Yasmeen N, Etienne M, Sharma PS, El-Kirat-Chatel S, Helú MB, Kutner W (2021) Molecularly imprinted polymer as a synthetic receptor mimic for capacitive impedimetric selective recognition of Escherichia coli K-12. Anal Chim Acta 1188:339177. https://doi.org/10.1016/j.aca.2021.339177
99. Moura SL, Fajardo LM, dos Anjos Cunha L, Sotomayor MDPT, Machado FBC, Ferrão LFA, Pividori MI (2018) Theoretical and experimental study for the biomimetic recognition of levothyroxine hormone on magnetic molecularly imprinted polymer. Biosens Bioelectron 107:203–210. https://doi.org/10.1016/j.bios.2018.01.028
100. Karami P, Gholamin D, Johari-Ahar M (2023) Electrochemical immunoassay for one-pot detection of thyroxin (T4) and thyroid-stimulating hormone (TSH) using magnetic and Janus nanoparticles. Anal Bioanal Chem 415:4741. https://doi.org/10.1007/s00216-023-04767-8
101. Afsarimanesh N, Mukhopadhyay SC, Kruger M (2018) Molecularly imprinted polymer-based electrochemical biosensor for bone loss detection. IEEE Trans Biomed Eng 65:1264–1271. https://doi.org/10.1109/TBME.2017.2744667
102. Moreira FTC, Rodriguez BAG, Dutra RAF, Sales MGF (2018) Redox probe-free readings of a B-amyloid-42 plastic antibody sensory material assembled on copper@carbon nanotubes. Sens Actuators B Chem 264:1–9. https://doi.org/10.1016/j.snb.2018.02.166
103. Vais RD, Yadegari H, Heli H, Sattarahmady N (2021) A β-amyloid(1–42) biosensor based on molecularly imprinted poly-pyrrole for early diagnosis of alzheimer's disease. J Biomed Phys Eng 11:215–228. https://doi.org/10.31661/jbpe.v0i0.1070
104. You M, Yang S, An Y, Zhang F, He P (2020) A novel electrochemical biosensor with molecularly imprinted polymers and aptamer-based sandwich assay for determining amyloid-β oligomer. J Electroanal Chem 862:114017. https://doi.org/10.1016/j.jelechem.2020.114017
105. Gomes RS, Gomez-Rodríguez BA, Fernandes R, Sales MGF, Moreira FTC, Dutra RF (2021) Plastic antibody of polypyrrole/multiwall carbon nanotubes on screen-printed electrodes for cystatin C detection. Biosensors 11. https://doi.org/10.3390/bios11060175
106. Aydin S, Ugur K, Aydin S, Sahin İ, Yardim M (2019) Biomarkers in acute myocardial infarction: current perspectives. Vasc Health Risk Manag 15:1–10. https://doi.org/10.2147/VHRM.S166157
107. Zhang G, Zhang L, Yu Y, Lin B, Wang Y, Guo M, Cao Y (2020) Dual-mode of electrochemical-colorimetric imprinted sensing strategy based on self-sacrifice beacon for diversified determination of cardiac troponin I in serum. Biosens Bioelectron 167:112502. https://doi.org/10.1016/j.bios.2020.112502
108. He S, Zhang P, Sun J, Ji Y, Huang C, Jia N (2022) Integrating potential-resolved electrochemiluminescence with molecularly imprinting immunoassay for simultaneous detection of dual acute myocardial infarction markers. Biosens Bioelectron 201:113962. https://doi.org/10.1016/j.bios.2022.113962
109. Carneiro LPT, Ferreira NS, Tavares APM, Pinto AMFR, Mendes A, Sales MGF (2021) A passive direct methanol fuel cell as transducer of an electrochemical sensor, applied to the detection of carcinoembryonic antigen. Biosens Bioelectron 175:112877. https://doi.org/10.1016/j.bios.2020.112877

110. Carneiro MCCG, Sousa-Castillo A, Correa-Duarte MA, Sales MGF (2019) Dual biorecognition by combining molecularly-imprinted polymer and antibody in SERS detection. Application to carcinoembryonic antigen. Biosens Bioelectron 146:111761. https://doi.org/10.1016/j.bios.2019.111761
111. Moreira FTC, Sales MGF (2019) Autonomous biosensing device merged with photovoltaic technology for cancer biomarker detection. J Electroanal Chem 855:113611. https://doi.org/10.1016/j.jelechem.2019.113611

Harnessing Molecularly Imprinted Polymers as Artificial Antibodies in Electrochemical Sensors for Disease Detection and Monitoring

7

Tamara Tasić, Vedran Milanković, Igor A. Pašti, and Tamara Lazarević-Pašti

Abstract

This chapter explores the integration of Molecularly Imprinted Polymers (MIPs) into electrochemical sensors, focusing on their applications as artificial antibodies in disease detection and monitoring. After discussing the basic principles of electrochemical sensors, we delve into the versatility of MIPs, which allow for the selective recognition of disease-related molecules. Their stability, cost-effectiveness, and non-immunogenicity make them ideal candidates for use in diagnostic technologies. The chapter showcases the synergy between MIPs and electrochemical transduction, converting target-binding events into measurable electrical signals. We highlight case studies where MIP-based electrochemical sensors have demonstrated enhanced sensitivity and specificity in detecting cancer biomarkers, autoimmune disease markers, and infectious agents. These examples underscore the potential of MIP-based sensors in revolutionizing disease management. The discussion covers challenges and future perspectives in the field, addressing the optimization of sensor performance and achieving the desired level of specificity. Additionally, we explore the potential of MIP-based electrochemical sensors as theranostic tools, combining diagnostics with targeted drug delivery and offering a glimpse into personalized medicine. In conclusion, MIP-based electrochemical sensors present a promising frontier in disease detection and monitoring, showcasing the innovative application of artificial antibodies. As technology advances, these biosensing platforms hold great

T. Tasić · V. Milanković · T. Lazarević-Pašti (✉)
VINČA Institute of Nuclear Sciences—National Institute of the Republic of Serbia, University of Belgrade, Belgrade, Serbia

I. A. Pašti
Faculty of Physical Chemistry, University of Belgrade, Belgrade, Serbia

S. Patra et al. (eds.), *Molecularly Imprinted Polymers: Path to Artificial Antibodies*, https://doi.org/10.1007/978-981-97-4379-7_7

potential to transform healthcare, enabling precise and accessible diagnostics for improved patient outcomes.

Keywords

Cancer biomarkers · Glucose detection · Autoimmune disease markers · Infectious agents monitoring · Neurological diseases biomarkers · Theranostic tools · Personalized medicine · Biosensing platforms · Healthcare transformation

7.1 Introduction

The relentless pursuit of innovative technologies has catalyzed groundbreaking advancements in sensor technology, focusing on disease detection and monitoring, capturing significant scientific attention. In the past few years, the requirement for sensitive and selective sensors has grown exponentially, driving efforts to revolutionize healthcare practices and improve patient outcomes. Among the remarkable discoveries, Molecularly Imprinted Polymers (MIPs) have emerged as a powerful class of synthetic materials with artificial antibody-like capabilities, captivating researchers and practitioners in the realm of electrochemical sensing [1–3].

MIPs are smart materials engineered to enhance pre-concentration and selectivity when extracting analytes from intricate matrices [4]. Responsive to external cues such as temperature, pH, biomolecules, and magnetic fields, these smart polymers enable advanced applications [5]. Termed stimuli-responsive or smart polymers, they mimic biological systems and can alter shape, size, and mechanical, optical, or electrical properties in response to signals, finding diverse applications in drug delivery, diagnostics, biotechnology, sensors, and more [3, 6]. MIPs are increasingly used in sensors, but conventional bulk polymerization methods have limitations. Bulk MIPs often lose binding sites, leading to slow rebinding or permanent trapping of the template. Nanotechnology has been integrated with molecular imprinting to overcome these challenges, resulting in nano MIPs. These nanomaterial-incorporated MIPs show improved rebinding, lower mass transfer resistance, and enhanced template accessibility. They offer extended lifetimes, storage stability, and cost-effectiveness compared to natural antibodies [7]. When we talk about bulk MIPs, it can be seen that the synthesis method is straightforward, resulting in irregular morphology, limited surface area, low adsorption capacity, heterogeneous distribution of binding sites, higher template leakage, and lower yield. Unlike bulk MIPs, nano MIPs have complex synthesis methods, uniform morphology, larger surface area, higher adsorption capacity, homogeneous binding sites distribution, less template leakage, and higher yield [8].

As mentioned, imprinting technology has gained a strong position in materials science, presenting challenges and opportunities from artificial receptors to specific catalysts. For example, graphene's unique properties enhance imprinted polymers, demonstrated in its use as an electrochemical sensor receptor layer and sorbent for various analytes. Various nanomaterial modifications consistently improve graphene's characteristics. For instance, introducing magnetic nanoparticles attracts non-magnetic layers toward external magnetic fields, leading to applications in separation and pre-concentration. The combination of MIP and graphene-based nanomaterials holds great promise for biomedical applications, offering a platform for researchers in need of sensitive and selective nano-sized materials [7].

The challenges in obtaining natural antibodies, ranging from ethical considerations to batch-to-batch variations and the risk of immunogenicity, have spurred the search for alternative recognition elements. Artificial antibodies have emerged as an attractive solution, bridging the gap between the limited supply of natural antibodies and the escalating demand for targeted, reproducible, and non-immunogenic recognition elements. The unyielding quest for precise and reliable disease diagnostics has motivated researchers to explore alternatives that emulate the exquisite specificity of natural antibodies. MIPs, also known as "plastic antibodies," have gained significant attention for their exceptional potential to bind to target molecules selectively, offering an innovative and versatile approach to disease detection and monitoring [9, 10]. By harnessing molecular imprinting principles, MIPs are meticulously crafted to possess tailored binding cavities that exclusively recognize and bind to specific analytes with precision akin to natural antibodies [11, 12].

Electrochemical sensors have emerged as stalwarts in analytical chemistry, revolutionizing how we detect and quantify various analytes [13]. Leveraging the redox reactions occurring at the electrode-electrolyte interface, electrochemical sensors offer numerous advantages, including high sensitivity, real-time detection, and the potential for miniaturization. The triad of working, reference, and counter electrodes, coupled with various electrochemical techniques, such as potentiometry, amperometry, and voltammetry, amplifies the versatility of these sensors. Integrating electrochemical sensors with innovative recognition elements, such as MIPs, augments their functionality, propelling them to the forefront of disease diagnostics [14, 15].

As with any emerging technology, MIP-based electrochemical sensors encounter challenges in their journey toward widespread implementation. Among the key hurdles are optimizing sensor performance to achieve the desired sensitivity and selectivity, addressing potential interference from complex biological matrices, and scaling up production processes [16]. However, the immense opportunities and ongoing research in nanotechnology, polymer science, and electrochemical transducers promise to overcome these obstacles, positioning MIP-based electrochemical sensors as powerful tools in the diagnostic arsenal.

The exceptional selectivity of MIPs as artificial antibodies and the sensitivity of electrochemical sensors offer a potent and transformative approach to disease detection and monitoring. This chapter aims to unravel the synergistic interplay between these cutting-edge technologies, highlighting their exceptional potential in revolutionizing healthcare and paving the way for a future of proactive and personalized diagnostics. Our exploration will encompass diverse medical conditions that stand to benefit from MIP-based electrochemical sensors. From cancer biomarker detection and infectious disease screening to autoimmune disease monitoring, diabetes management, and cardiovascular disease diagnostics, we will showcase the immense potential of this hybrid technology. As we delve into MIP-based electrochemical sensors, we embark on a journey of discovery and transformation that promises to redefine the landscape of disease diagnostics and healthcare management. Through the lens of MIP-based electrochemical sensors, we envision a future where precise and reliable disease detection becomes an integral part of personalized healthcare, empowering clinicians and patients alike with proactive and targeted strategies for improved health and well-being.

7.2 Electrochemical Sensors: Pioneers of Analytical Sensing

Widely utilized across diverse fields, including environmental monitoring, healthcare, and industrial process control, electrochemical sensors are analytical instruments that leverage electrochemical principles for detecting and quantifying chemical species within samples. These sensors offer numerous advantages, including high sensitivity, selectivity, and the ability to operate in real-time, making them essential tools for both research and practical applications. The heart of an electrochemical sensor is the electrode, a conductive material where the electrochemical reactions take place. Three main types of electrodes are used in electrochemical sensors: the working electrode, reference electrode, and counter electrode. The electrolyte, a solution providing ions necessary for the electrochemical reaction, interacts with the working electrode. The resulting redox reaction leads to changes in the electrode's potential or current, directly related to the concentration of the target analyte, allowing for quantitative analysis [17, 18].

Electrochemical sensors operate by detecting electrical signals generated by particular chemical reactions occurring at the surface of the electrode. Interaction of the target analyte with the working electrode initiates a redox reaction, altering the electrode's potential or current. These alterations directly correlate with the concentration of the target analyte, allowing quantitative analysis [17].

Types of electrochemical sensors are diverse, each tailored to specific detection requirements and applications. These sensors capitalize on the principles of electrochemistry, harnessing the unique redox reactions occurring at the electrode-electrolyte interface to detect and quantify analytes with exceptional sensitivity and selectivity. Understanding the various types of electrochemical sensors provides valuable insights into their working mechanisms and the broad scope of their applications [18, 19].

Potentiometric sensors (Fig. 7.1) stand as one of the most straightforward and extensively employed electrochemical sensors. Their operation involves measuring the potential disparity between a reference electrode and a working electrode when the analyte is present. The potential shift at the working electrode correlates directly with the concentration of the target analyte. The reference electrode ensures precise measurements as a persistent and precisely defined potential reference. Potentiometric sensors are particularly suitable for detecting ions, gases, and pH levels in solutions. These sensors find applications in environmental monitoring, such as measuring pH levels in soil and water quality assessments, as well as in

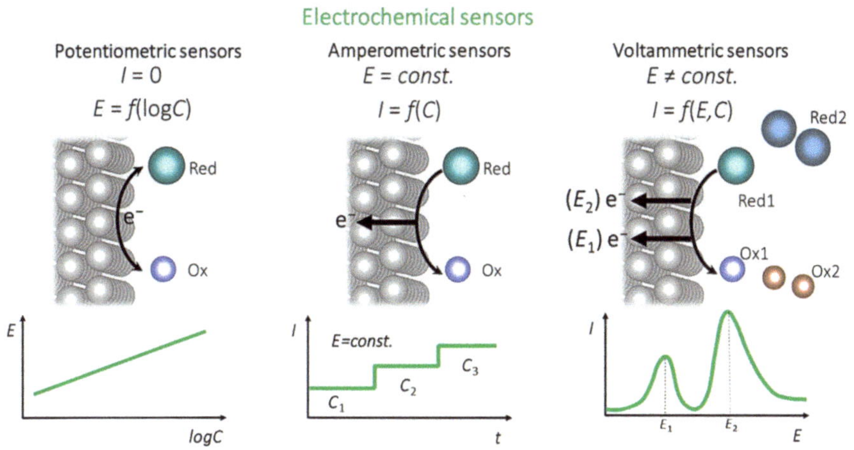

Fig. 7.1 Schematic representation of electrochemical sensors

clinical diagnostics, where they detect ions indicative of specific medical conditions [20].

Amperometric sensors, depicted in Fig. 7.1, measure the current produced by a redox reaction at the working electrode's surface. This current is proportionate to the analyte concentration in the sample. Upon interaction with the working electrode, the analyte undergoes a redox reaction, leading to electron transfer and the production of a measurable current. Renowned for their heightened sensitivity and swift response rates, amperometric sensors prove advantageous for real-time monitoring tasks. These sensors are widely employed in biomedical diagnostics, such as glucose monitoring for diabetes management, as well as in environmental monitoring to detect various pollutants and toxic substances [21].

Voltammetric sensors involve applying a varying potential to the working electrode and measuring the resulting current (Fig. 7.1). This dynamic approach enables the identification of different electrochemical reactions and the simultaneous detection of multiple analytes in a sample. The current response obtained during a voltammetric experiment provides valuable information about the analyte's electrochemical behavior, including its redox potential and diffusion coefficient. Voltammetric sensors are versatile and find applications in various fields, from chemical analysis to environmental monitoring and pharmaceutical research. They can be used to study complex reactions and characterize electroactive species in solution, making them invaluable tools in understanding chemical processes and identifying trace amounts of target analytes [22].

Electrochemical sensors are found to be extensively used in diverse applications. Firstly, they are employed for environmental monitoring, enabling the measurement of pollutants in air, water, and soil, thereby contributing to pollution control and environmental protection efforts [23–25]. Secondly, in biomedical and clinical diagnostics, electrochemical sensors play a crucial role in detecting biomarkers, glucose, cholesterol, and other analytes in blood or bodily fluids, facilitating rapid and cost-effective medical diagnostics [19]. Thirdly, these sensors are essential in ensuring food safety and quality control, as they detect contaminants, additives, and spoilage markers in food analysis [26]. Moreover, in industrial settings, electrochemical sensors are utilized to monitor chemical processes, ensuring safety and efficiency in manufacturing operations. For instance, monitoring of pesticide levels in food is essential [27–32]. Moreover, incorporating enzymes and antibodies into electrochemical biosensors has opened new avenues for specific and sensitive detection of biological analytes. Overall, electrochemical sensors are continuously evolving, driven by ongoing research and technological advancements, and their versatility in a wide range of applications makes them indispensable tools in modern analytical and diagnostic practices.

7.3 MIPs as Artificial Antibodies

MIPs are synthetic materials that emulate the specificity and selectivity of natural antibodies, making them invaluable as artificial antibodies for various applications. Engineered through the process of molecular imprinting, MIPs employ specific target molecules, known as templates, to create precisely tailored cavities or "imprints" within the polymer matrix. These imprints complement the target molecules' shape, size, and chemical characteristics, serving as effective binding sites [33, 34].

The key features of MIPs as artificial antibodies encompass exceptional selectivity and specificity, akin to natural antibodies. The molecular imprinting process ensures that the polymer selectively recognizes and binds to the template molecule or structurally similar molecules, effectively discriminating against other compounds [35]. Additionally, MIPs exhibit impressive stability, rendering them resistant to harsh environmental conditions and suitable for diverse applications across various settings. Their non-immunogenicity, stemming from their synthetic nature, is advantageous in medical and biological applications, as MIPs do not elicit immune responses in living organisms. Moreover, the versatility of MIPs allows for tailoring their binding capabilities to a wide range of molecules, making them indispensable tools in fields ranging from medicine to environmental monitoring [36].

Applications of MIPs as artificial antibodies span a broad spectrum of domains. Notably, they are commonly used in (bio)sensors for detecting specific analytes in intricate samples, playing pivotal roles in environmental, medical, and food monitoring. MIPs also find application in drug delivery as carriers for drug molecules, facilitating targeted release at the desired site, thereby enhancing drug delivery efficacy and reducing side effects. In chromatography and separation, MIPs act as powerful tools for selectively separating and purifying target molecules from intricate mixtures. Additionally, MIPs can mimic the function of enzymes, serving as biomimetic catalysts in catalyzing specific reactions [37]. Their specific binding capabilities also make them invaluable in biomedical research, providing insights into biological processes and interactions. Furthermore, MIPs are harnessed for affinity-based purification in biotechnology and pharmaceutical manufacturing, enabling the purification of therapeutic proteins and peptides [38].

While MIPs offer numerous advantages as artificial antibodies, challenges persist in enhancing their binding affinity and selectivity to approach the level of natural antibodies [39]. Nonetheless, advancements in nanotechnology and polymer science continue to drive improvements in MIP design and performance, propelling their application in cutting-edge research and industry.

MIPs exemplify promising artificial antibodies, offering a versatile, stable, and non-immunogenic alternative to their natural counterparts for a wide range of applications. As research progresses, their potential continues to expand, contributing to significant advancements in diagnostics, therapeutics, and various other fields, ultimately facilitating a myriad of innovative solutions in science and technology [39]. The exceptional capabilities of MIPs position them at the forefront of research and development, driving transformative changes in diverse domains and paving the way for a future of enhanced and personalized solutions.

7.4 MIP-Based Electrochemical Sensors

The utilization of MIPs in electrochemical sensors has attracted considerable attention owing to their distinct advantages, making them exceptionally valuable for the selective and sensitive detection of target analytes. By merging the specificity of MIPs with the sensitivity of electrochemical transducers, MIP-based electrochemical sensors are finding extensive applications across various fields, including environmental, medical, and food monitoring.

Key aspects of MIPs in electrochemical sensors include their high selectivity, as they are designed to exclusively recognize specific target molecules. During the molecular imprinting process, the polymer is imprinted with the structural features of the template molecule, resulting in binding sites that solely identify the target analyte amidst interfering species. Electrochemical sensors contribute to the sensitivity of MIP-based systems, detecting low analyte concentrations. The binding of the target analyte to the MIP triggers a measurable electrochemical response, enabling quantitative analysis even at trace levels. MIPs' robust and stable nature allows for their seamless integration into electrochemical sensors, ensuring long-term and reliable operation. Additionally, the real-time detection capabilities of electrochemical sensors enable rapid and continuous monitoring of analyte concentrations [19].

The applications of MIPs in electrochemical sensors are diverse and impactful. Environmental monitoring benefits from MIP-based electrochemical sensors for the detection of pollutants like heavy metals, pesticides, and toxins, enabling timely and accurate assessment of water and air quality [1]. In biomedical diagnostics, MIP-based electrochemical sensors detect biomarkers in clinical samples, facilitating early disease diagnosis and personalized medicine. For food safety, these sensors play a critical role in detecting foodborne pathogens, allergens, and contaminants, ensuring safety and quality control in the food industry [40]. Moreover, MIP-based electrochemical sensors are employed in pharmaceutical research to monitor drug levels in patients and assess the efficacy of drug delivery systems [41].

While MIP-based electrochemical sensors offer many benefits, some challenges persist, such as optimizing their performance to achieve the desired sensitivity and selectivity. Enhancing the affinity of MIPs towards target molecules and addressing potential interference from complex samples remain areas of active research. Looking ahead, ongoing developments in nanotechnology, MIP design, and electrochemical transducers hold the potential to further improve the performance of MIP-based electrochemical sensors. The amalgamation of MIPs' selectivity with the high sensitivity of electrochemical sensing promises the development of versatile, economical, and consistent sensors for a wide range of applications, revolutionizing the way we monitor and detect analytes across various fields [42].

7.5 MIP-Based Electrochemical Sensors in Disease Detection and Monitoring

MIP-based electrochemical sensors have shown great promise in disease detection and monitoring due to their unique capability for disease-specific biomarkers or pathogens recognition. These sensors combine the specificity of MIPs with the sensitivity of electrochemical transducers, enabling early and accurate diagnosis of various medical conditions, as well as providing valuable information for disease monitoring and management. Cancer biomarker detection is one of the notable applications of MIP-based electrochemical sensors, where they are employed to detect specific cancer biomarkers, such as proteins or nucleic acids, indicating the presence of cancer cells or providing insights into the disease's progression. By detecting these biomarkers in blood or other bodily fluids, these sensors aid in the early diagnosis of cancer, allowing for timely intervention and treatment [43]. Additionally, MIP-based electrochemical sensors can be engineered to recognize unique surface proteins or genetic markers of infectious agents, such as viruses, bacteria, or parasites. This enables fast and precise pathogen detection in patient samples, facilitating early identification and containment of infectious diseases [44, 45]. Monitoring autoimmune diseases is another significant application, where MIP-based electrochemical sensors can detect specific autoantibodies or disease-related biomolecules indicative of autoimmune conditions, helping assess disease activity and treatment response over time [46]. Furthermore, these sensors play a crucial role in diabetes management by monitoring glucose levels in patients. By specifically recognizing glucose molecules, MIP-based electrochemical sensors offer accurate and continuous glucose monitoring, helping individuals with diabetes manage their condition effectively [46]. These sensors could also be useful in studying and monitoring lung diseases, where the cholinergic system plays an important role [47]. Moreover, electrochemical sensors with MIPs targeting specific cardiovascular disease markers, such as cardiac troponins, enable rapid and sensitive detection of heart-related conditions, which is crucial for timely intervention and reducing the risk of further complications [48]. Beyond disease detection, MIP-based electrochemical sensors can be utilized for therapeutic drug monitoring, measuring drug concentrations in patients undergoing treatment, thereby optimizing drug dosages and ensuring therapeutic efficacy while minimizing adverse effects [49].

The advantages of MIP-based electrochemical sensors in disease detection and monitoring are numerous. Firstly, their high sensitivity enables the detection of disease biomarkers or pathogens even at low concentrations. Secondly, the specificity of MIPs allows for targeting specific molecules, reducing interference from other components in complex biological samples. Additionally, electrochemical sensors provide real-time or near-real-time results, enabling prompt decision-making in disease management. Cost-effectiveness is another advantage, as MIPs can be synthesized at a relatively low cost, making the sensors affordable for widespread use in healthcare settings. Moreover, MIP-based electrochemical sensors are amenable to point-of-care testing, allowing for convenient and accessible diagnostics in various healthcare settings [41, 42].

7.6 MIP-Based Electrochemical Sensors for Cancer Biomarker Detection

Prostate cancer is the third most prevalent cancer in men. Prostate-specific antigen (PSA) is a valuable biomarker for detecting this notorious disease. Detecting PSA is crucial for screening early prostate cancer and identifying disease recurrence after treatment.

To address this, Ertürk et al. developed a microcontact-PSA imprinted (PSA-MIP) biosensor chip. It can detect PSA accurately and quickly in real-time. They used methacrylic acid (MAA) and ethylene glycol dimethacrylate (EGDMA) to create the PSA-MIP electrodes through UV polymerization. They also made electrodes with Anti-PSA antibodies (Anti-PSA) for comparison. They tested the biosensor with different PSA concentrations from 10 fg mL^{-1} to 100 ng mL^{-1}. The PSA-MIP electrodes could detect PSA as low as 8.0×10^{-5} ng mL^{-1} (16×10^{-17} M), while Anti-PSA electrodes could detect it as low as 6.0×10^{-4} ng mL − 1 (12×10^{-16} M). They also checked if the biosensor could pick PSA among other substances like HSA and IgG. They successfully found PSA in diluted human serum samples and checked if the electrodes worked consistently. The study shows promising results, suggesting that combining the sensitivity of capacitive systems with the selectivity and reliability of microcontact-imprinting could help detect various substances in real-time, even at very low levels [50] (Fig. 7.2).

Karami et al. developed a novel dual-modality immunosensor capable of simultaneously detecting PSA and myoglobin (Myo) in human serum and urine samples. The sensor combines an MIP and a nanostructured biosensing layer. They first formed a self-assembled layer of 3,3′-dithiodipropionic acid di(N-hydroxysuccinimide ester) (DSP) on a gold screen-printed electrode (SPE) and covalently attached the target proteins (PSA and Myo) to it. Then, they synthesized the imprinted cocktail polymer (MIP(PSA, Myo)-SPE) on the electrode surface using specific templates for PSA and Myo. The researchers also created a nanocomposite (NCP) by decorating magnetite nanoparticles with multi-walled carbon

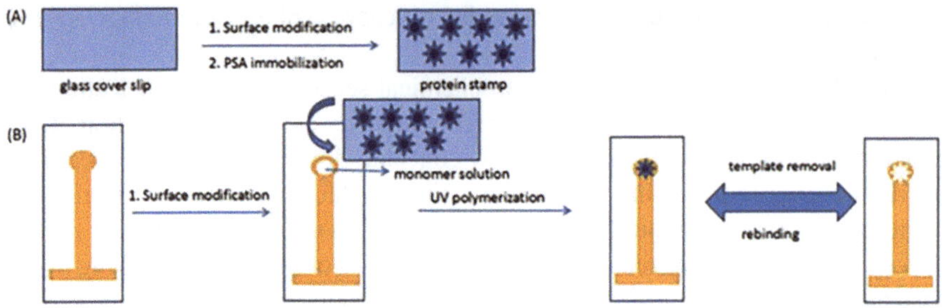

Fig. 7.2 Schematic representation of the microcontact imprinting of PSA onto the capacitive biosensor. (**a**) Preparation of the glass cover slips (protein stamps), (**b**) Preparation of the capacitive gold electrodes and microcontact imprinting via UV polymerization. Reprinted from [50]

nanotube, graphene oxide, and specific antibodies for PSA. The NCP was incubated with the MIP(PSA, Myo)-SPE. The dual-modality immunosensor demonstrated impressive LODs, with values of 5.4 pg mL^{-1} and 0.83 ng mL^{-1} and linear dynamic ranges of 0.01–100 ng mL^{-1} and 1–20,000 ng mL^{-1} for PSA and Myo, respectively. The sensor's capacity to accurately detect both PSA and Myo simultaneously and with exceptional sensitivity holds immense promise for advanced biosensor technology. This dual-analyte-specific receptor device is highly sought-after for integration into lab-on-chip kits, enabling the detection of a broad range of biomarkers present at extremely low levels during the initial phases of disease development [51].

Research published in 2016 presents the development and assessment of a novel synthetic receptor sensor by combining biomolecular recognition elements and molecular imprinting. The aim was to address challenges encountered in traditional protein imprinting methods. To achieve this, a thiolated DNA aptamer, known for its affinity for PSA, was complexed with PSA and immobilized on a gold electrode surface. Dopamine was electropolymerized around this complex to trap it in or near its binding conformation and localize PSA binding sites on the sensor surface. Upon removal of PSA, the resulting MIP cavity was expected to work synergistically with the embedded aptamer, forming a hybrid receptor (apta-MIP) with superior recognition properties compared to the aptamer alone. The PSA rebinding was evaluated to the apta-MIP surface using electrochemical impedance spectroscopy (EIS). The apta-MIP sensor exhibited high sensitivity, displaying a linear response to PSA concentrations ranging from 100 pg mL^{-1} to 100 ng mL^{-1}, and achieved LOD of 1 pg mL^{-1}. This limit was three times better than that of the aptamer-only sensor for PSA. Additionally, the sensor demonstrated low cross-reactivity with a closely related protein (human Kallikrein 2) and showed minimal response to human serum albumin (HSA). These results suggest that the sensor might possess resilience against the nonspecific binding of serum proteins [52].

A study published in 2014 demonstrates the creation of protein-specific plastic antibodies with tailored charged monomers around the binding site to enhance protein binding. These charged receptor sites are strategically placed on a neutral polymeric matrix to ensure proper protein reception at its binding site. The effectiveness of this approach is confirmed by preparing control materials using neutral monomers and a non-imprinted template. This work focuses on PSA, where serum levels above 10 ng mL^{-1} indicate a high likelihood of associated cancer. To produce the Protein Imprinted Materials with charged binding sites (C/PIM), graphene layers were used as a surface for the imprinting process. The protein was covalently attached to the graphene, and charged monomers, such as vinylbenzyl(trimethylammonium chloride) and vinyl benzoate, were introduced to label the binding site. These charged monomers self-organized around the protein, and radical polymerization of vinylbenzene was used to complete the process. Neutral PIM (N/PIM) without oriented charges and non-imprinted materials (NIM) without a template were prepared as control samples. The resulting materials were utilized to develop a simple and cost-effective potentiometric sensor for PSA. These materials were incorporated as ionophores in plasticized PVC membranes and tested on solid or liquid conductive contacts, including conductive carbon electrodes in syringes or inner reference solution in micropipette tips. The sensors with charged monomers displayed a more stable and sensitive response, with an average slope of −44.2 mV/decade and a low LOD of 5.8×10^{-11} mol L^{-1} (2 ng mL^{-1}). In contrast, the non-imprinted sensors showed lower sensitivity, with an average −24.8 mV/decade slope. The best sensors were successfully used for serum analysis, achieving recoveries between 96.9% and 106.1% and relative errors of 6.8% [53].

Alpha-fetoprotein (AFP) is a tumor marker used to diagnose and monitor certain types of cancer, particularly hepatocellular carcinoma (HCC), the most common form of liver cancer. Elevated AFP levels in the blood may indicate the presence of liver cancer, and monitoring AFP levels can be used to assess treatment response and recurrence.

Shen et al. developed an electrochemical sensor for detecting AFP cost-effectively and straightforwardly. The sensor was based on MIPs and was fabricated using a layer-by-layer coating approach on a glassy carbon electrode (GCE). Chitosan, glutaraldehyde, and the AFP antigen were sequentially applied, followed by the polymerization of acrylamide (AAM) as the functional monomer and N,N′-methylenebisacrylamide (MBAAm) as the cross-linker. The efficiency of the imprinting process was assessed using scanning electron microscopy (SEM), cyclic voltammetry (CV), and EIS. As a comparison, a non-imprinted polymer (NIP) film electrode was also prepared using the same steps but without AFP, and its selectivity coefficients were evaluated. Since AFP has poor electroactivity, its detection was achieved by monitoring the redox properties of $[Fe(CN)]^{3-/4-}$ as an electrochemical probe through differential pulse voltammetry (DPV). The method exhibited a linear detection range of 0.8–10,000 ng mL^{-1}, with a LOD of 0.096 ng mL^{-1}. Selectivity studies were conducted using bovine serum albumin (BSA), carcinoembryonic antigen (CEA), and the p24 antigen of the human immunodeficiency virus (HIV-p24). The reproducibility of the sensors was also tested. Additionally, the AFP imprinted sensor was successfully applied to determine AFP levels in a human serum sample [54].

A novel electrochemical sensor based on MIPs was developed for the sensitive detection of HIV-p24. The sensor was created on the surface of a GCE modified with multi-walled carbon nanotubes (MWCNTs) through surface polymerization. AAM was utilized as a functional monomer, N,N′-methylenebisacrylamide (MBA) as a cross-linking agent, and ammonium persulphate (APS) as an initiator (Fig. 7.3). The resulting MIPs electrochemical biosensor exhibited selective recognition capabilities for HIV-p24 and demonstrated a wide linear detection range from 1.0×10^{-4} to 2 ng cm^{-3}, along with a notably low LOD of 0.083 pg cm^{-3}. This performance outperforms many other HIV-p24 sensors employing alternative techniques. The sensor also demonstrated strong selectivity, repeatability, reproducibility, and stability. Furthermore, it successfully measured HIV-p24 in human serum samples, yielding satisfactory outcomes. The accuracy and reliability of the sensor's results were validated through comparison with enzyme-linked immunosorbent assay (ELISA) measurements. In summary, this research introduces an innovative electrochemical sensor leveraging MIPs to precisely detect HIV-p24. The sensor's remarkable performance characteristics make it a promising tool for practical applications [55].

In a study published in 2019, researchers developed an electrochemical MIP sensor for detecting the variable-format conformation protein gp120. To construct the sensor, they utilized a composite of graphene-like carbon nanofragments (CNF) and bismuth oxides (CNF-Bi) as the decorating material. Instead of using pure gp120, they employed small-molecule entry inhibitor NBD-556 and gp120 conjugates NBD-556@gp120 as templates. Pyrrole was selected as the electro-polymerization monomer. Results indicate that under optimized conditions, incorporating NBD-556

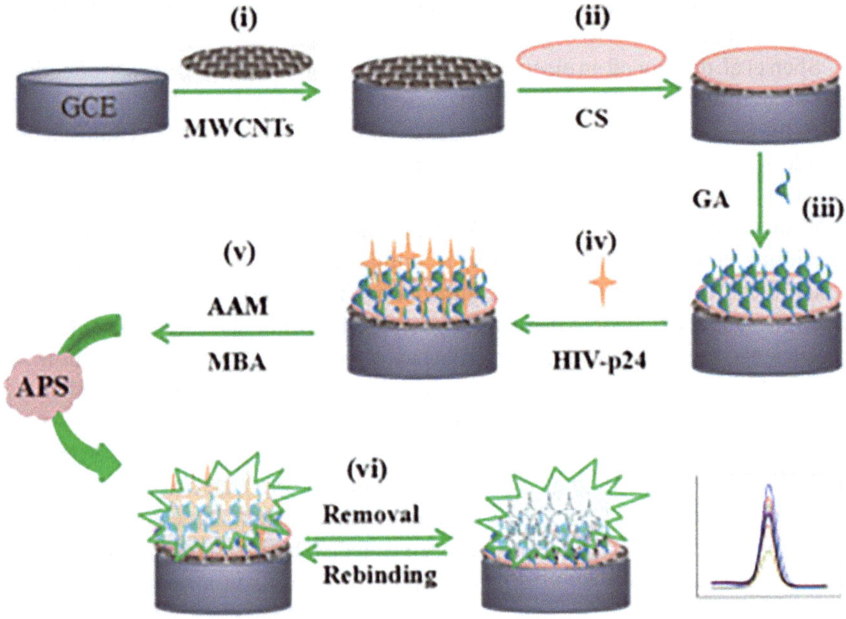

Fig. 7.3 Detailed procedure diagram for fabrication of the HIV-p24-MIPs/MWCNTs/GCE sensor. Reprinted from [55]

enhances the efficiency of specific recognition and analytical properties of the MIP sensor towards gp120. The sensor exhibited response currents proportional to NBD-556@gp120 concentrations within 0.0002–200 ng mL^{-1}, with an LOD of 0.0003 ng mL^{-1}. The NBD-556@gp120-based MIP sensor also demonstrated satisfactory stability and reproducibility. Notably, the sensor exhibited accurate results when applied to detecting gp120 in human plasma. This research concept holds significant promise for the early diagnosis of the HIV-1 virus and has potential applications for detecting other conformationally unstable proteins [56].

A biomimetic sensor was designed to identify an HIV-1, known as glycoprotein 41 (gp41). This protein, gp41, is a transmembrane component of HIV-1 responsible for facilitating the fusion of viruses with infected cells. It serves as a crucial marker for gauging the progression of HIV-1 infection and assessing the effectiveness of therapeutic measures. Lu et al. employed an epitope imprinting technique to construct the sensor in their study. The process involved using dopamine as a functional monomer, which was polymerized on the surface of a quartz crystal microbalance (QCM) chip in the presence of a synthetic peptide template. This template mimicked a specific portion of the gp41 protein, spanning amino acids 579–613. This polymerization formed a hydrophilic MIP film on the QCM chip. QCM measurements established that the MIP film exhibited high affinity towards the template peptide and demonstrated specific binding capability to the corresponding gp41 protein. Also, researchers determined the dissociation constant (K_d) of the MIP for the template peptide to be 3.17 nM. Developed MIP-based biomimetic sensor allowed direct and quantitative detection of the gp41 protein. It exhibited an LOD of 2 ng mL^{-1} for gp41, which is comparable to results obtained using the ELISA method, a widely accepted analytical technique. Moreover, they assessed the practical analytical performance of our sensor by evaluating its ability to detect gp41 in human urine samples, yielding satisfactory outcomes. This research showcases the potential of our biomimetic sensor as a sensitive and specific tool for the detection of gp41, contributing to the field of HIV-1 diagnostics [57].

In a study published in 2016, researchers developed an electrochemical sensor modified with a sensitive and selective MIP for detecting AFP. The sensor surface was specifically designed with Ag/AgCl nanoparticles of cube and spherical shapes. It was the first time that the effect of nanoparticle shape on the behavior of the MIP was studied. Cube-shaped nanoparticles were found to have high surface-to-volume ratios and superior electrocatalytic activity, making them a suitable platform for MIP synthesis. The researchers demonstrated how the change in nanomaterial morphology influenced the electrochemical and adsorption properties of the MIP towards AFP. They utilized cube-shaped nanoparticle@imprinted polymer to fabricate the electrochemical sensor, which exhibited good analytical performance for AFP detection using square wave stripping voltammetry. The sensor showed a linear current response for AFP concentrations ranging from 0.10 to 700.0 pg mL^{-1}, with an impressive ultra-trace LOD of 24.6 fg mL^{-1}. The proposed sensor exhibited high selectivity, sensitivity, simplicity, and clinical applicability for detecting AFP in human blood serum, plasma, and urine. Remarkably, the sensor achieved this without using antibodies or biological components, distinguishing it from previously reported systems. The sensor's potential lies in its ability to serve as a cost-effective and accessible alternative to commercially available and sophisticated ELISA kits for AFP determination [58].

Carcinoembryonic antigen (CEA) is a protein typically produced during fetal development, but its expression decreases significantly after birth. In healthy individuals, CEA levels in the blood are typically very low. CEA is considered a tumor marker, meaning it can be used as a biomarker to aid in diagnosing, prognosis, and monitoring certain types of cancer. It is most commonly associated with colorectal cancer but can also be elevated in other cancers, such as pancreatic, lung, breast, ovarian, and liver.

Moreira et al. created an inexpensive, disposable device using screen-printed cells with a three-electrode setup. They employed PCB technology to assemble electrodes (AgxO-working, AgxO-counter, and Ag/AgxO-reference) to detect CEA. The sensing material, a biomimetic matrix of polypyrrole (PPy), was synthesized through electro-polymerization of pyrrole in the presence of CEA, followed by proteolytic enzymatic removal of CEA. They prepared a non-imprinted film (NIPPy) simultaneously for comparison. The device demonstrated a linear response from 0.05 to 1.25 pg mL^{-1} using various electrochemical techniques. Selectivity studies and reproducibility testing confirmed its potential for point-of-care CEA screening due to its high sensitivity and stability, making it a promising tool for clinical use [59].

Wang et al. utilized surface molecular imprinting through self-assembled monolayers to create sensing components capable of identifying cancer biomarkers and other proteins. These components involve a silicon chip coated with gold, where hydroxyl-terminated alkanethiol molecules and the target biomolecules are simultaneously absorbed. The thiol molecules form ordered monolayers on the metal substrate, while the biomolecules can be extracted, leaving cavities resembling their shape in the monolayer. When the biomolecules are reintroduced to the sensor chip, their electrical potential changes, which can be measured through potentiometry. Researchers employed this technique to detect CEA in purified CEA solutions and the culture medium of a CEA-producing human colon cancer cell line. The accuracy of the CEA detection, verified against a standard immunoassay, demonstrated both sensitivity (detecting concentrations between 2.5 and 250 ng mL^{-1}) and specificity (no interference from hemoglobin; no response from a non-imprinted sensor). Comparable outcomes were observed for human amylase. Additionally, they achieved specific detection of poliovirus virions (no interference from adenovirus; no response from a non-imprinted sensor). These findings showcase the application of molecular imprinting principles in devising a novel approach for identifying protein-based cancer biomarkers and protein-related large structures such as viral capsids. This strategy holds promise for creating a versatile assay technique that could be highly sensitive, specific, straightforward, and potentially cost-effective [60].

Cancer antigen-125 (CA-125) is a biomarker linked to ovarian cancer. It is a protein biomarker. It was first identified in the early 1980s and has become an important tool for diagnosing, monitoring, and managing ovarian cancer. CA-125 is produced by normal tissues in the body, particularly in the lining of the chest and abdominal cavities, as well as in the lining of the lungs, uterus, and fallopian tubes. However, it is primarily known for being overexpressed in certain cancerous conditions, especially ovarian cancer. Elevated levels of CA-125 in the blood are an indication of ovarian cancer. However, it is not specific to this type of cancer and can also be elevated in other non-cancerous conditions.

Viswanathan et al. aimed to improve the detection of cancer biomarkers by developing a protein-imprinted polymer on a three-dimensional gold nanoelectrode ensemble (GNEE) for epithelial CA 125 detection. Initially, they integrated CA 125 template protein into a thin-film coating to create CA 125 imprinted GNEE. Subsequently, they removed the protein from the thin film's accessible surfaces, leaving CA 125 imprints. The developed CA 125 imprinted GNEE underwent analysis using CV, DPV, and EIS techniques. The thin, protein-imprinted sites on the GNEE surface facilitated the immunospecific capture of CA 125 molecules, with the bound CA 125 mass detected as a reduction in faradic current from the redox marker. The sensor exhibited good sensitivity, with a range of 0.5–400 U mL^{-1} and the lowest LOD of 0.5 U mL^{-1} under optimal conditions. They successfully tested spiked human blood serum and unknown real serum samples, with the assay's sensitivity remaining unaffected by non-specific proteins in the serum [61].

A study published in 2019 focuses on developing MIPs on a gold electrode surface for recognizing the CA-125 biomarker. The CA-125 imprinting was achieved by electro-polymerizing the pyrrole (Py) monomer on the gold electrode using CV, ensuring highly selective materials with excellent molecular recognition capabilities. The biosensor based on electrochemical transduction exhibited superior analytical parameters, demonstrating good selectivity and a low LOD of 0.01 U/mL, with a linear concentration range between 0.01 and 500 U mL^{-1}. This electrochemical biosensor was chosen for the study and successfully applied in analyzing CA-125 in artificial serum samples, yielding recovery rates ranging from 91 to 105% with an average relative error of 5.8% [62].

Nuclear matrix protein 22 (NMP22) is a protein that is usually found within the cell nucleus, specifically in the nuclear matrix, which is the framework that supports the organization of chromatin and various nuclear processes. NMP22 has gained significance as a biomarker for detecting and monitoring bladder cancer. When bladder cancer develops, NMP22 is released into the urine at higher levels than in healthy individuals, making it a useful indicator of bladder cancer presence and recurrence.

Research published in 2016 focuses on a sensing element composed of MIPs designed to detect NMP22 molecules. The authors increased MIP coating by growing zinc oxide (ZnO) nanorods on sensing electrodes. With a portable potentiostat, they monitored electrochemical reactions for NMP22 detection in urine samples. Phase 0 clinical trials then measured NMP22 levels in urine samples from bladder cancer patients using the integrated NMP22 MIP-coated ZnO nanorod electrodes. Results showed concentrations ranging from 128 ± 19 to 588 ± 53 ng mL^{-1} [63].

Neopterin is a biomarker molecule that plays a crucial role in the human immune system. It is a pteridine compound derived from guanosine triphosphate through a series of enzymatic reactions. Neopterin is primarily produced by activated human immune cells, particularly macrophages and dendritic cells, in response to various infections and inflammatory stimuli.

Sharma et al. developed a potentiometric sensor to detect neopterin using an electrochemically synthesized MIP. The MIP film was prepared using electrochemical co-polymerization of specific monomers with a neopterin template. They also synthesized a non-imprinted polymer (NIP) film for comparison. The sensor detected neopterin within the concentration range of 0.15–2.5 mM (38–633 μg mL^{-1}) with a limit of detection (LOD) of 22 μM (5.6 μg mL^{-1}). It effectively distinguished interferences from neopterin analogs and other substances like glucose and creatinine. Furthermore, the sensor successfully detected neopterin in serum samples [64].

Cancer Antigen 15-3 (CA 15-3) is a biomarker commonly used in managing breast cancer. It is a glycoprotein antigen produced by certain cancer cells, including breast cancer cells. CA 15-3 is a specific form of MUC1 (Mucin 1), a cell surface mucin glycoprotein overexpressed in many types of cancer.

To address this, Pacheco et al. developed an MIP-based electrochemical sensor for breast cancer monitoring by directly imprinting CA 15-3 on a screen-printed gold electrode (Au-SPE). They adsorbed CA 15-3 on the Au-SPE surface and then electro-polymerized 2-aminophenol around the protein. After extracting the imprinted protein, they conducted a voltammetric analysis using hexacyanoferrate(II/III) as the redox probe. The sensor showed a linear relationship between the peak current intensity of the redox probe and the logarithm of CA 15-3 concentration within the range of 5–50 U mL^{-1}, with a LOD of 1.5 U mL^{-1}. This MIP sensor offers quick (15 min) analysis, is cost-effective, easy to prepare, and has the potential for integration into small, portable point-of-care devices [65].

In a study published in 2018, researchers aimed to develop an MIP-based electrochemical sensor to quantify the human epidermal growth factor receptor two extracellular domain (HER2-ECD), a protein breast cancer biomarker. They used CV to electropolymerize the MIP on an Au-SPE using a solution containing phenol and HER2-ECD. The sensor's performance was assessed using CV and electrochemical impedance spectroscopy. The analysis of HER2-ECD was conducted using DPV with $[Fe(CN)_6]^{3-/4-}$ as the redox probe. The sensor exhibited a linear detection range from 10 to 70 ng mL^{-1} of HER2-ECD, with an LOD of 1.6 ng L^{-1} and a limit of quantification (LOQ) of 5.2 ng mL^{-1}. The MIP sensor demonstrated selectivity towards HER2-ECD over other protein biomarkers, and these proteins did not interfere with the analysis of HER2-ECD. When tested with spiked human serum samples, the sensor provided accurate and precise results, suggesting its potential as a non-invasive tool for analyzing HER2-ECD in breast cancer patients [66].

A study published in 2020 presents the development of an ultrasensitive electrochemical sensor using hybrid epitope imprinting and nanomaterial amplification to detect disease-associated molecules in trace amounts in human body fluids. The sensor was designed to specifically recognize neuron-specific enolase (NSE), a biomarker associated with neurological disorders. The hybrid epitope imprinting involved electro-polymerization in the presence of two computationally selected and cysteine-modified epitopes of NSE, as well as as-synthesized gold nanoparticles (AuNPs) and a functional monomer. The resulting hybrid MIPs, with and without AuNP decoration, were used to fabricate the electrochemical sensors to assess the impact of nanomaterial modification on biomarker sensing. The biomarker assay using the standard hybrid MIPs demonstrated 2.5-fold higher sensitivity than single epitope imprints. However, including AuNPs in the hybrid MIPs significantly enhanced the sensitivity, allowing the recognition of NSE in human serum over a concentration range of 25–4000 pg mL^{-1}, with a LOD of 25 pg mL^{-1}. Selectivity studies showed a high imprinting factor of 4.2, confirming the specificity of the AuNP-MIP cavities for the target NSE. The sensor exhibited negligible cross-reactivity with four reference molecules. Compared to conventional epitope imprinting strategies that rely on single epitopes, the hybrid epitope-MIPs, especially incorporating AuNPs, offered a more desirable sensing platform with high sensitivity, affinity, and specificity for NSE detection. This approach holds the potential for early and efficient diagnosis of neurological disorders and other diseases by detecting disease-associated molecules at very low concentrations in human body fluids [67].

An electrochemical sensor utilizing molecular imprinting was created by Wang et al. to detect NSE. This sensor was developed through the electrochemical polymerization of an ionic liquid, functionalized with a pyrrole group, within an arrangement of gold nanostructures. A precisely defined three-dimensional gold nanostructure was constructed on the surface of a GCE using a template-guided electrochemical deposition technique. The surface of the gold nanostructures was coated with 1-(3-mercaptopropyl)-3-vinyl-imidazolium tetrafluoroborate to generate active sites for attaching the molecularly imprinted film. Following this, an electrochemical polymerization process took place in an aqueous solution containing 1,3-di(3-N-pyrrolpropyl)imidazolium bromine ionic liquid and NSE. After removing NSE templates, a functional molecularly imprinted sensor was successfully fabricated. This sensor exhibited notable specificity and sensitivity towards NSE. It generated a linear response within the concentration range of 0.01–1.0 ng mL^{-1} and achieved a LOD of 2.6 pg mL^{-1}, requiring an incubation time of 15 min. The sensor's performance was validated by accurately determining NSE levels in clinical serum samples [68].

Moussa et al. introduced an innovative and environmentally friendly method for producing reduced graphene oxide adorned with silver nanoparticles (rGO-AgNPs) through a single-step green reduction process using an extract from *Allium sativum*. These rGO-AgNPs were combined with an MIP to create an electrochemical sensor with high sensitivity and selectivity for detecting lactic acid (LA), a critical component in screening cancer cells. The AuE/rGO-AgNPs/MIP setup indirectly detected LA via electrochemistry. The signal derived from $[Fe(CN)_6]^{3-/4-}$ redox peaks diminished consistently with increasing LA concentrations due to the disruption of electron transfer pathways, indicating a heightened sensitivity to the target molecule. The change in current ($\Delta I = I$ BLANK − I LA) was proportionate to LA concentration within the range of 10–250 μM, exhibiting an impressively low LOD of 0.726 μM (signal-to-noise ratio = 3). The developed LA sensor demonstrated excellent selectivity and remarkable stability. Furthermore, the sensor's resilience was tested across various pH conditions, yielding favorable outcomes. The practical applicability of the sensor was confirmed through successful tests on human serum samples, affirming its potential utility in cancer screening and monitoring [69].

Oral cancer ranks among the prevalent cancer types in Europe, necessitating the creation of cost-effective and dependable portable platforms for its diagnosis. Such platforms should exhibit high sensitivity and selectivity, and alongside prevention efforts, they are crucial for addressing the widespread nature of this condition. A study published in 2023 by Siciliano et al. presents the development and characterization of an electrochemical sensor based on MIPs for detecting transforming growth factor-beta 1 (TGF-β1), offering potential as an early screening tool for oral cancer. The biomimetic surface was engineered by electropolymerizing o-phenylenediamine (o-PD) onto platinum electrodes, employing TGF-β1 as the template molecule. The synthesis of MIPs, template removal, and the rebinding of TGF-β1 were tracked using DPV. The performance of the MIP sensor was assessed in both a buffer solution and saliva samples containing TGF-β1. The sensor demonstrated a linear response within the examined concentration range (from 20 ng mL^{-1} down to 0.5 ng mL^{-1}) and exhibited an exceptional LOD of 0.09 ng mL^{-1}. The sensor exhibited an affinity and selectivity for TGF-β1, even in the presence of interfering molecules. Additionally, the sensor's capability to detect the target molecule in saliva samples spiked with TGF-β1 was validated, yielding favorable recovery results. These findings suggest the potential for utilizing the proposed system in large-scale, rapid oral cancer screening procedures [70].

7.7 MIP-Based Electrochemical Sensors for Glucose Detection

Over the past few decades, there has been a growing interest in glucose bio-sensing technologies, primarily due to glucose metabolism's critical role in diseases such as diabetes. Caldara et al. introduced a novel approach known as dummy imprinting, focusing on synthesizing and optimizing a MIP specifically designed for glucose sensing. Micro-contact deposition was employed to integrate this polymer into a thermally conductive receptor layer. Essentially, MIP particles were pressed into a polyvinyl chloride adhesive layer using a polydimethylsiloxane stamp (Fig. 7.4). The prepared layer underwent evaluation through the heat-transfer method, enabling the determination of the receptor layer's specificity and sensitivity. Furthermore, the sensor's selectivity was assessed by analyzing its thermal response following exposure to increasing concentrations of various saccharide analogs in a phosphate-buffered saline (PBS) solution. The results reveal a linear detection range of 0.0194–0.3300 mM for glucose in PBS. Finally, the potential application of this sensor was demonstrated by subjecting the receptor layer to rising glucose concentrations in human urine samples, illustrating a linear detection range of 0.0444–0.3300 mM. The findings presented in this paper underscore the sensor's

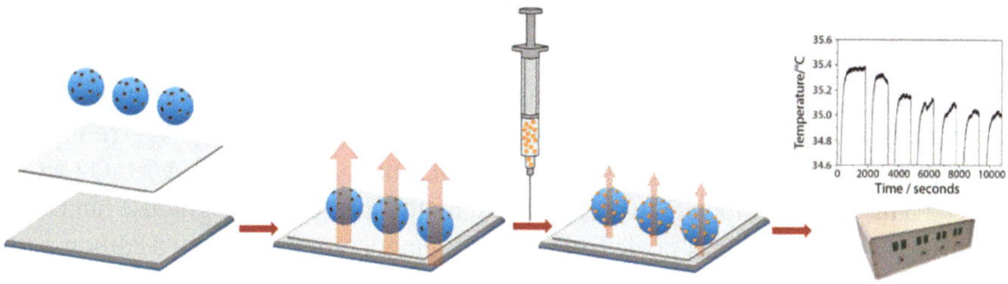

Fig. 7.4 Schematic representation of the deposition of the MIP particles and rebinding analysis with the HTM. Reprinted from [71]

practicality for non-invasive glucose monitoring and the analysis of food samples [71].

Widayani et al. synthesized an MIP using glucose as the template molecule via bulk polymerization. They explored its potential as an electrode in a glucose sensor. Results indicate that the MIP electrode's potential varies with glucose concentration, ranging from 72.5 to 295.8 mV for concentrations up to 7 mmol L^{-1}. Within the range of 0.02–5 mmol L^{-1}, the potential correlates linearly with glucose concentration, showing a sensitivity of approximately 43.7 ± 1.6 mV/mmol L^{-1}. Compared to a non-imprinted polymer electrode, the MIP electrode exhibits significantly higher sensitivity to glucose concentration changes [72].

Mirmosheni et al. introduced a new method for real-time glucose monitoring in water using a quartz crystal nanobalance (QCN) enhanced with an MIP, eliminating the need for enzymes. The MIP was created through free radical polymerization with various functional monomers, with acrylic acid proving the most effective. Testing within a flow cell setup showed that glucose concentration correlated linearly with frequency shifts, ranging from 10 to 200 mg L^{-1}, with an impressive correlation coefficient of R^2 = 0.9996. The sensor demonstrated a sensitivity of 1.20 Hz/mg L^{-1} and a detection limit of 4.4 mg L^{-1}. Importantly, the MIP-modified QCN sensor was unaffected by organic compound interference and ion exchange resins effectively removed interference from ionic compounds [73].

An electrochemical sensor was developed using Graphene Oxide-Molecular Imprinted Polymer (GO-MIP) for enzyme-free glucose determination. The creation of GO-MIP involved a sequence of functionalization, polymerization, and the introduction/removal of template molecules during the synthesis process. Under optimized conditions, the proposed GO-MIP-based electrode exhibited exceptional electrocatalytic activity for glucose oxidation and demonstrated a remarkable detection limit of 0.1 nM with a response time of approximately 2 min. The current response of the GO-MIP-based glucose sensor exhibited a linear correlation with the glucose concentration. Real-time testing of the electrodes in human blood yielded results that closely matched those from commercially available glucose monitors. Additionally, the material's reusability was assessed for up to eight cycles, and the interference of glucose with substances like ascorbic acid (AA), uric acid (UA), and dopamine (DA) was thoroughly examined. The results obtained underscore the promising potential of GO-MIP for highly sensitive and stable glucose sensing applications [74].

An enzyme-free glucose sensor was developed using the MIP technique. The process involved creating a novel electrode with polyvinyl acetate (PVA) reinforced by MnO_2/CuO loaded onto graphene oxide (GO) nanoparticles (PVA/MnO2@GO/CuO). This nanocomposite was modified in the presence of glucose and then imprinted. The sensor was made from this MIP nanocomposite using a carbon paste method combined with voltammetry. Results show that the response of the PVA/MnO2@GO/CuO MIP to glucose oxidation is significantly higher than that of the non-imprinted electrode. The sensor achieved a LOD of 53 μM, with linear responses observed for glucose concentrations from 0.5 to 4.4 mM. Both intra- and inter-day determinations had relative standard deviations below 6.0%, and the recoveries for various samples reached 96% [75].

A non-enzymatic potentiometric glucose sensor was developed for saliva using an MIP on a conducting polymer layer. It employs acrylamide and aminophenyl boronic acid as host molecules for glucose. The sensor, deposited on a screen-printed carbon electrode with gold nanoparticles, generates a signal by altering the polymer film's pKa value due to the formation of a boronate anion-glucose complex. It has a broad linear dynamic range from 3.2×10^{-7} to 1.0×10^{-3} M, with a low detection limit of 1.9 (±0.15) × 10^{-7} M. The sensor is highly selective for glucose and reliable in physiological fluids like saliva and finger-prick blood [76].

Zheng et al. introduced a novel molecularly imprinted electrochemical sensor (MIECS) constructed on a GCE that has been enhanced with carbon dots (CDs) and chitosan (CS). To fabricate the MIP film for glucose detection using differential pulse voltammetry, 3-aminobenzeneboronic acid was employed as a functional monomer, with glucose serving as the template. Several experimental parameters were optimized, including the molar ratio of functional monomer to template, volume ratio of CDs to CS, incubation time, and elution time. When tested with glucose as the model analyte, the MIECS demonstrated two assay ranges (0.5–40 μM and 50–600 μM) and a low LOD of 0.09 μM (S/N = 3) under optimized conditions. Additionally, the MIECS displayed remarkable selectivity, reproducibility, and stability. The sensor was successfully applied for preliminary glucose analysis in real human blood serum samples [77].

In a study published in 2021, a non-enzymatic MIECS sensor was developed for the detection of glucose. The sensor is constructed using a GCE that has been modified with hollow nickel nanospheres (HNiNS)/CDs–CS. These HNiNS and CDs–CS components are effective electrochemical materials, enhancing both the active surface area and electron-transport capabilities. In the preparation of the (MIP film, 3-aminobenzeneboronic acid and glucose were employed as the functional monomer and template, respectively. Various parameters were investigated, including the molar ratio of functional monomer to template, elution time, incubation time, and scan rates. The resulting HNiNS/CDs–CS/MIP/GCE sensor demonstrated strong electrochemical activity and remarkable recognition ability for glucose. Under optimal conditions, the sensor exhibited two broad linear response ranges for glucose concentrations, spanning 0.03–10 μM and 20–300 μM, with a relatively low LOD of 4.6 nM (S/N = 3). The sensor was successfully applied to detect glucose in human serum and fermentation samples [78].

In a proof-of-concept investigation published in 2020 by Bossard et al., the authors presented a study using laser-pyrolyzed paper as a substrate for rapid, enzyme-free detection of D-glucose via MIP technology. They imprinted the monomer 3-amino-4-hydroxybenzoic acid using electrochemical polymerization with D-glucose as the template molecule. The template removal was achieved through cyclic voltammetry in phosphate-buffered saline. The capacitance signal crucial for detection was extracted from the electrochemical impedance spectra after a short 60 s incubation with the analyte. Raman spectroscopy and scanning electron microscopy revealed highly porous nanographene on the electrodes. The sensor demonstrated strong linearity (R^2 = 0.97) within the clinically relevant range of 0–30 mmol dm^{-3}, with a limit of detection of 1.77 mmol dm^{-3}, excellent selectivity, and good reproducibility [79].

Karaman et al. developed a novel MIP electrochemical biosensor for glucose detection using a hierarchical N-rich carbon conductive-coated TNO structure (TNO@NC). They produced TNO@NC with minimal waste using a polypyrrole-chemical vapor deposition (PPy-CVD) method. Glucose imprinting was achieved via cyclic voltammetry. The biosensor showed selective, sensitive, and stable signals proportional to glucose concentrations, with a sensitivity of 18.93 μA μM^{-1} cm^{-2} (R^2 = 0.99) and a low LOD of 1.0 × 10^{-6} M, suggesting versatile applications in public health and food quality assessment [80].

Diabetes mellitus (DM) is commonly diagnosed through methods that can be invasive or involve enzymes, which are progressively seen as less desirable. If there is a correlation between salivary glucose (SG) levels and blood glucose (BG) levels, it could offer a useful avenue for early detection of DM. In their study, Diouf et al. employed an MIP approach to create an electrochemical non-enzymatic sensor for detecting SG at micro-molar concentrations. A MIP-based Au-SPE was developed by electropolymerizing Acrylamide/Bis-Acrylamide (AAM/NNMBA) in the presence of glucose as a template. Under optimal conditions, the sensor exhibited effective glucose detection while mitigating interferences from structurally similar substances like lactose and sucrose. Within a range of 0.5–50 μg mL^{-1}, the sensor demonstrated detection and quantification limits of 0.59 and 1.9 μg mL^{-1}, respectively. Moreover, real saliva glucose determinations were compared with finger prick blood measurements, yielding satisfactory results (R^2 = 0.99) through the partial least squares (PLS) statistical technique. This work has presented an economically viable, straightforward, and efficient sensing platform for non-enzymatic glucose detection. Consequently, this platform promises future advancements in accurate and dependable non-invasive DM diagnosis [81].

An electrochemical glucose sensor of remarkable sensitivity and selectivity was created by integrating gold nanoparticles-decorated molecularly imprinted polymers (AuNP-MIPs). This inventive approach involved the synthesis of AuNP-MIPs on a gold surface using a multistep amperometric technique. The process employed internally synthesized AuNPs alongside optimized conditions and glucose as the template molecule. Electrochemical tests on the AuNP-MIP sensor showcased its strong binding affinity to glucose, with a K_d of 3×10^{-8} M. In contrast, MIPs lacking AuNPs failed to detect the highest concentration within the investigative range (1.25 nM–2.56 μM). Comparative studies between AuNP-MIP and non-imprinted polymer (AuNP–NIP) underscored the remarkable selectivity of the former toward glucose. The sensor was challenged with interfering compounds (sucrose, dopamine, starch, and bovine serum albumin) to assess specificity, yielding minimal cross-reactivity except for sucrose. Innovatively, atomic force microscopy (AFM) was employed to investigate the behavior of imprinted polymers in fluid solvents, providing a new perspective. The sensor's capability to detect glucose in human serum was verified, showcasing an LOD of 1.25 nM and maintaining approximately 95% stability over a storage period of 40 days. This sensor's attributes, including its straightforward, cost-effective, and highly sensitive nature, hold significant promise for the detection of clinically relevant small molecules in the future [82].

An innovative voltammetric glucose sensor was produced using newly developed molecularly imprinted polymeric micelles (MIPMs) via a direct electrodeposition process. These MIPMs possessed characteristics of being photo-cross-linkable and nano-sized, resulting in a substantial specific surface area. The creation of MIPMs involved the self-assembly of an amphiphilic photo-cross-linkable copolymer combined with a molecular imprinting technique using glucose as the template molecule. An MIP film was established in real-time on the electrode surface by electrodepositing the MIPMs, and the subsequent photo-crosslinking produced a robust film that demonstrated notable resistance to solvent dissolution. This distinctive sensor exhibited a favorable response and selectivity toward glucose. The sensor's linear response ranged from 0.2 to 8 mM, with a relatively higher LOD of approximately 10 mM. This high limit indicated the presence of numerous effective recognition sites within the polymer matrix due to the extensive specific surface area of the MIPMs. Furthermore, the MIP sensor exhibited stability and reversibility [83].

Insulin, the polypeptide hormone responsible for regulating blood sugar levels, indicates both types of diabetes. An electrochemical sensor for insulin was created using a gold electrode customized with carboxylated f-MWCNTs and a MIP cryogel. This cryogel had specialized binding sites for insulin and large pores facilitating insulin movement. f-MWCNTs increased surface area and conductivity, reducing the energy needed for insulin oxidation. Insulin oxidation was directly measured using SWV in a flowing system. The resulting sensor had a linearity range of 0.050–1.40 pM, with a low LOD of 33 fM. It showed specificity and remained stable for 10 weeks when stored dry. When applied to human serum, it closely matched the results from the Elecsys insulin assay. This MIP sensor offers promise for diagnosing and managing diabetes [84].

Zhao et al. developed a novel electrochemical sensor to directly identify insulin using a molecularly imprinted technique known as epitope imprinting. In this process, a MIP was generated through electro-polymerization. The C-terminal polypeptide of insulin was utilized as the template molecule, and *o*-PD was employed as the functional building block. After removing template molecules using a NaOH solution, three-dimensional imprinting sites were formed, structurally matching the

C-terminal polypeptide of the insulin molecule. These imprinting sites could selectively recognize and bind to insulin molecules. By employing $K_3[Fe(CN)_6]$/$K_4[Fe(CN)_6]$ as an indicator, insulin could be indirectly detected. The response current displayed a linear correlation with insulin concentration, in the range of 1.0×10^{-14} to 5.0×10^{-13} M, and the LOD was 7.24×10^{-15} M. The sensor's development showcased strong specificity and stability. Furthermore, its successful application was demonstrated in determining insulin within serum samples, yielding satisfactory outcomes [85].

An electrochemical sensor utilizing silylated graphene oxide (Si-GO) was created for the specific and selective detection of the diabetes biomarker 2-Aminoadipic Acid (2-AAA). The sensor's production involved molecular imprinting technology combined with electrochemical sensing. Various MIPs were synthesized using different monomers, including 2-Acrylamido-2-methylpropanesulfonic acid (AMPS), MAA, and Vinyl pyridine (VPy), in different proportions. EGDMA was a cross-linker, while potassium peroxodisulphate (KPS) was an initiator. The modified electrode was created by modifying a GCE surface with the MIP (MIP/GCE) through a drop-casting method. The modified MIP electrode exhibited an amplified redox peak with a potential difference of 0.70 V and a redox current of 0.015 mA. This indicated the presence of complementary active sites of 2-AAA within the MIP. In contrast, the non-imprinted polymer (NIP) electrode displayed a weaker redox peak difference than the MIP. The optimal response time was 15 min, and the pH was 5. DPV analysis exhibited a good linear relationship with an R^2 value of 0.9932. The LOD and LOQ were calculated as 0.40×10^{-11} and 1.34×10^{-11} M, respectively [86].

Roy et al. introduced a novel flexible electrochemical sensor that monitors Diabetic Foot Ulcers (DFUs) by employing molecular imprinting on over-oxidized polypyrrole. DFUs are persistent wounds associated with diabetes that significantly impact the daily lives of affected individuals. Due to the notable increase in l-tyrosine levels in such cases, there's a pressing need for flexible sensing tools to quantify this biomarker. This helps assess the severity of foot ulcers to prevent complications, amputations, and the onset of serious symptoms. The flexible molecularly imprinted sensor was fabricated on polyethylene terephthalate (PET) substrates with carbon screen printing. This was achieved through the electro-polymerization of pyrrole followed by the incorporation of l-tyrosine into the polymer matrix. The developed sensor displayed a linear response over a wide concentration range of l-tyrosine, from 10 pM to 1 mM. The LOD was measured at 12.50 pM, and the sensor's sensitivity was calculated as 6.01 nA/pM/mm^2. Additionally, the sensor exhibited a rapid response time of under 1 min. The sensor's high selectivity for tyrosine was attributed to the over-oxidized nature of the templated poly-pyrrole. Furthermore, the sensor demonstrated a shelf life of 3 months, making it a suitable component for future connected health platforms designed for monitoring DFUs [87].

7.8 MIP-Based Electrochemical Sensors for Autoimmune Disease Marker Detection

Cytochrome c (Cyt c) serves as a crucial biomarker during the initial stages of apoptosis, contributing to the identification and treatment of various diseases, including cancer and autoimmune diseases. In their study, Campagnol et al. focus on an electrochemical sensor that employs an MIP to detect Cyt c with exceptional sensitivity. The MIP is developed through electro-polymerization of o-phenylenediamine in the presence of Cyt c, creating specific binding sites for Cyt c after subsequent solvent extraction. The voltammetric results reveal that the MIP-based sensor acts as a partially obstructed electrode surface. The obtained partition isotherm conforms to the Langmuir model, underscoring a strong affinity of the MIP for Cyt c, with a high association constant (K_a) of 5×10^{11} M^{-1}. By employing differential pulse voltammetry, remarkably heightened analytical sensitivity and an impressively low LOD in the femtomolar range are achieved, with minimal non-specific adsorption. Rigorous analytical recovery tests are conducted, encompassing potential interfering proteins and diluted human serum. These tests affirm the selectivity of the MIP-based sensor, reinforcing its potential for real sample analysis [88].

In a study published in 2022, a new type of functional monomer called pyrrole histidine was innovatively designed and synthesized by Çorman et al. This monomer was then used to create a specialized electrochemical sensor using molecular imprinting techniques to detect teriflunomide (TER). The process involved forming a molecularly imprinted thin film on a GCE by electropolymerizing a copolymer made up of pyrrole histidine (PyHis) and pyrrole in the presence of the TER template. The sensor, known as poly(pyrrole-co-pyrrole-histidine) [Poly(Py-co-PyHis)]@MIP/GCE, demonstrated a linear response to TER concentrations ranging from 0.1 to 1.0 pM, achieving a remarkably low LOD of 11.38 fM. This electrochemical sensor design successfully effectively recovered TER in synthetic serum samples and tablet dosage forms, with recovery rates of 97.56% and 100.35%, respectively. The developed [Poly(Py-co-PyHis)]@MIP/GCE sensor displayed exceptional electrochemical sensitivity to TER, a result attributed to the combined effect of the conducting polymer and the molecular imprinting techniques employed [89].

Interleukin-6 (IL-6) is a key cytokine that plays a crucial role in the immune response and inflammation, and its dysregulation has been implicated in various autoimmune diseases. Autoimmune diseases occur when the immune system mistakenly attacks healthy cells and tissues, leading to chronic inflammation and tissue damage. IL-6 is involved in several processes that contribute to autoimmune diseases:

- **Immune Cell Activation**: IL-6 is a potent activator of immune cells, including T and B cells. In autoimmune diseases, elevated IL-6 levels can drive the activation and proliferation of these immune cells, leading to an increased immune response against the body's tissues.

- **B Cell Differentiation**: IL-6 plays a crucial role in promoting the differentiation of B cells into plasma cells, which produce antibodies. In autoimmune diseases, this process can produce autoantibodies—antibodies targeting the body's tissues—resulting in tissue damage and inflammation.
- **Th17 Cell Differentiation:** IL-6 is a key factor in differentiating a subset of T helper cells known as Th17 cells. Th17 cells are involved in promoting inflammation and autoimmunity. Excessive Th17 cell activity has been implicated in autoimmune diseases such as rheumatoid arthritis, psoriasis, and multiple sclerosis.
- **Inflammatory Response:** IL-6 stimulates the production of other pro-inflammatory cytokines and chemokines, contributing to the overall inflammatory response in autoimmune diseases. This chronic inflammation can lead to tissue damage and exacerbate the autoimmune process.
- **Cytokine Storm**: In some autoimmune diseases, there can be a phenomenon called a "cytokine storm," with an overwhelming release of various cytokines, including IL-6, into the bloodstream. This excessive cytokine release can result in severe inflammation and tissue damage.
- **Systemic Effects:** Elevated IL-6 levels can have systemic effects on the body, impacting various organs and tissues. This can contribute to the multi-organ involvement often observed in autoimmune diseases.

In a study published in 2021, researchers presented an extremely sensitive biomimetic sensor to identify IL-6. Co-electropolymerizing pyrrole (Py) and carboxylated pyrrole alongside IL6 on an SPCE crafted the imprinted polymeric film. By utilizing oxalic acid, the IL6 biomarker molecule was subsequently eliminated to form recognition sites devoid of the biomarker. A control sample was prepared without the IL-6 biomarker as a comparative reference. The biomimetic sensor exhibited an LOD for IL-6 at 0.02 pg mL^{-1} in serum samples enriched with the biomarker. This biosensing device demonstrated notable sensitivity, reproducibility, accuracy, and swift response time [90].

Özcan et al., in their study, introduce a novel MIP biosensor utilizing a composite of graphene quantum dots (GQDs) and functionalized MWCNTs (f-MWCNTs) for the specific detection of IL-6 protein. Creating the IL-6 protein imprinted electrode involved depositing it onto the GQDs/f-MWCNTs composite while utilizing a solution containing 100.0 mM pyrrole and 25.0 mM IL-6 protein. The biosensor demonstrated a linearity range of 0.01–2.0 pg mL^{-1} and a remarkably low LOD of 0.0030 pg mL^{-1}, making it well-suited for analytical applications involving plasma samples. The biosensor's performance was further validated through stability, repeatability, and reproducibility assessments. In summary, this type of sensor is a promising tool for various analytical applications, particularly in plasma samples [91].

The following research involved the creation of an impedimetric biochip using molecularly imprinted polydopamine (MIP(pDa)) on peptide nanotube (PNT) functionalized screen-printed electrodes (SPEs). This biochip was a novel support matrix for constructing an electrochemical sensor to detect IL-6. IL-6 was utilized as a template molecule during the self-assembly polymerization process of dopamine. Dopamine functioned as a functional building block, initiating self-polymerization without requiring any initiators, enzymes, or cross-linkers. The designed electrode facilitated the selective binding of IL-6, allowing for the detection of IL-6 biomolecules within a concentration range of 1–200 pg mL^{-1}. This capability arose from the strong relationship between the redox pair response and the logarithm of IL-6 concentration. The MIP(pDa)/PNT electrode exhibited notable selectivity for IL-6, with high recovery rates observed in urine samples. This innovative method enabled the efficient, uncomplicated, and cost-effective detection of IL-6. The integration of MIP(pDa)/PNT with SPEs and the application of impedimetric detection presented a promising avenue for clinical diagnostics [92].

Afsarimanesh et al. introduced a method for selective detection of C-terminal telopeptide type I collagen (CTx-I). This is achieved using EIS. Artificial recognition sites for CTx-I were developed through precipitation polymerization to create a molecularly imprinted polymer. This involved using the CTx-I peptide as a template, MAA as a functional building block, and ethylene glycol methacrylate as a cross-linker. The sensor's performance showcased a linearity range from 0.1 to 2.5 ng mL^{-1}, with a LOD of 0.2 ng mL^{-1} [93].

Sundhoro et al. demonstrated the viability of such a device through the utilization of MIPs. The MIP-based sensor can detect soy-related allergen markers at concentrations as low as 100 ppb, two orders of magnitude below clinically significant levels. This capability was demonstrated in both controlled environments and intricate food samples. The efficacy of the sensor was cross-validated qualitatively using commercially accessible soy allergen detection lateral flow devices (LFDs). The results of this application tackle a longstanding analytical hurdle, paving the way for rapid, cost-efficient, and scalable techniques for the direct identification of allergen traces in food analysis [94].

A highly sensitive electrochemical sensor was developed for detecting β-lactoglobulin, which is crucial for cow's milk allergy safety. It utilized a carbon electrode modified with an MIP synthesized via a hydrothermal technique. Choline chloride served as the functional component, β-lactoglobulin as the template, and EGDMA as the cross-linking agent. To enhance specificity, the MIP was immobilized on a substrate comprising polyethyleneimine-reduced graphene oxide-gold nanoclusters (Au-NCs). Under optimized conditions, the sensor detected β-lactoglobulin with a linear range from 10^{-9} to 10^{-4} mg mL^{-1} and a LOD of 10^{-9} mg mL^{-1}. It performed consistently across four market-sourced milk samples, indicating practical applicability with real-world samples [95].

Artificial receptors designed to mimic their natural biological counterparts offer numerous benefits, including reduced production expenses, increased stability and versatility in storage, and the resolution of ethical concerns linked to animal-derived antibodies. In their study, Costa et al. custom-made MIP-allergen receptors were developed to match or even surpass the performance of biological antibodies. The objective was to create a specialized MIP as a synthetic antibody for detecting the hazelnut allergen Cor a 14. The target protein was integrated into a conducting polypyrrole receptor film on gold screen-printed electrodes (Au-SPE) to achieve this.

Electrochemical testing revealed a linear response within a dynamic range of 100 fg mL^{-1} to 1 μg mL^{-1}, with a LOD of 24.5 fg mL^{-1}, determined through SWV using calibration curves prepared with standards dissolved in phosphate buffer [96].

A study published in 2022 focuses on modifying surfaces using a combination of MIPs and SPIONs to detect gluten. SPIONs were synthesized through a chemical co-precipitation process and combined with MIPs via the polymerization of gluten and methyl methacrylate (MMA), serving as the template and monomer. This magnetic MIP (MMIP) was applied to modify a carbon-plate electrode. The MMIP electrode designed for gluten detection displayed a dynamic linear range of 5–50 ppm, boasting a correlation coefficient of 0.994. Furthermore, it exhibited a low LOD of 1.50 ppm, below the threshold the U.S. Food and Drug Administration (20 ppm) set. The sensor showcased exceptional selectivity, sensitivity, stability, and reproducibility qualities, making it a promising tool for gluten detection and analysis [97].

Lysozyme (LYZ) is a compact, positively charged protein with applications in medical treatment and the food industry as an antibacterial agent, although it can trigger allergic reactions. This research introduces high-affinity nanoMIPs specially designed for LYZ using a solid-phase methodology. These nanoMIPs were integrated into SPEs. The rapid measurement of LYZ at trace levels (pM) and the distinction between LYZ and structurally similar proteins (such as BSA and troponin-I) were facilitated by EIS within a short timeframe of 5–10 min [98].

7.9 MIP-Based Electrochemical Sensors for Infectious Agent Monitoring

The COVID-19 pandemic, caused by the novel coronavirus SARS-CoV-2, is a global health crisis that began in late 2019. Because of that, it is important for accurate and cost-effective portable sensors to aid in clinical diagnosis of COVID-19 [99].

In a study published in 2021, Raziq et al. introduce the pioneering development of an electrochemical sensor based on MIPs for detecting the nucleoprotein (ncovNP) of SARS-CoV-2. The sensor's core component is a disposable sensor chip—a thin film electrode—integrated with a MIP designed to exhibit specificity for ncovNP. This sensor chip is connected to a portable potentiostat. The resultant ncovNP sensor displayed a linear response to ncovNP present in the lysis buffer, with a range extending up to 111 fM. The sensor exhibited an LOD of 15 fM and a quantification limit of 50 fM. Significantly, the sensor successfully identified the presence of ncovNP in nasopharyngeal swab samples obtained from COVID-19-positive patients. This innovative approach opens up new avenues for rapidly developing diagnostic tools for COVID-19 [99].

A novel electrochemical sensor based on an MIP for detecting SARS-CoV-2-RBD was developed by Tabrizi et al. The sensor fabrication process involved the creation of a macroporous gold screen-printed electrode (MP-Au-SPE). The MIP was then synthesized onto the MP-Au-SPE surface through electro-polymerization of ortho-phenylenediamine. During this process, the SARS-CoV-2-RBD molecules were incorporated into the polymer matrix. Subsequent removal of the template molecules from the electrode was achieved using alkaline ethanol. The detection of SARS-CoV-2-RBD in saliva solutions was carried out using the EIS technique. The MIP-based sensor exhibited strong performance, showing a reliable response within a concentration range of 2.0–40.0 pg mL^{-1} of SARS-CoV-2-RBD, with an impressively low LOD of 0.7 pg mL^{-1}. These findings indicated that the developed MIP sensor displayed exceptional attributes in terms of selectivity, sensitivity, and stability for detecting SARS-CoV-2-RBD [100].

In a study published in 2021, Hashemi et al. introduce a novel 2D MIP platform built on graphene oxide decorated with an interconnected complex of polypyrrole-boronic acid. This innovative platform is engineered to detect SARS-CoV-2 antigen in aquatic biological samples with unparalleled sensitivity and specificity. The LOD achieved is 0.326 and 11.32 fg mL^{-1} using voltammetric and amperometric assays, respectively. Furthermore, the developed MIP platform showcases exceptional stability, selectivity, and accuracy in detecting the target antigen. These promising attributes open the door for creating highly specific and rapid diagnostic setups capable of detecting the antigen within just 1 min or 20 s using voltammetric or amperometric techniques. This advancement can revolutionize screening and diagnostic procedures, offering swift and precise detection of viral biomarkers [101].

Ayankojo et al., in their work, present the development of an electrochemical sensor that relies on a MIP synthetic receptor to accurately quantify the SARS-CoV-2 spike protein subunit S1 (ncovS1). The sensor's design leverages the covalent interaction between the 1,2-diols found in the heavily glycosylated protein and the boronic acid group of 3-aminophenyl boronic acid (APBA). The sensor exhibits reliable performance, offering a response time of 15 min, and it can detect ncovS1 in both phosphate-buffered saline and nasopharyngeal samples from patients. The LOD values are 15 and 64 fM, respectively [102].

Hepatitis C, caused by the Hepatitis C virus (HCV), is the most prevalent liver disease and can lead to severe health complications such as cirrhosis and hepatocellular carcinoma. A study published in 2022 focuses on creating a sensor for directly detecting HCV. The sensor employs a synthetic recognition element produced using molecular imprinting technology. This element is a MIP designed to have specific binding sites for the HCV envelope protein E2 (E2-MIP). By incorporating E2-MIP into an electrochemical sensor platform, the sensor can quantitatively assess the interaction between free E2 protein and HCV-mimetic particles (HCV-MPs) in human plasma. The sensor achieves a LOD of 4.6×10^{-4} ng mL^{-1} for HCV-MPs. This innovative electrochemical sensor for HCV detection offers a straightforward, rapid, cost-effective alternative to existing methods. It also opens the door to point-of-care diagnostics for HCV, providing a promising solution for addressing this pressing public health issue [103].

The following study introduces a novel approach to detecting HCV core antigens. The method employs a combination of aptasensing and molecular imprinting techniques, utilizing a multi-walled carbon nanotubes-chitosan nanocomposite (MWCNTs-Chit) as an enhanced immobilization interface. This interface enhances sensor conductivity and performance by enabling higher loading of the aptamer (Apt) sequence specific to the HCV core antigen. The sensor fabrication involves electro-polymerizing dopamine (DA) around the Apt [HCV core antigen] complex on a modified GCE coated with MWCNTs-Chit. This innovative sensor offers rapid response, cost-effectiveness, excellent sensitivity, and stability. Various electrochemical techniques assessed its performance, including CV, DPV, and EIS. The sensor's response demonstrates a linear detection range from 5.0 fg mL^{-1} to 1.0 pg mL^{-1}, with an impressively low LOD of 1.67 fg mL^{-1}. Notably, the proposed method was successfully applied to quantifying HCV core antigen in human serum samples, yielding satisfactory results. The developed sensor holds significant potential for practical applications in analyzing real samples [104].

Dengue is a mosquito-borne viral infection caused by the dengue virus (DENV), which belongs to the Flaviviridae family. It is a significant global public health concern, particularly in tropical and subtropical regions.

A novel impedimetric MIP sensor has been created to detect early-stage dengue infection. The approach involved modifying a screen-printed carbon electrode (SPCE) with electrospun polysulfone (PS) nanofibers, followed by a dopamine coating. During this process, non-structural protein 1 (NS1), a specific and sensitive biomarker for dengue virus infection, was used as a template. The self-polymerization of dopamine at room temperature allowed for the precise preservation of the NS1 template's structure, resulting in the formation of tailored imprinted sites that precisely match the target analyte. The impedimetric measurements exhibited a linear response within the 1–200 ng mL^{-1} concentration range. The newly created sensor exhibited the capacity to specifically identify NS1 concentrations as minimal as 0.3 ng mL^{-1}. Furthermore, the impedimetric sensor system was applied to determine NS1 levels in actual human serum samples. The results indicated satisfactory recoveries ranging from 95% to 97.14%, with less than 5% standard deviations. This advancement showcases a valuable tool for early dengue infection detection with its potential use in practical applications [105].

Buensuceso et al. demonstrated a direct and on-site method for dengue infection detection using a molecularly imprinted polyterthiophene film sensor. They imprinted a dengue nonstructural protein 1 (NS1) epitope, employing an epitope imprinting strategy with short peptide sequences as templates. Sensor fabrication involved cyclic voltammetry (CV) and real-time monitoring with electrochemical quartz crystal microbalance (EC-QCM). Calibration curves established linear detection ranges of 0.2–30 μg mL^{-1} and 0.2 to 10 μg mL^{-1}, with LODs of 0.073 μg mL^{-1} and 0.056 μg mL^{-1}, respectively. The resulting electropolymerized molecularly imprinted polyterthiophene (E-MIP) sensor exhibited excellent stability, sensitivity, and selectivity, suggesting precise cavity imprinting with tailored dimensions and functionalities [106].

Initially discovered in 1947, the Zika virus (ZIKV) didn't raise significant concern due to its limited impact. However, recent associations with microcephaly sparked increased research interest, notably in rapid detection methods. Tancharoen et al. introduced a novel ZIKV electrochemical biosensor using surface imprinted polymers and graphene oxide composites. The biosensor detects ZIKV by monitoring electrical signal changes as virus concentrations vary in buffer solutions and serum samples. It achieved a LOD of 2×10^{-4} PFU mL^{-1} (equivalent to 1 RNA copy mL^{-1}) in PBS and 2×10^{-3} PFU mL^{-1} to 5×10^{-2} PFU mL^{-1} (equivalent to 10–250 RNA copies mL^{-1}) in serum samples. SEM imagery estimated virus particle concentration, suggesting the biosensor's potential for accurate ZIKV detection, aiding in efficient and timely identification [107].

Nosocomial infections pose a significant challenge within hospital settings, directly impacting vulnerable patients or those with weakened immune systems. Among these infections, *Klebsiella pneumoniae (K. pneumoniae)* stands out as a primary culprit, causing a range of ailments in healthcare facilities. These include pneumonia, urinary tract infections, septicemias, and soft tissue infections. Notably, *K. pneumoniae* has developed heightened antibiotic resistance, further complicating treatment. In a study published by Pintavirooj et al., an innovative approach was taken to address the issue. A MIP-based electrochemical biosensor was developed for *K. pneumoniae* detection. This biosensor employed electrochemical techniques to gauge electrical signal changes across varying *K. pneumoniae* concentrations, ranging from 10 to 10^5 CFU mL^{-1}. The MIP-based *K. pneumoniae* sensor displayed a robust linear response, yielding a high correlation coefficient (R^2 value) of 0.9919. The sensor's sensitivity was assessed against bacteria sharing similar structural features with *K. pneumoniae*. Notably, the MIP-based *K. pneumoniae* biosensor, integrated with a gold electrode, exhibited superior sensitivity, registering a value of 7.51 (% relative current/log concentration). The sensitivity for the MIP sensors with *Pseudomonas aeruginosa* and *Enterococcus faecalis* was 2.634 and 2.226, respectively. Furthermore, the biosensor demonstrated remarkable performance with a LOD at 0.012 CFU mL^{-1} and a LOQ at 1.61 CFU mL^{-1}. This achievement can enhance *K. pneumoniae* detection accuracy within hospital environments, enabling early identification and management of nosocomial infections [108].

A study published in 2022 by Sharma et al. outlines the production of an MIP and its application in the electrochemical sensing of *K. pneumoniae* bacteria using an electrochemical method. The molecular imprinting process involved the polymerization of pyrrole in the presence of *K. pneumoniae*, followed by the removal of bacteria through ultrasonication to obtain the MIP. The electrode fabrication was achieved through electrophoretic deposition (EPD) of the MIP onto the surface of hydrolyzed ITO-coated glass. The DPV technique was employed for detection. The resulting sensor demonstrated excellent sensitivity (3 μA mL CFU-cm^{-2}) and LOD of 1.352 CFU mL^{-1} within a linear detection range from 1 to 105 CFU mL^{-1}. Selectivity studies were conducted against two other bacteria and various ions in healthy human urine. Spiked urine samples were tested to assess the sensor's applicability in real samples [109].

Pseudomonas aeruginosa, a bacterium known for causing prolonged chronic illnesses, can initiate infection even at low concentrations. Hence, it's crucial to promptly detect *Pseudomonas aeruginosa*. The following research from 2021 introduces a novel approach that combines molecular imprinting and aptasensing to quantify *Pseudomonas aeruginosa* (*P. aeruginosa*). The sensing performance is exceptional by harnessing the dual precise molecular recognition attributes of molecular imprinting polymers and aptamers. Gold nanoparticles are employed in this method as specific intermediaries and interfaces to enhance the loading efficiency of the aptamer sequence. The developed sensor exhibits an extensive linear dynamic range from 10^1 to 10^7 CFU mL^{-1} and boasts a LOD of 1 CFU mL^{-1}. Furthermore, the sensor's efficacy in identifying *P. aeruginosa* in blood samples was promising. This sensor offers several advantages, including cost-effectiveness, robust stability, rapid response time, and heightened sensitivity in measuring *P. aeruginosa* [110].

Liustrovaite et al. aimed to create an electrochemical sensor for detecting *Listeria monocytogenes* using modified platinum and screen-printed carbon electrodes (SPCEs) with an MIP application. They deposited non-imprinted polypyrrole (NIP-Ppy) and *Listeria monocytogenes*-imprinted polypyrrole (MIP-Ppy) layers on both SPCEs and Pt electrodes via electrochemical deposition. Various solutions (sulphuric acid, acetic acid, L-lysine, and trypsin) were used to extract bacteria from the Ppy-modified electrodes to optimize sensitivity and repeatability. The electrodes' performance was evaluated using pulsed amperometric detection (PAD). The most effective MIP-Ppy/SPCE sensor design involved using trypsin to remove bacteria. The LOD and LOQ for MIP-Ppy/SPCE were determined as 70 and 210 CFU mL^{-1}, respectively, with a linear detection range from 300 to 6700 CFU mL^{-1} [111].

Acinetobacter baumannii (A. baumannii), a troublesome source of nosocomial infections globally, poses a substantial threat to healthcare settings, leading to considerable illness and fatalities, particularly among intensive care unit patients. The presence of these bacterial strains in infections is a grave concern for healthcare professionals, driving changes in antibiotic strategies and hospital disinfection practices. Various procedures have been implemented to identify *A. baumannii*-colonized samples or hospital equipment, enabling appropriate decontamination measures. This study introduces a novel diagnostic test for A. *baumannii* using a MIP. The MIP was created on the surface of a GCE through the electro-polymerization of dopamine monomers alongside *A. baumannii* templates. Electrochemical techniques, including CV and impedance spectroscopy, were utilized to characterize the modified electrodes with non-imprinted polymer (NIP) and MIP in the context of the hexacyanoferrate redox system as an electrochemical probe. The sensor demonstrated a linear detection range from 10^2 to 10^7 CFU L^{-1} for *A. baumannii*, with an LOD of 30 CFU L^{-1}. The sensor effectively evaluated *A. baumannii* recovery from human blood serum samples. This represents the initial instance of a MIP sensor being used for *A. baumannii* detection [112].

Mycobacterium leprae is responsible for the endemic disease leprosy, which can become chronic if not diagnosed promptly. To facilitate early diagnosis, it was developed an epitope-imprinted sensor. Specifically, a MIP nanoparticles-based electrochemical QCM sensor was created to detect *Mycobacterium leprae* bacteria using its epitope sequence. To imprint this bacterial epitope, researchers employed multiple monomers, including 3-sulphopropyl methacrylate potassium salt, benzyl methacrylate, and 4-aminothiophenol. These imprinted nanoparticles were electropolymerized onto a gold-coated quartz electrode. Remarkably, the sensor exhibited specific binding capabilities to blood samples from infected patients, even in the presence of interfering substances such as 'matrix' and various plasma proteins like albumin and globulin. Furthermore, other peptide sequences closely resembling the epitope sequence but with two amino acid mismatches failed to exhibit any binding. The sensor underwent rigorous analytical testing, including selectivity, specificity, and matrix effects assessments, and demonstrated a low LOD of 0.161 nM and LOQ of 0.536 nM, along with excellent reproducibility (RSD 2.01%). Consequently, researchers successfully created a diagnostic tool for the bacterium responsible for leprosy straightforwardly, opening up the possibility for widespread clinical use and efficient population screening [113].

Typhoid disease, caused by *Salmonella typhi*, can be fatal if not diagnosed and treated promptly. In order to facilitate timely and accurate diagnosis, especially at an early stage, a biomimetic sensor was developed based on epitope imprinting technology for typhoid detection. Harijan et al. utilized immunoinformatic tools to predict the Salmonella invasive protein D (SipD) epitope sequence and employed it as the imprinting target. A molecularly imprinted matrix was created on the surface of a QCM chip through electropolymerization, using a synthetic 14-mer peptide that matched the predicted epitope sequence as the template. The sensor demonstrated selectivity by exhibiting a binding frequency shift of approximately −3000 Hz when exposed to blood plasma samples from infected individuals, indicating recognition of SipD proteins from the bacteria. The sensor proved capable of detecting the bacteria at an early stage, with a low LOD value of 0.50 ng mL^{-1}. Upon the extraction of the peptide sequence, the imprinted cavities selectively and specifically bound the targeted epitope sequences, both in laboratory samples and in 'real' samples from patients. This work presents an accurate, robust, cost-effective, and user-friendly diagnostic tool that has the potential to greatly assist the healthcare industry [114].

In a study published in 2021 by Wang et al. was developed an impedimetric sensor utilizing a conductive poly(3-thiopheneacetic acid) film that has been imprinted with bacteria (BICP), designed for the rapid, sensitive, and label-free detection of *Staphylococcus aureus* (*S. aureus*). The preparation of the BICP film was both convenient and environmentally friendly. It involved the in situ deposition of the BICP film onto a gold electrode surface, eliminating the need for toxic organic solvents and cross-linkers (Fig. 7.5). Electrochemical techniques and scanning electron microscopy were employed to characterize the imprinting and recognition process. Interestingly, the BICP exhibited a unique structure devoid of cocci-shaped cavities typically observed in poly(3-thiopheneacetic acid) (PTAA) matrices. Researchers investigated various factors influencing the imprinting and recognition processes to optimize sensing performance. Under the ideal conditions, the BICP film-based impedimetric sensor exhibited exceptionally rapid recognition within 10 min, an impressively low LOD of 2 CFU mL^{-1}, and a wide linear range from 10 to 10^8 CFU mL^{-1}. The sensor displayed high selectivity, universality, and repeatability. Additionally, we demonstrated the applicability of this sensor in analyzing real milk samples. This sensor offers a straightforward and versatile approach for the swift, sensitive, and specific detection of pathogenic bacteria, potentially making significant contributions to fields such as food safety [115].

Mugo et al. developed a versatile electrochemical sensor capable of detecting *Escherichia coli*, a pathogenic bacterium. This sensor is constructed using a

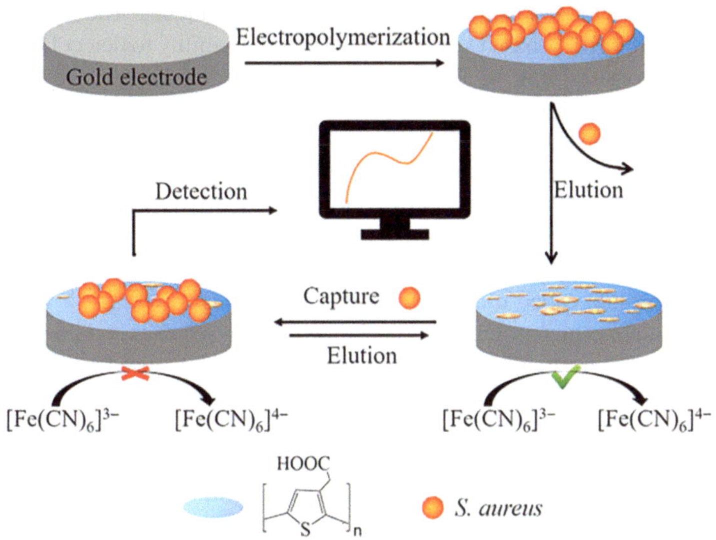

Fig. 7.5 Schematic illustration for the fabrication of BICP film-based impedimetric sensor for *S. aureus* detection. Reprinted from [115]

layer-by-layer assembly approach, combining MWCNTs, nanocellulose (CNC) films, and polyaniline (PANI) doped with phenylboronic acid (PBA). The sensing layer is a poly(methacrylic acid)-based pathogen microcontact imprinted polymer (PIP). The PIP@PBA/PANI@CNT/CNC sensor exhibits a strong affinity for *E. coli* even in complex biological matrices and can effectively distinguish *E. coli* from other bacterial species. Employing capacitance and impedance as transduction methods, the PIP sensor achieves a low LOD for *E. coli*, measuring at 8.7 ± 0.5 CFU mL^{-1}, with a rapid response time of ≤5 min [116].

7.10 MIP-Based Electrochemical Sensors for Neurological Disease Biomarker Detection

Parkinson's disease (PD) is a progressive and complex neurological disorder that primarily affects a person's motor functions. It is characterized by a range of motor and non-motor symptoms, and its exact cause is not yet fully understood [117].

Kumar et al. focused on PARK7/DJ-1, a protein known to be a highly sensitive biomarker for PD. They targeted the DJ-1 protein to create a specialized sensor using molecularly imprinted polypyrrole (MIPPy). The process involved the development of the MIPPy DJ-1 imprinted sensor through electro-polymerization of pyrrole (Py) monomer on an SPCE surface using CV. This method aimed to create nanostructures with high selectivity and molecular recognition capabilities. Additionally, the researchers compared this MIPPy sensor with an impedimetric immunosensor conjugated with AuNPs. The MIPPy and immuno sensors were thoroughly characterized using CV, EIS, and SEM techniques. The analytical performance of the MIPPy-based sensor demonstrated its ability to detect DJ-1 protein at concentrations as low as 1 nM, with a linear detection range from 1 to 500 nM. These results were comparable to those obtained with the DJ-1 immunosensor, with an LOD of 1 nM and a linear range from 1 to 500 nM. Importantly, these MIPPy and immuno sensors successfully analyzed DJ-1 protein levels in overexpressed and control N2A cells. Furthermore, the MIPPy-based sensor exhibited a strong correlation with the AuNP/PPy immunosensor and Western Blot analysis, confirming its reliability and accuracy in detecting DJ-1 protein [117].

In PD, a protein called α-Synuclein plays a significant role and is the primary component of abnormal protein deposits known as Lewy bodies and Lewy neurites, characteristic disease markers. Lee et al., in their study, focused on developing a highly sensitive electrochemical method for detecting α-Synuclein. Researchers used conductive polymers to synthesize three specific alpha-synuclein peptides as templates for molecular imprinting. These imprinted polymers were designed to selectively recognize and interact with alpha-synuclein, allowing for precise electrochemical measurements. All three peptide targets were successfully imprinted, although their effectiveness varied, likely due to differences in the carboxylic acid groups present in the target peptides. These carboxylic acids can bind to the aniline (AN) and m-aminobenzenesulfonic acid (MSAN) components of the MIP. Under optimized conditions, the imprinted electrode could detect alpha-synuclein from 1 to 1000 fg mL^{-1}, with a LOD of 10 fg mL^{-1}. As a result, this technology was applied to measure alpha-synuclein levels in the culture medium of human brain organoids derived from both normal individuals and those with idiopathic PD [118].

In their paper, Ma et al. synthesized a nanospherical conjugated microporous polymer (CMP) and used it with graphene nanosheets (GS) to enhance electrode performance. These materials were employed to create a molecularly imprinted electrochemical sensor for the highly sensitive detection of α-Synuclein. The MIP sensor was fabricated through the electro-polymerization of pyrrole and underwent characterization via SEM, CV, and DPV. Incorporating CMP-GS nanocomposites significantly improved the electron transport rate, signal intensity, and the electroactive surface of the GCE. Consequently, this enhancement led to improved sensitivity of the MIP sensor. It's worth highlighting that, when optimized, this sensor demonstrated an extensive linear detection range from 1×10^{-4} to 8 ng mL^{-1}, with a remarkably low LOD of 3.5×10^{-5} ng mL^{-1} at S/N = 3, compared to existing methods. The sensor exhibited high sensitivity, low susceptibility to interference, and excellent stability, making it a promising approach for detecting α-Synuclein in the blood of PD patients. Furthermore, this study introduces an innovative application of CMPs in developing MIP electrochemical sensors [119].

A new electrochemical sensor designed for detecting α-synuclein has been created. One well-documented cause of PD is a triplication in the α-synuclein (SNCA) gene, resulting in increased SNCA levels, which leads to the aggregation and subsequent neurotoxic effects responsible for the disease's progression. Consequently, the ability to detect SNCA levels is of critical diagnostic importance. This study employed an electrochemical approach to optimize the controlled assembly of poly(hydroxymethyl 3,4-ethylenedioxythiophene) nanotubes. These nanotubes were created using a peptide derived from SNCA as a template molecule, facilitating the formation of electrically conductive polymers imprinted with the peptide on specialized sensing electrodes. These tubular nanostructures may enhance the sensor's effective detection range. The sensor demonstrated the capability to detect SNCA peptide concentrations from 0.065 pM to 0.65 nM, boasting an impressively low LOD of 4.0 pM. Furthermore, this imprinted electrode was successfully applied to identify elevated levels of SNCA in a culture medium collected from midbrain organoids that were specific to human PD patients [120].

Ayankojo et al. (2023) introduce an electrochemical sensor combining an MIP biomimetic receptor with thin film metal electrodes for rapid detection of brain-derived neurotrophic factor (BDNF), a neurodegenerative disorder biomarker. The MIP-based sensor utilizes surface imprinting for specific binding sites. Under optimal conditions, it achieves an impressive LOD for BDNF, detecting concentrations as low as 9 pg mL^{-1}. It exhibits over fivefold selectivity against related neurotrophic factors and 30-fold selectivity against proteins with similar pI values. Competitive binding assays confirm its selectivity. Successful validation in fetal bovine serum highlights its potential for real sample analysis, offering a reliable, cost-effective platform for neurodegenerative disorder diagnosis [121].

Alzheimer's disease (AD) is a progressive and irreversible neurodegenerative disorder primarily affecting cognition, memory, and behavior. It is the most common cause of dementia among older adults and is characterized by symptoms that worsen over time [122, 123].

Pereira et al. introduced an innovative biosensor for electrochemical detection of amyloid β-42 (Aβ-42) peptide, a crucial AD biomarker. The biosensor utilizes paper as a platform, with a recognition layer based on an MIP created on a carbon ink electrode. The MIP, formed through electro-polymerization of Aβ-42 and O-phenylenediamine at neutral pH, retains the imprinted protein's shape after template molecule removal. Analytical performance was evaluated using square wave voltammetry, showing a linear response down to 0.1 ng mL^{-1} of Aβ-42. The biosensor offers simplicity, affordability, rapid response, and repeatability, potentially impacting early AD detection [122].

In AD, characterized by the accumulation of β-amyloid peptide in the brain, β-amyloid has emerged as a potentially valuable biomarker for diagnosing and monitoring the disease. Vais et al. utilized molecularly imprinted poly-pyrrole on screen-printed carbon electrodes through electro-polymerization in the presence of β-amyloid. The resulting imprinted poly-pyrrole acted as an artificial receptor with β-amyloid as the molecular template. The biosensor's performance was evaluated using cyclic voltammetry with a ferro/ferricyanide marker, optimizing parameters such as electro-polymerization cycles and β-amyloid binding duration to enhance sensitivity. The results showed a strong β-amyloid binding affinity with a Freundlich constant of 0.22 ng mL^{-1} and an impressive LOD of 1.2 pg mL^{-1}. The biosensor effectively detected β-amyloid in artificial cerebrospinal fluid, demonstrating its potential for early AD detection [124].

Özcan et al. present a rapid electrochemical biosensor for detecting amyloid-β 42-amino-acid-peptide (Aβ42) protein. The biosensor combines delaminated titanium carbide MXene (d-Ti3C2TX MXene) composite with MWCNTs incorporating MIPs. First, the d-Ti3C2TX/MWCNTs composite is prepared by mixing d-Ti3C2TX MXene with MWCNTs. Then, an Aβ42-imprinted d-Ti3C2TX/MWCNTs-modified glassy carbon electrode (GCE) is prepared by cyclic voltammetry using a mixture of pyrrole and Aβ42. The resulting electrode detects Aβ42 in a linear range of 1.0 to 100.0 fg mL^{-1}, with a LOD of 0.3 fg mL^{-1}. The biosensor's stability, repeatability, reproducibility, and reusability are evaluated. Finally, it is applied to analyze Aβ42 in plasma samples [125].

A highly sensitive and specific MIP sensor was developed to detect amyloid-β (1–42) (Aβ42). The sensor was created by modifying a GCE in a stepwise manner, first with electrochemically reduced graphene oxide (ERG) and then with poly(thionine-methylene blue) (PTH-MB). The MIPs were synthesized through electro-polymerization, with Aβ42 as the template and o-phenylenediamine (o-PD) and hydroquinone (HQ) as functional monomers. The sensor's preparation conditions were carefully examined, and under optimal conditions, the sensor exhibited a linear response in the concentration range of 0.12–10 μg mL^{-1}, with an impressive LOD of 0.018 ng mL^{-1}. The MIP-based sensor successfully detected Aβ42 in commercial fetal bovine serum (cFBS) and artificial cerebrospinal fluid (aCSF) [126].

A novel electrochemical biosensor has been created using a MIP for the impedimetric detection of Tau protein, a recognized biomarker associated with AD. Recent findings have linked AD symptoms to aggregated Tau proteins, specifically hyperphosphorylated Tau protein (Tangles), making it a promising biomarker for diagnosing Alzheimer's. The MIP was directly constructed on an SPCE through electro-polymerizing 3-aminophenol (AMP) in the presence of the protein template (p-Tau-441), utilizing cyclic voltammetry. The p-Tau-441 protein bound to the polymer framework was broken down by the enzymatic action of proteinase K in urea and subsequently removed, leaving behind vacant binding sites. The performance of both the imprinted and non-imprinted electrodes was assessed using electrochemical impedance spectroscopy. The MIP-based sensors demonstrated an impressive LOD of 0.02 pM in a pH 5.6 PBS buffer. The platform exhibited good selectivity and provided satisfactory results when tested with serum samples. This biosensor holds the potential to be a valuable tool for on-site Tau protein screening and serves as a cost-effective and readily manufacturable complement to established clinical methodologies, characterized by its rapid response time and affordability [127].

As mentioned, Tau protein holds great promise as a biomarker for the early detection of AD, emphasizing the critical need for a straightforward and efficient detection method. To address this need, an innovative sensing device was created using a SPEC coated with graphene oxide/Prussian Blue nanocubes (GO/PBNCs) to enable the selective and sensitive detection of Tau-441 protein. The MIP was constructed on the GO/PBNCs/C-SPE by electrochemically polymerizing 3-aminophenol (3-AMP) in the presence of the target protein using chronoamperometry. Subsequently, oxalic acid removed the template protein from the polymer matrix. In parallel, a non-imprinted material (NIP) was prepared without the target protein for comparison purposes. Under optimal conditions, this sensing platform displayed a linear detection range from 1.09 and 2.18 nmol L^{-1}, with an impressive LOD of 0.01 pmol L^{-1} in a spiked phosphate buffer solution (PBS). Importantly, the MIP sensor demonstrated minimal uric acid and bovine albumin interference. With its simplicity in production, affordability, and promising performance, this sensor stands as a potential strategic platform for the detection of both chemical and biological molecules [128].

Research published in 2017 by Wang et al. employed a novel approach to create a glial fibrillary acidic protein (GFAP)-imprinted polymer, mimicking biological processes. A carbon electrode was printed and modified with this imprinted polymer to develop a sensor for detecting GFAP. The sensor achieved a remarkable LOD of 0.04 μg mL^{-1} in an aqueous environment. To validate the methodology and assess the sensor's performance, GFAP detection was further explored in human blood serum samples [129].

7.11 Conclusion

The fusion of MIPs with electrochemical sensors marks a groundbreaking frontier in disease detection and monitoring, promising a transformative shift toward personalized and proactive healthcare. Beyond their current robust and efficient platform, these sensors envision a future where medical science integrates seamlessly with cutting-edge technologies, reshaping patient care and outcomes. The limitless potential of MIP-based electrochemical sensors sparks the imagination of a healthcare landscape where diseases are not only detected but predicted with unparalleled accuracy. The synergy of MIP selectivity and electrochemical sensitivity opens avenues for early interventions that can significantly alter disease trajectories. Envisaging a future of precision medicine, these sensors offer the prospect of tailoring diagnostic and therapeutic approaches to individual patients, dynamically adjusting treatments based on unique molecular signatures in real-time. Integrated into wearable devices, they enable continuous health monitoring, empowering individuals to participate actively in their healthcare. The global impact extends to rapid and cost-effective detection of infectious diseases, pollutants, and emerging health threats, fostering proactive public health strategies. The transformative journey of MIP-based electrochemical sensors transcends current achievements, propelling us into a realm where molecular precision and technological innovation redefine healthcare. As we stand at the crossover of this transformative era, the scientific promise of these sensors instigates collaborative efforts among scientists, healthcare professionals, and society to usher in a future where health is not merely managed but comprehensively mastered. This trajectory illuminates a path toward a healthcare landscape distinguished by unprecedented precision, accessibility, and efficacy.

References

1. Lazarević-Pašti T, Tasić T, Milanković V, Potkonjak N (2023) Molecularly imprinted plasmonic-based sensors for environmental contaminants—current state and future perspectives. Chemosensors 11:35
2. BelBruno JJ (2019) Molecularly imprinted polymers. Chem Rev 119:94–119
3. Thakur N, Das TR, Gupta S, Shukla SK, Patra S, Kumar D (2023) Chapter 5—Material selection for the fabrication of MIP-based devices: an overview. In: Singh M (ed) Molecularly imprinted polymers (MIPs). Elsevier, pp 101–114
4. Madikizela LM, Tavengwa NT, Tutu H, Chimuka L (2018) Green aspects in molecular imprinting technology: from design to environmental applications. Trends Environ Anal Chem 17:14–22
5. Morsi SMM, Abd El-Aziz ME, Mohamed HA (2023) Smart polymers as molecular imprinted polymers for recognition of target molecules. Int J Polym Mater Polym Biomater 72:612–635
6. Ober CK, Müllen K (2012) 8.01—Introduction—applications of polymers. In: Matyjaszewski K, Möller M (eds) Polymer science: a comprehensive reference. Elsevier, Amsterdam, pp 1–8
7. Roy E, Patra S, Tiwari A, Madhuri R, Sharma PK (2017) Introduction of selectivity and specificity to graphene using an inimitable combination of molecular imprinting and nanotechnology. Biosens Bioelectron 89:234–248
8. Orbay S, Kocaturk O, Sanyal R, Sanyal A (2022) Molecularly imprinted polymer-coated inorganic nanoparticles: fabrication and biomedical applications. Micromachines (Basel) 13:13
9. Hayden O, Lieberzeit PA, Blaas D, Dickert FL (2006) Artificial antibodies for bioanalyte detection—sensing viruses and proteins. Adv Funct Mater 16:1269–1278
10. Pan J, Chen W, Ma Y, Pan G (2018) Molecularly imprinted polymers as receptor mimics for selective cell recognition. Chem Soc Rev 47:5574–5587
11. Fresco-Cala B, Mizaikoff B (2020) Surrogate imprinting strategies: molecular imprints via fragments and dummies. ACS Appl Polym Mater 2:3714–3741
12. Piletsky S, Canfarotta F, Poma A, Bossi AM, Piletsky S (2020) Molecularly imprinted polymers for cell recognition. Trends Biotechnol 38:368–387
13. Kanoun O, Lazarević-Pašti T, Pašti I, Nasraoui S, Talbi M, Brahem A, Adiraju A, Sheremet E, Rodriguez RD, Ben Ali M, Al-Hamry A (2021) A review of nanocomposite-modified electrochemical sensors for water quality monitoring. Sensors 21:4131
14. Malitesta C, Mazzotta E, Picca RA, Poma A, Chianella I, Piletsky SA (2012) MIP sensors—the electrochemical approach. Anal Bioanal Chem 402:1827–1846
15. Kurbanoglu S, Yarman A, Zhang X, Scheller FW (2023) Electrochemical MIP sensors for environmental analysis. In: Regan F, Hansen P-D, Barceló D (eds) Biosensors for the marine environment: present and future challenges. Springer International Publishing, Cham, pp 139–164
16. Suryanarayanan V, Wu C-T, Ho K-C (2010) Molecularly imprinted electrochemical sensors. Electroanalysis 22:1795–1811
17. Antuña-Jiménez D, Díaz-Díaz G, Blanco-López MC, Lobo-Castañón MJ, Miranda-Ordieres AJ, Tuñón-Blanco P (2012) Chapter 1—Molecularly imprinted electrochemical sensors: past, present, and future. In: Li S et al (eds) Molecularly imprinted sensors. Elsevier, Amsterdam, pp 1–34
18. Bakker E, Telting-Diaz M (2002) Electrochemical sensors. Anal Chem 74:2781–2800
19. Wang Y, Xu H, Zhang J, Li G (2008) Electrochemical sensors for clinic analysis. Sensors 8:2043–2081
20. Sadak O (2023) 5—Potentiometric sensors. In: Barhoum A, Altintas Z (eds) Fundamentals of sensor technology. Woodhead Publishing, pp 99–121
21. Janata JI (2009) Amperometric sensors. In: Janata J (ed) Principles of chemical sensors, Springer, Boston, MA, pp 201–239
22. Edmonds TE (1988) Voltammetric and amperometric transducers. In: Edmonds TE (ed) Chemical sensors. Springer Netherlands, Dordrecht, pp 193–213

23. Liu Y, Xue Q, Chang C, Wang R, Liu Z, He L (2022) Recent progress regarding electrochemical sensors for the detection of typical pollutants in water environments. Anal Sci 38:55–70
24. Lazarević-Pašti TD, Bondžić AM, Pašti IA, Vasić VM (2012) Indirect electrochemical oxidation of organophosphorous pesticides for efficient detection via acetylcholinesterase test. Pestic Biochem Physiol 104:236–242
25. Lazarević-Pašti TD, Bondžić AM, Pašti IA, Mentus SV, Vasić VM (2013) Electrochemical oxidation of diazinon in aqueous solutions via electrogenerated halogens—Diazinon fate and implications for its detection. J Electroanal Chem 692:40–45
26. Rotariu L, Lagarde F, Jaffrezic-Renault N, Bala C (2016) Electrochemical biosensors for fast detection of food contaminants—trends and perspective. TrAC Trends Anal Chem 79:80–87
27. Milanković V, Tasić T, Pejčić M, Pašti I, Lazarević-Pašti T (2023) Spent coffee grounds as an adsorbent for malathion and chlorpyrifos—kinetics, thermodynamics, and eco-neurotoxicity. Food Secur 12:2397
28. Tasić T, Milanković V, Batalović K, Breitenbach S, Unterweger C, Fürst C, Pašti IA, Lazarević-Pašti T (2023) Application of viscose-based porous carbon fibers in food processing—malathion and chlorpyrifos removal. Food Secur 12:2362
29. Katnić Đ, Porobić SJ, Lazarević-Pašti T, Kojić M, Tasić T, Marinović-Cincović M, Živojinović D (2023) Sterilized plum pomace biochar as a low-cost effective sorbent of environmental contaminants. J Water Process Eng 56:104487
30. Katnić ĐB, Porobić SJ, Vujčić I, Kojić MM, Lazarević-Pašti T, Milanković V, Marinović-Cincović M, Živojinović DZ (2024) Irradiated fig pomace pyrochar as a promising and sustainable sterilized sorbent for water pollutant removal. Radiat Phys Chem 214:111277
31. Lazarević-Pašti T, Jocić A, Milanković V, Tasić T, Batalović K, Breitenbach S, Unterweger C, Fürst C, Pašti IA (2023) Investigating the adsorption kinetics of dimethoate, malathion and chlorpyrifos on cellulose-derived activated carbons: understanding the influence of physicochemical properties. C 9:103
32. Aničijević V, Tasić T, Milanković V, Breitenbach S, Unterweger C, Fürst C, Bajuk-Bogdanović D, Pašti IA, Lazarević-Pašti T (2023) How well do our adsorbents actually perform?—The case of dimethoate removal using viscose fiber-derived carbons. Int J Environ Res Public Health 20:4553
33. Chiappini A, Pasquardini L, Bossi AM (2020) Molecular imprinted polymers coupled to photonic structures in biosensors: the state of art. Sensors 20:5069
34. Ozcelikay G, Kaya SI, Ozkan E, Cetinkaya A, Nemutlu E, Kır S, Ozkan SA (2022) Sensor-based MIP technologies for targeted metabolomics analysis. TrAC Trends Anal Chem 146:116487
35. Cieplak M, Kutner W (2016) Artificial biosensors: how can molecular imprinting mimic biorecognition? Trends Biotechnol 34:922–941
36. Karimi Baker Z, Sardari S (2021) Molecularly imprinted polymer (MIP) applications in natural product studies based on medicinal plant and secondary metabolite analysis. Iran Biomed J 25:68–77
37. Mazzotta E, Malitesta C (2021) Enzyme-mimics molecularly imprinted polymers based on metal complexes: electropolymerization and electrocatalytic application. Methods Mol Biol 2359:233–240
38. Mostafa AM, Barton SJ, Wren SP, Barker J (2021) Review on molecularly imprinted polymers with a focus on their application to the analysis of protein biomarkers. TrAC Trends Anal Chem 144:116431
39. Parisi OI, Francomano F, Dattilo M, Patitucci F, Prete S, Amone F, Puoci F (2022) The evolution of molecular recognition: from antibodies to molecularly imprinted polymers (MIPs) as artificial counterpart. J Funct Biomater 13:p12-N.PAG
40. Kadhem AJ, Gentile GJ, Fidalgo de Cortalezzi MM (2021) Molecularly imprinted polymers (MIPs) in sensors for environmental and biomedical applications: a review. Molecules 26:6233
41. Yarman A, Kurbanoglu S, Jetzschmann KJ, Ozkan SA, Wollenberger U, Scheller FW (2018) Electrochemical MIP-sensors for drugs. Curr Med Chem 25:4007–4019

42. Gui R, Jin H, Guo H, Wang Z (2018) Recent advances and future prospects in molecularly imprinted polymers-based electrochemical biosensors. Biosens Bioelectron 100:56–70
43. Selvolini G, Marrazza G (2017) MIP-based sensors: promising new tools for cancer biomarker determination. Sensors 17:718
44. Singhal A, Parihar A, Kumar N, Khan R (2022) High throughput molecularly imprinted polymers based electrochemical nanosensors for point-of-care diagnostics of COVID-19. Mater Lett 306:130898
45. Hasseb AA, Abdel Ghani NdT, Shehab OR, El Nashar RM (2022) Application of molecularly imprinted polymers for electrochemical detection of some important biomedical markers and pathogens. Curr Opin Electrochem 31:100848
46. Çorman ME, Cetinkaya A, Armutcu C, Bellur Atici E, Uzun L, Ozkan SA (2022) A sensitive and selective electrochemical sensor based on molecularly imprinted polymer for the assay of teriflunomide. Talanta 249:123689
47. Milankovic V, Lazarevic-Pasti T (2021) The role of the cholinergic system in lung diseases. In: Dua K et al (eds) Targeting cellular signalling pathways in lung diseases. Springer Singapore, Singapore, pp 625–660
48. Phonklam K, Wannapob R, Sriwimol W, Thavarungkul P, Phairatana T (2020) A novel molecularly imprinted polymer PMB/MWCNTs sensor for highly-sensitive cardiac troponin T detection. Sensors Actuators B Chem 308:127630
49. Radi A-E, El-Naggar A-E, Nassef HM (2014) Molecularly imprinted polymer based electrochemical sensor for the determination of the anthelmintic drug oxfendazole. J Electroanal Chem 729:135–141
50. Ertürk G, Hedström M, Tümer MA, Denizli A, Mattiasson B (2015) Real-time prostate-specific antigen detection with prostate-specific antigen imprinted capacitive biosensors. Anal Chim Acta 891:120–129
51. Karami P, Bagheri H, Johari-Ahar M, Khoshsafar H, Arduini F, Afkhami A (2019) Dual-modality impedimetric immunosensor for early detection of prostate-specific antigen and myoglobin markers based on antibody-molecularly imprinted polymer. Talanta 202:111–122
52. Jolly P, Tamboli V, Harniman RL, Estrela P, Allender CJ, Bowen JL (2016) Aptamer–MIP hybrid receptor for highly sensitive electrochemical detection of prostate specific antigen. Biosens Bioelectron 75:188–195
53. Rebelo TSCR, Santos C, Costa-Rodrigues J, Fernandes MH, Noronha JP, Sales MGF (2014) Novel prostate specific antigen plastic antibody designed with charged binding sites for an improved protein binding and its application in a biosensor of potentiometric transduction. Electrochim Acta 132:142–150
54. Shen X, Ma Y, Zeng Q, Tao J, Huang J, Wang L (2016) Molecularly imprinted electrochemical sensor for advanced diagnosis of alpha-fetoprotein. Anal Methods 8:7361–7368
55. Ma Y, Shen X-L, Zeng Q, Wang H-S, Wang L-S (2017) A multi-walled carbon nanotubes based molecularly imprinted polymers electrochemical sensor for the sensitive determination of HIV-p24. Talanta 164:121–127
56. Ma Y, Liu C, Wang M, Wang L-S (2019) Sensitive electrochemical detection of gp120 based on the combination of NBD-556 and gp120. Talanta 196:486–492
57. Lu C-H, Zhang Y, Tang S-F, Fang Z-B, Yang H-H, Chen X, Chen G-N (2012) Sensing HIV related protein using epitope imprinted hydrophilic polymer coated quartz crystal microbalance. Biosens Bioelectron 31:439–444
58. Karfa P, Madhuri R, Sharma PK (2016) Retracted article: a battle between spherical and cube-shaped Ag/AgCl nanoparticle modified imprinted polymer to achieve femtogram detection of alpha-feto protein. J Mater Chem B 4:5534–5547
59. Moreira FTC, Ferreira MJMS, Puga JRT, Sales MGF (2016) Screen-printed electrode produced by printed-circuit board technology. Application to cancer biomarker detection by means of plastic antibody as sensing material. Sensors Actuators B Chem 223:927–935
60. Wang Y, Zhang Z, Jain V, Yi J, Mueller S, Sokolov J, Liu Z, Levon K, Rigas B, Rafailovich MH (2010) Potentiometric sensors based on surface molecular imprinting: detection of cancer biomarkers and viruses. Sensors Actuators B Chem 146:381–387

61. Viswanathan S, Rani C, Ribeiro S, Delerue-Matos C (2012) Molecular imprinted nanoelectrodes for ultra sensitive detection of ovarian cancer marker. Biosens Bioelectron 33:179–183
62. Rebelo TSCR, Costa R, Brandão ATSC, Silva AF, Sales MGF, Pereira CM (2019) Molecularly imprinted polymer SPE sensor for analysis of CA-125 on serum. Anal Chim Acta 1082:126–135
63. Lee M-H, Thomas JL, Chang Y-C, Tsai Y-S, Liu B-D, Lin H-Y (2016) Electrochemical sensing of nuclear matrix protein 22 in urine with molecularly imprinted poly(ethylene-co-vinyl alcohol) coated zinc oxide nanorod arrays for clinical studies of bladder cancer diagnosis. Biosens Bioelectron 79:789–795
64. Sharma PS, Wojnarowicz A, Sosnowska M, Benincori T, Noworyta K, D'Souza F, Kutner W (2016) Potentiometric chemosensor for neopterin, a cancer biomarker, using an electrochemically synthesized molecularly imprinted polymer as the recognition unit. Biosens Bioelectron 77:565–572
65. Pacheco JG, Silva MSV, Freitas M, Nouws HPA, Delerue-Matos C (2018) Molecularly imprinted electrochemical sensor for the point-of-care detection of a breast cancer biomarker (CA 15-3). Sensors Actuators B Chem 256:905–912
66. Pacheco JG, Rebelo P, Freitas M, Nouws HPA, Delerue-Matos C (2018) Breast cancer biomarker (HER2-ECD) detection using a molecularly imprinted electrochemical sensor. Sensors Actuators B Chem 273:1008–1014
67. Pirzada M, Sehit E, Altintas Z (2020) Cancer biomarker detection in human serum samples using nanoparticle decorated epitope-mediated hybrid MIP. Biosens Bioelectron 166:112464
68. Wang X, Wang Y, Ye X, Wu T, Deng H, Wu P, Li C (2018) Sensing platform for neuron specific enolase based on molecularly imprinted polymerized ionic liquids in between gold nanoarrays. Biosens Bioelectron 99:34–39
69. Ben Moussa F, Achi F, Meskher H, Henni A, Belkhalfa H (2022) Green one-step reduction approach to prepare rGO@AgNPs coupled with molecularly imprinted polymer for selective electrochemical detection of lactic acid as a cancer biomarker. Mater Chem Phys 289:126456
70. Siciliano G, Chiriacò MS, Ferrara F, Turco A, Velardi L, Signore MA, Esposito M, Gigli G, Primiceri E (2023) Development of an MIP based electrochemical sensor for TGF-β1 detection and its application in liquid biopsy. Analyst 148:4447–4455
71. Caldara M, Lowdon JW, Rogosic R, Arreguin-Campos R, Jimenez-Monroy KL, Heidt B, Tschulik K, Cleij TJ, Diliën H, Eersels K, van Grinsven B (2021) Thermal detection of glucose in urine using a molecularly imprinted polymer as a recognition element. ACS Sensors 6:4515–4525
72. Widayani, Yanti, Wungu TDK, Suprijadi (2017) Preliminary study of molecularly imprinted polymer-based potentiometric sensor for glucose. Proc Eng 170:84–87
73. Mirmohseni A, Pourata R, Shojaei M (2014) Application of molecularly imprinted polymer for determination of glucose by quartz crystal nanobalance technique. IEEE Sensors J 14:2807–2812
74. Alexander S, Baraneedharan P, Balasubrahmanyan S, Ramaprabhu S (2017) Highly sensitive and selective non enzymatic electrochemical glucose sensors based on graphene oxide-molecular imprinted polymer. Mater Sci Eng C 78:124–129
75. Farid MM, Goudini L, Piri F, Zamani A, Saadati F (2016) Molecular imprinting method for fabricating novel glucose sensor: polyvinyl acetate electrode reinforced by MnO2/CuO loaded on graphene oxide nanoparticles. Food Chem 194:61–67
76. Kim D-M, Moon J-M, Lee W-C, Yoon J-H, Choi CS, Shim Y-B (2017) A potentiometric non-enzymatic glucose sensor using a molecularly imprinted layer bonded on a conducting polymer. Biosens Bioelectron 91:276–283
77. Zheng W, Wu H, Jiang Y, Xu J, Li X, Zhang W, Qiu F (2018) A molecularly-imprinted-electrochemical-sensor modified with nano-carbon-dots with high sensitivity and selectivity for rapid determination of glucose. Anal Biochem 555:42–49
78. Wu H, Zheng W, Jiang Y, Xu J, Qiu F (2021) Construction of a selective non-enzymatic electrochemical sensor based on hollow nickel nanospheres/carbon dots–chitosan and molecularly imprinted polymer film for the detection of glucose. New J Chem 45:21676–21683

79. Bossard B, Grothe RA, Martins AB, Lobato A, Tasić N, Paixão TRLC, Gonçalves LM (2022) Nanographene laser-pyrolyzed paper electrodes for the impedimetric detection of d-glucose via a molecularly imprinted polymer. Monatshefte für Chemie—Chemical Monthly 153:1129–1135
80. Karaman C, Karaman O, Atar N, Yola ML (2021) A molecularly imprinted electrochemical biosensor based on hierarchical Ti2Nb10O29 (TNO) for glucose detection. Microchim Acta 189:24
81. Diouf A, Bouchikhi B, El Bari N (2019) A nonenzymatic electrochemical glucose sensor based on molecularly imprinted polymer and its application in measuring saliva glucose. Mater Sci Eng C 98:1196–1209
82. Sehit E, Drzazgowska J, Buchenau D, Yesildag C, Lensen M, Altintas Z (2020) Ultrasensitive nonenzymatic electrochemical glucose sensor based on gold nanoparticles and molecularly imprinted polymers. Biosens Bioelectron 165:112432
83. Yang Y, Yi C, Luo J, Liu R, Liu J, Jiang J, Liu X (2011) Glucose sensors based on electrodeposition of molecularly imprinted polymeric micelles: a novel strategy for MIP sensors. Biosens Bioelectron 26:2607–2612
84. Wardani NI, Kangkamano T, Wannapob R, Kanatharana P, Thavarungkul P, Limbut W (2023) Electrochemical sensor based on molecularly imprinted polymer cryogel and multiwalled carbon nanotubes for direct insulin detection. Talanta 254:124137
85. Zhao C-J, Ma X-H, Li J-P (2017) An insulin molecularly imprinted electrochemical sensor based on epitope imprinting. Chin J Anal Chem 45:1360–1366
86. Anirudhan TS, Mani A, Athira VS (2021) Molecularly imprinted electrochemical sensing platform for 2-Aminoadipic acid, a diabetes biomarker. React Funct Polym 168:105056
87. Roy S, Nagabooshanam S, Chauhan N, Kumar R, Wadhwa S, Mathur A (2021) Design and development of a novel flexible molecularly imprinted electroanalytical sensor for the monitoring of diabetic foot ulcers. Surf Interfaces 26:101310
88. Campagnol D, Karimian N, Paladin D, Rizzolio F, Ugo P (2022) Molecularly imprinted electrochemical sensor for the ultrasensitive detection of cytochrome c. Bioelectrochemistry 148:108269
89. Çorman ME, Cetinkaya A, Armutcu C, Atici EB, Uzun L, Ozkan SA Sensitive and selective recognition of teriflunomide based on molecularly imprinted electrochemical sensor via newly synthesized functionalized pyrrole monomer: pyrrole histidine. Available at SSRN: https://ssrn.com/abstract=4104464 or https://doi.org/10.2139/ssrn.4104464
90. Gonçalves MdL, Truta LAN, Sales MGF, Moreira FTC (2021) Electrochemical point-of care (PoC) determination of interleukin-6 (IL-6) using a pyrrole (Py) molecularly imprinted polymer (MIP) on a carbon-screen printed electrode (C-SPE). Anal Lett 54:2611–2623
91. Özcan N, Karaman C, Atar N, Karaman O, Yola ML (2020) A novel molecularly imprinting biosensor including graphene quantum dots/multi-walled carbon nanotubes composite for interleukin-6 detection and electrochemical biosensor validation. ECS J Solid State Sci Technol 9:121010
92. Tugce Yaman Y, Akbal Vural O, Bolat G, Abaci S (2022) Peptide nanotubes/self-assembled polydopamine molecularly imprinted biochip for the impedimetric detection of human Interleukin-6. Bioelectrochemistry 145:108053
93. Afsarimanesh N, Mukhopadhyay S, Kruger M, Marlena K (2017) Biosensors for the measurement of C-terminal telopeptide of type I collagen (CTX-I). J Osteoporos Phys Act 5:1000199
94. Sundhoro M, Agnihotra SR, Amberger B, Augustus K, Khan ND, Barnes A, BelBruno J, Mendecki L (2021) An electrochemical molecularly imprinted polymer sensor for rapid and selective food allergen detection. Food Chem 344:128648
95. Wang B, Hong J, Liu C, Zhu L, Jiang L (2021) An electrochemical molecularly imprinted polymer sensor for rapid β-Lactoglobulin detection. Sensors 21:8240
96. Costa R, Costa J, Moreira P, Brandão ATSC, Mafra I, Silva AF, Pereira CM (2022) Molecularly imprinted polymer as a synthetic antibody for the biorecognition of hazelnut Cor a 14-allergen. Anal Chim Acta 1191:339310

97. Limthin D, Leepheng P, Klamchuen A, Phromyothin D (2022) Enhancement of electrochemical detection of gluten with surface modification based on molecularly imprinted polymers combined with superparamagnetic iron oxide nanoparticles. Polymers 14:91
98. Singla P, Kaur S, Jamieson O, Dann A, Garg S, Mahon C, Crapnell RD, Banks CE, Kaur I, Peeters M (2023) Electrochemical and thermal detection of allergenic substance lysozyme with molecularly imprinted nanoparticles. Anal Bioanal Chem 415:4467–4478
99. Raziq A, Kidakova A, Boroznjak R, Reut J, Öpik A, Syritski V (2021) Development of a portable MIP-based electrochemical sensor for detection of SARS-CoV-2 antigen. Biosens Bioelectron 178:113029
100. Amouzadeh Tabrizi M, Fernández-Blázquez JP, Medina DM, Acedo P (2022) An ultrasensitive molecularly imprinted polymer-based electrochemical sensor for the determination of SARS-CoV-2-RBD by using macroporous gold screen-printed electrode. Biosens Bioelectron 196:113729
101. Hashemi SA, Bahrani S, Mousavi SM, Omidifar N, Behbahan NGG, Arjmand M, Ramakrishna S, Lankarani KB, Moghadami M, Firoozsani M (2021) Graphene-based femtogram-level sensitive molecularly imprinted polymer of SARS-CoV-2. Adv Mater Interfaces 8:2101466
102. Ayankojo AG, Boroznjak R, Reut J, Öpik A, Syritski V (2022) Molecularly imprinted polymer based electrochemical sensor for quantitative detection of SARS-CoV-2 spike protein. Sensors Actuators B Chem 353:131160
103. Antipchik M, Reut J, Ayankojo AG, Öpik A, Syritski V (2022) MIP-based electrochemical sensor for direct detection of hepatitis C virus via E2 envelope protein. Talanta 250:123737
104. Ghanbari K, Roushani M (2018) A nanohybrid probe based on double recognition of an aptamer MIP grafted onto a MWCNTs-Chit nanocomposite for sensing hepatitis C virus core antigen. Sensors Actuators B Chem 258:1066–1071
105. Arshad R, Rhouati A, Hayat A, Nawaz MH, Yameen MA, Mujahid A, Latif U (2020) MIP-based impedimetric sensor for detecting dengue fever biomarker. Appl Biochem Biotechnol 191:1384–1394
106. Buensuceso CE, Tiu BDB, Lee LP, Sabido PMG, Nuesca GM, Caldona EB, del Mundo FR, Advincula RC (2022) Electropolymerized-molecularly imprinted polymers (E-MIPS) as sensing elements for the detection of dengue infection. Anal Bioanal Chem 414:1347–1357
107. Tancharoen C, Sukjee W, Thepparit C, Jaimipuk T, Auewarakul P, Thitithanyanont A, Sangma C (2019) Electrochemical biosensor based on surface imprinting for Zika virus detection in serum. ACS Sensors 4:69–75
108. Pintavirooj C, Vongmanee N, Sukjee W, Sangma C, Visitsattapongse S (2022) Biosensors for Klebsiella pneumoniae with molecularly imprinted polymer (MIP) technique. Sensors 22:4638
109. Sharma R, Lakshmi G, Kumar A, Solanki P (2022) Polypyrrole based molecularly imprinted polymer platform for klebsiella pneumonia detection. ECS Sensors Plus 1:010603
110. Sarabaegi M, Roushani M (2021) Rapid and sensitive determination of Pseudomonas aeruginosa by using a glassy carbon electrode modified with gold nanoparticles and aptamer-imprinted polydopamine. Microchem J 168:106388
111. Liustrovaite V, Pogorielov M, Boguzaite R, Ratautaite V, Ramanaviciene A, Pilvenyte G, Holubnycha V, Korniienko V, Diedkova K, Viter R, Ramanavicius A (2023) Towards electrochemical sensor based on molecularly imprinted polypyrrole for the detection of bacteria—listeria monocytogenes. Polymers 15:1597
112. Roushani M, Sarabaegi M, Rostamzad A (2020) Novel electrochemical sensor based on polydopamine molecularly imprinted polymer for sensitive and selective detection of Acinetobacter baumannii. J Iran Chem Soc 17:2407–2413
113. Kushwaha A, Srivastava J, Singh AK, Anand R, Raghuwanshi R, Rai T, Singh M (2019) Epitope imprinting of Mycobacterium leprae bacteria via molecularly imprinted nanoparticles using multiple monomers approach. Biosens Bioelectron 145:111698
114. Harijan M, Shukla V, Singh AK, Raghuwanshi R, Nath G, Singh M (2022) Design of molecularly imprinted sensor for detection of typhoid using immunoinformatics and molecular imprinting. Biosens Bioelectron 10:100090

115. Wang R, Wang L, Yan J, Luan D, Tao S, Wu J, Bian X (2021) Rapid, sensitive and label-free detection of pathogenic bacteria using a bacteria-imprinted conducting polymer film-based electrochemical sensor. Talanta 226:122135
116. Mugo SM, Lu W, Dhanjai D (2020) A pathogen imprinted hybrid polymer capacitive sensor for selective Escherichia coli detection. Med Devices Sens 3:e10071
117. Dhinesh Kumar M, Karthikeyan M, Sharma N, Raju V, Vatsalarani J, Kalivendi SV, Karunakaran C (2022) Molecular imprinting synthetic receptor based sensor for determination of Parkinson's disease biomarker DJ-1. Microchem J 183:107959
118. Lee M-H, Thomas JL, Su Z-L, Yeh W-K, Monzel AS, Bolognin S, Schwamborn JC, Yang C-H, Lin H-Y (2021) Epitope imprinting of alpha-synuclein for sensing in Parkinson's brain organoid culture medium. Biosens Bioelectron 175:112852
119. Ma Y, Hu Q, Liu C, Wang L (2020) A nanospherical conjugated microporous polymer-graphene nanosheets modified molecularly imprinted electrochemical sensor for high sensitivity detection of α-Synuclein. J Electroanal Chem 862:113994
120. Lee M-H, Liu K-T, Thomas JL, Su Z-L, O'Hare D, van Wuellen T, Chamarro JM, Bolognin S, Luo S-C, Schwamborn JC, Lin H-Y (2020) Peptide-imprinted poly(hydroxymethyl 3,4-ethylenedioxythiophene) nanotubes for detection of α synuclein in human brain organoids. ACS Appl Nano Mater 3:8027–8036
121. Ayankojo AG, Boroznjak R, Reut J, Tuvikene J, Timmusk T, Syritski V. Electrochemical sensor based on molecularly imprinted polymer for rapid quantitative detection of brain-derived neurotrophic factor. Available at SSRN: https://ssrn.com/abstract=4501387 or https://doi.org/10.2139/ssrn.4501387.
122. Pereira MV, Marques AC, Oliveira D, Martins R, Moreira FTC, Sales MGF, Fortunato E (2020) Paper-based platform with an in situ molecularly imprinted polymer for β-amyloid. ACS Omega 5:12,057–12,066
123. Lazarevic-Pasti T, Leskovac A, Momic T, Petrovic S, Vasic V (2017) Modulators of acetylcholinesterase activity: from Alzheimer's disease to anti-cancer drugs. Curr Med Chem 24:3283–3309
124. Dehdari Vais R, Yadegari H, Heli H, Sattarahmady N (2021) A β-amyloid(1-42) biosensor based on molecularly imprinted poly-pyrrole for early diagnosis of Alzheimer's disease. J Biomed Phys Eng 11:215–228
125. Özcan N, Medetalibeyoglu H, Akyıldırım O, Atar N, Yola ML (2020) Electrochemical detection of amyloid-β protein by delaminated titanium carbide MXene/multi-walled carbon nanotubes composite with molecularly imprinted polymer. Mater Today Commun 23:101097
126. Ding M, Niu H, Guan P, Hu X (2023) Molecularly imprinted sensor based on poly-o-phenylenediamine-hydroquinone polymer for β-amyloid-42 detection. Anal Bioanal Chem 415:1545–1557
127. Ben Hassine A, Raouafi N, Moreira FTC (2021) Novel electrochemical molecularly imprinted polymer-based biosensor for tau protein detection. Chemosensors 9:238
128. Ben Hassine A, Raouafi N, Moreira FTC (2023) Novel biomimetic Prussian blue nanocubes-based biosensor for Tau-441 protein detection. J Pharm Biomed Anal 226:115251
129. Wang T, Fang Y, He Z (2017) Electrochemical quantitative detection of glial fibrillary acidic protein based on molecularly imprinted polymer sensor. Int J Electrochem Sci 12:7341–7350

8 Advancements in Molecularly Imprinted Polymer-Based Electrochemical Sensors: Fabrication, Functionalization, Applications, and Future Perspectives

Seema Maheshwari, Kuldeep Kaur, Simrat Kaur, and Ashok Kumar Malik

Abstract

MIPs are synthetic polymers with specific target binding sites created by the process of molecular imprinting. The process involves polymerization around a target molecule (template) which is embedded into a polymer matrix and subsequently removed, leaving behind molecular recognition sites whose internal cavity mirror the template's size, shape and active functional groups, allowing only the template to fit and interact effectively. Electrochemical sensors have emerged as one of the most popular sensor types due to several advantages such as rapid response, accuracy, ease of use, high sensitivity, portability, on-site analysis, and economic features. The selectivity and specificity of electrochemical sensors can be further enhanced by integrating them with molecularly imprinted polymers (MIPs). This enables them to distinguish between closely related substances in complex matrices making them useful for applications in fields like clinical diagnostics, disease monitoring, and food safety. For fabrication of MIP-based sensors, MIP layer can be directly synthesised on the surface of the electrode using techniques like electro-polymerization, or the MIP can be prepared separately and then deposited on the surface of the electrode. The detection can be carried out using different transduction methods such as amperometry, potentiometry, and impedimetry. This chapter critically reviews the recent advancements in MIP based electrochemical sensors covering fabrication of MIP along with electrochemical detection techniques. Further, the modification of MIP-based electrochemical sensors with diverse functional materials including nanomaterials, MOFs and ILs to improve the performance of MIP-based electrochemical sen-

S. Maheshwari · K. Kaur (✉) · S. Kaur
Department of Chemistry, Mata Gujri College, Fatehgarh Sahib, Punjab, India

A. K. Malik
Department of Chemistry, Punjabi University, Patiala, Punjab, India

S. Patra et al. (eds.), *Molecularly Imprinted Polymers: Path to Artificial Antibodies*, https://doi.org/10.1007/978-981-97-4379-7_8

sors has also been discussed. Moreover, the applications of MIP-based electrochemical sensors in healthcare including clinical diagnostics, pharmaceutical analysis and food safety have also been discussed citing examples from recent literature. The evaluation of the performance of these MIP-based electrochemical sensors in terms of performance parameters has also been carried out. Finally, the challenges and opportunities in the field of MIP-based electrochemical sensors are deliberated.

Keywords

Molecularly Imprinted Polymers · Electrochemical sensors · Disease detection MIP synthesis · Sensor fabrication

8.1 Introduction

MIPs are artificially synthesized materials that replicates the biological systems to recognize specific molecules such as substrate-enzyme or antigen-antibody. MIPs are often referred to as "synthetic receptors," "artificial receptors," or "plastic antibodies" because they can mimic the binding properties of natural receptors. MIPs are fashioned by polymerizing monomers in the presence of a template molecule. After polymerization, the template is removed, leaving behind cavities in the polymer matrix that are complementary in shape and functionality to the template molecule. When the target analyte is present in a sample solution, it selectively binds to the cavities in the MIPs due to the specific shape, size, and active functional groups of these cavities [1]. MIPs can be synthesized to recognize a wide range of targets, including small molecules to large proteins, and even cells [2, 3]. Molecularly imprinted polymers (MIPs) like biological recognition elements (enzymes/antibodies) show superior selectivity, but with the added advantage in terms of higher stability, lower cost, reusability, and tailorable design to any template [4]. Unlike natural receptors that demand pampering at body temperature, MIPs are more versatile in terms of operation conditions like extreme temperatures or pH and in organic solvents. Additionally, MIPs allow long term storage without the loss of performance, inexpensive synthesis and label-free selection [5]. By harnessing the incredible specificity and robustness of MIPs and combining them with real-time monitoring, we can build sensors capable of detecting the changes in targeted molecules, paving the way for non-invasive monitoring of biomarkers of health care, food adulterants, and clinical diagnosis [6].

There are different techniques/sensors which can be used to detect, various analytes such as Field-effect transistor (FET), Optical sensors, Chromatography, Colorimetric method, Surface-enhanced Raman scattering (SERS), and Electrochemical method. FET-based biosensor measures changes in electrical current due to analyte binding. FET-based sensors has certain limitations such as Debye length restrictions, subthreshold swing, and short-channel effects, impacting sensitivity and device performance [7]. Optical sensors, include fluorescence-based detection, which tracks changes in light emission after analyte reacts with a fluorescent probe. Fluorescence measurement, boast high sensitivity and environmental awareness, but come at a cost and demand sophisticated equipment [8]. Chromatography method, which separates and identifies analyte based on its interaction with a stationary phase. Chromatography-based sensors are a powerful analytical tool for various applications due to their unmatched sensitivity, selectivity, and ability to identify and quantify multiple components in complex samples, but has certain drawbacks such as time-consuming analysis, complex instrumentation and operation, less portability, require larger sample volumes, and has high operation costs [9]. SERS-based sensors identifies analyte molecules based on their unique vibrational response when illuminated by light. SERS based sensors also has certain fabrication and operational constraints such as stability issues, limited reusability, less selectivity and specificity, vacuum requirement, laser safety concerns [10]. Colorimetric method, which detects analyte through a color change caused by a chemical reaction. Colorimetric method has lower sensitivity compared to other instrumental methods, limited dynamic range and has susceptibility to interferences [7]. Electrochemical method, which measures the flow of electrons associated with analyte oxidation at an electrode. Electrochemical sensors offer several advantages over other types of sensors in various applications. One of the key advantages of electrochemical sensors is their high sensitivity as they allow for detection of low concentration of analytes. Electrochemical sensors can achieve low detection limits making them suitable for applications where trace amounts of analytes need to be detected [11]. Another advantage of electrochemical sensors is their fast response time which enable quick detection of changes in analyte concentrations making these sensors highly useful for real-time monitoring [12]. The ease of miniaturization of electrodes makes them suitable for use in portable and wearable devices. Their versatility and adaptability make them suitable for diverse analytical tasks [13]. Overall, the ease, speed, and affordability of electrochemical methods make them particularly attractive for analyte detection, especially in biological and medical applications [14].

Further points to consider, while electrochemical methods are very advantageous such as fast response, ease of use, high sensitivity, portability, possibility of on-site analysis, and economic features, but face certain challenges like interference from other biomolecules and the need for optimized electrodes for specific applications. The selectivity and specificity of electrochemical sensors can be further enhanced by integrating them with molecularly imprinted polymers (MIPs) to form MIP-based electrochemical sensors. In these sensors, the binding of the target molecule to the MIPs causes changes in the electrochemical properties of the sensor, such as conductivity, impedance, or voltametric signals [15]. The changes caused by MIP in electrochemical properties are then transduced into measurable electrical signals, providing a quantitative indication of the target molecule's concentration in the sample. MIP-based electrochemical sensors pose advantages such as elevated sensitivity, selectivity, and portability, making them a promising tool for several applications including environmental monitoring, biomedical applications, industrial process control, food industry, etc. [16]. MIP based electrochemical sensors have significant applications in healthcare sector. Specifically, they have found several applications in clinical diagnostics [17], food safety [18, 19], and drug analysis [20, 21]. MIP-based sensors can diagnose diseases by identification of key biomarkers associated with these diseases in body fluids. This can enable early diagnosis of diseases like cancer, diabetes, and cardiac disorders and also monitoring of disease progression. MIP-based sensors can also identify pathogens like bacteria and viruses, aiding in the rapid diagnosis of infectious diseases. MIP-based electrochemical sensors can also aid in monitoring drug levels in patients' blood, ensuring optimal dosages and minimizing side effects. MIP-based electrochemical sensors also find applications in food safety as they can aid in detection of toxins, allergens, and other contaminants in food products.

In this chapter, we explore the fundamental principles of MIP-based electrochemical sensors covering MIP synthesis and imprinting strategies along with design and fabrication of MIP-based electrochemical sensors. We present applications of MIP-based electrochemical sensors in healthcare sector covering the fields of clinical diagnostics, therapeutic drug monitoring, and food safety with examples from recent published literature. Further we provide a categorized list of various analytes detected using MIP based electrochemical sensors along with their detection limits.

8.2 Molecularly Imprinted Polymers (MIPs): Basic Principles

Molecularly imprinted polymers (MIPs) are synthetic polymeric materials with special binding cavities that perfectly fit to a specific target molecule, like natural receptors. The process of development of MIPs involves three main steps: (i) formation of a complex between the template molecule and the functional monomer(s) (ii) polymerization of the monomer(s) in the presence of a cross-linker to form the MIP (iii) extraction of the template molecule to create specific binding sites for the target molecule. The choice of monomers and crosslinkers is critical in obtaining desired MIP. Monomers are the building blocks of the polymeric matrix, and crosslinkers ensure the stability and long-term morphology of the binding sites. Hydrophilic monomers and crosslinkers are preferred for designing MIPs for applications in aqueous samples. After the polymerization process, the template molecule is removed from the polymeric matrix to create specific binding sites for the target molecule. The template extraction process is critical in determining the selectivity and sensitivity of the MIP material. Solvent extraction is the most commonly used technique, but thermal extraction and microwave-assisted extraction can also be used. MIPs can be synthesized using a variety of polymerization methods, including bulk polymerization, emulsion polymerization, suspension polymerization, precipitation polymerization and multi-step swelling polymerization. Bulk polymerization involves all the components mixed together in a single phase and polymerized simultaneously. After polymerization, the MIP is ground into particles and the template is removed. This is the simplest and most common method, but it may produce irregular-shaped particles. Suspension polymerization involves dispersing monomer droplets in an aqueous phase using a surfactant followed by polymerization. The resulting polymer particles can be isolated by filtration, washed, and the template removed [22]. The process produces spherical particles with uniform size and shape. Emulsion polymerization forms MIPs in the form of beads suspended in an aqueous solution. Surfactants are used to stabilize the monomer droplets in the aqueous phase. Polymerization arises within these droplets. The resulting polymer beads are collected, washed, and the template is removed. This is a more rapid and efficient method than suspension polymerization. In precipitation polymerization monomers are dissolved in solvent, then this solution was added dropwise into another solvent where the monomers are not soluble. Polymerization occurs as the monomers precipitate out of the solution due to insolubility. The precipitated particles are collected, washed, and the template is removed. This produces spherical particles with a homogeneous particle size distribution. Newer polymerization methods such as electro-polymerization, and photo-initiated polymerization are also being explored. Electro-polymerization involves the electrodeposition of polymers onto electrode surfaces in the presence of the template. This technique is commonly used for the preparation of MIP-coated electrodes for sensor applications. Photo-initiated polymerization uses light to initiate the polymerization reaction. Photopolymerizable monomers and cross-linkers are used, and the polymerization is initiated by exposure to UV light. This technique allows for spatial and temporal control over polymerization. Green polymerization approach uses green solvents such as water and ethanol as the porogen solvents. The choice of polymerization method depends on a number of aspects, including the desired properties of the MIPs, the complexity of the synthesis, and the cost. The most suitable method is selected based on these factors to create effective MIPs.

Several imprinting strategies have been employed for development of MIPs which include traditional imprinting strategies such as bulk imprinting and modern ones including surface imprinting, dummy template imprinting, epitope imprinting, and micro contact imprinting. In bulk imprinting technique, the template molecule is dispersed in a mixture of monomers and crosslinkers, and the polymerization is initiated. This approach is simple and widely used for small molecules. The surface imprinting, useful for macromolecules and proteins, involves immobilization of molecules on a surface, followed by polymerization is initiated on the surface. In dummy template imprinting, a structurally similar molecule (dummy template) is used as a template molecule during imprinting. After polymerization and template removal, specific binding sites for the target molecule are created built on the shape and functional groups of the dummy template. Epitope imprinting involves use of an epitope, which is short amino acid sequence (epitope) that is identified by an antibody is used as the template molecule making it useful for imprinting proteins [2]. In micro-contact imprinting, a micropatterned surface is used to create micro-sized cavities in the MIP. This approach is useful for imprinting small molecules and proteins [23]. In addition, living polymerization techniques, such as controlled radical polymerization (e.g., ATRP, RAFT), enables precise control over the polymer chain length and end functionality. These techniques can be employed to create MIPs with well-defined structures and binding sites.

The selection of the imprinting strategy depends on the characteristics of the template molecule and the intended application. The imprinting conditions can be optimized to create MIPs with high affinity and selectivity for the target molecule.

8.3 MIP-Based Electrochemical Sensors

The lack of selectivity of various electrodes used in electrochemical sensors has been of great concern. To solve this problem, researchers used materials that can specifically bind to the analyte, similar to the enzyme-substrate lock-and-key mechanism. These materials were used to modify the working electrode in an electrochemical setup. These modified electrodes were expected to increase the sensitivity of detection for the target analyte and reduce the interference from other molecules in the sample, leading to more accurate analysis. One of the most promising and innovative alternatives is to design and use artificial receptors such as molecularly imprinted polymers (MIPs). MIPs are synthetic materials that are created to have binding sites that are specifically tailored to a particular target molecule [3]. Electrochemical sensing combined with molecular imprinting can surpass conventional chemo sensing methodologies as it fulfils every necessary monitoring requirement. MIPs are a promising class of electrode modifiers that can provide high selectivity for target analytes. MIP-modified electrodes offer several advantages over non-MIP-modified electrodes, including high selectivity and sensitivity toward target analytes, reduced interference from other species, and improved repeatability [21]. Electrochemical sensors based on MIPs can offer high sensitivity and selectivity for a wide range of target analytes. MIPs can be designed to recognize specific target molecules, resulting in low detection limits and high accuracy.

8.3.1 Constitution and Working of MIP-Based Electrochemical Sensor

The electrochemical sensor consists of three electrodes: working, counter and reference electrodes. The main electrode is the working electrode, where the MIP is applied. It is where the target molecule interacts with the polymer and generates a signal. Counter electrode is also known as the auxiliary electrode. It completes the electrical circuit and prevents the buildup of changes on the working electrode and thus ensure the smooth flow of current. Reference electrode serves as a stable and reliable reference point for measuring the potential at the working electrode. These three electrodes are assembled together in a special container called an electrochemical cell as shown in Fig. 8.1. This cell holds the liquid sample and facilitates the exchange of ions between the electrodes. The MIP, specifically designed to bind the target molecule, is coated on the working electrode. The sample containing the target molecule is introduced into the cell. If the target molecule is present, it will bind to the specific cavities in the MIP on the working electrode. This binding event causes a change in the electrical properties of the working electrode, which is measured and translated into a signal [6]. Thus, in a MIP based electrochemical sensor, the MIP acts as the recognition element while the transducer is an electrode.

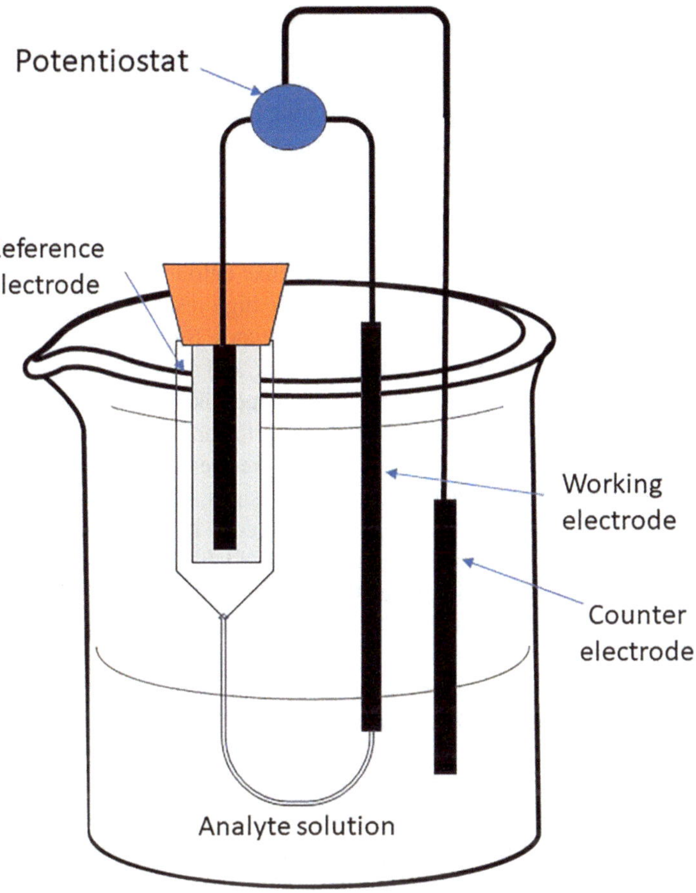

Fig. 8.1 Diagrammatic representation of electrochemical cell

8.3.1.1 Fabrication of Electrode

For fabrication of the electrode capable of converting chemical information into electrical signals, various materials and techniques are employed. Certain fabrication methods of electrochemical cell are electrodeposition, Chemical Vapor Deposition (CVD), Physical Vapor Deposition (PVD), screen printing [24], Inkjet Printing (IP), and micromachining. Electrodeposition is a technique of coating a substrate with a thin metal film by applying an electric current. It offers the advantage of precise control over film thickness and composition, good adhesion to substrate, suitable for complex electrode geometries but requires specialized equipment, limited to conductive substrates and do slow deposition for thick films [25]. PVD is the technique of depositing material from a source onto a substrate using physical processes like evaporation or sputtering. It provides high purity and density of deposited films, wide range of materials, good for large-area deposition. This method has disadvantages as it requires expensive equipment, it has limited control over film thickness and composition, and is not suitable for all materials [13]. CVD is another technique in which material is deposited from a gaseous precursor onto a substrate using chemical reactions. By this method conformal coating of complex structures can be achieved with high purity and density of films, and is good for insulating substrates, but requires careful control of temperature and gas flow, limited control over film thickness and composition, and is not suitable for all materials [26]. Screen printing is the technique of forcing a viscous ink through a patterned mesh onto a substrate to create a desired pattern. It offers the advantage of low cost, high throughput, suitable for large-area electrodes, good for thick films and has disadvantages such as limited resolution, difficulty in printing complex patterns, poor adhesion for some materials. Inkjet printing is a technique in which deposition of droplet-based conductive inks onto a substrate occurs using an inkjet printer. Printing by inkjet method provide high resolution and ability to print complex patterns, and is suitable for flexible substrates. But it has disadvantages, such as limited choice of conductive inks, high ink cost, and slow printing speed [27]. The technique of micromachining uses various techniques like photolithography and etching to fabricate microfluidic channels and other features on a substrate. It enables miniaturization of electrochemical cells, allows precise control over channel dimensions, and is suitable for complex microfluidic designs. However, it requires specialized equipment, expertise in handling, has high cost, and is limited to compatible materials. Spin coating is a versatile technique widely used in microfabrication to deposit thin films. It offers advantages like uniformity of thin films over large areas and film thickness can be controlled by adjusting spin speed and solution concentration etc. [28].

8.3.1.2 Electrochemical Detection

Electrochemical sensor works by measuring the electrical properties of a chemical reaction that occurs at the electrode surface. The electrical signals generated at the working electrode are often weak and subtle. The transducer acts as an amplifier and converts the chemical changes into quantifiable analytical signal and making them easier to detect and interpret. This amplification can involve changes in current (A), potential (V), charge (C), resistance (U), Impedance (I). On this basis there are several electrochemical detection methods including amperometry, potentiometry, voltammetry, conductometry, and impedimetry as shown in Fig. 8.2 [7, 9]. In amperometry, the electric current is measured at a constant potential, which is directly proportional to concentration of the analyte. In potentiometry, the potential difference is measured between electrodes at zero current, the monitoring of potential difference over time and under different conditions reveals specific potential behaviour of analyte [10]. In impedimetric method, Electrical Impedance Spectroscopy (EIS) is used to measure impedance which is a combination of capacitive, inductive, and resistive effects, and in conductometry, resistive devices can be used to measure the resistance of a system which is a measure of how easily electricity can flow through it. In voltammetry, the potential is varied linearly with time, and the current is measured at different applied potentials [11]. The information about analyte is gathered by analysing the current response. Voltametric techniques can be further classified into different categories based on the type of potential agitation used. The most common categories are:

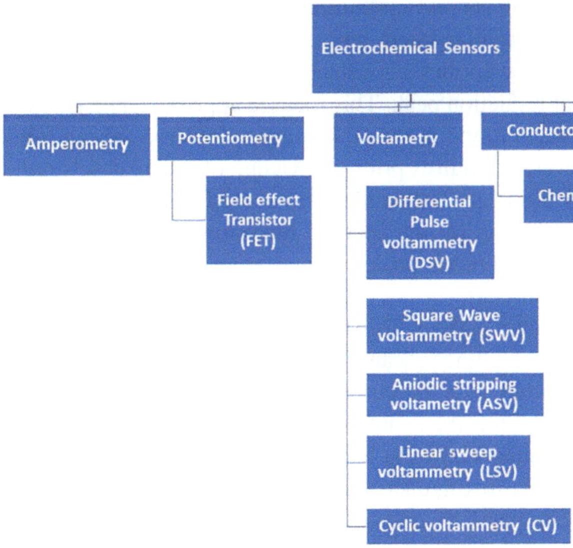

Fig. 8.2 Various methods of detection in electrochemical sensors

- Cyclic voltammetry (CV): The potential is swept to the working electrode, back and forth (increase or decrease) between two values, and the current is measured at each potential.
- Linear sweep voltammetry (LSV): The potential is swept linearly in one direction, and the current is measured at each potential.
- Differential pulse voltammetry (DPV): A series of short square-wave pulses are applied to the potential, and the pulses amplify and separate current peaks.
- Square-wave voltammetry (SWV): A square wave is applied to the potential, and the current is measured at two points during each cycle of the square wave.
- Anodic stripping voltammetry (ASV): The analyte is preconcentrated on the working electrode surface by employing at negative potential for a certain time. The potential is then scanned slowly in the positive direction, and the current is measured as the analyte is oxidized and desorbed from the electrode surface.

8.3.1.3 Sensitivity and Selectivity of Electrochemical Detection Techniques

The choice of specific electrochemical technique plays a crucial role in determining the sensitivity and selectivity of analyte detection. The sensitivity and selectivity characteristics of various detection techniques are summarized as follows:

(a) Cyclic voltammetry (CV): This technique scans the potential across a wide range, providing information about both oxidation and reduction peaks of the analyte. While it offers valuable insights into redox properties and reaction kinetics, its sensitivity for specific analytes can be moderate due to competing background currents. Regarding selectivity, the wider potential range scanned in CV can reveal peaks from interfering species, making selectivity a challenge.

This necessitates further analysis and optimization of electrode materials and supporting electrolytes to minimize interferences.

(b) Amperometry: This technique applies a constant potential at a value specific to the analyte's oxidation or reduction peak. This focused approach enhances sensitivity by minimizing background currents, leading to lower detection limits compared to CV. However, it requires prior knowledge of the analyte's redox potential for optimal performance. The fixed potential used in amperometry can improve selectivity by focusing on the specific redox potential of the target analyte. However, overlapping redox peaks of interfering species with similar potentials can still pose challenges.

(c) Differential pulse voltammetry (DPV): This technique applies a small, oscillating potential pulse on top of a linear scan. This modulation suppresses background currents and amplifies faradaic currents from the analyte, resulting in higher sensitivity than CV for trace amounts of analytes. DPV can offer improved selectivity due to their inherent background current suppression features. These techniques can distinguish analytes with closely spaced redox potentials more effectively than CV.

8.3.2 Modification of Electrode Surface with MIP

The MIP layer first needs to be integrated onto the electrode surface. This can be done in two ways: synthesis of MIP and immobilization separately (ex situ synthesis) or polymerizing of the MIP directly onto the electrode surface (in situ). Ex situ MIP sensors allow the completion of the synthesis and characterization of the synthesized MIP particles before their integration into the transducer. There are several methods for the preparation of ex situ MIP sensors including spin-coating, drop-coating, and Langmuir-Blodgett deposition. These methods involve the deposition of the MIP particles onto the transducer surface, followed by the transformation of these particles into thin film through solvent evaporation, chemical crosslinking or other means. Ex situ MIP sensors can also incorporate electro-catalytic materials that enable the improvement of the electrochemical response. For example, metal nanoparticles, carbon nanotubes, and graphene have been used to enhance the sensitivity and selectivity of MIP-based sensors. There are several approaches to the entrapment of MIP particles onto the electrode surface. One common approach is physical entrapment, which involves the use of a semipermeable membrane or polymeric material such as sol-gel or polystyrene to confine target molecule within its structure. Another approach is the integration of MIP particles directly onto the electrode by simply mixing the MIP particles with the electrode material. The modification of the transducer surface with MIP layer is shown in Fig. 8.3. This approach works great with carbon paste electrodes (CPEs) or graphite-epoxy composites (GECs). This approach is simple and the MIP has strong retention on electrode surface.

For the in situ generation of MIP sensors, two common approaches are: grafting and electro-polymerization. Chemical grafting involves the attachment of pre-formed MIP particles onto the electrode surface through covalent bonding or physical adsorption,

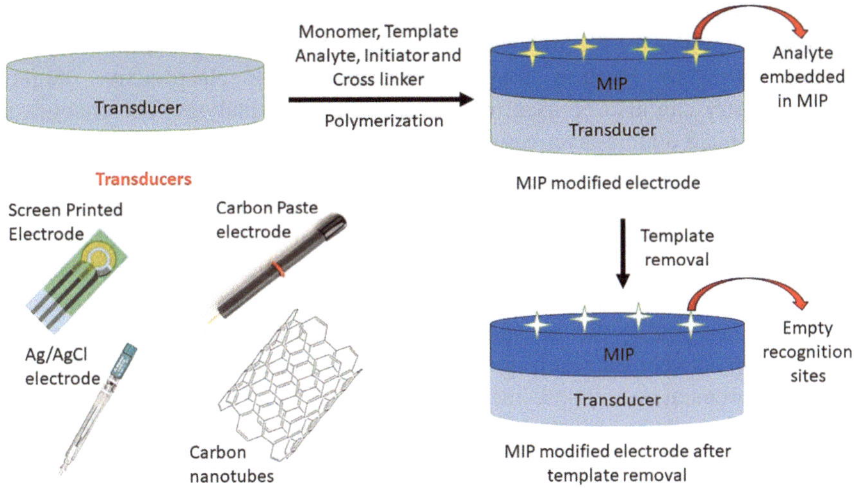

Fig. 8.3 Modification of transducer surface with MIP layer

followed by a polymerization step to form a thin film [29]. Electro-polymerization, on the other hand, involves the in situ synthesis of MIP particles on the electrode surface through electrochemical methods. This approach ensures the binding sites are closer to the sensor surface and better control of the film thickness [30]. Both approaches have been used successfully in the development of MIP-based sensors for various analytes, including drugs, proteins, and environmental pollutants [31]. The ex situ approach allows more precise control on each step individually, by frequently immobilize micro- or nanoparticles that enhance the electrochemical response in terms of sensitivity and selectivity [32]. On the other hand, the in situ methods are often one-step processes that result in strong adhesion of film to the transducer surface and control over growth of film thickness [33]. Achieving the ideal entrapment of analytes within the assembled polymer layer matrix is also possible through electrochemical techniques using conducting polymer-based imprinted layers. This is because CPs have several desirable properties including high electrical capacitance, diverse optical properties, electrical conductivity, good adhesion to electrode surfaces, and the ability to form chemically stable coatings [34]. Various conducting polymers made through electrochemical deposition can serve a polymeric backbone for building MIP formation. Polypyrrole (PPy), polyaniline (PANI), poly(3,4-ethylenedioxythiophene) (PEDOT), polythiophene (PTh), and ortho-phenylenediamine (OPDA) are all conducting polymers, e.g., polypyrrole imprinted with bovine leukemia virus glycoprotein and polydopamine layer for immunoglobulin G [35].

8.3.3 Measurement of Electrochemical Signals

Electrochemical sensors can utilize direct or indirect methods for determining target analytes. The direct method works for electroactive analytes within the potential range of analyte's electroactivity, i.e., where electrons directly transfer between the electroactive analyte and the electrode surface. Whereas indirect method is applied for electro inactive analytes. Voltammetry and amperometry directly measure electron transfer for quantifying electroactive targets. Potentiometry can also be used to determine charged analytes. The changes in electrochemical double-layer capacity reveals adsorption of electro inactive analytes on the transducer surface. The external redox probe such as hexacyanoferrate [Fe $(CN)_6]^{3-/4-}$, ferrocene, hexaammineruthenium (III) chloride, etc., are added in indirect analytical measurement to determine electro-inactive analytes. In this case, increased analyte concentration restricts the diffusion channels for the redox probe towards the electrode surface. This hinders electron transfer and leads to a decrease in the measured current, generating an inverse proportional relationship between current response and analyte concentration [12]. In the indirect measurement, we choose the potential range, in which the redox probe is electroactive, and the current changes are due to the interaction between probe and the analyte. The direct and indirect methods of measurement are represented in Fig. 8.4.

The electrodes used in electrochemical sensing can vary in terms of their shape, size, material, and modification procedures. Commonly used electrodes for electrochemical sensing are screen-printed electrodes (SPEs), carbon paste electrodes (CPE), pencil graphite electrode (PGE), gold electrodes, and glassy carbon electrodes (GCE). However, CPEs are among the most popular in electrochemical sensing due to their quiet background noise, excellent conductivity, and reliable accuracy [13]. Thus, the electrochemical sensors are relatively easy to use, relatively low cost, and offer a wide range of measurable analytes [14]. Electrochemical sensors employing different electrochemical techniques have been applied to detect various analytes as listed in Table 8.1.

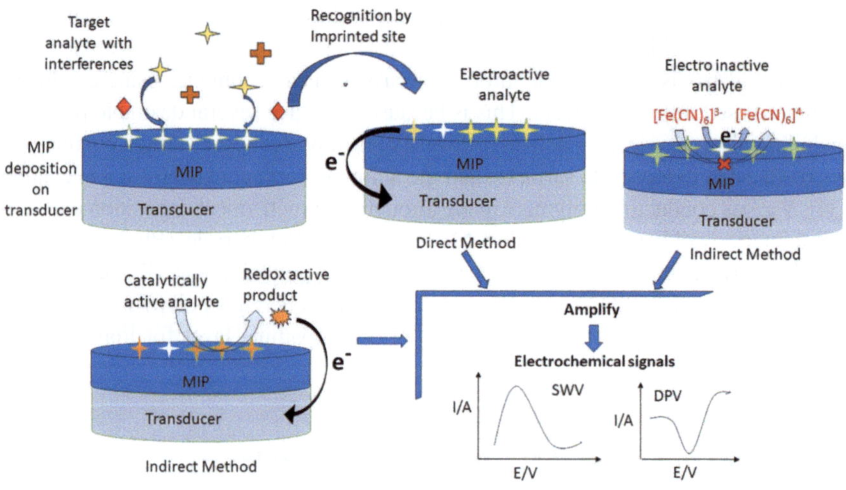

Fig. 8.4 Working of MIP-based electrochemical sensor

Table 8.1 Electrochemical sensors based on different electrochemical techniques to detect various analytes

S. No.	Electro-chemical technique	Fabrication method of electrode	Analyte	Receptor	Detection limit	Real sample analysis	Ref
1.	Amperometry	Electro-deposition	Ganciclovir	AuNPs/MIP/MWCNTs/GC	1.50×10^{-9} M	Human serum	[36]
2.	Cyclic Voltammetry (CV)	Screen printing	Insulin	MIP-SPCE	1.9×10^{-12} M	Pharmaceutical samples	[37]
3.	Differential Pulse voltammetry (DPV)	Dropping	Epinephrine	PABPA/MWCNTs/CS/GCE	2.9×10^{-8} M	Human blood serum	[38]
4.	CV	Spin coating	Dopamine	PABA/SiO_2-Au	0.018×10^{-6} M	Human serum	[39]
5.	DPV	Electrodeposition	Dopamine	ZnO-MIP array	–	Urine samples	[40]
6.	DPV	Drop casting	Paraquat	MIP-AuNC-CS	2.3×10^{-15} M	Tap water samples	[41]

8.4 Modified-MIP Based Electrochemical Sensors

The MIP can be modified with different functional materials such as nanoparticles, MOFs, and ionic liquids to improve their performance in electrochemical sensors. Modified electrochemical sensors gave a wide range of potential applications to detect pollutants in air and food to diagnose diseases. Some common modifications of MIP based electrochemical sensors are discussed below:

8.4.1 Nanoparticles Modified MIP-Based Electrochemical Sensors

The combination of nanoparticles with MIP can enhance the functioning of MIP based electrochemical sensors. Nanoparticles such as graphene oxide, carbon nanotubes, metal nanoparticles (e.g., AuNPs, AgNPs, PdNPs, PtNPs) or metal oxide nanomaterials (e.g., TiO_2, Fe_2O_3, etc.) can be used to increase the surface area of MIP cavities [42]. This allows the polymers to specifically capture the analyte to the active binding site, resulting in early detection of analyte. Nanoparticles also have good electrical properties and excellent conductivity, which can further enhance the working of MIP-based electrochemical sensors. Therefore, the combination of nanoparticles with MIP-based electrochemical sensors can lead to more sensitive and selective detection of metabolites, making them useful for various medical applications, including sensor design and point-of-care testing. Parate et al. developed cotinine sensor using screen-printed carbon electrodes modified with a molecularly imprinted polymer (MIP) made from ortho-phenylenediamine (oPD), further enhanced with graphene and PtNPs [43]. This sensor accurately measured cotinine level in saliva and tobacco products within 12 min. Another example is the detection of dopamine using a MIP-based electrochemical sensor modified with gold nanoparticles (AuNPs). The sensor showed high selectivity and sensitivity towards dopamine with a detection limit of 0.1 nM. Electrochemical MIP sensors can also be applied for chiral sensing. A new sensor developed by Pandey et al. employ the poly-aniline ferrocene sulfonic acid-carbon dot modified pencil graphite electrode precisely distinguish between D- and L- Ascorbic acid in pharmaceuticals and human serum [44]. In another study, the detection of 17β-estradiol was carried out using MIP layer made from monomers methacrylic acid (MAA) and divinylbenzene (DVB). Integration of MIP layer with Fe_3O_4 nanobeads-reduced graphene oxide nanostructure onto glassy carbon electrode resulted in enhanced electrochemical performance. The Fe_3O_4 nanobeads increased the surface area, while the reduced graphene oxide amplified the current response, leading to synergistic effects. In another study, Fe_3O_4-MIP@RGO-based biosensor was developed to successively test water samples with a LOD of 0.819 nM [45]. Martins et al. created a super sensitive electrochemical biosensor for detecting8-hydroxy-20-deoxyguanosine (8-OHdG) in urine samples. The sensor was built by layering polymer phenol on a gold electrode surface. An ingenious biosensor boost high selectivity and detect 8-OHdG with incredibly low detection limit of 0.74 pg mL^{-1} [46]. Thus, electrochemical MIP sensors integrated with nanomaterials are a promising new technology for extremely sensitive and selective recognition in a variety of samples, including pharmaceuticals, human serum, cerebrospinal fluid, and water [47].

8.4.2 Metal Organic Framework (MOF) Modified MIP-Based Electrochemical Sensors

MOFs have a huge surface area, porous structure and exposed active sites, which allows them to bind to target molecules more effectively [48]. MOFs are also relatively easy to synthesize and can be tailored to have specific properties, making them a versatile material for sensor applications [48]. These properties make MOFs attractive for use in MIP based electrochemical sensors (MIECs), as they can improve the sensitivity and selectivity of the sensors [49]. MOF@MIECs are a relatively new type of sensor. Few example, a novel MOF modified electrochemical sensor for the rapid and ultrasensitive detection of furosemide, a drug commonly used to treat chronic kidney diseases has been reported [50]. The sensor utilized a modified electrode incorporating C_3N_4 nanosheets, ZIF-8 (MOF), and MIP. This combination of materials enabled selective determination of furosemide within a linear range of 0.08–100 mM and a remarkably low limit of detection (LOD) of 8.0 nM. The sensor's effectiveness was demonstrated by successfully analyzing furosemide in tablet form [50].

8.4.3 Ionic Liquids (ILs) Modified MIP-Based Electrochemical Sensors

Ionic liquids and MIPs can improve the selectivity of electrochemical sensors in different ways. Ionic liquids can be incorporated into electrochemical sensors in various ways. They can act as building blocks (functional monomers), stabilizers (cross-linkers), reaction mediums (solvents), and surface tuners (organic surface modifiers) to increase the performance of molecularly imprinted polymers [51]. In MIP-based electrochemical sensors, ionic liquids can be used as a template for the formation of MIPs, which are highly selective for the target analyte due to their specific binding sites. Ionic liquids can also be used as an electrode modification agent, exhibiting upright electrocatalytic activity, elevated selectivity, accuracy, sensitivity, and reproducible electrochemical studies [52]. Functional monomers can act as ionic liquids in the synthesis of MIPs. Widely preferred functional monomers include aniline, 3-thiophene acetic acid, 3,4-ethylenedioxythiophene, pyrrole, and alkenyl ionic liquids, which can improve the selectivity and sensitivity of electrochemical sensors [53]. In addition, poly (ionic liquids) (PILs) can be formed by the polymerization of specially designed ionic liquid monomers with vinyl groups. These monomers have unique properties of both the polymers and the ionic liquids [54]. Hu et al. developed a method to detect Zearalenone, a toxic substance found in some food products [55]. The method uses ionic liquids to self-assemble a molecular imprinting layer on a gold nanoparticle decorated boron-doped mesoporous carbon composite. The resulting sensor is highly sensitive and selective, and can detect zearalenone at concentrations as low as 1×10^{-4} ng mL^{-1} [55].

8.5 Applications of MIP-Based Electrochemical Sensors for Health Care

MIP-based electrochemical sensors have been extensively studied for the detection of various analytes in biological samples. MIP-based electrochemical sensors have a wide range of applications, including food safety [19], clinical diagnostics [56, 57], and pharmaceutical analysis analysis [21]. The use of MIPs in electrochemical sensors has allowed for the development of highly sensitive and selective devices for the detection of these analytes in biological samples. One of the main advantages of MIP-based electrochemical sensors for biological samples is their ability to detect analytes in complex matrices such as blood, urine, and saliva. MIPs can be designed to selectively recognize and bind to specific biomolecules, even in the presence of other interfering molecules. This allows for the development of highly specific sensors that can detect analytes at low concentrations in complex biological matrices. The various applications of MIP-based electrochemical sensors in healthcare are discussed below:

8.5.1 MIP-Based Electrochemical Sensors for Clinical Diagnostics

Diagnosing a patient's condition is one of the most important steps in healthcare, as it determines the course of treatment. When a diagnosis is accurate and timely, the patient has the best chance of a good outcome because clinical decisions can be made based on a correct understanding of the problem. Applications of MIP based electrochemical sensors ranges from glucose sensing to precise detection in early signs of cancer or genetic diseases [56]. These sensors can work with different sample types, from solid tissues to various body fluids like blood, urine, and saliva, as long as they contain the relevant biomarker the sensor is looking for. Glucose biosensors have been the most popular type of biosensor since they were first commercialized. These sensors typically measure glucose levels by detecting the electrical current caused by a enzymatic oxidation of glucose by immobilized glucose oxidase on the sensor's surface. This reaction generates current proportional to glucose concentration, via the production of gluconic acid and hydrogen peroxide (H_2O_2). MIP based electrochemical sensors are also employed for detection of biomarkers related to various diseases. For example, a ratiometric MIP based electrochemical sensor has been developed for the selective and accurate detection of dopamine (DA). Nanoporous gold (NPG) was obtained via electrodeposition on a bare gold electrode, with enlarged surface for the subsequent immobilization of polythionine (pThi) and MIPs [58]. Cardiac troponin T (cTnT) is a biomarker for cardiac disease.

A novel cTnT-MIP sensor was constructed using a screen-printed carbon electrode (SPCE). The sensor utilized a polymethylene blue (PMB) redox probe immobilized on carbon nanotubes (CNTs) modified SPCE. The sensor response was detected using differential pulse voltammetry. The sensor presented a detection limit of 0.040 pg mL^{-1} and exhibited a brilliant binding affinity for determination of cTnT in human plasma [59].

Targeted metabolomics analysis can be used for biomarker discoveries by quantifying known metabolites with high accuracy and selectivity. This approach involves the measurement of a specific set of metabolites that are known to be associated with a particular disease or condition. By comparing the metabolite levels in samples from healthy individuals and those with the disease or condition, researchers can identify potential biomarkers that may be used for diagnosis, prognosis, or monitoring of the disease. Molecularly imprinted polymer (MIP)-based electro sensors have several advantages for metabolomics analysis. These sensors can be used for targeted metabolomics analysis to develop point-of-care test (POCT) systems, which patients can easily use. MIP-based electrochemical sensors can overcome the limitations of current analytical approaches, such as their complexity and limited availability, and provide improved sensitivity and accessibility of metabolite analysis in medicinal and clinical trials. Therefore, MIP technologies have the potential to revolutionize metabolomics analysis and enable the development of more accurate and efficient diagnostic tools [15]. Toxic substances produced by living cells or organisms are harmful to human body and may cause poisoning which could wreak biological damage or even death. The categorization of toxins encompasses low-molecular-weight chemicals, oligopeptides, and polypeptides, highlighting their diverse structural nature and potential biological impact. For example, Hu et al. constructed a signal on/off molecularly imprinted electrochemical sensor for sensitive detection of citrinin [60]. The authors used an electro polymerization method to synthesize MIPs with poly(thionine) serving dual function of molecular recognition element and reference signal. If citrinin exists, it will combine with MIPs and diffusion of $[Fe(CN)_6]^{3-/4-}$ towards the electrode was hindered, resulting in decreased signal. Conversely, in the absence of citrinin the signal increased. In this this strategy circumvents the need for an electrochemically active target molecule, and could be expanded to broader analyte detection that are not even electrochemically active. A list of various biomolecules/biomarkers detected using MIP based electrochemical sensors is provided in Table 8.2 below.

8.5.2 MIP-Based Electrochemical Sensors for Pharmaceutical Analysis

Various MIPs with diverse functionalities have been used for electrochemical sensing for the determination of drugs of a broad spectrum, encompassing analgesic, antibiotics, antivirals, cardiovascular, and central nervous system (CNS) acting agents. Few examples of these drugs are naproxen, histamine, S-propranolol, and ibuprofen [21]. MIPs have demonstrated significant potential for the detection and

Table 8.2 Application of MIP-based electrochemical sensors in clinical diagnostics

Biomarker	Disease/affect	Transducer	Monomer	Technique	Sample	Reference
Cholesterol	Clinical diagnostics of cholesterol	GCE	P-ATP, TBAP	CV & DPV	Urine	[61]
Progesterone	Reproductive hormone	Ag/AgCl	Aniline-co-metanilic acid	CV	Urine sample	[62]
Pyruvic acid	Cancer	CPE-CNT-CPE	MAA	ADPV	Plasma and Urine samples	[63]
Sarcosine	Prostate cancer	AU-SPE	p-ATP	IS	Human urine	[64]
Haemoglobin	Vital Protein	Ag-MOF@ MC-MIP/ AuE	APTES	DPV	Blood samples	[65]
Folic acid	B-complex vitamin, deficiency can cause cancer, megaloblastic anemia, cardiovascular disease, Alzheimer's disease, and psychiatric disorders	CPE Fe_3O_4@ MIP	MAA	SWV	Tablets, River water	[66]
Myoglobin	Protein, its level diagnosis muscle damage	Au-SPE Screen printed electrode	AA	EIS	Human serum and urine	[67]
Prostate specific antigen (PSA)	Tumor marker enzyme	PGE	Aniline	A	Tablet, Urine	[68]
Insulin	Anabolic hormone	Nano-MIP	BTSE	DPV	Human Plasma	[69]
Cortisol	Steroid hormone, diagnose adrenal gland disorders and chronic stress.	Au@MIP	o-PD	CV	Saliva	[70]

ADPV Anodic Differential Pulse Voltametry, *BTSE* 1,2-bis(triethoxysilyl)ethane, *PGE* Pencil graphite electrode, *o-PD* o-phenylenediamine, *CPE* Carbon paste electrode, *APTES* 3-Aminopropyl)triethoxysilane, *MAA* Methacrylic acid, *IS* Impedance spectrometry, *PGE* Pencil Graphite Electrode extraction of antibiotics from various food matrices, including meat, milk, fish, etc. Several reports document the development and application of MIP-based electrochemical sensors for the selective detection of antibiotics such as erythromycin [71], amoxicillin [72], tetracycline [73], and chloramphenicol. These sensors offer several advantages over conventional analytical methods, such as high selectivity, sensitivity, stability, low cost, and ease of fabrication. For instance, Mostafavi et al. developed a MIP sensor based on nanocomposite-incorporated carbon paste electrode sensor for the detection of diclofenac, a nonsteroidal anti-inflammatory drug used to treat fever, inflammation, and pain. It is available in several forms, including oral tablets, topical gels, creams, patches, and eye drops. Diclofenac works by blocking the production of prostaglandins. The polymerization of aniline was in the presence of triphenylamine, leading to the formation of a highly conductive three-dimensional aniline polymer network. To test, how well the sensor could distinguish between DCF and other molecules in binary solutions containing DCF/glucose, DCF/ascorbic acid, and DCF/urea. They found that the sensor could successfully be used to measure diclofenac in pharmaceutical and urine samples [74]. The potential of MIPs for antibiotic the detection and extraction for diverse samples (meat, milk, fish) is evident. For instance, a magnetic MIP nanoparticle (MMIP NP) modified carbon paste electrode (CPE) sensor was developed for sensitive and selective electrochemical determination of gemifloxacin (GM), a DNA gyrase inhibitor with broad-spectrum antibacterial activity [10]. Several studies have reported the use of MIP-based electrochemical sensors for antibiotics detection including amoxicillin, tetracycline, and chloramphenicol etc. For instance, magnetic molecularly imprinted polymer (MMIP) was fabricated for a selective electrochemical detection of amoxicillin [75]. Another study reported the detection of tetracycline in milk samples using MIP-based electrochemical sensor [73]. Therefore, MIP-based electrochemical sensors have great potential for the development of sensitive, rapid, and selective analytical methods for pharmaceutical analysis, and could be exploited further for healthcare applications [31]. A list of various pharmaceuticals analysed by using MIP-based electrochemical sensors is provided in Table 8.3.

8.5.3 MIP-Based Electrochemical Sensors for Food Safety

At each stage of food preparation, from production to retailing, food products are exposed to contaminants such as plasticizers, additives, preservatives, food colours, and artificial fragrances. Additionally, with environmental pollutants such as pesticides, insecticides, allergens, pathogens, and natural toxins. Consuming these hazardous substances and microorganisms can have adverse effects on consumers' health, such as high blood pressure, hypertension, allergic reactions, inflammation, hormonal dysfunction, and more [18]. Also, the monitoring of nutritional content in food samples is also important. MIP based electrochemical sensors have shown great potential for application in food samples. Molecularly imprinted electrochemical sensors could be used for testing pesticide residues, preservatives, toxins and other contaminants for ensuring food safety.

Table 8.3 Application of MIP-based electrochemical sensors in pharmaceutical analysis

Drug	Disease	Electrode	Monomer	Method	Sample	Reference
Diosgenin	Cardiovascular illness/ Diabetes	GCE	pABA	CV	Seed extract	[76]
Dopamine	Neurological	Au electrode	Thi	EIS & DPV	Artificial cerebrospinal fluids	[58]
Azithromycin	Antibiotic	GCE	3-TBA	CV & EIS	Urine, serum	[77]
Amoxicillin	Antibiotic	CPE	AAm	SWV	Milk and river water	[72]
Norfloxacin	Antibiotics	MIP/ CoFe-MOFs/ AuNP/ GCE	Pyrrole	DPV	Capsules & Milk	[78]
Cloxacillin	Antibiotic	Au/GO/ MIP SPCE	MAA	DPV	Milk	[79]
Diclofenac	Anti inflammation	RGO/ MIP	Aniline	DPV, CV	Tablet, Urine	[74]
Quetiapine	Anxiety disorder	Nano-MIP	MAA	CV	Urine	[80]
Ampicillin	Antibiotic	MIP-Fe_3N-Co_2N/CC	NNDMA	DPV	Milk	[81]
Tetracycline	Antibiotic	Fe_3O_4@ MIP	AA	SWV	Milk	[73]

GCE Glassy carbon electrode, *Thi* Thionine, *MAA* Methacrylic acid, *pABA* para Aminobenzoic acid, *AA* Acrylic acid, *NNDMA* N, N-Dimethylacrylamide, *AAm* Acrylamide, *CPE* Carbon paste electrode, *3-TBA* 3-Thienyl boronic acid, *RGO* Reduced graphene oxide, *SPCE* Screen printed carbon electrode

MIP-based electrochemical sensors can be used for the effective detection of pesticide residues, such as imidacloprid (IMI) and lindane in food products. These could also be used for monitoring of nutrients in food products. Foods with less nutritive value have been associated with deleterious effects on human health, including chronic diseases and deficiency diseases. For example, da Silva et al. developed a sensor for determination of lactose in milk samples. The amount of lactose in milk is an important indicator of the quality of milk and whether there are any problems with it, both in the food industry and for human and animal health. He reported the electro-polymerization of a MIP on a disposable graphite

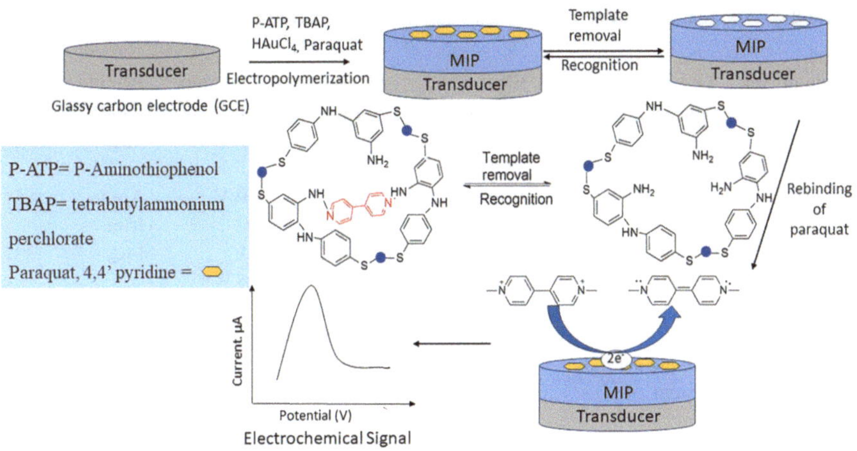

Fig. 8.5 Schematic representation of MIP modified electrochemical sensor for recognition of food adulterant (Paraquat) [41]

paper electrode. The MIP was made of polypyrrole (PPy) and electrosynthesized in the presence of lactose on a graphite paper electrode (PE). The sensor was able to detect lactose in whole and lactose-free milks with a detection limit of 0.88 nmol/L [82]. In another work, paraquat was determined using MIP modified glassy carbon electrode [41]. The MIP of p-aminothiophenol was assembled on the surface of Au nanoparticles modified electrode. The construction and working of the sensor is represented in Fig. 8.5. The limit of detection for determination of paraquat was found to be 2.3×10^{-15} mol L^{-1}.

Some fungi produce mycotoxins, which frequently contaminate food stuffs and animal feed, and pose significant health risks due to their ability to induce physiological abnormalities and tumorigenesis. Ratiometric electrochemical impedance immunosensors (RMIECs) leverage the high sensitivity of electrochemical techniques, superior selectivity of molecularly imprinted polymers (MIPs), and inherent accuracy of ratiometric detection mode, presenting potential as a highly effective platform for toxin analysis [58]. Hassan et al. created a MIP sensor for methyl parathion using a core-shell method. The selectivity of the magnetic-MIP-based sensor was evaluated using malathion, 4-nitrophenol, and phenol as potential interferents. The sensor demonstrated a high affinity towards methyl parathion in fish samples, exhibiting an exceptional pre-concentration capacity for this specific pesticide [83]. Table 8.4 provides a list of various food contaminants detected using MIP based electrochemical sensors.

Table 8.4 Application of MIP-based electrochemical sensors for detection of food contaminants

Food contaminant	Disease	Electrode	Monomer	Technique	Sample	Ref
Patulin	Mycotoxin, carcinogen	MIP/Au@ Cu-MOF/N--GQDs/GCE	Aniline	DPV	Apple juice	[84]
Diazinon	Toxic	MIP/CPE	MAA	SWV	Apple fruit	[85]
Methyl Parathion	Affects nervous system	MIP/ magneto-actuated electrode	MAA	CV &SWV	Fish	[83]
Malathion	Affects nervous system	MIP-au/SPE	AA	DPV	Olive fruit and oils	[86]
Chlorpyrifos	Affects nervous system	GCE	Thiophene derivatives	CV &DPV	Apple lettuces	[87]
Bisphenol A	Affects neuroendocrine system and reproductive system	CPE	HEMA	DPV, CV & EIS	Soft drinks	[88]
Paraquat	Toxic	MIP-AuNC-CS/ GCE	p-ATP	DPV	Tap water samples	[41]

CPE Carbon paste electrode, *HEMA* 2-Hydroxy ethyl methacrylate, *GCE* Glassy carbon electrode, *MAA* Methacrylic acid, *AA* Acrylic acid, *AuNCs-CS* Gold nanocrosses-Chitosan, *p-ATP* para-Aminothiophenol

8.6 Opportunities and Challenges

Electrochemical MIP sensors are a promising new technology for the development of real-time sensors for health screening. They are sensitive, have a long lifetime, and are less affected by the ambient environment than traditional electrochemical immunosensors. The widespread commercialization of these sensors necessitates addressing several challenges, including the requirement for pre-dilution of biological matrices such as human serum, plasma, saliva, and urine prior to analysis. This pre-dilution step limits the applicability and user friendliness of the sensors, research efforts are going on for developing strategies for direct analysis of undiluted biofluids. This is because these fluids contain numerous biomolecules that can interfere with the sensor signal. Future research should focus on developing MIP sensors that can be used to test undiluted body fluids. Another challenge is that most MIP sensors are currently only fabricated on a small scale. For these sensors to be commercialized, it will be necessary to develop mass production methods. Finally, many MIP sensors currently lack smart features such as the ability to automatically load samples, perform measurements, and clean the electrode surface. Future research should focus on developing MIP sensors with these smart features to make them more user-friendly and easier to integrate into clinical applications. Despite these challenges, the potential benefits of electrochemical MIP sensors are significant. These sensors can be used to develop real-time sensors for a wide range of diseases and health conditions. This would lead to earlier diagnosis and treatment, which could improve patient outcomes and reduce healthcare costs.

MIP-based electrochemical sensors harness the unique recognition properties of MIPs, to detect target molecules with high selectivity and sensitivity. MIP-based electrochemical sensors exhibit remarkable potential in the precise detection of analytes resulting in improved disease diagnosis, environmental protection, and food quality control. In healthcare, MIP-based electrochemical sensors enable rapid and accurate disease diagnosis and continuous monitoring of diseases. Their ability to offer accurate, selective, and rapid detection solutions positions them at the forefront of modern analytical techniques. As research continues to address challenges and explore new avenues, MIP-based electrochemical sensors are poised to shape the future of analytical chemistry, offering innovative solutions to real-world problems across various sectors. These developments hold the potential to revolutionize healthcare practices in the future.

References

1. Becskereki G, Horvai G, Tóth B (2021) The selectivity of molecularly imprinted polymers. Polymers 13(11):1781. https://doi.org/10.3390/polym13111781
2. Dar KK, Shao S, Tan T, Lv Y (2020) Molecularly imprinted polymers for the selective recognition of microorganisms. Biotechnol Adv 45:107640. https://doi.org/10.1016/j.biotechadv.2020.107640
3. Pan J, Chen W, Ma Y, Pan G (2018) Molecularly imprinted polymers as receptor mimics for selective cell recognition. Chem Soc Rev 47(15):5574–5587. https://doi.org/10.1039/C7CS00854F
4. Crapnell RD, Dempsey-Hibbert NC, Peeters M, Tridente A, Banks CE (2020) Molecularly imprinted polymer based electrochemical biosensors: overcoming the challenges of detecting vital biomarkers and speeding up diagnosis. Talanta Open 2:100018. https://doi.org/10.1016/j.talo.2020.100018
5. Pilvenyte G et al (2023) Molecularly imprinted polymer-based electrochemical sensors for the diagnosis of infectious diseases. Biosensors 13(6):6. https://doi.org/10.3390/bios13060620
6. BelBruno JJ (2019) Molecularly imprinted polymers. Chem Rev 119(1):94–119. https://doi.org/10.1021/acs.chemrev.8b00171
7. Achi F, Attar AM, Ait Lahcen A (2024) Electrochemical nanobiosensors for the detection of cancer biomarkers in real samples: Trends and challenges. TrAC Trends Anal Chem 170:117423. https://doi.org/10.1016/j.trac.2023.117423
8. Ahmad OS, Bedwell TS, Esen C, Garcia-Cruz A, Piletsky SA (2019) Molecularly imprinted polymers in electrochemical and optical sensors. Trends Biotechnol 37(3):294–309. https://doi.org/10.1016/j.tibtech.2018.08.009
9. Nsibande SA, Montaseri H, Forbes PBC (2019) Advances in the application of nanomaterial-based sensors for detection of polycyclic aromatic hydrocarbons in aquatic systems. TrAC Trends Anal Chem 115:52–69. https://doi.org/10.1016/j.trac.2019.03.029
10. Liu X et al (2020) Large-scale flexible surface-enhanced Raman scattering (SERS) sensors with high stability and signal homogeneity. ACS Appl Mater Interfaces 12(40):45,332–45,341. https://doi.org/10.1021/acsami.0c13691
11. Zhuang Y, Zhang M, Li X, Zhao M, Wang C, Chen S (2020) Design of high-performance electrochemistry sensors: elucidation of detection mechanism by DFT studies. J Electroanal Chem 860:113905. https://doi.org/10.1016/j.jelechem.2020.113905

12. Downs AM, Plaxco KW (2022) Real-time, in vivo molecular monitoring using electrochemical aptamer based sensors: opportunities and challenges. ACS Sens 7(10):2823–2832. https://doi.org/10.1021/acssensors.2c01428
13. Saputra HA (2023) Electrochemical sensors: basic principles, engineering, and state of the art. Monatshefte Für Chem Chem Mon 154(10):1083–1100. https://doi.org/10.1007/s00706-023-03113-z
14. Zhang L et al (2023) Electrochemical biosensors represent promising detection tools in medical field. Adv Sens Energy Mater 2(4):100081. https://doi.org/10.1016/j.asems.2023.100081
15. Ozcelikay G et al (2022) Sensor-based MIP technologies for targeted metabolomics analysis. TrAC Trends Anal Chem 146:116487. https://doi.org/10.1016/j.trac.2021.116487
16. Hasseb AA, Ghani N d TA, Shehab OR, El Nashar RM (2022) Application of molecularly imprinted polymers for electrochemical detection of some important biomedical markers and pathogens. Curr Opin Electrochem 31:100848. https://doi.org/10.1016/j.coelec.2021.100848
17. Mostafa AM, Barton SJ, Wren SP, Barker J (2021) Review on molecularly imprinted polymers with a focus on their application to the analysis of protein biomarkers. TrAC Trends Anal Chem 144:116431. https://doi.org/10.1016/j.trac.2021.116431
18. Ayerdurai V, Cieplak M, Kutner W (2023) Molecularly imprinted polymer-based electrochemical sensors for food contaminants determination. TrAC Trends Anal Chem 158:116830. https://doi.org/10.1016/j.trac.2022.116830
19. Elfadil D, Lamaoui A, Della Pelle F, Amine A, Compagnone D (2021) Molecularly imprinted polymers combined with electrochemical sensors for food contaminants analysis. Mol Basel Switz 26(15):4607. https://doi.org/10.3390/molecules26154607
20. Dauphin-Ducharme P et al (2019) Electrochemical aptamer-based sensors for improved therapeutic drug monitoring and high-precision, feedback-controlled drug delivery. ACS Sens 4(10):2832–2837. https://doi.org/10.1021/acssensors.9b01616
21. Afsharara H et al (2023) Molecularly imprinted polymer-modified carbon paste electrodes (MIP-CPE): a review on sensitive electrochemical sensors for pharmaceutical determinations. TrAC Trends Anal Chem 160:116949. https://doi.org/10.1016/j.trac.2023.116949
22. Elugoke SE et al (2021) Molecularly imprinted polymers (MIPs) based electrochemical sensors for the determination of catecholamine neurotransmitters—review. Electrochem Sci Adv 1(2):e2000026. https://doi.org/10.1002/elsa.202000026
23. Scheller FW, Zhang X, Yarman A, Wollenberger U, Gyurcsányi RE (2019) Molecularly imprinted polymer-based electrochemical sensors for biopolymers. Curr Opin Electrochem 14:53–59. https://doi.org/10.1016/j.coelec.2018.12.005
24. Gevaerd A, Silva LRG, Silva TA, Marcolino-Junior LH, Bergamini MF, Janegitz BC (2022) Screen-printed electrochemical sensors and biosensors for detection of biomarkers. In: Crespilho FN (ed) Advances in bioelectrochemistry volume 3: biosensors, wearable devices and biomedical applications. Springer International Publishing, Cham, pp 113–140. https://doi.org/10.1007/978-3-030-97921-8_5
25. Park S, Kang H, Kang CM, Kim Y, Kim EJ (2023) NiOx electrochemical sensor fabricated via electrodeposition and spin-coating. Electron Lett 59(6):e12761. https://doi.org/10.1049/ell2.12761
26. Kumunda C, Adekunle AS, Mamba BB, Hlongwa NW, Nkambule TTI (2021) Electrochemical detection of environmental pollutants based on graphene derivatives: a review. Front Mater 7. Accessed 21 Jan 2024 [Online]. https://www.frontiersin.org/articles/10.3389/fmats.2020.616787
27. Tortorich RP, Shamkhalichenar H, Choi J-W (2018) Inkjet-printed and paper-based electrochemical sensors. Appl Sci 8(2):2. https://doi.org/10.3390/app8020288
28. Kang Q, Takehara H, Ichiki T (2023) A polyaniline/polyvinyl acetate composite film electrode for highly sensitive electrochemical sensing of pH. Synth Met 297:117380. https://doi.org/10.1016/j.synthmet.2023.117380
29. Kidakova A, Reut J, Boroznjak R, Öpik A, Syritski V (2019) Advanced sensing materials based on molecularly imprinted polymers towards developing point-of-care diagnostics devices. Proc Est Acad Sci 68(2):158. https://doi.org/10.3176/proc.2019.2.07

30. Wang R et al (2021) Rapid, sensitive and label-free detection of pathogenic bacteria using a bacteria-imprinted conducting polymer film-based electrochemical sensor. Talanta 226:122135. https://doi.org/10.1016/j.talanta.2021.122135
31. Herrera-Chacón A, Cetó X, Del Valle M (2021) Molecularly imprinted polymers—towards electrochemical sensors and electronic tongues. Anal Bioanal Chem 413(24):6117–6140. https://doi.org/10.1007/s00216-021-03313-8
32. Mazzotta E, Di Giulio T, Malitesta C (2022) Electrochemical sensing of macromolecules based on molecularly imprinted polymers: challenges, successful strategies, and opportunities. Anal Bioanal Chem 414(18):5165–5200. https://doi.org/10.1007/s00216-022-03981-0
33. Liu T et al (2020) Review—in situ polymerization for integration and interfacial protection towards solid state lithium batteries. J Electrochem Soc 167(7):070527. https://doi.org/10.1149/1945-7111/ab76a4
34. Feldner A, Völkle J, Lieberzeit P, Fruhmann P (2023) Conductive molecularly imprinted polymers (cMIPs): rising and versatile key elements in chemical sensing. Chemosensors 11(5):5. https://doi.org/10.3390/chemosensors11050299
35. Barhoum A, Hamimed S, Slimi H, Othmani A, Abdel-Haleem FM, Bechelany M (2023) Modern designs of electrochemical sensor platforms for environmental analyses: principles, nanofabrication opportunities, and challenges. Trends Environ Anal Chem 38:e00199. https://doi.org/10.1016/j.teac.2023.e00199
36. Gholivand MB, Karimian N (2015) Fabrication of a highly selective and sensitive voltammetric ganciclovir sensor based on electropolymerized molecularly imprinted polymer and gold nanoparticles on multiwall carbon nanotubes/glassy carbon electrode. Sens Actuators B Chem 215:471–479. https://doi.org/10.1016/j.snb.2015.04.007
37. Zidarič T, Majer D, Maver T, Finšgar M, Maver U (2023) The development of an electropolymerized, molecularly imprinted polymer (MIP) sensor for insulin determination using single-drop analysis. Analyst 148(5):1102–1115. https://doi.org/10.1039/D2AN02025D
38. Thomas T, Mascarenhas RJ, Martis P, Mekhalif Z, Swamy BEK (2013) Multi-walled carbon nanotube modified carbon paste electrode as an electrochemical sensor for the determination of epinephrine in the presence of ascorbic acid and uric acid. Mater Sci Eng C Mater Biol Appl 33(6):3294–3302. https://doi.org/10.1016/j.msec.2013.04.010
39. Li J, Zhang N, Sun Q, Bai Z, Zheng J (2016) Electrochemical sensor for dopamine based on imprinted silica matrix-poly(aniline boronic acid) hybrid as recognition element. Talanta 159:379–386. https://doi.org/10.1016/j.talanta.2016.06.048
40. Wang H-H et al (2018) ZnO nanotubes supported molecularly imprinted polymers arrays as sensing materials for electrochemical detection of dopamine. Talanta 176:573–581. https://doi.org/10.1016/j.talanta.2017.08.083
41. Shan X et al (2019) A molecularly imprinted electrochemical sensor based on Au nanocross-chitosan composites for detection of paraquat. J Solid State Electrochem 23(4):1211–1220. https://doi.org/10.1007/s10008-018-04192-3
42. Gañán J, Martínez-García G, Morante-Zarcero S, Pérez-Quintanilla D, Sierra I (2023) Nanomaterials-modified electrochemical sensors for sensitive determination of alkaloids: recent trends in the application to biological, pharmaceutical and Agri food samples. Microchem J 184:108136. https://doi.org/10.1016/j.microc.2022.108136
43. Parate K, Karunakaran C, Claussen JC (2019) Electrochemical cotinine sensing with a molecularly imprinted polymer on a graphene-platinum nanoparticle modified carbon electrode towards cigarette smoke exposure monitoring. Sens Actuators B Chem 287:165–172. https://doi.org/10.1016/j.snb.2019.02.032
44. Pandey I, Jha SS (2015) Molecularly imprinted polyaniline-ferrocene-sulfonic acid-carbon dots modified pencil graphite electrodes for chiral selective sensing of D-ascorbic acid and L-ascorbic acid: a clinical biomarker for preeclampsia. Electrochim Acta 182:917–928. https://doi.org/10.1016/j.electacta.2015.10.005
45. Li Y, Zhao X, Li P, Huang Y, Wang J, Zhang J (2015) Highly sensitive Fe3O4 nanobeads/graphene-based molecularly imprinted electrochemical sensor for 17β-estradiol in water. Anal Chim Acta 884:106–113. https://doi.org/10.1016/j.aca.2015.05.022

46. Martins GV, Marques AC, Fortunato E, Sales MGF (2016) 8-hydroxy-2′-deoxyguanosine (8-OHdG) biomarker detection down to picoMolar level on a plastic antibody film. Biosens Bioelectron 86:225–234. https://doi.org/10.1016/j.bios.2016.06.052
47. Lin T, Xu Y, Zhao A, He W, Xiao F (2022) Flexible electrochemical sensors integrated with nanomaterials for in situ determination of small molecules in biological samples: a review. Anal Chim Acta 1207:339461. https://doi.org/10.1016/j.aca.2022.339461
48. Liang X-H, Yu A-X, Bo X-J, Du D-Y, Su Z-M (2023) Metal/covalent-organic frameworks-based electrochemical sensors for the detection of ascorbic acid, dopamine and uric acid. Coord Chem Rev 497:215427. https://doi.org/10.1016/j.ccr.2023.215427
49. Guo Z, Xu W, Xu G, Jia Q (2023) An updated overview of MOF@MIP in drug carrier and drug analysis: construction, application and prospective. TrAC Trends Anal Chem 167:117275. https://doi.org/10.1016/j.trac.2023.117275
50. Wang Y et al (2018) C3N4 nanosheets/metal–organic framework wrapped with molecularly imprinted polymer sensor: fabrication, characterization, and electrochemical detection of furosemide. ACS Sustain Chem Eng 6(12):16,847–16,858. https://doi.org/10.1021/acssuschemeng.8b04179
51. Cetinkaya A, Bilge S, Karadurmus L, Sınağ A, Ozkan SA (2022) The role and the place of ionic liquids in molecularly imprinted polymer-based electrochemical sensors development for sensitive drug assay. TrAC Trends Anal Chem 147:116512. https://doi.org/10.1016/j.trac.2021.116512
52. Kaur G, Kumar H, Singla M (2022) Diverse applications of ionic liquids: a comprehensive review. J Mol Liq 351:118556. https://doi.org/10.1016/j.molliq.2022.118556
53. Liu H, Jin P, Zhu F, Nie L, Qiu H (2021) A review on the use of ionic liquids in preparation of molecularly imprinted polymers for applications in solid-phase extraction. TrAC Trends Anal Chem 134:116132. https://doi.org/10.1016/j.trac.2020.116132
54. Lu Y et al (2020) Electrochemical determination of rutin based on molecularly imprinted poly (ionic liquid) with ionic liquid-graphene as a sensitive element. Sens Actuators B Chem 311:127911. https://doi.org/10.1016/j.snb.2020.127911
55. Hu X, Wang C, Zhang M, Zhao F, Zeng B (2020) Ionic liquid assisted molecular self-assemble and molecular imprinting on gold nanoparticles decorated boron-doped ordered mesoporous carbon for the detection of zearalenone. Talanta 217:121032. https://doi.org/10.1016/j.talanta.2020.121032
56. Wang L, Pagett M, Zhang W (2023) Molecularly imprinted polymer (MIP) based electrochemical sensors and their recent advances in health applications. Sens Actuators Rep 5:100153. https://doi.org/10.1016/j.snr.2023.100153
57. Ramanavicius S, Ramanavicius A (2022) Development of molecularly imprinted polymer based phase boundaries for sensors design (review). Adv Colloid Interf Sci 305:102693. https://doi.org/10.1016/j.cis.2022.102693
58. Yang J, Hu Y, Li Y (2019) Molecularly imprinted polymer-decorated signal on-off ratiometric electrochemical sensor for selective and robust dopamine detection. Biosens Bioelectron 135:224–230. https://doi.org/10.1016/j.bios.2019.03.054
59. Phonklam K, Wannapob R, Sriwimol W, Thavarungkul P, Phairatana T (2020) A novel molecularly imprinted polymer PMB/MWCNTs sensor for highly-sensitive cardiac troponin T detection. Sens Actuators B Chem 308:127630. https://doi.org/10.1016/j.snb.2019.127630
60. Hu X, Liu Y, Xia Y, Zhao F, Zeng B (2021) A novel ratiometric electrochemical sensor for the selective detection of citrinin based on molecularly imprinted poly(thionine) on ionic liquid decorated boron and nitrogen co-doped hierarchical porous carbon. Food Chem 363:130385. https://doi.org/10.1016/j.foodchem.2021.130385
61. Ji J, Zhou Z, Zhao X, Sun J, Sun X (2015) Electrochemical sensor based on molecularly imprinted film at Au nanoparticles-carbon nanotubes modified electrode for determination of cholesterol. Biosens Bioelectron 66:590–595. https://doi.org/10.1016/j.bios.2014.12.014

62. Lee M-H, O'Hare D, Guo H-Z, Yang C-H, Lin H-Y (2016) Electrochemical sensing of urinary progesterone with molecularly imprinted poly(aniline-co-metanilic acid)s. J Mater Chem B 4(21):3782–3787. https://doi.org/10.1039/C6TB00760K
63. Alizadeh T, Nayeri S (2020) An enzyme-free sensing platform based on molecularly imprinted polymer/MWCNT composite for sub-micromolar-level determination of pyruvic acid as a cancer biomarker. Anal Bioanal Chem 412(3):657–667. https://doi.org/10.1007/s00216-019-02273-4
64. Nguy TP, Van Phi T, Tram DTN, Eersels K, Wagner P, Lien TTN (2017) Development of an impedimetric sensor for the label-free detection of the amino acid sarcosine with molecularly imprinted polymer receptors. Sens Actuators B Chem 246:461–470. https://doi.org/10.1016/j.snb.2017.02.101
65. Mandani S, Rezaei B, Ensafi AA, Rezaei P (2021) Ultrasensitive electrochemical molecularly imprinted sensor based on AuE/Ag-MOF@MC for determination of hemoglobin using response surface methodology. Anal Bioanal Chem 413(19):4895–4906. https://doi.org/10.1007/s00216-021-03453-x
66. Garcia SM, Wong A, Khan S, Sotomayor MDPT (2021) A simple, sensitive and efficient electrochemical platform based on carbon paste electrode modified with Fe3O4@MIP and graphene oxide for folic acid determination in different matrices. Talanta 229:122258. https://doi.org/10.1016/j.talanta.2021.122258
67. Karami P, Bagheri H, Johari-Ahar M, Khoshsafar H, Arduini F, Afkhami A (2019) Dual-modality impedimetric immunosensor for early detection of prostate-specific antigen and myoglobin markers based on antibody-molecularly imprinted polymer. Talanta 202:111–122. https://doi.org/10.1016/j.talanta.2019.04.061
68. Vural T, Yaman YT, Ozturk S, Abaci S, Denkbas EB (2018) Electrochemical immunoassay for detection of prostate specific antigen based on peptide nanotube-gold nanoparticle-polyaniline immobilized pencil graphite electrode. J Colloid Interface Sci 510:318–326. https://doi.org/10.1016/j.jcis.2017.09.079
69. Garcia Cruz A et al (2020) Design and fabrication of a smart sensor using in silico epitope mapping and electro-responsive imprinted polymer nanoparticles for determination of insulin levels in human plasma. Biosens Bioelectron 169:112536. https://doi.org/10.1016/j.bios.2020.112536
70. Yeasmin S, Wu B, Liu Y, Ullah A, Cheng L-J (2022) Nano gold-doped molecularly imprinted electrochemical sensor for rapid and ultrasensitive cortisol detection. Biosens Bioelectron 206:114142. https://doi.org/10.1016/j.bios.2022.114142
71. Ayankojo AG, Reut J, Ciocan V, Öpik A, Syritski V (2020) Molecularly imprinted polymer-based sensor for electrochemical detection of erythromycin. Talanta 209:120502. https://doi.org/10.1016/j.talanta.2019.120502
72. López R, Khan S, Wong A, Sotomayor MDPT, Picasso G (2021) Development of a new electrochemical sensor based on mag-MIP selective toward amoxicillin in different samples. Front Chem 9:615602. https://doi.org/10.3389/fchem.2021.615602
73. Zeb S, Wong A, Khan S, Hussain S, Sotomayor MDPT (2021) Using magnetic nanoparticles/MIP-based electrochemical sensor for quantification of tetracycline in milk samples. J Electroanal Chem 900:115713. https://doi.org/10.1016/j.jelechem.2021.115713
74. Mostafavi M, Yaftian MR, Piri F, Shayani-Jam H (2018) A new diclofenac molecularly imprinted electrochemical sensor based upon a polyaniline/reduced graphene oxide nanocomposite. Biosens Bioelectron 122:160–167. https://doi.org/10.1016/j.bios.2018.09.047
75. Yang Y et al (2020) Magnetic molecularly imprinted electrochemical sensors: a review. Anal Chim Acta 1106:1–21. https://doi.org/10.1016/j.aca.2020.01.044
76. A new diosgenin sensor based on molecularly imprinted polymer of para aminobenzoic acid selected by computer-aided design—ScienceDirect. Accessed 25 Oct 2023. [Online]. https://www.sciencedirect.com/science/article/abs/pii/S0731708518327584
77. Stoian I-A et al (2020) Biomimetic electrochemical sensor for the highly selective detection of azithromycin in biological samples. Biosens Bioelectron 155:112098. https://doi.org/10.1016/j.bios.2020.112098

78. Ye C et al (2021) Study on the properties and reaction mechanism of polypyrrole@norfloxacin molecularly imprinted electrochemical sensor based on three-dimensional CoFe-MOFs/AuNPs. Electrochim Acta 379:138174. https://doi.org/10.1016/j.electacta.2021.138174
79. Jafari S, Dehghani M, Nasirizadeh N, Baghersad MH, Azimzadeh M (2019) Label-free electrochemical detection of Cloxacillin antibiotic in milk samples based on molecularly imprinted polymer and graphene oxide-gold nanocomposite. Measurement 145:22–29. https://doi.org/10.1016/j.measurement.2019.05.068
80. Motaharian A, Naseri K, Mehrpour O, Shoeibi S (2020) Electrochemical determination of atypical antipsychotic drug quetiapine using nano-molecularly imprinted polymer modified carbon paste electrode. Anal Chim Acta 1097:214–221. https://doi.org/10.1016/j.aca.2019.11.020
81. "Electrochemical assay of ampicillin using Fe3N-Co2N nanoarray coated with molecularly imprinted polymer | Request PDF." Accessed 25 Oct 2023. [Online]. https://www.researchgate.net/publication/342901543_Electrochemical_assay_of_ampicillin_using_Fe3N-Co2N_nanoarray_coated_with_molecularly_imprinted_polymer
82. da Silva J et al (2020) Non-enzymatic lactose molecularly imprinted sensor based on disposable graphite paper electrode. Anal Chim Acta 1143:53. https://doi.org/10.1016/j.aca.2020.11.030
83. Hassan AHA, Moura SL, Ali FHM, Moselhy WA, Taboada Sotomayor MDP, Pividori MI (2018) Electrochemical sensing of methyl parathion on magnetic molecularly imprinted polymer. Biosens Bioelectron 118:181–187. https://doi.org/10.1016/j.bios.2018.06.052
84. Hatamluyi B, Rezayi M, Beheshti HR, Boroushaki MT (2020) Ultra-sensitive molecularly imprinted electrochemical sensor for patulin detection based on a novel assembling strategy using Au@Cu-MOF/N-GQDs. Sens Actuators B Chem 318:128219. https://doi.org/10.1016/j.snb.2020.128219
85. Motaharian A, Motaharian F, Abnous K, Hosseini MRM, Hassanzadeh-Khayyat M (2016) Molecularly imprinted polymer nanoparticles-based electrochemical sensor for determination of diazinon pesticide in well water and apple fruit samples. Anal Bioanal Chem 408(24):6769–6779. https://doi.org/10.1007/s00216-016-9802-7
86. Aghoutane Y, Diouf A, Österlund L, Bouchikhi B, El Bari N (2020) Development of a molecularly imprinted polymer electrochemical sensor and its application for sensitive detection and determination of malathion in olive fruits and oils. Bioelectrochemistry 132:107404. https://doi.org/10.1016/j.bioelechem.2019.107404
87. Anirudhan TS, Athira VS, Nair SS (2022) Detection of chlorpyrifos based on molecular imprinting with a conducting polythiophene copolymer loaded on multi-walled carbon nanotubes. Food Chem 381:132010. https://doi.org/10.1016/j.foodchem.2021.132010
88. Metwally MG, Shehab OR, Ibrahim H, El Nashar RM (2022) Electrochemical detection of Bisphenol A in plastic bottled drinking waters and soft drinks based on molecularly imprinted polymer. J Environ Chem Eng 10(3):107699. https://doi.org/10.1016/j.jece.2022.107699

MIP-Based Optical Sensors: Exploring Principles and Applications as Artificial Antibodies in Disease Detection and Monitoring

9

Vedran Milanković, Tamara Tasić, Igor A. Pašti, and Tamara Lazarević-Pašti

Abstract

This chapter delves into the rapidly emerging field of molecularly imprinted polymers (MIPs) integrated into optical sensors, exploring their remarkable applications as artificial antibodies for disease detection and monitoring. We embark on the journey by elucidating the fundamental principles of optical sensors, emphasizing their unique ability to detect and quantify analytes through optical signals, including absorption, fluorescence, or refractive index changes. Unveiling the versatility of MIP synthesis and molecular imprinting, we underscore the exceptional selectivity of these synthetic receptors in recognizing target molecules. Presenting a compelling alternative to traditional antibodies and aptamers, MIPs exhibit enhanced stability, cost-effectiveness, and nonimmunogenicity, positioning them as promising candidates in disease diagnostics. The chapter further emphasizes the seamless integration of MIPs with optical transduction, where binding events between MIPs and targets induce measurable changes in light properties. The potential of MIP-based optical sensors for highly sensitive and real-time disease biomarker detection is explored, encompassing various medical conditions, from cancer to infectious diseases and autoimmune disorders. Through several illuminating case studies, we showcase the remarkable performance of MIP-based optical sensors in disease monitoring. Demonstrating their ability to detect disease-specific biomolecules and pathogens with exceptional specificity, MIPs pave the way for early diagnosis, personalized treatment, and improved patient outcomes. In addressing challenges and

V. Milanković · T. Tasić · T. Lazarević-Pašti (✉)
VINČA Institute of Nuclear Sciences—National Institute of the Republic of Serbia, University of Belgrade, Belgrade, Serbia

I. A. Pašti
Faculty of Physical Chemistry, University of Belgrade, Belgrade, Serbia

S. Patra et al. (eds.), *Molecularly Imprinted Polymers: Path to Artificial Antibodies*, https://doi.org/10.1007/978-981-97-4379-7_9

prospects in this exciting field, we delve into factors influencing sensor sensitivity and selectivity while highlighting cutting-edge advancements in MIP nanotechnology. Moreover, the chapter explores the potential synergy of MIP-based optical sensors with advanced imaging techniques, opening up new vistas for integrated diagnostic and therapeutic approaches. In conclusion, MIP-based optical sensors represent a promising frontier in disease detection and monitoring, harnessing the potential of artificial antibodies to revolutionize healthcare applications. As research progresses, these innovative biosensing platforms hold immense promise in reshaping medical diagnostics, propelling us toward a future of proactive and personalized healthcare.

Keywords

Prostate-Specific Antigen · Cancer Antigen 125 · Alpha-Fetorprotein · Allergens

9.1 Introduction

Optical sensing has long been recognized as a powerful analytical tool, harnessing the interaction between light and matter to detect and quantify various analytes. The versatility of optical sensing arises from a myriad of transduction mechanisms, including absorbance, fluorescence, refractive index changes, and surface plasmon resonance, offering a spectrum of detection capabilities suited to diverse analytical challenges. These optical sensing techniques have found applications in numerous fields, ranging from environmental monitoring and industrial process control to medical diagnostics and biotechnology [1].

Traditional sensing approaches often relied on natural antibodies or aptamers as the recognition elements to selectively bind target analytes. While effective, these natural receptors are not without limitations. Natural antibodies can exhibit batch-to-batch variations and may be sensitive to environmental conditions, leading to potential cross-reactivity with structurally similar compounds. Additionally, using natural antibodies can pose immunogenic risks, potentially eliciting immune responses when introduced into biological systems [2].

Molecularly imprinted polymers (MIPs) present an ingenious solution to these challenges, offering a powerful and versatile alternative to natural antibodies in optical sensing. MIPs are synthetic materials engineered through molecular imprinting, a process in which target molecules (known as templates) are used to create specific binding cavities or "imprints" in a polymer matrix. The resulting MIPs exhibit remarkable selectivity and affinity toward the template molecules, akin to the molecular recognition capabilities of natural antibodies [3]. Moreover, MIPs boast high stability and reproducibility, addressing the drawbacks associated with natural antibodies, and they are non-immunogenic, as they are chemically synthesized or produced through recombinant techniques [4].

Integrating MIPs into optical sensors creates a compelling synergy, as the selective recognition capabilities of MIPs align seamlessly with the sensitive and real-time detection capabilities of optical transduction. Binding events between MIPs and target analytes induce measurable changes in light properties, enabling the conversion of molecular recognition events into quantifiable optical signals. This amalgamation of MIPs and optical sensing opens new avenues for transformative applications, particularly in disease detection and monitoring [5].

In this chapter, we explore MIP-based optical sensors in-depth, peering into the principles that enable their exceptional selectivity, sensitivity, and stability. We begin by elucidating the fundamental principles of optical sensors, highlighting their ability to detect and quantify analytes through various optical signals. Absorbance measurements, fluorescence phenomena, refractive index changes, and surface plasmon resonance are among the optical transduction techniques that empower optical sensors to offer sensitive and real-time analyte detection. The applications of MIP-based optical sensors in disease detection and monitoring are vast and transformative. As we navigate through various medical conditions, from cancer and infectious diseases to autoimmune disorders, we explore how MIP-based sensors enable early and accurate diagnosis. Detecting disease-specific biomarkers, proteins, or nucleic acids in biological samples, MIP-based optical sensors open the gateway to personalized medicine, where tailored treatment strategies improve patient outcomes and prognosis.

9.2 Optical Sensors

Optical sensors are a class of analytical devices that utilize light to detect and quantify various analytes. By leveraging the interaction of light with target molecules or substances, optical sensors offer numerous advantages, including high sensitivity, noninvasiveness, and real-time monitoring capabilities. They find applications in diverse fields, ranging from environmental monitoring and biomedical diagnostics to industrial process control and beyond [6–9].

The key components of an optical sensor include the light source, sensing element, and optical transducer. Optical sensors utilize diverse light sources, such as light-emitting diodes (LEDs), lasers, or lamps, to produce the necessary illumination for the sensing process. The sensing element interacts with the target analyte and undergoes alterations in its optical properties as a response to the presence of the analyte. Subsequently, the optical transducer translates these modifications in the sensing element's optical properties into a quantifiable signal, often manifested as an electrical output [1].

There are diverse types of optical sensors (Table 9.1), each uniquely suited to particular purposes. Absorbance sensors, for instance, measure the extent to which the analyte absorbs light. As the concentration of the analyte increases, it causes a decrease in the intensity of transmitted light. These sensors play a vital role in various applications, each requiring precise monitoring and measurement of light absorption levels in response to varying analyte concentrations [10]. The working principle of absorbance sensors is shown in Fig. 9.1.

Table 9.1 Optical detection techniques and their advantages and disadvantages

Optical sensor type	Advantages	Disadvantages
Absorbance	• Selectivity • Reusability • Stability • Efficiency • Short response time • Cost effective	• Limited by mass transportation • Susceptible to steric hindrance • Simultaneous detection of multiple analytes is not possible
Fluorescence	• Selectivity • Reusability • Stability • Efficiency • Short response time	• Limited by mass transportation • Susceptible to steric hindrance • Simultaneous detection of multiple analytes is not possible • Expensive and complex equipment • Influenced by sample matrix
Refractometric	• Selectivity • Reusability • Stability • Efficiency • Short response time • Cost effective • Label-free detection	• Limited by mass transportation • Susceptible to steric hindrance • Simultaneous detection of multiple analytes is not possible
SPR	• Selectivity • Reusability • Stability • Efficiency • Real-time monitoring • Short response time • Label-free detection	• Limited by mass transportation • Susceptible to steric hindrance • Simultaneous detection of multiple analytes is not possible • Expensive and complex equipment • Influenced by sample matrix

Fluorescence sensors utilize the phenomenon of fluorescence. Here, the analyte absorbs light at a specific wavelength and subsequently emits light at a longer wavelength, generating a fluorescence signal that correlates directly with the concentration of the analyte. This unique property enables fluorescence sensors to provide highly sensitive and specific measurements across a range of applications, making them indispensable tools for analytical tasks requiring precise detection of analyte concentrations [11, 12]. The working principle of fluorescence sensors is shown in Fig. 9.2.

Refractive index sensors operate by detecting alterations in the refractive index of the sensing element due to the presence of analytes. These alterations modify the propagation of light within the sensor, ultimately leading to variations in the intensity or phase of transmitted light, which can then be directly linked to the concentration of the analyte. This method allows refractive index sensors to offer precise and specific measurements, making them valuable tools for applications that demand

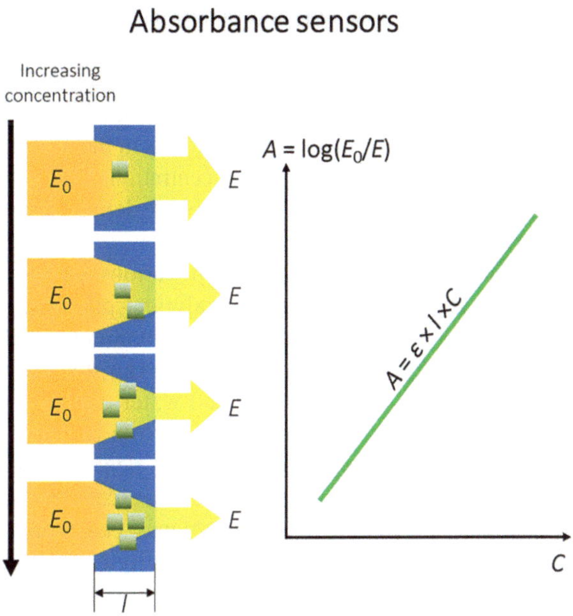

Fig. 9.1 Working principle of absorbance sensors

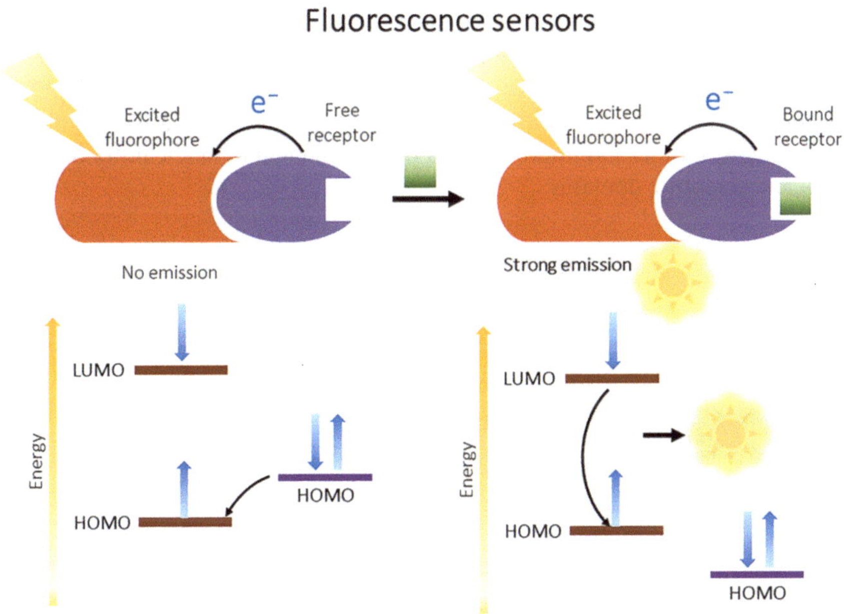

Fig. 9.2 Working principle of fluorescence sensors

accurate analyte concentration determination. The working principle of refractometric sensors is shown in Fig. 9.3 [13].

Finally, surface plasmon resonance (SPR) sensors (Fig. 9.4) operate by detecting alterations in refractive index near the surface of a metal film. These changes signal the binding of analytes and lead to a shift in the resonance angle, providing a quantitative measure of the analyte concentration. SPR sensors are widely used in various applications due to their ability to deliver real-time, label-free, and highly sensitive analyte detection [14, 15].

Optical sensors find diverse applications across various fields, making them indispensable tools in modern analytical practices. In environmental monitoring, they are crucial for detecting environmental pollutants such as gases, heavy metals, and organic compounds, ensuring the safety and sustainability of the environment by monitoring air and water quality [16, 17]. In biomedical diagnostics, optical sensors play a pivotal role in detecting biomarkers, pathogens, and disease-related molecules in biological samples, enabling early and accurate diagnosis and managing various medical conditions [18]. In the food industry, optical sensors are utilized to detect contaminants, allergens, and spoilage markers, ensuring food safety and quality control [19]. Moreover, in industrial settings, optical sensors are essential for monitoring and controlling diverse processes, such as chemical reactions and manufacturing, ensuring operational efficiency and product quality. Furthermore, these sensors serve as valuable tools in biochemical and biotechnological research, allowing scientists to study biomolecular interactions, enzymatic reactions, and cellular processes, driving advancements in various scientific disciplines [20, 21]. The versatility of optical sensors empowers researchers, industries, and professionals alike, propelling innovative solutions and improvements in environmental, medical, food, industrial, and scientific domains.

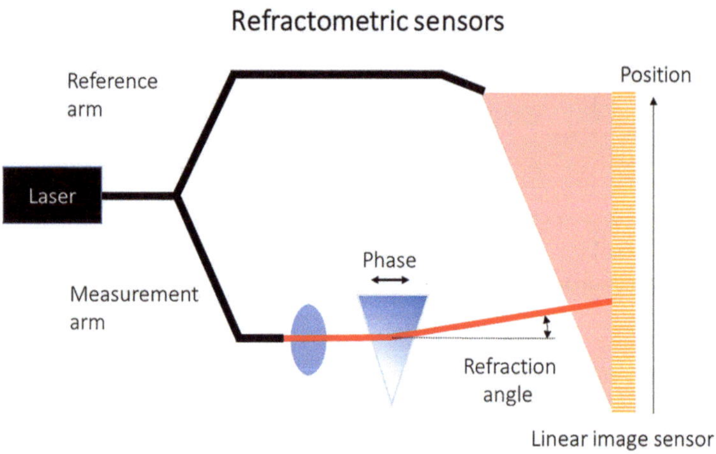

Fig. 9.3 Working principle of refractometric sensors

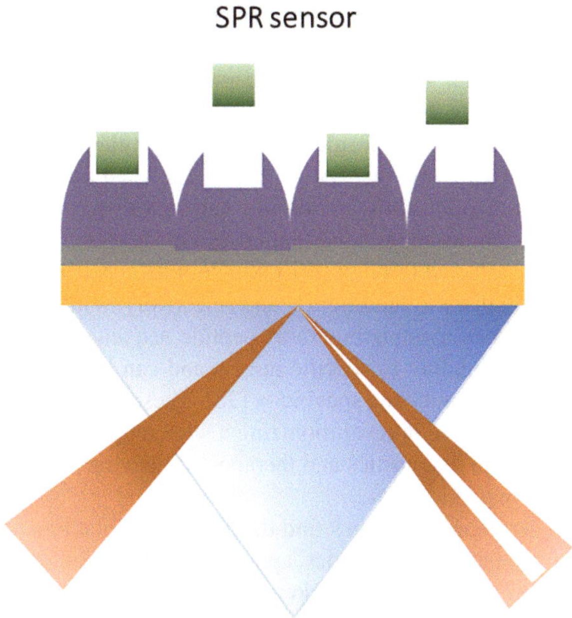

Fig. 9.4 Working principle of SPR sensors

9.3 Artificial Antibodies

Artificial antibodies, also known as synthetic antibodies or antibody mimetics, are engineered molecules designed to mimic natural antibodies' specificity and binding capabilities. These synthetic recognition elements have gained significant attention in various scientific and technological applications due to their unique advantages over traditional antibodies, offering enhanced stability, cost-effectiveness, and versatility [22].

Artificial antibodies possess key features that set them apart as powerful alternatives to natural antibodies. Like their natural counterparts, artificial antibodies demonstrate exceptional specificity and selectivity toward their target molecules, encompassing a wide range of analytes, from small molecules and peptides to proteins, nucleic acids, and even pathogens. In contrast to natural antibodies, which may be sensitive to environmental conditions and exhibit batch-to-batch variations, artificial antibodies are highly stable and reproducible, ensuring consistent and reliable performance across different applications [23]. Additionally, artificial antibodies offer the advantage of non-immunogenicity, as they are chemically synthesized or produced through recombinant techniques, eliminating the risk of immune responses commonly associated with natural antibodies. Furthermore, their cost-effectiveness is evident in their production, achievable through scalable chemical processes or standardized genetic engineering methods, making them accessible and appealing for research and industrial applications. These distinctive features position artificial antibodies as versatile and transformative tools with vast potential for advancing diverse fields, from biotechnology and medicine to research and diagnostics [24, 25].

Artificial antibodies encompass various types, each offering unique recognition capabilities. MIPs are synthetic materials generated through molecular imprinting, wherein target molecules are used as templates to create specific binding cavities within a polymer matrix. MIPs are renowned for their high selectivity and stability, finding applications in sensors, separation techniques, and drug delivery systems [26]. On the other hand, aptamers are short, single-stranded DNA or RNA molecules selected through SELEX, adopting specific three-dimensional structures that enable them to bind tightly and selectively to target molecules [27]. Peptide-based antibodies, comprising short peptide sequences, can be designed or selected for precise binding to specific targets and can be further engineered to enhance their binding affinity and specificity [28]. These diverse types of artificial antibodies hold great promise in revolutionizing a wide range of fields, from medicine and biotechnology to diagnostics and therapeutics, paving the way for innovative and transformative applications.

Artificial antibodies find diverse and pivotal applications across multiple fields. In biomedical diagnostics, aptamers and MIPs are instrumental in diagnostic assays, detecting disease biomarkers, pathogens, and various analytes in clinical samples, thus facilitating early and accurate disease diagnosis and enabling personalized and targeted therapies [29]. Moreover, artificial antibodies are promising as therapeutic agents, with aptamers engineered to inhibit specific disease-related proteins, offering targeted therapeutic interventions [30]. Environmental monitoring benefits from using artificial antibodies in detecting pollutants, toxins, and contaminants in air, water, and soil, thereby contributing to environmental safety and sustainability efforts [4]. In drug delivery, MIPs and aptamers are incorporated into drug delivery systems, ensuring targeted and controlled drug release, ultimately enhancing therapeutic efficacy and reducing side effects [31]. Additionally, artificial antibodies serve as valuable tools in biotechnology and research, enabling the study of protein–protein interactions, cellular processes, and biomolecular recognition, fostering advancements and breakthroughs in various scientific disciplines [32].

9.4 MIP-Based Optical Sensors

Molecularly imprinted polymers (MIPs) have found a prominent place within optical sensors, offering a robust and adaptable platform for the precise and selective detection of target substances. By uniting the extraordinary recognition capabilities of MIPs with optical transduction principles, MIP-based optical sensors deliver distinctive advantages across a spectrum of applications, ranging from environmental monitoring to cutting-edge medical diagnostics [33].

The principle of selective recognition and transduction of target analytes lies at the core of MIP-based optical sensors. The MIP acts as an artificial receptor, meticulously imprinted to identify a specific target molecule. During the imprinting process, the target analyte serves as a template, guiding the polymerization of functional monomers to create precise binding sites or "imprints" within the polymer matrix. These imprints exhibit a heightened affinity and selectivity toward the target analyte once it is extracted from the polymer. When the MIP-based sensor encounters a sample containing the target analyte, these analyte molecules selectively bind to the imprinted sites within the MIP. This binding event prompts a measurable alteration in the optical characteristics of the MIP, including absorbance, fluorescence, or refractive index. The optical transduction component of the sensor then translates this change into a detectable optical signal, which correlates with the concentration of the target analyte present in the sample [4, 34].

The advantages of MIP-based optical sensors make them invaluable tools in analytical applications. Demonstrating exceptional selectivity and specificity, MIPs exhibit a remarkable capacity to pinpoint the desired analyte while minimizing interference from other compounds within complex samples. Moreover, these sensors offer high sensitivity, enabling the detection of even minute concentrations of the target analyte. Their real-time or near-real-time detection capabilities facilitate swift and continuous monitoring and analysis [35]. Additionally, the inherent stability and reusability of MIPs enhance their cost-effectiveness and efficiency across diverse applications, establishing them as a promising and dependable choice in optical sensing [36].

MIP-based optical sensors find extensive use in various domains, underscoring their significance. Environmental monitoring is instrumental in detecting pollutants such as heavy metals, pesticides, and toxins in air, water, and soil samples, thereby safeguarding environmental well-being and ensuring quality control [4, 37]. In the field of biomedical diagnostics, these sensors play a pivotal role in the early diagnosis of diseases by identifying disease biomarkers, proteins, or nucleic acids in biological samples, thus facilitating personalized medicine [38, 39]. The food industry leverages MIP-based optical sensors to detect allergens, contaminants, and foodborne pathogens, assuring food safety and quality. It is important to monitor the concentration of environmental pollutants, especially pesticides in food [40–47]. Furthermore, in pharmaceutical analysis, these sensors aid in drug discovery and monitoring by detecting and quantifying drug molecules, thereby advancing scientific research and innovation. These sensors could also find application in studying and monitoring lung diseases where the cholinergic system plays an important role [48].

9.5 Applications of MIP-Based Optical Sensors in Disease Detection and Monitoring

MIP-based optical sensors play a critical role in disease detection and monitoring, finding applications in various medical conditions. In cancer biomarker detection, these sensors identify specific cancer biomarkers, such as proteins or nucleic acids, facilitating early cancer diagnosis and timely treatment [49]. Infectious disease screening benefits from engineered MIP-based sensors that recognize unique surface proteins or genetic markers of infectious agents, enabling rapid and sensitive detection of pathogens for early containment [50]. For autoimmune diseases, MIP-based sensors detect disease-related biomolecules, including autoantibodies, aiding in disease monitoring and treatment assessment. Diabetes management utilizes MIP-based sensors for accurate and continuous glucose monitoring, assisting patients in effectively managing their condition [35]. The versatility of MIP-based optical sensors in disease detection and monitoring showcases their transformative potential in advancing personalized healthcare and diagnostics. In Table 9.2, there is an overview of the most important characteristics of MIP-based optical sensors for disease detection and monitoring.

9.5.1 MIP-Based Optical Sensors for Cancer Biomarker Detection

9.5.1.1 PSA—Prostate-Specific Antigen

PSA is a protein produced by the prostate gland in males, and it is primarily known for its role in prostate health and cancer diagnosis. In its normal function, PSA is produced by the prostate and plays a role in liquefying semen. It helps to maintain the liquid state of semen, allowing sperm to swim freely. Elevated levels of PSA in the blood can be an indication of prostate cancer. However, it is important to note that PSA levels can also be elevated due to non-cancerous conditions such as benign prostatic hyperplasia (BPH), prostatitis (prostate inflammation), and even recent ejaculation or prostate procedures. A PSA blood test measures the level of PSA in the bloodstream. This test is commonly used to screen for prostate cancer or monitor the disease's progress in individuals who have already been diagnosed with prostate cancer. MIP-based optical sensors for PSA detection have the potential to offer highly specific and sensitive diagnostic tools for prostate cancer screening and monitoring. Ongoing research in this field aims to improve these sensors' performance and clinical applicability.

A method for enhancing the sensitivity of SPR biosensors is described by Sun et al. in their study published in 2020. In this approach, biotinylated phenylalanine (Biotin-Phe) monomers are rapidly self-assembled into nanoparticles under mild conditions. These self-assembled nanoparticles then serve as carriers for streptavidin–antibody complexes via the streptavidin-biotin interaction. The signal amplification occurs due to the substantial molecular weight of the nanoparticle–streptavidin–antibody conjugate. The sensor exhibited an impressive linearity range, spanning from as low as 1 pg/mL and extending up to 200 ng/mL, showcasing its versatility in quantifying PSA levels across a wide concentration range. Moreover, the sensor demonstrated a remarkable sensitivity with a limit of detection (LOD) as low as 1 ng/mL, making it highly suitable for precise and sensitive PSA biomarker detection [51].

In a study published in 2016, researchers have developed a highly sensitive and real-time detection method for PSA using a microcontact PSA-imprinted SPR sensor chip. The imprinted chip was created through UV polymerization using the microcontact imprinting technique, employing methacrylic acid (MAA) as the

Table 9.2 An overview of the most important characteristics of MIP-based optical sensors for disease detection and monitoring

Analyte	Type of detection	Functional monomer	LOD	Real sample	Reference
PSA	SPR	Biotin-Phe	1 ng/mL	Fresh male blood	[51]
PSA	SPR	MAA	91 pg/mL	Blood samples of healthy donors	[52]
PSA	SERS	Tannic acid	0.9 pg/mL	Serum samples	[53]
PSA	Fluorescence	Phenylboronic acid	1.67 ng/mL	/	[54]
CA-15-3	Fluorescence	4-Carboxyphenylboronic acid	50 µU/mL	Human serum samples	[55]
CA-125	Fluorescence	4-Carboxyphenylboronic acid	50 µU/mL	Human serum samples	[55]
CA-125	SPR	Py	0.01 U/mL	Human serum samples	[56]
AFP	Fluorescence	4-Vinylphenylboronic acid, and methyl methacrylate	48 ng/L	Human serum samples	[57]
AFP	SERS	MPBA	1 ng/mL	Human serum samples	[58]
AFP	Fluorescence	Phenylboronic acid	1.8 ng/mL	/	[54]
OTA	SPR	N-methacryloyl-(L)-phenylalanine and 2-hydroxyethyl methacrylate	0.028 ng/mL	Dried fig	[59]
AFB1	LSPR	/	0.3 ng/mL	Cereals	[60]
AFB1	SPR	N-methacryloyl-l-phenylalanine	1.04 pg/mL	Ground corn sample and fresh peanut	[61]
AFM1	SPR	MAPA	0.4 pg/mL	Raw milk samples	[62]
Endotoxins	SPR	Poly(hydroxyethyl methacrylate-N-methacryloyl-(L)-histidine methyl ester	0.023 ng/mL	Artificial plasma	[63]
Zearalenone	SPR	Pyrrole	0.3 ng/mL	Corn sample	[64]
DON	SPR	Polypyrrole	0.1 ng/mL	/	[65]
Enterococcus faecalis	SPR	MAH	1.05×10^2 CFU/mL	Sea water	[66]
Salmonella paratyphi	SPR	MAH	1.4×10^6 CFU/mL	Apple juice	[67]
Escherichia coli	SPR	MAH	0.57 CFU/mL	Urin	[68]

(continued)

Table 9.2 (continued)

Analyte	Type of detection	Functional monomer	LOD	Real sample	Reference
T4 bacteriophage	SPR	PVA	6×10^3 PFU/mL	Sea water and tap water	[69]
α-Casein	SPR	N-isopropyl acrylamide (NIPAm), N-tert-butylacrylamide (TBA), acrylic acid (Acc), and N-(3-aminopropyl) methacrylamide HCl (APM) were	127 ± 97.6 ng/mL	Waste water	[70]
Histamine	SPR	MAA	25 μg/L	Carp samples	[71]

functional monomer (Fig. 9.5). Under optimal conditions, they successfully detected PSA in standard solutions within a 0.1–50 ng/mL concentration range, achieving an impressive LOD of approximately 91 pg/mL (equivalent to 18×10^{-14} M). This research presents a promising approach for precise and reliable PSA detection in clinical applications [52].

Turan et al. introduced an innovative analytical approach for PSA detection by combining the exceptional selectivity of MIPs with the high sensitivity of SERS. They coated magnetic nanoparticles with an imprinted layer, using tannic acid as a functional monomer, diethylenetriamine as a cross-linker, and PSA as the template molecule. This innovative sensor boasts an impressive linearity range, spanning from an astonishingly low 0.5 pg/mL to a high of 1.0 μg/mL, demonstrating its ability to accurately quantify PSA levels over a wide concentration range. Furthermore, this sensor exhibits exceptional sensitivity with a remarkably low LOD of only 0.9 pg/mL, making it a valuable tool for the highly sensitive detection of PSA biomarkers in various applications [53].

Sunayama et al. presented a novel approach for the simultaneous detection of PSA biomarkers using a fluorescence-based sensing system with MIPs. This innovative sensor incorporates phenylboronic acid as the functional monomer, a strategic choice to enhance selectivity and sensitivity in capturing PSA biomarkers. One of the notable features of this sensor is its remarkable sensitivity, as evidenced by a low LOD of only 1.67 ng/mL. This outstanding sensitivity makes it exceptionally well-suited for the precise and ultrasensitive quantification of PSA levels, particularly in applications demanding the detection of minute concentrations with great accuracy [54].

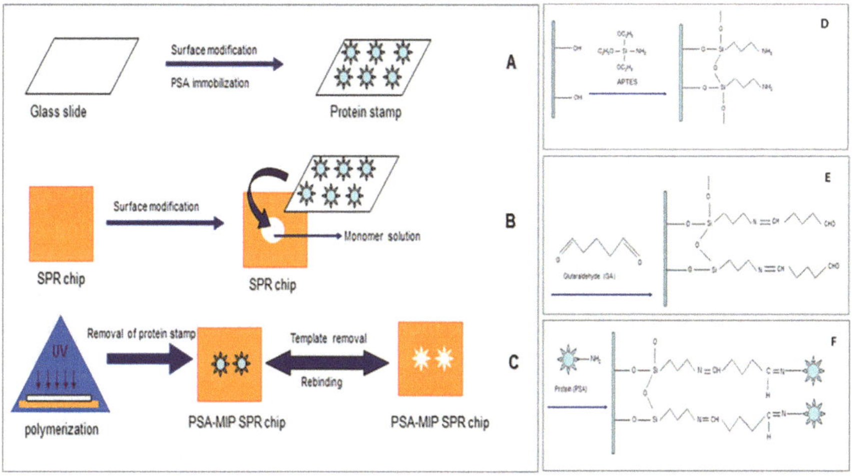

Fig. 9.5 Schematic representation of microcontact imprinting of PSA onto the SPR biosensor and surface modification of glass cover slips; (**a**) preparation of glass cover slips (protein stamps), (**b**) preparation of SPR chips, (**c**) microcontact imprinting of PSA via UV polymerization, (**d**) surface modification of glass cover slips with APTES, (**e**) activation of amino groups on glass cover slips with glutaraldehyde, and (**f**) PSA immobilization onto the glass cover slips. Reprinted from [52]

In summary, these studies present different methods for PSA detection, each with its own advantages. Sun et al. (2020) offer versatility with a wide linear range. The 2016 study by Ertürk et al. provides a good LOD and promises precise detection. Turan et al. achieve an exceptional sensitivity with an extremely low LOD, suitable for highly sensitive detection. Lastly, Sunayama et al. introduce a fluorescence-based method with high sensitivity, which is particularly useful for detecting low concentrations of PSA accurately. The choice of method would depend on the specific application and sensitivity requirements.

9.5.1.2 CA-15-3—Cancer Antigen 15-3 and CA-125—Cancer Antigen 125

CA-15-3 is a glycoprotein, and its levels in the blood are often elevated in individuals with breast cancer. It is primarily associated with breast carcinoma, particularly invasive ductal carcinoma, and metastatic breast cancer. It is important to note that CA 15-3 is not a definitive diagnostic test for breast cancer. Its levels can be elevated in other conditions, such as benign breast diseases, ovarian cancer, and lung cancer. Therefore, it is used in conjunction with other diagnostic and monitoring methods. CA-15-3 levels are typically measured through a blood test. Regular monitoring of CA-15-3 levels in breast cancer patients can help healthcare providers assess treatment efficacy and detect potential relapses.

CA-125 is a protein biomarker often associated with certain gynecological and abdominal cancers, particularly ovarian cancer. CA-125 is a glycoprotein, which means it is a protein molecule with sugar molecules attached. It is produced by various tissues in the body, including the lining of the uterus (endometrium), the fallopian tubes, and the lining of the chest and abdominal cavities (peritoneum). CA-125 is primarily used as a tumor marker, especially in the context of ovarian cancer. Elevated levels of CA-125 can indicate ovarian cancer, particularly in advanced-stage or recurrent ovarian cancer cases. It is less reliable as a diagnostic tool for early-stage ovarian cancer because CA-125 levels can be elevated in various non-cancerous conditions as well. In contrast, CA-125 can be elevated in noncancerous conditions such as endometriosis, uterine fibroids, pelvic inflammatory disease, and menstruation.

Optical sensors incorporating MIPs hold significant promise as cutting-edge diagnostic tools, demonstrating exceptional sensitivity and specificity in the detection of CA 125 and CA 15-3. These sensors represent a potential breakthrough in breast and ovarian cancer screening and monitoring. Current research endeavors are dedicated to enhancing the performance and expanding the clinical utility of these sensor platforms.

In a study published in 2021, Bahari et al. developed an efficient fluorescence immunosensor with 4-carboxyphenylboronic acid as a functional monomer designed to measure CA-15-3 concurrently. This sensor offers a wide linearity range, spanning from an impressively low 0.0005 U/mL to 40 U/mL, allowing for accurate quantification of CA-15-3 concentrations across a broad spectrum. Furthermore, the sensor exhibits sensitivity with an LOD of 50 μU/mL, making it suitable for precise detection of CA-15-3, even at relatively low concentrations [55].

Also, in the same study, Bahari et al. measure CA-125. Under the most favorable experimental conditions, the fluorescence intensity gradually increased in response to rising concentrations of CA125 antigens, spanning an impressive range of 0.0005–40 U/mL for each marker. The LOD achieved was a remarkable 50 μU/mL. [55]. Their developed method possesses several outstanding attributes, including a wide linear range, robust reproducibility, and straightforward operation, rendering it exceptionally suitable for the clinical diagnosis of CA125 and CA15-3 tumor markers.

The main goal for Rebelo et al. was to develop an SPR biosensor using MIPs for the specific recognition of CA-125. The imprinting of CA-125 through the electropolymerization of pyrrole (Py) as functional monomers on a gold electrode using cyclic voltammetry (CV) (Fig. 9.6). The biosensor employing electrochemical transduction exhibited superior analytical parameters, demonstrating excellent selectivity and a low detection limit (LOD) of 0.01 U/mL. It also offered a linear concentration range spanning from 0.01 to 500 U/mL. These results indicate its reliability and accuracy in practical applications [56].

In summary, Bahari et al.'s study in 2021 presented an efficient fluorescence immunosensor for both CA15-3 and CA-125 detection, with a wide linear range and a 50 μU/mL LOD for each marker. Rebelo et al. focused on CA-125 detection using an SPR biosensor with an even lower LOD of 0.01 U/mL and a broader linear range. Both studies demonstrated excellent selectivity and accuracy, providing valuable options for clinically diagnosing these tumor markers. The choice between these methods may depend on specific application requirements and the marker of interest.

9.5.1.3 AFP—Alpha-Fetorprotein

Alpha-fetoprotein (AFP) is a well-known cancer biomarker, primarily associated with certain types of liver cancer and certain germ cell tumors. AFP is most notably associated with hepatocellular carcinoma (HCC), the most common type of liver cancer. In individuals with HCC, elevated AFP levels are often observed. AFP serves as a diagnostic and prognostic marker for HCC. AFP levels can be elevated

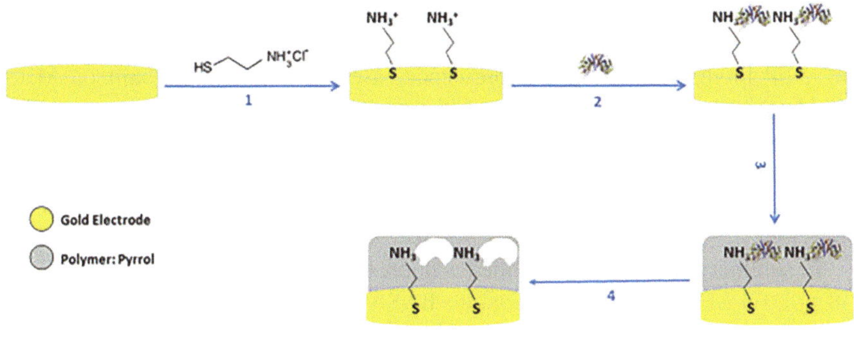

Fig. 9.6 Graphical representation of the different stages of MIP assembly. Reprinted from [56]

in some cases of testicular cancer, especially non-seminomatous germ cell tumors. It can be used as a marker for diagnosis, monitoring, and assessing treatment response in these cases. It may also be elevated in some ovarian tumors, such as ovarian teratomas. It is used in cancer diagnosis, monitoring treatment response, and assessing disease progression. However, its clinical utility is limited to certain cancers and is not a standalone diagnostic tool. Other diagnostic tests and imaging studies are often used alongside AFP measurements to evaluate a patient's condition comprehensively. MIP optical sensors have the potential to be developed for the specific detection of alpha-fetoprotein (AFP). MIP-based optical sensors for AFP detection have the potential to offer highly specific and sensitive diagnostic tools for cancer screening, monitoring, and management. Ongoing research aims to improve the performance and clinical applicability of these sensors.

In 2015, Tan et al. published an article that presents a novel fluorescent MIP with the ability to selectively recognize and trigger fluorescence in response to the tumor biomarker AFP. The methodology employs amino-modified Mn-doped ZnS quantum dots (QDs) as a stable substrate, 4-vinylphenylboronic acid, and methyl methacrylate as functional monomers, γ-methacryloxypropyl trimethoxysilane as a grafting agent, and AFP as the template molecule. A graft imprinting process is carried out on the surface of the QDs. The resultant MIP-QDs exhibit a strong linear response to α-fetoprotein within the concentration range of 50 ng/L to 10 μg/L, with a remarkably low LOD at 48 ng/L. Importantly, this approach is versatile and has the potential to be adapted for the recognition of various other glycoproteins, widening its applicability [57].

Ye et al., in their study, introduced a novel approach called the boronate-affinity sandwich assay (BASA) designed for the precise and sensitive detection of trace glycoproteins within complex samples. BASA operates on the principle of forming sandwiches involving boronate-affinity MIPs, target glycoproteins, and boronate-affinity SERS probes. In this type of sensor, 4-mercaptophenylboronic acid (MPBA) was used as a functional monomer. This sensor system exhibits an impressive linearity range, spanning from as low as 1 ng/mL to a high of 10 mg/mL, enabling precise quantification of AFP concentrations across a wide spectrum. The sensor demonstrates remarkable sensitivity, featuring a low LOD of just 1 ng/mL. This exceptional sensitivity makes it exceptionally well-suited for the precise and ultra-sensitive quantification of AFP levels, even at extremely low concentrations [58].

Sunayama et al. presented a groundbreaking method for simultaneously detecting AFP biomarkers using a fluorescence-based sensing system built on MIPs. Within this system, phenylboronic acid was used as a functional monomer. This sensor system offers a sensitivity with a remarkable LOD of 1.80 ng/mL. This innovative approach holds the potential to serve as a novel platform for the simultaneous detection of multiple proteins, offering significant promise in various applications [54].

In summary, Tan et al. developed a fluorescent MIP with a moderate LOD of 48 ng/L for AFP. Ye et al. introduced the BASA method with remarkable sensitivity, achieving an impressive 1 ng/mL LOD for AFP. Sunayama et al. employed a fluorescence-based sensing system with phenylboronic acid. They achieved an

LOD of 1.80 ng/mL for AFP, showcasing the potential for simultaneous detection of multiple proteins. Each method has its strengths, and the choice between them may depend on specific application requirements and the level of sensitivity needed.

9.5.2 MIP-Based Optical Sensors for Infectious Disease Screening

MIP-based optical sensors represent a cutting-edge technology with a wide range of applications, particularly in the field of infectious disease detection. These sensors are specifically designed to detect microorganisms, such as bacteria and viruses, offering significant promise in several key areas. From clinical diagnostics to pathogen monitoring and public health surveillance, MIP-based optical sensors are poised to revolutionize disease detection. Moreover, their versatility extends to vital domains like healthcare, food safety, and environmental monitoring, further underscoring their potential to enhance various aspects of our lives [72].

For detecting ochratoxin A (OTA) contamination in dried figs, Akgönüllü et al. developed a label-free and highly selective SPR sensor. In their approach, researchers designed a MIP film on the SPR sensor chip using a light-initiated polymerization process involving N-methacryloyl-(L)-phenylalanine and 2-hydroxyethyl methacrylate as functional monomers in the presence of OTA as a template molecule. The OTA-imprinted sensor demonstrated a wide linear detection range from 0.1 to 20 ng/mL, with an impressively low LOD calculated at 0.028 ng/mL. Overall, their results demonstrate the successful application of the OTA-imprinted SPR sensor to accurately determine OTA levels in dried fig samples. This sensor holds promise for addressing OTA contamination concerns in agricultural products [59].

In a study published in 2022, researchers presented an innovative sensor platform designed to detect aflatoxin B1 (AFB1), a prevalent food toxin often found in cereals. Their approach relies on a molecularly imprinted polymer film that offers specific binding to AFB1. Additionally, it leverages the fluorescence signal generated by the analyte molecule, which is intensified by the localized electric field induced in close proximity to the surface of silver nanoparticles. These nanoparticles are excited at the wavelength corresponding to localized surface plasmon resonance (LSPR). The sensor's remarkable performance includes an LOD for AFB1 as low as 0.3 ng/mL, significantly surpassing the sensitivity of a fluorescent sensor lacking silver nanoparticles. The plasmon-enhanced fluorescence factor stands at an impressive 33, and the sensor's linear dynamic range spans from 0.3 to 25 ng/mL. This platform holds great promise for the precise and selective detection of AFB1 in cereals and related food products [60].

A novel and highly sensitive plasmonic sensing approach has been devised by Akgönüllüet al. for the detection of AFB1. This method capitalizes on enhanced SPR nanosensors. Initially, AFB1 is precomplexed with N-methacryloyl-l-phenylalanine, serving as the template molecule and functional monomer. The AFB1-imprinted nanosensor demonstrates an impressively broad linear detection range, spanning from 0.0001 to 10.0 ng/mL, and it achieves an incredibly low LOD at 1.04 pg/mL. This innovative approach holds great promise for highly sensitive, selective, and practical detection of AFB1, especially in the analysis of food samples, mitigating potential health risks associated with aflatoxin contamination [61].

Akgönüllüet et al. developed an SPR sensor based on molecular imprinting to directly detect trace amounts of aflatoxin M1 (AFM1) in raw milk samples. Researchers created a polymer nanofilm integrated with gold nanoparticles (AuNPs) on a gold plasmonic sensor chip coated with allyl mercaptan to achieve this. N-methacryloyl-l-phenylalanine (MAPA) was selected as the functional monomer. This method enables the highly sensitive detection of AFM1, with a remarkably low LOD of 0.4 pg/mL. It exhibits good linearity across a wide concentration range from 0.0003 to 20.0 ng/mL under optimized conditions. This research holds significant promise for enhancing the detection of AFM1 in raw milk, contributing to improved food safety and quality assurance in the dairy industry [62].

In research published in 2021, researchers developed a SPR sensor for the rapid, sensitive, and selective detection of endotoxins. They employed a molecular imprinting approach to create this sensor, which offers a straightforward and efficient means of endotoxin recognition. To construct the sensor, nanofilms were synthesized using poly(hydroxyethyl methacrylate-N-methacryloyl-(L)-histidine methyl ester) as a functional monomer on the surfaces of SPR chips through ultraviolet polymerization. This sensor system offers a substantial linearity range from 0.5 to 100 ng/mL. Remarkably, the sensor exhibits exceptional sensitivity, featuring a low LOD of 0.023 ng/mL. In summary, our study presents a simple, fast, sensitive, and selective SPR sensor based on molecular imprinting for the detection of endotoxins, showcasing its potential for various analytical and diagnostic purposes [63].

Zearalenone, a mycoestrogen, was successfully detected using an SPR sensor functionalized with Pyrrole as a functional monomer. The MIPPy-SPR sensor displayed a linear response within the 0.3–3000 ng/mL concentration range and the LOD of 0.3 ng/mL. These findings suggest that combining SPR sensing with a MIPPy film represents a promising and effective alternative method for the sensitive and selective detection of zearalenone [64].

The objective of the study published in 2011 by Choi et al. was to explore the applicability of the MIP technique for detecting mycotoxin deoxynivalenol (DON) using an SPR transducer. To achieve this, polypyrrole was used as a functional monomer. The sensor demonstrated a linear response in the concentration range of 0.1–100 ng/mL for DON detection. The sensor exhibits exceptional sensitivity, with an LOD of less than 0.1 ng/mL. In summary, these findings suggest that the combination of SPR sensing with a MIPPy film as a synthetic receptor holds promise as a method for the specific detection of DON [65].

The detection of *Enterococcus faecalis* was accomplished using an SPR sensor functionalized with N-methacryloyl-(L)-histidine-methylester (MAH) as the functional material. This sensor system offers a linearity range spanning from 2×10^4 CFU/mL to 1×10^8 CFU/mL, enabling the precise quantification of *Enterococcus faecalis* concentrations across this range. The sensor exhibits sensitivity with a LOD as low as 1.05×10^2 CFU/mL. This novel method is a surface modification technique that relies on the biotarget itself as a recognition element. It offers versatility to adapt to various biotargets with differing molecular structures, sizes, and physicochemical properties. Such a dependable and adaptable platform holds potential applications ranging from microbiome characterization to forensic analysis, revitalizing outdated detection strategies [66].

This study presents the creation of an SPR sensor featuring a microcontact imprinted sensor chip designed for the detection of *Salmonella paratyphi*. The process involved the preparation of stamps representing the target microorganism, followed by the development of microcontact *S. paratyphi*-imprinted SPR chips employing the functional monomer N-methacryloyl-L-histidine methyl ester (MAH). Real-time detection of *Salmonella paratyphi* was achieved over a concentration range of 2.5×10^6—15×10^6 CFU/mL. Real sample experiments were conducted using apple juice, where *Salmonella paratyphi* recognition was successfully achieved using these SPR sensors, with an LOD of 1.4×10^6 CFU/mL. In conclusion, the SPR sensor demonstrates promise as an excellent tool for monitoring *Salmonella paratyphi* in food products or contaminated water, facilitating the development of rapid and effective control strategies [67].

Escherichia coli was detected using an SPR sensor functionalized with N-methacryloyl-(L)-histidine methyl ester (MAH) as the functional material. This sensor system offers a linearity range spanning from 1.5×10^1 CFU/mL to 1.5×10^6 CFU/mL, enabling the precise quantification of *Escherichia coli* concentrations across this range. The LOD for this biomimetic SPR-based sensor was determined to be 0.57 CFU/mL. Furthermore, the feasibility of the sensor was assessed in a simulated urine mimic environment. This research demonstrates the potential of these tailor-made receptors and the biomimetic sensor for sensitive *E. coli* detection, particularly in urine samples [68].

In the study by Özgecan et al., the detection of T4 bacteriophage was successfully accomplished using an SPR nanosensor functionalized with polyvinyl alcohol (PVA) as a functional monomer. This sensor system offers a linearity range spanning from 1×10^4 PFU/mL to 4×10^6 PFU/mL, allowing for precise quantification of T4 bacteriophage concentrations across this range. The sensor exhibits sensitivity with a LOD as low as 6×10^3 PFU/mL. This remarkable sensitivity makes it highly suitable for the precise and ultrasensitive quantification of T4 bacteriophage, even when present at concentrations as low as 6×10^3 PFU/mL [69].

These studies demonstrate the versatility and potential of SPR sensors for detecting various analytes, including mycotoxins, bacteria, and bacteriophages. Each study highlights specific sensor designs and their capabilities, such as low LODs, wide linear ranges, and high sensitivity, catering to different detection needs in diverse applications.

9.5.3 MIP-Based Optical Sensors for Allergens

Among allergens, casein proteins, the most prevalent group of proteins in milk, represent a prominent category of milk protein allergens in food ingredients. Consequently, these proteins serve as an ideal target for cleaning validation. In a study published in 2018 by Ashly et al., MIP nanoparticles (nanoMIPs) were synthesized with a high affinity for bovine α-casein using a solid-phase imprinting method. N-isopropyl acrylamide (NIPAm), N-tert-butylacrylamide (TBA), acrylic acid (Acc), and N-(3-aminopropyl) methacrylamide HCl (APM) were used as functional monomers. Subsequently, these nanoMIPs were subjected to thorough characterization and incorporated into a label-free SPR-based sensor. This sensor system provides a linearity range spanning from 0.5 to 8 ppm, achieving an impressive detection limit of 127 ± 97.6 ng/mL (equivalent to 0.127 ppm). The sensor developed in their study holds the potential to empower food manufacturers to effectively monitor and manage the risk associated with food allergens in food processing environments. Doing so ensures that the food produced meets safety standards and is safe for consumption by consumers [70].

The detection of histamine is accomplished using an SPR sensor functionalized with methacrylic acid (MAA) as a functional monomer. Jiang et al. innovative sensing technique allows the detection of histamine within an extensive concentration range spanning from 25 to 1000 μg/L. Remarkably, the sensor exhibits high sensitivity with a low LOD of 25 μg/L, making it well-suited for accurate detection of histamine, even at relatively low concentrations. The results clearly demonstrated that this sensor platform exhibits a high degree of specificity, specifically designed for the detection of histamine. This development holds significant promise in meeting the demand for advanced and cost-efficient sensor systems with enhanced sensitivity [71].

In summary, both studies focus on the development of SPR-based sensors for specific analyte detection. The first study targets the detection of casein proteins, which are common milk protein allergens, and achieves a low detection limit, making it valuable for allergen monitoring in food processing. The second study focuses on histamine detection and offers a wide linear detection range and high sensitivity, catering to the need for accurate histamine quantification. Both sensor platforms exhibit specificity for their respective target analytes and have potential applications in food safety and quality control.

9.6 Current Challenges

Navigating the frontier of MIP-based optical sensors presents a spectrum of challenges that warrant careful consideration. While the continuous evolution of nanotechnology, material science, and signal processing has significantly improved the performance of optical sensors, inherent complexities that demand strategic solutions exist. One prominent challenge lies in optimizing the overall performance of MIP-based optical sensors, requiring a delicate balance between sensitivity, selectivity, and response time. Achieving enhanced sensitivity often involves intricate adjustments to the molecular imprinting process and nanomaterial integration, addressing issues of binding site accessibility and affinity. Broader applicability across a diverse range of analytes necessitates ongoing refinement of MIP design, considering the dynamic nature of target molecules. Striking a balance between rapid response times and selectivity is a nuanced task, demanding innovative approaches to sensor architecture and signal processing.

Furthermore, challenges persist in the integration of MIPs with optical transduction methods. Achieving uniformity and stability of imprinted sites across the sensor surface is essential for consistent and reliable performance. The potential for nonspecific binding and interference from complex sample matrices adds an additional layer of complexity, requiring advanced strategies to mitigate background signals and enhance the sensors' specificity in real-world applications. Another challenge emerges in ensuring the scalability and reproducibility of these innovative biosensing platforms. Standardizing the fabrication processes while maintaining the unique imprinting characteristics poses a significant hurdle. Addressing the variability in sensor performance and ensuring robustness across different manufacturing environments are critical aspects that necessitate ongoing research and development efforts.

The interdisciplinary nature of MIP-based optical sensors calls for collaborative efforts to overcome these challenges. A convergence of expertise in chemistry, nanotechnology, material science, and optics is imperative to develop comprehensive solutions. Collaborative initiatives may also facilitate the integration of advanced imaging techniques with MIP-based sensors, paving the way for synergistic diagnostic and therapeutic approaches.

While MIP-based optical sensors hold tremendous promise in disease detection and monitoring, addressing the multifaceted challenges is pivotal for unlocking their full potential. Ongoing research activities spanning diverse scientific disciplines are crucial in refining these sensors and ensuring their seamless integration into analytical and diagnostic landscapes, ultimately contributing to advancing scientific research, healthcare practices, and environmental sustainability.

9.7 Conclusion

Advances in nanotechnology, material science, and signal processing are continuously improving optical sensor performance, making them more sensitive, selective, and suitable for a wider range of applications. Nevertheless, optical sensors have emerged as indispensable tools for sensitive and non-destructive analytical measurements. Their versatility, real-time capabilities, and noninvasiveness make them valuable assets in diverse fields. As technology continues to advance, optical sensors are expected to play an increasingly critical role in advancing scientific research, enhancing healthcare practices, and promoting environmental sustainability.

While MIP-based optical sensors offer numerous advantages, challenges remain in optimizing their performance, including enhancing sensor sensitivity, selectivity, and response time. Ongoing research in MIP design, nanotechnology, and optical signal processing holds promise for overcoming these challenges and expanding the applications of MIP-based optical sensors. MIPs integrated into optical sensors represent a cutting-edge and versatile selective and sensitive analyte detection platform. Their ability to mimic the recognition properties of natural antibodies, combined with the benefits of optical transduction, positions them as powerful tools in various fields, from environmental monitoring to healthcare diagnostics. As research and technology continue to advance, MIP-based optical sensors hold tremendous potential for revolutionizing analytical sensing and driving innovation in diverse applications.

References

1. Regtien P, Dertien E (2018) 7—Optical sensors. In: Regtien P, Dertien E (eds) Sensors for mechatronics, 2nd edn. Elsevier, pp 183–243
2. Voskuil J (2014) Commercial antibodies and their validation. F1000Res 3:232
3. Turiel E, Esteban AM (2020) 8—Molecularly imprinted polymers. In: Poole CF (ed) Solid-phase extraction. Elsevier, pp 215–233
4. Lazarević-Pašti T, Tasić T, Milanković V, Potkonjak N (2023) Molecularly imprinted plasmonic-based sensors for environmental contaminants—current state and future perspectives. Chemosensors 11:35
5. Díaz-Álvarez M, Martín-Esteban A (2021) Molecularly imprinted polymer-quantum dot materials in optical sensors: an overview of their synthesis and applications. Biosensors (Basel):11
6. Balasurya S, Kokilavani S, Janani B, Sudheer Khan S (2023) 5—Recent advances in bioelectronic noses based on olfactory receptors. In: Gupta RK et al (eds) Nanotechnology-based e-noses. Woodhead Publishing, pp 125–141
7. Lippitsch ME, Riegler H, Aussenegg FR, Hermann G, Müller E (1988) Picosecond absorption and fluorescence studies on large phytochrome from rye. Biochemie und Physiologie der Pflanzen 183:1–6
8. Posch HE, Wolfbeis OS (1989) Optical sensor for hydrogen peroxide. Microchim Acta 97:41–50
9. Lazarević Pašti T, Momić T, Onjia A, Vujisić L, Vasić V (2010) Myeloperoxidase-mediated oxidation of organophosphorus pesticides as a pre-step in their determination by AChE based bioanalytical methods. Microchim Acta 170:289–297
10. Burgess LW (1995) Absorption-based sensors. Sens Actuators B 29:10–15
11. Pickup JC, Hussain F, Evans ND, Rolinski OJ, Birch DJS (2005) Fluorescence-based glucose sensors. Biosens Bioelectron 20:2555–2565
12. Schäferling M (2012) The Art of Fluorescence Imaging with Chemical Sensors. Angew Chem Int Ed 51:3532–3554
13. Maru K (2019) Design of transmission-type refractive index sensor, based on silica planar lightwave circuit using combination of refractive angle and phase measurements. Sensors 19:4095
14. Prabowo BA, Purwidyantri A, Liu KC (2018) Surface plasmon resonance optical sensor: a review on light source technology. Biosensors (Basel):8
15. Shrivastav AM, Cvelbar U, Abdulhalim I (2021) A comprehensive review on plasmonic-based biosensors used in viral diagnostics. Commun Biol 4:70
16. Sabri N, Aljunid SA, Salim MS, Ahmad RB, Kamaruddin R (2013) Toward optical sensors: review and applications. J Phys Conf Ser 423:012064
17. Yan X, Li H, Su X (2018) Review of optical sensors for pesticides. TrAC Trends Anal Chem 103:1–20
18. Pirzada M, Altintas Z (2020) Recent progress in optical sensors for biomedical diagnostics. Micromachines 11:356
19. Sun R, Li Y, Du T, Qi Y (2023) Recent advances in integrated dual-mode optical sensors for food safety detection. Trends Food Sci Technol 135:14–31
20. Sharma AK, Jha R, Gupta BD (2007) Fiber-optic sensors based on surface plasmon resonance: a comprehensive review. IEEE Sens J 7:1118–1129

21. Haake HM, Schütz A, Gauglitz G (2000) Label-free detection of biomolecular interaction by optical sensors. Fresen J Anal Chem 366:576–585
22. Chen J, Garcia ES, Zimmerman SC (2020) Intramolecularly cross-linked polymers: from structure to function with applications as artificial antibodies and artificial enzymes. Acc Chem Res 53:1244–1256
23. Hayden O, Lieberzeit PA, Blaas D, Dickert FL (2006) Artificial antibodies for bioanalyte detection—sensing viruses and proteins. Adv Funct Mater 16:1269–1278
24. Whitcombe M, Alexander C, Vulfson E (2000) Imprinted polymers: versatile new tools in synthesis. Synlett 2000:0911–0923
25. Cieplak M, Kutner W (2016) Artificial biosensors: how can molecular imprinting mimic biorecognition? Trends Biotechnol 34:922–941
26. Becskereki G, Horvai G, Tóth B (2021) The selectivity of molecularly imprinted polymers. Polymers (Basel):13
27. Byun J (2021) Recent progress and opportunities for nucleic acid aptamers. Life (Basel):11
28. Mohit E, Hashemi A, Allahyari M (2014) Breast cancer immunotherapy: monoclonal antibodies and peptide-based vaccines. Expert Rev Clin Immunol 10:927–961
29. Borrebaeck CAK (2000) Antibodies in diagnostics—from immunoassays to protein chips. Immunol Today 21:379–382
30. Parisi OI, Francomano F, Dattilo M, Patitucci F, Prete S, Amone F, Puoci F (2022) The evolution of molecular recognition: from antibodies to molecularly imprinted polymers (MIPs) as artificial counterpart. J Funct Biomater 13:12
31. Eniola AO, Hammer DA (2003) Artificial polymeric cells for targeted drug delivery. J Control Release 87:15–22
32. Chang TMS (1984) Artificial cells in medicine and biotechnology. In: Chang TMS (ed) Microencapsulation and artificial cells. Humana Press, Totowa, NJ, pp 5–24
33. Keitel B, Batista AD, Mizaikoff B, Fresco-Cala B (2022) Molecularly imprinted polymer sensors in environmental analysis. Encyclopedia of Sensors and Biosensors, 4, Elsevier B.V. 851–867
34. Yang W, Ma Y, Sun H, Huang C, Shen X (2022) Molecularly imprinted polymers based optical fiber sensors: a review. TrAC Trends Anal Chem 152:116608
35. Saylan Y, Akgönüllü S, Yavuz H, Ünal S, Denizli A (2019) Molecularly imprinted polymer based sensors for medical applications. Sensors 19:1279
36. Ahmad OS, Bedwell TS, Esen C, Garcia-Cruz A, Piletsky SA (2019) Molecularly imprinted polymers in electrochemical and optical sensors. Trends Biotechnol 37:294–309
37. Fang L, Jia M, Zhao H, Kang L, Shi L, Zhou L, Kong W (2021) Molecularly imprinted polymer-based optical sensors for pesticides in foods: recent advances and future trends. Trends Food Sci Technol 116:387–404
38. Kadhem AJ, Gentile GJ, Fidalgo de Cortalezzi MM (2021) Molecularly imprinted polymers (MIPs) in sensors for environmental and biomedical applications: a review. Molecules 26:6233
39. Lipovka A, Fatkullin M, Averkiev A, Pavlova M, Adiraju A, Weheabby S, Al-Hamry A, Kanoun O, Pašti I, Lazarevic-Pasti T, Rodriguez RD, Sheremet E (2024) Surface-enhanced raman spectroscopy and electrochemistry: the ultimate chemical sensing and manipulation combination. Crit Rev Anal Chem 54:110–134
40. Tasić T, Milanković V, Batalović K, Breitenbach S, Unterweger C, Fürst C, Pašti IA, Lazarević-Pašti T (2023) Application of viscose-based porous carbon fibers in food processing—malathion and chlorpyrifos removal. Foods 12:2362
41. Tasić T, Lazarević-Pašti T (2022) Organophosphates and depression. In: Lazarević-Pašti T (ed) Organophosphates: detection, exposure and occurrence. Volume 1: Impact on health and the natural environment. Nova Science Publishers, pp 95–116
42. Milanković V, Tasić T, Pejčić M, Pašti I, Lazarević-Pašti T (2023) Spent coffee grounds as an adsorbent for malathion and chlorpyrifos—kinetics, thermodynamics, and eco-neurotoxicity. Foods 12:2397

43. Lazarević-Pašti T, Aničijević V, Baljozović M, Aničijević DV, Gutić S, Vasić V, Skorodumova NV, Pašti IA (2018) The impact of the structure of graphene-based materials on the removal of organophosphorus pesticides from water. Environ Sci Nano 5:1482–1494
44. Katnić Đ, Porobić SJ, Lazarević-Pašti T, Kojić M, Tasić T, Marinović-Cincović M, Živojinović D (2023) Sterilized plum pomace biochar as a low-cost effective sorbent of environmental contaminants. J Water Process Eng 56:104487
45. Lazarević-Pašti T, Jocić A, Milanković V, Tasić T, Batalović K, Breitenbach S, Unterweger C, Fürst C, Pašti IA (2023) Investigating the adsorption kinetics of dimethoate, malathion and chlorpyrifos on cellulose-derived activated carbons: understanding the influence of physicochemical properties. C 9:103
46. Katnić ĐB, Porobić SJ, Vujčić I, Kojić MM, Lazarević-Pašti T, Milanković V, Marinović-Cincović M, Živojinović DZ (2024) Irradiated fig pomace pyrochar as a promising and sustainable sterilized sorbent for water pollutant removal. Radiat Phys Chem 214:111277
47. Aničijević V, Tasić T, Milanković V, Breitenbach S, Unterweger C, Fürst C, Bajuk-Bogdanović D, Pašti IA, Lazarević-Pašti T (2023) How well do our adsorbents actually perform?—the case of dimethoate removal using viscose fiber-derived carbons. Int J Environ Res Public Health 20:4553
48. Milankovic V, Lazarevic-Pasti T (2021) The role of the cholinergic system in lung diseases. In: Dua K et al (eds) Targeting cellular signalling pathways in lung diseases. Springer Singapore, Singapore, pp 625–660
49. Bhakta S, Mishra P (2021) Molecularly imprinted polymer-based sensors for cancer biomarker detection. Sensors Actuators Rep 3:100061
50. Singh KRB, Natarajan A (2023) Molecularly imprinted polymer-based optical immunosensors. Luminescence 38:834–844
51. Sun T, Zhang Y, Zhao F, Xia N, Liu L (2020) Self-assembled biotin-phenylalanine nanoparticles for the signal amplification of surface plasmon resonance biosensors. Microchim Acta 187:473
52. Ertürk G, Özen H, Tümer MA, Mattiasson B, Denizli A (2016) Microcontact imprinting based surface plasmon resonance (SPR) biosensor for real-time and ultrasensitive detection of prostate specific antigen (PSA) from clinical samples. Sens Actuators B 224:823–832
53. Turan E, Zengin A, Suludere Z, Kalkan NÖ, Tamer U (2022) Construction of a sensitive and selective plasmonic biosensor for prostate specific antigen by combining magnetic molecularly-imprinted polymer and surface-enhanced Raman spectroscopy. Talanta 237:122926
54. Sunayama H, Takamiya K, Takano E, Horikawa R, Kitayama Y, Takeuchi T (2021) Simultaneous detection of two tumor marker proteins using dual-colored signaling molecularly imprinted polymers prepared via multi-step post-imprinting modifications. Bull Chem Soc Jpn 94:525–531
55. Bahari D, Babamiri B, Salimi A (2021) Ultrasensitive molecularly imprinted fluorescence sensor for simultaneous determination of CA125 and CA15–3 in human serum and OVCAR-3 and MCF-7 cells lines using Cd and Ni nanoclusters as new emitters. Anal Bioanal Chem 413:4049–4061
56. Rebelo TSCR, Costa R, Brandão ATSC, Silva AF, Sales MGF, Pereira CM (2019) Molecularly imprinted polymer SPE sensor for analysis of CA-125 on serum. Anal Chim Acta 1082:126–135
57. Tan L, Chen K, Huang C, Peng R, Luo X, Yang R, Cheng Y, Tang Y (2015) A fluorescent turn-on detection scheme for α-fetoprotein using quantum dots placed in a boronate-modified molecularly imprinted polymer with high affinity for glycoproteins. Microchim Acta 182:2615–2622
58. Ye J, Chen Y, Liu Z (2014) A boronate affinity sandwich assay: an appealing alternative to immunoassays for the determination of glycoproteins. Angew Chem Int Ed 53:10,386–10,389
59. Akgönüllü S, Armutcu C, Denizli A (2022) Molecularly imprinted polymer film based plasmonic sensors for detection of ochratoxin A in dried fig. Polym Bull 79:4049–4067
60. Sergeyeva T, Yarynka D, Lytvyn V, Demydov P, Lopatynskyi A, Stepanenko Y, Brovko O, Pinchuk A, Chegel V (2022) Highly-selective and sensitive plasmon-enhanced fluorescence sensor of aflatoxins. Analyst 147:1135–1143

61. Akgönüllü S, Yavuz H, Denizli A (2020) SPR nanosensor based on molecularly imprinted polymer film with gold nanoparticles for sensitive detection of aflatoxin B1. Talanta 219:121219
62. Akgönüllü S, Yavuz H, Denizli A (2021) Development of gold nanoparticles decorated molecularly imprinted-based plasmonic sensor for the detection of aflatoxin M1 in milk samples. Chemosensors 9:363
63. Çimen D, Aslıyüce S, Tanalp TD, Denizli A (2021) Molecularly imprinted nanofilms for endotoxin detection using an surface plasmon resonance sensor. Anal Biochem 632:114221
64. Choi S-W, Chang H-J, Lee N, Kim J-H, Chun HS (2009) Detection of mycoestrogen zearalenone by a molecularly imprinted polypyrrole-based surface plasmon resonance (SPR) Sensor. J Agric Food Chem 57:1113–1118
65. Choi S-W, Chang H-J, Lee N, Chun HS (2011) A surface plasmon resonance sensor for the detection of deoxynivalenol using a molecularly imprinted polymer. Sensors 11:8654–8664
66. Erdem Ö, Saylan Y, Cihangir N, Denizli A (2019) Molecularly imprinted nanoparticles based plasmonic sensors for real-time Enterococcus faecalis detection. Biosens Bioelectron 126:608–614
67. Perçin I, Idil N, Bakhshpour M, Yılmaz E, Mattiasson B, Denizli A (2017) Microcontact imprinted plasmonic nanosensors: powerful tools in the detection of Salmonella paratyphi. Sensors 17:1375
68. Özgür E, Topçu AA, Yılmaz E, Denizli A (2020) Surface plasmon resonance based biomimetic sensor for urinary tract infections. Talanta 212:120778
69. Erdem Ö, Cihangir N, Saylan Y, Denizli A (2020) Comparison of molecularly imprinted plasmonic nanosensor performances for bacteriophage detection. New J Chem 44:17,654–17,663
70. Ashley J, Shukor Y, D'Aurelio R, Trinh L, Rodgers TL, Temblay J, Pleasants M, Tothill IE (2018) Synthesis of molecularly imprinted polymer nanoparticles for α-casein detection using surface plasmon resonance as a milk allergen sensor. ACS Sens 3:418–424
71. Jiang S, Peng Y, Ning B, Bai J, Liu Y, Zhang N, Gao Z (2015) Surface plasmon resonance sensor based on molecularly imprinted polymer film for detection of histamine. Sens Actuators B 221:15–21
72. Rajpal S, Mishra P (2022) Next generation biosensors employing molecularly imprinted polymers as sensing elements for in vitro diagnostics. Biosensor Bioelectron 11:100201

Molecularly Imprinted Polymers-Based Optical Sensors for Biomedical Diagnostics

10

Simrat Kaur, Kuldeep Kaur, and Seema Maheshwari

Abstract

Molecularly imprinted polymers (MIPs) are custom-designed artificial receptor materials that have the ability to identify and selectively attach with specific target molecules such as proteins, small organic molecules, or ions. The molecular imprinting process used to create MIPs entails synthesizing the polymer from its monomers with cross-linking agents while the target molecule is present in the matrix. This results in the formation of precise binding sites in the polymer network that mirror the shape and chemical properties of the template molecule. MIPs have gained significant attention in various fields, including chemical sensing, biosensing, environmental monitoring, and healthcare monitoring. MIPs can be integrated into optical sensors as a thin film or coating on a solid substrate and used to detect specific target molecules. MIP-based optical sensors may use optical absorbance/transmission, fluorescence, or surface plasmon resonance (SPR) phenomenon as the signal transduction mechanism. Consequently, variations in the absorbance, fluorescence, or refractive index of the sensor material, resulting from the binding of the target molecule, can be identified and employed for the quantification of the target analyte's concentration. MIP-based optical sensors offer many advantages, such as good specificity, sensitivity, and reusability. As a result, they have been applied for the recognition of a diverse analytes, encompassing drugs, toxins, and biomolecules. Optical sensors utilizing molecularly imprinted polymers (MIPs) have shown promise in disease detection and monitoring due to their ability to provide highly specific and sensitive detection of biomolecules associated with various diseases making them useful for applications in healthcare and diagnostics. In the present chapter, the fundamental principles MIP-based optical sensors have been discussed. The signal transduction mechanisms including absorbance, fluorescence, and SPR-based techniques

S. Kaur · K. Kaur (✉) · S. Maheshwari
Department of Chemistry, Mata Gujri College, Fatehgarh Sahib, Punjab, India

S. Patra et al. (eds.), *Molecularly Imprinted Polymers: Path to Artificial Antibodies*, https://doi.org/10.1007/978-981-97-4379-7_10

have been presented. In addition, the fabrication process involving integration of MIP into optical sensor devices and functionalization of MIP has been discussed. In this chapter, we delve into the fundamental principles, working mechanisms, and fabrication of MIP-based optical sensors. We converse about practical applications of these sensors in disease detection and monitoring, drawing examples from recent research findings.

Keywords

Molecularly imprinted polymers · Optical sensors · Molecular imprinting Disease detection and monitoring · Chemical sensing

10.1 Introduction

Synthetic polymer materials known as molecularly imprinted polymers (MIPs) can recognize and selectively bind specific molecules of interest such as proteins, small organic molecules, or ions. The selective recognition sites in MIPs are designed to mimic the binding by natural antibodies or receptors. These recognition sites are tailored to bind with high affinity to a specific target analyte. The preparation of MIPs involves molecular imprinting process which entails the synthesis of the polymer from its monomers using cross-linking agents with the participation of the target molecule, leading to formation of precise binding spots within the polymer matrix that mirror the shape and chemical properties of the template molecule. The recognition mechanism of MIP-based biosensors involves generating distinct binding spots within a polymer matrix through template-assisted polymerization, enabling the selective and high-affinity capture of target analytes, such as pathogenic bacteria, and their subsequent conversion into measurable signals through integration with various transducer platforms for bacterial detection [1]. Molecularly imprinted polymers (MIPs) offer a range of remarkable advantages, including straightforward synthesis [2], exceptional reusability [3], a wealth of customizable binding sites [4], cost-effectiveness [5], design flexibility [6], and remarkable stability under challenging chemical and physical conditions [7]. The application of MIPs as sensor platforms enables the creation of highly selective sensors capable of identifying specific molecules with remarkable accuracy and high sensitivity [8, 9]. These attributes have generated a growing interest in the application of sensors utilizing MIPs in several areas including environmental monitoring [10], food safety [7, 11], biomarker detection [12], disease diagnosis [13], pesticide residue detection [14], and detection of toxic pollutants [15].

Optical sensor technology has gained immense importance due to its ability to provide precise, rapid, and no-invasive measurements across a vast array of applications [16]. As technology continues to advance, optical sensors are destined to play an increasingly crucial role across several fields. The integration of optical sensor technology with molecularly imprinted polymers (MIPs) has resulted in development of sensors utilizing MIPs. Sensors utilizing MIPs combine the specificity of molecular imprinting with the sensitivity of optical detection. These have garnered significant interest in the past few years as highly selective and sensitive sensors for a diverse array of analytes [17]. Sensors utilizing MIPs have been investigated for diverse healthcare applications [18]. One area where MIP-based biosensors have shown great promise is the diagnosis and monitoring of diseases [19, 20]. These sensors provide a rapid and specific detection method for cancer [21], cardiovascular [22, 23], and skeletal diseases [24]. MIP-based optical sensors have also been employed for rapid identification of diseases like COVID-19, where MIPs not only provide accurate disease detection but also offer long-term performance and reusability, making them a preferred choice for creating highly sensitive disease detectors [25]. MIPs can be designed to recognize specific proteins or other biomarkers associated with various infectious diseases, including HIV-1, COVID-19, and Dengue virus [26]. MIP-based optical sensors have also been testified for continuous monitoring of diseases, therapeutic drug monitoring, and environmental monitoring [27].

In the present chapter, the fundamental principles MIP-based optical sensors have been discussed. The signal transduction mechanisms including absorbance, fluorescence, and surface plasmon resonance (SPR)-based techniques have been presented. In addition, the fabrication process involving integration of MIP into optical sensor device and functionalization of MIP has been discussed. Further, the uses of MIP-based sensors in detection of disease and monitoring have been discussed citing examples from recent literature. Additionally, we deliberate the challenges inherent in the realm of optical sensors utilizing MIPs for disease detection and monitoring. Beyond merely highlighting their real-world applications and advantages, we address the prospects of these sensors, alongside ongoing efforts and strategic developments. This review aims to pave the way for the broader adoption of MIP-based optical sensors in healthcare, encompassing disease monitoring, clinical diagnostics, and various other domains.

10.2 Fundamentals of Molecularly Imprinted Polymers

Molecularly imprinted polymers (MIPs) are man-made materials that have been developed as synthetic receptors or "plastic antibodies" [28]. They are designed to mimic the recognition mechanisms of natural receptors, such as antigen–antibody interactions, and work on this principle [28]. The fundamental principles underlying molecularly imprinted polymers (MIPs) encompass several key elements involving template molecules, cross-linking agents, and functional monomers as shown in Fig. 10.1. The process of molecular imprinting involves polymerization of functional monomers and cross-linking agents in the presence of the template molecule. The components of MIPs include template molecule, functional monomers, and cross-linking agents the choice of which plays a crucial role in determining the characteristics of MIP. The components are briefly explained below:

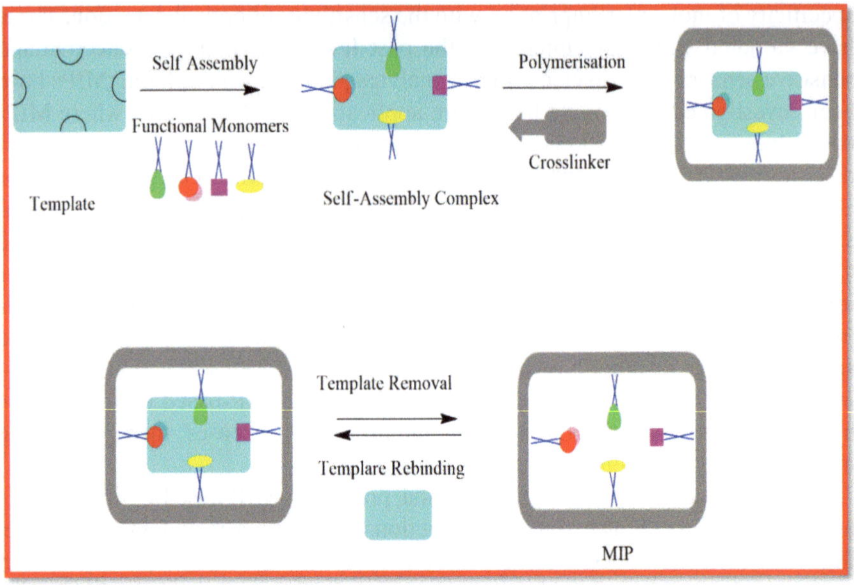

Fig. 10.1 Process of molecular imprinting (MIP) and template recognition

(a) Template Molecule: The development of MIPs commences with the selection of a template molecule that serves as the specific target for which the polymer is devised to exhibit selective binding. This template can encompass a variety of substances, such as pharmaceuticals, small organic compounds, proteins, or other substances of interest.
(b) Functional Monomers: Functional monomers are carefully chosen based on their capacity to engage in non-covalent interactions with the template molecule. These interactions might encompass hydrogen bonding, van der Waals forces, or electrostatic attractions. Typically, these monomers are selected to emulate the chemical features pertaining to the template molecule.
(c) Cross-Linking Agents: To establish a three-dimensional polymer network, cross-linking agents are employed. These agents play a key role in preserving the structure and stability of the MIP. Cross-linkers are typically compounds with bifunctional or multifunctional properties that react with the functional monomers.

The polymerization process involves the combination of functional monomers and cross-linking while the template molecule is present. Polymerization methods can encompass bulk polymerization, precipitation polymerization, and suspension polymerization. Throughout this process, the monomers self-assemble around the template molecule, generating cavities or binding sites with a shape and functionality that correspond to the template. After polymerization, the template molecule is extracted from the polymer matrix, resulting in the retention of cavities that are tailored to the size, shape, and functional groups of the template. The resulting MIP comprises binding sites that exhibit specificity for the template molecule or molecules bearing a similar structural likeness. These binding sites enable the MIP to selectively detect and bind the target analyte even within complex mixtures (Fig. 10.1). When the MIP encounters the target analyte, it engages in selective binding, akin to the lock-and-key principle.

The analyte can subsequently be released by altering environmental conditions, such as adjusting pH, temperature, or ionic strength. The resulting MIP comprises binding sites that exhibit specificity for the template molecule or molecules bearing a similar structural likeness. These binding sites enable the MIP to selectively detect and bind the target analyte even within complex mixtures [29]. MIPs offer notable benefits, including elevated selectivity, stability, and reusability when contrasted with alternative recognition elements like antibodies or natural receptors [4].

Regarding the manner of interaction between the template and functional monomer molecules, the self-assembly process is generally categorized into three main types: covalent, non-covalent, and semi-covalent imprinting species [30]. In covalent imprinting, the template–monomer complex is stabilized by reversible covalent bonds between the template molecule and the functional monomer. This results in a more homogeneous distribution of binding sites within the polymer, significantly enhancing sensitivity. However, template removal can be challenging due to the strong interaction [31]. In contrast, non-covalent imprinting relies on weak non-covalent interactions, such as hydrogen bonding, electrostatic interactions, and Van der Waals forces, like biological recognition processes [31]. The use of weak non-covalent bonds simplifies the process and allows for a broader selection of functional monomers for various templates. While this approach offers benefits like easy template removal, the main limitation is lower binding site selectivity, which requires improvement [31]. The semi-covalent imprinting method combines features of covalent and non-covalent binding. During the imprinting process, the template–monomer complex is initially stabilized through covalent interactions, but the rebinding of targets occurs through non-covalent interactions [31]. This approach offers impressive template binding and efficient rebinding properties, but it often demands strong covalent bonding between the monomer and the template to withstand polymerization conditions.

In terms of the polymerization techniques used with the template–monomer complex and cross-linking monomers, the polymerization process can be of different types including bulk polymerization, precipitation polymerization, emulsion and suspension polymerization, multistep swelling polymerization, sol–gel polymerization, electro-polymerization, and surface imprinting, among others [31]. Bulk polymerization is the most widely accepted method for MIP preparation due to its efficiency and cost-effectiveness. However, it has some drawbacks, such as a low number of binding sites near the surface, nonuniform morphology, low binding capacity, and slow mass transfer [31]. Precipitation polymerization is a quick and cost-effective method, yielding MIPs with uniform morphology, but it requires a significant amount of organic porogen solvents [31]. Emulsion and suspension polymerization techniques can produce high-yield MIP particles with uniform sizes, improved binding site homogeneity, and accessibility [31]. However, they often require substantial amounts of organic solvents or auxiliary reagents, leading to more complex washing procedures and potentially compromising the purity and stability of the template–monomer complex. Multistep swelling polymerization is suitable for producing monodisperse imprinted beads with high yield, but it is time-consuming. Sol–gel polymerization offers advantages such as ease of operation at ambient temperature without chemical or thermal decomposition. It uses environmentally friendly reaction solvents and results in MIPs of suitable size with a highly porous structure, high chemical and thermal stability, long lifetime, and good repeatability [31]. Electro-polymerization allows for the direct formation of a thin MIP film with a controlled thickness on an electrode's surface, providing a simple, rapid, and cost-effective way to prepare MIPs. Surface imprinting, which involves imprinting MIPs on the surface of supporting substrates like silica (SiO_2), magnetic nanoparticles (MNPs), carbon nanotubes (CNTs), quantum dots (QDs), gold and silver NPs (AuNPs and AgNPs), and graphene oxide (GO), is an attractive technique to enhance the efficiency of MIP preparation [3]. It offers advantages such as high binding capacities, homogeneous distribution of binding sites, improved mass transfer, stability, and dispersion [3]. Lately, there has been significant attention in the research field toward nanostructured modified molecularly imprinting polymers, owing to their outstanding properties such as a high surface-to-volume ratio, cost-effectiveness, and easy preparation and handling.

The innovative concept of smart imprinting combines molecular imprinting and nanotechnology. There is compelling evidence supporting the superiority of nano-MIPs over bulk MIPs across various applications such as pesticide residue detection [32], protein target analysis [33], drug delivery systems [34], selective enrichment techniques [35], sensing methods including electrochemical sensing [36], and photoelectrochemical sensors [37], among others. Particularly, among nanostructured materials, carbonaceous materials like "graphene" have captured the interest of researchers due to their remarkable electrical, thermal, and physical characteristics [38]. The integration of molecularly imprinted technology and nanotechnology has significantly improved the selectivity, specificity, and sensitivity of MIPs for target molecules, leading to enhanced interest in the application of MIP-based optical sensors for drug delivery and theragnostic applications [39, 40]. Smart MIPs, designed with a specific focus, exhibit intelligence by responding to outward stimuli of the likes of temperature, pH, biomolecules, and magnetic fields, leading to enhanced applications for these polymers [41, 42].

Each synthesis method has its own advantages and disadvantages. The selection of the synthesis method for MIP is based on factors such as the nature of the analyte, the desired sensor format, and the sensor's intended application. After synthesis, the MIPs are integrated into the optical sensor platform to enable selective and sensitive identification of target analytes.

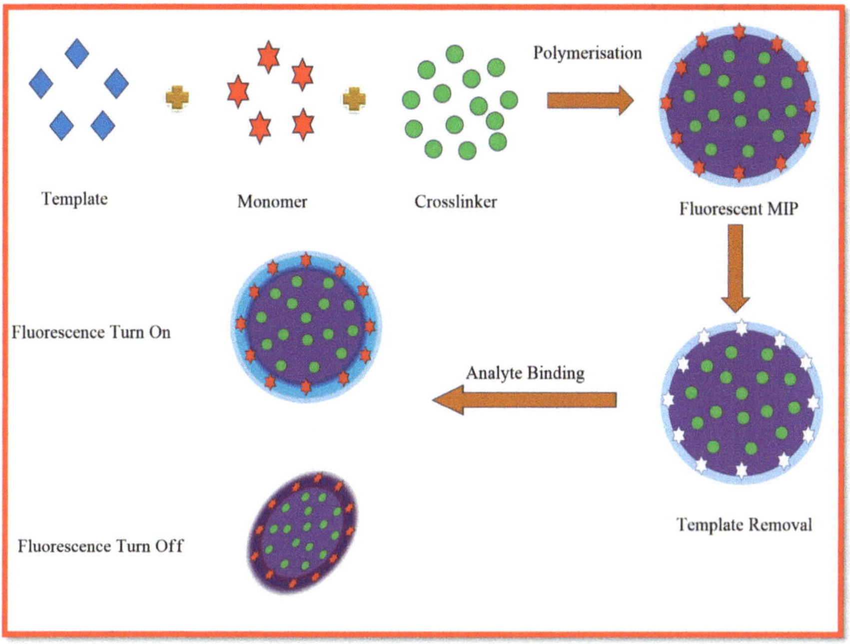

Fig. 10.2 Illustration of working of fluorescence-based MIP sensor

10.3 Design and Fabrication of MIP-Based Optical Sensors

The MIP-based optical sensors comprise two key elements: a recognition unit and a transducer element. The MIP serves as the recognition unit that interacts specifically with target analytes. The interplay of the MIP and the target molecule initiates a quantifiable optical reaction. The transducer component converts this binding event into a measurable optical signal. This response can manifest in diverse ways, encompassing alterations in fluorescence, absorbance, refractive index, or surface plasmon resonance (Fig. 10.2). By harnessing these alterations, MIP-based sensors have the capacity to identify and quantify analytes across a range of fields, including environmental monitoring [43], healthcare [9], pharmaceuticals [8], and food safety [7], among others. The design and assembly of MIP-based optical sensors involve selection of the most appropriate optical setup, immobilization of the MIP layer on the transducer element, functionalization of the MIP layer, and integration into a device.

10.3.1 Selection of Optical Set-Up

Based on the type of optical signal being measured, we can have different types of sensors, including absorbance-based sensors, fluorescence-based sensors, and SPR-based sensors, the working principle of which is explained in subsequent paragraphs.

10.3.1.1 Absorbance-Based Sensors

Absorbance-based optical sensors are designed to measure the absorption of light by a sample. The principle behind these sensors is based on the Beer–Lambert Law, which relates the absorbance of light to the amount of the absorbing substance in the sample. Absorbance-based optical sensors provide accurate and reliable measurements, making them indispensable tools in research, quality control, and industrial processes [44]. Their flexibility and applicability across different fields make them a crucial part of modern analytical instrumentation. Optical sensors utilizing molecularly imprinted polymers (MIPs) incorporate MIPs as the recognition element, designed for specific interactions and binding with the target analyte, and incorporate a transducer component to relay the bonding occurrence. These sensors are primarily concerned with detecting alterations in the absorbance signals generated by the transducer when a complex forms between the target analyte and MIP. This approach offers an efficient, swift, and cost-effective means of sensitively detecting analytes through the analysis of absorbance signals [45].

10.3.1.2 Fluorescence-Based Sensors

Fluorescence-based optical sensors make use of fluorescence signals to detect and measure the presence and concentration of specific molecules or analytes in various samples. They leverage the fluorescence properties of certain compounds, which emit light at a longer wavelength when excited by a shorter-wavelength light source. The binding of a specific molecule to MIP recognition element results in changes in the fluorescence signal being measured, which can be enhanced or reduced resulting in fluorescence turn-on or fluorescence turn-off sensors as illustrated in Fig. 10.2. Fluorescence detection has gained prominence as the favored optical method owing to its ability to achieve low limits of detection (LOD), simplified design, and practicality. Fluorescence-based optical sensors are highly versatile, offering excellent sensitivity and specificity, and have a significant impact in diverse fields for detecting and quantifying a wide range of analytes [46]. They have emerged as the preferred optical sensing platforms for sensing minute concentrations of diverse analytes, including mycotoxins, pesticides, and viruses [47–50].

10.3.1.3 SPR-Based Sensors

A potent spectroscopic method—surface plasmon resonance (SPR) observes subtle alterations in the effective refractive index at the metal–dielectric interface employed for live and unlabelled detection of various biomolecular interactions [51]. This is achieved by exploiting physical and optical phenomena arising from the interaction of light with thin metal films. In essence, it relies on the principles of physical optics and occurs at the interface between a metal surface, typically modified with explicit recognition elements, and a surrounding medium, which can be either vacuum/air or a liquid. When a particular molecule recognized by the receptor on this modified surface binds, it induces alterations in the refractive index on the metal's surface, along with an angular shift [52]. The SPR rotation offset is extraordinarily responsive to changes in refractive index or mass density within a single layer near the metal's surface, allowing for the real-time monitoring of interactions between targets and their receptors [53]. These characteristics bestow SPR with numerous advantages, including easy fabrication, rapid responsiveness, high sensitivity, no need for sample purification, label-free detection, real-time monitoring, high-volume capacity, minimal reagent usage, robust resistance to interference, and non-destructive analysis [52, 54]. The integration of molecularly imprinted polymers (MIPs) into SPR sensors enhances their specificity. This integration capitalizes on the strengths of SPR and the high selectivity of MIPs, thereby significantly contributing to the field of optical sensing. It finds applications in areas like biochemistry, pharmaceuticals, environmental monitoring, and material science [55]. A summary of various MIP-based optical detection techniques highlighting the advantages and disadvantages is given in Table 10.1.

10.3.2 Immobilization of MIP Layer

Coating the optical transducer with a molecularly imprinted polymer (MIP) layer is a crucial step in the crafting of MIP-based optical sensors. The MIP layer is designed to selectively bind to a target analyte, making it an integral part of the sensor's recognition and detection system. The choice of coating technique can significantly influence the performance and effectiveness of the MIP-based sensor. The popular coating techniques, including spin-coating, dip-coating, and electropolymerization, are discussed further.

Spin-coating is a deposition technique that entails applying a thin and uniform layer of molecularly imprinted polymer (MIP) onto the optical transducer by spinning the substrate at high speeds, leveraging centrifugal force to evenly spread the MIP solution and create a thin film. This process involves first depositing a small quantity of the MIP solution onto the center of the substrate, after which the

Table 10.1 Summary of advantages and disadvantages of various MIP-based optical detection techniques

Detection technique	Principle	Advantages	Disadvantages
Absorbance-based sensors	Utilizes absorbance of light by MIPs	Simple setup, cost-effective, suitable for qualitative analysis	Limited sensitivity, susceptible to interference from other molecules
Fluorescence-based sensors	MIPs with fluorescent labels emit fluorescence upon binding	High sensitivity, wide dynamic range, quantitative analysis	Complex instrumentation, potential photobleaching of fluorophores, costlier compared to absorbance sensors.
Surface plasmon resonance	Detects alterations in refractive index occurring on a sensor surface as a result of binding occurrences	Real-time or continuous observation, detection without labeling, enhanced specificity	Expensive setup, limited to certain types of samples, sensitivity affected by sample properties.

substrate is rapidly spun at high speeds, typically reaching thousands of revolutions per minute. As a result of this centrifugal force, the MIP solution effectively spreads outward, forming a thin and even coating on the surface. Spin-coating offers advantages such as precise control over the thickness and uniformity of the MIP layer and is particularly well-suited for generating thin, uniform films on a broad range of substrate types. It finds applications in the advancement of sensors based on MIPs for various analytes, including applications in biosensors and chemical sensors.

Dip coating is a straightforward, yet highly effective technique used to coat substrates. In this process, the substrate is gradually immersed in a molecularly imprinted polymer (MIP) solution, and it is then carefully withdrawn at a controlled speed to achieve the formation of a uniform coating. This uniformity is attained through the combined actions of capillary forces and solvent evaporation. Dip coating offers several advantages; it is adaptable to various substrate shapes and sizes, making it a versatile choice, and it is particularly well suited for creating thicker MIP coatings [56]. In recent years, there has been an enhanced utilization of the method to incorporate nanoMIPs as artificial receptors to enhance the functionality of thermocouples. These modified surfaces have been applied as heat-sensitive devices for identifying a range of substances, including small molecules and proteins [57].

Electropolymerization is an electrochemical technique employed in the direct synthesis of a polymer film on the surface of an optical transducer. This process involves dissolving the molecularly imprinted polymer (MIP) monomer in a solvent to create an electrochemical solution. The substrate functions as the operational electrode and is immersed in the monomer solution. An electrical potential is applied among the working electrode (the substrate) and a counter electrode, leading to the oxidation of the monomers at the working electrode. This oxidation process results in the polymerization of the monomers and the formation of the MIP layer. Electropolymerization is advantageous for its ability to provide precise control over the thickness of the MIP layer and is particularly suitable for in situ MIP layer formation on the sensor surface [58]. This technique is commonly employed in the development of MIP-based sensors where control over the layer thickness is crucial, such as in electrochemical sensors and the creation of molecularly imprinted electrodes [58]. The choice of coating technique depends on factors like the required layer thickness, substrate material, and the specific properties of the target analyte. Each method has its advantages and limitations, making it important to carefully select the appropriate technique based on the needs of the MIP-based optical sensor application.

10.3.3 Functionalization of the MIP Layer

Functionalization is a crucial step in the crafting of molecularly imprinted polymers (MIPs) based optical sensors to enhance their interaction with target molecules. It involves the modification of the surface of the MIP to improve its recognition and binding capabilities. Functionalization typically involves adding functional groups to the polymer matrix to tailor its selectivity, specificity, and affinity toward the target molecule. Within the context of MIPs, functionalization refers to the deliberate modification of the polymer matrix to impart it with specific molecular recognition properties. The primary objective of functionalization in MIPs is to create a polymer matrix that can specifically recognize and attach to specific target molecules, known as template molecules. These templates serve as a molecular "blueprint" for the desired binding sites within the polymer. Some key aspects of functionalization include:

1. Covalent Attachment: Functional groups can be covalently attached to the MIP's polymer matrix. For instance, the introduction of specific functional groups, such as amino (-NH_2) or carboxyl (-COOH) groups, can provide sites for chemical bonding with the target molecules.
2. Imparting Chemical Affinity: By introducing functional groups capable of forming chemical interactions with the target molecule (e.g., hydrogen bonding, electrostatic interactions, or van der Waals forces), the MIP's
3. binding sites become more specific to the target.

The functionalization is usually followed by integrating the MIP into a functional device for its ease of practical application. A key advantage of MIP-based optical sensors is their capacity to offer remarkable specificity, enabling the differentiation of closely related molecules. Furthermore, these sensors can be customized for various analytes by adjusting the MIP composition during synthesis. This adaptability has spurred the development of a wide array of MIP-based optical sensor technologies, applicable in diverse settings, from clinical diagnostics to the identification of environmental contaminants [9, 59]. This introductory glimpse only scratches the surface of the potential inherent in MIP-based optical sensor technologies. Ongoing research in this domain promises to bring forth innovative materials and techniques, expanding the scope of their utility and establishing them as an exciting area of exploration and application within the realm of chemical and biological sensing.

10.4 Applications in Disease Detection and Monitoring

Integrating MIPs with optical sensor platforms has a diverse array of applications across healthcare including disease detection, disease monitoring, and drug analysis. The present section explores the utility of MIP-based optical sensors in healthcare including disease detection, disease monitoring, therapeutic drug monitoring, environmental exposure monitoring, and increasing the therapeutic efficiency (Fig. 10.3).

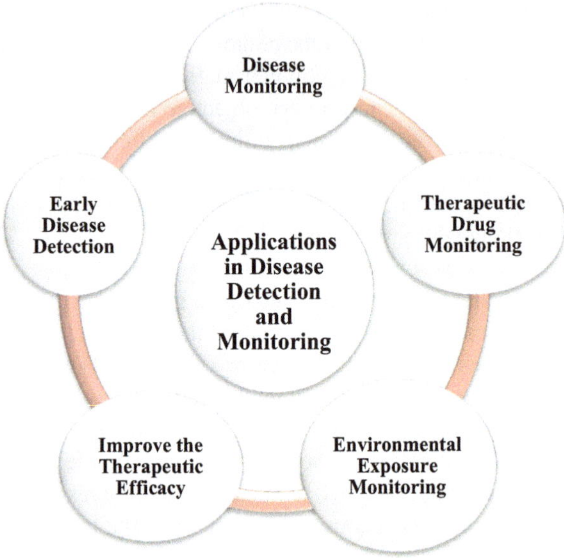

Fig. 10.3 Applications of sensors based on MIPs in disease detection and monitoring

10.4.1 Early Disease Detection

The detection of diseases is of paramount importance in healthcare as it can lead to early diagnosis and tailored treatments for specific diseases. Early detection of certain diseases like cancer is imperative to curtail their spread. Cancer is a widespread disease worldwide and is among the primary contributors to mortality and a significant obstacle to extending lifespan [60]. If unchecked and unmitigated, the progression of the disease can be swift and challenging to treat, possibly resulting in the loss of the affected individual's life. Hence, there is a pressing need for the development and implementation of innovative and expeditious methods for the detection of cancer biomarkers, facilitating earlier cancer diagnosis. Biomarkers, which are distinct attributes indicative of typical biological functions, disease processes, or responses to exposures or interventions, including therapeutic treatments, play a pivotal role in this pursuit. Numerous cancer biomarkers, including prostate-specific antigen (PSA), sarcosine, neuron-specific enolase, and α-fetoprotein, etc. have been previously documented [23]. MIP-based sensors have become as a vital analytical methodology for the determination of cancer biomarkers. For instance, Xing et al. (2019) synthesized the dual MIP-based plasmonic immunosandwich assay, or duMIP-PISA, for the specific extraction of neuron-specific enolase (NSE), a biomarker for small cell lung cancer (SCLC), from complex biotic samples [61]. A plasmonic substrate, consisting of a self-assembled gold nanoparticle monolayer coated on a glass slide with a C-terminal epitope imprint, was employed to selectively extract the target protein [61]. Additionally, nanotags composed of Raman-responsive $Ag@SiO_2$ nanoparticles imprinted with an N-terminal epitope were utilized for the accurate labeling of the captured protein [61]. The sandwich-like complexes comprising MIP-protein-MIP exhibited notably heightened surface-enhanced Raman scattering (SERS) signals, facilitating the highly sensitive detection of the target protein [61]. The combination of dual MIP-based recognitions guaranteed the assay's exceptional specificity, complemented by SERS detection for unparalleled sensitivity [61]. Compared to traditional antibody-based assays like commercial kit-based ELISA, the plasmonic immunosandwich assay utilizing dual MIPs offered several significant advantages including less time-consuming and simpler procedure, reduced sample volume, and expanded linear range [61].

Several other diseases can also be detected using specific biomarkers associated with these diseases. The diagnosis of brain diseases using some ocular biomarkers present in the cerebrospinal fluid (CSF) has also been demonstrated by Jinn et al. (2019). The authors described a visually distinctive hydrogel sensor embedded with light-interfering bio-responsive avidin imprinted silica nanoparticles (SNPs) [62]. MIP hydrogels can precisely attach to a target molecule, and this is a "compact and straightforward label-free system that does not need any additional operations." Additionally, MIP hydrogels are biocompatible and can be used for continuous monitoring by a patient. Compared to other methods that necessitate an expert with advanced skills and intricate, complex equipment instruments, MIP technology is a promising alternative for creating a sensor of the insertion type which can be observed in a continuous manner by a patient [62]. This hydrogel was affixed to an intraocular lens (IOL) and was capable of detecting eye abnormalities through a smartphone application. When avidin binds to the hydrogel, it initiates a change in the hydrogel's volume, resulting in a modification of SNP spacing. This alteration conforms to Bragg's Law for regular SNP arrays, inducing a change in the light wavelength that translates the change in hydrogel volume into an amplified optical signal in the form of reflected light, yielding a lower detection limit [62]. This molecular-imprinted SNP hydrogel sensor demonstrated remarkable sensitivity and selectivity [62]. In another study by Resende et al. (2020), a three-dimensional silica photonic crystal (PhC) sensor based on MIP was designed for the detection of fibrinopeptide B (FPB), a vital biomarker for venous thromboembolism [63]. The reduction in intensity of the MIP reflectance peak was found to be directly correlated with the FPB concentration. The reduction could be attributed to alterations in refractive index initiated by the presence of FPB [63]. Conversely, the intensity of the nonimprinted polymer (NIP) reflectance peak displayed irregular changes due to the absence of specific recognition sites tailored for FPB. When tested in synthetic human urine samples, the MIPP sensor exhibited a similar response to FPB, achieving a low limit of detection (LOD) of 0.13 ng/mL [63].

Urinary tract infections (UTIs) are prevalent bacterial infections that, if left untreated, can lead to severe consequences, including organ failure and fatalities. Consequently, the diagnosis of UTIs holds significant importance for public health. In this study by Özgür et al., 2020 [64], an endeavor was made to create a biomimetic sensor for UTI diagnosis by combining the molecular imprinting technique with an SPR-based sensor [64]. In this approach, methacrylic acid (MAH) was chosen as the functional monomer because of its specific functional groups, which play a crucial role in designing receptors for *E. coli* detection on the surface of MIP sensor. *E. coli* receptors were customized onto a gold surface of the SPR chip utilizing the MAH monomer with specific functional groups. Prior to creating the polymeric film, AgNPs were added to the polymer blend to boost the sensitivity of the biomimetic sensor. This approach allowed to monitor the kinetic analysis of the target microorganism in real time and without the need for labels in both aqueous solutions and a urine mimic [64]. The biomimetic SPR-based sensor exhibited high selectivity in detecting *E. coli*. Notably, the limit of detection (LOD) and analysis time for this sensor were determined to be 0.57 CFU/mL and approximately 20 min, respectively, in aqueous solutions, both of which were lesser than the UTI threshold values and quicker than conventional methods [64].

10.4.2 Disease Monitoring

MIP-based optical sensors play a vital role in continuous disease monitoring, particularly in conditions such as diabetes and cardiovascular diseases. Their high sensitivity and selectivity make them well suited for detecting specific disease-related biomarkers. These sensors can offer real-time, nonintrusive, or minimally intrusive surveillance, ensuring that patients receive timely information about their health [65]. For instance, in diabetes, MIP-based sensors can provide continuous glucose monitoring, helping patients manage their blood sugar levels more effectively. In cardiovascular diseases, they can detect cardiac biomarkers, allowing for immediate intervention in case of abnormal elevations. The long-term stability of MIP sensors ensures reliable data collection over extended periods. By reducing the need for frequent laboratory testing, these sensors offer convenience to patients and promote early intervention, ultimately improving disease management and patient outcomes. Several studies have proved the utility of MIP-based optical sensors in continuous monitoring of diseases some of which are presented here. Brown et al. [66] presented a minimally intrusive method for measuring glucose levels through the skin, i.e., transdermal glucose (TG) using a fluorescent biosensor. The authors developed an alternative method using microneedles (MN) for TG collection. They performed diffusion studies on Yucatan miniature pig skin affixed to an in-line diffusion flow cell, as well as on various skin sites of human participants. The researchers used a mutant variant of the glucose-binding protein (GBP) from *E. coli* coupled with a fluorescent probe as the biosensor for TG measurement. They found that the biosensor consisting of the GBP with a "his tag" exhibited high selectivity and micromolar sensitivity to glucose, making it suitable for TG measurements. In both animal and human trials, they observed increased skin permeability and transdermal glucose diffusion facilitated by microneedle application. Importantly, there was a considerable positive linear correlation between TG measured using the fluorescence-based biosensor and blood glucose levels (BG). This correlation held true for intact human skin as well as MN-treated human skin. The micromolar sensitivity of GBP minimized the necessary volume for analyzing interstitial fluid glucose, enabling shorter microneedle application durations compared to previous research [66].

Cardiovascular diseases encompass a spectrum of conditions that impact the heart and blood vessels, including ailments like coronary heart disease, cerebrovascular disease, peripheral arterial disease, rheumatic heart disease, congenital heart defects, deep vein thrombosis, and pulmonary embolism [67]. Troponin serves as a primary biomarker for cardiovascular diseases. Notably, Baldoneschi et al. developed an MIP-based optical sensor for troponin I, achieving an impressive limit of detection at 8.9 nM [22]. Additionally, SPR was employed to detect cardiac troponin T by employing dopamine as a monomer to craft MIP on a bare gold chip [68]. MIP-based optical sensors have also found application in the detection of myoglobin, showcasing their versatility in the realm of cardiovascular disease diagnostics [69, 70].

10.4.3 Therapeutic Drug Monitoring

Therapeutic drug monitoring (TDM) is an essential clinical procedure that entails measuring and assessing drug concentrations in a patient's bloodstream or other biological fluids to ensure the safe and effective administration of medications. MIP-based optical sensors have been successfully utilized for therapeutic drug monitoring [71]. For example, Ren and Li (2020) developed a flower-like Ag-embedded MIP material to create a high-performance SERS sensor for detecting glibenclamide, an oral antidiabetic medication prescribed aimed at the management of type II diabetes [72]. Glibenclamide is pivotal in triggering the release of insulin from pancreatic B-cells, thereby reducing serum glucose levels. Nevertheless, excessive glibenclamide usage can result in various side effects making it important to monitor the levels of glibenclamide in blood. In flower-shaped Ag-embedded MIP material, the imprinted coating efficiently shielded the flower-shaped Ag from oxidation, thus enhancing the stability of the SERS substrate. The flower-shaped Ag@MIPs served as an efficient and ultra-sensitive SERS platform allowing for the quantitative detection of glibenclamide, even at trace concentrations as low as 1 ng mL^{-1} [72]. Similarly, a study by Ahmadpour and Hosseini, explored a luminescence sensor on solid-phase, utilizing molecularly imprinted polymer-CdSeS/ZnS quantum dots, designed for the specific extraction and detection of sulfasalazine in biological samples [73]. Sulfasalazine (SSZ) is a drug commonly used for managing inflammatory bowel disease and rheumatoid arthritis which needs monitoring. The MIP-CdSeS/ZnS quantum dots sensor harnessing the superior selectivity of MIPs along with the favorable fluorescence characteristics of quantum dots (QDs), to allow highly selective and sensitive sensing of SSZ. The method's limits of detection (LOD), determined at signal-to-noise (S/N) ratios of 3 and 10, were found to be 0.0071 and 0.024 μmol L-1, respectively [73]. Therefore, MIP-based optical sensors hold significant promise in therapeutic monitoring of various drugs in biological samples.

10.4.4 Environmental Exposure Monitoring

Environmental exposure monitoring is of paramount importance for several reasons, as it plays a crucial role in safeguarding public health and the environment. It is a fundamental component of environmental protection and public health. It provides the data necessary for informed decision making, regulatory compliance, and the safeguarding of human and environmental well-being. It is an essential tool for managing and mitigating the impact of human activities on the environment. Histamine poisoning linked to the consumption of seafood is a worldwide public health concern. Ingesting elevated levels of histamine can lead to a spectrum of illnesses, ranging from mild symptoms like tingling sensations to potentially fatal outcomes. Due to the elevated levels of free amino acids, like histamine seafood is susceptible to swift deterioration caused by both microbial and chemical processes. Rahtuvanoğlu et al., 2020 centered a study on creating a biomimetic biosensor tailored for the detection of histamine in both cheese and canned tuna samples [74]. This biosensor was engineered by employing histamine-imprinted nanoparticles (MIP), which were applied to coat the surface of an SPR chip. Subsequently, these modified chips underwent testing to assess their capability in identifying histamine within food samples. The biosensor displayed exceptional sensitivity, a remarkably low detection threshold, and an extensive range for linear detection, rendering it an efficient tool for histamine detection [74]. In a separate investigation, a MIP-based ultrasensitive fluorescence-based sensor has been developed for detecting thiamphenicol in milk [75]. The fluorescence sensor utilized a hybrid sensing probe consisting of mesoporous carbon and quantum dots embedded within an imprinted polymer. The MIP-based fluorescence sensor provided a sensitive and selective detection method with fast response time and high accuracy for detection of thiamphenicol. The optosensor underwent testing on spiked milk samples, demonstrating a linear response to thiamphenicol concentrations ranging from 0.01 to 10 ng/mL, with a limit of detection of 0.003 ng/mL [75]. The optosensor also demonstrated high selectivity for thiamphenicol over other antibiotics and food additives and was found to be stable and reusable for up to 10 cycles [75]. Similarly, a MIP-based SERS sensor was developed to detect enrofloxacin in food matrices [76]. The AgNPs-MIPs showed specific identification of enrofloxacin in complex environments, with a minimum detection limit of 0.25 ng/mL. The method offered a reliable, cost-effective, time-saving, highly selective, and sensitive approach for detecting enrofloxacin residues in real food samples [76].

10.4.5 Improve the Therapeutic Efficacy

Achieving precise targeting of cancer cells while enhancing the effectiveness of various therapies remains a challenging endeavor. To address this, nanoparticles featuring dual-template imprinting and a core–shell structure were synthesized by Hui Peng et al. (2020). The core of these nanoparticles comprised fluorescent silica nanoparticles (f-SiO_2), while the outermost shell consisted of the imprinted polymer layers. The design of the imprinted layer was achieved through a free-radical precipitation method on the surface of $FSiO_{2.}$ At the same time, this layer encased gadolinium-doped silicon quantum dots and photosensitizers (Ce6). Two template molecules were introduced during the polymerization process: one was the CD59 protein epitope (YNCPNPTADCK), overexpressed on the surface of numerous solid cancers, and the other was the chemotherapeutic agent doxorubicin (DOX) [77]. Additionally, the embedded Ce6 had the capability to generate toxic 1O2 when subjected to 655 nm laser irradiation, contributing to the eradication of cancer cells. This combined effect, in conjunction with the presence of loaded DOX, resulted in a synergistic approach to cancer therapy [77]. Furthermore, the inclusion of gadolinium-doped silicon quantum dots, Ce6, and DOX provided targeted fluorescence imaging (FI) and magnetic resonance imaging (MRI) capabilities to the MIPs [77]. In depth in vitro and in vivo experiments validated the exceptional targeting precision and desirable treatment outcomes, with minimal harm to healthy tissues and organs [77]. Consequently, these designed MIPs have the potential to advance targeted biomarker recognition and precision-guided treatments, aided by cell imaging tools. Additionally, it discusses the in vivo anticancer effects of the MIPs and their potential as drug delivery systems [77].

In summary, MIP-based optical sensors are emerging as a powerful tool across healthcare, with applications in disease detection (including cancer and UTI biomarkers), continuous disease monitoring (like diabetes), therapeutic drug monitoring (to ensure medication effectiveness and avoid side effects), environmental exposure monitoring (detecting harmful substances in food), and even hold promise for targeted cancer therapies. Molecularly imprinted polymers (MIPs) in optical sensors hold an advantage for disease detection and monitoring compared to other sensors as they only bind strongly to the target molecule, reducing interference from other molecules present in the sample (like urine or blood). This high selectivity leads to more accurate disease detection. Table 10.2 shows a

Table 10.2 A list of representative MIP-based optical sensors for the detection of analytes with detection limits

Optical detection techniques	Fabrication methods	Analytes	Receptor components	Detection limit	Real sample Analyses	References
Fluorescence	Vertical deposition	Fibrinopeptide B (FPB)	Molecularly-imprinted photonic polymers	0.13 ng/mL	Human urine	[63]
SPR	Microcontact imprinting method	*E. coli*	N-methacryloyl-(L)-histidine methyl ester (MAH) for tailor-made *E. coli* receptor	0.57 CFU/mL	Urine	[64]
SPR	Self-assembling adhesive imprinted method	Troponin I	Dopamine in alkaline aqueous solutions via the imprinting of short peptides derived from TnT (10–15 amino acids).	15.4 nM	Human serum	[68]
Fluorescence spectroscopy	Solid-phase extraction method	Sulfasalazine	MIP conjugated with immobilized CdSeS/ZnS alloyed QDs	0.0071 $\mu molL^{-1}$	Human plasma and urine	[73]
SPR	Mini emulsion polymerization technique	Histamine	Histamine imprinted nanoparticles	0.58 $ng.mL^{-1}$	Fish and cheese samples	[74]
Optosensor	Surface modification method	Thiamphenicol	Mesoporous carbon and CdTe*CdS*ZnS quantum dots embedded in a molecularly imprinted polymer	0.04 $\mu g\ L^{-1}$	Milk	[75]
Surface-enhanced Raman spectroscopy	Reduction method	Enrofloxacin	Ag NPs–MIPs	0.25 ng/mL	Pork	[76]

representative list of optical detection techniques for analyte detection along their detection limits. As seen in the table, MIP-based sensors can achieve very low detection limits, so they can identify even tiny amounts of the target molecule. This is crucial for early disease detection when the biomarker concentration might be low. The table also highlights the versatility of MIPs in targeting various disease markers.

10.5 Challenges and Future Perspectives

Molecularly imprinted polymers (MIPs) incorporated into optical sensors confront several challenges in their current applications. One significant obstacle is achieving the desired selectivity and sensitivity, especially when dealing with complex sample matrices. Enhancing MIPs' specificity toward target analytes without succumbing to cross-reactivity remains a critical challenge for accurate and reliable detection. For this, optimization of the parameters for imprinting is required to attain MIPs with superior performance with desired properties. Research efforts are needed to optimize the imprinting process and improve the recognition capabilities of MIPs. The choice of polymer material and imprinting methodology can significantly affect the recognition ability of MIPs. The kinetics of binding events between MIPs and target molecules also pose a challenge, influencing the response times of sensors. Establishing standardized protocols for MIP synthesis is imperative to enhance reproducibility, as variations in polymerization conditions can affect sensor performance. Designing MIPs that can be regenerated and reused multiple times without losing their binding affinity is a challenge. For real-time sensing of analytes, the MIP-based optical sensors need to be integrated into portable and miniaturized devices for onsite and point-of-care applications which pose challenges.

Despite the challenges, the future outlook for molecularly imprinted polymers in optical sensors is promising, driven by innovative approaches and emerging technologies. Multifunctionality stands out as a key area for development, with the potential for MIPs to detect multiple analytes simultaneously or integrate seamlessly with other sensing technologies. The integration of nanotechnology into MIPs holds great promise, offering opportunities to enhance sensitivity, selectivity, and overall sensor functionality. As advancements continue, the utilization of optical sensors based on MIPs in real-world scenarios, such as medical diagnostics and environmental monitoring, is expected to expand. Further developments in smart materials, enabling MIPs to dynamically adapt to changing conditions, will contribute to increased versatility and applicability in diverse fields.

10.6 Conclusion

MIPs hold great potential in sensor technology due to their unique recognition properties. They offer advantages like high selectivity, sensitivity, stability, low cost, and ease of fabrication. MIP-based sensors show significant potential in biomedical fields, such as disease diagnosis, drug delivery, and monitoring biomarkers. Further research in these areas can lead to significant advancements in healthcare. MIP-based optical sensors represent a transformative technology in the realm of selective and sensitive detection. Recent literature underscores the versatility and potential of these sensors in diverse applications, from environmental monitoring to healthcare diagnostics. With ongoing advancements and interdisciplinary collaborations, the future of MIP-based optical sensors promises even greater precision, miniaturization, and accessibility, ultimately revolutionizing how we perceive and address analytical challenges.

References

1. Zhang J, Wang Y, Lu X (2021) Molecular imprinting technology for sensing foodborne pathogenic bacteria. Anal Bioanal Chem 413(18):4581–4598. https://doi.org/10.1007/s00216-020-03138-x
2. Yarman A, Kurbanoglu S, Zebger I, Scheller FW (2021) Simple and robust: the claims of protein sensing by molecularly imprinted polymers. Sens Actuators B Chem 330:129369. https://doi.org/10.1016/j.snb.2020.129369
3. Dong C, Shi H, Han Y, Yang Y, Wang R, Men J (2021) Molecularly imprinted polymers by the surface imprinting technique. Eur Polym J 145:110231. https://doi.org/10.1016/j.eurpolymj.2020.110231
4. Polymers | Free full-text | Molecularly imprinted polymer-based sensors for protein detection. Accessed 28 Oct 2023 [Online]. https://www.mdpi.com/2073-4360/15/3/629
5. Soufi GJ, Iravani S, Varma RS (2021) Molecularly imprinted polymers for the detection of viruses: challenges and opportunities. Analyst 146(10):3087–3100. https://doi.org/10.1039/D1AN00149C
6. Approaches to the rational design of molecularly imprinted polymers developed for the selective extraction or detection of antibiotics in environmental and food samples—Jamieson—2021—physica status solidi (a)—Wiley Online Library. Accessed 29 Oct 2023 [Online]. https://onlinelibrary.wiley.com/doi/full/10.1002/pssa.202100021
7. Cao Y, Feng T, Xu J, Xue C (2019) Recent advances of molecularly imprinted polymer-based sensors in the detection of food safety hazard factors. Biosens Bioelectron 141:111447. https://doi.org/10.1016/j.bios.2019.111447
8. Ramanavicius S, Samukaite-Bubniene U, Ratautaite V, Bechelany M, Ramanavicius A (2022) Electrochemical molecularly imprinted polymer based sensors for pharmaceutical and biomedical applications (review). J Pharm Biomed Anal 215:114739. https://doi.org/10.1016/j.jpba.2022.114739
9. Saylan Y, Akgönüllü S, Yavuz H, Ünal S, Denizli A (2019) Molecularly imprinted polymer based sensors for medical applications. Sensors 19(6):1279. https://doi.org/10.3390/s19061279
10. Zarejousheghani M, Rahimi P, Borsdorf H, Zimmermann S, Joseph Y (2021) Molecularly imprinted polymer-based sensors for priority pollutants. Sensors 21(7):2406. https://doi.org/10.3390/s21072406
11. Ozcelikay G et al (2022) Novel sensor approaches of aflatoxins determination in food and beverage samples. Crit Rev Anal Chem:1–20. https://doi.org/10.1080/10408347.2022.2105136
12. Lahcen AA et al (2021) Laser-scribed graphene sensor based on gold nanostructures and molecularly imprinted polymers: application for Her-2 cancer biomarker detection. Sens Actuators B Chem 347:130556. https://doi.org/10.1016/j.snb.2021.130556
13. Prabhu SN, Gooneratne CP, Mukhopadhyay SC (2021) Development of MEMS sensor for detection of creatinine using MIP based approach—a tutorial paper. IEEE Sensors J 21(20):22170–22181. https://doi.org/10.1109/JSEN.2021.3077060
14. Fang L et al (2021) Molecularly imprinted polymer-based optical sensors for pesticides in foods: recent advances and future trends. Trends Food Sci Technol 116:387–404. https://doi.org/10.1016/j.tifs.2021.07.039

15. Deng D, Fang X, Duan D, Li K (2022) A gel fluorescence sensor based on CDs@SiO2/FeS2@MIPs for the visual detection of p-chlorophenol. Anal Methods 14(17):1721–1729. https://doi.org/10.1039/D1AY01849C
16. Uniyal A, Srivastava G, Pal A, Taya S, Muduli A (2023) Recent advances in optical biosensors for sensing applications: a review. Plasmonics 18(2):735–750. https://doi.org/10.1007/s11468-023-01803-2
17. Rico-Yuste A, Carrasco S (2019) Molecularly imprinted polymer-based hybrid materials for the development of optical sensors. Polymers 11(7):1173. https://doi.org/10.3390/polym11071173
18. Dehdari Vais R, Yadegari H, Heli H, Sattarahmady N (2021) A β-amyloid(1-42) biosensor based on molecularly imprinted poly-pyrrole for early diagnosis of Alzheimer's disease. J Biomed Phys Eng 11(2):215–228. https://doi.org/10.31661/jbpe.v0i0.1070
19. Ramajayam K, Ganesan S, Ramesh P, Beena M, Kokulnathan T, Palaniappan A (2023) Molecularly imprinted polymer-based biomimetic systems for sensing environmental contaminants, biomarkers, and bioimaging applications. Biomim Basel Switz 8(2):245. https://doi.org/10.3390/biomimetics8020245
20. Wang Y et al (2022) Application progress of magnetic molecularly imprinted polymers chemical sensors in the detection of biomarkers. Analyst 147(4):571–586. https://doi.org/10.1039/d1an01112j
21. Matsumoto H, Sunayama H, Kitayama Y, Takano E, Takeuchi T (2019) Site-specific post-imprinting modification of molecularly imprinted polymer nanocavities with a modifiable functional monomer for prostate cancer biomarker recognition. Sci Technol Adv Mater 20(1):305–312. https://doi.org/10.1080/14686996.2019.1583495
22. Baldoneschi V, Palladino P, Banchini M, Minunni M, Scarano S (2020) Norepinephrine as new functional monomer for molecular imprinting: an applicative study for the optical sensing of cardiac biomarkers. Biosens Bioelectron 157:112161. https://doi.org/10.1016/j.bios.2020.112161
23. Lamaoui A, Amine A (2022) Key advances in MIP-based sensors applied for cancer and cardiovascular biomarkers detection. Curr Top Med Chem 22(7):529–548. https://doi.org/10.2174/1568026622666220307124003
24. Ostrovidov S et al (2023) Molecularly imprinted polymer-based sensors for the detection of skeletal- and cardiac-muscle-related analytes. Sensors 23(12):12. https://doi.org/10.3390/s23125625
25. Tomichan R et al (2023) Insight of smart biosensors for COVID-19: a review. Lumin J Biol Chem Lumin 38(7):1102–1110. https://doi.org/10.1002/bio.4430
26. Pilvenyte G et al (2023) Molecularly imprinted polymer-based electrochemical sensors for the diagnosis of infectious diseases. Biosensors 13(6):620. https://doi.org/10.3390/bios13060620
27. Bräuer B, Unger C, Werner M, Lieberzeit PA (2021) Biomimetic sensors to detect bioanalytes in real-life samples using molecularly imprinted polymers: a review. Sensors 21(16):16. https://doi.org/10.3390/s21165550
28. Chen H, Guo J, Wang Y, Dong W, Zhao Y, Sun L (2022) Bio-inspired imprinting materials for biomedical applications. Adv Sci 9(28):2202038. https://doi.org/10.1002/advs.202202038
29. El-Schich Z et al (2020) Molecularly imprinted polymers in biological applications. BioTechniques 69(6):406–419. https://doi.org/10.2144/btn-2020-0091
30. Mohiuddin I, Grover A, Aulakh JS, Lee S-S, Malik AK, Kim K-H (2020) Porous molecularly-imprinted polymer for detecting diclofenac in aqueous pharmaceutical compounds. Chem Eng J 382:123002. https://doi.org/10.1016/j.cej.2019.123002
31. Zhou T, Ding L, Che G, Jiang W, Sang L (2019) Recent advances and trends of molecularly imprinted polymers for specific recognition in aqueous matrix: preparation and application in sample pretreatment. TrAC Trends Anal Chem 114:11–28. https://doi.org/10.1016/j.trac.2019.02.028
32. Farooq S et al (2018) Molecularly imprinted polymers' application in pesticide residue detection. Analyst 143(17):3971–3989. https://doi.org/10.1039/c8an00907d

33. Rebelo TSCR, Costa R, Brandão ATSC, Silva AF, Sales MGF, Pereira CM (2019) Molecularly imprinted polymer SPE sensor for analysis of CA-125 on serum. Anal Chim Acta 1082:126–135. https://doi.org/10.1016/j.aca.2019.07.050
34. Omranipour HM, Sajadi Tabassi SA, Kowsari R, Rad MS, Mohajeri SA (2015) Brimonidine imprinted hydrogels and evaluation of their binding and releasing properties as new ocular drug delivery systems. Curr Drug Deliv 12(6):717–725. https://doi.org/10.2174/1567201812666150316110838
35. Zhao X, Mai Y, Chen D, Zhang M, Hu H (2019) Selective enrichment of clenbuterol onto molecularly imprinted polymer microspheres with tailor-made structure and oxygen functionalities. Polymers 11(10):1635. https://doi.org/10.3390/polym11101635
36. Pacheco JG et al (2019) Electrochemical sensing of the thyroid hormone thyronamine (T0AM) via molecular imprinted polymers (MIPs). Talanta 194:689–696. https://doi.org/10.1016/j.talanta.2018.10.090
37. Liu M, Ding X, Yang Q, Wang Y, Zhao G, Yang N (2017) A pM leveled photoelectrochemical sensor for microcystin-LR based on surface molecularly imprinted TiO2@CNTs nanostructure. J Hazard Mater 331:309–320. https://doi.org/10.1016/j.jhazmat.2017.02.031
38. Roy E, Patra S, Tiwari A, Madhuri R, Sharma PK (2017) Introduction of selectivity and specificity to graphene using an inimitable combination of molecular imprinting and nanotechnology. Biosens Bioelectron 89:234–248. https://doi.org/10.1016/j.bios.2016.02.056
39. Bărăian A-I, Iacob B-C, Bodoki AE, Bodoki E (2022) In vivo applications of molecularly imprinted polymers for drug delivery: a pharmaceutical perspective. Int J Mol Sci 23(22):22. https://doi.org/10.3390/ijms232214071
40. Kang MS, Cho E, Choi HE, Amri C, Lee J-H, Kim KS (2023) Molecularly imprinted polymers (MIPs): emerging biomaterials for cancer theragnostic applications. Biomater Res 27:45. https://doi.org/10.1186/s40824-023-00388-5
41. Thakur N, Das TR, Gupta S, Shukla SK, Patra S, Kumar D (2023) Chapter 5—Material selection for the fabrication of MIP-based devices: an overview. In: Singh M (ed) Molecularly imprinted polymers (MIPs). Elsevier, pp 101–114. https://doi.org/10.1016/B978-0-323-91925-8.00006-5
42. Morsi S, Abd El-Aziz M, Mohamed H (2022) Smart polymers as molecular imprinted polymers for recognition of target molecules. Int J Polym Mater Polym Biomater 72:1–24. https://doi.org/10.1080/00914037.2022.2042287
43. Rebelo P et al (2021) Molecularly imprinted polymer-based electrochemical sensors for environmental analysis. Biosens Bioelectron 172:112719. https://doi.org/10.1016/j.bios.2020.112719
44. Khonina SN, Kazanskiy NL, Butt MA (2023) Optical fibre-based sensors—an assessment of current innovations. Biosensors 13(9):835. https://doi.org/10.3390/bios13090835
45. Ahmad OS, Bedwell TS, Esen C, Garcia-Cruz A, Piletsky SA (2019) Molecularly imprinted polymers in electrochemical and optical sensors. Trends Biotechnol 37(3):294–309. https://doi.org/10.1016/j.tibtech.2018.08.009
46. Sargazi S et al (2022) Fluorescent-based nanosensors for selective detection of a wide range of biological macromolecules: a comprehensive review. Int J Biol Macromol 206:115–147. https://doi.org/10.1016/j.ijbiomac.2022.02.137
47. Sergeyeva T et al (2019) Development of a smartphone-based biomimetic sensor for aflatoxin B1 detection using molecularly imprinted polymer membranes. Talanta 201:204–210. https://doi.org/10.1016/j.talanta.2019.04.016
48. Tang Y et al (2018) Upconversion particle@Fe3O4@molecularly imprinted polymer with controllable shell thickness as high-performance fluorescent probe for sensing quinolones. Talanta 181:95–103. https://doi.org/10.1016/j.talanta.2018.01.006
49. Xu J, Prost E, Haupt K, Bui BTS (2018) Direct and sensitive determination of trypsin in human urine using a water-soluble signaling fluorescent molecularly imprinted polymer nanoprobe. Sens Actuators B Chem 258:10–17. https://doi.org/10.1016/j.snb.2017.11.077
50. Yang J, Feng W, Liang K, Chen C, Cai C (2020) A novel fluorescence molecularly imprinted sensor for Japanese encephalitis virus detection based on metal organic frame-

works and passivation-enhanced selectivity. Talanta 212:120744. https://doi.org/10.1016/j.talanta.2020.120744
51. Capelli D, Scognamiglio V, Montanari R (2023) Surface plasmon resonance technology: recent advances, applications and experimental cases. TrAC Trends Anal Chem 163:117079. https://doi.org/10.1016/j.trac.2023.117079
52. Patil PO et al (2019) Graphene-based nanocomposites for sensitivity enhancement of surface plasmon resonance sensor for biological and chemical sensing: a review. Biosens Bioelectron 139:111324. https://doi.org/10.1016/j.bios.2019.111324
53. Zhou J et al (2019) Surface plasmon resonance (SPR) biosensors for food allergen detection in food matrices. Biosens Bioelectron 142:111449. https://doi.org/10.1016/j.bios.2019.111449
54. Mahmoudpour M, Dolatabadi JEN, Torbati M, Tazehkand AP, Homayouni-Rad A, de la Guardia M (2019) Nanomaterials and new biorecognition molecules based surface plasmon resonance biosensors for mycotoxin detection. Biosens Bioelectron 143:111603. https://doi.org/10.1016/j.bios.2019.111603
55. Anand U et al (2022) Recent advances in the potential applications of luminescence-based, SPR-based, and carbon-based biosensors. Appl Microbiol Biotechnol 106(8):2827–2853. https://doi.org/10.1007/s00253-022-11901-6
56. Fang M et al (2019) Preparation of highly conductive graphene-coated glass fibers by sol-gel and dip-coating method. J Mater Sci Technol 35(9):1989–1995. https://doi.org/10.1016/j.jmst.2019.05.027
57. Caldara M et al (2023) Deposition methods for the integration of molecularly imprinted polymers (MIPs) in sensor applications. Adv Sensor Res 2(7):2200059. https://doi.org/10.1002/adsr.202200059
58. "The development of an electropolymerized, molecularly imprinted polymer (MIP) sensor for insulin determination using single-drop analysis—analyst (RSC Publishing). doi: https://doi.org/10.1039/D2AN02025D." Accessed 29 Oct 2023 [Online]. https://pubs.rsc.org/en/content/articlehtml/2023/an/d2an02025d
59. Ayivi RD, Adesanmi BO, McLamore ES, Wei J, Obare SO (2023) Molecularly imprinted Plasmonic sensors as Nano-transducers: an effective approach for environmental monitoring applications. Chemosensors 11(3):3. https://doi.org/10.3390/chemosensors11030203
60. Bray F, Laversanne M, Weiderpass E, Soerjomataram I (2021) The ever-increasing importance of cancer as a leading cause of premature death worldwide. Cancer 127(16):3029–3030. https://doi.org/10.1002/cncr.33587
61. Xing R, Wen Y, Dong Y, Wang Y, Zhang Q, Liu Z (2019) Dual molecularly imprinted polymer-based plasmonic immunosandwich assay for the specific and sensitive detection of protein biomarkers. Anal Chem 91(15):9993–10000. https://doi.org/10.1021/acs.analchem.9b01826
62. Jinn WS et al (2019) A visually distinguishable light interfering bioresponsive silica nanoparticle hydrogel sensor fabricated through the molecular imprinting technique. J Mater Chem B 7(45):7120–7128. https://doi.org/10.1039/C9TB01579E
63. Resende S, Frasco MF, Sales MGF (2020) A biomimetic photonic crystal sensor for label-free detection of urinary venous thromboembolism biomarker. Sens Actuators B Chem 312:127947. https://doi.org/10.1016/j.snb.2020.127947
64. Özgür E, Topçu AA, Yılmaz E, Denizli A (2020) Surface plasmon resonance based biomimetic sensor for urinary tract infections. Talanta 212:120778. https://doi.org/10.1016/j.talanta.2020.120778
65. Wang M et al (2022) A wearable electrochemical biosensor for the monitoring of metabolites and nutrients. Nat Biomed Eng 6(11):11. https://doi.org/10.1038/s41551-022-00916-z
66. Brown S et al (2018) Minimally invasive technique for measuring transdermal glucose with a fluorescent biosensor. Anal Bioanal Chem 410(27):7249–7260. https://doi.org/10.1007/s00216-018-1336-8
67. Olvera Lopez E, Ballard BD, Jan A (2023) Cardiovascular disease. In: StatPearls, Treasure Island (FL): StatPearls Publishing. Accessed 29 Oct 2023 [Online]. http://www.ncbi.nlm.nih.gov/books/NBK535419/

68. Palladino P, Minunni M, Scarano S (2018) Cardiac troponin T capture and detection in real-time via epitope-imprinted polymer and optical biosensing. Biosens Bioelectron 106:93–98. https://doi.org/10.1016/j.bios.2018.01.068
69. Piloto AM, Ribeiro DSM, Rodrigues SSM, Santos C, Santos JLM, Sales MGF (2018) Plastic antibodies tailored on quantum dots for an optical detection of myoglobin down to the femtomolar range. Sci Rep 8(1):1. https://doi.org/10.1038/s41598-018-23271-z
70. Piloto AML, Ribeiro DSM, Rodrigues SSM, Santos JLM, Sampaio P, Sales G (2021) Imprinted fluorescent cellulose membranes for the on-site detection of myoglobin in biological media. ACS Appl Bio Mater 4(5):4224–4235. https://doi.org/10.1021/acsabm.1c00039
71. Garzón V, Pinacho DG, Bustos R-H, Garzón G, Bustamante S (2019) Optical biosensors for therapeutic drug monitoring. Biosensors 9(4):132. https://doi.org/10.3390/bios9040132
72. Ren X, Li X (2020) Flower-like Ag coated with molecularly imprinted polymers as a surface-enhanced Raman scattering substrate for the sensitive and selective detection of glibenclamide. Anal Methods 12(22):2858–2864. https://doi.org/10.1039/D0AY00575D
73. Ahmadpour H, Hosseini SMM (2019) A solid-phase luminescence sensor based on molecularly imprinted polymer-CdSeS/ZnS quantum dots for selective extraction and detection of sulfasalazine in biological samples. Talanta 194:534–541. https://doi.org/10.1016/j.talanta.2018.10.053
74. Rahtuvanoğlu A, Akgönüllü S, Karacan S, Denizli A (2020) Biomimetic nanoparticles based surface plasmon resonance biosensors for histamine detection in foods. ChemistrySelect 5(19):5683–5692. https://doi.org/10.1002/slct.202000440
75. Sa-nguanprang S, Phuruangrat A, Bunkoed O (2022) An optosensor based on a hybrid sensing probe of mesoporous carbon and quantum dots embedded in imprinted polymer for ultra-sensitive detection of thiamphenicol in milk. Spectrochim Acta A Mol Biomol Spectrosc 264:120324. https://doi.org/10.1016/j.saa.2021.120324
76. Neng J, Wang Y, Zhang Y, Chen P, Yang K (2023) MIPs–SERS sensor based on ag NPs film for selective detection of enrofloxacin in food. Biosensors 13(3):3. https://doi.org/10.3390/bios13030330
77. Epitope molecularly imprinted polymer nanoparticles for chemo-/photodynamic synergistic cancer therapy guided by targeted fluorescence imaging | ACS Applied Materials & Interfaces. Accessed 21 Oct 2023 [Online]. https://pubs.acs.org/doi/10.1021/acsami.0c00468

MIP-Based Chip Sensors 11

Fulden Ulucan-Karnak and Cansu İlke Kuru

Abstract

Recent years have seen a significant increase in interest in the ground-breaking technology of a commercial chip platform for the detection of molecules. Various materials, including molecular imprinted surfaces, have been developed for recognition surfaces in chip sensors. The process of molecular imprinting is used to prepare affinity-based polymers for analytical targets. Molecularly imprinted polymers are an exciting new kind of sensing element that offers excellent stability, cheap cost, strong affinity for target molecules, and increased selection options for monomers and polymerization protocols. The combination of these recognition components with materials at the nanoscale improves binding selectivity as well as sensitivity, allowing for the construction of very versatile diagnostic devices that can be utilized directly in analysis. The purpose of this review article is to identify many ways for molecular imprinting molecules and incorporation with various chip technologies to explain the most effective techniques for creating molecularly imprinted polymer-based sensor systems.

Keywords

Molecular imprinting technologies · Chip · Sensors

F. Ulucan-Karnak
Medical Biochemistry Department, Institute of Health Sciences, Ege University, Bornova, Izmir, Türkiye

C. İ. Kuru (✉)
Department of Biotechnology, Graduate School of Natural and Applied Sciences, Ege University, İzmir, Türkiye

Buca Municipality Buca Science and Art Center, Buca, Izmir, Türkiye

S. Patra et al. (eds.), *Molecularly Imprinted Polymers: Path to Artificial Antibodies*, https://doi.org/10.1007/978-981-97-4379-7_11

11.1 Introduction

The polymeric matrices created using the imprinting technology, known as "molecularly imprinted polymers" (MIPs), are strong molecular recognition elements that can mimic natural recognition components like antibodies and biological receptors. The ultimate objective of molecular imprinting is to produce MIPs that can perform in a manner akin to that of biological receptors, giving them the ability to take the place of these naturally occurring components in some applications. Three crucial reagents are used in the molecular imprinting procedure to produce MIPs with high affinity and specificity to a particular target molecule: template molecules, functional monomers, and cross-linkers. The degree of affinity and specificity attained in the finished product is significantly influenced by the choice of the proper reagents [1, 2].

In terms of size, form, and spatial arrangement of the functional groupings, these artificial recognition sites should complement the template. The template molecule and functional monomers can interact in covalent, non-covalent, or semi-covalent ways to produce these spatial configurations. Then, together with the template molecule, co-polymerization of the cross-linking and functional monomers that make up the pre-complex is accomplished. Following polymerization, the template molecule is eliminated, leaving complemental gaps that will serve as artificial recognition sites. The non-covalent strategy, however, has been the one that is most frequently employed (Fig. 11.1) [3–5].

MIPs are prepared using a variety of processes, including precipitation polymerization, bulk polymerization, suspension polymerization, and in situ polymerization. Each of these techniques has their advantages and disadvantages (Table 11.1). However, using MIP as a component of recognition in the manufacture of electrochemical and optical sensors has yielded encouraging results, there are numerous disadvantages to using standard methods for their fabrication. For instance, the binding sites located deep within the material are frequently destroyed when the bulk polymerization technique is employed. Long diffusion distances, particularly

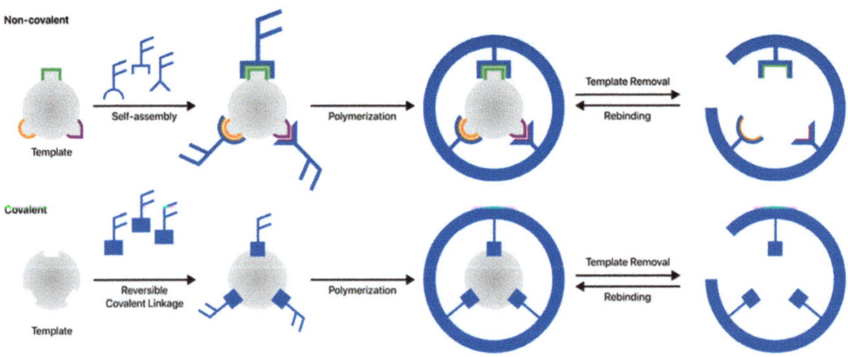

Fig. 11.1 Non-covalent and covalent-based MIP procedure, retrieved from open access journal licensed under CC BY 4.0 [3]

Table 11.1 Comparation of different preparation methods of MIP

Preparation methods	Advantages	Disadvantages
Bulk polymerization	Easy to produce, low cost, robust, resistant to harsh environments	Irregular particles, low yield, template leakage
Emulsion polymerization	Water-soluble, narrow particle distribution, high specific surface area, good dispersity, and regular particle size	Low binding capacity
Suspension polymerization	Regular particles	Low recognition and polydisperse size distribution
Surface imprinting	Particle size is uniform and controlled, with high adsorption capacity, strong stability and selectivity, good reproducibility, quick binding kinetics, and mass transfer	Restricted surface areas
Precipitation polymerization	There are no stabilizers, the process is easy, the yields are high, the time is short, the size is small and consistent, and diverse compounds can be imprinted	High dilution conditions, precise synthetic parameter adjustments, and a high porogen volume

when dealing with big templates, cause the template to either become permanently imprisoned inside the matrix or cause their rebinding to become extremely sluggish. To address these issues and enhance MIP's performance, molecular imprinting has been coupled with nanotechnology [6–10].

Molecularly imprinted polymers (MIPs), when combined with the benefits of rapid synthesis, resilience, regeneration (and thus cost-effectiveness), as well as low initial production costs, offer an alluring substitute for traditional biological receptors. When compared to enzymes or antibodies, they were far easier to make and more economically made. They exhibit both storage and operational stability and have a superior rate of regeneration after use. They exhibit selectivity and binding strength in particular. MIPs are thought of as plastic antibodies or antibodies because of their excellent sensitivity and selectivity, and they perform even better than commercial antibodies. The capacity of MIPs to alter the entire polymer structure or only its functional sites while utilizing various monomers and functional monomers in various ratios is another intriguing feature. Therefore, even when subjected to extremes of pH, organic medium, or autoclave conditions, They have very good stability, repeatability, and biocompatibility [11–13]. Imprinted technologies can be also combined with graphene, graphene oxide, QDs, etc. to enhance their superior properties like selectivity, specificity, and conductivity [14, 15].

Molecularly imprinted polymers (MIPs) have a molecular complementary recognition process identical to natural interactions between receptors and ligands. They have advantages such as chemical, mechanical and thermal stability, ease of manufacture, low cost, and receptor-like affinity for target molecules. MIPs are used in various applications, including separation methods, artificial enzymes, immunoassays, biosensors, antibody mimics, catalysis, drug delivery, organic synthesis. Despite their potential for cell recognition, early research focused on adsorption, separation, sensing, and catalysis. The search for new MIPs for cell recognition continues due to their promise for basic cell-based biomedical applications, as well as ongoing efforts to improve artificial receptors for cellular functions [16].

A search of the PubMed database for the phrase "molecular imprinted polymer sensor" reveals a growing trend in the publication counts since the 2000s. Senay et al. provided some insight into MIP-based sensors and determined that the most researched MIP sensors are electrochemically based, with optical and mass-sensitive sensors coming in second and third, respectively [17]. Capacitance sensors, chemiresistors, electrochemical impedance spectroscopy, and field effect transistors are the methods used the most fundamental measurement techniques for using MIP recognition polymers in sensors. Electrochemical sensors, optical sensors such as surface plasmon resonance (SPR), and quartz crystal microbalance (QCM) sensors are other implementations for MIPs. Although the ultimate format selection for a given application can be random, the target molecule and its source are typically the determining factors. Also computational methods could be used for designing novel MIP materials, interactions, and MIP-based sensors [18].

11.2 Sensor Design Using Molecular Imprinting

Molecular imprinted polymers can be used as the recognition layers of the biosensor surfaces. This layer is coated onto the transducer element of a biosensor construction. The coating step is the most important task in sensor setup. Thin films and prefabricated micro- and nanoparticles are the two physical forms on which MIPs are often interfaced. There are numerous approaches to interface micro- or nanoparticles, including embedding them into polymer matrices with core–shell designs and covalent bonding to the surface or drop-casting spin-coating, and spray-coating them onto the surface directly [19–22]. As an alternative, electropolymerization could be used for MIP-based sensor designs. By altering electrochemical parameters, it offers clear benefits over other kinds of polymerization techniques for controlling layer thickness and porosity. This approach is also quick, simple, and inexpensive. The template is removed from the matrix of polymer during polymerization to create complementary holes [23]. These designs are illustrated in Fig. 11.2.

MIPs are used in biochips to detect biomolecules, providing label-free, adaptable, and low-cost sensors ideal for lab-on-chip devices. This chip-based sensing technology can be used for point-of-care diagnostics and wireless transmission of analysis findings in remote areas. MIPs are quick, inexpensive, portable, and can be integrated into smart responsive biomedical systems, making them a promising alternative to artificial receptors in microfluidic-based lab-on-a-chip devices [24, 25].

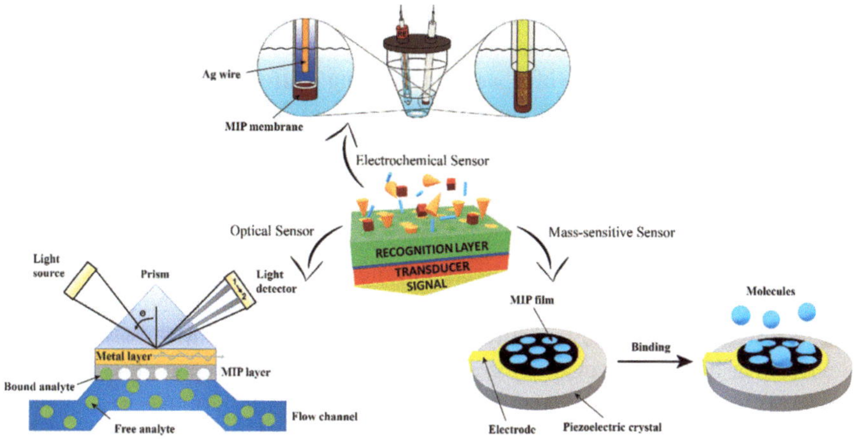

Fig. 11.2 MIP-based sensor designs, retrieved from open access journal licensed under CC BY 4.0 [19]

11.3 Applications of MIP-Based Chip Sensors

In recent decades, sensor-related scientific research has become very active. The term "sensor" refers to a functionalized surfaces that transmits signals from imperceptible chemical or physical interactions occurring at a particular interface to discernible reactions. Due to their distinct capacity to identify analytes, sensors containing a layer of polymeric molecularly compatible cavities, also known as MIP sensors, have attracted a lot of attention. MIP sensors have been used in conjunction with a wide range of analyte determination techniques, including surface plasmon resonance, fluorescence, electrochemical, and surface-enhanced Raman scattering [26].

Increasing interest has been shown in using MIPs as the molecular recognition component in sensors through their selective binding, as a result of the requirement for dependable and cost-effective materials for integrated devices for point-of-care testing (POCT). Binding between target analyte and MIP can be detected by electrochemical impedance spectroscopy (EIS) in electrochemical sensors or a change in refractive index in surface plasmon resonance sensors. Many analytes with none optical or electrochemical characteristics can be detected by using MIPs, which can generate electrochemical or optical signals in response to the template interacting with the functional chemical groups at the imprinting site [27].

11.3.1 Point of Care Approach for Clinical Diagnostics

According to numerous papers, electrochemical sensors combined with microfluidic technology are seen as a point-of-care sensing platform thanks to advantages like adjustable sample volume, affordability, and portability. It is undeniable that microfluidic components have an incredibly high level of autonomy in terms of sampling due to internal capillary action [26].

This benefit was paired with a wearable sensing platform by Wang et al. to realize onsite, individualized health monitoring. Through delicately imprinted cavities, the sensing platform permitted different metabolites in sweat to be detected. Laser-engraved graphene (LEG) and redox-active nano-reporters (RARs) worked together to produce electrochemical responses. The results of the sensing were transmitted through Bluetooth to a piece of software that was designed and shown on a mobile device. Additionally, the research team unveiled a smartwatch that was ready for market and had the crucial wearable MIP sensing patch name "NutriTrek." This product was more technologically sophisticated than other MIP-based sensing devices that had been previously described. The previously reported sensing platforms were typically hampered by strenuous washing procedures to regenerate imprinted polymeric receptors, hence the work introduced the first wearable sensing system based on an MIP in real-time detecting mechanism. This adaptable microfluidic MIP sensor made it possible to successfully monitor branched chain and certain amino acids [28].

Today, there is a growing imperative for the rapid and precise detection of cancer biomarkers. The device's own effectiveness in terms of cost, specificity, analysis time, and detection limit is significantly influenced by the design of devices appropriate for use [27]. A disposable, low-cost device for CEA detection was created by Moreira et al. based on a screen-printed cell with a three-electrode configuration. The biomimetic material was made of a polypyrrole (PPy) imprinted matrix which was electro-polymerized on the surface of working electrode while CEA was present. The protein was then extracted from the sample following proteolytic enzymatic action. By exposing the sensor to various compounds, selectivity studies were carried out. The sensor's repeatability was then examined. The findings demonstrated that this polymeric antibody-based sensor is an effective and stable instrument for CEA screening in POCT [29].

The Rebelo et al.'s paper describes synthesis of molecular imprinting technology-based artificial receptors for detecting the stress biomarker -amylase in human saliva. The gold screen-printed electrodes (AuSPEs), which are inexpensive and disposable, are the main components of the portable electrochemical devices for monitoring -amylase. The enzyme was taken out of the polymer film after electropolymerization to create the target enzyme's precise recognition sites (Fig. 11.3). Limit of detection was attained (LOD 3×10^{-4} mg/mL) and the MIP biosensor demonstrated a very broad linear concentration range. In this study, an effective method for creating an electrochemical system to gather salivary proteomic data in clinical setting was revealed. Specifically recognizing the stress biomarker amylase through the construction and electrosynthesis of artificial receptors at the sensor surface to gauge patients' physiological and psychological stress was based on molecular imprinting technology and electrochemical techniques. Furthermore, detection was carried out simply. The detection is made easy, affordable, and ideal for PoC application by the use of disposable chips, AuSPEs, and portable and tiny potentiostats. The sample matrix had no

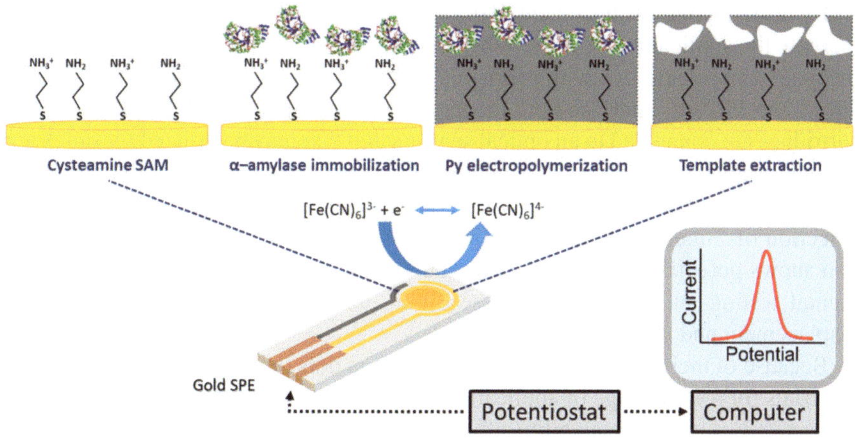

Fig. 11.3 An illustration of the synthetic procedure used to create the MIP film on the AuSPE surface. The following steps are included in the whole procedure: the imprinting process by electrosynthesis of a poly(pyrrole) thin film in the presence of template molecules; the (i) cysteamine adsorption on the working region of electrode, (ii) immobilization of biomolecule templates, and (iii) target biomolecules are extracted from the polymeric matrix. [30]

discernible impact on the biosensor's function, and excellent recoveries were attained. So, quick and precise screening assays for clinical diagnosis can be performed using MIP-based biosensors [30].

Wearable MIP-based electrochemical sensors combine several benefits while avoiding many of the frequently noticed drawbacks of other MIP-based sensing technologies of the time. Clinical experiments with untreated biological fluids can be done using the suggested sensor, for example, because the imprinted polymeric traps can be electrochemically replenished after each type of analyte rebinding. In multicycle investigations, it has great reusability as well. This sensor's capillary effect—which comes from microfluidic accessories—allows for micro-volume sampling. It is once again shown that the combination of an electrochemical MIP sensor and a microfluidic system results in highly practical sensing for point-of-care and in situ applications [26].

11.3.2 Lab-on-a-Chips for Environmental Detection

As an alternative biosensing approach, lab-on-a-chip (LOC) combines the techniques of multiple laboratory operations, including chemical synthesis, DNA sequencing, biochemical analysis, and more, on tiny microfluidic single chips. Its advantages include cost-effectiveness, parallelization, ergonomics, rapid diagnostic speed, and sensitivity. In the meanwhile, LOC platforms, which are environmentally friendly sensing for both onsite detection and a substitute for the commercial rival, are based on molecularly imprinted biosensing methods and enable their use in a number of sectors [24].

By utilizing dye-displacement MIPs, Goossens's study presents a new method to increase the application of MIPs. When exposed to an aqueous target, these artificial receptors can release dye with precision. A cheap, disposable lab-on-chip cartridge allows the MIPs to interface with the target for dye displacement. Additionally, the cartridges are connected to an optical sensor setup that may be utilized for quick and simple sample screening. Lastly, an application proof demonstrates the utilization of our sensing system for the dose-responsive characterization (4PL)-based indirect detection of 2-methoxphenidine utilizing dye-loaded MIPs. Additionally, this system shows potential for several additional sensing applications, including environmental testing and food quality assessment, due to the universality of the dye displacement and sensing principle [31].

Because of their remarkable properties, carbon dots (or CDs) are attractive nanomaterials for a wide range of applications, including fluorescent nanosensors. These characteristics include scalability, affordability, tunability, stability, water solubility, and biocompatibility. These features when combined with the durability, high affinity, selectivity, and reusability of MIPs enable the creation of new types of fluorescent nanosensors that are more selective and sensitive to their target molecules. Recent advancements in the integration of CDs and MIPs have led to the creation of green CDs/MIPs-based fluorescent nanosensors that are sensitive enough to detect a variety of contaminants in environmental samples [32].

Insecticides like parathion-methyl (PM) must be quickly detected, which calls for techniques with great sensitivity and selectivity. In Li et al.'s study, the silver terephthalate metal–organic frameworks (Fe_3O_4/C-dots@Ag-MOFs) and Fe_3O_4 nanozyme-supported carbon quantum dots used here as amplification elements were combined to create a dual catalytic amplification method. A unique electrochemical microfluidic paper-based chip for PM detection was created based on this methodology. The MIP collected PM from the sample that was inserted into the reaction site and produced a reduction current response at about 0.53 V in a three-electrode system integrated into the chip. Ag-MOFs' and Fe_3O_4/C-dots simultaneous catalysis greatly improved the signal. The chip, which had a detection limit of 1.16 × 1011 mol/L, was effectively employed to identify the presence of PM in agricultural products and environmental samples, with recovery rates ranging from 82.7 to 109% and a relative standard deviation (RSD) of less than 5.0%. With the use of a microfluidic paper-based chip, this approach significantly increased the chips' selectivity and sensitivity by combining a dual catalytic amplification strategy with an MIP [33].

The goal of Zhang et al.'s study was to create new fluorescent microfluidic paper chips that would specifically recognize and detect the pesticide through a mechanism called electron transfer-induced fluorescence quenching. Imprinted paper chips (paper@QDs@MIPs) were produced by depositing CdTe quantum dots (QDs) onto cellulose paper. The molecularly imprinted fluorescence sensor can be transferred using this technique from the liquid phase to the paper-based phase for portable and quick analysis. The paper chips with imprints were successfully analyzed, and they performed an optimal ordered spatial network structure, fluorescence quality, and chemical stability. In less than 18 minutes, the paper@QDs@MIPs demonstrated that 2,4-D binding considerably decreased the intensity of fluorescence. It also demonstrated high detectability of 90 nM and good linearity in the range of 0.83–100 M. Additionally, the recoveries of the spiked bean sprouts ranged between 94.2% and 107.0% at three concentration levels, with less than 5.9% relative standard deviation. The tool offered a powerful platform for the quick identification, practicality, and detection of minute food contaminants in intricate matrices, safeguarding food, and advancing surface imprinting investigations [34].

Karrat et al. devised a novel strategy to address the problem of ion extraction and molecularly imprinted polymers; two polymers, one for rutin and the other for hexavalent chromium (Cr(VI)), were made. Through the use of a diphenylcarbazide solution, Cr (VI) was removed from the polymer quickly. The aluminum solution facilitated the rapid and simple removal of rutin molecules. Since the removal processes need extracting with development of coloration, the solutions might be used simultaneously for the analyte's extraction and detection (Fig. 11.4). An analytical kit that can be 3D printed has been created to detect Cr (VI) in cellphones. The

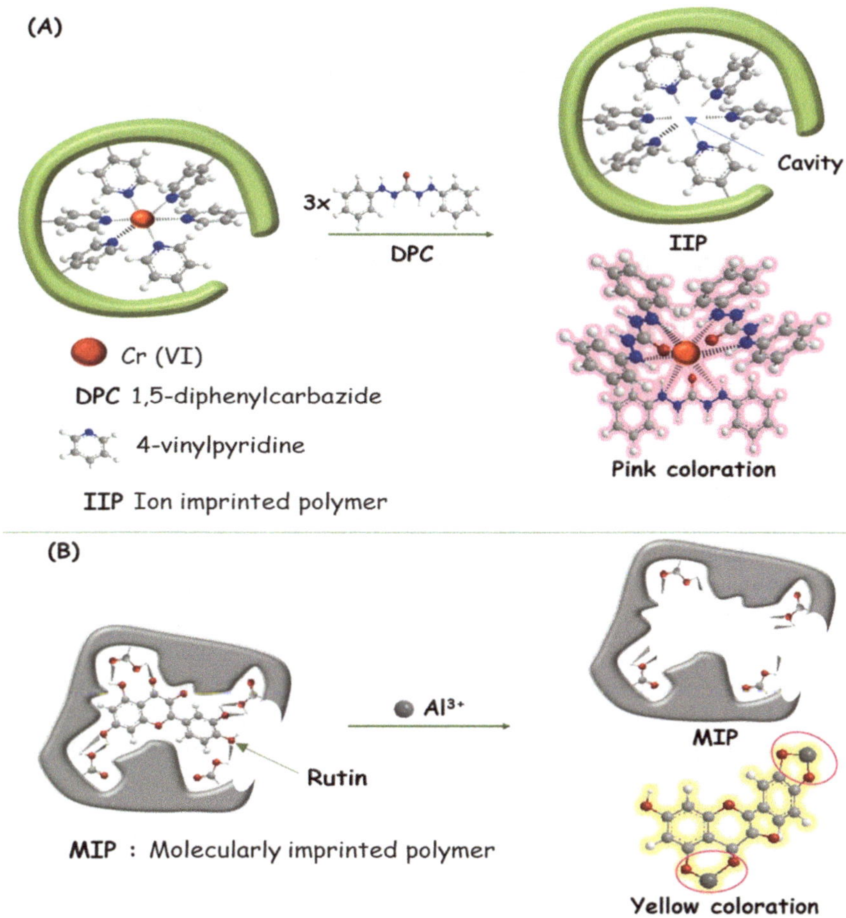

Fig. 11.4 Schematic demonstration for the removal of Cr(VI) from ion imprinted polymer (**a**) and rutin from MIP (**b**). A swift and gentle breaking of the covalent connections between the 4-vinylpyridine and Cr(VI) was made possible by the strong interaction between Cr(VI) and DPC (Scheme 1A). Scheme 1B shows that the interaction between Al(III) and rutin produces a complex with a yellow hue that makes the extraction visible to the bare eye [35]

sorbent in this instrument was the ionic imprinted polymer (IIP) column, and the sorbent was the diphenylcarbazide solution [35].

The microfluidics LOC system has adopted MIPs, another effective separation technique. MIPs can create specific cavities with functional groups that are complementary to the structure of analyte molecules by using conjugation and interaction between templates, cross-linking agents, and functional monomers. This produces highly selective molecular-binding sites. MIPs, aptamers, and antibodies are currently three effective components that can be used with microfluidic LOC devices to form molecule-specific capture [36].

11.4 Conclusion

In this chapter, combinations of molecular imprinting and microfluidic-based chip technologies were given to implement them in different fields. Some studies were mentioned in detail but of course the literature is not limited to them. In Table 11.2, studies about MIP-based chip sensors were given as summarized with their target molecules, fabrication technologies, detection methods, detection limit, and real sample analysis informations.

11.5 Discussion and Future Perspectives

In recent years, biosensing devices have emerged as a promising substitute by combining micro-total analysis systems (TAS) and lab-on-a-chip (LOC) with benchmark designs to provide a starting point for chemical analysis, environmental monitoring, and medical diagnostics in terms of ergonomics, cost-effectiveness, speed, and sensitivity of the diagnostic process [24].

Due primarily to their great selectivity in comparison to other phases, molecularly imprinted polymers (MIPs) have been used in a growing number of analytical methods. As alternatives for natural receptors in sensors and chromatography, MIPs have drawn a lot of interest. This is because molecular imprinting is likely the most universal and economical method for creating artificial receptors. Comparing imprinted polymers to natural receptor molecules reveals numerous significant advantages: In addition to being created for nearly any substance, polymers offer several other advantages as well. They can function with organic solvents, are stable at low and high pH levels, pressure, and temperatures, and have an affinity that is similar to that of natural biomolecules [52].

MIPs, or molecularly imprinted polymers, are renowned for their ability to specifically recognize a large range of various analytes. A substantial obstacle is presented by the intricate immobilization procedure of MIPs onto sensor electrodes. The difficulty of integrating MIP sensors with transducers is one of their limitations; the most straightforward and clever method to get around this problem is electropolymerization, which allows the MIP to be produced in place at an electrode surface by simply adjusting the layer's thickness in relation to the charge transferred.

Table 11.2 Applications of MIP-based chip sensors in the literature

Target molecule	Fabrication technology	Detection method	Detection limit	Real sample analysis	Ref.
Anesthetic propofol	Electropolymerization	Electrical detection	0.1 μg/mL	Human serum	Hong et al. [37]
Carcinoembryonic antigen (CEA)	Electropolymerization	Electrochemical	0.08 ng/mL	Human serum	Carneiro et al. [38]
SARS-CoV-2 spike protein	Surface imprinting strategy	Electrochemical	64 fM	Nasopharyngeal samples	Ayankojo et al. [39]
Carcinoembryonic antigen (CEA)	Surface imprinting technology	Electrochemical	0.32 ng/mL	Human serum	Qi et al. [40]
Domoic acid (DA)	Direct photo-grafting	Surface plasmon resonance	5 μg/L	–	Lotierzo et al. [41]
Glyphosate	Coating	Optical, chemical	0.04 μg/L	Environmental waters	Alberti et al. [42]
Aminoacids	Artificial antibody	Wearable	–	Human serum and sweat	Wang et al. [28]
Carcinogenic embryonic antigen (CEA)	Surface imprinting	Electrochemical	pg CEA per mL	–	Moreira et al. [29]
Amylase	Surface imprinting	Electrochemical	3.0 10^{-4} mg/mL	Saliva	Rebelo et al. [30]
2,4-Dichlorophenoxyacetic acid (2,4-D)	Surface imprinting	Microfluidic	90 nM	Bean sprout	Zhang et al. [34]
SARS-CoV-2 nucleoprotein (ncovNP)	Surface imprinting	Electrochemical	15 fM	Nasopharyngeal swab samples	Raziq et al. [43]
Hepcidin-25	Epitope imprinting	Optical	3.5 fM	Human serum	Cenci et al. [44]
Lysozyme	Coating	Optical	0.015 nM	–	Bakhshpour-Yucel [45]
Zearalenone (ZEN)	Surface imprinting	Mass-based (QCM)	0.30 ng/L	Rice samples	Çapar et al. [46]

(continued)

Table 11.2 (continued)

Target molecule	Fabrication technology	Detection method	Detection limit	Real sample analysis	Ref.
Creatinine	Precipitation polymerization	Electrochemical	50 ppm	Human serum	Prabhu et al. [47]
Salbutamol	Electropolymerization	Electrochemical	100 pM	Swine meat and feed samples	Dechtrirat et al. [48]
2-Furaldehyde	Surface imprinting	Optical	12 μg/L	Milk samples	Alberti et al. [49]
Cortisol	Surface imprinting	Wearable	0.35 nM	Sweat	Mei et al. [50]
Microcystin (MC-RR)	Epitope imprinting	Fluorescent	0.43 μg/L	Real water samples	Li et al. [51]

Making compact devices for clinical diagnostics, environmental management, and the pharmaceutical sectors using this method is particularly appealing [27].

In the future respects, additional focus must be given to the development of MIPs for nanomedicine in the future. Protein corona development must be controlled if protein imprints are used for in vivo targeting, hence some methods must be created and put into practice. By integrating MIPs with inorganic nanoparticles as hybrid MIPs, it appears possible to create materials that combine targeting, imaging, and drug delivery. In the upcoming years, a variety of intriguing combinations will be produced, offering a wide range of patient care options. Specifically, MIP/porous polymers and carbon nanomaterials will mark a major breakthrough in the biotech sensor industry. A further ideal goal that is always being pursued and could soon be realized is artificializing detection systems through the use of MIP in combination with different analytical techniques [53, 54].

By incorporating smart technologies like the Internet of Medical Things (IoMT) and artificial intelligence (AI), the futuristic application of MIP-based biosensors could be further developed. Later, the medical, scientific, and research communities would be able to share the data that was obtained through POC testing of biologic molecules using IoT integrated MIP-based biosensors [55].

The process of commercializing MIPs has been sluggish. Rather than sensors, materials for separation are used in the majority of commercial applications. The largest challenge to the creation of MIP-based materials for sensors, or for chromatographic applications, may be the incidence of nonspecific binding to the templated material. This happens as a result of the polymer's inherent adsorbent properties, which make some interaction—even a slight one—with the target molecule inevitable. Furthermore, a variety of rebinding constants brought forth by nonhomogeneous binding sites complicate the intended application of the polymers. The requirement to separate the target from the as-produced polymer complicates the use of MIPs as sensors. In certain sensing applications, the rebinding sense is hampered by any unextracted target presence. Given the abundance of MIP sensing findings in the chemical literature, it is reasonable to predict that commercial applications of MIPs to sensors will arise shortly [18].

References

1. Ansari S, Masoum S (2019) Molecularly imprinted polymers for capturing and sensing proteins: current progress and future implications. TrAC Trends Anal Chem 114:29–47
2. Kang MS, Cho E, Choi HE, Amri C, Lee JH, Kim KS (2023) Molecularly imprinted polymers (MIPs): emerging biomaterials for cancer theragnostic applications. Biomater Res 27(1):45
3. Akgönüllü S, Kılıç S, Esen C, Denizli A (2023) Molecularly imprinted polymer-based sensors for protein detection. Polymers 15(3):629. https://doi.org/10.3390/polym15030629
4. Hasanah AN, Safitri N, Zulfa A, Neli N, Rahayu D (2021) Factors affecting preparation of molecularly imprinted polymer and methods on finding template-monomer interaction as the key of selective properties of the materials. Molecules 26(18):5612. https://doi.org/10.3390/molecules26185612
5. Li S, Cao S, Whitcombe MJ, Piletsky SA (2014) Size matters: challenges in imprinting macromolecules. Prog Polym Sci 39(1):145–163

6. Sajini T, Mathew B (2021) A brief overview of molecularly imprinted polymers: highlighting computational design, nano, and photo-responsive imprinting. Talanta Open 4:100072
7. Baker ZK, Sardari S (2021) Molecularly imprinted polymer (MIP) applications in natural product studies based on medicinal plant and secondary metabolite analysis. Iran Biomed J 25(2):68
8. Kamaruzaman S, Nasir NM, Faudzi SMM, Yahaya N, Hanapi NSM, Ibrahim WNW (2021) Solid-phase extraction of active compounds from natural products by molecularly imprinted polymers: synthesis and extraction parameters. Polymers 13, no. 21:3780. https://doi.org/10.3390/polym13213780
9. He JX, Pan HY, Xu L, Tang RY (2021) Application of molecularly imprinted polymers for the separation and detection of aflatoxin. J Chem Res 45(5–6):400–410
10. Shao Y, Duan J, Wang M, Cao J, She Y, Cao Z, Li G, Jin F, Wang J, Abd El-Aty AM (2023) Application of molecularly imprinted electrochemical biomimetic sensors for detecting small molecule food contaminants. Polymers 15(1):187. https://doi.org/10.3390/polym15010187
11. Yarman A, Kurbanoglu S, Zebger I, Scheller FW (2021) Simple and robust: the claims of protein sensing by molecularly imprinted polymers. Sensors Actuators B Chem 330:129369
12. Bedwell TS, Whitcombe MJ (2016) Analytical applications of MIPs in diagnostic assays: future perspectives. Anal Bioanal Chem 408:1735–1751
13. Ye K, Xu S, Zhou Q, Wang S, Xu Z, Liu Z (2023) Advances in molecular imprinting technology for the extraction and detection of quercetin in plants. Polymers 15(9):2107
14. Roy E, Patra S, Tiwari A, Madhuri R, Sharma PK (2017) Introduction of selectivity and specificity to graphene using an inimitable combination of molecular imprinting and nanotechnology. Biosens Bioelectron 89:234–248
15. Chen JL (2017) Determination of tetracycline using imprinted polymethacrylates along with fluorescent CdTe quantum dots on plastic substrates. Microchim Acta 184:1335–1343. https://doi.org/10.1007/s00604-017-2118-2
16. Pan J, Chen W, Ma Y, Pan G (2018) Molecularly imprinted polymers as receptor mimics for selective cell recognition. Chem Soc Rev 47(15):5574–5587
17. Senay RH, Ulucan F, Kuru CI, Akgol S (2018) Ch. Biosensors designed by molecular imprinted polymers, smart biosensing technologies: past, present and future. In: Smart biosensing technologies: past, present and future. Hacettepe University, pp 25–68
18. BelBruno JJ (2019) Molecular imprinted polymers. Chem Rev 119(1):94–119. https://doi.org/10.1021/acs.chemrev.8b00171
19. Leibl N, Haupt K, Gonzato C, Duma L (2021) Molecularly imprinted polymers for chemical sensing: a tutorial review. Chemosensors 9(6):123. https://doi.org/10.3390/chemosensors9060123
20. Schmidt RH, Mosbach K, Haupt K (2004) A simple method for spin-coating molecularly imprinted polymer films of controlled thickness and porosity. Adv Mater 16(8):719–722
21. Mazzotta E, Turco A, Chianella I, Guerreiro A, Piletsky SA, Malitesta C (2016) Solid-phase synthesis of electroactive nanoparticles of molecularly imprinted polymers. A novel platform for indirect electrochemical sensing applications. Sensors Actuators B Chem 229:174–180
22. Jambhulkar S, Ravichandran D, Zhu Y, Thippanna V, Ramanathan A, Patil D et al (2023) Nanoparticle assembly: from self-organization to controlled micropatterning for enhanced functionalities. Small 20:2306394
23. Pilvenyte G, Ratautaite V, Boguzaite R, Ramanavicius S, Chen C-F, Viter R, Ramanavicius A (2023) Molecularly imprinted polymer-based electrochemical sensors for the diagnosis of infectious diseases. Biosensors 13(6):620. https://doi.org/10.3390/bios13060620
24. Karasu T, Özgür E, Uzun L (2023) MIP-on-a-chip: artificial receptors on microfluidic platforms for biomedical applications. J Pharm Biomed Anal 226:115257
25. Akgönüllü S, Denizli A (2022) Sensors & lab-on-a-chipmolecular imprinting-based sensors: lab-on-chip integration and biomedical applications. J Pharm Biomed Anal 115213
26. Wang L, Pagett M, Zhang W (2023) Molecularly imprinted polymer (MIP) based electrochemical sensors and their recent advances in health applications. Sensors Actuators Rep 5:100153
27. Selvolini G, Marrazza G (2017) MIP-based sensors: promising new tools for cancer biomarker determination. Sensors 17(4):718

28. Wang M, Yang Y, Min J, Song Y, Tu J, Mukasa D et al (2022) A wearable electrochemical biosensor for the monitoring of metabolites and nutrients. Nat Biomed Eng 6(11):1225–1235
29. Moreira FT, Ferreira MJM, Puga JR, Sales MGF (2016) Screen-printed electrodes are produced by printed-circuit board technology. Application to cancer biomarker detection using a plastic antibody as sensing material. Sensors Actuators B Chem 223:927–935
30. Rebelo TS, Miranda IM, Brandão AT, Sousa LI, Ribeiro JA, Silva AF, Pereira CM (2021) A disposable saliva electrochemical MIP-based biosensor for detection of the stress biomarker α-amylase in point-of-care applications. Electrochem 2(3):427–438
31. Goossens J, Vandenryt T, Lowdon JW, Van Grinsven B, Eersels K, Deferme W, Thoelen R (2023) Design of a lab-on-chip cartridge for the optical detection of small molecules based on dye-displacement MIPs. IEEE Trans Instrum Meas 72:1
32. Keçili R, Hussain CG, Hussain CM (2023) Emerging trends in green carbon dots coated with molecularly imprinted polymers for sensor platforms. TrAC Trends Anal Chem 166:117205
33. Li S, Pang C, Ma X, Zhang Y, Xu Z, Li J et al (2021) Microfluidic paper-based chip for parathion-methyl detection based on a double catalytic amplification strategy. Microchim Acta 188:1–8
34. Zhang Z, Ma X, Jia M, Li B, Rong J, Yang X (2019) Deposition of CdTe quantum dots on microfluidic paper chips for rapid fluorescence detection of pesticide 2, 4-D. Analyst 144(4):1282–1291
35. Karrat A, García-Guzmán JJ, Palacios-Santander JM, Amine A, Cubillana-Aguilera L (2023) New strategies for the removal of the template from the ion and molecularly imprinted polymers: application to the fast and on-site Cr (VI) detection with a smartphone. Sensors Actuators B Chem 386:133751
36. Guo L, Feng J, Fang Z, Xu J, Lu X (2015) Application of microfluidic "lab-on-a-chip" for the detection of mycotoxins in foods. Trends Food Sci Technol 46(2):252–263
37. Hong CC, Lin CC, Hong CL, Lin ZX, Chung MH, Hsieh PW (2016) Handheld analyzer with on-chip molecularly-imprinted biosensors for electrical detection of propofol in plasma samples. Biosens Bioelectron 86:623–629
38. Carneiro LP, Ferreira NS, Tavares AP, Pinto AM, Mendes A, Sales MGF (2021) A passive direct methanol fuel cell as transducer of an electrochemical sensor, applied to the detection of carcinoembryonic antigen. Biosens Bioelectron 175:112877
39. Ayankojo AG, Boroznjak R, Reut J, Öpik A, Syritski V (2022) Molecularly imprinted polymer based electrochemical sensor for quantitative detection of SARS-CoV-2 spike protein. Sensors Actuators B Chem 353:131160
40. Qi J, Li B, Zhou N, Wang X, Deng D, Luo L, Chen L (2019) The strategy of antibody-free biomarker analysis by in-situ synthesized molecularly imprinted polymers on movable valve paper-based device. Biosens Bioelectron 142:111533
41. Lotierzo M, Henry OYF, Piletsky S, Tothill I, Cullen D, Kania M et al (2004) Surface plasmon resonance sensor for domoic acid based on grafted imprinted polymer. Biosens Bioelectron 20(2):145–152
42. Alberti G, Spina S, Arcadio F, Pesavento M, De Maria L, Cennamo N et al (2023) MIP-assisted 3-hole POF chip faced with SPR-POF sensor for glyphosate detection. Chemosensors 11(7):414
43. Raziq A, Kidakova A, Boroznjak R, Reut J, Öpik A, Syritski V (2021) Development of a portable MIP-based electrochemical sensor for detection of SARS-CoV-2 antigen. Biosens Bioelectron 178:113029. https://doi.org/10.1016/j.bios.2021.113029. Epub 2021 Jan 23
44. Cenci L, Andreetto E, Vestri A et al (2015) Surface plasmon resonance based on molecularly imprinted nanoparticles for the picomolar detection of the iron regulating hormone Hepcidin-25. J Nanobiotechnol 13:51. https://doi.org/10.1186/s12951-015-0115-3
45. Bakhshpour-Yucel M (2023) SPR-based sensing of lysozyme using Lyz-MIP-modified graphene oxide surfaces. Chem Pap 77:2671–2678. https://doi.org/10.1007/s11696-022-02656-5
46. Çapar N, Polat İ, Yola BB, Atar N, Yola ML (2023) A novel molecular imprinted QCM sensor based on MoS2NPs-MWCNT nanocomposite for zearalenone determination. Microchim Acta 190(7):262

47. Prabhu SN, Gooneratne CP, Mukhopadhyay SC (2021) Development of MEMS sensor for detection of creatinine using MIP based approach–a tutorial paper. IEEE Sensors J 21(20):22170–22181
48. Dechtrirat D, Sookcharoenpinyo B, Prajongtat P, Sriprachuabwong C, Sanguankiat A, Tuantranont A, Hannongbua S (2018) An electrochemical MIP sensor for selective detection of salbutamol based on a graphene/PEDOT: PSS modified screen printed carbon electrode. RSC Adv 8(1):206–212
49. Alberti G, Arcadio F, Pesavento M, Marzano C, Zeni L, Zeid NA, Cennamo N (2022) Detection of 2-furaldehyde in milk by MIP-based POF chips combined with an SPR-POF sensor. Sensors 22(21):8289
50. Mei X, Yang J, Yu X, Peng Z, Zhang G, Li Y (2023) Wearable molecularly imprinted electrochemical sensor with integrated nanofiber-based microfluidic chip for in situ monitoring of cortisol in sweat. Sensors Actuators B Chem 381:133451
51. Li B, Qi J, Liu F et al (2023) Molecular imprinting-based indirect fluorescence detection strategy implemented on paper chip for non-fluorescent microcystin. Nat Commun 14:6553. https://doi.org/10.1038/s41467-023-42244-z
52. Uludag Y, Esen E, Kokturk G, Ozer H, Muhammad T, Olcer Z et al (2016) Lab-on-a-chip based biosensor for the real-time detection of aflatoxin. Talanta 160:381–388
53. Garnier M, Sabbah M, Ménager C, Griffete N (2021) Hybrid molecularly imprinted polymers: the future of nanomedicine? Nanomaterials (Basel) 11(11):3091. https://doi.org/10.3390/nano11113091
54. Gao M, Gao Y, Chen G, Huang X, Xu X, Lv J et al (2020) Recent advances and future trends in the detection of contaminants by molecularly imprinted polymers in food samples. Front Chem 8:616326
55. Yadav AK, Verma D, Dalal N, Kumar A, Solanki PR (2022) Molecularly imprinted polymer-based nanodiagnostics for clinically pertinent bacteria and virus detection for future pandemics. Biosens Bioelectron 12:100257

Importance of MIPs as Artificial Antibodies for the Detection of Bacteria and Viruses

12

Abdulhakim Umar Toro, Abdulaziz Umar Kurya, Parveen Bansal, and Sudheesh K. Shukla

Abstract

In light of the persistent threat of pandemics, researchers are diligently advancing innovative technologies to diagnose and treat infections caused by a variety of bacteria and viruses. Molecularly imprinted polymers (MIPs) stand out as synthetic polymeric materials featuring specific recognition sites that mirror template molecules in shape, size, and functional groups. Their stability, specificity, low cost, ease of preparation, and potential for miniaturization position them as promising alternatives to natural bioreceptors in sensor technology. Notably, imprinting of bacteria and viruses have garnered significant attention due to the high selectivity, simplicity, rapidity, excellent stability, low cost, and eco-friendliness of the process. This comprehensive review aims to assess key accomplishments in MIP-based sensors, encompassing electrochemical, optical,

A. U. Toro (✉)
Department of Biomedical Engineering, Shobhit Institute of Engineering and Technology (Deemed to-be-University), Modipuram, Meerut, Uttar Pradesh, India

A. U. Kurya
Department of Biotechnology, School of Life and Allied Health Sciences, Glocal University, Saharanpur, Uttar Pradesh, India

P. Bansal
University Centre of Excellence in Research, Baba Farid University of Health Sciences, Faridkot, Punjab, India

S. K. Shukla

Centre for Nanoscience and Nanotechnology, School of Chemical Engineering and Physical Sciences, Lovely Professional University, Phagwara, Punjab, India

Department of Applied Chemistry, University of Johannesburg, Johannesburg, South Africa

S. Patra et al. (eds.), *Molecularly Imprinted Polymers: Path to Artificial Antibodies*, https://doi.org/10.1007/978-981-97-4379-7_12

fluorescence, and piezoelectric approaches, particularly in detecting and quantifying host proteins, gene transcripts, and cellular processes relevant to the identification of bacteria and viruses. Additionally, the review outlines pivotal advancements in MIP-based sensor technologies and delineates future trends, underscoring their potential applications in detecting biomarkers associated with bacteria and viruses.

Keywords

Molecularly imprinted polymer · Artificial antibodies · MIPs · Virus · Bacteria Infection · Point-of-care · Biosensor

12.1 Introduction

The domain of scientific inquiry has been captivated by the promising discipline of molecular imprinting. This innovative technique has swiftly ascended as a beacon of hope, promising the creation of artificial entities akin to natural receptors, known as molecularly imprinted polymers (MIPs). Distinguished from their organic counterparts, MIPs showcase unparalleled advantages, boasting traits of economic feasibility, steadfast stability, and remarkable adaptability. Consequently, these bespoke synthetic marvels have found their indispensable niche across a plethora of domains, spanning from intricate separation methodologies and meticulous sample pretreatment to sensors, catalysis, and drug delivery [1]. To ensure complete polymerization, using the right proportions of template molecules, functional monomers, cross-linkers, and initiators is crucial. Once polymerization is complete, the resulting matrix forms specific recognition cavities perfectly suited for the template molecules as depicted in Fig. 12.1.

Several cost-effective biosensing techniques have been engineered to detect bacteria, utilizing antibodies or enzymes tethered to electrochemical or optical transducers. These methods, geared toward point-of-care applications, offer affordability and user-friendliness. However, they demand stringent regulation of working conditions such as pH, temperature, and pressure due to the inherent characteristics of the biological recognition elements employed [2]. The remarkable specificity exhibited by molecularly imprinted polymers (MIPs) toward their target analytes renders them highly attractive. Through the molecular imprinting technique, specific molecular recognition sites are meticulously crafted, founded on the principle of complementarity between these imprinted sites and the analyte. Consequently, MIPs possess the unique ability to selectively bind the desired

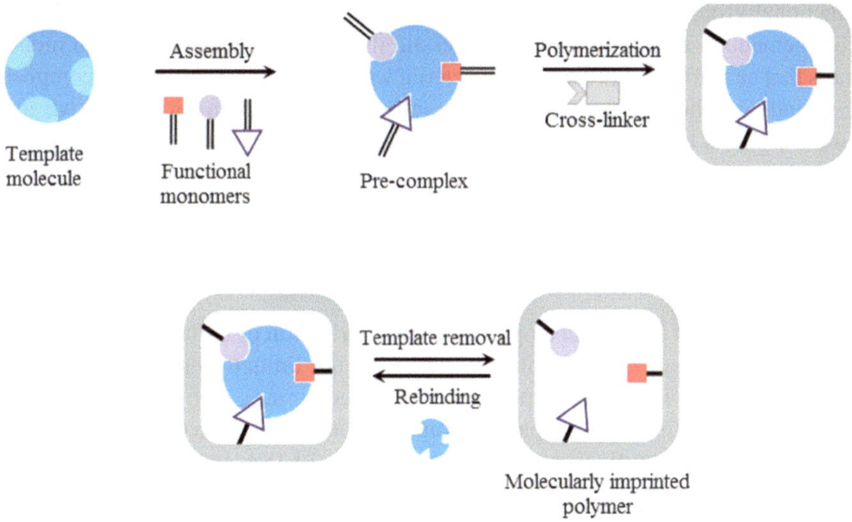

Fig. 12.1 Elucidate the schematic representation of the foundational principles inherent in the preparation of molecularly imprinted polymers

analytes, which in turn act as templates guiding their development [3]. Liustrovaite et al. explored the challenges associated with imprinting entire bacterial cells within polymers, a task rendered exceptionally difficult due to their size. Numerous studies have investigated the applicability of MIP-based sensors for bacterial detection. Two crucial factors influencing bacterial recognition need consideration:

(i) Differentiation based on bacterial cell shape (e.g., round or rod-shaped bacteria like *Staphylococcus aureus* or *Escherichia coli*) and,
(ii) Chemical recognition, involving interactions between functional groups in polymers and those localized on the cell surface, such as cis-diol groups of glycan chains.

Addressing these emerging issues, Piletsky et al. raised pertinent questions regarding the selection of materials suitable for electrode modification using MIP-based layers [1]. Creating artificial receptors for bacteria poses a greater challenge compared to designing receptors for small or even large molecules like proteins. This complexity stems from the substantial size of bacteria and the multitude of functional groups present in their cell walls, making it arduous to create selective complementary recognition sites. Successful imprinting requires functional monomers that can strongly interact with bacterial walls. However, these walls' composition and morphology can vary significantly, as observed in Gram-positive and Gram-negative bacteria, or even within the same bacteria under different environmental conditions [3].

12.2 MIPs Design and Application

Various polymers have been employed in the fabrication of molecularly imprinted polymer (MIP)-based biosensors, including but not limited to polypyrrole, polyscopoletin, polythiophene, poly(m-phenylenediamine), poly(o-phenylenediamine), polyaniline, polydopamine, among others. MIP-based assays have demonstrated competitiveness with assays, such as ELISA, in terms of specificity, sensitivity, and compatibility with technologies, such as lateral flow lab-on-a-chip, the commercial potential of these assays has not yet been completely realized [4].

12.3 MIP-Based Electrochemical Sensors for Bacteria and Virus Detection

Surface imprinting of whole organism is attributed to the unique characteristics of microbes, which can often be identified based on their morphology and surface properties. By leveraging these distinctive features, the whole virus surface imprinting technique offers a straightforward means of virus detection [5]. Viruses can be detected by a whole virus, to detect different types of viruses such as influenza virus, hepatitis A virus, hepatitis B virus, apple stem pitting virus, tobacco mosaic virus, and HIV virus [6].

The identification of specific biomarkers plays a pivotal role in the detection of various bacterial and viral entities. These include, but are not limited to, biomarkers associated with viral infections such as influenza virus neuraminidase inhibitor peramivir, HIV-gp41, HIV-gp120, Dengue-NS1, SARS-CoV-2 antigen, SARS-CoV-2 spike protein, SARS-CoV-2 nucleocapsid protein, C-reactive protein, hepatitis C virus core antigen, and E2 protein. Additionally, biomarkers indicative of nosocomial infections encompasses *K. pneumoniae*, *P. aeruginosa*, *L. monocytogenes*, and *A. baumannii* [7]. Table 12.1 summarizes electrochemical sensors based on molecularly imprinted polymers for the detection of bacteria and viruses.

Table 12.1 Electrochemical sensors based on molecularly imprinted polymers for the detection of bacteria and viruses

Microbe/biomarker	Electrochemical analysis methods	Electrodes	Sensitivity	Polymers	Reference
SARS-CoV-2 nucleocapsid protein	DPV, EIS	SPCE	LOD 3.0 fM	P-Arg, gold/graphene nanohybrids	[8]
HCV surface protein E2	DPV	SPE	LOD 0.46 pg/mL	PmPD	[9]
K. pneumoniae	CV, DPV	ITO	LOD 1.352 CFU/mL	Ppy	[10]
L. monocytogenes	DPV, EIS	GCE	LOD 2 CFU/mL	Polythionine, MXenes nanoribbon (Ti3C2TxR)	[11]
SARS-CoV-2	CV, SWV	Gold-based thin-film metal electrodes (Au-TFME)	15 CFU mL^{-1}	ncovS1-MIP film	[7]
E. coli O157:H7	ECL	Nitrogen-doped graphene quantum dots (N-GQDs)	8 CFU mL^{-1}	PDA-SIPs	[12]
Zika virus	CV/EIS	SPGE//[Fe(CN)6]3−/4−	10–3 PFU	Prepolymer-GO composites under UV light	[13]
Bacillus cereus	CV	CPE/[Fe(CN)6]3−/4−	102 CFU mL^{-1}	Pyrrole	[14]
Acinetobacter baumannii	CV, EIS, DPV	GCE	LOD (CFU/mL; LR 102–107 CFU/mL	Polydopamine	[15]
Pseudomonas aeruginosa	CV, EIS, DPV	GCE	LOD 1 CFU/mL; LR 10–107 CFU/mL	Polydopamine	[16]
K. pneumoniae	CV	AuSPE	LOD 0.012 CFU/mL; LOQ 1.61 CFU/mL; LR 101–105 CFU/mL	Poly(MAM:AAM:NVP), graphene oxide	[17]

12.4 MIP-Based Optical Sensors for Bacteria and Virus Detection

In recent decades, the global landscape of infectious diseases has prompted a paradigm shift in the development of advanced sensing technologies for the rapid and precise detection of pathogenic microorganisms. Among the various approaches, molecularly imprinted polymer (MIP)-based optical sensors have emerged as a pioneering strategy, offering unparalleled selectivity and sensitivity in the detection of bacteria and viruses. This introduction provides an in-depth exploration of the rationale behind the utilization of MIPs in optical sensor platforms, the significance of addressing bacterial and viral infections, and the current state of research in this evolving field [18].

Molecularly imprinted polymers (MIPs) represent a class of synthetic materials designed with a remarkable capability to recognize and selectively bind to specific target molecules. The synthesis of MIPs involves a template polymerization process where a polymer is formed around a template molecule, creating cavities with molecularly imprinted sites that are structurally complementary to the template. Upon removal of the template, these cavities become highly selective recognition sites for the target molecule. The versatility of MIPs in terms of tailoring their chemical composition, size, and morphology has led to their integration into various sensing platforms [19].

12.5 Optical Sensors: Principles and Applications

Optical sensors, harnessing the principles of light-matter interactions, have garnered significant attention in the field of biosensing. The fundamental basis lies in the ability of these sensors to detect changes in optical properties, such as absorbance, fluorescence, or refractive index, upon the interaction with target analytes. The advantages of optical sensors include real-time monitoring, noninvasiveness, and high sensitivity. MIPs, when incorporated into optical sensor designs, enhance these advantages by providing a molecularly imprinted recognition layer with a high affinity for specific target analytes [20].

12.6 Rationale for Bacteria and Virus Detection

The detection of bacteria and viruses is of paramount importance in healthcare, environmental monitoring, and food safety. Rapid and accurate identification of pathogenic microorganisms is critical for timely clinical intervention, outbreak control, and ensuring the safety of food and water supplies. Conventional methods for bacterial and viral detection often involve time-consuming culture techniques or expensive molecular assays. The development of MIP-based optical sensors addresses the need for rapid, cost-effective, and highly specific detection methods, holding tremendous potential for revolutionizing disease diagnostics and biosafety measures [21].

12.7 Significance of MIP-Based Optical Sensors

The integration of MIPs into optical sensors for bacteria and virus detection presents a multifaceted approach to overcome the limitations of traditional sensing methods. MIPs offer inherent stability, low cost, and the ability to operate in diverse environmental conditions. The molecular imprinting process imparts a high degree of specificity, enabling the discrimination of closely related bacterial strains or viral species. Moreover, the tailorable nature of MIPs allows for the customization of recognition sites to target specific biomarkers associated with infectious agents [22].

12.8 MIP-Based Optical Sensors for Detection of Bacterial and Viral Infection

COVID-19: Coronaviruses are a group of enveloped, single-stranded RNA viruses belonging to the Coronaviridae family. These viruses are known for their distinctive crown-like appearance under a microscope, which is due to the presence of spike proteins on their surface. Coronaviruses can infect a variety of animals, including bats, camels, and birds, and some of them have the potential to jump from animals to humans, leading to zoonotic infections [2]. Utilizing the localized surface plasmon resonance (LSPR) principle, Funari et al. have innovatively crafted an opto-microfluidic sensing platform featuring gold nanospikes created through electrodeposition. This platform demonstrates efficacy in identifying and quantifying antibodies specifically targeting the SARS-CoV-2 spike protein. This development presents a potential point-of-care testing solution that complements conventional serological assays, promising a more accessible, cost-effective, and rapid approach to quantitative diagnostics for SARS-CoV-2 [23]. Considering the urgent demand for swift and uncomplicated SARS-CoV-2 detection techniques, a novel SPR-POF-MIP sensor has been developed. This sensor enables the direct identification of SARS-CoV-2 virions in liquid solutions. It incorporates a molecularly imprinted polymer nano-layer, precisely tailored to recognize the S1 subunit of the SARS-CoV-2 spike protein, as illustrated in Fig. 12.2 [24].

Bacterial Detection: Electrochemiluminescence biosensor was developed for *Escherichia coli* quantitative detection based on polydopamine surface imprinted polymer and nitrogen-doped graphene quantum dots. Interestingly, the sensing system had high selectivity to the target analyte and provided new opportunities and speed up disease diagnosis, treatment, and prevention with pathogen [12].

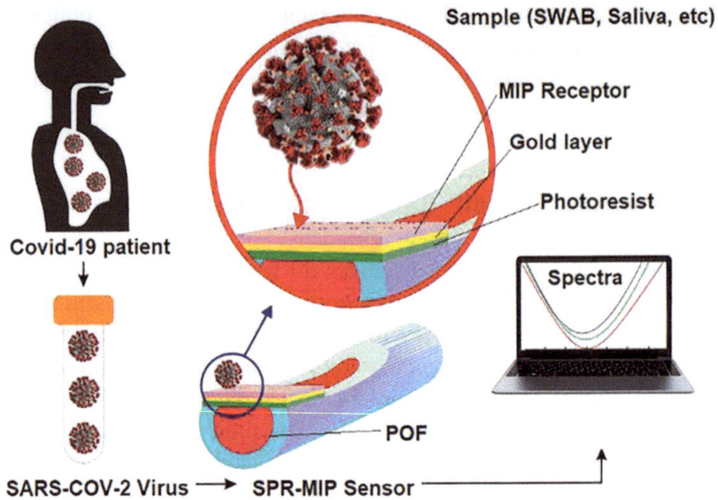

Fig. 12.2 Comprehensive delineation of a sensor architecture designed for the detection of SARS-CoV-2 across diverse matrices

12.9 MIP-Based Fluorescence Sensors for Bacteria and Virus Detection

In the dynamic landscape of biosensor development, the fusion of fluorescence technology with molecularly imprinted polymers (MIPs) has emerged as a formidable frontier, offering unprecedented opportunities for highly specific and sensitive detection of target analytes. This synergy capitalizes on the unique advantages of both fluorescence-based detection and MIPs, paving the way for innovative solutions in diverse fields such as medical diagnostics, environmental monitoring, and pharmaceutical research [25].

12.10 Fundamental Principles of Fluorescence Biosensors

Fluorescence biosensors harness the intrinsic properties of fluorophores to facilitate the detection of specific biological interactions. At the core of their functionality lies the ability of certain molecules to absorb light at a particular wavelength and subsequently emit light at a longer wavelength. This phenomenon, known as fluorescence, serves as a robust basis for biosensor design. The exquisite sensitivity of fluorescence detection, coupled with real-time monitoring capabilities and minimal background interference, has propelled fluorescence biosensors to the forefront of analytical techniques [26].

12.11 Molecularly Imprinted Polymers (MIPs): Design, Synthesis, and Fusion with Fluorescence Sensors

MIPs, on the other hand, represent a cutting-edge approach to molecular recognition. The process of molecular imprinting involves polymerization in the presence of a template molecule, resulting in the creation of cavities within the polymer matrix that mimic the size, shape, and chemical functionality of the template. The subsequent removal of the template leaves behind highly selective binding sites, rendering MIPs tailor made for the recognition of specific target molecules. This inherent selectivity, along with the stability of MIPs under harsh conditions, positions them as ideal candidates for integration into biosensing platforms [27].

The marriage of fluorescence technology with MIPs introduces a novel dimension to biosensor design. MIPs can be strategically engineered to incorporate fluorophores directly or be paired with fluorescent tags. This strategic integration endows the biosensor with the ability to undergo a detectable change in fluorescence intensity upon the binding of the target analyte. The resulting sensor exhibits heightened sensitivity and selectivity, providing a powerful tool for the discrimination of closely related molecules and the accurate quantification of target analytes [28].

12.12 MIP-Based Fluorescence Sensors for Detection of Relevant Diseases

Detection of Viruses: A specialized fluorescent sensor, utilizing virus-molecular imprinted polymers (virus-MIPs), has been meticulously engineered for the precise recognition and exceptionally sensitive detection of the Japanese encephalitis virus (JEV). This innovative biosensor incorporates virus-MIPs, strategically affixed to the surface of silica microspheres that have been modified with a fluorescent dye, namely pyrene-1-carboxaldehyde (PC). The fluorescence intensity of PC undergoes enhancement through the mechanism of fluorescence resonance energy transfer (FRET), wherein the virus serves as the energy donor and PC as the energy acceptor. This pioneering technology is poised to redefine virus detection methodologies, owing to its heightened sensitivity, streamlined operational simplicity, remarkable stability, and cost-effectiveness. The integration of virus-MIPs with the FRET-based fluorescent system presents a novel and robust approach, promising significant advancements in the realm of virus detection [6].

As depicted in Fig. 12.3, by incorporating a fluorescence source whose fluorescence properties alter upon target recognition during the conventional molecularly imprinted polymers (MIPs) fabrication process, the envisioned molecular imprinting-based fluorescence sensors (MI-FL sensors) adeptly fulfil the need for precise and sensitive detection of target substances within complex samples. These sensors exhibit distinct fluorescence signatures or even discernible color changes upon binding to specific analytes, obviating the necessity for additional time-consuming extraction of templates from the MIPs framework [29].

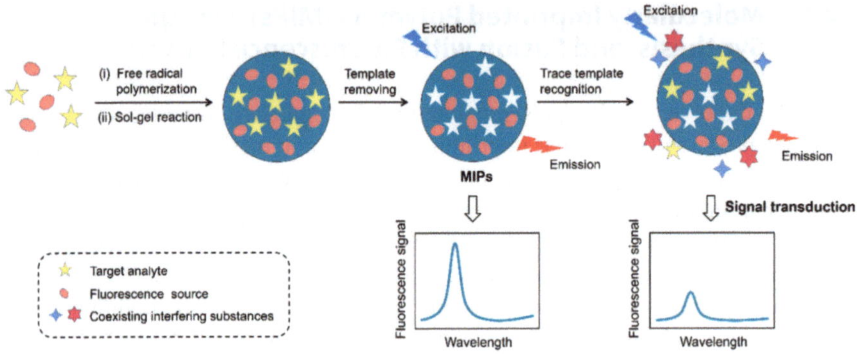

Fig. 12.3 Refinement of the preparation procedure for molecularly imprinted fluorescence sensors, involving the incorporation of fluorescent materials into the matrix of molecularly imprinted polymers (MIPs)

Detection of Bacteria: Bacterial infections persist as a significant contributor to global mortality and illness. The majority of contagious diseases stem from a limited set of pathogens—approximately 20 out of the 1400 human pathogens identified. While ongoing research emphasizes vaccines, novel antibacterial treatments, and pharmaceuticals, an additional approach to curb infectious outbreaks involves prioritizing the establishment and enhancement of early warning systems. These systems would indicate the presence of pathogens in both qualitative and quantitative terms. Given that toxin-contaminated food and beverages pose a considerable health risk and wield substantial socioeconomic impact worldwide, the detection of infectious bacteria remains a formidable scientific challenge and an urgent practical concern [30].

A highly sensitive fluorescence nano-biosensors were designed for detection of Shigella species. For achieved this purpose, DNA probes and gold nanoparticles (AuNPs) were designed and synthesized, respectively. Then, two DNA probes as signal reporter were immobilized on surface of AuNPs. On the other hand, iron nanoparticles (MNPs) were synthesized and modified with sulfosuccinimidyl 4-nmaleimidomethyl cyclohexane-1-carboxylate (SMCC). The third DNA probe was immobilized on surface of MNPs for separation of target DNA. The MNP-DNA probe and DNA probe-AuNP-fluorescence DNA probe were added to target DNA. The MNP-DNA probe-target DNA-DNA probe-AuNP-fluorescence DNA probe complex was isolated by a magnet. The specificity and sensitivity of the designed fluorescent DNA biosensor were determined which were good enough for the precise detection of Spa gene in four *Shigella* spp. at low concentrations bacteria (CFU mL^{-1}). Therefore, the current method will provide a highly sensitive detection method for high-risk pathogens in the future [31].

Liu et al. have devised and synthesized a fluorescent conjugated polymer featuring a phenylboronic acid group. This polymer is capable of covalently binding with cis-diol-containing compounds, such as sialic acid (SA), through the formation of a cyclic ester. The resulting imprinted nanoparticles (NPs) exhibit a consistent size of approximately 30 nm and display a more robust fluorescence compared to the unmodified polymer NPs. Through a cell imaging assay, it was clearly demonstrated that SA-imprinted NPs selectively bind with SA that is over-expressed in DU 145 cancer cells. The favorable characteristics of these SA-imprinted NPs, including their small and uniform size, intense fluorescence, and low cytotoxicity, suggest their promising potential as an effective fluorescent probe for targeted cancer cell imaging [32].

12.13 MIP-Based Electrochemical Sensors for Bacteria and Virus Detection

In the relentless pursuit of advancing diagnostics for infectious diseases, the fusion of molecular imprinting technology with electrochemical sensing has emerged as a transformative paradigm. This chapter embarks on a comprehensive exploration of molecularly imprinted polymer-based electrochemical sensors specifically tailored for the detection of bacterial and viral infections. This marriage of molecular specificity and electrochemical precision holds promise not only for the rapid identification of pathogens but also for addressing the pressing need for sensitive, selective, and real-time diagnostic tools in the realm of infectious diseases [7].

The urgency of combatting bacterial and viral infections underscores the limitations of traditional diagnostic methods, often marred by delays, complexity, and a compromise between sensitivity and specificity. Against this backdrop, molecularly imprinted polymer-based electrochemical sensors emerge as innovative solutions that strive to revolutionize infectious disease diagnostics. This chapter navigates the intricate landscape where molecularly imprinted polymers, designed to selectively recognize pathogens of interest, seamlessly integrate with electrochemical transduction methods to enable swift and reliable detection [33]. A comprehensive outline is illustrated in Fig. 12.4 that illustrate the assembly process of voltametric sensors and biosensors that necessitate direct bacterial involvement for detection,

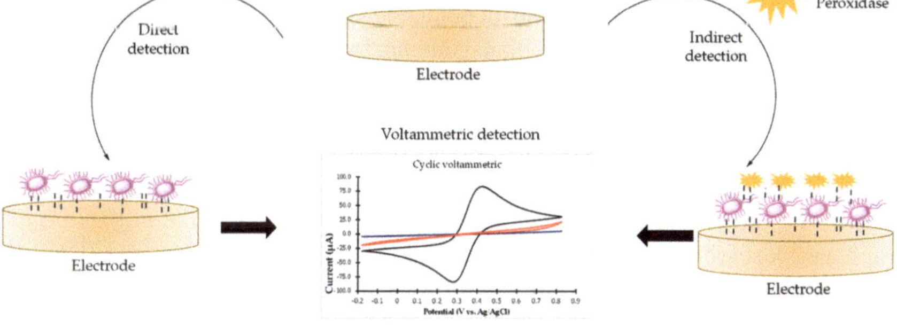

Fig. 12.4 Comprehensive schematic for the development of biosensors employing bacteria directly in the context of voltametric detection

with a pivotal focus on electrode modification. Various materials, such as nanoparticles, carbon-based substances, molecularly imprinted polymers, antibodies, and bacteriophages, have been utilized in electrode construction [34].

12.14 MIP-Based Electrochemical Sensors for the Detection of Viral Infection

Molecularly imprinted polymers (MIPs) offer a means to detect and identify infectious diseases through imprinting with entire viruses or specific proteins, known as biomarkers. The straightforward detection of viruses can be accomplished through surface imprinting of the entire virus, capitalizing on the ease of recognition based on their distinct morphology and surface characteristics [5].

SARS-CoV-2: The emergence of severe acute respiratory syndrome coronavirus-2 (SARS-CoV-2) in 2019 ignited the global COVID-19 pandemic. Despite its recent appearance, this novel virus has garnered significant scrutiny from the scientific community. Over the past few years, there has been a noteworthy surge in research activity, with an escalating number of studies leveraging molecularly imprinted polymers (MIPs) for the purpose of detecting SARS-CoV-2. This heightened interest underscores the urgent need for innovative and effective diagnostic approaches in the ongoing battle against the pandemic [35–37].

In a recent advancement, a compact electrochemical sensor has been created, incorporating a molecularly imprinted polymer (ncovNP-MIP) as an artificial recognition element designed for the specific detection of the SARS-CoV-2 antigen (ncovNP). The synthesis conditions for ncovNP-MIP were meticulously optimized, showcasing the sensor's adeptness in selectively re-binding to the ncovNP. Encouragingly, the validation of the sensor's performance using clinical samples from patients' nasopharynx swabs yielded promising results. This affirmed the sensor's efficacy in detecting ncovNP within intricate biological matrices, signifying its potential for practical applications in real-world diagnostic scenarios [33].

HIV: The introduction of molecularly imprinted polymers (MIPs) marks a significant leap forward in the realm of electrochemical sensors, providing an exceptionally precise and sensitive platform for detecting diverse biomolecules, including viruses like the human immunodeficiency virus (HIV). The global health challenge posed by HIV demands swift and dependable diagnostic instruments to enable timely intervention and effective management. In this regard, MIP-based electrochemical sensors have emerged as promising contenders, demonstrating their capacity to potentially transform HIV detection methodologies [38].

Ma et al. innovatively engineered an electrochemical sensor based on molecularly imprinted polymers (MIPs) for the detection of the variable-format conformations of the protein gp120. The sensor's construction involved the utilization of a graphene-like carbon nanofragment (CNF) and bismuth oxide composites (CNF-Bi) as decorating materials, with the small-molecule entry inhibitor NBD-556 and gp120 conjugates NBD-556@gp120 serving as the template in place of gp120. Pyrrole was employed as the electropolymerization monomer. The MIP sensor exhibited enhanced recognition capabilities, a broader linear range, and a lower limit of detection (LOD), surpassing current techniques for gp120 detection significantly. This methodology not only expands the application scope of molecular imprinting technology by providing flexibility in the selection of template molecules but also contributes to advancing the molecularly imprinted electrochemical sensor's utility, particularly in the early diagnosis of HIV-1 virus infections [39].

12.15 MIP-Based Electrochemical Sensors for the Detection of Bacterial Infection

In recent years, the development of rapid, sensitive, and cost-effective techniques for the detection of bacterial infections has become imperative for effective healthcare management. Electrochemical sensors have emerged as powerful tools in this domain, offering advantages such as high sensitivity, low detection limits, and ease of miniaturization. Molecularly imprinted polymers (MIPs), a class of synthetic polymers with tailor-made recognition sites for specific target molecules, have garnered significant attention for their application in electrochemical sensors for the detection of bacteria [3]. The prevalence of bacterial infections and the associated public health challenges underscore the need for innovative diagnostic technologies. Traditional methods often involve time-consuming processes, skilled personnel, and are prone to false positives or negatives. Electrochemical sensors, with their ability to convert biochemical interactions into electrical signals, have proven to be promising platforms for sensitive and specific bacterial detection [40].

Recently, electrochemical biosensors employing the molecularly imprinted polymer (MIP) technique have been applied for the detection of *K. pneumoniae*. This study critically evaluates the effectiveness of cyclic voltammetry (CV) as an electrochemical biosensor-based detection method, relying on the observation of current variations occurring at the sensor surface due to biochemical interactions on the electrode. The specificity of the MIP-based sensor was rigorously assessed, particularly targeting rod-type bacteria.

The results unequivocally demonstrated that the MIP exhibits a diminished sensitivity response toward other rod-type bacteria, underscoring its exceptional specificity for *Klebsiella pneumoniae*. The MIP-based sensor designed for *K. pneumoniae* detection boasts several advantageous features. Notably, its analytical process is straightforward, enabling ease of analysis. Furthermore, it ensures swift detection for diagnostic purposes, contributing to timely interventions. Importantly, the sensor exhibits a remarkable level of accuracy and specificity specifically tailored for the precise detection of *K. pneumoniae*, solidifying its potential as a robust tool in biosensing applications [17].

A polypyrrole (PPy)-based molecularly imprinted polymer (MIP) platform was developed to enable the electrochemical sensing of Klebsiella pneumonia, facilitating real-time monitoring of its concentration levels. The electrode fabrication process involved employing electrophoretic deposition (EPD) to apply the MIP onto the hydrolyzed indium tin oxide (ITO)-coated glass surface. The subsequent detection of *K. pneumonia* was carried out utilizing the electrochemical differential pulse voltammetry (DPV) technique. Following the synthesis, the resulting product underwent comprehensive characterization using Fourier transform infrared spectroscopy (FTIR) to elucidate its structural attributes and validate the successful synthesis of the intended MIP. To assess the MIP's specificity, selectivity studies were conducted, involving comparisons against two other bacterial strains and various ions commonly found in healthy human urine. This multifaceted approach ensures the robustness and reliability of the developed MIP platform for the targeted electrochemical detection of *K. pneumonia*, holding promise for enhanced diagnostic applications in microbial monitoring [10].

12.16 Conclusion

The use of MIPs in biosensors represents a promising field in scientific research, offering economic feasibility, stability, and adaptability across diverse applications, from separation methodologies to sensors and drug delivery. Addressing challenges in imprinting entire bacterial cells, MIPs have significantly advanced both electrochemical and optical detection methods for pathogens, including SARS-CoV-2 and HIV. MIP-based optical sensors, particularly in the context of diseases like COVID-19 and bacterial infections, show the potential for revolutionizing diagnostics and biosafety measures by discerning specific biomarkers. The integration of fluorescence technology with MIPs enhances the specificity and sensitivity of biosensors, contributing to the evolving landscape of biosensor development and disease diagnostics. This synergy between molecular imprinting techniques and sensor technologies is poised to play a pivotal role in shaping the future of biosensing and disease diagnostics.

References

1. Piletsky S, Canfarotta F, Poma A, Bossi AM, Piletsky S (2020) Molecularly imprinted polymers for cell recognition. Trends Biotechnol 38(4):368–387. https://doi.org/10.1016/j.tibtech.2019.10.002
2. McClements J, Bar L, Singla P, Canfarotta F, Thomson A, Czulak J, Johnson RE, Crapnell RD, Banks CE, Payne B, Seyedin S, Losada-Pérez P, Peeters M (2022) Molecularly imprinted polymer nanoparticles enable rapid, reliable, and robust point-of-care thermal detection of SARS-CoV-2. ACS Sens 7(4):1122–1131. https://doi.org/10.1021/acssensors.2c00100
3. Liustrovaite V, Pogorielov M, Boguzaite R, Ratautaite V, Ramanaviciene A, Pilvenyte G, Holubnycha V, Korniienko V, Diedkova K, Viter R, Ramanavicius A (2023) Towards electrochemical sensor based on molecularly imprinted polypyrrole for the detection of bacteria—listeria monocytogenes. Polymers 15(7):1597. https://doi.org/10.3390/polym15071597
4. Lowdon JW, Diliën H, Singla P, Peeters M, Cleij TJ, van Grinsven B, Eersels K (2020) MIPs for commercial application in low-cost sensors and assays—an overview of the current status quo. Sensors Actuators B Chem 325:128973. https://doi.org/10.1016/j.snb.2020.128973
5. Cui F, Zhou Z, Zhou HS (2020) Molecularly imprinted polymers and surface imprinted polymers based electrochemical biosensor for infectious diseases. Sensors (Switzerland) 20(4):996. https://doi.org/10.3390/s20040996

6. Liang C, Wang H, He K, Chen C, Chen X, Gong H, Cai C (2016) A virus-MIPs fluorescent sensor based on FRET for highly sensitive detection of JEV. Talanta 160:360–366. https://doi.org/10.1016/j.talanta.2016.06.010
7. Pilvenyte G, Ratautaite V, Boguzaite R, Ramanavicius S, Chen CF, Viter R, Ramanavicius A (2023) Molecularly imprinted polymer-based electrochemical sensors for the diagnosis of infectious diseases. Biosensors 13(6):620. https://doi.org/10.3390/bios13060620
8. Ferreira MDP, Yamada-Ogatta SF, Teixeira Tarley CR (2023) Electrochemical and bioelectrochemical sensing platforms for diagnostics of COVID-19. Biosensors 13(3):336. https://doi.org/10.3390/bios13030336
9. Rodrigues PD, Tierno PFGMM, Carvalho GAC, Santos JMMG (2023) Molecularly imprinted polymer-based electrochemical sensing in protein detection. Path Sci 9(9):4001–4018. https://doi.org/10.22178/pos.96-7
10. Sharma R, Lakshmi GBVS, Kumar A, Solanki P (2022) Polypyrrole based molecularly imprinted polymer platform for Klebsiella pneumonia detection. ECS Sens Plus 1(1):010603. https://doi.org/10.1149/2754-2726/ac612c
11. Jiang X, Lv Z, Ding W, Zhang Y, Lin F (2022) Pathogen-imprinted polymer film integrated probe/Ti3C2Tx MXenes electrochemical sensor for highly sensitive determination of listeria monocytogenes. J Electrochem Sci Technol 13(4):431–437. https://doi.org/10.33961/jecst.2022.00269
12. Chen S, Chen X, Zhang L, Gao J, Ma Q (2017) Electrochemiluminescence detection of Escherichia coli O157:H7 based on a novel polydopamine surface imprinted polymer biosensor. ACS Appl Mater Interfaces 9(6):5430–5436. https://doi.org/10.1021/acsami.6b12455
13. Tancharoen C, Sukjee W, Thepparit C, Jaimipuk T, Auewarakul P, Thitithanyanont A, Sangma C (2019) Electrochemical biosensor based on surface imprinting for Zika virus detection in serum. ACS Sens 4(1):69–75. https://doi.org/10.1021/acssensors.8b00885
14. Ait Lahcen A, Arduini F, Lista F, Amine A (2018) Label-free electrochemical sensor based on spore-imprinted polymer for Bacillus cereus spore detection. Sensors Actuators B Chem 276:114–120. https://doi.org/10.1016/j.snb.2018.08.031
15. Roushani M, Sarabaegi M, Rostamzad A (2020) Novel electrochemical sensor based on polydopamine molecularly imprinted polymer for sensitive and selective detection of Acinetobacter baumannii. J Iran Chem Soc 17(9):2407–2413. https://doi.org/10.1007/s13738-020-01936-9
16. Sarabaegi M, Roushani M (2021) Rapid and sensitive determination of Pseudomonas aeruginosa by using a glassy carbon electrode modified with gold nanoparticles and aptamer-imprinted polydopamine. Microchem J 168:106388. https://doi.org/10.1016/j.microc.2021.106388
17. Pintavirooj C, Vongmanee N, Sukjee W, Sangma C, Visitsattapongse S (2022) Biosensors for Klebsiella pneumoniae with molecularly imprinted polymer (MIP) technique. Sensors 22(12):4638. https://doi.org/10.3390/s22124638
18. Fang L, Jia M, Zhao H, Kang L, Shi L, Zhou L, Kong W (2021) Molecularly imprinted polymer-based optical sensors for pesticides in foods: recent advances and future trends. Trends Food Sci Technol 116:387–404. https://doi.org/10.1016/j.tifs.2021.07.039
19. Pathak AK, Limprapassorn P, Kongruttanachok N, Viphavakit C (2022) Molecularly imprinted polymer-based optical sensor for isopropanol vapor. J Sens Actuat Netw 11(3):46. https://doi.org/10.3390/jsan11030046
20. Huang X, Xia L, Li G (2023) Recent progress of molecularly imprinted optical sensors. Chemosensors 11(3):168. https://doi.org/10.3390/chemosensors11030168
21. El-Schich Z, Zhang Y, Feith M, Beyer S, Sternbæk L, Ohlsson L, Stollenwerk M, Wingren AG (2020) Molecularly imprinted polymers in biological applications. Biotechniques 69(6):407–420. https://doi.org/10.2144/btn-2020-0091
22. Akgönüllü S, Kılıç S, Esen C, Denizli A (2023) Molecularly imprinted polymer-based sensors for protein detection. Polymers 15(3):629. https://doi.org/10.3390/polym15030629
23. Funari R, Chu KY, Shen AQ (2020) Detection of antibodies against SARS-CoV-2 spike protein by gold nanospikes in an opto-microfluidic chip. Biosens Bioelectron 169:112578. https://doi.org/10.1016/j.bios.2020.112578

24. Cennamo N, D'agostino G, Perri C, Arcadio F, Chiaretti G, Parisio EM, Camarlinghi G, Vettori C, Di Marzo F, Cennamo R, Porto G, Zeni L (2021) Proof of concept for a quick and highly sensitive on-site detection of sars-cov-2 by plasmonic optical fibers and molecularly imprinted polymers. Sensors 21(5):1–17. https://doi.org/10.3390/s21051681
25. Qu H, Fan C, Chen M, Zhang X, Yan Q, Wang Y, Zhang S, Gong Z, Shi L, Li X, Liao Q, Xiang B, Zhou M, Guo C, Li G, Zeng Z, Wu X, Xiong W (2021) Recent advances of fluorescent biosensors based on cyclic signal amplification technology in biomedical detection. J Nanobiotechnol 19(1):1–28. https://doi.org/10.1186/s12951-021-01149-z
26. Koveal D (2023) Functional principles of genetically encoded fluorescent biosensors for metabolism and their quantitative use. J Neurochem 168:496–505. https://doi.org/10.1111/jnc.15878
27. Yemiş F (2013) Molecularly imprinted polymers and their synthesis by different methods. Polym Polym Comp 21(3):145–150
28. Ansari S, Masoum S (2021) Recent advances and future trends on molecularly imprinted polymer-based fluorescence sensors with luminescent carbon dots. Talanta 223:121411. https://doi.org/10.1016/j.talanta.2020.121411
29. Yang Q, Li J, Wang X, Peng H, Xiong H, Chen L (2018) Strategies of molecular imprinting-based fluorescence sensors for chemical and biological analysis. Biosens Bioelectron 112:54–71. https://doi.org/10.1016/j.bios.2018.04.028
30. Yadav AK, Verma D, Dalal N, Kumar A, Solanki PR (2022) Molecularly imprinted polymer-based nanodiagnostics for clinically pertinent bacteria and virus detection for future pandemics. Biosens Bioelectron X 12:100257. https://doi.org/10.1016/j.biosx.2022.100257
31. Elahi N, Kamali M, Baghersad MH, Amini B (2019) A fluorescence Nano-biosensors immobilization on iron (MNPs) and gold (AuNPs) nanoparticles for detection of Shigella spp. Mater Sci Eng C 105:110113. https://doi.org/10.1016/j.msec.2019.110113
32. Liu R, Cui Q, Wang C, Wang X, Yang Y, Li L (2017) Preparation of sialic acid-imprinted fluorescent conjugated nanoparticles and their application for targeted cancer cell imaging. ACS Appl Mater Interfaces 9(3):3006–3015. https://doi.org/10.1021/acsami.6b14320
33. Raziq A, Kidakova A, Boroznjak R, Reut J, Öpik A, Syritski V (2021) Development of a portable MIP-based electrochemical sensor for detection of SARS-CoV-2 antigen. Biosens Bioelectron 178:113029. https://doi.org/10.1016/j.bios.2021.113029
34. Lopez-Tellez J, Ramirez-Montes S, Ferreira TA, Santos EM, Rodriguez JA (2022) Application of voltammetric sensors for pathogen bacteria detection: a review. Chemosensors 10(10):424. https://doi.org/10.3390/chemosensors10100424
35. Kurya AU, Aliyu U, Gusau MA (2020a) Understanding Covid-19 based on clinical insights and recent advancements in therapeutic discovery. Asian J Biol Life Sci 9(2):251–259. https://doi.org/10.5530/ajbls.2020.9.39
36. Kurya AU, Toro AU, Bello UR, Sharma DC (2020b) A review study on efficacy of immunomodulatory agents used for the treatment of Covid-19. Asian J Biol Life Sci 9(2):246–250. https://doi.org/10.5530/ajbls.2020.9.38
37. Ou X, Liu Y, Lei X, Li P, Mi D, Ren L, Guo L, Guo R, Chen T, Hu J, Xiang Z, Mu Z, Chen X, Chen J, Hu K, Jin Q, Wang J, Qian Z (2020) Characterization of spike glycoprotein of SARS-CoV-2 on virus entry and its immune cross-reactivity with SARS-CoV. Nat Commun 11(1):1620. https://doi.org/10.1038/s41467-020-15562-9
38. Silwana B, Matoetoe MC (2022) Review—nanostructured electrochemical sensors for determination of the first generation of the NNRTIs for HIV-1. ECS Adv 1(4):046502. https://doi.org/10.1149/2754-2734/ac9323
39. Ma Y, Liu C, Wang M, Wang LS (2019) Sensitive electrochemical detection of gp120 based on the combination of NBD-556 and gp120. Talanta 196:486–492. https://doi.org/10.1016/j.talanta.2018.12.062
40. Gnimatin JP, Weyori EW, Agossou SM, Adokiya MN (2022) Bacterial infections epidemiology and factors associated with multidrug resistance in the northern region of Ghana. Sci Rep 12(1):22069. https://doi.org/10.1038/s41598-022-26547-7

Important of MIPs as Artificial Antibodies in Cancer Detection and Monitoring

13

Nur Nadhirah Mohamad Zain, Nor Munira Hashim, Noorfatimah Yahaya, Nik Nur Syazni Nik Mohamed Kamal, Kavirajaa Pandian Sambasevam, and Muggundha Raoov Ramachandran

Abstract

Early cancer detection plays a crucial role in successful treatment, with biomarkers serving as vital components in this process. The emergence of molecularly imprinted polymers (MIPs) as artificial antibiotics, or receptor-like materials, has opened up new possibilities in cancer detection and monitoring as targeted tools. MIPs, synthetic materials with molecular recognition capabilities, selectively bind to molecules of interest, including cancer biomarkers. Despite recent advancements in molecular imprinting technology, the application of MIPs as artificial antibiotics remains a significant challenge. This chapter highlights recent advancements (2016–2023) and provides an overview of the current state of MIPs in the selective extraction/microextraction of cancer biomarkers. It discusses the design, fabrication, and analytical performance of MIPs, including the limit of detection and quantification, recovery, and imprinting factors. The focus is on the design and synthesis of MIPs specifically tailored for cancer detection, emphasizing their enhanced specificity and selectivity compared to conventional detection methods. Furthermore, the article explores the applications of MIPs in cancer biomarker analysis and their potential as diagnostic tools for early cancer

N. N. M. Zain (✉) · N. M. Hashim · N. Yahaya · Nik Nur Syazni Nik Mohamed Kamal
Department of Toxicology, Advanced Medical and Dental Institute, Universiti Sains Malaysia, Kepala Batas, Pulau Pinang, Malaysia
e-mail: nurnadhirah@usm.my

K. P. Sambasevam
Advanced Materials for Environmental Remediation (AMER) Research Group, Faculty of Applied Sciences, Universiti Teknologi MARA, Cawangan Negeri Sembilan, Kampus Kuala Pilah, Kuala Pilah, Negeri Sembilan, Malaysia

M. R. Ramachandran
Department of Chemistry, Faculty of Science, Universiti Malaya, Kuala Lumpur, Malaysia

S. Patra et al. (eds.), *Molecularly Imprinted Polymers: Path to Artificial Antibodies*, https://doi.org/10.1007/978-981-97-4379-7_13

detection. Lastly, the challenges and future perspectives of MIPs in the field of cancer monitoring are addressed, highlighting the promising advancements and potential impact of this technology on improving cancer diagnostics and therapies.

Keywords

Molecularly imprinted polymers · Artificial antibiotics · Biomarkers · Cancer

Abbreviations

Acr	Acrylamide
AFP	Alpha-fetoprotein
AIBN	Azobisisobutyronitrile
APMA	3-Aminopropylmethacrylate
APMS	2-Acrylamido-2-methyl-1-propanesulfonic acid
APS	Ammonium persulfate
APTES	Aminopropyltriethoxysilane
AuNP	Gold nanoparticles
AuNP SAM coated arrays	Gold nanoparticle self-assembled monolayer
Au-SPE	Gold screen-printed gold electrode
Bio-MIP-ePADs	Antibody-free biomarker analysis by in situ synthesized molecularly imprinted polymers on movable valve microfluidic paper-based electrochemical device
Bis-Acr	Bis-acrylamide
BnTES	Benzyltriethoxysilane
BTVP	*N*-3,5-113 bis(Trifluoromethyl)-phenyl-*N*′-4-vinylphenylurea
CA 15-3	Carbohydrate antigen 15-3
CA	Cancer antigen
CB with PtRu	Carbon black with platinum-ruthenium alloy
CD	Cluster of differentiation
CDS@ZnO/ITO	Cadmium sulfide nanoparticle-sensitized zinc oxide immobilized on indium tin oxide
CEA	Carcinoembryonic antigen
c-Kit	Tyrosine-protein kinase kit
CME	Cysteine-modified epitope
CMSM	Core-molecule-shell-molecule-nanoparticles
CNE	Carbon nanotube electrode
CP1/CP2	Amphiphilic conjugated polythiopenes
CV	Cyclic voltammetry
Cys	Cysteine
Cys-EP1/Cys-EP2	Cysteine-modified epitopes of NSE
Cyss	Cystine
Cyt c	Cytochrome c

DMFC	Direct methanol fuel cell
DMIP	Dual-template rectangular nanotube molecularly imprinted polypyrrole
DPV	Differential pulse voltammetry
DR	Degradation product
DSP	3,3′-Dithiodipropionic acid di(*N*-hydroxysuccinimide ester)
EAMA.HCl	*N*-(2-aminoethyl) methacrylamide hydrochloride
EC	Electrochromic cell
EDOT	3,4-Ethylenedioxythiopene
EGDMA	Ethylene glycol dimethacrylate
EGFR	Epidermal growth factor receptor
EpCAM	Epithelial cell adhesion molecule
EtOH	Ethanol
FFPE	Formalin-fixed paraffin embedded
FIA	Flow injection analysis
FTO/PEDOT	Fluorine-doped tin oxide/poly(3,4-polyethylene disoxythiopene)
GO	Graphene oxide
GSH	Glutathione
GSSG	Glutathione disulfide
H_2SO_4	Sulfuric acid
HAc	Acetic acid
HE4	Human epididymis secretory protein 4
HER-2 ECD	Human epidermal growth factor receptor 2
HER-2	Human epidermal growth factor 2
IBTES	Isobutylltriethoxysilane
KCl	Potassium chloride
LC-MS/MS	Liquid chromatography-Tandem mass spectrometry
$LiClO_4$	Lithium chloride
MAA	Methacrylic acid
MBAA	*N,N′*-methylene-bis-acrylamide
MeCN	Methyl cyanide/Acetonitrile
MeOH	Methanol
MES	2-Morpholinoethanesulfonic acid
MIP/CNT CPE sensor	Molecularly imprinted polymer/carbon nanotube
MIP	Molecularly imprinted polymer
MIP@CPE	Molecularly imprinted polymer@ carbon paste electrode
MIPPy	Pyrrole-based molecularly imprinted polymer
MIP-SERS biosensor	Molecularly imprinted polymer-surface enhanced Raman spectroscopy
MISPE	Molecularly imprinted solid phase extraction
mMIP microsphere	Magnetic molecularly imprinted polymer microsphere
MO	Methyl orange

MTCL-MDSPE	Multiplex tags chemical isotope labeling combined with magnetic dispersive solid phase extraction
MWCNT	Multiwalled carbon nanotube
$NADP^+$	Nicotinamide adenine dinucleotide phosphate
NADPH	Coenzyme II reduced tetrasodium salt
$NaNO_3$	Sodium nitrate
NanoMIP	Nano-molecularly imprinted polymer
NaOH	Sodium hydroxide
NDKB	Nucleoside diphosphate kinase
NH_4OH	Ammonium hydroxide
$Ni(OH)_2$	Nickel (II) hydroxide
NIPAm	*N*-isopropylacrylamide
NMP	Nucleophosmin
NNMBA	*N*,*N*′-methylenebisacrylamide
NSE	Neuron-specific enolase
odMIP-PISA	Orthogonal dual molecularly imprinted polymer-based plasmonic immunosandwich assay
OPD	*O*-phenylenediamine
o-PD	o-phenylenediamine
OVA1	Ovarian 1
OXA	Oxalic acid
PA	Pyruvic acid
PB	Phosphate buffer
PBS	Phosphate-buffered saline
PEGDA	Polyethylene glycol diacrylate (Mn 575)
Poly(TB) MIP	Poly(toluidine blue) molecularly imprinted polymer
ProGRP	Progastrin-releasing peptide
PSA	Prostate-specific antigen
Py	Pyrrole
QCM	Quartz crystal microbalance
ROMA	Risk of ovarian malignancy algorithm
SAM	S-Adenosyl methionine
SAR	Sarcosine
SDS	Sodium dodecyl sulfate
SMIP-SERS	Surface molecularly imprinted polymer-surface-enhanced Raman spectroscopy
TBAm	*N*-tertbutylacrylamide
TCEP buffer	Tris carboxy ethyl phosphene
TEMED	Tetramethylethylenediamene
TEOS	Tetraethyl orthosilicate
THF	Tetrahydrofuran
DVB-80	Divinylbenzene-80
TPVU	*N*-3,5–133 bis(Trifluoromethyl)-phenyl-*N*′-4-vinylphenylurea
TRIM	Trimethylolpropane trimethacrylate

triMIP-PISA	Molecularly imprinted polymer-based plasmonic immunosandwich assay
UHPLC-MS/MS	Ultrahigh-performance liquid chromatography-tandem mass spectrometry
UP water	Ultrapure water
UPTES	3-ureidopropyltriethoxysilane
UV polymerization	Ultraviolet polymerization
VEGF	Vascular endothelial growth factor
VOC	Volatile organic compounds
VPA	Vinyl phosphoric acid
$(TBA)ClO_4$	Tetrabutylammonium perchlorate
$[Fe(CN)6]^{3-/4-}$	Ferricyanide
4-VP	4-Vinyl pyridine
5-HIAA	5-Hydroxyindoleacetic acid

13.1 Introduction

Cancer is becoming a global issue due to the increasing number of fatalities worldwide. According to the American Cancer Society (ACS), it ranks as the second leading cause of death in the United States, with projections of 1,918,030 new cancer and 609,360 cancer-related deaths in 2022. These data clearly indicate the severity of cancer and undoubtably cause tremendous pressure on society. Unfortunately, the survival rate for cancer is often dismal, primarily due to the absence of affordable diagnostic methods, delayed diagnosis, and inadequate access to necessary resources. Early detection of cancer, prompt treatment, and subsequently monitoring can significantly diminish mortality rates associated with this deadly disease [1]. Efficient early diagnosis and treatment can have a significantly impact on reducing cancer incidence and mortality rates. In this context, in addition to the current array of complex clinical methods used for cancer diagnosis—such as X-ray MRI, computerized tomography (CT), magnetic resonance imaging or MRI, positron emission tomography, endoscopy, sonography, thermography, cytology, and biopsy—detecting cancer biomarkers presents a promising, cost effective, and highly accurate approach for early diagnosis. Most of these complex clinical methods are time consuming, require experienced technicians, and sometimes become very costly [2]. In this regard, there is an urgent need for designing, evolving, and finding new designs for novel diagnostic tools. Newly developed devices have to provide as an option for solving the existing problems within classical clinical diagnostics that require sophisticated and expensive instrumentation.

Biomarkers are defined by both the Food and Drug Administration (FDA) and the European Union National Institute of Health as a characteristic that are quantitatively measured and assessed as indicators of normal biological processes, pathogenic processes, or pharmacologic responses to therapeutic intervention [3]. In the relentless pursuit of innovative solutions to combat the complex challenges posed by cancer, the convergence of multiple scientific disciplines has led to remarkable breakthroughs. One such groundbreaking approach involves the utilization of molecularly imprinted technology (MIT) or molecularly imprinter polymers (MIPs) as artificial antibiotics in the realm of cancer detection and monitoring. As conventional diagnostic methodologies often fall short in terms of accuracy, speed, and reliability, researchers have turned their attention to MIPs as a cutting-edge tool that offers the promise of revolutionizing cancer diagnostics. MIPs are synthetic analogs of natural biological antibody–antigen systems and can selectively bind molecules that have been templated during the synthesis process through a "lock and key" mechanism. Generally, the MIPs synthesis process consists of prepolymerization, polymerization, and template molecule elution. After the template molecules are eluted., the recognition cavity complementary to the template molecule in shape, size, or special structure formed in the highly cross-linked polymer matrix. MIPs, distinguished by their tailor-made receptor sites that are meticulously designed to bind with specific target molecules, emulate the principles of molecular recognition seen in natural antibodies [4]. This ability to selectively capture target molecules makes MIPs an ideal candidate for detecting cancer biomarkers with remarkable precision.

By harnessing this molecular imprinting technique, researchers aim to transform the landscape of cancer detection, enabling earlier diagnosis, effective monitoring of treatment responses, and ultimately, improved patient outcomes. The specific recognition properties, as well as the chemical and physical stability of MIPs, make them attractive for cancer detection and monitoring applications. MIPs, or MIT, offer several advantages over other detection methods such as immunoassays, flow cytometry, ELISA, Western blot, and RT-PCR, which have limited sensitivity and accuracy. Moreover, these approaches are often costly, intricate, and time-consuming, requiring sizable sample sizes. The current scenario emphasizes the need for a practical approach to early cancer diagnosis. In this context, highly selective nanomaterials incorporated with detection techniques tools are emerging as game-changers in the clinical landscape, with ever-increasing expectations [5]. To elaborate, the following advantages of MIPs can be categorized into four important parts. First, MIPs show high selectivity toward distinct target molecules, enabling precise identification and quantification of different analytes. Second, the polymer is stable in various conditions and exhibits wide pH tolerance with minimal nonspecific binding. Furthermore, by adjusting the polymerization conditions and choosing the appropriate functional monomer, MIPs can be designed to detect a wide range of analytes, showcasing their versatility. Importantly, the cost of their synthesis and preparation is relatively low compared to other biological factors [6].

This chapter delves into the multifaceted role of MIPs as artificial antibiotics in the context of cancer detection and monitoring. It explores the underlying principles of MIPs synthesis and their potential to enhance the sensitivity and specificity of diagnostic assays. Furthermore, it elucidates the implications of this technology in the broader scope of personalized medicine, highlighting the prospects of tailoring treatments based on real-time data obtained through MIPs-based monitoring systems. As we embark on this journey through the realm of innovative cancer diagnostics, it becomes evident that MIPs hold the key to unlocking a new era of precision medicine, where early intervention and accurate monitoring converge to redefine the future of cancer care.

13.2 Cancer Biomarkers

13.2.1 Definition

A biomarker has been defined as a measurable substance or characteristic in the human body that could indicate of a physiological or pathological process or pharmacological response to a therapeutic intervention [7, 8]. Changes in blood pressure and lactate levels following exercise and the concentration of blood glucose in diabetes mellitus are some examples of classical biomarkers [9, 10]. A biomarker can also be referred to as a biochemical indicator that can track the changes or possible changes in the structure and function of cells and subcellular structures of systemic organs and tissues [11]. In this context, biomarkers can be categorized based on (i) molecular (chemicals, proteins, or genes), (ii) cellular (cell type, cell morphology, and tissue histology), or (iii) imaging (X-ray, CT, PET, or MRI features) [12]. Figure 13.1 illustrates types of biomarkers that are classified based on molecular characteristic. The choice of a biomarker type depends on the condition being studied and the biological system under investigation. As a result, a diverse variety of biomarkers is used in many different disciplines ranging from clinical medicine to drug testing and toxicology and even to environmental monitoring and soil science [13–15].

In clinical medicine, the main purpose of biomarkers is to diagnose, prognose, or predict disease. In addition, biomarkers can be used to distinguish between benign and malignant conditions or one type of malignancy from another, as well as to monitor the status of the disease (e.g., to detect cancer recurrence). These biomarkers are also used as tools to monitor the clinical response and the effect of therapy [8, 16]. Figure 13.2 illustrates (a) terminology and (b) clinical uses of cancer biomarkers.

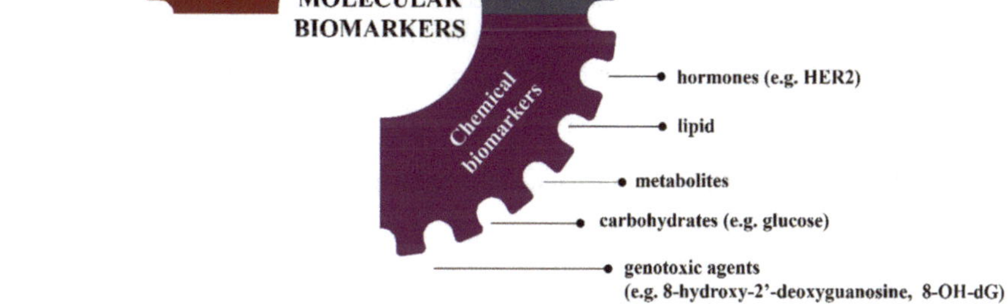

Fig. 13.1 Schematic illustration of classification of various biomarkers based on molecular characteristic

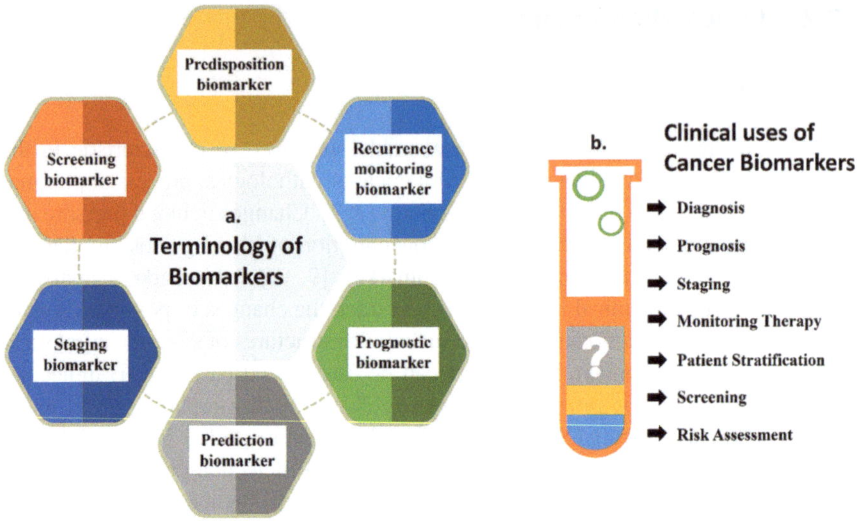

Fig. 13.2 Schematic illustration of the (**a**) terminology and (**b**) clinical uses of biomarkers. (Adapted from [16])

According to the National Cancer Institute, a biomarker is defined as "a biological molecule found in blood, other body fluids, or tissues that is a sign of a normal or abnormal process, or of a condition or disease." Biomarkers for cancer are also referred to as "cancer biomarkers" or "tumor markers," substances found in higher-than-normal levels in the circulation (whole blood, plasma, or serum), in secretions (urine, stools, sputum, or nipple discharge), or in tissue-derived of some people with cancer [17]. These biomarkers could be the product of cancerous cells or as response to cancer, and they can be either genes, RNAs, proteins, metabolites, or extracellular vesicles [11, 17]. For example, the immune system may react to the cancer by producing substances that can be detected in the blood. These substances may indicate the presence of a cancer, which is not actually produced by cancer cells [18]. Nevertheless, noncancerous conditions may also cause an increment in some cancer biomarkers to be higher than normal, while not every cancer patient may share similar expression of these biomarkers [19]. In this sense, knowledge about cancer biomarkers and development of reliable, accurate, and sensitive detection tools are important for both early diagnosis and monitoring the disease progression. Figure 13.3 illustrates sample types of noninvasive biomarkers that can be obtained from cancer patients.

13.2.2 Examples of Cancer Biomarkers and FDA-Approved Protein Tumor Markers

The type of biomarker used depends both on the application and the availability of appropriate assay methods. Table 13.1 lists the Food and Drug Administration (FDA) approved protein biomarkers in current clinical practice. The use of

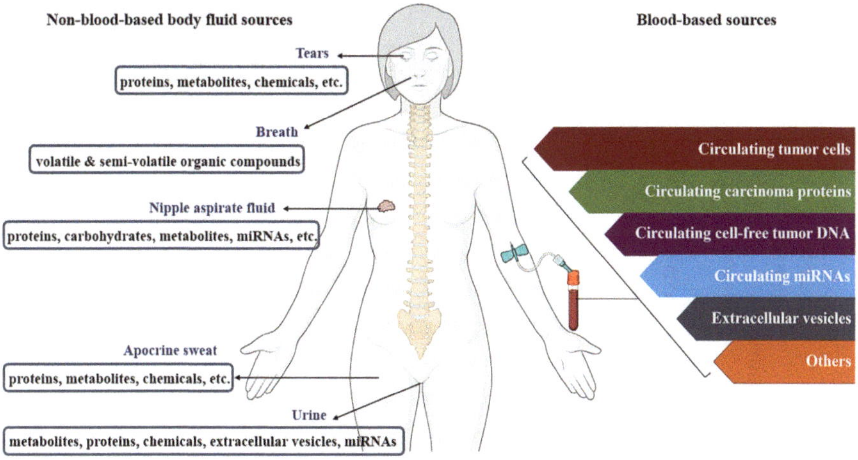

Fig. 13.3 Schematic illustration of sample types of noninvasive biomarkers. (Adapted from [20]. Representative of human anatomy of female adult was created with BioRender.com (https://biorender.com/))

FDA-approved biomarkers has played a significant role in facilitating the appropriate selection of the type of biomarker. This, in turn, will enable effective strategies in terms of disease diagnosis, prognosis, detection of residual disease or recurrence, and provide precision treatment for cancer patients [21]. Under the FDA's Medical Device Law, biomarkers have been approved as an in vitro diagnostic devices (IVDs). According to FDA regulations, biomarkers have been categorized into one of three classes (Classes I, II, or III). This classification depends on the intended use of the particular assay and its associated risk to the patient. An intended use with a higher risk is classified under Class III. This class requires a premarket application (PMA) to the FDA. Class II is for the moderate-risk intended uses of biomarkers, which in most cases requires a 510(k) submission. A 510(k) is referred to as a premarket submission made to the FDA to demonstrate that the device to be marketed is as safe and effective. This is substantially equivalent to a legally marketed device (section 513(i)(1)(A) FD&C Act) [22]. On the other hand, biomarkers categorized under Class I would offer little or no potential for the unreasonable risk of injury or illness, and these types of biomarkers are generally exempt from FDA premarket review [23, 24].

In cancer, molecular biomarkers are commonly genes or gene products, such as proteins. Examples include CA15-3 and CA27-29. Both biomarkers are carbohydrate-containing protein antigens of the transmembrane glycoprotein MUC-1, which is capable of inhibiting tumor cell lysis and reducing cell–cell interaction [25]. Various studies have demonstrated that increased serum CA15-3 levels at diagnosis are correlated with higher breast cancer stage, tumor size, and worse overall survival and disease-free survival [26–28]. Besides the above-mentioned biomarkers, assessment of HER-2/neu status by immunohistochemistry is further routine laboratory testing for predicting the response to anti-HER2 therapy. The

Table 13.1 List of FDA-approved protein tumor markers currently used in clinical[a] (Source: [43])

Biomarker	Clinical use	Specimen	Methodology	Submission type	Year first approved or cleared	Device class	Product code
Breast cancer							
Circulating tumor cells • EpCAM, • CD45, • Cytokeratins 8, • Cytokeratins 18+, • Cytokeratins 19+	Prediction of cancer progression and survival	Whole blood	Immunomagnetic capture/ immune-fluorescence	510(k)	2005	2	NQI
Estrogen receptor (ER)	Prognosis, response to therapy	FFPE tissue	Immunohistochemistry	510(k)	1999	2	MYA
Progesterone receptor (PR)	Prognosis, response to therapy	FFPE tissue	Immunohistochemistry	510(k)	1999	2	MXZ
HER-2/neu	Assessment for therapy	FFPE tissue	Immunohistochemistry	510(k)	1998	3	MVC
CA-15-3	Monitoring disease response to therapy	Serum plasma	Immunoassay	510(k)	1997	2	MOI
CA27.29	Monitoring disease response to therapy	Serum	Immunoassay	510(k)	1997	2	MOI
Prostate cancer							
Pro2PSA	Discriminating cancer from benign disease	Serum	Immunoassay	PMA	2012	3	OYA
p63 protein	Aid in differential diagnosis	FFPE tissue	Immunohistochemistry	510(k)	2005	1	NTR
Free PSA	Discriminating cancer from benign disease	Serum	Immunoassay	510(k)	1997	3	MTG
Total PSA	Prostate cancer diagnosis and monitoring	Serum	Immunoassay	PMA	1986	2	LTJ, MTF

Testicular cancer							
Alpha-fetoprotein (AFP)[b]	Management of cancer	Serum, plasma, amniotic fluid[b]	Immunoassay	PMA	1992	3	LOK
Ovarian cancer							
ROMA • HE4 + CA-125	Prediction of malignancy	Serum	Immunoassay	510(k)	2011	2	ONX
OVA1 • Multiple proteins	Prediction of malignancy	Serum	Immunoassay	510(k)	2009	2	ONX
HE4	Monitoring recurrence or progression of disease	Serum	Immunoassay	510(k)	2008	2	OIU
CA-125	Monitoring disease progression, response to therapy	Serum plasma	Immunoassay	510(k)	1997	2	LTK
Colorectal cancer							
Fibrin/fibrinogen degradation product (DR-70)	Monitoring progression of disease	Serum	Immunoassay	510(k)	2008	2	NTY
Human hemoglobin • Fecal occult blood	Detection of fecal occult blood (home use)	Feces	Lateral flow immunoassay	510(k)—CLIA waived	1976	2	KHE
Hepatocellular cancer							
AFP-L3%	Risk assessment for development of disease	Serum	HPLC, microfluidic capillary electrophoresis	510(k)	2005	2	NSF

(continued)

Table 13.1 (continued)

Biomarker	Clinical use	Specimen	Methodology	Submission type	Year first approved or cleared	Device class	Product code
Gastrointestinal stromal tumors							
c-Kit	Detection of tumors, aid in selection of patients	FFPE tissue	Immunohistochemistry	PMA	2004	3	NKF
Pancreatic cancer							
CA19-9	Monitoring disease status	Serum, plasma	Immunoassay	510(k)	2002	2	NIG
Thyroid cancer							
Thyroglobulin	Aid in monitoring	Serum, plasma	Immunoassay	510(k)	1997	2	MSW
Bladder cancer							
Nuclear mitotic apparatus protein • NuMA • NMP22	Diagnosis and monitoring of disease (professional and home use)	Urine	Lateral flow immunoassay	PMA	1996	3	NAH
Not specified							
Carcino-embryonic antigen	Aid in management and prognosis	Serum, plasma	Immunoassay	510(k)	1985	2	DHX

[a] While hCG is commonly used as a tumor marker, it has not been cleared/approved for this application by the FDA
[b] AFP is a Class III analyte because of its non-cancer intended use (aid in prenatal diagnosis of birth defects)

HER-2/neu oncogene is a member of the epidermal growth factor receptor (EGFR) or HER family, which is involved in cell–cell and cell–stromal communication via a process known as signal transduction. Overexpression of the *HER2* gene is associated with aggressiveness and an increased risk of breast cancer recurrence [29, 30]. In this regard, the use of anti-HER2 therapy (e.g., trastuzumab, lapatinib, pertuzumab, and ado-trastuzumab emtansine) has significantly improved the outcome of breast cancer patients with HER2 overexpression [31–33]. These data also indicate the usefulness of clinical application of HER-2/neu as an important gene for molecular targeting treatment of breast cancer.

Several other biomarkers can be used for various cancers, including PSA as a biomarker to diagnose and monitor prostate cancer [34]; CA19-9 as a biomarker to monitor disease status of pancreatic cancer [35], serum thyroglobulin (Tg) as a biomarker for the detection of residual thyroid tissue, persistent disease, and relapse [36], serum fibrinogen-fibrin degradation products (DR-70) as predictive and prognostic biomarkers in colorectal cancer [37, 38], OVA1 as a biomarker for the early detection of ovarian cancer [39], NMP22 as a urine protein-based biomarker for the detection, surveillance and treatment response prediction of bladder cancer [40], and c-Kit as a predictive biomarker for gastrointestinal cancer [41]. Carcinoembryonic antigen, on the other hand, is a nonspecific serum biomarker that is highly expressed in various cancers, including colorectal, ovarian, cervical, lung, breast, gastric, pancreatic, urinary bladder, prostate, and lymphoma [42].

13.3 Design and Synthesis of Molecularly Imprinted Polymers (MIPs)

Designing and synthesizing MIPs involves highly selective and specific polymer matrices that can recognize and bind selectively to a target molecule. MIPs are synthetic materials that have gained significant attention in various fields, including chemical analysis, separation sciences, drug delivery, and sensors, due to their ability to mimic natural receptors and exhibit target-specific binding [44]. Generally, the design of MIPs involves (a) target molecule, (b) functional monomers, (c) cross-linker, and (d) solvent.

(a) The target molecule (template) that the MIPs will be imprinted for. This could be a small organic molecule, a biomolecule, or even a metal ion.
(b) Appropriate functional monomers can form complementary interactions (hydrogen bonding, van der Waals, electrostatic, etc.) with the target molecule. These monomers will be polymerized to create the polymer matrix.
(c) Cross-linker molecule that will help in forming a stable three-dimensional network of polymers. The ratio of cross-linker to monomers will affect the porosity and binding sites of the MIPs.
(d) The solvent system that will be used for polymerization should be compatible with the monomers and target molecules.

A smart polymer is a type of material that can be used to respond consistently to an external stimulus. There are a wide range of systems that are designed to make use of this type of material. Each of these is built to be specific to a single potential stimulus. Through the integration of smart material technology, a variety of smart MIPs have been created and designed to provide self-control capabilities [45, 46].

In recent years, the landscape of MIPs design has undergone a transformative shift, courtesy of computational methods. These methods have introduced an efficient and environmentally conscious alternative to traditional trial-and-error approaches. Computational designs have emerged as potent tools for screening functional monomers in molecular imprinting [44]. The crux of functional monomer selection hinges on its capability to interact with the template. These simulations are frequently conducted in the gas phase, facilitating robust interactions between the template and functional monomer by eliminating the presence of other prepolymerization reagents such as porogens, initiators, and cross-linking monomers. This gas-phase simulation environment offers the utmost purity, devoid of any interferences. A noteworthy aspect is the binding energy, which attains its peak when robust interactions are observed between the functional monomer and the template. Essentially, the functional monomer exhibiting the highest binding energy to the template molecule is generally an optimal fit for MIPs synthesis [47].

The extensive utilization of computational tools in MIPs design for diverse organic compounds is evident [44]. Notably, density functional theory emerges as the prevailing computational method for the selection of functional monomers, while molecular dynamics predominantly serves to validate interactions between monomers and templates. These computational tools provide significant advantages by reducing the need for trial-and-error synthetic approaches, thereby resulting in cost efficiencies and waste reduction. A pivotal aspect is the congruence between the outcomes of computational methods and experimental data [48].

Although MIPs have shown promising results when utilized as a recognition element for fabricating optical and electrochemical sensors, their fabrication through conventional techniques can lead to various issues. For instance, when a bulk polymerization process is utilized, the binding sites are usually lost in the resulting material. Due to the long diffusion distances between the matrix and the template, the rebinding process can take a long time. This is why a combination of molecular imprinting and nanotechnology has been used to improve the MIP's performance. Moreover, compared to natural antibodies, MIPs are more durable and reusable. They can also be stored at ambient temperatures and are 200 times cheaper [49].

13.3.1 Synthesis of MIPs

The fundamental "three-element concept" pertains to the components of MIPs, which include a basic monomer, a template, a cross-linker, a polymerization initiator, and a porogen [46]. Figure 13.4 illustrates the MIPs synthesis process [50].

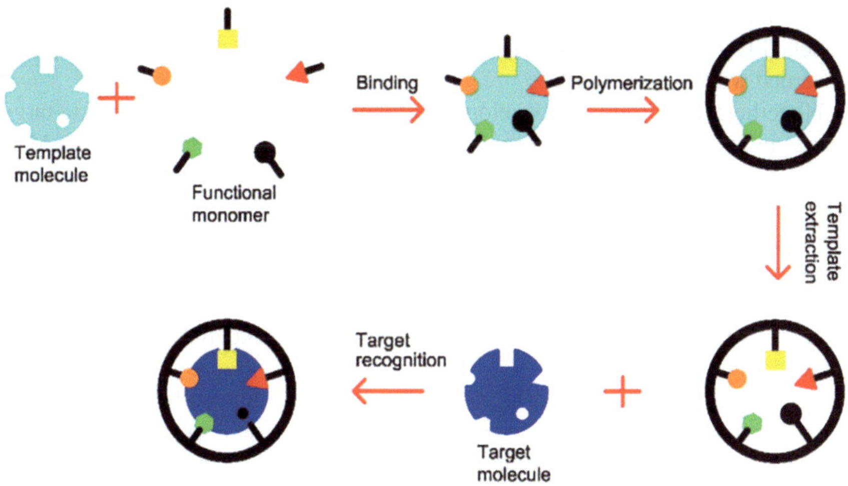

Fig. 13.4 Schematic diagram of the molecularly imprinting process [50]

Conventionally, the process of imprinting begins with the formation of a complex between the monomer and the template molecule. Following this, a cross-linker is introduced, maintaining the integrity of the monomer–template complex, and facilitating the development of an artificially engineered three-dimensional polymer framework [51]. Various synthetic strategies for producing MIPs, including covalent, semi-covalent, and noncovalent approaches, have been extensively explored and reviewed [52].

13.3.1.1 The Utilization of Green Prepolymerization Reagents in the Synthesis of MIPs

In contemporary practices, there is a notable shift toward embracing the principles of green chemistry. This shift extends to the synthesis of MIPs, where the incorporation of benign or environmentally favorable reagents is pursued. This strategic choice underscores the paramount importance of ensuring both user safety and environmental well-being. The core tenet of green chemistry is the reduction of waste generation and the avoidance of toxic or hazardous reagents and solvents across the spectrum of chemical product development and application [53].

13.3.1.2 MIPs Templates

The contemporary trend is oriented toward the integration of green chemistry principles, which poses challenges and opportunities in the synthesis of MIPs. Incorporating environmentally friendly prepolymerization reagents poses a challenge, particularly due to the frequent use of organic compounds as template molecules, which is hard to circumvent in molecular imprinting. In the context of green synthesis, a notable challenge in MIPs formation revolves around the necessity to employ an equimolar quantity of the template molecule. This is essential for generating the complementary polymer cavities responsible for selectively binding the analyte molecule [54]. Unfortunately, this equivalent template molecule is typically detached from the MIPs, and regardless of its environmental or user toxicity, it is discarded with organic waste.

Nonetheless, a promising development involves the substitution of toxic target compounds (organic) with environmentally friendly "dummy templates" in MIPs synthesis. This substitution gains significance because, once templates are removed to create complementary cavities, they are usually not recovered from the eluting solvent. Notably, the utilization of "dummy molecules" as green templates has been explored for various compounds. For example, in the creation of a MIPs for acrylamide extraction from heat-processed food, propionamide was used as a dummy template due to its reduced toxicity compared to the target molecule [55].

However, challenges persist in identifying suitable compounds as dummy templates for many target compounds, often requiring the synthesis of such templates before generating a MIP. This can extend synthesis times, delaying MIP applications [56, 57]. Conversely, the concept of multitemplates has gained ground, where a single MIP is developed for multiple target compounds. This minimizes reagent consumption, aligning with greener practices [58].

13.3.1.3 Functional Monomers of MIPs

Extensive efforts have been directed toward using less toxic substances as functional monomers in MIPs. Particularly, ionic liquids have gained significant attention as potential substitutes for traditional organic functional monomers during MIPs synthesis [59–64]. The attraction toward utilizing ionic liquids in molecular imprinting stems from their properties that align with the principles of green chemistry. These properties encompass low vapor pressure, high thermal stability, compatibility with a broad spectrum of organic solvents, effective extractability for various organic, inorganic, and organometallic materials, high viscosity, and substantial ionic conductivity [65, 66]. While ionic liquids are currently classified as green solvents due to these attributes, their toxicity, and potential environmental dispersion, predominantly determined by their water solubility, warrant further evaluation [66].

In contrast, conventional functional monomers like methacrylic acid and acrylic acid, known for their toxicity to freshwater and marine organisms [67], have not been extensively studied for their potential release from MIPs during prolonged usage in environmental waters. As a progressive approach, the utilization of ionic liquids as functional monomers in MIP synthesis is being favored over such compounds [44].

13.3.1.4 Cross-Linking Monomers and Initiators of MIPs

Ethylene glycol dimethacrylate has consistently maintained its status as the preferred cross-linking monomer across a multitude of molecular imprinting studies. Nonetheless, the growing emphasis on environmentally mindful imprinting methodologies has triggered an exploration of alternative options. In this pursuit, calcium chloride, an inorganic compound, has been harnessed as a cross-linking monomer to craft imprinted composite membranes for d-tryptophan, epitomizing a more ecologically conscious approach [68]. Concerning radical initiators, the common place 2,2′-azobisisobutyronitrile has primarily served as a polymerization activator, employed in modest quantities. Consequently, its ecological impact has garnered relatively diminished attention.

13.3.1.5 Porogenic Solvents for MIPs Synthesis

Recent advancements in research underscore a notable shift toward harnessing environmentally friendly solvents as effective porogens in the synthesis of MIPs [69]. Porogenic solvents, occasionally employed excessively during polymerization reactions, can result in their disposal through evaporation, often exacerbated by heating procedures, leading to air contamination. This trend is particularly conspicuous in suspension and precipitation polymerization methodologies. Notably, the field of molecular imprinting has witnessed the emergence of eco-conscious porogenic solvents, encompassing water and ionic liquids [69, 70].

The incorporation of ionic liquids as porogenic solvents in MIPs synthesis is exemplified by the creation of an MIPs with selectivity for testosterone, where 1-butyl-3-methylimidazolium tetrafluoroborate assumed the role of the porogen [70]. These strides are of paramount significance, given that traditional organic solvent frequently employed in this realm exhibit elevated vapour pressures, thus posing risks to operator safety. Consequently, the ongoing strides in this realm hold considerable potential for elevating both environmental sustainability and occupational well-being within the ambit of MIPs synthesis.

13.3.2 Polymerization Techniques for MIPs

Three widely employed polymerization techniques are bulk, precipitation, and suspension polymerization. These methodologies have been extensively documented and applied in the literature, with a specific focus on elucidating the process of each polymerization technique and the resultant particle morphology for each approach [71].

Bulk polymerization yields a complete conversion of all reagents into a solid polymer, subsequently requiring the removal of excess solvent through evaporation. In contrast, both precipitation and suspension polymerization involve the utilization of surplus porogenic solvent during synthesis. Consequently, these two techniques (precipitation and suspension) yield solid polymers within a significantly larger volume of organic solvent. As viewed from an environmental standpoint, both precipitation and suspension polymerization methods contribute to substantial liquid waste generation. Hence, in this context, the bulk polymerization approach emerges as relatively more environmentally conscious. However, it's important to note that bulk polymerization is not without drawbacks, as it often necessitates subsequent milling of the polymer. This milling procedure can lead to the formation of irregularly shaped particles and potential disruption of binding sites [71, 72].

13.3.3 Template Removal

The pivotal role of this phase in MIPs synthesis cannot be overstated, primarily due to the critical necessity for the thorough removal of template molecules. Insufficient elimination of template molecules from the MIPs structure results in a diminished availability of binding cavities for subsequent rebinding interactions, inevitably leading to reduced extraction efficiency. Moreover, the persistence of template molecules could lead to their release during analytical applications, introducing errors and potentially causing an overestimation of analytical results [73].

As a consequence, the pursuit of achieving complete template removal takes on paramount significance. This endeavor harmonizes seamlessly with the principles of green analytical chemistry, as it aims not only to minimize solvent costs but also to shorten the time required for thorough template elimination. A notable technique that has emerged in this context is pressurized hot water extraction, which presents an innovative approach by utilizing a non-organic solvent for template removal. Throughout molecular imprinting procedures, substantial volumes of organic solvents enriched with template molecules are generated and subsequently discarded post-template removal. In response to this, the adoption of water as a green solvent within a pressurized liquid extraction system has been investigated [74]. Beyond its environmentally friendly attributes, the use of water for template removal offers an advantage in terms of cost-effectiveness.

13.4 Characterization of MIPs

Nowadays, MIPs are widely synthesized via conventional chemical method or electropolymerization. Since each method directed and tailored in different manner, sufficient characterizations are necessary to validate the formation of MIPs. In this section, the analysis of the surfaces of MIPs using techniques involving electron microscopy, determination of particle and pore sizes, is discussed. Moreover, MIPs have gained extensive use in the real-time monitoring of cancer biomarkers due to their cost-effective, fast, and highly sensitive detection capabilities. Consequently, the immobilization of particles onto the MIPs matrix holds a vital significance, as it can be examined using atomic force microscopy (AFM) methods. More recent studies have placed significant emphasis on modifying MIPs to enhance their ability for recognition, focusing on sensitivity and selectivity. Consequently, techniques such as nuclear magnetic resonance (NMR) and infrared (IR) spectroscopies have gained importance for revealing the MIPs' structures. Moreover, the utilization of MIPs in challenging environments has prompted researchers to explore the thermal characteristics of these synthesized MIPs.

13.4.1 Surface Morphological Characterization

MIPs often find useful applications in the solid-phase extraction (SPE) techniques as column or surface modifiers for the preconcentration and eventually detection of cancer biomarkers [75]. Hence, understanding the optimal size range of MIPs particles is crucial to ensure their efficient operation, preventing issues like clogging and flow disruption. Employing surface analysis methods like electron microscopy and particle size analysis becomes instrumental in comprehending the particle's geometry and dimensions. These factors significantly influence the re-binding affinity between the MIPs and the template molecule [76]. Theoretically, the smaller particle size would possess higher surface area which create bigger opportunities for binding with the target molecules. Typically, precipitation polymerization creates the opportunity to modify the particle size into micro-nano scale sizes. In theory, smaller particle sizes inherently offer a greater surface area, thus presenting amplified binding possibilities with target molecules. This capacity for surface manipulation, however, varies across different synthesis pathways. Notably, research indicates that MIPs size generated through bulk polymerization aligns with the dimensions of the employed sieves. Suspension polymerization methods typically yield MIPs droplets within the 50–500 μm diameter range. Conversely, emulsion polymerization has the potential to produce even smaller droplets, ranging from 1 to 100 μm. Additionally, precipitation polymerization stands out for its capability to facilitate particle size modification, potentially achieving dimensions at the micro-nano scale [76].

13.4.1.1 Electron Microscopy

Electron microscopy serves as an umbrella term encompassing analytical techniques such as scanning electron microscopy (SEM), field-emission scanning electron microscopy (FESEM), and transmission electron microscopy (TEM). These methodologies operate based on the fundamental principle of electron beam interaction with the surface of MIPs [77, 78]. In a typical scenario, SEM functions on the principle wherein an electron beam is systematically scanned across the surface of MIPs. This action results in the ejection of secondary electrons, consequently generating images that depict the morphology of the MIPs surface. SEM analysis usually helpful in determining various particle size or geometrical distribution of the MIPs. Certain investigations have even highlighted the utility of SEM in drawing logical comparisons between imprinted and nonimprinted polymers. Notably, Lu and colleagues have conducted research on the label-free detection of nucleoside diphosphate kinase NM23-H2 (NDKB), a significant biomarker for diagnosing colorectal cancer, using MIPs. The MIP-based sensor's structure was analyzed at various stages of fabrication. Particularly, in Fig. 13.5a and b, a lack of notable alterations was observed between the MIPs and the MIPs from which the template had been removed. This suggests that a stable MIPs structure was established through the binding-removal process, ensuring a dependable response during biomarker detection. Furthermore, in Fig. 13.5d, the nonimprinted polymer (NIP) illustrates that Au@Ag nanoparticles were scarcely coated by polydopamine (PDA), resulting in the hindrance of the reporter molecule, thereby yielding no signal for detection purposes [77].

Apart from SEM, transmission electron microscopy (TEM) is increasingly becoming favored within the MIPs research community. TEM offers a more detailed view of MIPs surfaces in comparison to SEM. Unlike SEM, which employs an ejection mechanism, TEM involves transmitting an electron beam through the sample,

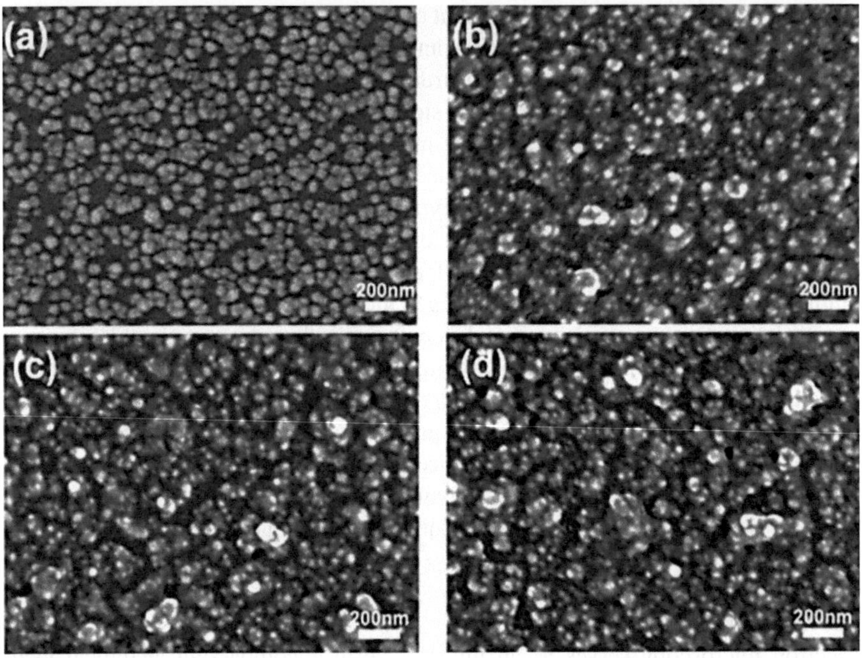

Fig. 13.5 SEM images of (**a**) Au@Ag nanoparticles assembled on silicon wafer, (**b**) The MIPs biosensor with NDKB protein template, (**c**) The MIPs biosensor after the removal of the NDKB protein template, (**d**) The NIPs biosensor

allowing for deeper insights into surface interactions. Consequently, TEM provides information on particle size, geometry, and can effectively differentiate between hollow and compact material structures. Hashkavayi et al. presented TEM images of magnetic MIPs (MMIPs), displaying spherical particles with a size of 30 nm. The study further showcased noticeable voids within the MMIPs structure following template removal, consistent with SEM findings illustrated in Fig. 13.6. It is important to note that while SEM offers qualitative insights into MMIP's spherical shape, TEM images provide greater clarity, enabling precise calculation of the spherical size with enhanced accuracy [79].

13.4.1.2 Particle Size Analysis

Electron microscopy furnishes qualitative evaluations, whereas particle size analysis yields empirical quantitative outcomes for assessing the particle dimensions of MIPs. Typically, two methodologies are utilized for particle size analysis: laser diffraction (LD) and dynamic light scattering (DLS), both of which aid in comprehending the consistency of particle size distribution. LD functions by illuminating particles with a laser beam, leading to scattering effects, while DLS involves the illumination of suspended particles in a compatible solvent with laser light to generate data on size distribution. Despite its capacity to provide quantitative data, particle size analysis is less frequently employed compared to electron microscopy. This technique was predominantly utilized by researchers deeply engaged in particle size

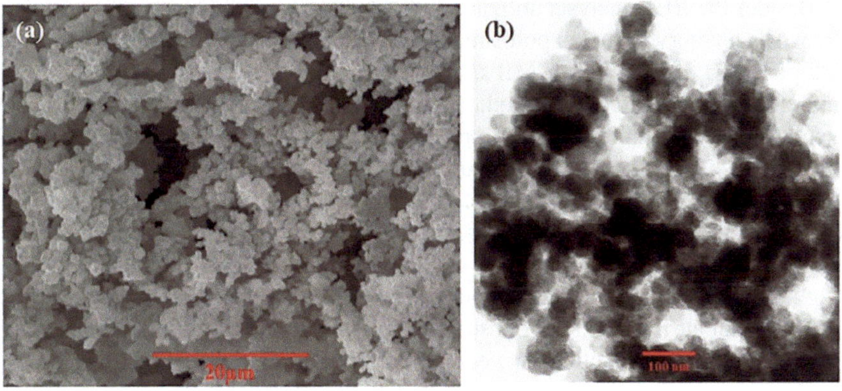

Fig. 13.6 (**a**) SEM images of Fe_3O_4@MIPs, (**b**) TEM image of MIPs

modification, as it served as a key avenue to attain specific objectives during the advancement of MIPs. Mousazadeh et al. employed the DLS technique in conjunction with their FESEM analysis. Specifically, they utilized the DLS method to ascertain the hydrodynamic size of MIPs, resulting in a reported size of 175.4 nm and a polydispersity index (PDI) of 0.073. In contrast, the MIPs particle size determined through FESEM was found to be 104 nm. The authors deduced that these measurements aligned with the DLS analysis. Furthermore, the authors concluded that the diminutive size of the MIPs could be attributed to the compact dimensions of the template molecule [80].

13.4.1.3 Surface Area and Pore Size Determination

Porosity, or particle porosity, refers to the presence of openings or pores on a material's surface, or the proportion of open pores to the total volume of the particle. This phenomenon holds significant importance in the production of MIPs, as these pore characteristics directly influence the binding capacity of template molecules to the MIPs' surface. Two extensively researched techniques are employed for determining porosity: nitrogen gas adsorption and mercury intrusion porosimetry. Typically, a nitrogen adsorption isotherm is utilized to extract details like the overall pore volume, surface area, and distribution of pore sizes [81]. To investigate pore volume, researchers rely on two critical factors: the adsorption of nitrogen gas onto particle pores and the volume characteristics of the particles themselves. Additionally, the specific surface area of MIPs particles is determined using the Brunauer, Emmett, and Teller (BET) method, which calculates a uniform coverage of pore volumes. Calculating parameters like average pore diameter, total pore volume (cm^3/g), and specific surface area (m^2/g) is a common practice. In this context, Kandi et al. observed an increase in total pore volume and specific surface area upon MIPs formation compared to NIPs. The reported MIPs displayed a total pore volume of 0.0186 cm^3/g, whereas NIPs exhibited 0.0063 cm^3/g. Additionally, the specific surface area of the MIPs was reported as 7.538 m^2/g, while the NIPs showed 1.887 m^2/g [82].

Besides the BET analysis utilizing nitrogen adsorption, an alternative technique known as mercury intrusion porosimetry can be employed to ascertain surface area and pore size distribution of MIPs. This method involves introducing the sample onto a holder, subsequently introducing mercury, and increasing pressure to drive mercury into the sample's pores. The correlation between pressure increments and the volume of mercury intruded into the pores is pivotal in determining pore size within this technique. Nevertheless, in comparison to nitrogen adsorption-based BET analysis, mercury intrusion porosimetry is less frequently employed for MIPs analysis. In certain cases, the pore size distribution derived from adsorption data might differ from the information obtained through mercury intrusion. This discrepancy is often attributed to factors such as grain compression during high-pressure intrusion and the distinction between pore body and pore throat distribution data furnished by mercury intrusion [83].

13.4.1.4 Atomic Force Microscopy

Atomic force microscopy (AFM) has significantly contributed to molecular biology by facilitating the identification of uneven surfaces like proteins and ribonucleic acids. Interestingly, the technique's utility expanded when material scientists started manipulating material surface chemistry through processes like immobilization or incorporation of supplementary fillers [84]. In AFM, a scanning cantilever equipped with a tip is employed to image a surface. As the cantilever interacts with the sample's surface, forces are generated, causing a reflected laser beam to be directed toward a detector. The deflection of the reflected beam occurs as the cantilever's tip moves vertically across the surface, and this motion is subsequently detected as shown in Fig. 13.7.

As for MIPs composites, surface roughness would have been additional validation, since cancer biomarkers were used to get immobilized as template molecules on the surface of MIPs [85, 86]. In the context of immobilizing the cancer biomarker 5-hydroxyindole-3-acetic acid (5-HIAA) onto molecularly imprinted polypyrrole (MIPPy), the optimization process, including the number of applied cyclic voltammetry (CV) cycles, was verified using AFM. In a study by Moncer et al., the optimization of MIP sensor thickness was accomplished through varying CV cycles (ranging from 2 to 8). The impact on sensitivity was assessed through differential pulse voltammetry (DPV) current and surface roughness analysis via AFM, as depicted in Fig. 13.8. Results revealed an incremental current response and sensitivity enhancement up to 3 CV cycles. However, beyond 3 cycles, the current declined due to increased MIPs thickness on the working electrode. This excessive thickness hindered rebinding of template biomarkers to MIPPy's recognition sites. This trend was also evident in AFM analysis, which exhibited a significant thickness increase after the fourth cycle. Additionally, roughness measurements confirmed that above 3 CV cycles, the MIPPy film became notably rougher, indicating greater incorporation of 5-HIAA molecules into MIPPy's cavities [85]. Given that immobilization processes often lead to the creation of rougher surfaces, multiple research studies support the idea that AFM can be effectively employed to visualize the successful coating and deposition of MIPs particles onto surfaces.

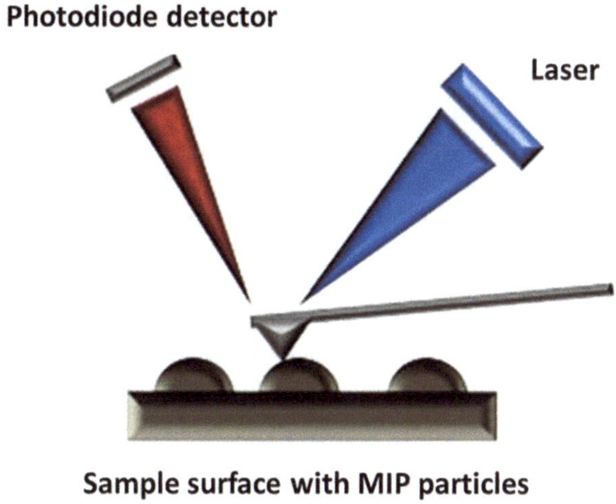

Fig. 13.7 AFM imaging works by using a cantilever tip to reflect a laser beam, which is then detected as the tip moves across a surface. (Adapted with permission from [76])

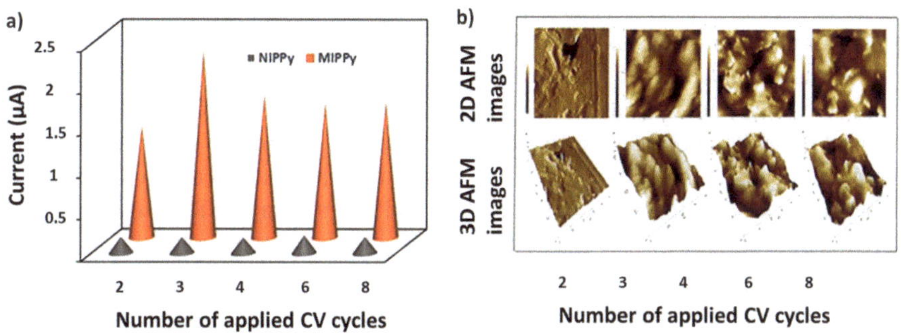

Fig. 13.8 (**a**) Impact of CV cycles on the sensitivity MIPPy against 5-HIAA; (**b**) AFM analysis on the surface roughness with respect to the number of CV cycles

13.4.2 Chemical Characterization

Analyzing chemical modifications on the molecular structure of MIPs demands thorough investigation using spectroscopic techniques. Notably, nuclear magnetic resonance (NMR) and infrared (IR) spectroscopies stand out as effective methods for chemical characterization. Typically, when subjected to a magnetic field, the atomic nuclei of MIPs yield NMR spectra. Researchers have harnessed both solid-state and liquid-state NMR techniques to delve into the chemical structure of MIPs. NMR characterization relies on the magnetic properties of atomic nuclei, often involving analysis of hydrogen (^{1}H) and carbon-13 (^{13}C) [87, 88]. Certain researchers employ a mapping technique that involves 2D NMR COSY (Correlation Spectroscopy) between ^{1}H and ^{13}C NMR. This technique facilitates detailed analysis of molecular interactions and correlations within the molecular structure of the sample [89]. This method makes it possible to investigate complex molecular interactions and correlations within the sample's molecular structure. Researchers acquire a better understanding of the chemical alterations and structural changes that take place in MIPs by mapping the links between various atomic nuclei (MIPs). This method advances our knowledge of the intricate structure of MIPs and helps to improve both their design and functional attributes.

Infrared (IR) spectroscopy is a widely used chemical characterization technique that involves irradiating an MIPs containing specific functional groups. The radiation is absorbed by these functional groups, resulting in distinct absorption bands within the fingerprint region of the spectrum. By analyzing the obtained IR spectrum, the presence of functional groups in MIPs can be deduced. This technique often involves comparing MIPs spectra with NIPs spectra to ensure structural similarity once the template is removed. During the assembly of MIPs, weak non-covalent interactions facilitate easy removal and rebinding of template molecules.

A study by Oliveira et al. demonstrated the effectiveness of IR spectroscopy in differentiating between MIPs based on the tumor biomarker CA 15-3 (associated with breast cancer) and their corresponding NIPs. The respective IR fingerprint regions exhibited significant differences, indicating the presence of unique interactions in the MIPs. Moreover, the results suggested the absence of significant interactions between CA 15-3 and the NIPs, highlighting the specificity of the MIPs–template interactions [90].

In conclusion, it can be said that a wide variety of approaches are available for accurately characterizing MIPs in cancer biomarker detection. These methods can be used in tandem because they are complementary. It should be highlighted, nonetheless, that the effective application of a particular characterization technique depends greatly on the particular template and MIPs compositions used within a certain research topic.

13.5 Applications of MIPs in Cancer Biomarker Analysis

MIPs have found various applications in cancer detection, utilizing both non-covalent and covalent synthesis methods as shown in Table 13.2 [75, 77, 85, 86, 90–120]. These synthetic polymers are designed to recognize specific biomolecules and are widely used for the detection of cancer biomarkers. Some commonly detected cancer biomarkers include ProGastrin-releasing peptide (ProGRP), neuron-specific enolase (NSE), α-fetoprotein (AFP), sarcosine (SAR), carcinoembryonic antigen (CEA), and carbohydrate antigen 15-2 (CA 15-3), each associated with different cancer types [75, 90, 92–97, 99–101, 108, 109, 119]. In addition to protein biomarkers, volatile organic compounds like nonanal and hexanal have been identified as biomarkers for disease monitoring, particularly in lung cancer patients. These compounds are present in various bodily fluids and are detected in human exhalation, feces, urine, blood, sweat, saliva, and milk [91].

Non-covalent MIPs synthesis methods include precipitation polymerization, often utilizing acetonitrile as the porogen solvent. This technique involves mixing templates, monomers, initiators, and cross-linkers [75, 91–93, 95, 98, 99, 101, 102,

121]. Other polymerization methods such as bulk and solvothermal polymerization have also been employed [94, 96, 97]. Meanwhile, there was a unique polymerization method of MIPs which was in situ polymerization where monomers, initiator, and crosslinker were added to cell lines which already have natural biomarkers as templates [103]. Covalent MIPs synthesis has been employed in the production of MIPs-based sensors, particularly in electrochemical detections. This approach enhances sensor characteristics, including selectivity, response time, surface area, sensitivity, and stability. Covalent MIPs are synthesized through electropolymerization of templates onto electrode surfaces, such as silver and gold surfaces, without the use of initiators or crosslinkers [77, 85, 86, 90, 104–119, 122]. However, Patra et al. (2015) used another method called controlled radical polymerization in synthesizing MIP based- electrochemical sensor in detecting PSA from blood and urine samples [120].

The limit of detection (LOD) values for different biomarkers vary across the applications mentioned in Table 13.2. For instance, the LOD for nonanal, a volatile organic compound used as a lung cancer biomarker, is 4.5 mg L^{-1} [91]. NSE and ProGRP, which are biomarkers for various cancers, exhibit LODs of 0.11 nM and 0.06 nM, respectively [75]. SAR, a biomarker for prostate cancer, has an LOD of 0.02 nM, while CEA shows varying LOD values, with one method yielding LODs of 5.6×10^{-14} M for immobilized AuNPs SAM-coated arrays and 3.5 fg mL^{-1} for fluorescent molecularly imprinting conjugated polythiophenes (FMICPs) [94, 99, 100]. In other cases, AFP detection using various methods results in LODs of 0.38 μM, 3.3 pg mL^{-1}, and 15 fg mL^{-1} in urine, saliva, and serum matrices [95, 100, 107]. Similarly, CA 15–3 detection showcases a range of LOD values, including 1.16 U mL^{-1}, 0.10 U mL^{-1}, and 1.5 U mL^{-1} for various MIP based-biosensors [90, 108, 109]. The LODs for other cancer biomarkers like nucleoside diphosphate kinase NM23-H2 (NDKB), human epidermal growth factor receptor (HER2-ECD), cytochrome c (Cyt c), and Neopterin vary from 0.82 pg mL^{-1} to 22 μM [77, 110, 116, 118]. These LOD values reflect the sensitivity of the detection methods and the potential for early cancer diagnosis. In summary, the LOD values reported in the table provide insights into the sensitivity of MIP-based methods for cancer biomarker detection, suggesting that MIPs hold promise for enhancing the accuracy of cancer diagnosis by detecting biomarkers at extremely low concentrations.

In the realm of cancer detection, MIPs exhibit great promise, showcased through both non-covalent and covalent synthesis methods. These synthetic polymers, designed to selectively recognize specific cancer biomarkers, are opening new avenues for early diagnosis and monitoring. By employing a range of techniques such as precipitation polymerization and electrochemical sensing, MIPs offer impressive sensitivities, as evident from the LOD values reported in Table 13.2. With the potential to detect biomarkers at extremely low concentrations, MIPs are poised to revolutionize cancer diagnostics, contributing to improved patient outcomes and fostering advancements in personalized healthcare strategies

Table 13.2 Types of cancer biomarkers and their respective MIPs, synthesis conditions of MIPs, analytical methods, and validations in different matrices

Cancer biomarker	Analytical method	Matrix	Template/monomer/progen solvent/initiator/cross-linker	Polymerization condition (imprinting approach, solvents, temperature/time)	Template eluting (solvents/methods)	LOD/LOQ	Recovery (%)	Imprinting factor	Matrix effect (%)	Ref.
Non-covalent MIPs										
Nonanal (VOC)	MIPs-based chemiresistive sensor/potentiostat	Human plasma	Nonanal/MAA/MeCN/AIBN/EDGMA	Precipitation polymerization in MeCN at 65–70 °C/16 h	Soxhlet extraction using MeCN	4.5 mg L^{-1}/–	–	–	–	[91]
NSE and ProGRP	On-line MISPE/LC–MS/MS	Serum	NLLGLIEA and ELPYR/EAMA.HCl and TPVU/MeCN and THF/AIBN/DVB-80	Precipitation polymerization/MeCN and THF/60 °C/48 h	MeCN ∀→ 0.1 M MeOH → 10% HCl → MeOH/wash sequentially	0.11 nM/0.37 nM (NSE), 0.06 nM/0.2 nM (ProGRP)	54.3–94.1	28	–	[75]
ProGRP	Online MISPE/LC-MS	Serum	Z-NLLGLIEA[Nle]/EAMA HCl/MeCN/DVB-80	Precipitation polymerization/sonicated (10 min) → ice-bath temperature (10 min) → 60 °C (48 h)	MeCN → MeOH/0.1 M HCl (90/10, v/v) → MeOH	17.2 pM/–	–	–	–	[92]
			Z-NLLGLIEA[Nle]/EAMA HCl and *N*-3,5-*bis*(trifluoromethyl)-phenyl-*N′*-4-vinylphenylurea/MeCN/DVB-80							

ProGRP	mMIP microsphere	Serum	NLLGLIEAK/DVB-80 and BTPV/MeCN: Toluene (75:25, v/v)/AIBN/MAA and EAMA.HCl	Two-step precipitation polymerization: 1. Co-polymerization of DVB-80 and MAA to produced poly(MAA-co-DVB-80) microspheres with presence of AIBN and MeCN:Toluene → chemical precipitation of poly(MAA-co-DVB-80) microspheres with presence of Fe_3O_4 precursor and alkalinization step to produced magnetic poly(MAA-co-DVB-80) 2. Delayed addition of NLLGLIEAK into mixture of magnetic poly(MAA-co-DVB-80), EAMA.HCl, and BTPV to produce MMIP microsphere	Washing with 50 mM ammonium bicarbonate	39 pM/129 pM	25	6.1	–	[93]
	mMIP polymer microsphere			Precipitation polymerization with template presence in the first step for 24 h followed by magnetization procedure						
SAR	MTCL-MDSPE/ UHPLC-MS/MS	Urine	MTCL373-SAR/	Solvothermal polymerization	Washed by anhydrous etOH and deionized water in turn for three times	0.02 nM	96.1–107.4	–	–	[94]
SAR	MIP@CPE/CV and DPV	Urine	SAR/MAA/H_2O: MeCN, (1/4; v/v)/AIBN/EGDMA	Precipitation polymerization in an oil bath at 60 °C for 20 h	Eluted several times with MeOH:HAc (9/1; v/v)	0.38 μM	97.4–103.8	–	8.1	[95]
SAR	MISPE with electrochemical determination at a MWCNT/Nafion®/ $Ni(OH)_2$-modified screen-printed electrode/FIA	Urine	SAR/APMS in MeCN:MeOH (1:1, v/v) and 4-VP/AIBN/TRIM	Bulk polymerization at 60 °C in an oil bath for 24 h	Eluted with acetone, MeCN, 1 M HCl and UP water	0.96 μmol L^{-1}/3.20 μmol L^{-1}	95–108	–	–	[96]
SAR	DMFC powered biosensor/EC	Serum	SAR/CB with PtRu, 40:20 wt% and VPA/MES buffer/APS and TEMED/ Acr and Bis-Acr	Bulk radical polymerization at 30 °C for 1 h	Stirring in 0.5 M H_2SO_4 for 30 min and washing with UP water and MeOH several times	0.04 μM	–	–	–	[97]

(continued)

Table 13.2 (continued)

Cancer biomarker	Analytical method	Matrix	Template/monomer/progen solvent/initiator/ cross-linker	Polymerization condition (imprinting approach, solvents, temperature/ time)	Template eluting (solvents/ methods)	LOD/LOQ	Recovery (%)	Imprinting factor	Matrix effect (%)	Ref.
PA	MIP/CNT-CPE sensor/voltammetry	Plasma and urine	PA/MAA/MeCN/AIBN/ EGDMA	Precipitation polymerization in MeCN using oil bath at 60 °C for 12 h	Washing the polymer with hot etOH and water	0.048 μM	96.7–104.0 (plasma) and 97.8–103.2 (urine)	–	–	[98]
CEA	odMIP-PISA/ Raman spectroscopy	Serum	CEA glycated epitope-immobilized AuNPs SAM coated arrays/, APTES and TEOS/ anhydrous EtOH/IBTES/ UPTES	Precipitation polymerization in anhydrous EtOH	Rinsed with MeCN: H2O: HAc (v/v; 50: 49: 1) for 20 min at room temperature → washing with anhydrous EtOH and water for three times each	5.6×10^{-14} M/5.6×10^{-12} M	–	12.3	–	[99]
AFP and CEA	FMICPs/LC-MS/ MS	Saliva and serum	AFP or CEA/CP1 or CP2/ PBS/1-hydroxycyclohexyl phenyl ketone (photo-initiator)/PEGDA	Ultrasonication of CP1/CP2 and VPBA in PBS → AFP/CEA stock solution was added and shaken at room temperature for 1 h → added PEGDA and 1-hydroxycyclohexyl phenyl ketone → UV polymerization	Dispersed and shaken for 2 h at room temperature in HAc: MeOH (1:9) mixture	15 fg mL^{-1} (AFP) and 3.5 fg mL^{-1} (CEA)	96.9–110.0 (saliva, AFP), 97.6–110.0 (serum, AFP) and 97.0–110 (serum, CEA)	8.4 (AFP) and 9.8 (CEA)	–	[100]
AFP	triMIP-PISA/ Raman spectrography	Serum	AFP-anchored array/ APTES and BnTES/ anhydrous etOH containing 3% NH_4OH (28%)/IBTES/TEOS	Precipitation polymerization in anhydrous etOH containing 3% NH_4OH (28%) at 25 °C for 50 min	Wash with etOH and water for three times each	–	–	14.4	–	[101]
N-phenylgly- cine	*N*-phenylglycine imprinted polymer	Urine	*N*-phenylglycine/ acrylamide and trimethylpropane triacylate/MeCN/AIBN	Precipitation polymerization in MeCN for 24 h at 60 °C	Soxhlet extraction with MeOH:HAc (9:1, v/v) → removed residual acid with MeOH	0.016 mg L^{-1}/0.055 mg L^{-1}	84.7–100.0 0%	–	–	[102]
Lung cancer markers (protein epitopes)	NanoMIP/LC–MS/ MS	NCI-H23, NCI-H460, NCI-H522 and A549 cell lines	Protein epitopes/NIPAm, TBAm and APMA/PBS/ potassium persulfate, and TEMED/MBAA	In situ polymerization: Mixture of functional monomers (NIPAm, TBAm, and APMA) and cross-linker (MBAA) was prepared in PBS and purged with N_2 gas for 20 min → added to flask of cells with contained no media → addition of initiator (potassium persulfate and TEMED) → polymerization at room temperature for 1 h	Template was not removed. NanoMIPs were collected from trypsinized cells and wash with PBS for further sequencing analysis	–	–	–	–	[103]

Covalent MIPs										
CA-125	MIP@AuSPE/EIS	Artificial serum	CA-125/Py/PB	Electropolymerization/PB/–/25 °C/–	25 × 10^{-3} mol L^{-1} SDS in PB and MeOH (10:1, v/v)/ incubated overnight	0.01 U/mL/–	91–105	–	–	[104]
5-HIAA	Electrochemical sensor based on MIPPy/potentiostat/ galvanostat	Serum, urine and plasma	5-HIAA/Py/0.1 M $LiClO_4$	Electropolymerization/0.1 M $LiClO_4$/–	0.05 M NaOH/ overoxidation mechanism	5 × 10^{-5} M/–	–	–	–	[85]
PSA	MIP based-electrochemical nanobiosensor/ potentiostat/ galvanostat	Serum	PSA/Py/0.1 M KCl	Electropolymerization/0.1 M KCl/4 °C/24 h	0.5 M H_2SO_4 and 0.005 M oxalic acid/incubated overnight	2.0 pg mL^{-1}/–	–	–	–	[105]
GSSG, GSH, Cys, Cyss, $NADP^+$ and NADPH	MIP-decorated multiplex electrochemical sensors/redox potential	Blood	Cyss, Cys, NADPH and $NADP^+$/OPD/$[Fe(CN)_6]^{3-/4-}$	Electropolymerization/ $[Fe(CN)_6]^{3-/4-}$/	0.1 M NaOH/immersion and cyclic voltammetry	20 pmol L^{-1}/–	92–112	–	–	[106]
AFP and CEA	DMIP based one-by-one approach sensor/ Potentiostat/ Galvanostat	Serum	AFP and CEA/Py and MO/ PBS	Electropolymerization between 0.0 and 0.7 V at a 50 mV/s scan rate/ PBS	0.1 M NaOH/rinsed	3.3 pg mL^{-1} (AFP) and 1.6 pg mL^{-1} (CEA)	92.0–98.8	–	–	[107]
CA 15-3	Disposable electropolymerized molecularly imprinted electrochemical sensor/CV	Serum	CA 15-3/CNE/AuNP/oPD	Electropolymerization between 0.2 and 1.0 V in the oPD solution at scan rate of 50 mV s^{-1}/oPD	PBS → deionized water → dried under N_2 stream	1.16 U mL^{-1}/3.87 U mL^{-1}	101.8–104.3	–	–	[90]
CA 15-3	Poly(toluidine blue) MIP/Potentiostat	Serum	CA 15-3/protein imprinted poly(TB) films which were grown in TB tailed SAM/0.1 M PBS TB-SAM	Electropolymerization in 0.1 M PBS between −0.50 and 0.70 V, at The scan rate of 20 mV s^{-1}	Electropolymerization at −0.55 or −0.65 to 0.35 V, at 50 mV s^{-1}, in 0.1 Mol L^{-1} PBS (at pH 7.4) containing 0.1 mol L^{-1} $NaNO_3$ (dopant)	0.10 U mL^{-1}	99–102	–	–	[108]

(continued)

Table 13.2 (continued)

Cancer biomarker	Analytical method	Matrix	Template/monomer/progen solvent/initiator/ cross-linker	Polymerization condition (imprinting approach, solvents, temperature/ time)	Template eluting (solvents/ methods)	LOD/LOQ	Recovery (%)	Imprinting factor	Matrix effect (%)	Ref.
CA 15–3	Molecularly imprinted electrochemical sensor	Serum	CA 15–3/Au-SPE/10 mM 2-aminophenol	Electropolymerization where CA 15-3 was put onto activated Au-SPE surface for 20 min → rinsed with PBS → added 2-aminophenol	Immersed in 0.5 M oxalic acid (12 h), and 0.1 M PBS (1 h)	1.5 U mL^{-1}	–	–	–	[109]
NDKB	Label-free MIP-SERS biosensor/Raman spectroscopy	Serum	NDKB/SERS active substrate and dopamine/10 mM Tris-HCl buffer	Oxidative polymerization by immersing template directly with SERS active substrate as monomer in Tris-HCl buffer for 9 h at room temperature	Wash several times with ultrapure water and dried	0.82 pg mL^{-1}	–	–	–	[77]
HER2-ECD	Electrochemical MIP sensor	Serum	HER2-ECD/ AuSPEs/20 mM phenol and phosphate buffer	Electropolymerization by (CV) in phenol and PB	Rinsed with ultra-pure water and 40 μL of an SDS (0.5%) and acetic acid (0.5%) solution for 30 min → UP water (10 min) → PB (5 min)	1.6 ng mL^{-1}/5.2 ng mL^{-1}	80.6–90.2	–	–	[110]
EGFR and VEGF	Au-SPE)-based biosensor modified with an MIP	Serum	EGFR and VEGF/DSP modified AuSPE/10 mM Acr and 70 mM NNMBA	Incubate DSP with EGFR and VEGF → polymerized with using persulfate by adding AAM and NNMBA → add APS in PBS → stop polymerization using hydroquinone solution in etOH	MIP(s)-SPE was incubated in the dilute solution of OXA for 12 h	0.01 pg mL^{-1} (EGFR) and 0.005 pg mL^{-1} (VEGF)	–	–	–	[111]
CEA	Self-powered and self-signaled autonomous electrochemical biosensor	Urine	CEA/Py on FTO/PEDOT/ PB	Electropolymerizing by CV in PB	Washed with PB and incubated in proteinase K for 5 h	0.14 ng mL^{-1}	–	–	–	[112]
CEA	Bio-MIP-ePADs	Serum	CEA/surface imprinting consists of GO, chitosan, and glutaraldehyde/PBS and dopamine	Surface imprinting using GO, chitosan, and glutaraldehyde → washed with PBS → addition of CEA → addition of PBS containing dopamine → electropolymerization	Removed thoroughly by SDS/HAc solution	0.32 ng mL^{-1}	–	–	–	[113]

CEA	SMIP/SERS	Blood	CEA/CMSM-coated circular arrays/PBS	Boronate affinity-based controllable oriented Surface imprinting between CEA in PBS and CMSM-coated circular arrays	Immersion in 0.1 M HAc containing 10% SDS (wt/vol)	0.064 pg mL^{-1}	98.4–106.3	17.65	–	[114]
CEA	DMFC sensor/potentiostat/galvanostat	Serum	CEA/Py and EDOT/PEDOT	Py and EDOT were electropolymerized onto PEDOT in presence of CEA	Incubation of electrode in proteinase K	0.08 ng mL^{-1}	–	–	2.65	[115]
Neopterin	Potentiometricche-mosensor/	Synthetic serum	Neopterin/2,2′-bithiophene-5-boronic acid and 2-(cytosin-1-yl)ethyl p-bis(2,2′-bithien-5-yl) methylbenzolate/0.1 M $(TBA)ClO_4$ and 0.1 M NaOH in MeCN (1:9)/2,4,5,2′,4′,5′-hexa (thiophene-2-yl)-3,3′-bithiophene	Electropolymerization in 0.1 M $(TBA)ClO_4$	Electropolymerization in 0.1 M carbonatebuffersolution	22 μM	–	~ 6	–	[116]
NSE	MIP-based synthetic receptors/cyclic voltammetry	Serum	CME/QCM gold surface/scopoletin/TCEP buffer	Immobilization of CME onto QCM gold surface based on self-assembled monolayer (SAM) formation (3 h) → addition of scopoletin in 0.1 M NaCl → electropolymerization	Anodic desorption with the multiple pulse amperometry technique at 0.9 V for 30 s	2 ng mL^{-1}	–	8.8	–	[117]
			Aqueous mixture of histidine containing peptide/scopoletin/gold wire surface	Electropolymerization of the aqueous mixture of the histidine containing peptide and scopoletin on the gold wire surface	Alkaline treatment in 0.1 M NaOH solution (40 min) → thorough washing with distilled water	1 ng mL^{-1}	–	11.0	–	
NSE	AuNPs decorated epitope-mediated hybrid MIPs electrochemical sensor/square wave voltammetry	Serum	Cys-Ep1 and Cys-Ep2/gold working electrode/PBS–TCEP	Incubation of gold working electrode in mixed solution of epitopes	Anodic potential application of gold working electrode in water	200 pg mL^{-1}	–	7.3	–	[86]

(continued)

Cancer biomarker	Analytical method	Matrix	Template/monomer/progen solvent/initiator/ cross-linker	Polymerization condition (imprinting approach, solvents, temperature/ time)	Template eluting (solvents/ methods)	LOD/LOQ	Recovery (%)	Imprinting factor	Matrix effect (%)	Ref.
Cytochrome c (Cyt c)	Molecularly imprinted electrochemical sensor/potentiostats	Serum	Cyt c/gold electrode and o-PD/0.1 M acetate buffer	Electropolymerization of Cyt and o-PD onto gold electrode in 0.1 M acetate buffer	The gold electrode was dipped into 0.1 M H_2SO_4	4.2×10^{-2} pg mL^{-1}	98.5–131	133	–	[118]
AFP	Molecularly imprinted photochemical and electrochemical sensor	Serum	AFP/CdS@ZnO/ITO/PBS	Electropolyrization of AFP in PBS onto CdS@ZnO/ITO	CdS@ZnO/ITO was immersed into 0.1 M NaOH for 20 min	0.38 pg mL^{-1}	99.2–105.2	–	–	[119]
PSA	Molecularly imprinted electrochemical sensor	Blood and urine	PSA/itaconic acid/DMSO/ EDGMA	Controlled radical polymerization	Not stated	0.25 pg L-1 (SWSV) and 3.04 pg L-1 (DPSV)	99.2–101.2 (blood) and 99.6–101.0 (urine)	–	–	[120]

13.6 Comparison of Method-Based MIPs to Conventional Methods in Cancer Detection

Measurements of cancer progression and diagnosis are very important for successful monitoring of cancer. Early-stage cancer can be efficiently treated using the available therapeutic interventions. However, the lack of symptoms during these initial stages often resulted in delayed diagnosis in the more advanced stages, where the treatments may be less effective. Thus, cancer biomarkers play a very important role in cancer diagnosis since these markers exhibit unique genomic changes in cancer cells as the disease advances [123]. In general, there are several advantages and drawbacks to each strategy for the determination and quantification of cancer. Some of the bioassays that are used include immunohistochemistry, radioimmunoassay, fluorescent spectrometry, and enzyme-linked immunosorbent assay (ELISA) [124]. However, these approaches have several drawbacks, such as complicated procedures and equipment, that are expensive, laborious, difficult, and time consuming [125]. Other techniques were also applied, such as nuclear magnetic resonance (NMR), X-ray, tissue biopsy, electrical signal monitoring, and scanning techniques based on positron and computed emission tomography. However, because most of the methods are size-based processes, the techniques are limited to a huge number of cells. Techniques such as mass spectrometry, electrophoresis, surface plasmon, and polymerase chain reaction (PCR) have been developed for the early determination of cancer [124]. These methods suffer from low sensitivity and selectivity due to the production of unnecessary results, which require implementing the whole procedure again. One of the main disadvantages of the methods is the limited examination of the response of patients toward therapeutic actions.

Rapid and accurate tools for the determination of cancer-specific biomarkers are one of the ways to improve early cancer diagnosis [126]. The specific biomarkers are present in the blood and serum specimens of the patients. Thus, the quest to explore analytical diagnostic methods such as biosensors is significant for identifying and determining biomarkers. A biosensor is an analytical device composed of biological elements that lead to direct or indirect interaction with the analyte as well as a transducer to transduce a signal. While immunosensors are a category of biosensors in which antibodies are employed as biological components. Some of the advantages of biosensors include high sensitivity and selectivity, being portable, and compact, and requiring the minimum amount of sample or specimen [127–130].

The majority of the developed biosensors used natural biomolecules as the recognition elements, which raises concerns pertaining to the stability of these sensors under harsh conditions. In addition, the relatively high cost of natural biomolecules like antibodies adds an extra cost to cover the overall expense of the developed sensors. To overcome these shortcomings, MIPs are proposed as low-cost, stable, and sensitive alternatives to natural antibodies. Compared to other available methods for cancer detection, the MIP-based methods offer various advantages that include: i) the method requires minimal cost as compared to natural receptors since the method is based on the polymerization step, ii) the methods can be reusable, and iii) cost-effectiveness compared to other similar products [2].

The recent development in incorporating MIPs onto magnetic nanoparticles has shown significant advancement due to their wide applicability. These magnetic MIPs can be used to simultaneously identify, preconcentrate, and isolate important biomarkers for various applications. The magnetism properties have simplified the overall process to possibly separate the particles in less than a minute and significantly reduce the washing steps of MIPs preparation [131]. Moreover, the development of new polymerization methods with the use of custom-made monomers can improve the efficiency and stability of MIPs [2]. The MIPs can be synthesized by using multiple monomers that can assist toward the amino acid-like binding sites, which will ultimately improve the binding nature of the imprinted sites [132, 133]. Other than that, multimodal biomarker imprinting strategies also offer accurate and reliable strategies for cancer detection [134]. Through comprehensive research and the assistance of modern technology, there is potential for the development of affordable and cost-effective MIP-based sensors that are accessible to all.

The current biosensor platforms, which are comprised of electrical and optical systems, detect cancer-related changes through electrical property alteration, and optical biosensors employ light interaction for sensitive cancer biomarker detection. While MIP-based biosensors used synthetic receptors to selectively bind to cancer-specific molecules [135]. In general, electrical biosensors offer label-free detection, optical biosensors provide real-time monitoring, and MIPs biosensors win in specificity and stability. Nevertheless, there is an urgency for these MIP-based sensors to be rapidly developed for cancer detection, with improved stability and cost-effectiveness that could resolve the financial strain of medical costs.

13.7 Challenges and Future Perspective of MIPs in Cancer Monitoring

The early, rapid, and accurate detection of cancer biomarkers plays a pivotal role in the successful treatment of this life-threatening disease. In recent decades, the development of MIPs has brought hope to the field of cancer detection, and the progress achieved thus far is indeed remarkable. However, it is important to acknowledge that MIPs face several significant challenges that require ongoing research efforts to overcome. First, one of the key challenges with MIPs is their selectivity, which can be lower when multiple templates (targets) are involved. This is often due to the complex interplay of different target molecules and the heterogeneity of imprinting sites within the polymer framework. Achieving high selectivity in such cases remains a substantial hurdle. Second, the recognition of biomacromolecules by MIPs remains a formidable challenge. Biomacromolecules are characterized by their large molecular size and the presence of numerous diverse functional groups, making it difficult for MIPs to precisely recognize bind to these complex structures. Third, the issue of template bleeding is a concern, particularly when identifying multiple trace targets of interest. Preventing cross-contamination or interference between templates is essential for accurate detection and quantification. Fourth, the development of efficient and environmentally friendly polymerization methods for MIPs is an ongoing requirement. Green chemistry principles are increasingly important, and finding sustainable synthesis routes for MIPs is an essential aspect of their future development.

13.8 Conclusion

In conclusion, early cancer detection is a critical factor in improving the success rates of cancer treatment. Biomarkers have emerged as essential components in this process, facilitating the identification and monitoring of cancer. The introduction of MIPs as artificial antibodies or receptor-like materials has expanded the horizons of cancer detection and monitoring. MIPs, with their molecular recognition capabilities, exhibit selective binding to molecules of interest, including cancer biomarkers. This chapter has provided an overview of the recent advancements in MIPs technology from 2016 to 2023, with a particular focus on their role in selectively extracting and microextracting cancer biomarkers. The discussion encompassed the design, fabrication, and analytical performance of MIPs, evaluating their detection limit of detection and quantification, recovery, and imprinting factor. Notably, MIPs designed specifically for cancer detection have demonstrated enhanced specificity and selectivity when compared to conventional detection methods. Furthermore, this article delved into the practical applications of MIPs in cancer biomarker analysis, highlighting their potential as diagnostic tools for early cancer detection. By offering improved precision and sensitivity, MIPs have the potential to revolutionize cancer diagnostics. There are challenges to be overcome, including the refinement of MIPs as artificial antibiotics, the future of MIPs in cancer monitoring appears promising. These advancements hold significant potential for enhancing cancer diagnostics and therapies, ultimately contributing to better patient outcomes and improved cancer care.

References

1. Suzaei FM, Batista AD, Mizaikoff B, Rahimi S, Daryanavard SM, Abdel-Rehim M (2022) Molecularly imprinted polymers for selective extraction/microextraction of cancer biomarkers: a review. Microchim Acta 189(7):255. https://doi.org/10.1007/s00604-022-05356-9
2. Bhakta S, Mishra P (2021) Molecularly imprinted polymer-based sensors for cancer biomarker detection. Sens Actuat Rep 3:100061. https://doi.org/10.1016/j.snr.2021.100061
3. Ben Moussa F (2023) Molecularly imprinted polymers meet electrochemical cancer chemosensors: a critical review from a clinical and economic perspective. Microchem J 191:108838. https://doi.org/10.1016/j.microc.2023.108838
4. Wang S, Zhang L, Zeng J, Hu X, Wang X, Yu L et al (2023) Multi-templates molecularly imprinted polymers for simultaneous recognition of multiple targets: from academy to application. TrAC Trends Anal Chem 166:117173. https://doi.org/10.1016/j.trac.2023.117173
5. Gkika DA, Tolkou AK, Lambropoulou DA, Bikiaris DN, Kokkinos P, Kalavrouziotis IK et al (2024) Application of molecularly imprinted polymers (MIPs) as environmental separation tools. RSC Appl Polymers. https://doi.org/10.1039/D3LP00203A
6. Sengar MS, Kumari P, Sengar N, Singh SK (2024) Molecularly imprinted polymer technology for the advancement of its health surveillances and environmental monitoring. ACS Appl Polymer Mater 6:1086–1105. https://doi.org/10.1021/acsapm.3c02185

7. Ahmad A, Imran M, Ahsan H (2023) Biomarkers as biomedical bioindicators: approaches and techniques for the detection, analysis, and validation of novel biomarkers of diseases. Pharmaceutics 15(6):1630. https://doi.org/10.3390/pharmaceutics15061630
8. Jain KK (2010) The handbook of biomarkers, 1st edn. Humana, Totowa, NJ
9. Haller N, Behringer M, Reichel T, Wahl P, Simon P, Kruger K et al (2023) Blood-based biomarkers for managing workload in athletes: considerations and recommendations for evidence-based use of established biomarkers. Sports Med 53(7):1315–1333. https://doi.org/10.1007/s40279-023-01836-x
10. Tiongco R-E, Bituin A, Arceo E, Rivera N, Singian E (2018) Salivary glucose as a non-invasive biomarker of type 2 diabetes mellitus. J Clin Exp Dent 10(9):e902. https://doi.org/10.4317/jced.55009
11. Tang S, Yuan K, Chen L (2022) Molecular biomarkers, network biomarkers, and dynamic network biomarkers for diagnosis and prediction of rare diseases. Fundam Res 2(6):894–902. https://doi.org/10.1016/j.fmre.2022.07.011
12. Wishart DS, Bartok B, Oler E, Liang KYH, Budinski Z, Berjanskii M et al (2021) MarkerDB: an online database of molecular biomarkers. Nucleic Acids Res 49(D1):D1259–D1267. https://doi.org/10.1093/nar/gkaa1067
13. He Z, Gao K, Dong L, Liu L, Qu X, Zou Z et al (2023) Drug screening and biomarker gene investigation in cancer therapy through the human transcriptional regulatory network. Comput Struct Biotechnol J 21:1557–1572. https://doi.org/10.1016/j.csbj.2023.02.005
14. Kadim MK, Risjani Y (2022) Biomarker for monitoring heavy metal pollution in aquatic environment: an overview toward molecular perspectives. Emerg Contam 8:195–205. https://doi.org/10.1016/j.emcon.2022.02.003
15. Califf RM (2018) Biomarker definitions and their applications. Exp Biol Med (Maywood) 243(3):213–221. https://doi.org/10.1177/1535370217750088
16. Kamel HFM, Al-Amodi HSB (2016) Cancer biomarkers. In: Role of biomarkers in medicine. IntechOpen
17. Bodaghi A, Fattahi N, Ramazani A (2023) Biomarkers: promising and valuable tools towards diagnosis, prognosis and treatment of Covid-19 and other diseases. Heliyon 9(2):e13323. https://doi.org/10.1016/j.heliyon.2023.e13323
18. Gonzalez H, Hagerling C, Werb Z (2018) Roles of the immune system in cancer: from tumor initiation to metastatic progression. Genes Dev 32(19–20):1267–1284. https://doi.org/10.1101/gad.314617.118
19. Nagpal M, Singh S, Singh P, Chauhan P, Zaidi MA (2016) Tumor markers: a diagnostic tool. Natl J Maxillofac Surg 7(1):17–20. https://doi.org/10.4103/0975-5950.196135
20. Li J, Guan X, Fan Z, Ching LM, Li Y, Wang X et al (2020) Non-invasive biomarkers for early detection of breast cancer. Cancers (Basel) 12(10):2767. https://doi.org/10.3390/cancers12102767
21. Sarhadi VK, Armengol G (2022) Molecular biomarkers in cancer. Biomol Ther 12(8):1021. https://doi.org/10.3390/biom12081021
22. FDA (2022) Premarket Notification 510(k). https://www.fda.gov/medical-devices/premarket-submissions-selecting-and-preparing-correct-submission/premarket-notification-510k. Accessed 25 Aug 2023
23. Beaver JA, Tzou A, Blumenthal GM, McKee AE, Kim G, Pazdur R et al (2017) An FDA perspective on the regulatory implications of complex signatures to predict response to targeted therapies. Clin Cancer Res 23(6):1368–1372. https://doi.org/10.1158/1078-0432.CCR-16-1098
24. Scherf U, Becker R, Chan M, Hojvat S (2010) Approval of novel biomarkers: FDA's perspective and major requests. Scand J Clin Lab Invest Suppl 242:96–102. https://doi.org/10.3109/00365513.2010.493415
25. Seale KN, Tkaczuk KHR (2022) Circulating biomarkers in breast cancer. Clin Breast Cancer 22(3):e319–e331. https://doi.org/10.1016/j.clbc.2021.09.006

26. Fakhari A, Gharepapagh E, Dabiri S, Gilani N (2019) Correlation of cancer antigen 15-3 (CA15-3) serum level and bony metastases in breast cancer patients. Med J Islam Repub Iran 33:142. https://doi.org/10.34171/mjiri.33.142
27. Tang Z, Li L, Shen L, Shen X, Ju S, Cong H (2018) Diagnostic value of serum concentration and integrity of circulating cell-free DNA in breast cancer: a comparative study with CEA and CA15-3. Lab Med 49(4):323–328. https://doi.org/10.1093/labmed/lmy019
28. Chu WG, Ryu DW (2016) Clinical significance of serum CA15-3 as a prognostic parameter during follow-up periods in patients with breast cancer. Ann Surg Treat Res 90(2):57–63. https://doi.org/10.4174/astr.2016.90.2.57
29. Pupa SM, Ligorio F, Cancila V, Franceschini A, Tripodo C, Vernieri C et al (2021) HER2 signaling and breast cancer stem cells: the bridge behind HER2-positive breast cancer aggressiveness and therapy refractoriness. Cancers (Basel) 13(19):4778. https://doi.org/10.3390/cancers13194778
30. Miligy IM, Toss MS, Gorringe KL, Lee AHS, Ellis IO, Green AR et al (2019) The clinical and biological significance of HER2 over-expression in breast ductal carcinoma in situ: a large study from a single institution. Br J Cancer 120(11):1075–1082. https://doi.org/10.1038/s41416-019-0436-3
31. Bradley R, Braybrooke J, Gray R, Hills R, Liu Z, Peto R et al (2021) HER2-positive breast cancer: a meta-analysis of 13 864 women in seven randomised trials. Lancet Oncol 22(8):1139–1150. https://doi.org/10.1016/S1470-2045(21)00288-6
32. Dieci MV, Miglietta F, Griguolo G, Guarneri V (2020) Biomarkers for HER2-positive metastatic breast cancer: beyond hormone receptors. Cancer Treat Rev 88:102064. https://doi.org/10.1016/j.ctrv.2020.102064
33. Choong GM, Cullen GD, O'Sullivan CC (2020) Evolving standards of care and new challenges in the management of HER2-positive breast cancer. CA Cancer J Clin 70(5):355–374. https://doi.org/10.3322/caac.21634
34. Duffy MJ (2020) Biomarkers for prostate cancer: prostate-specific antigen and beyond. Clin Chem Lab Med 58(3):326–339. https://doi.org/10.1515/cclm-2019-0693
35. Fahrmann JF, Schmidt CM, Mao X, Irajizad E, Loftus M, Zhang J et al (2021) Lead-time trajectory of CA19-9 as an anchor marker for pancreatic cancer early detection. Gastroenterology 160(4):1373–1383 e1376. https://doi.org/10.1053/j.gastro.2020.11.052
36. Garo ML, Campenni A, Petranovic-Ovcaricek P, D'Aurizio F, Giovanella L (2023) Evolution of thyroid cancer biomarkers: from laboratory test to patients' clinical management. Clin Chem Lab Med 61(5):935–945. https://doi.org/10.1515/cclm-2022-1087
37. Cai L, Tu M, Yin X, Zhang S, Zhuang W, Xia Y et al (2022) Combination of serum CST4 and DR-70 contributes to early diagnosis of colorectal cancer. Clin Chim Acta 531:318–324. https://doi.org/10.1016/j.cca.2022.04.1000
38. Saridemir S, Güven HE, Aksel B, Doğan L (2020) Serum AMDL DR-70 levels: a new concept in the diagnosis and follow-up of colorectal carcinoma. Biomark Med 14(8):621–628. https://doi.org/10.2217/bmm-2020-0004
39. Zhang R, Siu MKY, Ngan HYS, Chan KKL (2022) Molecular biomarkers for the early detection of ovarian cancer. Int J Mol Sci 23(19):12041. https://doi.org/10.3390/ijms231912041
40. Chakraborty A, Dasari S, Long W, Mohan C. (2019) Urine protein biomarkers for the detection, surveillance, and treatment response prediction of bladder cancer. (2156–6976 (Print))
41. Koch C, Trojan J (2015) Established and potential predictive biomarkers in gastrointestinal cancer—c-Kit, Her2, ras and beyond. Digestion 91(4):294–302. https://doi.org/10.1159/000376573
42. Hall C, Clarke L, Pal A, Buchwald P, Eglinton T, Wakeman C et al (2019) A review of the role of carcinoembryonic antigen in clinical practice. Ann Coloproctol 35(6):294–305. https://doi.org/10.3393/ac.2019.11.13
43. Füzéry AK, Levin J, Chan MM, Chan DW (2013) Translation of proteomic biomarkers into FDA approved cancer diagnostics: issues and challenges. (1542–6416 (Print))
44. Madikizela LM, Tavengwa NT, Tutu H, Chimuka L (2018) Green aspects in molecular imprinting technology: from design to environmental applications. Trends Environ Analyt Chem 17:14–22. https://doi.org/10.1016/j.teac.2018.01.001

45. Morsi SMM, Abd El-Aziz ME, Mohamed HA (2023) Smart polymers as molecular imprinted polymers for recognition of target molecules. Int J Polym Mater Polym Biomater 72(8):612–635. https://doi.org/10.1080/00914037.2022.2042287
46. Thakur N, Das TR, Gupta S, Shukla SK, Patra S, Kumar D (2023) Chapter 5—Material selection for the fabrication of MIP-based devices: an overview. In: Singh M (ed) Molecularly imprinted polymers (MIPs). Elsevier, pp 101–114
47. Ahmadi F, Yawari E, Nikbakht, M (2014) Computational design of an enantioselective molecular imprinted polymer for the solid phase extraction of S-warfarin from plasma. (1873–3778 (electronic))
48. Gholivand MB, Karimian N, Malekzadeh G (2012) Computational design and synthesis of a high selective molecularly imprinted polymer for voltammetric sensing of propazine in food samples. Talanta 89:513–520. https://doi.org/10.1016/j.talanta.2012.01.001
49. Roy E, Patra S, Tiwari A, Madhuri R, Sharma PK (2017) Introduction of selectivity and specificity to graphene using an inimitable combination of molecular imprinting and nanotechnology. Biosens Bioelectron 89:234–248. https://doi.org/10.1016/j.bios.2016.02.056
50. Kadhem AJ, Gentile GJ, Fidalgo de Cortalezzi MM (2021) Molecularly imprinted polymers (MIPs) in sensors for environmental and biomedical applications: a review. Molecules 26(20). https://doi.org/10.3390/molecules26206233
51. Martins N, Carreiro EP, Locati A, Ramalho JPP, Cabrita MJ, Burke AJ et al (2015) Design and development of molecularly imprinted polymers for the selective extraction of deltamethrin in olive oil: an integrated computational-assisted approach. J Chromatogr A 1409:1–10. https://doi.org/10.1016/j.chroma.2015.07.025
52. Mayes AG, Whitcombe MJ (2005) Synthetic strategies for the generation of molecularly imprinted organic polymers. Adv Drug Deliv Rev 57(12):1742–1778. https://doi.org/10.1016/j.addr.2005.07.011
53. Sheldon RA (2000) Atom utilisation, E factors and the catalytic solution. Comptes Rendus de l'Acad des Sci Ser IIC Chem 3(7):541–551. https://doi.org/10.1016/S1387-1609(00)01174-9
54. Danylec B, Schwarz LJ, Harris SJ, Boysen RI, Hearn MT (2015) The application of template selectophores for the preparation of molecularly imprinted polymers. (1420–3049 (electronic))
55. Ning F, Qiu T, Wang Q, Peng H, Li Y, Wu X et al (2017) Dummy-surface molecularly imprinted polymers on magnetic graphene oxide for rapid and selective quantification of acrylamide in heat-processed (including fried) foods. Food Chem 221:1797–1804. https://doi.org/10.1016/j.foodchem.2016.10.101
56. Hu J-H, Feng T, Li W-L, Zhai H, Liu Y, Wang L-Y et al (2014) Surface molecularly imprinted polymers with synthetic dummy template for simultaneously selective recognition of nine phthalate esters. J Chromatogr A 1330:6–13. https://doi.org/10.1016/j.chroma.2014.01.008
57. Ji W, Ma X, Zhang J, Xie H, Liu F, Wang X (2015) Preparation of the high purity gingerols from ginger by dummy molecularly imprinted polymers. J Chromatogr A 1387:24–31. https://doi.org/10.1016/j.chroma.2015.02.002
58. Lu W, Wang X, Wu X, Liu D, Li J, Chen L et al (2017) Multi-template imprinted polymers for simultaneous selective solid-phase extraction of six phenolic compounds in water samples followed by determination using capillary electrophoresis. J Chromatogr A 1483:30–39. https://doi.org/10.1016/j.chroma.2016.12.069
59. Fan J-P, Tian Z-Y, Tong S, Zhang X-H, Xie Y-L, Xu R et al (2013) A novel molecularly imprinted polymer of the specific ionic liquid monomer for selective separation of synephrine from methanol–water media. Food Chem 141(4):3578–3585. https://doi.org/10.1016/j.foodchem.2013.06.040
60. Guo L, Deng Q, Fang G, Gao W, Wang S (2011) Preparation and evaluation of molecularly imprinted ionic liquids polymer as sorbent for on-line solid-phase extraction of chlorsulfuron in environmental water samples. J Chromatogr A 1218(37):6271–6277. https://doi.org/10.1016/j.chroma.2011.07.016
61. Wang Y, Han M, Liu G, Hou X, Huang Y, Wu K et al (2015) Molecularly imprinted electrochemical sensing interface based on in-situ-polymerization of amino-functionalized ionic

liquid for specific recognition of bovine serum albumin. Biosens Bioelectron 74:792–798. https://doi.org/10.1016/j.bios.2015.07.046
62. Xiang H, Peng M, Li H, Peng S, Shi S (2017) High-capacity hollow porous dummy molecular imprinted polymers using ionic liquid as functional monomer for selective recognition of salicylic acid. J Pharm Biomed Anal 133:75–81. https://doi.org/10.1016/j.jpba.2016.11.007
63. Yuan S, Deng Q, Fang G, Wu J, Li W, Wang S (2014) Protein imprinted ionic liquid polymer on the surface of multiwall carbon nanotubes with high binding capacity for lysozyme. J Chromatogr B 960:239–246. https://doi.org/10.1016/j.jchromb.2014.04.021
64. Yuan Y, Liang S, Yan H, Ma Z, Liu Y (2015) Ionic liquid-molecularly imprinted polymers for pipette tip solid-phase extraction of (Z)-3-(chloromethylene)-6-flourothiochroman-4-one in urine. J Chromatogr A 1408:49–55. https://doi.org/10.1016/j.chroma.2015.07.028
65. Kagimoto J, Taguchi S, Fukumoto K, Ohno H (2010) Hydrophobic and low-density amino acid ionic liquids. J Mol Liq 153(2):133–138. https://doi.org/10.1016/j.molliq.2010.02.002
66. Ruiz-Aceituno L, Sanz ML, Ramos L (2013) Use of ionic liquids in analytical sample preparation of organic compounds from food and environmental samples. TrAC Trends Anal Chem 43:121–145. https://doi.org/10.1016/j.trac.2012.12.006
67. Sverdrup LE, Källqvist T, Kelley AE, Fürst CS, Hagen SB (2001) Comparative toxicity of acrylic acid to marine and freshwater microalgae and the significance for environmental effects assessments. Chemosphere 45(4):653–658. https://doi.org/10.1016/S0045-6535(01)00044-3
68. Zhou Z, He L, Mao Y, Chai W, Ren Z (2017) Green preparation and selective permeation of d-tryptophan imprinted composite membrane for racemic tryptophan. Chem Eng J 310:63–71. https://doi.org/10.1016/j.cej.2016.10.070
69. Wang X-H, Tang L, Yang F-F, Ying L-L, Huang Y-P, Liu Z-S (2017) Green synthesis of water-compatible and thermo-responsive molecularly imprinted nanoparticles. Eur Polym J 92:174–184. https://doi.org/10.1016/j.eurpolymj.2017.05.012
70. He C, Long Y, Pan J, Li K, Liu F (2008) Molecularly imprinted silica prepared with immiscible ionic liquid as solvent and porogen for selective recognition of testosterone. Talanta 74(5):1126–1131. https://doi.org/10.1016/j.talanta.2007.08.009
71. Gama MR, Bottoli CBG (2017) Molecularly imprinted polymers for bioanalytical sample preparation. J Chromatogr B 1043:107–121. https://doi.org/10.1016/j.jchromb.2016.09.045
72. Zhang H, Dramou P, He H, Tan S, Pham-Huy C, Pan H (2012) Molecularly imprinted stationary phase prepared by reverse micro-emulsion polymerization for selective recognition of gatifloxacin in aqueous media. J Chromatogr Sci 50(6):499–508. https://doi.org/10.1093/chromsci/bms028
73. Lorenzo RA, Carro AM, Alvarez-Lorenzo C, Concheiro A (2011) To remove or not to remove? The challenge of extracting the template to make the cavities available in molecularly imprinted polymers (MIPs). Int J Mol Sci 12(7):4327–4347. https://doi.org/10.3390/ijms12074327
74. Batlokwa BS, Mokgadi J, Nyokong T, Torto N (2011) Optimal template removal from molecularly imprinted polymers by pressurized hot water extraction. Chromatographia 73(5):589–593. https://doi.org/10.1007/s10337-010-1884-3
75. McKitterick N, Bicak TC, Switnicka-Plak MA, Cormack PAG, Reubsaet L, Halvorsen TG (2021) On line duplex molecularly imprinted solid-phase extraction for analysis of low-abundant biomarkers in human serum by liquid chromatography-tandem mass spectrometry. J Chromatogr A 1655:462490. https://doi.org/10.1016/j.chroma.2021.462490
76. De Middeleer G, Dubruel P, De Saeger S (2016) Characterization of MIP and MIP functionalized surfaces: current state-of-the-art. TrAC Trends Anal Chem 76:71–85. https://doi.org/10.1016/j.trac.2015.11.007
77. Lu Y, Liu Q, Fu B, Li P, Xu W (2023) Label-free MIP-SERS biosensor for sensitive detection of colorectal cancer biomarker. Talanta 258:124461. https://doi.org/10.1016/j.talanta.2023.124461
78. Silva CF, Menezes LF, Pereira AC, Nascimento CS (2021) Molecularly imprinted polymer (MIP) for thiamethoxam: a theoretical and experimental study. J Mol Struct 1231:129980. https://doi.org/10.1016/j.molstruc.2021.129980

79. Hashkavayi AB, Alizadeh A, Azimi R, Peyrovi M, Raoof JB, Chun H (2023) Development of core-shell magnetic molecularly imprinted polymer-based electrochemical sensor for sensitive and selective detection of ezetimibe. Heliyon 9(6):e17169. https://doi.org/10.1016/j.heliyon.2023.e17169
80. Mousazadeh M, Jahangiri-Manesh A, Nikkhah M, Abbasian S, Moshaii A, Masroor MJ et al (2022) Detection of hexanal gas as a volatile organic compound cancer biomarker using a nanocomposite of gold nanoparticles and selective polymers. J Electroanal Chem 905:115962. https://doi.org/10.1016/j.jelechem.2021.115962
81. Asadi E, Azodi-Deilami S, Abdouss M, Kordestani D, Rahimi A, Asadi S (2014) Synthesis, recognition and evaluation of molecularly imprinted polymer nanoparticle using miniemulsion polymerization for controlled release and analysis of risperidone in human plasma samples. Korean J Chem Eng 31(6):1028–1035. https://doi.org/10.1007/s11814-013-0287-1
82. Kandi MTS, Meshkat SS, Hosseinzadeh S, Behroozsarand A (2023) Synthesis & characterization of novel MIP with RAFT polymerization of (2-hydroxy ethyl methacrylate)/chitosan as a nanocarrier for drug delivery applications. Int J Biol Macromol 250:126052. https://doi.org/10.1016/j.ijbiomac.2023.126052
83. Clarkson CR, Solano N, Bustin RM, Bustin AMM, Chalmers GRL, He L et al (2013) Pore structure characterization of North American shale gas reservoirs using USANS/SANS, gas adsorption, and mercury intrusion. Fuel 103:606–616. https://doi.org/10.1016/j.fuel.2012.06.119
84. Gulati K, Adachi T (2023) Profiling to probing: atomic force microscopy to characterize nano-engineered implants. Acta Biomater 170:15–38. https://doi.org/10.1016/j.actbio.2023.08.006
85. Moncer F, Adhoum N, Catak D, Monser L (2021) Electrochemical sensor based on MIP for highly sensitive detection of 5-hydroxyindole-3-acetic acid carcinoid cancer biomarker in human biological fluids. Anal Chim Acta 1181:338925. https://doi.org/10.1016/j.aca.2021.338925
86. Pirzada M, Sehit E, Altintas Z (2020) Cancer biomarker detection in human serum samples using nanoparticle decorated epitope-mediated hybrid MIP. Biosens Bioelectron 166:112464. https://doi.org/10.1016/j.bios.2020.112464
87. Steinke JHG, Dunkin IR, Sherrington DC (1996) Molecularly imprinted anisotropic polymer monoliths. Macromolecules 29(1):407–415. https://doi.org/10.1021/ma9505033
88. Long Y, Philip JY, Schillén K, Liu F, Ye L (2011) Insight into molecular imprinting in precipitation polymerization systems using solution NMR and dynamic light scattering. J Mol Recognit 24(4):619–630. https://doi.org/10.1002/jmr.1097
89. Shevchenko N, Pankova G, Laishevkina S, Iakobson O, Koshkin A, Shabsels B (2019) Core-shell polymer particles containing derivatives of 1,3-diphenyl-β-diketonate boron difluoride: synthesis and spectroscopic investigation of toluene vapor sorption. Colloids Surf A Physicochem Eng Asp 562:310–320. https://doi.org/10.1016/j.colsurfa.2018.11.051
90. Oliveira AEF, Pereira AC, Ferreira LF (2023) Disposable electropolymerized molecularly imprinted electrochemical sensor for determination of breast cancer biomarker CA 15-3 in human serum samples. Talanta 252:123819. https://doi.org/10.1016/j.talanta.2022.123819
91. Jahangiri-Manesh A, Mousazadeh M, Nikkhah M, Abbasian S, Moshaii A, Masroor MJ et al (2022) Molecularly imprinted polymer-based chemiresistive sensor for detection of nonanal as a cancer related biomarker. Microchem J 173:106988. https://doi.org/10.1016/j.microc.2021.106988
92. Rossetti C, Switnicka-Plak MA, Gronhaug Halvorsen T, Cormack PA, Sellergren B, Reubsaet L (2017) Automated protein biomarker analysis: on-line extraction of clinical samples by molecularly imprinted polymers. Sci Rep 7:44298. https://doi.org/10.1038/srep44298
93. McKitterick N, Braathen F, Switnicka-Plak MA, Cormack PAG, Reubsaet L, Halvorsen TG (2020) Magnetic synthetic receptors for selective clean-up in protein biomarker quantification. J Proteome Res 19(8):3573–3582. https://doi.org/10.1021/acs.jproteome.0c00258
94. Chen SE, Hu J, Yan P, Sun J, Jia W, Zhu S et al (2021) 12-Plex UHPLC-MS/MS analysis of sarcosine in human urine using integrated principle of multiplex tags chemical iso-

tope labeling and selective imprint enriching. Talanta 224:121788. https://doi.org/10.1016/j.talanta.2020.121788
95. Sheydaei O, Khajehsharifi H, Rajabi HR (2020) Rapid and selective diagnose of Sarcosine in urine samples as prostate cancer biomarker by mesoporous imprinted polymeric nanobeads modified electrode. Sensors Actuators B Chem 309:127559. https://doi.org/10.1016/j.snb.2019.127559
96. de Cássia Mendonça J, da Rocha LR, Capelari TB, Prete MC, Angelis PN, Segatelli MG et al (2020) Design and performance of novel molecularly imprinted biomimetic adsorbent for preconcentration of prostate cancer biomarker coupled to electrochemical determination by using multi-walled carbon nanotubes/Nafion®/Ni(OH)2-modified screen-printed electrode. J Electroanal Chem 878:114582. https://doi.org/10.1016/j.jelechem.2020.114582
97. Ferreira NS, Carneiro LPT, Viezzer C, Almeida MJT, Marques AC, Pinto AMFR et al (2022) Passive direct methanol fuel cells acting as fully autonomous electrochemical biosensors: application to sarcosine detection. J Electroanal Chem 922:116710. https://doi.org/10.1016/j.jelechem.2022.116710
98. Alizadeh T, Nayeri S (2020) An enzyme-free sensing platform based on molecularly imprinted polymer/MWCNT composite for sub-micromolar-level determination of pyruvic acid as a cancer biomarker. Anal Bioanal Chem 412(3):657–667. https://doi.org/10.1007/s00216-019-02273-4
99. Zhou L, Wang Y, Xing R, Chen J, Liu J, Li W et al (2019) Orthogonal dual molecularly imprinted polymer-based plasmonic immunosandwich assay: a double characteristic recognition strategy for specific detection of glycoproteins. Biosens Bioelectron 145:111729. https://doi.org/10.1016/j.bios.2019.111729
100. Tawfik SM, Elmasry MR, Sharipov M, Azizov S, Lee CH, Lee YI (2020) Dual emission nonionic molecular imprinting conjugated polythiophenes-based paper devices and their nanofibers for point-of-care biomarkers detection. Biosens Bioelectron 160:112211. https://doi.org/10.1016/j.bios.2020.112211
101. Pang J, Li P, He H, Xu S, Liu Z (2022) Molecularly imprinted polymers outperform lectin counterparts and enable more precise cancer diagnosis. Chem Sci 13(16):4589–4597. https://doi.org/10.1039/d2sc01093c
102. Feng L, Liang X, Mao X, Wan H, Wu Y, Han Q (2021) Study on the preparation of molecular imprinted polymer for analysis of N-phenylglycine in human urine. J Chromatogr B Analyt Technol Biomed Life Sci 1182:122918. https://doi.org/10.1016/j.jchromb.2021.122918
103. Piletska E, Magumba K, Joseph L, Garcia Cruz A, Norman R, Singh R et al (2022) Molecular imprinting as a tool for determining molecular markers: a lung cancer case. RSC Adv 12(28):17747–17754. https://doi.org/10.1039/d2ra01830f
104. Rebelo T, Costa R, Brandao A, Silva AF, Sales MGF, Pereira CM (2019) Molecularly imprinted polymer SPE sensor for analysis of CA-125 on serum. Anal Chim Acta 1082:126–135. https://doi.org/10.1016/j.aca.2019.07.050
105. Yazdani Z, Yadegari H, Heli H (2019) A molecularly imprinted electrochemical nanobiosensor for prostate specific antigen determination. Anal Biochem 566:116–125. https://doi.org/10.1016/j.ab.2018.11.020
106. Liu J, Wang Y, Liu X, Yuan Q, Zhang Y, Li Y (2019) Novel molecularly imprinted polymer (MIP) multiple sensors for endogenous redox couples determination and their applications in lung cancer diagnosis. Talanta 199:573–580. https://doi.org/10.1016/j.talanta.2019.03.018
107. Taheri N, Khoshsafar H, Ghanei M, Ghazvini A, Bagheri H (2022) Dual-template rectangular nanotube molecularly imprinted polypyrrole for label-free impedimetric sensing of AFP and CEA as lung cancer biomarkers. Talanta 239:123146. https://doi.org/10.1016/j.talanta.2021.123146
108. Ribeiro JA, Pereira CM, Silva AF, Sales MGF (2018) Disposable electrochemical detection of breast cancer tumour marker CA 15-3 using poly(Toluidine Blue) as imprinted polymer receptor. Biosens Bioelectron 109:246–254. https://doi.org/10.1016/j.bios.2018.03.011
109. Pacheco JG, Silva MSV, Freitas M, Nouws HPA, Delerue-Matos C (2018) Molecularly imprinted electrochemical sensor for the point-of-care detection of a breast cancer bio-

marker (CA 15-3). Sensors Actuators B Chem 256:905–912. https://doi.org/10.1016/j.snb.2017.10.027
110. Pacheco JG, Rebelo P, Freitas M, Nouws HPA, Delerue-Matos C (2018) Breast cancer biomarker (HER2-ECD) detection using a molecularly imprinted electrochemical sensor. Sensors Actuators B Chem 273:1008–1014. https://doi.org/10.1016/j.snb.2018.06.113
111. Johari-Ahar M, Karami P, Ghanei M, Afkhami A, Bagheri H (2018) Development of a molecularly imprinted polymer tailored on disposable screen-printed electrodes for dual detection of EGFR and VEGF using nano-liposomal amplification strategy. Biosens Bioelectron 107:26–33. https://doi.org/10.1016/j.bios.2018.02.005
112. Tavares APM, Truta L, Moreira FTC, Carneiro LPT, Sales MGF (2019) Self-powered and self-signalled autonomous electrochemical biosensor applied to cancinoembryonic antigen determination. Biosens Bioelectron 140:111320. https://doi.org/10.1016/j.bios.2019.111320
113. Qi J, Li B, Zhou N, Wang X, Deng D, Luo L et al (2019) The strategy of antibody-free biomarker analysis by in-situ synthesized molecularly imprinted polymers on movable valve paper-based device. Biosens Bioelectron 142:111533. https://doi.org/10.1016/j.bios.2019.111533
114. Lin X, Wang Y, Wang L, Lu Y, Li J, Lu D et al (2019) Interference-free and high precision biosensor based on surface enhanced Raman spectroscopy integrated with surface molecularly imprinted polymer technology for tumor biomarker detection in human blood. Biosens Bioelectron 143:111599. https://doi.org/10.1016/j.bios.2019.111599
115. Carneiro LPT, Ferreira NS, Tavares APM, Pinto A, Mendes A, Sales MGF (2021) A passive direct methanol fuel cell as transducer of an electrochemical sensor, applied to the detection of carcinoembryonic antigen. Biosens Bioelectron 175:112877. https://doi.org/10.1016/j.bios.2020.112877
116. Sharma PS, Wojnarowicz A, Sosnowska M, Benincori T, Noworyta K, D'Souza F et al (2016) Potentiometric chemosensor for neopterin, a cancer biomarker, using an electrochemically synthesized molecularly imprinted polymer as the recognition unit. Biosens Bioelectron 77:565–572. https://doi.org/10.1016/j.bios.2015.10.013
117. Tchinda R, Tutsch A, Schmid B, Sussmuth RD, Altintas Z (2019) Recognition of protein biomarkers using epitope-mediated molecularly imprinted films: histidine or cysteine modified epitopes? Biosens Bioelectron 123:260–268. https://doi.org/10.1016/j.bios.2018.09.010
118. Campagnol D, Karimian N, Paladin D, Rizzolio F, Ugo P (2022) Molecularly imprinted electrochemical sensor for the ultrasensitive detection of cytochrome c. Bioelectrochemistry 148:108269. https://doi.org/10.1016/j.bioelechem.2022.108269
119. Zang Y, Zhang Y, Wei R, Xue H, Jiang J (2023) Difunctional molecularly imprinted polymers and heterostructured CdS nanoparticle-sensitized ZnO nanorod arrays for antibody-free photoelectrochemical alpha-fetoprotein sensor. J Electroanal Chem 944:117631. https://doi.org/10.1016/j.jelechem.2023.117631
120. Patra S, Roy E, Madhuri R, Sharma PK (2015) Nano-iniferter based imprinted sensor for ultra trace level detection of prostate-specific antigen in both men and women. Biosens Bioelectron 66:1–10. https://doi.org/10.1016/j.bios.2014.10.076
121. Chen W, Wang Z, Gu S, Wang J (2019) Detection of hexanal in humid circumstances using hydrophobic molecularly imprinted polymers composite. Sensors Actuators B Chem 291:141–147. https://doi.org/10.1016/j.snb.2019.04.065
122. Iwanowska A, Yusa S, Nowakowska M, Szczubialka K (2016) Selective adsorption of modified nucleoside cancer biomarkers by hybrid molecularly imprinted adsorbents. J Sep Sci 39(15):3072–3080. https://doi.org/10.1002/jssc.201600132
123. Ullah MF, Aatif M (2009) The footprints of cancer development: cancer biomarkers. Cancer Treat Rev 35(3):193–200. https://doi.org/10.1016/j.ctrv.2008.10.004
124. Tandon S, George SM, McIntyre R, Kandasubramanian B (2020) Polymeric immunosensors for tumor detection. Biomed Phys Eng Express 6(3):032001. https://doi.org/10.1088/2057-1976/ab8a75
125. Hassanpour S, Hasanzadeh M (2021) Label-free electrochemical-immunoassay of cancer biomarkers: recent progress and challenges in the efficient diagnosis of cancer employing

electroanalysis and based on point of care (POC). Microchem J 168:106424. https://doi.org/10.1016/j.microc.2021.106424
126. Hassanpour S, Baradaran B, Hejazi M, Hasanzadeh M, Mokhtarzadeh A, de la Guardia M (2018) Recent trends in rapid detection of influenza infections by bio and nanobiosensor. TrAC Trends Anal Chem 98:201–215. https://doi.org/10.1016/j.trac.2017.11.012
127. Wang X-Y, Feng Y-G, Wang A-J, Mei L-P, Yuan P-X, Luo X et al (2021) A facile ratiometric electrochemical strategy for ultrasensitive monitoring HER2 using polydopamine-grafted-ferrocene/reduced graphene oxide, Au@Ag nanoshuttles and hollow Ni@PtNi yolk-shell nanocages. Sensors Actuators B Chem 331:129460. https://doi.org/10.1016/j.snb.2021.129460
128. Chen Y, Wang X-Y, Wang A-J, Mei L-P, Yuan P-X, Luo X et al (2021) Ultrasensitive ratiometric electrochemical immunoassay of N-terminal pro-B-type natriuretic peptide based on three-dimensional PtCoNi hollow multi-branches/ferrocene-grafted-ionic liquid and Co N C nanosheets. Sensors Actuators B Chem 326:128794. https://doi.org/10.1016/j.snb.2020.128794
129. Wang XY, Feng YG, Wang AJ, Mei LP, Luo X, Xue Y et al (2021) Facile construction of ratiometric electrochemical immunosensor using hierarchical PtCoIr nanowires and porous SiO(2)@Ag nanoparticles for accurate detection of septicemia biomarker. Bioelectrochemistry 140:107802. https://doi.org/10.1016/j.bioelechem.2021.107802
130. Gao F, Du L, Tang D, Lu Y, Zhang Y, Zhang L (2015) A cascade signal amplification strategy for surface enhanced Raman spectroscopy detection of thrombin based on DNAzyme assistant DNA recycling and rolling circle amplification. Biosens Bioelectron 66:423–430. https://doi.org/10.1016/j.bios.2014.12.006
131. Ansari S, Karimi M (2017) Recent configurations and progressive uses of magnetic molecularly imprinted polymers for drug analysis. Talanta 167:470–485. https://doi.org/10.1016/j.talanta.2017.02.049
132. Ye L (2016) Molecularly imprinted polymers with multi-functionality. Anal Bioanal Chem 408(7):1727–1733. https://doi.org/10.1007/s00216-015-8929-2
133. Luo L, Zhang F, Chen C, Cai C (2019) Visual simultaneous detection of hepatitis A and B viruses based on a multifunctional molecularly imprinted fluorescence sensor. Anal Chem 91(24):15748–15756. https://doi.org/10.1021/acs.analchem.9b04001
134. Xie X, Ma X, Guo L, Fan Y, Zeng G, Zhang M et al (2019) Novel magnetic multi-templates molecularly imprinted polymer for selective and rapid removal and detection of alkylphenols in water. Chem Eng J 357:56–65. https://doi.org/10.1016/j.cej.2018.09.080
135. Hong R, Sun H, Li D, Yang W, Fan K, Liu C et al (2022) A review of biosensors for detecting tumor markers in breast cancer. Life (Basel) 12(3):342. https://doi.org/10.3390/life12030342

Advancements in Molecularly Imprinted Polymers for Selective Recognition of Cancer Biomarkers

14

Abraham Ulises Chávez-Ramírez, Vanessa Vallejo-Becerra, Juan de Dios Galindo-de-la-Rosa, Soane Fernández-Puig, Jannu Ricardo Casanova-Moreno, Apurva Rohokale, Goldie Oza, and Arístides Camilo Valdés-González

Abstract

The artificial antibody acting as molecular armada for detecting and monitoring Cancer. The quest for cancer detection has revolutionized the diagnostic world by identifying a plethora of signature markers that can be used as a representative of different kinds of cancers. In these lines, several biosensors have been developed that use different biomolecules as recognition systems for their detection. However, the high cost and stability issues associated with these biomacromolecules used as recognition elements have led researchers to explore novel alternatives. Molecularly imprinted polymers (MIPs) are self-engineered artificial antibodies that can interact with the targeted ligands within the targeting cavities without the requirement of any other linker. Overall, they exhibit augmented stability, selectivity, multiplexing capacities, and versatility in comparison to natural antibodies. A detailed explanation for applications of MIPs in cancer biomarker detection is explored in this chapter. It encompasses an overview of classical recognition elements, highlighting their limitations, and introduces MIPs as an attractive solution. The synthesis strategies of MIPs are discussed,

A. U. Chávez-Ramírez (✉) · S. Fernández-Puig · J. R. Casanova-Moreno
A. Rohokale · G. Oza
Centro de Investigación y Desarrollo Tecnológico en Electroquímica S. C., Queretaro, Mexico
e-mail: achavez@cideteq.mx

V. Vallejo-Becerra · J. de Dios Galindo-de-la-Rosa
Universidad Autónoma de Querétaro, Queretaro, Mexico

A. C. Valdés-González
Instituto Nacional de Higiene, Epidemiología y Microbiología, Ministerio de Salud Pública de Cuba, Havana, Cuba

S. Patra et al. (eds.), *Molecularly Imprinted Polymers: Path to Artificial Antibodies*, https://doi.org/10.1007/978-981-97-4379-7_14

and the challenges encountered in their implementation are also addressed. This chapter will exemplify different applications of MIPs in monitoring cancer by detecting a range of biomarkers such as alpha-fetoprotein, B-cell immunoglobulins, catecholamines, and prostate-specific antigens.

Keywords

Moleculary imprinted polymers · Cancer biomarkers · Sensor · Cancer detection
Molecular recognition

14.1 Introduction

Cancer pathology is the most important challenges in modern medicine, accounting for a large number of deaths worldwide each year. To overcome the mortality and morbidity of this disease, it is very pertinent that there is early detection and precise monitoring of the ailment. Traditional diagnostic methods, such as imaging techniques and biomarker analysis, have made considerable progress in detecting different types of cancer. However, there is still a growing need for more sensitive, specific, and cost-effective diagnostic tools [1–3].

In recent years, various materials have emerged for the determination of cancer biomarkers, including molecularly imprinted polymers (MIPs). These materials have become technology that has the potential to revolutionize the detection and monitoring of different stages of cancer. MIPs are synthetic materials that behave as biomimetic receptors, capable of selectively recognizing the analyte for which they have been prepared. These materials are designed to act as artificial antibodies (plastic antibodies) by mimicking the binding capabilities of natural antibodies found in the immune system [4].

The operation of MIPs is based on the principle of molecular imprinting, where the desired analyte is used as a template during the polymerization process (Fig. 14.1). This leads to the formation of cavities or "binding sites" in the polymer that are complementary in shape and chemical functionality to the target molecule. Once the template is removed, these cavities remain, providing the MIP with the ability to selectively recognize and capture the target analyte [6]. These characteristics give the materials obtained advantages over natural receptors, such as their versatility and stability, and can be used in the detection of cancer biomarkers, disease monitoring, and controlled drug release. Other advantages include high sensitivity, which allows the detection of cancer biomarkers even at low concentrations; relatively low costs, which makes them more accessible for widespread use in clinical settings; and MIP-based assays have been used in the manufacture of selective phases for the development of chemical sensors that can be easily integrated into existing diagnostic platforms, facilitating their adoption by healthcare professionals [4].

MIPs are now being designed to become a crucial component of the molecular armada used in the battle against cancer, providing a powerful tool to detect cancer

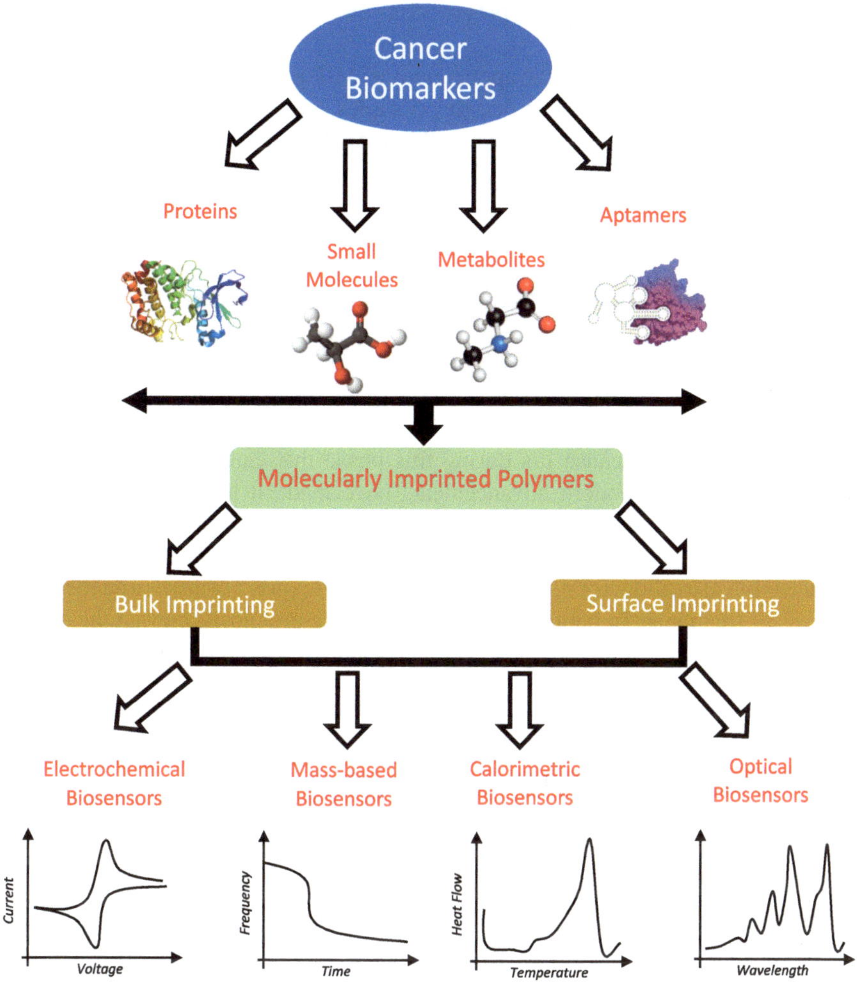

Fig. 14.1 Schematic representation of molecularly imprinted polymers (MIPs) in the application of different cancer biomarker detection. Reproduced from [5]

at early stages, monitor disease progression, and facilitate targeted therapies to improve patient outcomes [7].

In this chapter, we explore the applications of MIP in cancer detection and monitoring. We will delve into their design principles, synthesis methods, and recent advances in the field. In addition, we will discuss the challenges and prospects of MIPs as artificial antibodies to revolutionize cancer diagnostics and personalized medicine.

14.2 Cancer Detection and Diagnostic Challenges

The primary objective of cancer screening and early identification is to achieve a cure by identifying cancer, whether it be the malignancy itself or its initial abnormality, at the earliest possible stage, well before any symptoms emerge. This early detection is crucial because it's during this phase that cancer treatment is most effective [8]. While modern advancements have made it normal to detect solid tumors, more sophisticated technologies are essential for detecting cancer even before symptoms appear. These advanced tools not only enable the early detection of cancer but also allow for monitoring disease progression and assessing the effectiveness of treatments in patients in order to enhance prognosis, elevate the quality of life, reduce healthcare expenses, and increase the chances of recovery for individuals battling cancer.

14.2.1 Imaging Techniques

Considerable effort has been directed to the development of diagnostic probes or biomarkers and strategies that aim to target specific molecular and genetic irregularities, coupled with advanced analytical techniques known for their high sensitivity. Imaging is crucial in cancer protocols, providing morphological, structural, metabolic, and functional data. Integration with tools like in vitro analysis aids clinical decisions. Early cancer detection via imaging-based screening significantly reduces mortality for specific cancers [9]. Traditional high-resolution anatomical imaging methods like plain film X-ray, computed tomography (CT), and magnetic resonance imaging (MRI) are proficient in revealing information about lesion characteristics, including location, size, morphology, and structural alterations to nearby tissues. Nevertheless, these techniques offer limited insights into tumor physiology. Positron emission tomography (PET) is commonly employed to offer an understanding into the biological functions of a tumor and the microenvironment surrounding it [10].

Positron emission tomography (PET) is a noninvasive imaging technique that furnishes physiological details about tissue and organ function. It achieves this by administering radioactive compounds (radiotracers), detecting emitted radiation, and reconstructing the radiotracer's. Detection of various types of cancer can be done by imaging the temporal distribution of labeled biological molecules, to create "physiologic maps" of the functions or processes relevant to the labeled molecules in various areas of a specific organ. This method aids in detecting overlooked cancerous lesions and provides a more precise analysis of lymph node metastasis compared to conventional imaging. Drawbacks include radiation exposure from intravenous radiopharmaceutical administration, high cost, and limited availability due to the need for specialized personnel for reliable imaging [11, 12].

Computed tomography (CT) is a noninvasive and painless imaging technique utilizing specialized X-ray equipment to generate detailed scans of internal body areas. Studies have shown that CT can be effective in both colorectal cancer screening (including screening for large polyps) and lung cancer screening [13, 14]. Concerns related to CT scans involve the potential risks associated with ionizing radiation and potential reactions to the intravenous contrast agent, which enhances visualization. Ionizing radiation exposure may slightly elevate the lifetime risk of cancer. This is especially worrisome for pediatric patients, as their higher sensitivity to radiation and longer lifespan increase the potential for radiation-induced cancer over time [15, 16].

Magnetic resonance imaging (MRI) scanners assume a crucial role in cancer diagnosis, staging, and treatment planning. They yield highly detailed images that effectively distinguish between normal and abnormal tissue. This precision enables the radiology team to accurately assess the growth of cancerous cells within the body, and it is valuable for detecting metastases, which indicate the spread of cancer to new body regions. MRI excels in providing superior contrast for soft tissues when compared to CT scans. Consequently, radiologists frequently utilize MRI for medical imaging involving the brain, spinal cord, muscles, ligaments, blood vessels, and even the interior of bones. MRI is not recommended when internal or external metal objects are present due to potential interference with magnetic fields. MRI is not able to produce clear images of bone matter and blood. This means they typically can't be used for bone cancer and hematologic malignancies [17].

Magnetic resonance spectroscopy (MRS) leverages the varying resonant frequencies of atomic nuclei in different electric environments to identify chemical groups and molecules, enabling noninvasive in vivo measurement of numerous metabolites. However, it faces challenges, such as the proximity of metabolite frequencies, requiring a uniform magnetic field for resolution. Difficulties arise near bone and air interfaces, leading to incomplete data coverage. Weak signals, due to low metabolite concentrations relative to water in tissue, result in limited spatial resolution clinically, with only millimolar-concentration metabolites detectable [18].

14.2.2 Molecular Diagnostic Techniques

Lately, molecular cancer diagnosis involves measuring gene and protein-related information in cancer patients for diagnostic and prognostic purposes. Swift diagnosis of infectious pathogens is vital to mitigate significant cancer outbreaks. Below are concise descriptions of key molecular diagnostic techniques used for cancer detection [19].

Polymerase chain reaction (PCR) and reverse transcriptase PCR (RT-PCR) are highly sensitive methods for detecting tumor marker-expressing cells, particularly in patients with localized or metastatic cancer [20]. PCR can be highly sensitive, but it may also produce false-positive or false-negative results, the accuracy of PCR depends on the quality and purity of the DNA or RNA extracted from the sample. It is not well-suited for detecting large-scale chromosomal rearrangements or structural variations in the genome [21]. PCR involves multiple temperature cycles, which can be time-consuming, leading to delays in obtaining results. It can be expensive due to the need for reagents, equipment, and skilled personnel. There's a risk of sample cross-contamination during the PCR amplification process. These limitations may be addressed by combining PCR with other molecular techniques for a more comprehensive analysis of cancer-related genetic alterations.

Fluorescence in situ hybridization (FISH) detects chromosomal abnormalities, gene amplifications, and translocations in cancer cells by fluorescently labeling specific DNA sequences. However, FISH may not consistently enable low clonal cell detection for minimal residual disease (MRD). Multiple tests are necessary, given that commercial labs often conduct single-gene analyses requiring prior knowledge. Additionally, FISH probes may have lower sensitivity for unbalanced translocations, and adjusting cutoffs for different probes poses a technical challenge [22].

Immunohistochemistry (IHC) employs both monoclonal and polyclonal antibodies to identify the tissue distribution of a target antigen in both healthy and diseased states. This technique utilizes specific antibodies for detecting the presence and localization of particular proteins in cancer tissues [23]. This helps in identifying tumor markers and assessing their expression levels.

Liquid biopsies present an appealing option for diagnosing and prognosing cancer in patients. The vast array of analytes circulating in the blood makes it a promising technique for managing oncological patients clinically [24]. This method entails analyzing circulating tumor DNA (ctDNA), circulating tumor cells (CTCs), or exosomes in various bodily fluids like blood and urine. Liquid biopsy, though promising, faces challenges hindering its integration into routine clinical practice. An inherent drawback is the potential oversight of crucial genetic alterations in early disease stages due to extremely low concentrations of ctDNA from specific tumors. This delay could impact timely diagnosis and essential therapeutic interventions. Age-related clonal hematopoiesis of indeterminate potential (CHIP) in some cases may impede ctDNA testing, leading to misinterpretation and inappropriate therapeutic decisions for patient management [25].

DNA microarrays can simultaneously analyze the expression levels of thousands of genes, enabling the profiling of gene expression patterns in cancer cells and the identification of biomarkers [26]. DNA methylation patterns can be altered in cancer cells, methylation analysis helps in identifying epigenetic changes associated with cancer development and progression [27]. Proteomic techniques allow the analysis of protein profiles in cancer cells and tissues, aiding in the discovery of potential biomarkers and therapeutic targets [28]. MicroRNAs (miRNAs) are small RNA molecules that play a role in cancer development. Analyzing the levels of specific miRNAs in bodily fluids can provide insights into cancer detection and prognosis. Comprehensive genomic and transcriptomic profiling of cancer tissues helps identify mutations, gene expression patterns, and pathways that are specific to different cancer types [29].

These molecular diagnostic techniques offer a precise understanding of genetic and molecular cancer alterations, facilitating early diagnosis, personalized treatments, and treatment response monitoring. They revolutionize cancer care by guiding targeted therapies and improving outcomes. Challenges include the time and cost of analysis. The lack of molecular diagnostics regulation poses risks, as clinical validity studies are not mandatory. While some manufacturers conduct these studies, others may not. If a molecular diagnostic lacks sufficient study, its benefits remain uncertain, potentially leading to suboptimal treatments based on test results [30].

14.3 Classical Recognition Elements in Biosensors

Biosensors are devices that use a biological recognition element to detect the presence of specific analytes and that have a transducer attached that may be capable of producing a measurable signal whose intensity depends on the nature of the analyte providing a rapid, specific, and sensitive biochemical signal translation [31–37].

The first biosensor of which there is a report used glucose oxidase (GOx) immobilized on a semipermeable dialysis membrane using an amperometric electrode quantifying the oxygen whose concentration is related to the concentration of glucose in the sample [38]. This work was described by Clark and Lyons in 1962 who also described how "to make electrochemical sensors (pH, polarographic, potentiometric, or conductometric) more intelligent" by adding "enzyme transducers as membrane-enclosed sandwiches" [39, 40].

Biosensors have a wide range of applications and have helped detect various analytes of interest such as polluting agents, detection of heavy metals in soil, important biomarkers in biomedical engineering, pesticides in water, drugs, hormones, and viruses. These devices present advantages over conventional analytical detection techniques and present an improvement in response time, lower cost, miniaturization, automation, and in some cases increased detection sensitivity and the possibility of being used in medical laboratories, quality laboratories of food, microbial detection, facilitating and making timely the detection of analytes with high precision [41].

Classical recognition elements in biosensors share some common characteristics:

(a) *Specificity:* It is the ability of the device to selectively bind and respond only to the desired analyte in the presence of other molecules, where an analyte of interest can be specifically discriminated among interferents that may be present in the sample, guaranteeing the high specificity of the biosensor and the reliability of the detection.

(b) *Stability*: It is the ability of the biosensor to maintain its precision and performance over time under certain storage and operating conditions, considering the device's resistance to degradation and variability in its responses.
(c) Sensitivity: It is the ability to accurately detect and quantify analytes at small concentrations; High sensitivity implies being able to identify very low levels of analytes of interest.
(d) *Ease of regeneration:* It is the ability of the biosensor to restore its initial detection capacity after being used, where molecules attached to the biological recognition element or the sensor surface can be selectively eliminated, allowing multiple analyzes to be performed [42].
(e) *Linearity:* Linearity is a very important characteristic since it contributes to the detection pressure, where the greater the linearity, the better the detection of the analyte [43].

In the medical field, specifically in cancer research, biosensors have shown great potential, and the number of investigations has increased year after year, with the early detection of cancer being fundamental and the one that is presented with the greatest interest since a diagnosis Timely treatment can ensure that patients are given appropriate treatment and improve their quality of life. We are still seeking to develop these devices with a more precise approach, making them more economical and able to generate information about the etiology of the disease [44].

There is a wide variety of biological recognition molecules used in biosensors that interact selectively with the analytes of interest (Fig. 14.2), which are listed below:

(a) *Enzymes:* Enzymes belong to the protein family and have great specificity toward specific substrates. They participate in most reactions in life including metabolic pathways and cell growth [45]. In the area of biosensors, they are

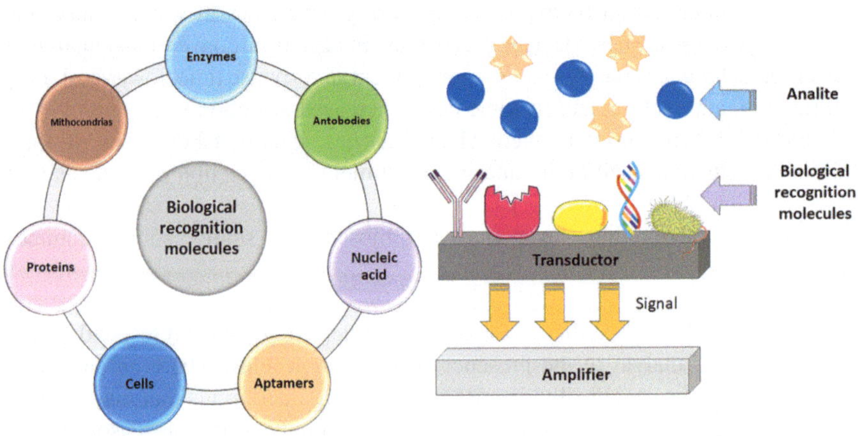

Fig. 14.2 Schematic representation of the main receptor biomolecules used in the development of biosensors

presented as one of the most used recognition elements due to their high selectivity and sensitivity and by using an adequate biomolecule immobilization technique, the detection stability can even be improved [46].

(b) *Antibodies:* Immunosensors can be defined as biosensors that use antibodies as a recognition molecule, whose performance depends on the type of antibody and its immobilization approach. Antibodies can recognize biomarkers due to their high sensitivity and specificity, performing recognition through tuning between the antibody and the antigen [47].

(c) *Nucleic acids:* A biosensor can also use nucleic acids that are fundamental biological components for genetic storage and transfer processes [48]. An oligonucleotide with a known RNA or DNA sequence is used that can detect DNA/RNA fragments or chemical or biological species, called a genosensor [49].

(d) *Aptamers:* Aptamers are RNA or DNA oligonucleotides that have great selectivity and the ability to fold to target molecules such as peptides, drugs, proteins, metal ions, amino acids, whole cells, becoming tools to be used in the development of biosensors [50].

(e) *Cells:* Living cells coupled to transducers have been used to produce a quantifiable and readable signal that depends on the transduction mechanism and the type of cell used, such as bacteria, fungi, yeast, algae, animal cells of fish, rats and even humans [51].

(f) Proteins or peptides. Proteins have become molecular bioreceptors due to the specificity of binding to the target analyte, producing a change in the output signal [52]. Peptides have the same basic component as proteins; Therefore, it can be used for biological recognition, where it can be produced by screening and optimizing artificial peptide libraries, where short-chain peptides have better conformational and chemical stability than proteins [53].

(g) Mitochondria. Mitochondria are complex energy-producing organelles present in eukaryotic cells that play an important role in the regulation of many biological processes and are a source of reactive oxygen species (ROS) [54]. Immobilized mitochondria have been used in piezoresonators, allowing the piezometric and voltammetric translation of analytic signals, generated in response to external stimuli and mitochondrial processes [55].

14.3.1 Antibodies as Gold Standard Recognition Elements

Antibodies comprise of two heavy and two light peptide chains connected by disulfide bonds that have a "Y"-shaped structure of immunoglobulins, classifying them into five important classes IgG, IgM, IgA, IgD, and IgE depending on the differences between the heavy chains [43]. Antibodies have two main regions, the Fc region that interacts with components of the immune system such as phagocytes to promote the elimination of antigen and an antigen-binding region (Fab) where geometric coupling with the antigen is carried out, composed of a variable domain and constant of each light and heavy chain, giving the shape to the parapet at the amino terminal end of the monomer [56]. Biosensors based on the antigen–antibody union are defended as immunosensors and various transducers such as thermal, magnetic, piezoelectric, electrochemical, fluorescent, and optical have been used [57].

Immunosensors are classified as labeled and nonlabeled. The first do not require the use of markers or chemical labels for the detection of the analyte, taking advantage of the intrinsic capacity of the antibodies to bind specifically to their antigens, becoming a powerful detection tool due to its simplicity because it does not require additional marking processes generating a detectable signal change, produces physical changes that are directly detected by the transducer. They have the advantage of simplifying procedures, reducing costs associated with the manufacturing of the devices and greater sensitivity due to the interference that can be generated due to the label [58]. Labeled immunosensors use labels or marker substances for the detection of specific analytes, which generate detectable signals. Among the labels used are enzymes, fluorophores, nanoparticles, or radioactive isotopes giving great versatility, which produce a signal change when the analyte binds to the labeled antibody. This type of detection has high selectivity, high sensitivity, and specificity, giving a wide range of applications [59, 60].

Immunosensors play an important role in the diagnosis of diseases due to their ability to quickly and accurately detect biomarkers, achieving early detection of diseases, essential for effective treatment and avoiding the progression of diseases, managing to differentiate similar diseases due to their high detection specificity and minimize false positives and negatives and these devices also allow the monitoring of medical treatments because levels of biomarkers can be measured during the course of the disease [61].

Immunosensors participate in the accurate, timely, and early detection of cancer, improving survival, diagnosis, and prognosis rates of patients. Proteins, peptides, or other tumor-related metabolites that are released into the blood or isolated in tumor cells are the analytes most used for cancer detection. There are immunosensors that use urine, blood, and saliva as samples that can be a less invasive strategy compared to more invasive procedures such as biopsies. This detection is difficult because the identification of more than one biomarker is needed to diagnose cancer-causing diseases, requiring the design of efficient multi-detection devices, becoming a great challenge today [62]. This type of device is presented as a valuable tool for effective diagnosis and treatment and for improving medical care and the fight against cancer.

14.3.2 Enzymes and Aptamers in Biosensor as Recognition Elements

Enzymatic biosensors are the most used in scientific research due to their high specificity and good catalytic power under mild reaction conditions of the enzymes [63–65]. To improve the stability of enzymes, as well as their recovery, reuse, increase in stability, loss of activity, enzyme immobilization techniques are used, helping to design biosensors and being able to use a wide range of transducers depending on the type of biosensors that are used [66, 67].

Enzymes function as key mediators for the regulation of cellular metabolism, being one of the macromolecules with diverse catalytic functions, carrying out a large series of reactions, from hydrolysis, oxidation–reduction, union of molecules, etc. [35]. Enzymes offer excellent characteristics to biosensors including selectivity and specificity that ease the biosensing process by skipping purification stages or matrix effects. Redox enzymes, among all types of enzymes, are particularly well suited to being coupled with electrochemical methods because electrons can be transferred from and to the enzymes via electroactive surfaces. Since the beginning of time, several sensors have been invented and improved with the goal of enhancing knowledge and enabling better decision making. To deliver relevant information, sensors have been integrated into almost every aspect of life and have engaged a huge array of processes and systems. No matter where they are used, some similar properties are sought to define efficient sensors: simplicity, dependability, and quick reaction are the most desirable traits.

Biosensors that make use of aptamers (synthetic single-stranded nucleic acids) have attracted attention in recent years due to the great selectivity of binding to target molecules and that they can acquire various structures because they can fold in two dimensions (2D) and three dimensions (3D) presenting less spatial blocking and greater surface density, increasing the binding performance with the analyte [68–70]. Due to their chemical nature, aptamers (characteristics of nucleic acids) have stable structural stability under various environmental conditions such as temperature, pH range from 2 to 12 and storage. They have the capacity for regeneration unlike antibodies and can be synthesized with the capacity for thermal refolding and can be chemically modified depending on the detection criteria of the target molecule [43]. Aptamers have been used as small molecule therapeutic agents and can be combined with molecular techniques, flow cytometry, or nanoparticle-based detection for diagnosis. Aptamers act as antagonists because they inhibit protein–protein interactions and as antagonists by increasing the capacity of target proteins [71]. However, these biomolecules present various disadvantages that limit their therapeutic potential in vivo, such as the degradation of the nuclease causing a decrease in the lifetime of the aptamers, where several efforts have been made to avoid this, where modifications are made at the sites of nucleic acids, such as the sugar rings of nucleosides, the ends of nucleic acid chains, and phosphodiester bonds [71]. There is also another drawback because aptamers can be excreted through renal filtration due to their mass (6–30 kDa) and short diameter (less than 5 nm), which causes larger molecules such as cholesterol or proteins to bind to the end increasing its size [72, 73].

The aptamer selection technology has been improved, for example, next-generation sequencing (NGS) used at the beginning of the SELEX process for the analysis of random oligonucleotide libraries has been shown [74–76].

Sensitivity, limit of detection (LOD), specificity, dynamic range, and reproducibility indicate the performance of a clinical biosensor that allows understanding the capacity and technological limitations to analyze whether the device could be applied; therefore, it is important have a clinical purpose before developing the biosensor. In some cases, the detection of multiple biomarkers is required due to the complexity of the disease to give a timely diagnosis where it must be defined whether it is necessary to provide the concentration of the biomarkers as an output value or indicate only the presence or absence of these [77].

The use of electrochemical transducers for the design and development of biosensors has attracted a lot of attention due to the great detection sensitivity that can be obtained, and which can help improve miniaturization by generating portable devices. The aforementioned biorecognition elements have great performance and reliability of recognition of specific analytes. They also present great disadvantages such as their short useful work life, relatively high cost, low stability, and durability, which is why several efforts have been made in genetic engineering to overcome these limitations and also the use of nanomaterials such as nanozymes (nanomaterials that emulate catalyzed reactions). by enzymes), however, when dealing with larger molecules, such as nucleic acids and proteins, problems arise such as nonspecific adsorption of biomolecules on the electrode surfaces [78].

Each biomarker may require different clinical cutoff values, dynamic range of performance, advanced signal processing, novel detection methods, integration of machine learning algorithms to improve accuracy due to the physical, chemical, and biological properties of each biomarker [77].

14.3.3 Limitations and Challenges of Traditional Recognition Systems

Biomolecules used as bioreceptors in biosensors present certain important limitations and challenges. The sensitivity of recognition systems varies depending on the type of analyte and sometimes the necessary detection ranges cannot be achieved, requiring extremely high sensitivity. Furthermore, biomolecules such as antibodies and enzymes can have a limited shelf life and their production, extraction, and purification can be very expensive and time-consuming, limiting their availability.

Many biomolecules can be denatured because they are very unstable to environmental variations. The denaturation of biomolecules is a process in which the three-dimensional structures of these molecules are altered, causing the loss of their biological activity. Thermal denaturation can severely affect the activity of enzymes, causing their inactivation due to the breaking of weak bonds, such as hydrogen bonds and hydrophobic interactions. Also raising the temperature can separate DNA strands. Extreme changes in pH can affect the electrical charge of protein functional groups, which alters their interactions and three-dimensional structure [79].

The immobilization of biomolecules on surfaces continues to be one of the most important tools used in the area of biosensors, where efficient and stable confinement of the biomolecule on the surface of the transducer is required, as well as the chemical inertness and biocompatibility of the support, and that the analyte can interact with the immobilized biomolecule. The retention of the activity of the immobilized biomolecule is a great challenge, such as antigen–antibody binding, hybridization of complementary oligonucleotides, enzymatic catalysis, using different physical or chemical immobilization strategies such as adsorption, attachment to polymeric supports or membranes, cross-linking and covalent bonding [79].

Sometimes immobilization can negatively affect the activity of biomolecules due to the disruption of interactions between the subunits of a protein, changes in the electrical charge of the functional groups or the exposure of active sites, therefore the choice of the immobilization method is important for the performance of biosensors. Over time, biomolecules can lose their activity, limiting their use in long-term or implantable detection devices [80]. The orientation of immobilized proteins on the surface of supports such as antibodies plays a fundamental role in immunosensors, which requires a surface modification of the support, increasing costs and if not done, it can cause random immobilization of the antibodies, causing with a loss of sensitivity and loss of detection efficiency.

14.4 Molecularly Imprinted Polymers (MIPs): An Attractive Solution

Over the years, the scientific community has been inspired by nature, by the processes and phenomena that occur in it, trying to understand them and seeking to imitate them for use in different areas of science. This is the case of molecular recognition and biomimetics. Molecular recognition refers to the process by which molecules interact specifically with each other due to the complementary properties of their chemical structures. This interaction can be highly selective and is fundamental in many biological and chemical functions such as where a molecule is selectively recognized by a suitable receptor binding to it; enzymes recognize and bind to specific substrates to catalyze biochemical reactions; and receptors on cells recognize and bind to signaling molecules to initiate cellular responses and interactions where antibodies recognize antigens. Biomimetics is an interdisciplinary field that involves the study and emulation of biological systems, processes, and structures to develop innovative solutions to various technological, engineering, and design challenges. In biomimetic design, researchers, and engineers are inspired by nature's solutions to complex problems that have evolved over millions of years [81, 82].

By observing how organisms have adapted and optimized their forms and functions, scientists seek to replicate these strategies to create more efficient, sustainable, and resilient technologies. Some examples include:

- **Sensor design:** Biomimetics can be applied to create sensors that mimic biological molecular recognition systems. For example, the design of odor and taste sensors inspired by human olfactory and taste receptors [83, 84].
- **Biosensors:** Biosensors are devices that use biomolecules to detect and quantify specific analytes, mainly molecules of medical interest. Biomimetics is applied here to mimic the interaction between biomolecules and analytes, improving the sensitivity and specificity of biosensors [5].

- **Therapy and medicine:** In the medical field, biomimetics can influence the design of drugs and therapies that rely on the specific interaction between molecules, such as the design of drugs that bind to receptors on cancer cells [85].

14.4.1 Definition and Features of MIPs

Molecular printing involves the creation of stable, highly cross-linked polymers with selective molecular recognition properties tailored to specific analytes. This recognition is achieved by generating complementary cavities in terms of size, shape, and chemical interaction sites during the polymerization process. Molecularly imprinted polymers (MIPs) are artificial materials meticulously crafted to possess highly specific recognition sites for targeted molecules. The synthesis of these materials employs "molecular imprinting," wherein polymerization occurs around a target molecule or template, followed by the removal of the template. This process results in the creation of sites within the polymer structure that mirror the shape and functionality of the target molecule. MIPs are engineered materials designed to emulate the selectivity and specificity observed in biological molecular recognition systems [86].

The concept of MIPs is based on the idea of creating three-dimensional cavities in the polymer structure that can recognize and selectively binding to a specific molecule. These cavities are designed to be highly specific for the target molecule in terms of size, shape, and chemical properties. Once the template is removed, the resulting polymer has a three-dimensional structure containing recognition sites that can interact with the target molecule by mimicking the binding sites of biological entities such as antibodies and enzymes, the interactions that occur in molecularly imprinted materials are mainly host–host (as in enzymes) where a lock-key complex is formed [87].

MIPs are prepared by contacting the analyte (target) or a dummy template [88] (similar structure to the target) with a functional monomer having complementary functional groups in an appropriate solvent, so that a stable prepolymerization complex is formed. Polymerization then takes place in the presence of a cross-linking agent. Once the polymer is obtained, the template molecule is removed by washing the MIP, with a solvent capable of removing high dielectric constant (high polarity). The use of solvents of higher polarity than the porogen favors the removal of the template molecule thus leaving "complementary cavities" available in the polymer matrix capable of recognizing and binding the analyte selectively (Fig. 14.3). It is important to highlight that MIPs have advantages such as low cost, ease of preparation in addition to their low toxicity, they are biocompatible and generally biodegradable [89].

Before being used, the MIPs are characterized both physically and chemically and validated by evaluating different parameters: evaluation of the binding kinetics and the binding properties with the analyte, which includes the determination of several parameters such as the cross-linking of the template molecule with the polymer; the limit of detection and quantification; and the selectivity and the application to selected samples.

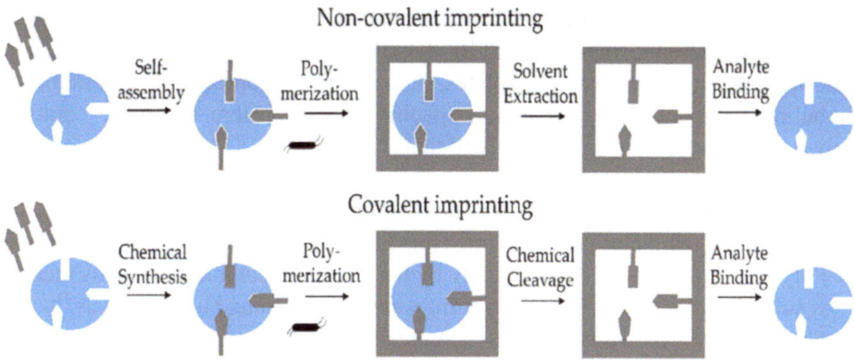

Fig. 14.3 Schematic representation of covalent and non-covalent mechanisms of molecularly imprinted procedures. Reproduced from Lusina et al. [89]

14.4.2 Synthesis Strategies of MIPs

There are different strategies to obtain MIPs depending on the type of interaction involved, which are based on covalent and noncovalent interactions between the template molecule and the functional monomer.

- **Covalent imprinting** is a specific molecular imprinting technique used in the synthesis of MIPs. In this strategy, the binding sites within the polymer matrix are created through the formation of strong covalent bonds between the template molecule and the functional monomers during the polymerization process. It offers several advantages, including high binding specificity and stability. This results in highly selective and stable recognition sites within the MIP. However, it can be more challenging and time consuming than some other imprinting techniques because of the need to find suitable functional monomers that can react covalently with the template molecule (Fig. 14.3) [90].
- **Noncovalent imprinting** is another molecular imprinting technique. In this strategy, the binding sites within the polymer matrix are created through reversible, noncovalent interactions (often involving links hydrogen, π–π, dipole–dipole, ionic, and/or hydrophobic interactions) between the template molecule and functional monomers during the polymerization process. This results in recognition sites that are complementary in shape and functionality to the template molecule. Noncovalent imprinting is more straightforward and less time-consuming than covalent imprinting because it does not require the selection and synthesis of functional monomers with specific covalent reactivity. However, it may offer slightly lower binding affinities and stability compared to covalent imprinting, but it is still highly effective. This molecular printing strategy is the most widely used today (Fig. 14.3) [90].

14.4.2.1 Polymerization Techniques for MIP Fabrication

The polymerization techniques used for the fabrication of MIPs depend on the specific application to the chosen templates, such as alpha-fetoprotein, B-cell immunoglobulins, catecholamines, or cancer-specific antigens. Here are some common polymerization techniques used in MIP fabrication:

- **Bulk polymerization**: In bulk polymerization, the monomers, cross-linkers, initiator, and template molecule are mixed in a single homogeneous phase. Polymerization is initiated, typically by heat or chemical initiators. This method is suitable for producing MIPs in various forms, including bulk materials, beads, powders, and films. It is relatively simple and widely used [91].
- **Emulsion polymerization (EP)** involves creating a stable emulsion of water and an organic solvent. The template molecule, monomers, cross-linkers, and initiator are incorporated into this emulsion. Polymerization occurs in the water phase, leading to the formation of MIP microspheres or nanoparticles with a controlled size distribution. This method is particularly suitable for producing MIPs with a narrow particle size distribution. However, mini emulsion polymerization is a variation of EP involving smaller droplets of monomer and initiator within the emulsion. This technique is used for producing MIP nanoparticles with well-defined sizes and properties [92, 93].
- **Precipitation polymerization** involves dissolving the monomers, cross-linkers, initiator, and template molecule in a good solvent and then adding a poor solvent to induce polymer precipitation. This results in the formation of MIP nanoparticles or microparticles that can be easily collected and purified. Precipitation polymerization is often used for producing MIP nanoparticles [94].
- **Surface imprinting** is a technique used to create MIPs with selective binding sites on the surface of a preexisting substrate. The template molecule is adsorbed onto the substrate, and then a thin layer of polymer is polymerized on top of it. After polymerization, the template molecule is removed, leaving behind selective binding sites on the substrate's surface. This approach is often used for creating MIP-coated sensors or solid-phase extraction materials as described in works of [95, 96], the polymerization of PSA on the graphene surface in which the MIP-graphene-based sensor for PSA, was prepared in three steps (1) the preparation of graphene oxide, (2) the synthesis of protein imprinted material, and (3) the assembly of the potentiometric sensors. In the case of the analyte cancer antigen 15-3 (CA 15-3), the MIP-based voltammetric sensor to monitor breast cancer was imprinting on direct surface in a screen-printed gold electrode (Au-SPE) (Fig. 14.4). This printing is performed by absorbing the analyte on the Au-SPE surface and then electropolymerizing the 2-aminophenol functional monomer around the CA 15-3. Subsequently, the analyte is extracted and monitored by voltammetric analysis (measuring the signal before and after the template removal) with hexacyanoferrate (II/III as redox probe, to determine if all the analyte was extracted from the structure.

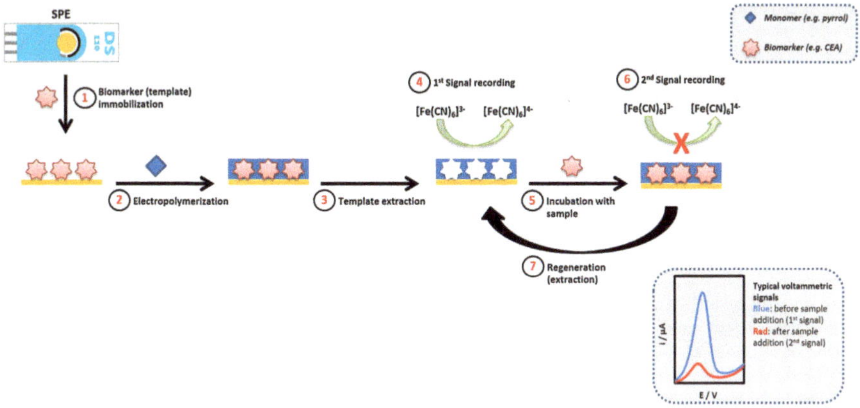

Fig. 14.4 Schematic illustration of the construction and operation of the MIP/Au-SPE. Reproduced from [95]

There are other polymerization techniques to obtain MIP in different configurations such as sheets [97] and over Si (core@shell) as core and MIP as shell [98] that allow for better interaction of the analyte with the material.

14.4.2.2 Template, Functional Monomers, and Cross-Linker Selection and Design

Synthesis of MIPs for cancer biomarkers involves several key steps: template selection, functional monomer selection, cross- = linker selection, polymerization, and template removal. Here's a general overview of each step, focusing on template, functional monomers, and cross-linker selections [99, 100]:

- **Template selection:** The template or target is the molecule that can be a part of the molecule (epitope) or a dummy you want the MIP to recognize and bind to. Thus, the molecule can have a functional group or chemical feature that allows for specific interactions with the chosen functional monomers, which in this case would be an antibody, b- catecholamine or a cancer biomarker (Fig. 14.5) [101].
- **Functional monomer selection:** Functional monomers are the building blocks of MIPs, and they are responsible for interacting with the template molecule. They must be selected that can form specific interactions with the template, such as hydrogen bonding, electrostatic interactions, or hydrophobic interactions. Common functional monomers for MIPs include acrylic acid, methacrylic acid, acrylamide, and various other vinyl monomers in the case of organic MIPs. The choice depends on the nature of interactions required with the template [102].
- **Cross-linker selection:** Cross-linkers are molecules that create a three-dimensional network within the polymer matrix, providing structural integrity and stability to the MIP. The most common cross-linkers include ethylene glycol dimethacrylate (EGDMA) and divinylbenzene (DVB). The choice of cross-linker influences the porosity and rigidity of the MIP, which can affect binding

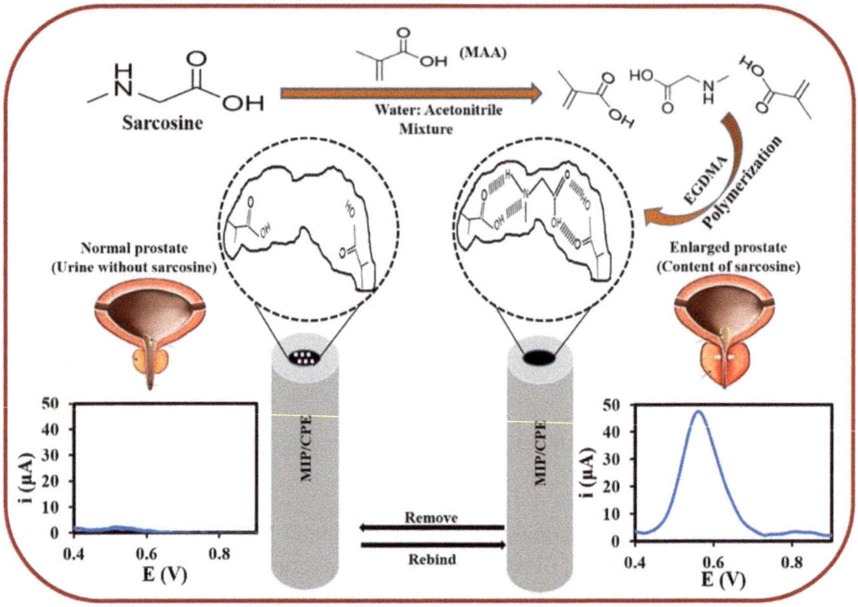

Fig. 14.5 The schematic representation of sensor and SAR MIP nanostructures synthesis. Reproduced from [101]

kinetics and capacity. The cross-linker-to-monomer ratio should be carefully optimized to achieve the desired polymer properties [103].

- **Polymerization:** For the synthesis take place: Mix the template, functional monomers, cross-linker, and a solvent (typically a nonpolar solvent like toluene or a polar solvent like acetonitrile) to form a pre-polymerization mixture. Polymerization could be initiated through different methods like thermal, UV, or chemical initiation. This leads to the formation of a polymer network with specific binding sites imprinted to match the template's shape and chemical functionality. One important step is the optimization conditions in polymerization that involves temperature or light, time and the initiator) for the specific MIP design [104].
- **Template removal:** After polymerization, the template molecule needs to be removed from the MIP to form cavities or binding sites that recognize the target biomarker. This is accomplished by washing the MIP with a suitable solvent or buffer that dissolves the template while preserving the MIP structure. The choice of solvent is crucial to achieve the maximum extraction efficiency of the molecule from the structure [105].

Optimizing the synthesis conditions and thoroughly characterizing the MIP are crucial for developing an effective cancer biomarker recognition system. It may require iterative adjustments in the selection of template, functional monomers, and cross-linkers to achieve the desired specificity and sensitivity [106].

14.4.2.3 Selectivity and Sensitivity of MIPs in Biomarker Detection

MIPs have significant attention nowadays in biomarker detection, and there are some parameters that may affect their practical functions. Selectivity and sensitivity are critical factors to consider during the application of this materials. Selectivity is defined by IUPAC [107] as the extent to which a method can be used to determine analytes in matrices without interference from other similar components. In the case of MIP, the best method must be sought to measure selectivity in real samples [108].

Selectivity:

- **Molecular recognition:** MIPs are designed to selectively bind to a specific target molecule. This selectivity is achieved through the molecular imprinting process, where the polymer is synthesized in the target molecule's presence.
- **Cross-reactivity:** One challenge is ensuring that the MIPs do not cross-react with structurally similar molecules. Achieving high selectivity can be challenging, and it often requires careful optimization of the polymerization conditions and template molecule selection [109].

Sensitivity refers to the capability of an assay to identify extremely low concentrations of a specific substance. In analytical terms, this sensitivity is commonly denoted as the limit of detection (LOD). The LOD is influenced by numerous factors that can impact the sensitivity of molecularly imprinted polymers (MIP).

Sensitivity:

- **Binding affinity:** The sensitivity of MIPs in biomarker detection depends on their binding affinity for the target biomolecule. Higher affinity typically results in better sensitivity.
- **Optimization:** To enhance sensitivity, researchers may modify the MIP synthesis process, such as increasing the concentration of the target molecule during polymerization or using different functional monomers.
- **Detection methods:** The choice of detection method also influences sensitivity. Common detection methods include fluorescence, electrochemical, and mass spectrometry-based techniques. The choice of method should be compatible with the specific MIP and target biomarker [109].

There are some factors that influence the selectivity and sensitivity of MIPs in biomarker detection most of them are related to the material synthesis and design. The typical factors are the target selection, polymerization conditions, MIP morphology, binding and reusability conditions, interference, and validation, etc.

Template molecule selection: Choosing the appropriate template molecule (the biomarker of interest) is crucial. The template molecule should be structurally and chemically like the target biomarker.

Polymerization conditions: The choice of monomers, cross-linkers, and polymerization conditions (e.g., temperature, solvent) during MIP synthesis can significantly affect the selectivity and sensitivity of the resulting MIPs.

MIP morphology: The morphology of the MIPs, such as particle size, porosity, and surface area, can impact binding kinetics and sensitivity. Controlling these properties can be essential for achieving high sensitivity.

Regeneration and reusability: The ability to regenerate and reuse MIPs is vital for practical applications. If MIPs can be easily regenerated without losing their selectivity and sensitivity, they can be more cost effective.

Interference and matrix effects: Real-world samples often contain various interfering substances and matrix effects that can affect biomarker detection. Ensuring that MIPs maintain their selectivity and sensitivity in complex sample matrices is important.

Analytical method validation: Rigorous validation of the analytical method using MIPs is necessary to determine its sensitivity, specificity, and limit of detection. This includes calibration, accuracy, precision, and robustness assessments [108].

14.4.2.4 Advantages over Natural Antibodies and Other Recognition Elements

As mentioned above, MIPs are synthetic materials that are designed to mimic the molecular recognition capabilities of natural antibodies [84] and other recognition elements. Some of the advantages of MIPs over natural antibodies are describe above:

- **Stability:** MIPs are highly stable and can withstand a wide range of harsh conditions, including extreme temperatures, pH levels, and chemical environments. In contrast, natural antibodies can be sensitive to these conditions and may lose their activity.
- **Cost-effectiveness:** MIPs can be produced at a lower cost compared to natural antibodies, which require expensive and time-consuming processes like animal immunization and purification.
- **Durability:** MIPs have a longer shelf life compared to natural antibodies, which can degrade over time. This makes them more suitable for long-term storage and use in various applications.
- **Tailored specificity:** MIPs can be precisely designed and engineered to recognize specific target molecules with high selectivity. This allows for the customization of MIPs for a wide range of applications, from drug delivery to environmental sensing.
- **Wide target range:** MIPs can be created to recognize a broad spectrum of target molecules, including small organic compounds, proteins, and even cells. Natural antibodies may have limitations in terms of the types of targets they can recognize.
- **Ease of production:** MIPs can be synthesized using straightforward polymerization techniques, making them relatively easy to produce in large quantities and with consistent quality.
- **Resistance to denaturation:** MIPs do not undergo denaturation, as natural antibodies can when exposed to extreme conditions. This ensures their functionality is maintained over time.

- **Nonimmunogenic:** MIPs are synthetic materials and do not elicit an immune response when used in biological systems, whereas natural antibodies can trigger immune reactions in certain applications [109–111].

Despite these advantages, MIPs also have some limitations [84, 108]. For example, they may not always achieve the same level of sensitivity and specificity as natural antibodies, and their design and optimization can be complex. Additionally, they typically lack the versatility and adaptability of natural antibodies, which can evolve to recognize new antigens over time.

14.5 Challenges in Implementing MIPs for Cancer Biomarker Detection

The utilization of molecularly imprinted polymers (MIPs) in cancer theragnostic is still in its initial phases, facing various challenges. A significant obstacle arises from the intricate and diverse nature of cancer cell subtypes and tissues, leading to a multitude of potential MIP targets. This intricate landscape complicates the design of MIPs capable of selectively recognizing and binding these targets within complex biological environments, impeding the progress of target-specific diagnostics and cancer nanomedicines. Additionally, the absence of standardized synthesis and characterization methods for MIPs necessitates meticulous optimization of various parameters, including monomers and template molecules. Printing large molecules is often impractical due to the inherent demands and nature of this printing technique. The homogeneity of binding sites in the polymerization process, especially in block polymerization, is generally lacking. Furthermore, the absence of a standardized assessment method for MIP selectivity and binding affinity hinders the comparative analysis of MIP-based sensors. Finally, inadequate research on toxicity, biodegradability, compatibility with biological systems, and fluid dispersion presents practical hurdles for the in vivo applications of MIPs.

For smaller molecules, developing these materials is somewhat simpler, yet challenges persist in achieving a final product suitable for practical medical applications [112]. However, by studying the advantages that these materials have and considering their optimization at all stages, a useful product could be reached for diagnosis and why not for competitive marketing with natural molecule sensors [113].

Multiple articles describe obtaining MIP for the determination of biomarkers in several types of cancer such as lung, breast, prostate, among others, and the advantages that these have with respect to natural molecules, as reflected in this chapter. However, even if we try to have the greatest selectivity in plastic bodies, there is still a long way to go in relation to their natural analogues. Another challenge that arises in the case of MIPs is to take them beyond the laboratory and industrialize their production. All this is related to the fact that natural molecules such as enzymes and antibodies are still more trusted to make a biosensor than synthetic materials [5].

However, MIPs have consistently demonstrated their capability to address key challenges in biosensor manufacturing. These challenges include ensuring the stability of detection probes and optimizing the reaction time of the sensing mechanism, achieving remarkable sensitivity and selectivity despite the inherent complexity of various bio-fluids.

14.6 Applications of MIPs in Cancer Biomarker Detection

In the last decades, a large number of articles have been publishes that uses MIPs in cancer biomarkers detection and as we are mentioned before, this material is an attractive approach mainly for sensing, detection and drug-controlled liberation of different analytes Fig. 14.6 [3, 114].

MIPs can be designed to recognize and bind to specific cancer biomarkers such as proteins, small molecules, metabolites, and genes. In early detection of cancer biomarkers, MIPs are using as recognition elements in biosensors or diagnostic assays that can detect these biomarkers even at very low concentrations. On the other hand, MIPs can be used to create drug delivery systems that release therapeutic agents specifically at the tumor site [100]. The mechanism for this approach is imprinting the polymer with a molecule that is overexpressed in cancer cells, the drug delivery system can target cancer cells while minimizing damage to healthy tissues [89].

Other used of MIPs in biomarker detection is in separation and purification of biomarkers [106] from complex biological samples, this approach is particularly important in diagnostic and biomarker quantification.

We could mention some other applications of MIPs in different types of cancer diseases:

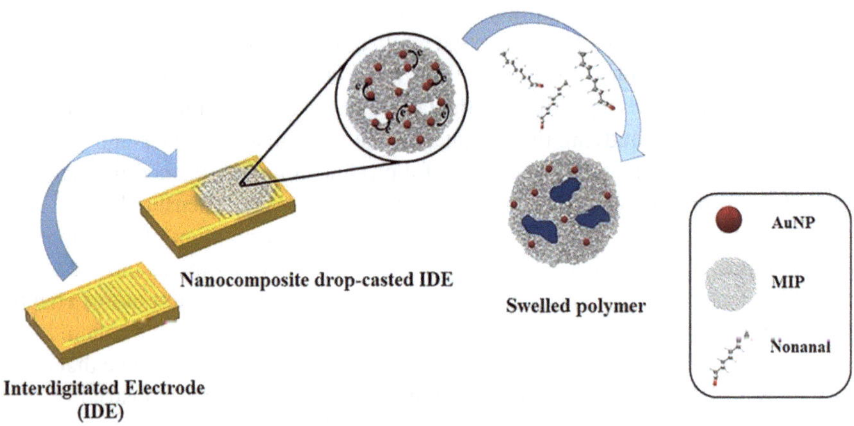

Fig. 14.6 Schematic representation of the nonanal chemiresistive sensor. When nonanal is adsorbed in the cavities of MIP, the polymer swells and the electrical resistance increases [114]

- In separation and purification of biomarkers [106] from complex biological samples, this approach is particularly important in diagnostic and biomarker quantification.
- MIP-based biosensors and assays can be developed for point-of-care testing, enabling rapid and onsite detection of cancer biomarkers. This is especially valuable in resource-limited settings where access to advanced diagnostic equipment is limited [115, 116].
- As contrast agents in imaging techniques, such as MRI or PET scans, to enhance the visualization of cancer cells or tumors. By designing MIPs with specific targeting capabilities, they can improve the accuracy of cancer diagnostics [117].
- To monitor the effectiveness of cancer treatment by tracking changes in biomarker levels over time. This helps healthcare providers adjust treatment plans and assess the response to therapy [118].

To describe MIPs approaches in cancer diseases the following Table 14.1 have a list of these materials in different medical applications.

Despite the growing interest and extensive research carried out on the development of polymers for various applications, they are still focused on laboratory-level research. However, their implementation in medicine is being studied due to the advantage of these materials over natural molecules, and MIPs can provide a sustainable alternative to solve these problems. Therefore, we can say that MIPs are a viable and more economical alternative (although still under study) to obtain devices that are used as an alternative in the diagnosis of cancer diseases and that are responsible for processing and analyzing patient samples with these conditions.

14.7 Multisensing Systems on Microfluidic Platforms

Obtaining information from multiple sensors in a single sample has always been attractive from an analytical perspective. Most of the time, however, this requires integrating separate dedicated sensors for each analyte in a single platform. If the goal is to have a portable point-of-care (POC) or even wearable device, or if the available sample volume is small, miniaturization of both the sensors and the platform is critical. Microfluidic systems, defined as platforms handling low volumes (microliters or less) for chemical purposes, have proven to be a suitable platform to integrate several sensors in a small space.

Early examples of these implementations took the form of chips focused on the measurement of levels of relevant substances for the evaluation of the general state of the patient (e.g., glucose, lactate, urea, creatinine, and pH) [125]. Some of these evolved into commercial multisensing platforms, the most famous one being the i-STAT developed by Abbott. This handheld instrument can analyze up to 26 different analytes depending on the cartridges (i.e., chips) employed. Recently, research efforts are focused on more complex analytes, such as the multiplexed diagnosis of infectious diseases [59, 60].

Table 14.1 MIPs approaches in cancer diseases

MIPs	Target biomarker	Limit of detection	Sensor type	Application	Ref.
MIPs-gold nanoparticles (AuNps)	Gaseous nonanal	4.5 mg/L	Chemiresistor	Detection of nonanal a cancer-related biomarker	[114]
MIP-nanoparticles	Methotrexate		Drug delivery systems	Dossification of methotrexate anti-breast cancer	[100]
MIP-graphene	Prostate-specific antigen (PSA)	2 ng/mL	Potentiometric	Testing the variability of PSA expression by different human prostate cancer cell lines	[96]
MIP-screen-printed gold electrode (au-SPE)	Cancer antigen 15-3 (CA 15-3)	1.5 U mL^{-1}	Electrochemical	Detection (point-of-care) of a breast cancer biomarker	[95]
Supermagnetic Fe_3O_4@ZIF-8@ MIP	Sarcosine	0.4 pM	Electrochemical	Detection of sarcosine in urine	[119]
MIP immunosensor based on AuNPs/PTh-modified glass carbon electrode	CEA	0.2589 pg/ mL	Electrochemical	Detection of carcinoma embryonic antigen	[120]
MIP immunosensor based on AuNPs/PTh-modified glass carbon electrode	Alpha-fetoprotein (AFP)	0.8138 pg/ mL	Immunosensor	Detection of AFP and future perspectives of detection hepatocellular carcinoma and germinal tumors	[121]
MIP three-dimensional gold nanoelectrode ensemble (GNEE)	CA-125	0.5 U mL^{-1}	Electrochemical	Detection of ovarian cancer marker	[122]
Nano-iniferter platform for MIP and multiwalled carbon nanotubes (MWCNTs) decorated with manganese nanoparticles	PSA	0.25 pg L^{-1}	Electrochemical	Detection of PSA (men and woman prostate, breast cancer, and forensic studies)	[123]
Difunctional polydopamine-based MIPs (PDA-MIP) on CdS nanoparticle-decorated ZnO nanorod arrays (CdS@ZnO NAs)	AFP	0.38 pg mL^{-1}	Photoelectrochemical	Sensor for AFP detection (hepatocellular carcinoma, testicular tumor, and yolk sac tumor)	[124]

As microfluidic platforms expand from the traditional chip-based systems, interesting applications emerge. Microfluidic multisensing devices are currently being developed that continuously measure biologically relevant analytes through wearable patches. By being adhered to the skin, these devices can sample sweat directly from the patient. For example, integration of single analyte sensors has been reported for glucose and cortisol [126]. Furthermore, Kim et al. integrated potentiometric sensors for Ca^{2+}, K^+, and Na^+ in a single patch [57].

14.7.1 Microfluidics for Cancer Detection: Advantages and Applications

The use of microfluidic devices has also been exploited in cancer study and diagnosis [127]. A wide variety of devices and strategies have been implemented to detect, and sometimes quantify, tumor markers such as cells, exosomes, nucleic acids, and proteins [128].

An important fraction of the literature on this field deals with the fabrication of chips for separation and capture of circulating tumor cells (CTCs) [129]. These are thought to be involved in the spreading of the cancer as they travel to different organs through the bloodstream. A main challenge in detecting these cells is their low abundance compared to other cells. Therefore, selective bioreceptors such as aptamers and antibodies have been immobilized on the chips to selectively capture the CTCs [130]. Even when whole cells are not released from the tumor, they can release other targets of clinical importance. Exosomes, along with circulating tumor DNA and RNA, have been used for early cancer detection. Therefore, microfluidic systems have been developed to separate and detect these entities, leading to the so-called "microfluidic liquid biopsy" [131]. Also, several devices have been designed to study DNA methylation. These epigenetic alterations are very relevant in cancer study, as they can silence or activate genes, influencing cancer progression, metastasis, and drug resistance [132]. Proteins and glycoproteins are also common tumor markers. Carcinoembryonic antigen (CEA), for example, has been reported in anomalous levels in gastrointestinal, pancreatic, lung, breast, and ovarian cancers. Zhou et al. developed a plasmonic immunochip that contains gold nanowells modified with antibodies [133]. By measuring a wavelength shift in the optical response, they were able to detect concentrations of CEA as low as 5 ng/mL.

Due to the relative simplicity in integrating miniaturized electrodes in microfluidic devices, electrochemical biosensors have been widely employed in cancer POC platforms [1]. However, electronic components for optical measurements have been continuously decreasing in size and price, making the integration of optical sensors more and more frequent. Many of the detection schemes include nanomaterials such as nanoparticles, nanotubes, graphene, and metal–organic frameworks incorporated into the microfluidic devices for cancer-related assays [134]. The main advantage of these materials is their smaller scale compared with the microchannels, therefore increasing the surface area available for the assays. As well, some of the materials have intrinsic magnetic or optical properties that aid in the separation or detection processes.

Finally, microfluidic cell cultures are suitable to study tumor interactions with other tissues and with drugs. In particular, complex three-dimensional (3D) cell cultures involving more than one cell type have been developed to mimic human organs. A variety of these so-called organs-on-a-chip have been reported, to model the heart, gut, lung, liver, and kidney, among others [135].

14.7.2 Integration of MIPs into Microfluidic Devices

MIPs have been integrated into microfluidic platforms for both separation and sensing applications [136]. In sensing, the inclusion of MIPs in microfluidic platforms is very desirable since the better stability of MIPs compared to antibodies [137] could result in better shelf life for point-of-care applications. Indeed, MIPs have been integrated into a variety of lab-on-a-chip platforms, resulting in acoustic, optical, electrochemical, and even thermal sensors [138].

Traditionally, molecularly imprinted polymers have been synthesized in bulk form, yielding blocks typically a few millimeters in size. These blocks are then broken down into smaller pieces by milling and then sieved to the desired particle size. Although this strategy has been used to prepare miniaturized sensors [139], it is not particularly useful for inclusion into microfluidic devices. Instead, microfabrication scientists prefer patterning techniques that they are familiar with to fabricate the microfluidic chip themselves. As such, a variety of methods have been adapted to synthesize the MIPs in ways more compatible with microfabrication. These strategies include photolithography, stereolithography, inkjet printing, dip-pen nanolithography, and micro-contact printing [140].

The most common synthesis method of molecularly imprinted polymers is free radical polymerization. An initiator is included that can be activated thermally or through light. Photopolymerization is particularly useful for patterning as light can be directed selectively to certain regions of a substrate. Several types of photolithography, stereolithography, and near-field assisted optical lithography have been used in the structuring of MIPs [141] Huang et al., for example, developed an MIP photoresist based on acrylic copolymers with functional groups suitable for the detection of albuterol (Fig. 14.7a) [142]. This resist was spin-coated and exposed through a mask to modify electrodes 30 × 30 μm in size. Alternatively, Hong prepared a molecularly imprinted polymer to detect the anesthetic propofol [143]. Instead of spinning the mixture on the substrate, they used a spacer between the substrate and the mask, essentially making a mold and filling the gap with the MIP precursor solution (Fig. 14.7b).

Electropolymerization of electroactive precursors can also be used to prepare MIPs [144]. Although most electrodeposited MIPs are made of conducting polymers, some electroactive precursors result in nonconductive MIPs [145]. Electropolymerization offers spatial control since the polymer is only deposited at the electrode [146]. Furthermore, MIPs of different thickness can be obtained by controlling the charge passed during the electrodeposition reaction (Fig 14.6c) [147]. For example, MIPs have been deposited on traditional wire-in-glass disk

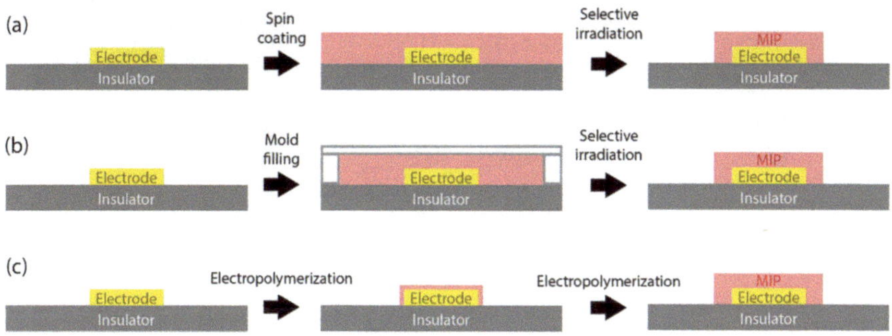

Fig. 14.7 Common strategies for patterning of MIPs on electrodes for microfluidic systems. (**a**) A viscous prepolymeric mixture is used as photoresist. (**b**) A mold is filled with a less viscous prepolymeric mixture and a selected region is irradiated. (**c**) The polymer is electrodeposited on the electrode surface

microelectrodes with diameters of 25 µm or less [148–150]. As well, point-of-care devices have been developed by in situ electropolymerization of MIPs. Weng et al., for example, reported a device with a poly(3,4-ethylenedioxythiophene) (PEDOT) MIP deposited on Pt structures to detect morphine [151]. Nagabooshanam et al. deposited polypyrrole on gold electrodes in a commercial microfluidic chip to detect the pesticide chlorpyrifos [152]. Ge et al., on the other hand, electropolymerized a poly-o-phenylenediamine MIP on a paper-based device [153]. The selectivity of MIPs has been exploited in multisensing microfluidic platforms. Huang et al., for example, developed a device in which three different MIPs were included in the same chamber for detection of cholesterol, testosterone, and progesterone [154].

Although most examples of MIP detection in the literature are using small molecules as templates, there are some reports of them being used for macromolecules like proteins [155]. It is expected that, in the future, MIPs are exploited for detection of even larger analytes, like pathogenic bacteria [156, 157]. Finally, it must be pointed out that microfluidic devices have also been used to produce monodisperse MIP spheres [158].

14.7.3 Multi-Biomarker Sensing for Enhanced Accuracy

The detection of multiple biomarkers coupled to advanced computing techniques has shown promising results to better diagnose life-threatening conditions. In particular, sets of biomarkers have been reported for a variety of cancer types [128, 129, 159]. In comparison with traditional proteomic or genomic analysis which uses expensive and complex equipment (e.g., liquid chromatography and mass spectrometry), multisensing point-of-care devices can provide faster and simpler measurements [160, 161]. As such, there have been several attempts to integrate multiple assays in microfluidic chips. Gao et al., for example, developed a chip

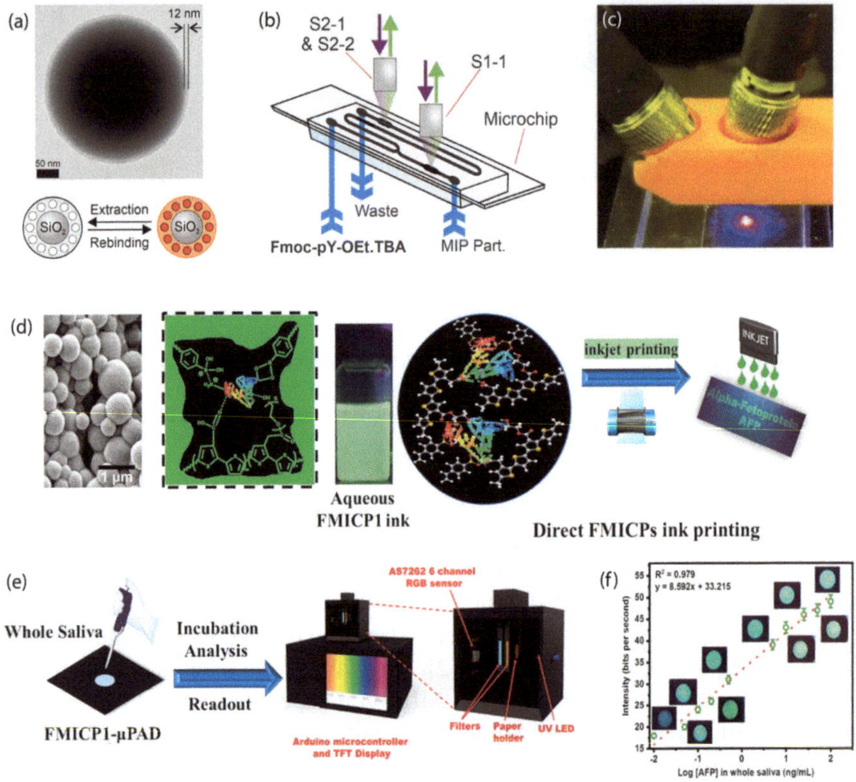

Fig. 14.8 Examples of cancer biomarker detection in MIP-based microfluidic platforms. SiO_2/MIP core–shell nanoparticle (**a**) for the detection of the phosphorylated tyrosine derivative Fmoc-pY-OEt.TBA flow through the microchannels (**b**) and get mixed with the analyte to get detected through optic fibers (**c**) coupled with an illumination source and a USB spectrophotometer. (**d**) Fluorescent MIP nanoparticles sensitive to α-fetoprotein (AFP) are used to create an ink that can be printed to create micro paper-based analytical devices (μPADs), which can be measured in a low-cost equipment (**e**) to quantitatively determine AFP concentration (**f**). Subfigures **a–c** reproduced from Lab Chip, 2023, 23, 466 with permission from the Royal Society of Chemistry. Subfigures **d–f** adapted from Biosensors and Bioelectronics, 160, Tawfik et al. Dual emission nonionic molecular imprinting conjugated polythiophenes-based paper devices and their nanofibers for point-of-care biomarkers detection, 112211, Copyright (2020), with permission from Elsevier

based on giant magnetoresistance to simultaneously detect twelve tumor markers [162].

Molecularly imprinted polymers to recognize molecules that can act as biomarkers for cancer have been employed either for diagnostic or therapeutic (i.e. directed drug delivery) applications [5, 85, 163]. Some common analytes are nuclear matrix protein 22, prostate-specific antigen, alpha-fetoprotein, carcinoembryonic antigen, and cancer antigen 125. Such MIPs have been implemented inside microfluidic platforms. Burnage et al. used a thin (12 nm) MIP shell to coat SiO_2 nanoparticles (Fig. 14.7a). These particles flowed through microchannels and chambers for extraction, derivatization, and detection of phosphorylated peptides (Fig. 14.8b and c) [164]. Tawfik et al. developed fluorescent imprinted polymers to recognize α-fetoprotein (AFP) and carcinoembryonic antigen (CEA) [165]. These polymers can take the shape of nanoparticles (Fig. 14.8d) or fibers and were included in paper-based POC assays where fluorescence intensity could be measured with either a smartphone or a purposely built low-cost instrument (Fig. 14.8e and f).

14.8 Prospects and Future Directions

14.8.1 Nanotechnology and MIPs in Cancer Diagnostics

To generate MIP nanoparticles, microprinting methods, current achievements have shifted to the epitope imprinting technique. Solid-phase MIP synthesis has addressed several MIP production challenges while focusing on the uses of the produced materials in the drug administration process, diagnostics, imaging, and tissue engineering [166]. In the early history of the activities of antigen–antibodies and enzymes–substrate interaction, the molecular imprinting technique was carried out [167].

From a general perspective, there is an increasing demand for biosensors that accounts for a $1 billion market worldwide. They have become well known due to their numerous benefits. The system is made up of immobilized receptors (enzymes, genetic material, antibodies, antigens, cells, or tissues) that can recognize the analyte. The most often employed substance is an antibody. However, they are pricey, unstable, and only useful under physiological circumstances. The transducer transforms the biological event's response into a quantifiable signal [168].

For theragnostic methods, molecularly imprinted polymers are one of the most important biomaterials that act as artificial receptors that can recognize substances with higher affinity and greater selectivity [85]. NIPAm based can be imprinted using an imprinting technique with an epitope (EGFR) allowing researchers to target the cancer cells that can overexpress the protein (EGFR). The MIPs were able to distinguish the cells that expressed EGFR at various intensities. Before polymerization, the template peptide was initially immobilized on glass beads to prepare these MIPs. Low-affinity polymers and monomers can be removed by washing step and high-affinity MIPs separated from the template molecules. This made it simple by the solid-phase method [166].

Popularity was gained quickly by the glycoprotein-imprinted polymers as superior receptors for diagnostics, inhibition, nanomedicines, and cancer targeting as they can accurately target the glycosides and glycans overexpressed in different tumors. Glycoprotein-imprinted polymers can target the specific glycosylated cell receptors and glycoprotein biomarkers in various malignancies. They could be used as nanomedicines in a variety of cancer treatment modalities, including targeted NanoMIP therapies, drug administration, photothermal therapy, and photodynamic therapy for eliminating tumor cells.

MIPs were created by combining hydrogen bonding, van der Waals, covalent bonding, and hydrophobic in various cross-linkers and monomers. Integrated with complementing analytical techniques like chromatographic columns or microextraction systems since these cross-linkers and monomers are fixed to build on to the specific recognition sites in nanoscale imprinted cavities. In solution pretreatment, MIPs have frequently been employed as high affinity materials. MIPs have been used before electrochemical methods, surface-enhanced resonance, Raman scattering, fluorescence, and colorimetry are all gaining popularity recently.

Boronic acids as fascinating ligand applications targeting specifically many cancer locations, drug transport, biosensing, separation, and self-assembly are enabled by this overexpressed most uniquely [110]. For the detection of cancer biomarkers, systems made up of MIP-based detection are versatile and have demonstrated favorable outcomes, enabling high sensitivity and selectivity. Even though aptamer-based and classic antibody targeting materials have high specificity, they are created by time-consuming and labor-intensive techniques such as systematic evolution of ligands or antibody screening. On the other hand, MIPs molecular requires only a target molecule of high-purity sample and at a reduced enables the simple synthesis of specific targeting materials [85].

To detect various target analytes in a single complicated solution, the multitemplate approach allows molecularly imprinted polymer-based biosensors to be developed. Various nanomaterials, such as nanocomposites or gold having two or more nanoparticles, can be employed to change the electrode surface. Therefore, conductivity, surface area, and reaction sensitivity rise. "MIP-based electrochemical biosensors" show excellent specificity, selectivity, and sensitivity to target molecules to detect metastasis the breast cancer-CA 27.29 and in the diagnosis, and response to treatment using ovarian cancer-CA-125. In the diagnosis, assess the response to treatment of the prostate cancer-prostate-specific antigen, and in diagnosis and response to treatment of germ cell tumors using the liver cancer-fetoprotein [169].

Integrating the magnetic graphene oxide, boronate affinity, and surface imprinting to extract and identify glycoproteins selectively by creating the magnetic molecularly imprinted polymers (MGO/MIPs). It has a remarkable significance, cost toward the template because of MGO/MIPs, and simple operation as the separation of the Fe_3O_4 nanoparticles can performed by an external magnetic field shows a sufficient level of selectivity, a fast elution rate, and a low cost [106]. For cancer biomarkers, MIPs during polymerization the molecule should not contain any unreactive groups, it should have nice chemical stability, and for better binding the functional groups in the template must complementarily integrate with the polymeric matrix [5].

MIP-based electrochemical sensors are incredibly strong and have demonstrated the ability to analyze solutions in all stages. Using the nanotubes made up of carbon and metallic nanoparticles for implementing this property. On the measurement substrate, the direct creation of receptors is allowed because conductive polymers like polypyrrole have shown to be particularly appealing for inclusion into MIP-based electrochemical sensors allowing for mass-scale chip manufacture. As a result, sensors for numerous targets such as hormone disruptors, medicines, diagnostics, and even DNA molecules have been developed [170].

New biosensing platform that enhances sensitivity and selectivity with nanostructured gold and molecularly imprinted polymer by using LSG electrodes. The Her-2 protein (human epidermal growth factor receptor 2), a key breast tumor biomarker, is detected using this biomimetic sensing technology. LSG electrodes are created by employing a CO_2 laser to irradiate a polyimide sheet. Onto the LSG, the nanostructured gold is electrodeposited to improve the immobilization of Her-2 and sensitivity on the sensor surface. The created LSG-AuNS-MIP sensor is integrated with a built open-source electrochemical analyzer capable of detecting Her-2, the detection of various disease biomarkers, and shows potential applicability as a point-of-care device. MIPs are seen as a technique for enhancing the selectivity of electrochemical-based sensors using manufactured analogs to natural antigen–antibody pairs. The new sensing approach was effectively employed for detecting Her-2 in the serum samples under ideal conditions [171].

An electropolymerized dopamine molecularly imprinted polymer (MIP) electrochemiluminescence (ECL) biosensor to detect squamous cell carcinoma antigen (SCCA) has been developed. Enabling the loading of additional the Fe (III)-MIL-88B-NH2 metal–organic framework (MOF) and ZnSeQDs (ZnSe quantum dots) that aid in the conversion of S_2O_8 2 into SO_4%. Due to the powerful ECL emission, a three-way amplification of the signal approach for ZnSeQDs emission has been developed. This biosensor is used to enhance the selectivity in SCCA detection and has biosensor various advantages, including high selectivity, specificity, and stability [172].

MIP-coated inorganic nanoparticles platform for a wide range of significances in biomedical science, including drug cargoes and imaging contrast agents. The combination of MIPs and inorganic nanomaterials such as silica (SiO_2), gold (Au), silver (Ag), iron oxide (Fe_3O_4), and quantum dots (QDs) yields highly versatile materials that combine many properties from both components. These materials, which have a multicomponent structure that is made up of an inorganic core and an imprinted polymer shell, have improved qualities and new functionalities. Given the growing importance of point-of-care testing devices, MIPs with microfluidic systems develop innovative devices that can be used in point-of-care applications. The variability of their analyte binding location and the limited accessibility of internal binding sites hinder the overall effectiveness of big and macroscopic bulk material-based MIPs.

In addition, removing the template to obtain efficient MIPs is tough. Large monolithic constructions may also be challenging to combine with a sensor's devices or signal transducer. These issues are typically handled by mechanically reducing large monolithic materials to microparticles, such as grinding. MIPs of nanoform have various advantages over bulk and microscale MIPs that constitutes a larger surface-to-volume ratio, improved detection capabilities, and decreased mass-transfer resistance. These nano/MIPS can be integrated with nanoparticles for effective distribution of the nanocomplex. To facilitate the surface imprinting of the template molecule on their polymeric shell, gold (Au), silica (SiO_2), QDs, silver (Ag), and iron oxide (Fe_3O_4) nanoparticles have been extensively used as core material. This combination uses the intrinsic capabilities of the inorganic core and the customized qualities of the polymeric shell to create materials with unique functional properties suitable for a variety of biomedical applications [173].

They exhibit promising stability, specific binding, and loading in a variety of hostile environments, flexibility, and antibody-like recognition as drug-delivery vehicles. At the site of interest with an extremely powerful external magnetic field, allowing for higher selectivity and efficient treatment, magnetic drug targeting (MDT) entails enriching SPIONs. In patients, magnetic drug targeting of drug-loaded SPIONs is getting closer to being used.

Further, adding an imprinted polymer shell (thin) to the MNPs surface improves the physicochemical properties of the sought-after MDT by improving binding kinetics, rising uniform circular shape, and binding capacity in the aqueous blood circulatory. Local heat amplification because MNPs are present in their structure can cause magnetic hyperthermia when exposed to an external magnetic field following the drug release when MMIP is loaded [174].

Grafting or surface polymerization is used to create core–shell MIP particles. Some carrier NP with an MIP envelope include the bound template as well as additional nanomaterials in a core–shell shape. After polymerization, the template is eluted, leaving epitopes for the analyte. A sandwich immunoassay could benefit from such a nanocomposite. Other approaches involve depositing MIP thin layers onto cantilever or electrode surfaces for analyte binding. Before polymerization begins, on the prepared beads surface all the MIP materials are adsorbed such as thin or porous silica, the bead has been removed, resulting in a circular particle covered with a MIP (thin layer) when the free-radical polymerization is completed. The approach has the advantages of having a stable coated layer, high surface density of polymer chains, and onto the same substrate the capacity to integrate multiple polymers. Furthermore, to be used as a sensor, an MIP must be linked with a transducer or reading method capable of measuring the amount of rebinding the target to the MIP following test sample exposure [175, 176].

Molecular imprinting has been employed to trap tumor cells by employing fixed cells as templates successfully. Might create nanocavities created theoretically by molecular imprinting that closely matches the chemical characteristics of exosomes by employing exosomes as templates. These nanocavities may function like antibodies, capturing exosomes from biofluids. Nevertheless, it is difficult to collect enough pure exosomes for use as templates because of the limits of current exosome capture technology. As a result, we believe that nanoparticle synthesis with similar size distribution and surface features to exosomes as templates should be a nice way to produce fake antibodies for exosome capture. There is a negatively charged exosome membrane capable of interacting with positively charged molecules. Exosomes were successfully captured using anion exchange resin. This approach, however, necessitates the collection of distinct effluent fractions, and pollutants and exosomes may co-elute, lowering the recovery rate and purity. When cationic molecules are introduced into the molecular imprinting process, electrostatic interaction may be expected to give an additional increase in exosome-specific capturing sensitivity.

Therefore, a molecule imprinting polymer (synthetic antibody) for exosome capturing was developed using imprinting the template and electrostatic interaction.

The nanocavities on the MIP exactly match the chemical binding and shape of the target exosomes. By performing NTA on MIP-isolated exosomes from urine and HeLa-CCM, the distribution of the size of the particles by the MIP from urine and HeLa-CCM was analyzed [177].

Proteome maps of cell surfaces to discover cell surface markers and epitopes for diagnostic and therapeutic purposes are generated by the snapshot imprinting method. The selective targeting of three discovered epitopes of epidermal growth factor receptors utilizing molecularly imprinted polymer nanoparticles and two cancer cell lines, MDA-MB-468 and HN5, were studied. This method discovered 438 proteins, 283 of which were transmembrane or extracellular. The capacity to examine the cell surface proteins, labeling, and affinity separation is the main significance of the molecular imprinting approach [178].

A universal technique to produce an antibody-free hydrogel with micro interfaces acquired by molecular recognition and cell imprinting imparted by the affinity of boronate. Manmade antibody captures human hepatocarcinoma SMMC-7721 cells with excellent efficiency. 99% of these collected cells are released with no loss of proliferative potential after incubating SMMC-7721 cells on a 4.5 cm^2 hydrogel. When 1000 times the number of non-target cells, particularly leukemia Jurkat cells, the SMMC-7721 cells caught with an enrichment factor that is as high as 13.5 3.2 (n = 3), confirming the artificial antibody's better selectivity for the capture of the targeted CTCs. Notably, the SMMC-7721 cells may be successfully caught after they have been spiked into whole blood, indicating that this technique holds considerable promise for further molecular characterization of CTCs [179].

A hydrogel layer (thermo-responsive) imprinted with sialic acid (SA) was created for the capturing and releasing of cancer cells. The sialic acid imprinting technique was carried out at 37 °C utilizing a functional monomer(thermo-responsive), resulting in adjustable sialic acid-recognition sites with sialic acid binding at 37 °C and weak binding at lower temperatures (e.g., 25 °C). The sialic acid-imprinted hydrogel layer is used to selectively recognize cancer cells because sialic acid can frequently be found at the glycan terminals of cell membrane proteins or lipids. Furthermore, by lowering the temperature, noninvasively trapped cells are released. With high capture efficiency, cell selectivity, and more reliable and long-lasting sialic acid-imprinted sites compared to natural receptors or antibodies, this hydrogel layer (thermo-responsive) has the potential to be highly effective for cancer diagnosis [180].

By grafting an-SA-imprinted hydrogel layer onto mesoporous silica nanoparticles (MSNs), sialic acid-imprinted mesoporous nanocarriers were created. Because SA is typically overexpressed in cancer cells due to the terminal of glycoproteins and lipids on the cell membrane, SA-imprinted mesoporous nanocarriers (SIMNs) could show targeted drug delivery and cancer selectivity. SIMNs were loaded with doxorubicin, also known as DOX, to form DOX-laden SIMNs. SIMNs@DOX nanoparticles selectively target cancer cells with SA-overexpression due to their surface recognition sites for SA molecules. Endocytosis by cancer cells results in the release of DOX due to low pH levels in endo–/lysosomes, allowing for drug

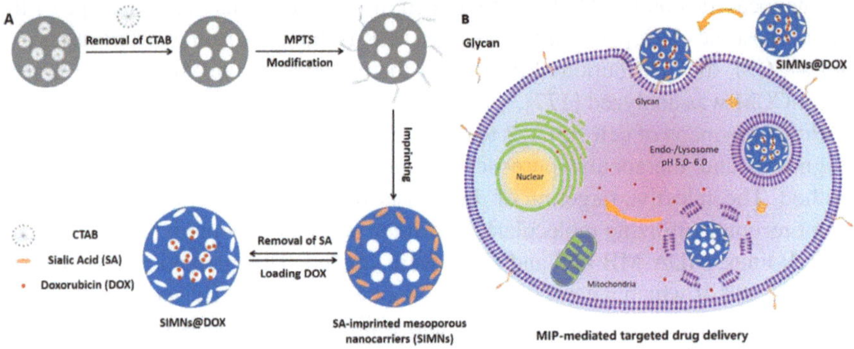

Fig. 14.9 (**a**) Strategy to prepare SA-imprinted sites on MSNs for the loading of antitumor drug DOX. (**b**) Schematic illustration of the SIMNs@DOX with SA recognition ability to target SA-overexpressed cancer cells and control release of antitumor drug [181]

delivery targeting cancer cells without harming normal cells. Please see Fig. 14.9 for more information [181].

Sialic acid-imprinted fluorescent core–shell molecularly imprinted polymer particles (SA-MIPs) were used on the cell surface of cancer cell lines. Sialic acid is recognized. When compared to previously employed core–shell SA-MIPs, improved scattering properties and susceptibility. The cancer cell lines are screened using the lectins Maackia amurensis Lectin I (MAL I, 2,3-SA) and Sambucus Nigra Lectin (SNA, 2,6-SA) even though sialic acid-imprinting was done without regard for the 2,3- and 2,6-SA forms. The cancer cell lines used in this investigation had varying binding behaviors with the SA-MIPs, and the lectin-binding pattern was also revealed by this. Furthermore, two distinct sialic acid conjugates were employed to prevent the sialic acid-MIPs' binding to lung cancer and breast, skin, and cell lines, respectively, demonstrating the SA-MIPs' specificity in confocal fluorescence microscopy and flow cytometr [182].

14.9 Conclusion and Perspectives

In this chapter, we reviewed the current state of the design and synthesis of molecular imprinting polymers (MIPs) for their use in recognizing and identifying cancer biomarkers. Because of their excellent properties as artificial recognition elements and their importance in early cancer detection, MIPs will soon have an important commercial market around the world.

As stated, the excellent properties of MIPs such as their high recognition capacity, their stability in both organic and inorganic media, the simplicity of their synthesis, and preparation as well as their low cost make these artificial polymeric materials an important tool for the detection of cancer biomarkers.

The fact that MIPs can be synthesized for almost any analyte, using a reduced number of monomers, both organic (such as methacrylic acid) and inorganic (sol–gel type MIPs) make them ideal for the preparation of recognition systems and extraction of different analytes, ranging from very simple molecules such as sarcosine to molecules as complex as proteins, which gives them excellent properties for use in biosensors, among many other fields of chemical analysis and cancer treatment. During their preparation, MIPs not only acquire cavities in shape, size, and binding sites for the template molecule, but they can also be synthesized combined with other nanomaterials (NpMg, NpAu, QDs, graphene, etc.) that greatly improve their properties and possibilities of application. All of the above allow systems based on MIPs to compete with advantages over the use of antibodies in the detection and treatment of cancer.

As is known, the approach to cancer detection and treatment is a complex process that involves numerous biomarkers and factors. Therefore, the possibility of imprinting MIPs in the presence of multiple templates makes it possible to use them for the recognition and differential detection of multiple cancer biomarkers in complex samples, allowing the development of electronic tongues and noses for this purpose. Although the development and application of MIPs is in its initial phases and there is still much ground to investigate, future development directions must follow this path.

In our view, the combination of the unique recognition properties of MIPs, their simplicity, and low preparation cost as well as the possibility of combining them with other nanomaterials in microfluidic systems holds a promising future for the development of clinical analysis devices that can be used by the patients themselves. To this end, some problems associated with the accessibility of the analyte to the binding and recognition sites in the polymer must still be resolved, its obtaining in particle sizes suitable for use in nano- and microfluidic systems, making this interaction reversible or easy to regenerate, and improve the mass transfer of the analyte to the cavities, all with an adequate cost–benefit ratio. Likewise, efficient and sensitive signaling of recognition and the reduction of nonspecific interactions will be important points to improve in future developments.

The fact that MIPs can be easily integrated into numerous analytical techniques exponentially expands their applications in numerous fields of cancer biomarker extraction and separation. An important aspect to keep in mind is that, similar to antibodies, their use as recognition elements must be linked to a transducer capable of signaling its interaction with the analyte. The selected transducer must be simple and reliable, without the need for external devices or supplies, so this is an important aspect that must be worked on in future developments and applications of MIPs.

While short-term tests have shown promising results in low toxicity and good biocompatibility for many molecularly imprinted polymers (MIPs), further research is necessary to assess their long-term toxicity before clinical use. To advance MIP-based nanomedicine clinically, evaluating biocompatibility, biodegradability, and selecting appropriate in vivo animal models are crucial. Despite early-phase obstacles, MIPs are progressing and hold immense promise in early cancer detection and nanomedicine against cancer. As MIP advancements continue, their unique characteristics position them to set a standard in cancer detection and therapy.

Acknowledgments The authors acknowledge the National Council of Humanities, Science, and Technology (CONAHCYT) for supporting this work through the Frontier Science project 2096211.

References

1. Cui F, Zhou Z, Zhou HS (2019) Measurement and analysis of cancer biomarkers based on electrochemical biosensors. J Electrochem Soc 167:037525
2. Popat K, McQuuen K, Feeley TW (2013) The global burden of cancer. Best Pract Res Clin Anaesthesiol 27:399–408
3. Troyer DA, Mubiru J, Leach RJ, Naylor SL (2004) Promise and challenge: markers of prostate cancer detection, diagnosis and prognosis. Dis Markers 20:117–128
4. Xu D, Zhu W, Wang C, Tian T, Cui J, Li J, Wang H, Li G (2014) Molecularly imprinted photonic polymers as sensing elements for the creation of cross-reactive sensor arrays. Chem A Eur J 20:16620–16625
5. Bhakta S, Mishra P (2021) Molecularly imprinted polymer-based sensors for cancer biomarker detection. Sens Actuat Rep 3:100061
6. Wulff G (1995) Molecular imprinting in cross-linked materials with the aid of molecular templates—a way towards artificial antibodies. Angew Chem Int Ed Engl 34(17):1812–1832
7. Zhang Y, Li M, Gao X, Chen Y, Liu T (2019) Nanotechnology in cancer diagnosis: progress, challenges and opportunities. J Hematol Oncol 12(1):1–13
8. Sun Y, Haglund TA, Rogers AJ, Ghanim AF, Sethu P (2018) Review: microfluidics technologies for blood-based cancer liquid biopsies. Anal Chim Acta 1012:10–29
9. Fass L (2008) Imaging and cancer: a review. Mol Oncol 2(2):115–152
10. Histed SN, Lindenberg ML, Mena E, Turkbey B, Choyke PL, Kurdziel KA (2012 Apr) Review of functional/anatomical imaging in oncology. Nucl Med Commun 33(4):349–361
11. Griffeth LK (2005) Use of PET/CT scanning in cancer patients: technical and practical considerations. Proc (Bayl Univ Med Cent) 18(4):321–330
12. Pulumati A, Pulumati A, Dwarakanath BS, Verma A, Papineni RVL (2023) Technological advancements in cancer diagnostics: improvements and limitations. Cancer Rep (Hoboken) 6(2):e1764
13. Berrington de González A, Mahesh M, Kim K-P et al (2009) Projected cancer risks from computed tomographic scans performed in the United States in 2007. Arch Intern Med 169(22):2071–2077
14. Mira JG, Potter JL, Fullerton GD, Ezekiel J (1982) Advantages and limitations of computed tomography scans for treatment planning of lung cancer. Int J Radiat Oncol Biol Phys 8(9):1617–1623
15. Frush DP, Donnelly LF, Rosen NS (2003) Computed tomography and radiation risks: what pediatric health care providers should know. Pediatrics 112(4):951–957
16. Pearce MS, Salotti JA, Little MP et al (2012) Radiation exposure from CT scans in childhood and subsequent risk of leukaemia and brain tumours: a retrospective cohort study. Lancet 380(9840):499–505
17. Mayerhoefer ME, Archibald SJ, Messiou C, Staudenherz A, Berzaczy D, Schöder H (2020) MRI and PET/MRI in hematologic malignancies. J Magn Reson Imaging 51(5):1325–1335
18. Kauppinen RA, Peet AC (2011) Using magnetic resonance imaging and spectroscopy in cancer diagnostics and monitoring: preclinical and clinical approaches. Cancer Biol Ther 12(8):665–679
19. Thenrajan T, Alwarappan S, Wilson J (2023) Molecular diagnosis and cancer prognosis-A consice review. Diagnostics (Basel) 13:4
20. Ganesh V, Raj MD, Jose G, Moreno MD, Leonard G, Gomella MD (1998) Utilization of polymerase chain reaction technology in the detection of solid tumors. Cancer 82(8):1419–1608
21. Wong ML, Medrano JF (2018) Real-time PCR for mRNA quantitation. BioTechniques 39–1:75

22. Ikbal Atli E, Gurkan H, Onur Kirkizlar H, Atli E, Demir S, Yalcintepe S, Kalkan R, Demir AM (2021) Pros and cons for fluorescent in situ hybridization, karyotyping and next generation sequencing for diagnosis and follow-up of multiple myeloma. Balkan J Med Genet 23(2):59–64
23. Duraiyan J, Govindarajan R, Kaliyappan K, Palanisamy M (2012) Applications of immunohistochemistry. J Pharm Bioallied Sci 4(Suppl 2):S307–S309
24. Armakolas A, Kotsari M, Koskinas J (2023) Liquid biopsies, novel approaches and future directions. Cancers (Basel) 15(5):1579
25. Di Capua D, Bracken-Clarke D, Ronan K, Baird AM, Finn S (2021) The liquid biopsy for lung cancer: state of the art, limitations and future developments. Cancers (Basel) 13(16):3923
26. Kim IJ, Kang HC, Park JG (2004) Microarray applications in cancer research. Cancer Res Treat 36(4):207–213
27. Lakshminarasimhan R, Liang G (2016) The role of DNA methylation in cancer. Adv Exp Med Biol 945:151–172
28. Kwon YW, Jo HS, Bae S, Seo Y, Song P, Song M, Yoon JH (2021) Application of proteomics in cancer: recent trends and approaches for biomarkers discovery. Front Med (Lausanne) 8:747333
29. Jo SY, Hong N, Lee S, Jeong JJ, Won J, Park J, Kim GJ, Kim SK, Kim S, Rhee Y (2023) Genomic and transcriptomic profiling reveal molecular characteristics of parathyroid carcinoma. Exp Mol Med 55(5):886–897
30. Choe H, Deirmengian CA, Hickok NJ, Morrison TN, Tuan RS (2015) Molecular diagnostics. J Am Acad Orthop Surg 23 Suppl(0):S26–S31
31. Andryukov BG, Besednova NN, Romashko RV, Zaporozhets TS, Efimov TA (2020) Label-free biosensors for laboratory-based diagnostics of infections: current achievements and new trends. Biosensors 10:690041
32. Das P, Das M, Chinnadayyala SR, Singha IM, Goswami P (2016) Recent advances on developing 3rd generation enzyme electrode for biosensor applications. Biosens Bioelectron 79:386–397
33. Homola J (2008) Surface plasmon resonance sensors for detection of chemical and biological species. Chem Rev 108:462–493
34. Honeychurch KC, Piano M (2018) Electrochemical (bio) sensors for environmental and food analyses. Biosensors 8:57
35. Mehrotra P (2016) Biosensors and their applications—a review. J Oral Biol Craniof Res 6:153–159
36. Morales MA, Halpern JM (2018) Guide to selecting a biorecognition element for biosensors. Bioconjug Chem 29:3231–3239
37. Wang J (2008) Electrochemical glucose biosensors. Chem Rev 108:814–825
38. Clark LC Jr, Lyons C (1962) Electrode systems for continuous monitoring in cardiovascular surgery. Ann N Y Acad Sci 102:29–45
39. Nambiar S, Yeow JTW (2011) Conductive polymer-based sensors for biomedical applications. Biosens Bioelectron 26:1825–1832
40. Sassolas A, Blum LJ, Leca-Bouvier BD (2011) Immobilization strategies to develop enzymatic biosensors. Biotechnol Adv 30(3):489–571
41. Jazib A, Jawayria N, Muhammad A, Muhammad F, Ali R (2017) Biosensors: the fundamentals, designs, types and Most recent impactful applications: a review. J Biosens Bioelectron 8(1):1–9
42. Njagi JI, Kagwanja SM (2011) The Interface in biosensing: improving selectivity and sensitiviy. Interfaces Interphases Analyt Chem 11:225–247
43. Naresh V, Lee N (2021) A review on biosensors and recent development of nanostructured materials-enabled biosensor. Sensors 21:1109
44. Li YCE, Lee IC (2020) The current trends of biosensors in tissue engineering. Biosensors 10:88
45. Gao F, Liu G, Qiao M, Li Y, Yi X (2022) Biosensors for the detection of enzymes based on aggregatio-induced emission. Biosensor 12:953

46. Sadani K, Nag P, Thian XY, Mukherji S (2022) Enzymatic optical biosensors for healthcare applications. Biosens Bioelectron X 12:100278
47. Sharma S, Byrne H, O'Kennedy RJ (2016) Antibodies and antibody-derived analytical biosensors. Essays Biochem 60:9–18
48. Yang Z, Li J, Wang Y, Gao F, Su J, Liu Y, Wang H (2023) Metal/covalent-organic framework-based biosensor for nuclei acid detection. Coord Chem Rev 491:215249
49. Palchetti I, Mascini M (2008) Nucleic acid biosensor for environmental pollution monitoring. Analyst 7:846–854
50. Song S, Wang L, Li J, Fan C, Zhao J (2008) Aptamer-based biosensor. TrAC Trend Anal Chem 27:108–117
51. Gupta N, Renugopalakrishan V, Liepmann D, Paulmurugan R, Malhotra BD (2019) Cell-based biosensors: recent trends, challenges and future perspectives. Biosens Bioelectron 143:111435
52. Alexandrov K, Vickers CE (2022) In vivo protein-based biosensor: seeing metabolism in real time. Trends Biotechnol 41:19–26
53. Liu Q, Wang J, Boyd BJ (2015) Peptide-based biosensors. Talanta 136:114–127
54. De Michele R, Carimi F, Fromeer WB (2014) Mitochondrial biosensor. Int J Biochem Cell Biol 48:39–44
55. Stobiecka M, Jakiela S, Chalupa A, Bednarczyk P, Dworakowska B (2017) Mitochondria–based biosensors with piezometric and RELS transduction for potassium uptake and release investigations. Biosens Bioelectron 88:114–121
56. Morea V, Lesk AM, Tramontano A (2000) Antibody modeling: implications for engineering and design. Methods 20:267–279
57. Kim T, Hoang E, Esfandyarpour R (2021) A 3D printed wearable bioelectronic patch for multi-sensing and in situ sweat electrolyte monitoring. Adv Mater Technol 6:2001021
58. Filk H, Avan AA (2019) Nanostructures for nonlabeled and labeled electrochemical immunosensors: simultaneous electrochemical detection of cancer markers: a review. Talanta 205:120153
59. Chen F, Hu Q, Li H, Xie Y, Xiu L, Zhang Y, Guo X, Yin K (2023a) Multiplex detection of infectious diseases on microfluidic platforms. Biosensor 13:410
60. Chen H, Zhang J, Huang R, Wang D, Deng D, Zhang Q, Luo L (2023b) The applications of electrochemical Immunosensors in the detection of disease biomarkers: a review. Molecules 28:3605
61. Pollap A, Świt P (2022) Recent advance in sandiwch SERS immunosensor for cancer detection. Int J Mol Sci 23:4740
62. Karunakaran C, Pandiaraj M, Santharaman P (2005) Immunosensor. Biosens Bioelectron:205–245
63. Nguyen HH, Lee SH, Lee UJ, Fermin CD, Kim M (2019) Immobilized enzymes in biosensor applications. Materials 12:121
64. Rocha TG, Pedro PH, de Souza MCM, Monteiro RRC, dos Santos JCS (2020) Lipase cocktail for optimized biodiesel production of free fatty acids from residual chicken oil. Catal Lett 151:1155–1166
65. Ronkainen NJ, Halsall HB, Heineman WR (2010) Electrochemical biosensors. Chem Soc Rev 39:1747–1763
66. Monteiro RRC, Lima PJM, Pinheiro BB, Freire TM, Dutra LMU, Fechine PBA, Gonçalves LRB, de Souza MCM, dos Santos JCS, Fernandez-Lafuente R (2019) Immobilization of lipase a from Candida Antarctica onto chitosan-coated magnetic nanoparticles. Int J Mol Sci 20:4018
67. Nguyen HH, Kim M (2017) An overview of techniques in enzyme immobilization. Appl Sci Converg Technol 26:157–163
68. Dhiman A, Kalra P, Bansal V, Bruno JG, Sharma TK (2017) Aptamer-based point-of-care diagnostic platforms. Sens Actuat B Chem 246:535–553
69. Hong P, Li W, Li J (2012) Applications of aptasensors in clinical diagnostics. Sensors 12:1181–1193

70. Tombelli S, Minunni M, Mascini M (2005) Analytical applications of aptamers. Biosens Bioelectron 20:2424–2434
71. Hwang SD, Midorikawa N, Punnarak P, Kikuchi Y, Kondo H, Hirono I, Aoki T (2012) Inhibition of hirame rhabdovirus growth by RNA aptamers. J Fish Dis 35:927–934
72. Ayodele OO, Adesina AO, Pourianejad S, Averritt J, Ignatova T (2021) Recent advances in nanomaterial-based aptasensors in medical diagnosis and therapy. Nano 11:932
73. Zhuo Z, Yu Y, Wang M, Li J, Zhang Z, Liu J, Wu X, Lu A, Zhang G, Zhang B (2017) Recent advances in SELEX technology and aptamer applications in biomedicine. Int J Mol Sci 18:2142
74. Blind M, Blank M (2015) Aptamer selection technology and recent advances. Mol Ther Nucleic Acids 4:e223
75. Heo K, Min SW, Sung HJ, Kim HG, Kim HJ, Kim YH, Choi BK, Han S, Chung S, Lee ES (2016) An aptamer-antibody complex (oligobody) as a novel delivery platform for targeted cancer therapies. J Control Release 229:1–9
76. Lee CH, Lee SH, Kim JH, Noh YH, Noh GJ, Lee SW (2015) Pharmacokinetics of a cholesterol-conjugated aptamer against the hepatitis C virus (HCV) NS5B protein. Mol Ther Nucleic Acids 4:e254
77. Brilian AP, Cabral PC, Freitas P, Fernades E (2021) The challenges of developing biosensors for clinical assessment: a review. Chemosensors 9:299
78. Daniel RM, Dines M, Petach HH (1996) The denaturation and degradation of stable enzymes at high temperatures. Biochem J 317:1–11
79. Cosnier S (2003) Biosensors based on electropolymerized films: new trends. Anal Bioanal Chem 377:507–520
80. Welch NG, Scoble JA, Muir BW, Pigram PJ (2017) Orientation and characterization of immobilized antibodies for improved immunoassays (review). Biointerphases 12:02D301
81. Al-Kindy S, Badía R, Suárez-Rodríguez JL, Díaz-García ME (2000) Molecularly imprinted polymers and optical sensing applications. Crit Rev Anal Chem 30(4):291–309
82. Fernández-González A, Guardia L (2007) Reconocimiento molecular mediante materiales biomiméticos: impresión molecular. An Quím RSEQ 103(2):14–14
83. Dickert FL, Lieberzeit P, Tortschanoff M (2000) Molecular imprints as artificial antibodies—a new generation of chemical sensors. Sens Actuat B Chem 65(1–3):186–189
84. Xu J, Miao H, Wang J, Pan G (2020) Molecularly imprinted synthetic antibodies: from chemical design to biomedical applications. Small 16(27):1906644
85. Kang MS, Cho E, Choi HE, Amri C, Lee JH, Kim KS (2023) Molecularly imprinted polymers (MIPs): emerging biomaterials for cancer theragnostic applications. Biomater Res 27:1
86. He S, Zhang L, Bai S, Yang H, Cui Z, Zhang X, Li Y (2021) Advances of molecularly imprinted polymers (MIP) and the application in drug delivery. Eur Polym J 143:110179
87. Pichon V, Chapuis-Hugon F (2008) Role of molecularly imprinted polymers for selective determination of environmental pollutants—a review. Anal Chim Acta 622(1–2):48–61
88. Yang LL, Li YJ, Jiang ZF, Li QY, Ma RR, He JY, Yuan CS (2021) Investigating two distinct dummy templates molecularly imprinted polymers as paclitaxel adsorbent in synthesis system and releaser in biological samples. Microchem J 165:106042
89. Lusina A, Michał C (2022) Molecularly imprinted polymers as state-of-the-art drug carriers in hydrogel transdermal drug delivery applications. Polymers 14:640
90. Pérez MCC (2015) Síntesis y caracterización de polímeros de impresión molecular para aplicaciones analíticas (Doctoral dissertation, Universidade da Coruña)
91. Ertürk G, Mattiasson B (2017) Molecular imprinting techniques used for the preparation of biosensors. Sensors 17(2):288
92. Vaihinger D, Landfester K, Kräuter I, Brunner H, Tovar GE (2002) Molecularly imprinted polymer nanospheres as synthetic affinity receptors obtained by miniemulsion polymerisation. Macromol Chem Phys 203(13):1965–1973
93. Zeng Z, Hoshino Y, Rodriguez A, Yoo H, Shea KJ (2010) Synthetic polymer nanoparticles with antibody-like affinity for a hydrophilic peptide. ACS Nano 4(1):199–204

94. Li P, Fei R, Yuan C (2003) Morphologies and binding characteristics of molecularly imprinted polymers prepared by precipitation polymerization. Polymer Int 52(12):1799–1806
95. Pacheco JG, Silva MS, Freitas M, Nouws HP, Delerue-Matos C (2018) Molecularly imprinted electrochemical sensor for the point-of-care detection of a breast cancer biomarker (CA 15-3). Sensors Actuators B Chem 256:905–912
96. Rebelo TS, Santos C, Costa-Rodrigues J, Fernandes MH, Noronha JP, Sales MGF (2014) Novel prostate specific antigen plastic antibody designed with charged binding sites for an improved protein binding and its application in a biosensor of potentiometric transduction. Electrochim Acta 132:142–150
97. Mamipour Z, Nematollahzadeh A, Kompany-Zareh M (2021) Molecularly imprinted polymer grafted on paper and flat sheet for selective sensing and diagnosis: a review. Microchim Acta 188:1–21
98. Fernández-Puig S, Lazo-Fraga AR, Korgel BA, Oza G, Dutt A, Vallejo-Becerra V et al (2022) Molecularly imprinted polymer-silica nanocomposite based potentiometric sensor for early prostate cancer detection. Mater Lett 309:131324
99. He JY, Li Q, Xu HX, Zheng QY, Zhang QH, Zhou LD et al (2023) Recognition and analysis of biomarkers in tumor microenvironments based on promising molecular imprinting strategies with high selectivity. TrAC Trends Anal Chem:117033
100. Yoosefi S, Esfandyari-Manesh M, Ghorbani-Bidkorpeh F, Ahmadi M, Moraffah F, Dinarvand R (2022) Novel biodegradable molecularly imprinted polymer nanoparticles for drug delivery of methotrexate anti-cancer; synthesis, characterization, and cellular studies. DARU J Pharmac Sci 30(2):289–302
101. Sheydaei O, Khajehsharifi H, Rajabi HR (2020) Rapid and selective diagnose of sarcosine in urine samples as prostate cancer biomarker by mesoporous imprinted polymeric nanobeads modified electrode. Sensors Actuators B Chem 309:127559
102. Alizadeh T, Nayeri S (2020) An enzyme-free sensing platform based on molecularly imprinted polymer/MWCNT composite for sub-micromolar-level determination of pyruvic acid as a cancer biomarker. Anal Bioanal Chem 412:657–667
103. Ansell RJ (2015) Characterization of the binding properties of molecularly imprinted polymers. Molecularly imprinted polymers in biotechnology, pp 51–93
104. Ye L (2016) Molecularly imprinted polymers with multi-functionality. Anal Bioanal Chem 408:1727–1733
105. García-Calzón JA, Díaz-García ME (2007) Characterization of binding sites in molecularly imprinted polymers. Sensors Actuators B Chem 123(2):1180–1194
106. Suzaei FM, Batista AD, Mizaikoff B, Rahimi S, Daryanavard SM, Abdel-Rehim M (2022) Molecularly imprinted polymers for selective extraction/microextraction of cancer biomarkers: a review. Microchim Acta 189:255
107. Vessman J, Stefan RI, Van Staden JF, Danzer K, Lindner W, Burns DT et al (2001) Selectivity in analytical chemistry. Pure Appl Chem 73(8):1381–1386
108. Becskereki G, Horvai G, Tóth B (2021) The selectivity of molecularly imprinted polymers. Polymers 13(11):1781
109. Spivak DA (2005) Optimization, evaluation, and characterization of molecularly imprinted polymers. Adv Drug Deliv Rev 57(12):1779–1794
110. Ali MM, Zhu S, Amin FR, Hussain D, Du Z, Hu L (2022) Molecular imprinting of glycoproteins: From preparation to cancer theranostics. Theranostics 12:2406–2426
111. Ansell RJ, Ramström O, Mosbach K (1996) Towards artificial antibodies prepared by molecular imprinting. Clin Chem 42(9):1506–1512
112. Cecchini A, Raffa V, Canfarotta F, Signore G, Piletsky S, MacDonald MP, Cuschieri A (2017) In vivo recognition of human vascular endothelial growth factor by molecularly imprinted polymers. Nano Lett 17(4):2307–2312
113. Verheyen E, Schillemans JP, Van Wijk M, Demeniex MA, Hennink WE, Van Nostrum CF (2011) Challenges for the effective molecular imprinting of proteins. Biomaterials 32(11):3008–3020

114. Jahangiri-Manesh A, Mousazadeh M, Nikkhah M, Abbasian S, Moshaii A, Masroor MJ, Norouzi P (2022) Molecularly imprinted polymer-based chemiresistive sensor for detection of nonanal as a cancer related biomarker. Microchem J 173:106988
115. Park R, Jeon S, Jeong J, Park SY, Han DW, Hong SW (2022) Recent advances of point-of-care devices integrated with molecularly imprinted polymers-based biosensors: from biomolecule sensing design to intraoral fluid testing. Biosensors 12(3):136
116. Tarannum N, Kumar D, Surya SG, Dramou P (2023) Point-of-care detection assay based on biomarker-imprinted polymer for different cancers: a state-of-the-art review. Polym Bull 80(1):1–46
117. Haupt K, Medina Rangel PX, Bui BTS (2020) Molecularly imprinted polymers: antibody mimics for bioimaging and therapy. Chem Rev 120(17):9554–9582
118. Parisi OI, Ruffo M, Malivindi R, Vattimo AF, Pezzi V, Puoci F (2020) Molecularly imprinted polymers (MIPs) as theranostic systems for sunitinib controlled release and self-monitoring in cancer therapy. Pharmaceutics 12(1):41
119. Tang P, Wang Y, He F (2020) Electrochemical sensor based on super-magnetic metal–organic framework@ molecularly imprinted polymer for Sarcosine detection in urine. J Saudi Chem Soc 24(8):620–630
120. Lai Y, Deng Y, Yang G, Li S, Zhang C, Liu X (2018) Molecular imprinting polymers electrochemical sensor based on AuNPs/PTh modified GCE for highly sensitive detection of carcinoma embryonic antigen. J Biomed Nanotechnol 14(10):1688–1694
121. Lai Y, Zhang C, Deng Y, Yang G, Li S, Tang C, He N (2019) A novel α-fetoprotein-MIP immunosensor based on AuNPs/PTh modified glass carbon electrode. Chin Chem Lett 30(1):160–162
122. Viswanathan S, Rani C, Ribeiro S, Delerue-Matos C (2012) Molecular imprinted nanoelectrodes for ultra sensitive detection of ovarian cancer marker. Biosens Bioelectron 33(1):179–183
123. Patra S, Roy E, Madhuri R, Sharma PK (2015) Nano-iniferter based imprinted sensor for ultra trace level detection of prostate-specific antigen in both men and women. Biosens Bioelectron 66:1–10
124. Zang Y, Zhang Y, Wei R, Xue H, Jiang J (2023) Difunctional molecularly imprinted polymers and heterostructured CdS nanoparticle-sensitized ZnO nanorod arrays for antibody-free photoelectrochemical alpha-fetoprotein sensor. J Electroanal Chem 944:117631
125. Satoh W, Hosono H, Yokomaku H, Morimoto K, Upadhyay S, Suzuki H (2008) Integrated electrochemical analysis system with microfluidic and sensing functions. Sensors (Basel) 21:1111–1127
126. Naik AR, Zhou Y, Dey AA, González-Arellano DL, Okoroanyanwu U, Secor EB, Hersam MC, Morse J, Rothstein JP, Carter KR, Watkins JJ (2022) Printed microfluidic sweat sensing platform for cortisol and glucose detection. Lab Chip 22:156–169
127. Akgönüllü S, Bakhshpour M, Piskin AS, Denizli A (2021) Microfluidic Systems for Cancer Diagnosis and Applications. Micromachines 12:1349
128. Hong R, Sun H, Li D, Yang W, Fan K, Liu C, Dong L, Wang G (2022) A review of biosensors for detecting tumor markers in breast cancer. Life 12:342
129. Zhang F, Chen J, Wang M, Drabier R (2013) A neural network approach to multi-biomarker panel discovery by high-throughput plasma proteomics profiling of breast cancer. BMC Proc 7:S10
130. Farshchi F, Hasanzadeh M (2021) Microfluidic biosensing of circulating tumor cells (CTCs): recent progress and challenges in efficient diagnosis of cancer. Biomed Pharmacother 134:1111153
131. Bargahi N, Ghasemali S, Jahandar L, Nazari A (2022) Recent advances for cancer detection and treatment by microfluidic technology, review and update. Biol Proced Online 24:5
132. Adampourezare M, Dehghan G, Hasanzadeh M, Feizi M (2021) Application of lateral flow and microfluidic bio-assay and biosensing towards identification of DNA-methylation and cancer detection: recent progress and challenges in biomedicine. Biomed Pharmacother 141:11118445

133. Zhou J, Tao F, Zhu J, Lin S, Wang Z, Wang X, Ou J, Li Y, Liu QH (2019) Portable tumor biosensing of serum by plasmonic biochips in combination with nanoimprint and microfluidics. Nanophptonics 8:0173
134. Xiang Y, Hu C, Wu G, Xu S, Li Y (2023) Nanomaterial-based microfluidic systems for cancer biomarker detection: recent applications and future perspectives. TrAC Trends Anal Chem 158:116835
135. Wu Q, Liu J, Wang X, Feng L, Wu J, Zhu X, Wen W, Gong X (2020) Organ-on-a-chip: recent breakthroughs and future prospects. Biomed Eng Online 19:9
136. Saylan Y, Denizli A (2019) Molecularly imprinted polymer-based microfluidic systems for point-of-care applications. Micromachines 10:766
137. Guo L, Feng J, Fang Z, Xu J, Lu X (2015) Application of microfluidic "lab-on-a-chip" for the detection of mycotoxins in foods. Trends Food Sci Technol 46:252–263
138. Karasu T, Özgür E, Uzun L (2023) MIP-on-a-chip: artificial receptors on microfluidic platforms for biomedical applications. J Pharm Biomed Anal 226:115257
139. Wackers G, Putzeys T, Peeters M, de Cauter LV, Cornelis P, Wübbenhorst M, Tack J, Troost F, Verhaert N, Doll T, Wagner P (2020) Towards a catheter-based impedimetric sensor for the assessment of intestinal histamine levels in IBS patients. Biosens Bioelectron 158:112152
140. Tokonami S, Shiigi H, Nagaoka T (2009) Review: micro- and nanosized molecularly imprinted polymers for high-throughput analytical applications. Anal Chim Acta 641:7–13
141. Fuchs Y, Soppera O, Haupt K (2012) Photopolymerization and photostructuring of molecularly imprinted polymers for sensor applications—a review. Anal Chim Acta 717:7–20
142. Huang H, Huan S, Lin C, Lee Y (2007) A multi-array sensor via the integration of acrylic molecularly imprinted photoresists and ultramicroelectrodes on a glass chip. Anal Chim Acta 582:137–146
143. Hong C, Chan P, Lin C, Hong C (2010) A disposable microfluidic biochip with on-chip molecularly imprinted biosensors for optical detection of anesthetic propofol. Biosens Bioelectron 25:2058–2064
144. El Sharif HF, Denninson SR, Tully M, Crossley S, Mwangi W, Bailey D, Graham SP, Reddy SM (2022) Evaluation of electropolymerized molecularly imprinted polymers (E-MIPs) on disposable electrodes for detection of SARS-CoV-2 in saliva. Anal Chim Acta 1206:339777
145. Sharma PS, Pietzyk A, D'Souza F, Kutner W (2012) Electrochemically synthesized polymers in molecular imprinting for chemical sensing. Anal Bioanal Chem 402:3177–3204
146. Crapnell RD, Hudson A, Foster CW, Eersels K, Grinsven B, Cleij TJ, Banks CE, Peeters M (2019) Recent advances in electrosynthesized molecularly imprinted polymer sensing platforms for bioanalyte detection. Sensors 19:1204
147. Gonçalves LM (2021) Electropolymerized molecularly imprinted polymers: perceptions based on recent literature for soon-to-be world-class scientists. Curr Opin Electrochem 25:100640
148. Glasscott MW, Vannoy KJ, Kazemi R, Verber MD, Dick JE (2020) μ-MIP: molecularly imprinted polymer-modified microelectrodes for the ultrasensitive quantification of GenX (HFPO-DA) in river water. Enviorn Sci Technol Lett 7:489–495
149. Gómez-Caballero A, Aranzazu Goicolea M, Barrio RJ (2005) Paracetamol voltammetric microsensors based on electrocopolymerized–molecularly imprinted film modified carbon fiber microelectrodes. Analyst 130:1012–1018
150. Tsai T, Han H, Cheng C, Chen L, Chang H, Jason Chen J (2012) Modification of platinum microelectrode with molecularly imprinted over-oxidized polypyrrole for dopamine measurement in rat striatum. Sens Actuat B Chem. 171–172:93–101
151. Weng C, Yeh W, Ho K, Lee G (2007) A microfluidic system utilizing molecularly imprinted polymer films for amperometric detection of morphine. Sens Actuat B Chem. 121:576–582
152. Nagabooshanam S, Roy S, Deshmukh S, Wadhwa S, Sulania I, Mathur A, Krishnamurthy S, Bharadwaj LM, Roy SS (2020) Microfluidic affinity sensor based on a molecularly imprinted polymer for ultrasensitive detection of Chlorpyrifos. ACS Omega 5:31765–31773

153. Ge L, Wang S, Yu J, Li N, Ge S, Yan M (2013) Molecularly imprinted polymer grafted porous au-paper electrode for an microfluidic electro-analytical origami device. Adv Funct Mater 23:3115–3123
154. Huang S, Lee G, Chien F, Chen S, Chen W, Yang M (2006) A microfluidic system with integrated molecular imprinting polymer films for surface plasmon resonance detection. J Mixeomech Microeng 16:1251
155. Ramanavicius S, Samukaite-Bubniene U, Rataitaite V, Bechlany M, Ramanaviciues A (2022) Electrochemical molecularly imprinted polymer based sensors for pharmaceutical and biomedical applications (review). J Pharm Biomed Anal 215:114739
156. Ashley J, Shahbazi M, Kant K, Chidambara VA, Wolff A, Bang DD, Sun Y (2017) Molecularly imprinted polymers for sample preparation and biosensing in food analysis: progress and perspectives. Biosens Bioelectron 91:606–615
157. Mi F, Hu C, Wang Y, Wang L, Peng F, Geng P, Guan M (2022) Recent advancements in microfluidic chip biosensor detection of foodborne pathogenic bacteria: a review. Anal Bioanal Chem 414:2883–2902
158. Orbay S, Sanyal A (2023) Molecularly imprinted polymeric particles created using droplet-based microfluidics: preparation and applications. Micromachines 14:763
159. Kang KN, Koh EY, Jang JY, Kim CW (2022) Multiple biomarkers are more accurate than a combination of carbohydrate antigen 125 and human epididymis protein 4 for ovarian cancer screening. Obstet Gynecol Sci 65:346–354
160. Sardesai AU, Tanak AS, Krishnan S, Striegel DA, Schully KL, Clarck DV, Muthukumar S, Prasad S (2021) An approach to rapidly assess sepsis through multi-biomarker host response using machine learning algorithm. Sci Rep 11:16905
161. Tothill IE, Biosensors for cancer markers diagnosis (2009) Semin Cell Dev Biol 20:55–62
162. Gao Y, Huo W, Zhang L, Lian J, Tao W, Song C, Tang J, Shi S, Gao Y (2019) Multiplex measurement of twelve tumor markers using a GMR multi-biomarker immunoassay biosensor. Biosens Bioelectron 123:204–210
163. Selvolini G, Marrazza G (2017) MIP-based sensors: promising new tools for cancer biomarker determination. Sensor 17:718
164. Burnage SC, Bell J, Wan W, Kislenko E, Rurack K (2023) Combining a hybrid chip and tube microfluidic system with fluorescent molecularly imprinted polymer (MIP) core–shell particles for the derivatisation, extraction, and detection of peptides with N-terminating phosphorylated tyrosine. Lab Chip 23:466–474
165. Tawfik SM, Elmasry MR, Sharipov M, Azizov S, Lee CH, Lee Y (2020) Dual emission nonionic molecular imprinting conjugated polythiophenes-based paper devices and their nanofibers for point-of-care biomarkers detection. Biosens Bioelectron 160:112211
166. Piletsky S, Canfarotta F, Poma A, Maria Bossi A (2020) Title: molecularly imprinted polymers for cell recognition. Trends Biotechnol. 38(4):368–387
167. Sajini T, Mathew B (2021) A brief overview of molecularly imprinted polymers: highlighting computational design, nano and photo-responsive imprinting. Talanta Open 4:100072
168. Louadj L, Pagani A, Benghouzi P, Sabbah M, Griffete N (2023) How molecularly imprinted polymers can be used for the diagnostic and treatment of tropical diseases? Chem Afr 6(1):3–14
169. Pilvenyte G, Ratautaite V, Boguzaite R, Ramanavicius A, Viter R, Ramanavicius S (2023) Molecularly imprinted polymers for the determination of cancer biomarkers. Int J Mol Sci 24(4):4105
170. Lowdon JW, Diliën H, Singla P, Peeters M, Cleij TJ, van Grinsven B, Eersels K (2020) MIPs for commercial application in low-cost sensors and assays—an overview of the current status quo. Sensors Actuators B Chem 325:128973
171. Lahcen AA, Rauf S, Aljedaibi A, de Oliveira Filho JI, Beduk T, Mani V, Alshareef HN, Salama KN (2021) Laser-scribed graphene sensor based on gold nanostructures and molecularly imprinted polymers: application for Her-2 cancer biomarker detection. Sensors Actuators B Chem 347:130556

172. Mo G, Qin D, Jiang X, Zheng X, Mo W, Deng B (2020) A sensitive electrochemiluminescence biosensor based on metal-organic framework and imprinted polymer for squamous cell carcinoma antigen detection. Sensors Actuators B Chem 310:127852
173. Orbay S, Kocaturk O, Sanyal R, Sanyal A (2022) Molecularly imprinted polymer-coated inorganic nanoparticles: fabrication and biomedical applications. Micromachines 13(9):1464
174. Sanadgol N, Wackerlig J (2020) Developments of smart drug-delivery systems based on magnetic molecularly imprinted polymers for targeted cancer therapy: a short review. Pharmaceutics 12:1–31
175. Denmark DJ, Mohapatra S, Mohapatra SS (2020) Point-of-care diagnostics: molecularly imprinted polymers and nanomaterials for enhanced biosensor selectivity and transduction. Eurobiotech J 4:184–206
176. Kadhem AJ, Gentile GJ, de Cortalezzi MMF (2021) Molecularly imprinted polymers (Mips) in sensors for environmental and biomedical applications: a review. Molecules 26:20
177. Liu L, Liu J, Zhou W, Sui Z, Liu J, Yang K, Zhang L, Liang Z, Zhang Y (2022) An artificial antibody for exosome capture by dull template imprinting technology. J Mater Chem B 10:6655–6663
178. Piletsky SS, Piletska E, Poblocka M, Macip S, Jones DJL, Braga M, Cao TH, Singh R, Spivey AC, Aboagye EO, Piletsky SA (2021) Snapshot imprinting: rapid identification of cancer cell surface proteins and epitopes using molecularly imprinted polymers. Nano Today 41:101304
179. Liu L, Dong C, Li X, Li S, Ma B, Zhao B, Li X, Liang Z, Yang K, Zhang L, Zhang Y (2020) Antibody-free hydrogel with the synergistic effect of cell imprinting and boronate affinity: toward the selective capture and release of undamaged circulating tumor cells. Small 16(2020):1904199
180. Ma Y, Yin Y, Ni L, Miao H, Wang Y, Pan C, Tian X, Pan J, You T, Li B, Pan G (2021) Thermo-responsive imprinted hydrogel with switchable sialic acid recognition for selective cancer cell isolation from blood. Bioact Mater 6(5):1308–1317
181. Yin Y, Guan L, Wang Y, Ma Y, Pan J, Peng Y, Pan G (2021) Sialic acid-imprinted mesoporous nanocarriers for tumor cell targeted drug delivery. Coll Interf Sci Commun 42:100421
182. Beyer S, Kimani M, Zhang Y, Verhassel A, Sternbæk L, Wang T, Persson JL, Härkönen P, Johansson E, Caraballo R, Elofsson M, Gawlitza K, Rurack K, Ohlsson L, El-Schich Z, Wingren AG, Stollenwerk MM (2022) Fluorescent molecularly imprinted polymer layers against sialic acid on silica-coated polystyrene cores—assessment of the binding behavior to cancer cells. Cancers 14(8):1875

Molecularly Imprinted Polymers as Artificial Antibodies in Therapeutic Applications

15

Parmida Harirchi, Kiarash Bastani, Kiana Sakhaee, and Payam Zahedi

Abstract

In recent years, the fields of tissue engineering and drug delivery have witnessed remarkable progress, driven by the pursuit of enhancing therapeutic outcomes and addressing a wide range of medical challenges. Among the cutting-edge technologies that have captured considerable attention in this domain are Molecularly Imprinted Polymers (MIPs). These sophisticated biomaterials have undergone significant advancements in both their applications and synthesis processes since their initial proposal five decades ago. Molecular imprinting, a highly effective synthetic technique, lies at the core of MIPs' ability to create durable materials with predetermined molecular selectivity. Often referred to as "artificial antibodies," MIPs possess unique qualities that render them intriguing candidates for various applications, particularly in tissue engineering and targeted drug delivery. The incorporation of novel capabilities into MIPs has further augmented their potential, enabling responsiveness to diverse environmental conditions and facilitating selective molecular binding. This adaptability allows for precise adjustments in binding and controlled release of various molecular targets, while also streamlining the process of affinity separation. While the molecular imprinting of various substances has proved successful, cells have emerged as particularly challenging targets. Early research predominantly focused on microprinting techniques, but recent advancements have steered away from this approach, opting for epitope imprinting to develop MIP nanoparticles (NPs). Moreover, the obstacles encountered in MIP manufacture during the early stages have been effectively resolved through the development of inno-

P. Harirchi · K. Bastani · K. Sakhaee · P. Zahedi (✉)
Nano-Biopolymers Research Laboratory, School of Chemical Engineering, College of Engineering, University of Tehran, Tehran, Iran
e-mail: phdzahedi@ut.ac.ir

S. Patra et al. (eds.), *Molecularly Imprinted Polymers: Path to Artificial Antibodies*, https://doi.org/10.1007/978-981-97-4379-7_15

vative techniques such as solid-phase MIP synthesis. The present study highlights the pivotal developments in the creation and utilization of molecularly imprinted polymers that possess the distinctive ability to specifically recognize physiologically active substances. The underlying concepts supporting the design of these highly effective and template-specific molecularly imprinted polymers take center stage in this exploration. Notably, the emphasis is placed on the imprinting of intricate macromolecular and supramolecular templates, showcasing the versatility and precision of these synthetic materials. In the realm of chemistry and molecular biology, considerable attention is given to the current and potential applications of molecularly imprinted polymers. Their promising uses span a broad spectrum, ranging from advanced imaging techniques and targeted medication delivery to diagnostics and the field of tissue engineering, where MIPs demonstrate immense potential. In conclusion, this article provides a succinct summary of numerous cell imprinting techniques, underscoring the significance of molecularly imprinted polymers in pushing the boundaries of tissue engineering and drug delivery. As research in this area continues to evolve, MIPs remain at the forefront of groundbreaking technologies, poised to revolutionize the medical landscape and improve the lives of countless individuals worldwide.

Keywords

Molecularly imprinted polymers · Tissue engineering · Drug delivery Artificial antibody · Biomaterials

15.1 Introduction

Molecularly imprinted polymers (MIPs) represent a category of synthetic polymers recognized for their capacity to selectively identify and attach to specific molecules, guided by factors like shape, dimensions, and functional groups. Often referred to as "artificial antibodies" or "plastic antibodies," MIPs emulate the role of natural antibodies and possess unique properties such as the ability to interact with cells, induce specific cell signaling, intracellular cascades, and modulate cell behavior. These properties distinguish them from other natural systems [1]. MIPs have several advantages over other biological receptors such as low production cost, high physical strength, and less sensitivity to changes in environmental conditions making them a suitable alternative for these receptors. Additionally, they exhibit resistance to elevated pressure and can withstand various chemicals such as organic solvents, acids, bases, and metal ions [2]. Moreover, they are a versatile material choice due to their unlimited storage ability, robustness, and stability at high temperatures, which makes them applicable in a wide temperature range for various purposes. To synthesize these materials, a template molecule, one or two cross-linkers, and one or two functional monomers are used. Initially, a complex is formed between the template and functional monomers, which is then surrounded and fixed by cross-linkers to create a three-dimensional polymer network. Finally, the template molecule is removed from the polymer matrix, resulting in a cavity with the same shape and size as the template molecule [3]. Multiple polymerization techniques are employed to create MIPs, with noteworthy methods being free radical polymerization (FRP), atom transfer radical polymerization (ATRP), nitroxide-mediated radical polymerization (NMP), and reversible addition-fragmentation chain transfer (RAFT) polymerization. MIPs have found diverse applications, such as in drug delivery systems (DDSs), owing to their exceptional selectivity, stability, and cost-effectiveness [4–7]. MIPs have emerged as promising tools for effecting targeted drug transport to specific cellular entities or tissues within the domain of DDSs. Additionally, MIPs offer the capacity to modulate drug release kinetics, thereby augmenting therapeutic efficacy and ensuring treatment safety. Compared to conventional drug delivery modalities, MIP-based DDS presents several advantages, including ameliorated adverse effects, promoted drug loading capacities, and attributes such as biocompatibility, biodegradability, reduced toxicity, cost-effective synthesis, and versatile routes of administration. Nonetheless, challenges still need to be addressed before MIP-based DDS can be widely adopted. These encompass the enhancement of MIP cytotoxicity profiles and bolstering biocompatibility, while concurrently improving strategies for different administration routes [6–10]. Moreover, MIPs have also been investigated for use in tissue engineering (TE) and regenerative medicine, particularly, they have been studied for their potential use in creating scaffolds for tissue regeneration. MIP-based scaffolds can be designed to mimic the extracellular matrix (ECM) of natural tissues, providing a suitable environment for cell growth and differentiation. Recent investigations underscore the utility of MIP-based scaffolds in regenerating diverse tissues such as bone, cartilage, and nerve tissue (Fig. 15.1) [11, 12].

In this chapter, different methods of synthesis and design of MIPs are first introduced. Then, by examining the mechanism of these materials as DDSs and their use in TE, conclusions are drawn about the advantages and challenges of using these items in the mentioned applications. All the discussed cases are discussed with the approach of using MIPs as artificial antibodies.

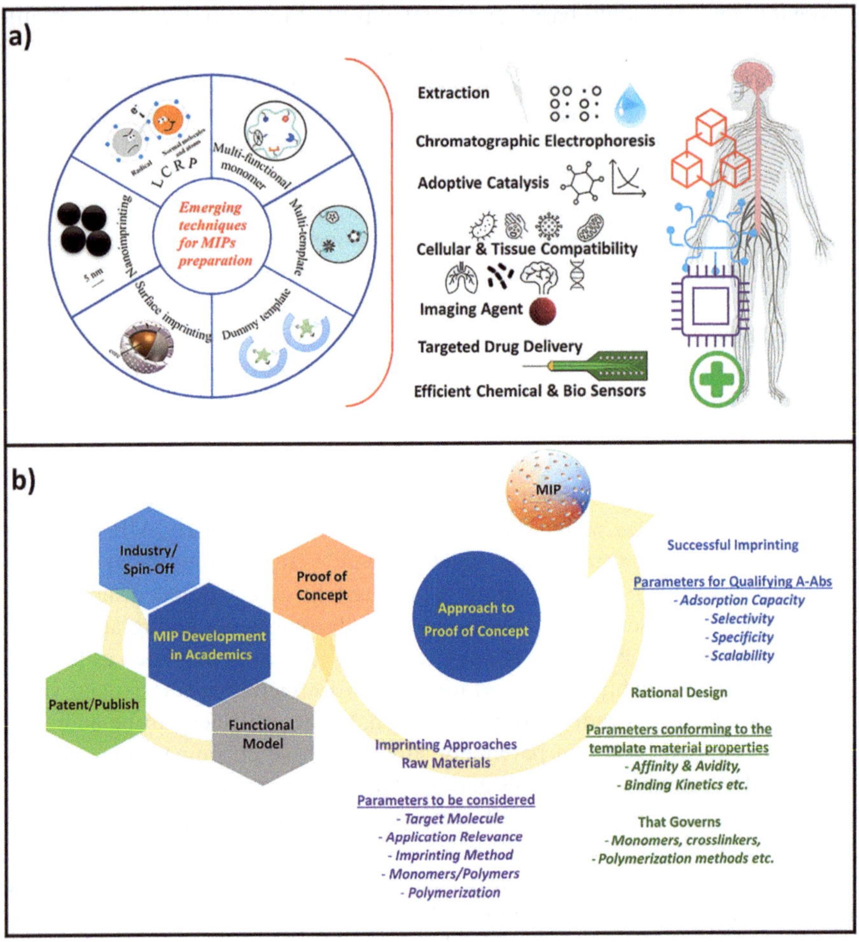

Fig. 15.1 (**a**) Depiction of MIP progress aligned with specific biological models and the potential utilization of high-performance MIPs in cutting-edge biomedical contexts, (**b**) the overview of the key determinants crucial to the prosperous creation of MIPs, along with a step-by-step guide to their development within a standard research framework [10]

15.2 MIP Synthesis and Design Strategies Toward Artificial Antibody

Molecularly imprinted polymers (MIPs) are polymeric matrices that can selectively bind to a specific type of molecule [13]. Within the realm of molecular imprinting, monomers undergo polymerization in the presence of a target molecule that acts as a template throughout the polymer synthesis process [14]. The formation of complementary sites, mirroring the template molecule, leads to products with an affinity akin to that seen in antibody–antigen complexes [15]. In 1931, Polyakov first proposed the concept of molecular imprinting and observed that when polymerization was carried out using benzene to produce silica, the resulting silica exhibited remarkable adsorption properties toward the additive [16]. Scientists attributed this observation to the existence of silica pores within the polymeric structures. Polyakov, in 1949, replicated the same synthesis as described earlier, with the addition of methyl orange dye as an extra component [17]. Consequently, the resulting silica exhibited selectivity toward methyl orange dye, and a similar outcome was achieved when Dickey repeated the process with other dyes. Since Dickey's method, numerous chemists have adopted a similar approach to craft specific silica-based sorbents. However, interest in imprinted silica diminished due to concerns regarding the materials instability and lack of reproducibility [18, 19]. In 1972, Wulff and Sarhan introduced the covalent approach to the field of molecular imprinting, while in the 1980s, Mosbach pioneered the noncovalent approach. These contributions significantly revitalized the development of molecular imprinting [20].

15.2.1 MIPs Synthesis

There are three main phases in the synthesis procedure of MIPs [21] (Fig. 15.2a). During this process, a stable polymer is formed, which can be customized in terms of its properties like toughness, porousness, size, and how it reacts to different conditions or stimuli [22, 23]. In molecular imprinting, there are two main approaches that depend on the interactions involved in forming the initial complex before polymerization [24].

The first approach uses a covalent bond between template molecules and functional monomers to form reversible covalent bonds [25]. After the template has been polymerized, the chemical cleavage of the corresponding covalent bonds takes place, facilitating the removal of the template from the polymeric matrix [14]. The covalent approach offers a significant benefit in terms of the robustness of covalent interactions. This robustness plays a crucial role in achieving a consistent distribution of recognition sites and minimizing nonspecific interactions [26]. In addition to the difficulties that come with achieving a thermodynamic equilibrium due to the strength of the covalent bonds, this approach can only be used to recognize a few target molecules, such as alcohols, aldehydes, ketones, amines, and carboxylic acids, because of the strength of the covalent bonds [27, 28].

The second method relies on relatively subtle noncovalent interactions between the template and functional monomers. These noncovalent interactions encompass a range of forces, including hydrogen bonds, ionic interactions, van der Waals forces, and dipole–dipole interactions. These forces are formed during both the polymerization and rebinding phases [29]. Compared to the covalent approach, this method offers numerous benefits. Noncovalent imprinting procedures are known for their simplicity, allowing for the utilization of a variety of monomers, including methacrylic acid, 4-vinyl benzoic acid, acrylamide, vinyl pyrrolidone, 2-hydroxyethyl methacrylate, and others. Furthermore, this approach provides the capability to target a wide range of molecules [30]. Noncovalent methods come with certain

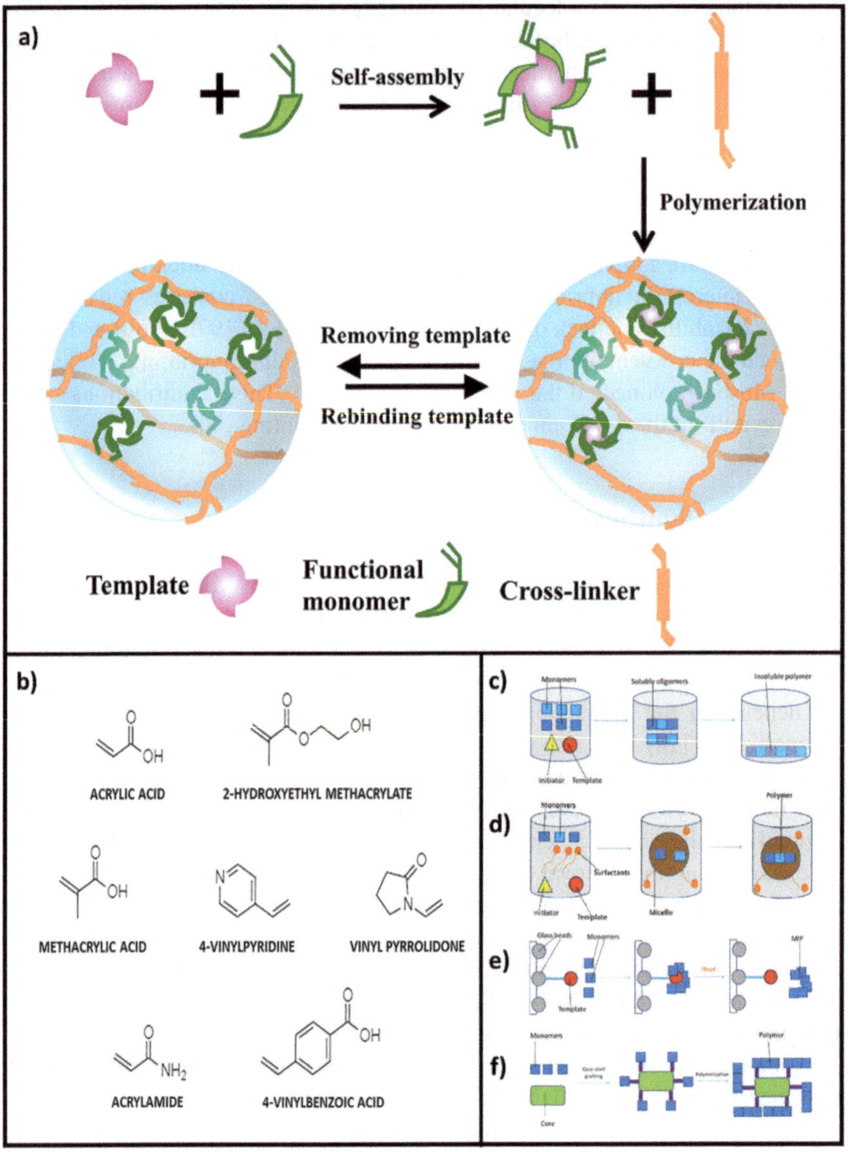

Fig. 15.2 (**a**) Schematic diagram of MIP formation, (**b**) chemical structure of some of the monomers used for MIPs synthesis, (**c**) diagram illustrating precipitation polymerization: monomers, the template, and the initiator are all soluble; the polymer chain continues to grow until it reaches a size at which it becomes insoluble, leading to the precipitation of the polymer, (**d**) emulsion polymerization: monomers are included in micelles with surfactants on their surface; the polymerization occurs inside the micelles, (**e**) solid-phase approach: Polymerization takes place in the vicinity of the template, which is immobilized on glass beads. Detachment is achieved through heat, (**f**) core–shell grafting followed by polymerization [6, 38]

drawbacks, including the necessity for a substantial surplus of functional monomers. This is due to the requirement to shift the equilibrium toward the associated form of the prepolymerization complex, leading to the creation of nonspecific binding sites during polymerization. After a comprehensive examination of both covalent and noncovalent techniques, Whitcombe devised a semi-covalent intermediary approach. In this approach, template and functional monomers engage in covalent interactions during polymerization, while noncovalent interactions come into play during the rebinding phase [31, 32]. As a result of this approach, it becomes feasible to merge the heightened affinity observed in the early stages of polymerization with the swift binding kinetics experienced in the subsequent rebinding phase [33]. Creating MIPs involves considering several critical factors, including the choice of functional monomers, cross-linking agents, initiators, and solvents, the proportions among these reagents, the polymerization method employed, and the imprinting technique used [34, 35]. A few monomers that are used in molecular imprinting are shown in Fig. 15.2b, which provides a brief overview of their structures.

Several polymerization techniques have been used to synthesize MIPs over the years [36]. Free radical polymerization stands as the quintessential approach for polymer production, necessitating the application of either heat or light to initiate the process [37]. Because of the challenges in controlling the propagation and termination phases of this process, coupled with a high polydispersity index, various monomers are compatible with this method. However, this approach often results in the creation of highly branched, atactic MIPs characterized by reduced binding specificity and selectivity [38]. Consequently, controlled radical polymerization has gained prominence over free radical polymerization. This shift is primarily attributed to its adaptability as a reversible deactivation radical polymerization method, encompassing techniques such as atom transfer radical polymerization, reversible addition-fragmentation chain transfer, nitroxide-mediated polymerization, and iniferter-mediated polymerization. These methods enable precise control over molecular weight distribution and stereochemistry. However, it's important to note that they do not provide control over the size of the synthesized nanoparticles [39, 40]. Today, there exists a variety of material polymerization techniques, all of which can be harnessed for the production of nanoMIPs. These methods encompass precipitation polymerization, emulsion polymerization, and core–shell polymerization, often followed by grafting processes [41, 42] (Fig. 15.2c–e). In 1999, Ye et al. introduced precipitation polymerization as a novel method. This technique allows for the straightforward, one-step synthesis of uniform and spherical MIP nanoparticles [43]. The method involves the use of a substantial amount of solvent. The solvent dissolves the monomers, initiator, and template, but not the polymer that will eventually form. Polymer chains start to form from monomers and oligomers until they reach a size that causes them to precipitate. After a washing and centrifugation step, MIPs are collected [44]. Various factors such as solvent polarity, temperature, and stirring rate can be adjusted to control the process [45]. In contrast to certain alternative methods, this approach eliminates the necessity for employing stabilizer molecules. Nevertheless, it comes with its own set of limitations, such as the requirement for a relatively larger quantity of template material and an extended processing duration [46]. Emulsion polymerization stands as another vital method used in the crafting of nanoMIPs. This technique enables the creation of uniform MIP nanoparticles with exposed binding sites on their surfaces, facilitating interactions with surface molecules [47]. The polymerization procedure is commonly conducted in both water-in-oil and oil-in-water emulsions, facilitated by the presence of a surfactant. However, it's worth noting that the utilization of water-in-oil emulsion is less common [48]. Through the employment of this method, it is possible to polymerize small emulsions, either in the form of mini-emulsions or micro-emulsions. This yields nanoparticles with varying diameters, spanning from 30 to 500 nm for mini-emulsions and from 5 to 50 nm for micro-emulsions [49]. For mini-emulsion polymerization, a co-surfactant serves to stabilize the monomers, after which homogenization is achieved through either sonication or stirring. The key distinction between the two polymerization approaches lies in the requirement for a greater surfactant concentration in the latter, aimed at producing smaller particles [50, 51]. Emulsion polymerization offers numerous benefits, including its high yield and applicability for protein imprinting. However, it often poses challenges in terms of polymer purification, primarily because of the presence of surfactants, which can lead to substantial time and effort being expended [52]. Moreover, it's possible to create core–shell MIP nanoparticles through the technique of coreshell grafting followed by polymerization [53]. Diverse preformed particles serve as the central core components of these nanoparticles. This assortment includes organic polymers, silica, superparamagnetic iron oxides, quantum dots, upconversion nanophosphors, carbon dots, and gold/silver particles. These particles are connected to the MIPs, typically forming the outer shell of the nanoparticles [54–57]. A fine control over the size of the MIP can be achieved using this technique [58]. One of the latest methodologies in nanoMIP synthesis is the solid-phase approach. It involves polymerization subsequent to the immobilization of the template molecule on glass beads that have been activated with NaOH to expose their -OH groups, facilitating the silanization process [59] (Fig. 15.2f). After that, nanoMIPs need to be purified to be ready for use [56]. Solid-phase synthesis offers the prospect of creating MIPs with exceptionally strong binding affinity and a uniform dispersion of recognition sites [60]. This enables the template to be recycled as well and the template can be removed from the polymer without undergoing the washing phase. Nevertheless, this approach has its limitations, including a reduced yield and its limited efficiency in generating templates with substantial structures [35]. In addition, the detachment of MIPs happens through the application of heat, which makes it impossible to increase the yield of the reaction by performing multiple synthesis on the same solid phase using thermo-sensitive templates such as proteins [61]. Moreover, materials created with surface imprinting techniques showcase heightened selectivity and sensitivity. This enhancement makes them valuable for enhancing recognition performance, thanks to the inclusion of surface-based recognition sites. In these procedures, the template is situated on the surface of the polymer, with the most commonly employed methods being solid-phase polymerization and emulsion polymerization [56]. When dealing with a challenging template, especially one of substantial size, a viable approach is to employ a fragment imprinting strategy, often referred to as segment imprinting. This method utilizes a segment or a portion of the target molecule as a pseudo-template to craft MIPs [62]. This method of imprinting is called "dummy template imprinting," and it uses substances that have chemical properties similar to those of the target molecules [63]. Hence, this approach can prove exceptionally valuable, particularly when certain molecules are costly or pose toxic risks [54]. Similar to natural proteins, MIPs can also undergo post-imprinting chemical modifications. These modifications encompass the addition, transformation, or conjugation of functional groups, either by incorporating existing groups or by introducing new ones, after the imprinting process [51]. Beyond elevating the specificity and sensitivity of MIPs, this method holds the potential to augment various parameters. To achieve success with this approach, it is essential to initiate the polymerization process using functional monomers that possess functional groups amenable to chemical modification. Examples include amino or carboxyl groups, as well as reversible bonds like disulfide and imine bonds [64]. There are several ways to modify monomer residues after molecular imprinting, including hydrolysis and other methods of reaction. It is possible to either eliminate or add functional groups that are either unwanted or desired for the application of MIPs without compromising their selectivity and specificity in molecular recognition processes through post-imprinting modifications [65]. These modifications can include the incorporation of fluorescent dyes at specific sites related to the imprinting process. For instance, Saeki et al. developed a method for detecting prostate-specific antigen (PSA) using an MIP [66]. Following the molecular imprinting stage, the polymer underwent two subsequent modifications. Initially, they introduced a novel modification agent known as PIR-C. This agent interacted with thiol groups within the polymer, establishing molecularly imprinted nanocavities. This process resulted in orthogonal dual interaction sites, as PIR-C bound to the polymer's thiol groups [67]. In particular, they employed 3-fluorophenyl boronic acid for the recognition of PSA sugar chains and carboxyphenyl groups for protein binding. These selections played a pivotal role in enhancing the specificity and selectivity of the MIP's interactions with its target. Subsequently, a second post-imprinting modification was implemented on PIR-C's secondary amino groups to incorporate the fluorescent dye Alexa Fluor 647 [68]. Moreover, to address a limitation linked to molecular imprinting, specifically the variability in recognition site affinity, an additional post-imprinting adjustment has been introduced, referred to as "capping." To block functional monomer residues within cavities exhibiting lower affinity, a sequential approach is adopted. Initially, the residues within higher-affinity cavities are blocked. Subsequently, tailored post-imprinting techniques are employed to modify the residues within lower-affinity cavities. An illustrative instance of this method can be observed in the creation of a fluorescent MIP designed for lysozyme recognition [69]. To protect the functional monomer residues located in the high-affinity binding sites of the MIP film, a diluted solution of lysozyme was introduced to the MIP film post-creation of the lysozyme-imprinted polymer. During this step, a capping procedure was employed to address the residues within the binding sites that demonstrated lower affinity for the protein. This involved a reaction between the amino groups on the functional monomer units and p-isothiocyanatophenyl α-D-mannopyranoside. This reaction effectively prevented these moieties from binding with the fluorescent molecule, diverting the binding to the high-affinity cavities instead. According to the experimental findings, the capping process significantly boosted the selectivity of the MIPs [38].

15.2.2 Design Strategies

The success of MIPs in various applications depends on careful design considerations. This section aims to provide a brief overview of key design strategies to create MIPs that are specific, stable, and have high affinity.

15.2.2.1 Selection of Template Molecule

The template molecule is the target molecule that dictates the shape and functionality of MIP's binding sites. It is essential that the template used in the MIP be carefully selected, as it will determine whether the MIP will be able to recognize and bind to specific analytes. The template can be a small molecule, a peptide, a protein, or even a complex biomolecule. Among the factors that should be considered are the template's structural characteristics, its functional groups, as well as its relevance to the intended application [70–72].

15.2.2.2 Selection of Functional Monomers

Functional monomers are the building blocks that interact with the template molecule during polymerization, creating complementary binding sites. The choice of functional monomers is pivotal to achieving strong and specific interactions. Monomers should possess functional groups that can form hydrogen bonds, electrostatic interactions, or hydrophobic interactions with the template. Rational selection of monomers based on their compatibility with the template's chemical structure enhances binding site affinity [73, 74].

15.2.2.3 Cross-Linker Type and Density

Cross-linkers link the functional monomers together, forming the polymer matrix. Cross-linker type and density influence polymer porosity, rigidity, and stability. An optimal cross-linker density ensures that binding sites are accessible to the polymer without compromising the stability of the polymer. The cross-linker should be chosen to prevent steric hindrances that might hinder template binding and analyte recognition [61, 75, 76].

15.2.2.4 Polymerization

Various polymerization techniques can be utilized for crafting MIPs, each offering distinct advantages. Bulk polymerization, for instance, entails the polymerization of monomers and cross-linkers in the presence of a template molecule. Precipitation polymerization results in imprinted nanoparticles. Surface imprinting involves the creation of binding sites on particle surfaces. In order to select the right polymerization method, one needs to consider the application, the desired particle size, and the ease of removal of the template [77].

15.2.2.5 Template Removal Techniques

After polymerization, the template molecule needs to be removed, leaving behind cavities that match the template's shape and functional groups. Traditional methods like solvent extraction and thermal degradation are common but can damage the polymer matrix or be inefficient. The modern techniques of extraction, such as microwave-assisted extraction, supercritical fluid extraction, and enzymatic degradation, are gentler and more efficient than the traditional techniques, preserving the stability of the MIP in the process [78].

15.2.2.6 Computational Modeling for Rational Design

Within the realm of MIP design, computational modeling assumes a critical role by aiding in the prediction of optimal interactions between the template molecule and functional monomers. Molecular docking, quantum mechanical calculations, and molecular dynamics simulations assist in understanding binding site formation and affinity. These simulations help select suitable monomers, estimate binding energies, and guide experimental design. The design process is made faster as a result of the synergy between the computational predictions and experimental validation [46, 79].

15.2.2.7 Surface Imprinting and Nano-Structuring

Surface imprinting involves creating binding sites specifically on particle surfaces, resulting in increased accessibility and binding kinetics. Utilizing nanomaterials as templates, this technique introduces nano-structuring, enhancing surface area and affording greater control over binding site geometry. As a result of these strategies, they are particularly useful in sensor applications where it is vital to have rapid binding and release of analytes [80].

15.2.2.8 Post-Imprinting Modifications

Post-imprinting modifications enable the fine-tuning of MIP properties. Surface functionalization with additional recognition elements, such as aptamers or enzymes, can enhance binding selectivity. Furthermore, stimuli-responsive polymers can also be incorporated into MIPs to create "smart" MIPs that change their conformation or affinity to bind to specific environmental cues as a result of specific environmental signals [81].

15.2.2.9 Cross-Selective MIPs

Cross-selective MIPs are designed to recognize multiple structurally similar target molecules. This approach involves optimizing the balance between affinity and selectivity for different analytes. For applications involving complex samples with multiple potential targets, cross-selectivity can be a challenge, but it is of utmost importance for achieving successful results [82].

15.2.2.10 Template Analog Approach

When dealing with challenging or rare template molecules, synthesizing MIP molecules can be aided by using analogs of the template molecule that have similar structural features. This method ensures that the produced binding sites are designed to interact with the desired analyte while avoiding the difficulties of using templates [83].

15.3 MIPs as Artificial Antibody in DDSs

Advances in nanotechnology have significantly transformed the landscape of DDSs, introducing the ability to precisely target and control the release of therapeutic agents within the body [84]. This innovation has paved the way for more effective treatments with reduced side effects. Central to the success of these systems is the recognition element, which is responsible for identifying and interacting with specific molecules of interest [85]. Traditional antibodies have high specificity and affinity for antigens, making them primary tools for detecting antigens [86]. However, MIPs have emerged as a compelling alternative to traditional antibodies in drug delivery applications [87]. They are synthetic materials designed with highly specific and customizable binding affinity for a target molecule [38]. One way to achieve this is through a process called molecular imprinting. This involves synthesizing the polymer while the target molecule is present, creating spaces or binding sites that are specifically designed to match the size, shape, and chemical properties of the target molecule [6]. MIPs have the ability to recognize molecules as one of their fundamental properties; combined with their resistance to a wide range of chemical and physical conditions (pH, temperature, organic solvents, pressure), these synthetic macromolecules could be used for a wide range of applications [23]. Numerous pharmaceutical agents, such as peptides, gene-based medications, and small molecules characterized by limited solubility and/or lipophilicity, face challenges in conventional delivery methods. This is primarily because of their molecular size, charge, and difficulties in absorbing into the systemic circulation, often due to enzyme degradation and insufficient absorption [88]. Within a DDS, the administration of a therapeutic agent is meticulously orchestrated to achieve specific goals. This includes delivering the agent at the intended rate and time while mitigating adverse effects, enhancing patient adherence, prolonging the duration of pharmacological effects, and potentially achieving targeted delivery, all while ensuring an appropriate dosage [89]. The loading capacities of DDS can have a substantial impact on the effectiveness of treatment and the frequency of dosing. This influence primarily stems from their effects on pharmacokinetics and pharmacodynamics [90]. DDSs have significantly improved with the use of polymers as vehicles for delivering drugs [88]. Polymer-based DDS employs the strategy of distributing drugs within a polymer matrix or encapsulating them behind a polymer membrane. This approach is designed to achieve controlled release of the drugs over time [91]. This can potentially reduce or eliminate various side effects, making the drug safer [92]. However, conventional polymeric vehicles cause burst releases of drugs, resulting in potential toxicity issues and undesirable side effects for patients [93]. Despite the progress made in polymeric delivery systems, certain critical areas remain insufficiently explored. These include concerns related to toxicity, physicochemical instability, and the responsiveness of these systems to specific stimuli

[94]. Approximately one out of every 5000 polymer-based drug development strategies enters preclinical studies and becomes a market-approved dosage form after 10 years [88, 95].

15.3.1 Drug Delivery Based on a Rate-Program

A rate-programmed drug delivery approach entails the controlled release of a drug over an extended duration at a predetermined rate [96]. In this mechanism, MIPs are crafted to release the drug at a precise rate. This rate can be finely tuned by manipulating factors like pH, temperature, and exposure to light, all with the aim of attaining the intended therapeutic outcome [97]. The rate of drug release can be regulated by altering either the drug concentration within the MIPs or by making structural modifications to the MIPs themselves [6]. Numerous investigations have proposed that MIPs can function as sustained drug delivery carriers for the controlled release of theophylline. This approach has the potential to mitigate side effects and enhance the drug's bioavailability [98, 99].

15.3.1.1 Factors Influencing Drug Release Kinetics in MIPs

The rate at which drugs are released from MIPs is dependent on several factors, including the size and shape of the binding sites they contain. These binding sites play a crucial role in determining drug release rates, with smaller sites resulting in slower diffusion and larger ones promoting faster release [100].

Environmental Conditions

It is a fundamental aspect of rate programming that MIPs are responsive to environmental factors, such as temperature and humidity [101]. For instance, alterations in pH levels can activate the discharge of certain medications, enabling them to be directed toward particular areas in the body where pH levels vary, including the stomach and intestines. These body parts have different pH levels. Additionally, it is feasible to make use of temperature fluctuations or exposure to specific light wavelengths to stimulate or regulate the discharge of drugs based on the requirements of the system [46].

Concentration of Drug in MIPs

The concentration of the drug loaded into the MIPs is one of the most crucial parameters in the rate programming process [14]. To achieve a more accurate control of the drug release process, you can adjust the drug concentration in the polymer matrix. Increasing the drug concentration generally leads to a longer and more consistent release of the drug [66].

MIP Structure Modification

It has been found that the rate of drug release can be greatly influenced by tailoring MIPs to fit the structure of the drug molecules. Studies show that researchers can accurately manage the kinetics of drug release by adjusting factors such as the density of binding sites and the size of imprinted cavities. This guarantees that drugs are released for therapeutic needs [88].

15.3.1.2 Diffusion-Controlled Release Vs. Affinity-Controlled Release

In the realm of MIPs, understanding the mechanisms behind drug release is crucial for tailoring the rate of release to specific therapeutic needs. The release of drugs from MIPs can be attributed to two primary mechanisms. The controlled release of drugs from MIPs relies on concentration gradients within the material [102]. The rate at which drugs are released in a controlled manner is contingent on the level of affinity between the drug and the MIP. The fundamental principle that allows for diffusion-controlled release is molecular diffusion [103]. In this mechanism, the rate at which drugs are released from MIPs is primarily determined by the concentration gradients established within the polymer matrix. In the initial stages of loading drug molecules into MIPs, they are present at higher concentrations within the polymer than in the surrounding environment [36]. As the drug molecules gradually diffuse from regions with higher concentrations of the drug molecules to the regions with lower concentrations of the drug molecules, the polymer matrix gradually begins to lose these drug molecules. To achieve diffusion-controlled release, it's crucial to take into account the following factors: Firstly, the dimensions and configuration of drug compounds. The size and shape of the drug molecules have a significant impact on how quickly they diffuse through the matrix of the MIP [104]. Smaller, more compact drug molecules typically diffuse more rapidly. Second, the porosity of the polymer structure. The porosity of the polymer structure has an impact on the pathways available for the drug to diffuse [105]. As a result of a more porous polymer, a faster release of the drug is possible. The diffusion-controlled release of drugs is commonly used to achieve sustained release, but it may not be highly targeted. The controlled release method can be particularly useful for drugs that do not have a strong binding affinity for MIPs but still require controlled release to maintain therapeutic concentrations. In contrast, affinity-controlled release is controlled by specific binding interactions between the drug and the MIP that determine how much of a drug is released [69]. Regarding the speed at which drugs are released within the body, the binding affinity, which signifies the intensity of these interactions, holds a pivotal role in dictating the rate of drug release. Affinity-controlled release occurs when drug molecules are selectively captured by imprinted cavities or binding sites within the MIP. There is a correlation between the strength of these interactions and how tightly the drug is held within the polymer, which affects the release rate of the drug. A heightened binding affinity leads to a decelerated release rate of the drug [106, 107]. When it comes to MIP-based drug delivery, the choice between diffusion-controlled release and affinity-controlled release depends on the specific therapeutic objectives and the nature of the drug being delivered. By understanding these mechanisms, researchers can create tailored MIPs to achieve desired drug release profiles and optimize the treatment process. Affinity-controlled release is influenced by binding site density, which affects overall binding capacity and the rate of release. Higher site density can result in prolonged drug retention. Additionally, binding strength, determined by molecular recognition sites created during MIP synthesis, is an important factor to consider in achieving desired drug release profiles [56].

15.3.1.3 Mathematical Models and Equations for Rate Programming

Mathematical models and equations can be beneficial in forecasting and managing the release of drugs from MIPs. Higuchi's model, which is based on Fick's law of diffusion, is utilized to explain drug release from matrixes like MIPs. The Ritger–Peppas model is more fitting for systems that exhibit abnormal (non-Fickian) diffusion, while the Weibull model is used to represent diverse release mechanisms in different systems [108].

Higuchi's Model for Drug Release

Higuchi's model is often used to describe drug release from matrix-based delivery systems like MIPs [109]. The model is founded on Fick's law of diffusion, which is a fundamental principle of mass transport is obtained by Eq. 15.1.

$$Q = k{\cdot}t^{0.5}, \tag{15.1}$$

where Q represents the cumulative amount of drug released at time t and k is a constant related to the diffusivity of the drug within the polymer matrix. The model assumes that drug release from MIPs is primarily governed by Fickian diffusion.

Ritger–Peppas Model for Anomalous (Non-Fickian) Diffusion

As an alternative to the Fickian diffusion model, it is possible to use the Ritger–Peppas model in situations where drug release from MIPs exhibits anomalous (non-Fickian) diffusion [109]. This model is often represented as Eq. 15.2.

$$Q = k{\cdot}t^{n}, \tag{15.2}$$

where Q is the cumulative drug release at time t, k is a kinetic constant, and n is a diffusion exponent that characterizes the release mechanism. Different values of n correspond to different types of drug release mechanisms, including Fickian (n = 0.5), case II transport (n = 1), and other non-Fickian behaviors. A Ritger–Peppas model can be used to characterize complex drug release mechanisms beyond simple diffusion, making it suitable for situations in which MIP-based delivery systems exhibit unconventional release kinetics that challenge simple diffusion-based models.

Weibull Model for Different Release Mechanisms

A Weibull model is a mathematical tool that is capable of representing the release of drugs in a variety of delivery systems, including MIPs [110]. In order to accommodate different release mechanisms within the same system, the Weibull model typically takes the form of Eq. 15.3:

$$Q = Q_0 \cdot \left[1 - \exp\left(-\left(\frac{t}{\alpha} \right)^{\beta} \right) \right], \tag{15.3}$$

where Q represents the cumulative drug release at time t, Q_0 is the initial amount of drug in the system, and α and β are shape parameters that determine the release profile. The flexibility of this tool allows for the analysis and representation of various release mechanisms across multiple systems.

It can be concluded that mathematical models and equations, including Higuchi's, Ritger-Peppa's, and Weibull's, provide researchers with essential tools to understand, predict, and control drug release kinetics from MIP-based DDSs. As a result of these models, it is possible to optimize drug delivery profiles in order to achieve desired therapeutic outcomes.

15.3.2 Administration by Transdermal Route

MIP-based DDSs have found application in transdermal formulations due to their ability to provide controlled release, increased drug loading, and potential for enantioselectivity [111]. Consequently, there has been a reduction in dosing frequency, a decrease in side effects, and an enhancement in therapeutic outcomes. Nicotine stands out as an excellent candidate for transdermal drug delivery due to its high lipophilicity, which facilitates easy skin penetration [112, 113]. An underlying hypothesis is that continuous nicotine release into the skin can promote smoking abstinence [114]. This approach enables the maintenance of a low drug concentration in the bloodstream. Nevertheless, nicotine's tendency to volatilize from the reservoir poses limitations on its application in transdermal formulations [115]. To overcome this drawback, Ruela's [116] group designed nicotine-based MIP particles dispersed in mineral oil (a nonpolar vehicle). The nicotine-loaded particles were chemically and thermally stable after loading. MIP and NIP particles exhibited maximum absorption capacities of approximately 60 μg mg^{-1} and 38 μg mg^{-1}, respectively. This highlights the significant impact of selective cavities in MIP. In skin penetration studies conducted on a porcine model, the MIP formulation demonstrated skin penetration rates similar to the Nicopatch® commercial nicotine patch. Moreover, it exhibited zero-order kinetics for a period of 48 h. It's worth noting that commercial nicotine patches, such as Nicopatch® and Nicorette®, provided longer durations of action compared to MIP [117]

Table 15.1 The formulations of nicotine patches

Nicotine formulation types	Nicotine content (mg)	Surface area (cm^2)	Application duration (h)	In vitro release amount (μg cm^{-2})
Nicopatch®	17.5	10.0	24	700
Nicorette®	8.3	10.0	16	500
Nicotine MIP	10.0	1.8	48	2200

(Table 15.1).

Although both MIPs and NIPs regulate the release and skin penetration rate of nicotine, MIPs excel in drug loading capacity and achieving sustained release when compared to conventional polymers. This makes MIPs a more favorable choice for transdermal DDSs.

15.3.3 Delivering Drugs in a Stimuli-Responsive Manner

An optimal stimuli-responsive DDS has the capacity to respond to internal triggers such as reactive oxygen species, alterations in pH and temperature, and overexpressed proteins and enzymes. This responsiveness enables the precise and controlled local delivery of drugs in terms of dosage and release rate. Polymer-based DDSs that respond to both internal and external stimuli have successfully achieved controlled drug release, notably for antitumor agents. While traditional polymer-based DDSs offer advantages in terms of affordability, low toxicity, and controlled release, they may not be the most cost-effective solution [88]. The concept of stimuli-responsive drug delivery refers to the development of smart DDSs that are capable of sensing changes in their environment and responding to specific triggers. Triggers, also known as "stimuli," include external or internal factors such as pH levels, temperature, light, enzymes, or even specific molecules. These systems are characterized by their ability to release drugs on demand or in a highly controlled manner in response to external stimuli [118]. Stimuli-responsive drug delivery offers precise control over the release of medication, which is a significant advantage. Researchers can develop delivery systems that respond to specific triggers, ensuring that drugs are administered at the exact time and location where they are most needed. This level of precision enables effective treatment of various medical conditions, including cancer, inflammation, and chronic illnesses [37]. The pH-sensitive DDSs utilize pH variations in the body, such as the acidic environment of the stomach or the neutral pH of healthy tissues. When the delivery system encounters a specific pH level, it triggers drug release. Anticancer drugs encapsulated in pH-responsive nanoparticles may remain stable in the bloodstream, but release their cargo once they reach the slightly acidic tumor microenvironment [39]. Another notable category of DDSs is temperature-sensitive systems. These systems exploit temperature changes, either in the body or externally applied, to control drug release. As an example, hyperthermia treatments for cancer involve raising the temperature of the tumor region, which activates thermoresponsive drug carriers, resulting in localized drug release and increased therapeutic effectiveness [26]. Light-responsive drug delivery involves using light as a trigger for drug release. This method is very flexible because different wavelengths of light can be used to activate drug delivery systems (DDSs). Light-responsive systems are especially beneficial in photodynamic therapy because they allow for precise drug release in specific areas [20]. Finally, enzyme-responsive DDSs are designed to react to specific enzymes present in the body. In the context of inflammation or diseases characterized by high enzyme levels, these systems can release drugs selectively in response to enzymatic activity, which ensures a timely and targeted therapeutic response [119].

15.3.4 The MIP-Based Anti-Cancer DDS

The (MIP)-based DDSs for cancer treatment can react to various internal stimuli, but pH changes have been particularly effective. This is primarily due to the fact that the typical physiological pH, which is around 7.4, exceeds the acidic pH range of 6–7 typically observed within tumor environments. Another prospective therapeutic trigger involves the elevated levels of glutathione (GSH) within tumor cells, generating a reduction potential within the local milieu. This responsiveness to reduction can also be harnessed in MIP-based DDSs that are sensitive to stimuli [88]. Because 5-FU undergoes swift metabolism within the human body, characterized by a short plasma half-life of merely 6 min, it often necessitates a substantial dosage to maintain its therapeutic efficacy. However, the administration of high doses can lead to the accumulation of toxic metabolites, resulting in severe side effects [120]. The advantages offered by MIPs can be synergistically combined with stimuli-responsiveness to DDs. In recent studies, a reduction-responsive MIP was developed using poly(2-isopropenyl-2-oxazoline) (PiPOx), synthesized through ring-opening polymerization of iPOx. This MIP incorporated 3,3-dithiodipropionic acid (DTDPA) as a cross-linker, containing disulfide bonds that are sensitive to reduction, allowing for a reduction-triggered drug release mechanism. The PiPOx-MIPs demonstrated a superior drug-carrying capacity compared to PiPOx-NIPs, with a loading capacity of 85 mg g^{-1} as opposed to 66 mg g^{-1}. Furthermore, PiPOx-MIPs exhibited drug release rates of 30% at pH 7.4 and 20% at pH 2.0 in the absence of a reducing agent. However, the addition of tris(2-carboxyethyl) phosphine (TCEP) as a reducing agent led to a substantial increase in drug release, surpassing 50%. In contrast, PiPOx-NIPs released the drug irrespective of the presence of reducing environments (Fig. 15.3a_1–a_3) [121]. In comparison to conventional polymers, PiPOx-MIP DDS may improve 5-FU bioavailability and reduce its toxicity. Nonetheless, both MIP and NIP showed an initial burst release, probably due to specific surface adsorption. Moreover, in vivo studies and an evaluation of the administration route are needed [88]. The PiPOx-MIPs delivered 5-FU more effectively to tumor cells than conventional polymers in vitro, however, based on this type of testing. Several studies have indicated that MIP-based DDSs can control and/or target drug release in response to endogenous stimuli. Drugs with narrow therapeutic windows may benefit from this. However, in vivo, experiments are required to substantiate whether imprinting technology offers a unique advantage over conventional polymers in achieving sustained and controlled drug release, potentially improving drug stability and therapeutic efficacy in harsh conditions [88, 122].

15.3.5 An Exogenous Stimulus-Responsive MIP-DDS Based

MIP-based DDS can also be activated through external stimuli, particularly magnetic nanoparticles that react to magnetic fields [88, 121]. Molecular imprinting technology and magnetic nanoparticles can be combined to create DDSs that have

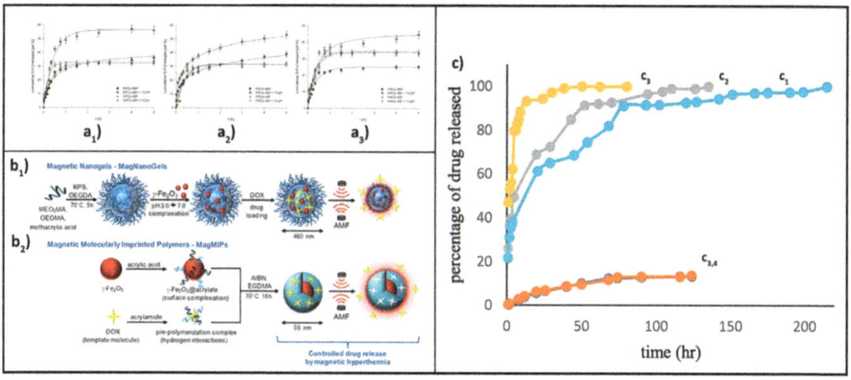

Fig. 15.3 Release profiles of 5-FU from drug-loaded PiPOx-MIPs and NIPs at (**a**$_1$) pH 7.4; (**a**$_2$) pH 6.5; (**a**$_3$) pH 2.0, Schematic illustration of the synthesis of (**b**$_1$) MagNanoGels are created using precipitation radical copolymerization, followed by the assembly of MNPs inside the nanogels. and (**b**$_2$) MagMIPs are produced through the subsequent grafting of an acrylic acid compound onto the surface of MNPs, followed by polymer growth in the presence of DOX to facilitate imprinting polymerization. Investigations were conducted on the loading and release of DOX in the presence of an AMF, evaluating the in vitro release profiles of MMIP, MIP NPs, and MNIP (**c1**, **c2**, and **c3**, pH 5), as well as MMIP and MIP NPs (**c3** and **c4**, pH 7.4) [88]

superior biocompatibility, are responsive to stimuli, and have stable drugs. This can reduce the severity of side effects that may be caused by drug action on non-target tissues. It was reported by Cazares-Cortes et al. [123] that two magnetic nanosystems were compared for the controlled release of DOX: magnetic nanogels containing thermosensitive polymers and core–shell magnetic MIP (MMIP) NPs (Fig. 15.3b$_1$ and b$_2$). The magnetic nanosystem acts as a "hot spot," inducing localized hyperthermia upon the application of an alternative magnetic field (AMF), thereby facilitating the release of DOX. DOX released from MMIPs after 4 h of exposure to AMF has increased from 10% to 60% of the total payload due to disrupted hydrogen bonds between DOX and the functional monomer at in vitro drug release studies performed at 37 °C. After exposure to AMF, DOX release from magnetic nanogels increased from 24% to 45%, but in this case, conformational changes in the gel network explained the increase. While MMIPs released less DOX overall (7 m) than nanogels (16.7 m), they released more DOX after 4 h. Furthermore, both the MMIPs and nanogels showed a decrease in cancer cell viability (from 88% to 60% and from 54% to 30%, respectively) after exposure to the AMF [124]. Since MMIPs have a relatively low loading capacity, larger doses of MMIPs may be needed in clinical practice compared to nanogels [125]. While nanogels rely on physical entrapment for DOX release, MMIPs achieve specific binding between the drug and the imprinted cavities. By reducing DOX leakage in the systemic circulation, MMIPs may reduce side effects. Hassanpour et al. [126] synthesized MMIPs for azidothymidine-controlled release. This indicates that MMIPs have a significantly higher adsorption capacity (~170 mg g^{-1}) than the corresponding MNIPs (~38 mg g^{-1}), MIP NPs (~46 mg g^{-1}), and NIP NPs (~18 mg g^{-1}). A pH 7.4 drug release study showed that MMIPs released 14% of the loaded drug, but mostly at pH 5 (Fig. 15.3c_1–c_4). Also, the MMIPs were 50 times more toxic to MCF-7 cancer cells than free azidothymidine while having no effect on healthy MCF-10 cells. MMIPs exhibited potential for pH-responsive drug delivery despite not assessing the effect of AMFs on drug release [105].

15.3.6 Delivering Drugs to Their Target with Active Targeting

DDSs transport medications to both healthy and diseased tissues, even if the drugs have the potential to harm healthy cells, resulting in significant side effects, particularly when dealing with toxic drugs like anti-cancer agents. In addition to their active targeting capabilities, controlled DDSs can also precisely govern the spatial release of drugs. This spatial control helps mitigate the side effects that arise from damage to healthy cells, ultimately enhancing the overall therapeutic profile. Activated polymeric DDSs achieve precise drug delivery to specific target sites by recognizing distinctive cell markers [118]. Typically, to target particular cells, traditional polymeric DDSs need to be linked with receptor ligands. Ligands can encompass proteins, peptides, carbohydrates, nucleic acids, and small molecules. However, these conjugation processes for polymeric DDSs can be labor-intensive, expensive (e.g., the RGD peptide targeting sequence costs 145 USD per 10 mg), and may not provide adequate control over drug release. Over the past 5 years, various research studies have demonstrated significant potential for MIP-based DDSs, offering a more promising alternative [127]. Efforts to improve drug delivery for anti-tumor agents have been numerous due to the narrow therapeutic window of these drugs. Recent studies have shown significant promise for MIP-based DDSs. However, in some cases, MIP-based DDSs may fail to load drugs adequately, such as in contact lenses, which can lead to ineffective drug release. Molecular imprinting can enhance drug loading in MIP-based DDSs with insufficient loading, suggesting potential benefits over NIP-based DDSs for anti-tumor drug delivery [88, 96]. It is possible to increase loading capacity by employing fewer cross-linkers, but this strategy must take into account the trade-off between drug loading and cavity stability. Additionally, some MIPs may release a small burst of drug molecules upon adsorption to their surface. In some cases (such as anti-tumor drugs), this can be extremely toxic. However, burst release may be beneficial in other cases (e.g., at the beginning of an antibiotic treatment). The burst effect can perhaps be modulated more effectively through improved purification strategies. While it is extremely challenging to design a suitable release profile, MIPs can be easily modified to implement a variety of release profiles to suit the needs of the drug under consideration. As a result, the combination of implantable microchips and MIPs may prove to be an effective avenue for drug delivery within the reach of clinical trials within the next decade [128].

15.4 MIPs as Artificial Antibody in TE

Biomaterial researchers are presently encountering the dilemma of creating and manufacturing new materials that demonstrate enhanced biological performance, including improved compatibility with cells and tissues, as well as the ability to mimic antibodies. A noteworthy research trend involves the development of synthetic receptors that display affinity and specificity comparable to those found in biological systems [129]. To produce novel biomaterials with enhanced intelligence and customization capabilities for recognizing, binding, or interacting with toxins, cytokines, or growth factors (GFs), researchers can transfer molecular or structural information from a target substance to a synthetic polymer [130]. A novel approach for treating several protein-mediated diseases may be developed through this method to study physiological interactions [131]. In recent years, there has been a surge of interest in MIPs due to their potential to develop polymeric systems capable of selectively recognizing double-stranded DNAs, viruses, bacteria, or even cells [132]. MIPs, synthetic receptors conceived in the 1970s, demonstrate a specific binding affinity for template molecules. They mimic biological interactions in synthetic materials, making them comparable to DNA. MIPs have an advantage over natural antibodies because of their highly specific binding sites. Additionally, they are stable and can withstand harsh conditions such as temperature and pH [12]. In addition, MIPs can bind to a wide range of targets, making them a cheaper and more versatile alternative to natural antibodies that must be specifically made for each antigen [133, 134]. Additionally, MIPs can be prepared with greater ease and exhibit extended storage stability, remaining effective for several months or even years without performance degradation. Due to these inherent advantages, MIPs have gained widespread application across a multitude of domains, including biosensors, DDSs, and analyte extraction techniques [135]. Nonetheless, this technology is not without its limitations, which encompass a relatively low imprinting factor (the ratio of template binding in the imprinted polymer compared to binding in the non-imprinted polymer) and a restricted selection of available functional monomers. Despite its success in small molecule imprinting, MI technology is limited in its application to larger biomolecules. MIPs are used in the field of TE [107]. An approach based on cells, biomaterials (e.g., scaffolds), and GFs is used to develop artificial tissues using TE principles. TE aims to create biological replacements with the objective of restoring, preserving, or enhancing tissue functionality, primarily for transplantation purposes. Utilizing TE MIPs presents a promising avenue for incorporating cell-stimulating cues into scaffolds within tissue engineering, enhancing the engineered constructs' capacity to emulate the intricate features of native tissue and, subsequently, promoting better integration post-implantation [12].

15.4.1 Molecular Imprinting for TE Applications

Molecular recognition is critical at many scale levels in natural systems. In the context of the immune system, foreign substances within the body, termed antigens, are detected through the interaction with globular proteins called antibodies. These antibodies have specific antigen-binding sites that enable them to recognize and respond to foreign materials within the body [136]. There are both fibrillar and amorphous components of the extracellular matrix (ECM), which interact with the cells normally through their surface receptors (integrins) from the ECM. The interactions between cells and the ECM can influence cell differentiation, cell growth, orientation, molecule production, and ECM remodeling [137]. Additionally, growth

Table 15.2 The translation of MI into TE, along with the clinical application of the specific case

Description in brief	Strategy for fabrication	Clinical application	Ref
This is a type of nanoparticle that is capable of binding an antibacterial drug in order to release it on a timer	Precipitation polymerization	Antibacterial wound dressing	[139]
There are nanoparticles imprinted with matrix metallopeptidase 9 (MMP-9), whose scavenging may contribute to better regeneration of cardiac tissue in the future	Precipitation polymerization	Cardiac tissue regeneration	[103]
Biomimetic sponges are functionalized with particles selective for laminin and fibronectin	Precipitation polymerization	Cardiac tissue regeneration	[140]
The surface of the PDMS has been imprinted with osteoblast-like cells. The surface has been seeded with stem cells, which have differentiated to the imprinted phenotype as a result of the imprinting process	Micro-contact printing	Bone tissue regeneration	[141]

factors (GFs) engage in binding interactions with insoluble components of the extracellular matrix (ECM), affording them the ability to intricately modulate their concentration and functional impact. Consequently, scaffolds are required to offer more than just mechanical and structural support; they must also actively guide and regulate cell attachment, migration, proliferation, and differentiation [138]. Therefore, it is crucial to expand the capabilities of scaffolds to include biological signals that facilitate cell interactions and allow for the controlled delivery of GFs. The following sections will delve into the utilization of MIPs in tissue engineering applications, as detailed in Table 15.2.

15.4.2 Introducing Chemical Cues into a Scaffold by Molecular Imprinting

A critical challenge in the field of TE is the development of functionalized scaffolds that can effectively guide cell behavior and promote tissue formation by facilitating interactions between cells and biomaterials. For such interactions to occur, cells must have the ability to recognize and engage with biomolecular materials. Achieving this recognition can be accomplished by incorporating chemical cues into the scaffold. It is noteworthy to highlight that MI has been employed as a technique to introduce carriers for these biomolecules within this context.

Based on the previous work, Criscenti et al. [142] developed bioactive scaffolds for TE by combining Soft-MI with electrospinning and combining these two techniques. The surfaces of electrospun mats underwent imprinting with different proteins and growth factors, including FGF-2 and BMP-2, before being used for cell culture. The imprinting technique was observed to exert a notable impact on cell proliferation, cell count, and metabolic activity. In another study, Rosellini et al. [143] suggested that the combination of MI and scaffolding could be utilized for applications in TE. A peptide sequence derived from fibronectin was employed in precipitation polymerization to imprint MAA-based microgel beads with an identical peptide sequence. Through a straightforward deposition method, these beads were subsequently employed to functionalize films, leading to the creation of an innovative bioactive scaffold. Furthermore, in addition to their specific binding affinity for both the individual peptide sequence and the complete fibronectin molecule, the researchers demonstrated that the MI beads could enhance cell adhesion and proliferation when utilized for scaffold functionalization. It was reported by Pan et al. [144] that thermo-responsive cell culture substrates designed for cell sheet tissue engineering were developed through the imprinting of PNIPAAm surfaces. The tetrapeptide RGDS (Arg-Gly-Asp-Ser), known for its ability to enhance cell adhesion, was employed to imprint the hydrogel with a peptide adhesive. This method allowed the cultivation of a single layer of cells by immobilizing the peptide on the surface and incubating it at 37 °C for 24 h. In the realm of cell culture, scientists discovered that by reducing the temperature to a certain level, the surface properties of the substrate underwent a transformation from hydrophobic to hydrophilic. This change triggered the cells to detach naturally, allowing the researchers to harvest both the cells and the ECM without causing any harm. In a more recent investigation conducted by the same research team, they employed microcontact printing to imprint RGD onto a surface. This innovative approach enabled the researchers to showcase the surface's capability to bind and re-bind the peptide, ultimately leading to the development of a platform with customizable cell adhesion properties [145].

15.4.3 Molecular Imprinting for Controlling Cellular Activities through Physical Cues

It's also viable to influence cellular behavior by incorporating physical cues within the scaffold, simulating the natural ECM environment to steer and regulate their activity [146]. Researchers have extensively investigated MIPs as a valuable tool for guiding cellular behavior by employing them as templates to influence cell growth. Among the various methods for MIP fabrication, micro-contact imprinting using cells has emerged as a popular and widely adopted approach [147]. The method reported by DePorter et al. [148] is simple and inexpensive, requiring the culture of mammalian cells onto a glass slide, followed by the injection of AAM into the cavity left by the mammalian cells. A hydrogel surface imprinted with cell features was found to act as a cue for cell adhesion and proliferation. Mahmoudi et al. [149] reported that stem cells were differentiated using MIPs. This research focused on chondrocytes that were shaped like a sphere (matured) as well as those that were shaped like a spindle (dedifferentiated). The chondrocyte cells were poured with a PDMS solution, and at the end of the curing process, this solution was removed from the chondrocyte cells. To investigate the molecular attributes, specifically gene expression, of the template cell types, adipose-derived mesenchymal stem cells were seeded onto the PDMS surface. Notably, these cells not only altered the shape of the cavities but also expressed their molecular characteristics within these

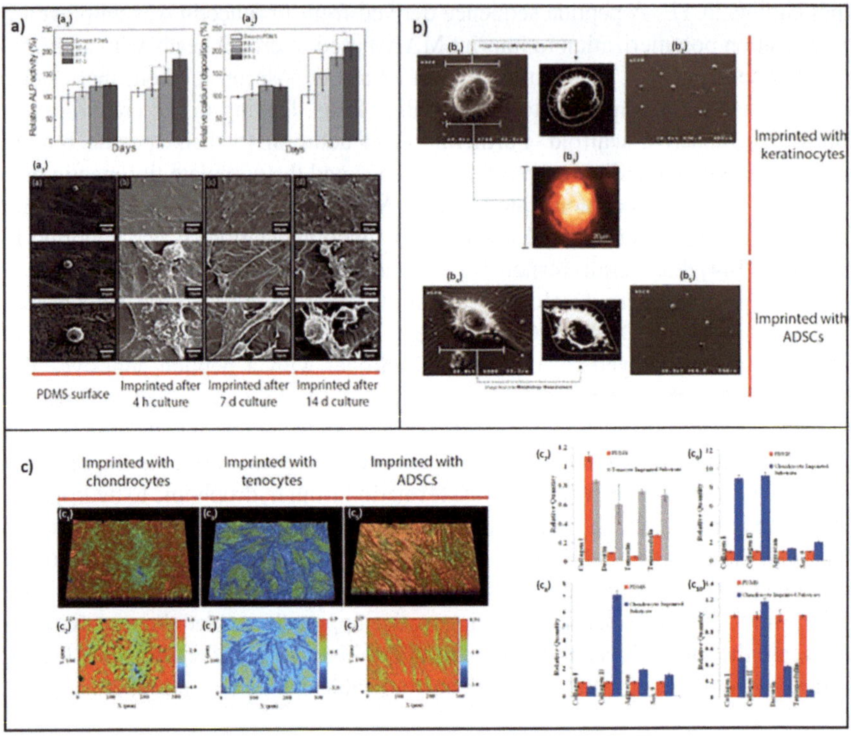

Fig. 15.4 Cell MI to create physical cues able to guide cell activities. In (**a**) ALP activity and calcium deposition were measured at 7 and 14 days for osteoblast-like cell culture on smooth PDMS surface and on PDMS surface imprinted with the same cell type after different culturing times (4 h, 7 days, and 14 days). Image reprinted with permission from. In (**b**) atomic force microscope images of ADSCs cells cultured on a keratinocytes-imprinted and ADSCs-imprinted silicone substrate. Image reprinted with permission from. In (**c**) profilometry images of the PDMS substrates imprinted with different cell types (chondrocytes, tenocytes, and ADSCs), alongside the gene expression results of different cell cultures on the substrates (specifically, (c_7, c_8) ADSCs, (c_9) fibroblasts and (c_{10}) tenocytes). (**d**) Examples of different types (ceramic, polymeric, metallic) nanoparticles which can be utilized for various applications in TERM including tissue targeting and imaging, bioactive agent delivery, modulating mechanical properties of scaffolds, and providing antimicrobial and antitumor properties [12]

cavities. Another noteworthy study, conducted by Mashinchian et al. [150], showcased that when stem cells were cultured on a surface imprinted with mature keratinocytes, they underwent a transformation, adopting keratinocyte-like appearances, and exhibited the expression of keratinocyte marker genes and proteins during their in vitro cultivation, as illustrated in Fig. 15.4b. In a related study, Bonakdar et al. [151] delved into experiments involving adipose-derived mesenchymal stem cells subjected to imprinting with cell-like structures. A thorough analysis of cell morphology and the modulation of specific genes conclusively demonstrated the induction of shape alteration and cellular differentiation, as illustrated in Fig. 15.4c. In a conclusive study, Jeon and collaborators [141] conducted pioneering research where they successfully crafted a patterned surface model using an innovative casting technique. This breakthrough allowed them to faithfully replicate the shape and patterns obtained from osteoblast-like cells cultured over varying durations (4 h, 7 days, and 14 days). They subsequently re-cultured the same cell type on these negative-patterned surfaces, meticulously monitoring several biological activities, including cell viability, alkaline phosphatase activity, and calcium deposition. Remarkably, the results revealed a substantial enhancement in all these activities when compared to the previous time period, as illustrated in Fig. 15.4a.

15.4.4 Drug Delivery and Sequestering Using Molecular Imprinting

Due to their capacity to selectively bind template molecules, MIPs stand as ideal candidates for incorporation into DDSs [128]. DDSs merit special attention as they possess the unique capability to deliver specific medications to targeted regions, ensuring localized delivery while safeguarding surrounding tissues from damage [152]. Additionally, they can serve the purpose of gradually releasing drugs over an extended timeframe, maintaining controlled dosage levels to attain the highest therapeutic effectiveness without causing excessive or insufficient drug delivery [13]. Mao et al. [153] successfully crafted MIP nanospheres using precipitation polymerization, intending to utilize them in Drug Delivery Systems (DDS). These MIP nanospheres exhibited a remarkable ability to bind and release vancomycin over extended periods, surpassing 18 days, with accelerated release rates observed under lower pH conditions. This research holds promise for the development of DDS strategies aimed at preventing bacterial infections [154]. MIPs also were used in the delivery of cancer drugs in another recent study by Gu et al. [155]. In an in vivo tumor mouse model, the researchers detailed the utilization of a MIP-based prodrug delivery system (proDDS). This proDDS exhibited the remarkable ability to accumulate and selectively bind to the tumor site while locally releasing the drug dosage. Notably, fluorescence imaging results indicated that the imprinted proDDS demonstrated a superior targeting capability and a prolonged retention time when contrasted with the imprinted control. This innovative approach resulted in a significant reduction in the tumor's growth rate [156]. In the realm of TE, DDS plays a pivotal role in delivering GFs and other therapeutic agents to developing tissues. This not only has the potential to enhance tissue regeneration but also offers versatile applications. For instance, an intriguing approach was reported in an article where vancomycin, an antibiotic, was employed as a template for wound healing applications, leveraging its unique properties to promote and expedite the wound healing process [157]. To develop an antibacterial wound dressing using the MIP, a crucial step involved encapsulating the produced MIP within an alginate matrix. This encapsulation process was essential to transform the MIP into an effective wound dressing [157]. Through in vitro drug release and biological activity assessments against Gram-positive bacteria, it was observed that (i) the release of antibiotics was notably decelerated post-encapsulation in the alginate matrix, and (ii) the quantity of drug released within the initial 24 h proved to be adequate in inhibiting bacterial growth [158]. As a result of the MIP-based DDS, antibiotics are released steadily and effectively, obviating the need for frequent dressing changes. This novel application of MIPs in crafting a DDS for skin wound healing was recently presented in a paper authored by Koudehi et al. [139]. To provide a more detailed account, the researchers crafted MIP nanoparticles using gentamicin as the template molecule. These nanoparticles were subsequently electrospun together with polyvinyl alcohol (PVA) and gelatin to create a composite material. Electrospun fiber mats derived from this composite material were then subjected to in vitro cytotoxicity assessments and drug release studies. Furthermore, they were implanted in a rat model for in vivo evaluation. Remarkably, in comparison to electrospun PVA/gelatin fibers and an untreated control group, the novel composite biomaterials exhibited an accelerated wound healing response within a 14-day period. This outcome underscores the significant potential of these composite materials in enhancing wound healing processes [159]. Additionally, an equally significant application of MIPs in TE scaffolding is their capacity to capture and retain biomolecules, effectively enhancing the process of tissue regeneration [160]. As GFs are naturally abundant near injury sites, scaffolds can be thought of as tools that harness this natural process. They capture these biomolecules, creating an environment conducive to tissue regeneration, and effectively leveraging the body's own mechanisms for healing [161]. Because of the specificity of MIPs, they have the potential to be a promising solution to this challenge [162]. MI was used, for example, by Teixeira et al. [163] in order to create imprinted nanoparticles that could sequester endogenous GFs (TGF-β3). In the study's outcomes, it became evident that these nanoparticles possessed a unique capability to selectively bind with the GF, not only in controlled laboratory assays but also in complex human fluids such as platelet lysate. Notably, when human adipose stem cells were exposed to the imprinted nanoparticles along with platelet lysate, they demonstrated a notably increased deposition of collagen matrix compared to the control group, which had been incubated with non-imprinted nanoparticles. This discovery holds great promise for applications in tissue engineering and regeneration [62, 163]. Beyond GFs, it's crucial to recognize that a multitude of other molecules may exist at the injury site, potentially hindering the regenerative process [164]. MIP that possesses the capacity to selectively capture these molecules from the surrounding environment can significantly accelerate the regeneration process [165]. Cristallini et al. [103] applied the Molecular Imprinting (MI) technique to craft particles using poly(ethylene glycol) ethyl ether methacrylate along with matrix metallopeptidase 9 (MMP-9), with a specific imprint of MMP-9 itself. Multiple investigations have indicated that the removal of MMP-9 from the myocardium in the early stages of a stroke may promote more effective tissue regeneration. These studies revealed that the nanoparticles effectively captured the template MMP-9. Furthermore, they demonstrated that the entire bio-engineered structure, which exhibited characteristics at the nano, micro, and millimeter scales, successfully replicated a cardiac ECM-like environment upon applying the MIPs to the scaffold.

15.5 Advantages and Challenges of MIPs in Drug Delivery and TE

MIPs are a remarkable type of synthetic material. What makes them truly fascinating are their molecularly specific cavities, like tiny, custom-made hideaways. These cavities have opened up exciting new frontiers in the worlds of drug delivery and tissue engineering [166]. These remarkable nanoparticles have a special talent that sets them apart from natural antibodies. They can identify molecules at a molecular level, a skill that holds the promise of revolutionizing medical treatments due to their incredible molecular recognition abilities. But it's important to remember that crafting and using MIPs can be quite tricky. There are hurdles to overcome when it comes to designing, making, and using them [35]. This section explores the complex world of MIPs, including their benefits and the challenges of integrating them into drug delivery and TE [167].

The hallmark advantage of MIPs in drug delivery lies in their ability to achieve precise targeting [46]. MIPs are designed to bind selectively to specific molecules, similar to how a key and lock mechanism binds to different molecules. This specificity enables the delivery of therapeutic agents directly to disease sites, thereby minimizing collateral damage to healthy tissues as the therapeutic agents are delivered to the disease sites directly [157]. One application of MIPs is their ability to detect cancer-specific markers. This feature allows them to concentrate chemotherapy agents at the tumor site, which can lead to improved effectiveness and reduced toxicity throughout the body. Additionally, MIPs offer an advantage in controlled drug release kinetics. By using a precisely calibrated polymer structure, drug molecules can be released in a controlled and sustained manner, maintaining optimal drug concentrations for an extended period of time [13]. With this controlled release profile, it is not only possible to improve therapeutic outcomes but also to reduce the frequency at which a drug needs to be administered [168]. In chronic conditions where maintaining stable drug concentrations can be crucial, patients benefit from consistent drug levels, which can be especially helpful in managing chronic conditions with the need for consistent drug levels. Moreover, MIPs are also very effective in protecting labile compounds from degradation. It is well known that many drugs are susceptible to degradation and enzymatic breakdown. The MIPs are capable of encapsulating these compounds, shielding them from harsh environments and the activity of the enzymes [46]. As a result of this protective effect, the drug remains intact during its transit through the body, preserving its therapeutic strength until it reaches the target site where it is needed. It is important that DDSs be biocompatible, and MIPs often exhibit a reduced immunogenicity. As opposed to other drug carriers that may trigger immune responses, the biocompatibility of MIPs means that they are less likely to cause adverse reactions in the patient. In the case of long-term treatment, where immune system involvement may compromise the effectiveness of the treatment, this property is particularly advantageous.

There is no doubt that MIPs are a key component of TE that facilitates enhanced cellular integration. By emulating the ECM, MIPs provide an environment conducive to cell adhesion, proliferation, and differentiation [169]. It is through this process that tissue regeneration and tissue integration can be promoted, which facilitates the growth of functional tissues. The MIPs provide a platform that can be used to tailor surface properties to orchestrate specific cell behaviors [170]. There are a number of ways in which bioactive molecules, peptides, or physical features can be incorporated onto MIP surfaces in order to guide cellular responses. The customization of the tissue enhances the possibility of desirable cellular interactions, thereby influencing the growth and function of the tissue [171]. During the process of TE, scaffolds play a crucial role in providing structural support for tissue growth [172]. It has been shown that MIPs have a high degree of mechanical stability and are capable of being customized, allowing them to be used as scaffold materials in the development of tissue-engineered constructs [173]. In order to ensure that engineered tissues can withstand physiological forces and develop properties similar to those of native tissues, this support must be provided.

15.5.1 Challenges of MIPs in Drug Delivery

In recent years, numerous intriguing research endeavors have explored the potential of MIP-based DDSs, pointing toward promising advancements in the near future. A substantial portion of these studies has concentrated on enhancing DDSs specifically for antitumor agents. These agents hold significance as prime candidates for improved DDSs, chiefly owing to their narrow therapeutic margins [88, 174]. For the advancement of MIP-based DDSs, methacrylic monomers, and cross-linkers, such as HEMA, MAA, MBA, EGDMA, and EDMA, have been extensively used as monomers and cross-linkers [45, 175, 176]. Despite the fact that these monomers are considered to have acceptable toxicity and excellent biocompatibility, there has not been yet any detailed evaluation of their long-term toxicity and metabolic pathways. Moreover, additional materials such as POSS, β-CD coated surfaces, CNTs, CS, silicon, as well as stimuli-sensitive materials have also been used to improve the performance of MIPs in terms of drug release behavior, biodegradability, and biocompatibility [113, 177]. There is also a major obstacle when it comes to the development of most MIP-based DDS that are based on organic solvents (e.g., toluene, chloroform). In this way, the noncovalent interactions between the template molecules and the functional monomers are made easier and maintained. Despite this, in terms of medical translation and application, the presence of residual organic solvents can cause damage to healthy cells, thus resulting in serious side effects [178]. The use of organic solvents during the synthesis may also result in a marked difference in the drug-release behavior of MIPs in aqueous media, as well as potentially increasing manufacturing costs. The use of supercritical carbon dioxide ($scCO_2$) technology in creating polymers offers numerous benefits. These include controlled morphology, non-toxicity, absence of solvent residues, and the ability to avoid purification and drying processes. As a result, $scCO_2$ technology serves as an environmentally friendly and safe alternative [88]. Additionally, $scCO_2$ has the ability to stabilize hydrogen bonds between templates and functional monomers. However, it's crucial to acknowledge that this technology isn't widely utilized or readily available, even on a laboratory scale. One reason for the limited use of $scCO_2$ technology could be the need for specialized and costly equipment, as well as extreme operating conditions related to temperature and pressure. Nevertheless, $scCO_2$ holds great potential for creating MIP-based DDSs that are highly pure and compliant with GMP standards [179].

It's important to mention that there's a challenge related to the examples of literature discussed above. Specifically, only a few studies have reported in vivo data for the assessment of MIP-based DDSs. Whenever a potential DDS is being evaluated, it is absolutely essential that in vivo research is conducted as part of the evaluation process. Since most studies have only evaluated toxicity and drug release in vitro, it is crucial to gather additional data and conduct further investigations. This is necessary to ensure that the positive behavior of MIPs observed in vitro translates into in vivo models, where safety and adequate drug release are demonstrated. MIPs cannot be adequately evaluated in terms of their potential to improve clinical outcomes unless they are evaluated in vivo. There is also a concern in some cases that MIP-based DDSs (e.g., contact lenses) are inadequately loaded with drugs. The result may be a sub-par, ineffective release of the drug [180]. In situations where MIP-based DDSs are not loaded to their full potential, they still tend to outperform NIP-based DDSs. This suggests that molecular imprinting has the potential to improve drug loading in DDSs. By using fewer cross-linkers, the loading capacity of the imprinted cavity can be increased, but the trade-off between drug loading and cavity stability must be considered. The adsorption of drug molecules onto the surface of some MIPs may also result in a small initial burst release due to the adsorption of those molecules onto the MIP surface. In some drugs (e.g., antitumor agents), this can have serious toxic effects as a result of this reaction. It is also possible, though, for burst release to be beneficial in other instances (e.g., at the beginning of a course of antibiotics) [181]. There might be a way to improve the modulation of the burst effect by using more effective purification strategies. It is a tremendous challenge to design a suitable release profile, but MIPs offer ample opportunity for modifications to achieve a variety of different release profiles that can be customized to fit the needs of the drug of interest, despite the fact that this can be extremely challenging.

15.5.2 Challenges of MIPs in TE

To attain tissue regeneration, TE represents an innovative strategy that necessitates expertise spanning multiple domains, including engineering, biology, biomaterial science, and medicine. The scaffolding approach in TE is meticulously crafted to facilitate the generation of living tissues that seamlessly fuse with the implantation site and stimulate tissue regeneration. This approach combines three essential components: biomaterials, cells, and GFs [182]. To ensure a seamless integration, it is crucial that the engineered structure replicates the intricate nature of the natural ECM environment and maintains control over cellular activities. In this context, MI is striving to elevate the scaffold's capacity to emulate biological characteristics by integrating synthetic receptors that operate in a manner akin to those found in natural systems. This advancement will ultimately enhance the scaffold's ability to mimic biological processes. As previously discussed in earlier sections, MI has already found successful application in TE scaffolds, introducing chemical and physical cues. Furthermore, this technology has also been harnessed for the controlled GFs/drugs and the sequestration of endogenous biomolecules, thereby amplifying the tissue regeneration process. Beyond its role in tissue regeneration and repair, MI holds promise as a solution for creating in vitro models that faithfully replicate complex tissue environments. This enables the study of physiological and pathological aspects as well as the efficient testing of new drug formulations [183].

In addition to the previously mentioned applications, there are several significant challenges in TE that remain to be addressed. One such challenge is the insufficient availability of active cells for seeding scaffolds, especially when creating scaffolds for implantation. MI can play a pivotal role in addressing these challenges through its data analysis capabilities [184]. Although harvesting adult cells directly from the patient might seem straightforward in theory, it is generally avoided in practice. This is primarily due to the invasiveness of the procedure and the potential harm it can inflict on other tissues in the patient's body. On the other hand, utilizing stem cells presents its own set of challenges, including issues related to incorrect differentiation and the maintenance of phenotype. In this context, the use of MI substrates in TE offers an attractive solution. It provides a cost effective and straightforward method for expanding and differentiating stem cells, with reduced differentiation potential compared to GFs. Kashani et al. [185] in a recent report introduced an innovative approach that combines MIPs with microfluidic devices through a two-step process. First, they fabricated a chondrocyte-imprinted substrate using a microfluidic device, and then they cultured stem cells on this substrate using a second microfluidic device. After a 14-day culture period, the cells successfully differentiated into chondrocytes, exhibiting the characteristic spherical morphology of chondrocytes. A microfluidic device, when used independently, offers superior control over the microenvironment in both temporal and spatial dimensions. It also allows for the stimulation of multiple cells at a high rate through the utilization of multiple microfluidic channels. The incorporation of MI into this system can be argued to enhance its efficiency by providing precise control over cell placement. Another crucial aspect of TE is the capability to monitor the development of tissue in real time. This real-time monitoring can offer valuable insights into the regeneration process, and scaffold degradation after implantation, and contribute to a better understanding of the physiology or pathology of the target tissue within an in vitro model. Notably, MI has demonstrated its potential to address this challenge by enabling noninvasive fluorescent labeling. In recent literature, there has been extensive research on fluorescent labeling, particularly in the context of tumor cells. Cancer cells tend to overexpress specific membrane molecules, setting them apart from normal cells and making them easily distinguishable through fluorescent labeling techniques. For example, Demir et al. [186] harnessed the power of carbon nanodots (CDs) in combination with a MIP shell that exhibited a specific affinity for glucuronic acid. They tailored this MIP shell to target a specific segment of hyaluronan, a polymer commonly found in the ECM. Interestingly, this hyaluronan segment is found in higher concentrations in certain cancerous cells, including those in the colon, breast, prostate, lung, bladder, and cervix. The remarkable selectivity of the MIP shell enabled the precise labeling of cancer cells. Upon exposure to UV light, the core of the CDs emitted a powerful and unmistakable fluorescence, bestowing upon it a valuable capability to identify these cancer cells. Another example is the work of Panagiotopoulou et al. [187] conducted research employing fluorescently labeled MI nanoparticles. Their goal was to visualize both fixed and living human keratinocytes to investigate whether indicators of cell malignancy, such as sialic acid (SA) and hyaluronan, were present within these cells. The remarkable aspect of their work was the high level of selectivity observed between these nanoparticles and their respective templates, all while ensuring they did not harm the cells. What's intriguing is that these nanoparticles could be tailored to various sizes and were capable of binding to both the extracellular membrane portions (for particles measuring 400 nm) and the internal membranes of cells (for particles measuring 150 nm). This versatility in size allowed for a comprehensive examination of cell structures and potential malignancy indicators. Furthermore, MIPs have been harnessed for direct therapeutic applications, facilitating the precise binding and elimination of specific cell types, along with enabling cell imaging. In these direct therapies, MIPs are employed to selectively target and eliminate particular cell populations. Recent research has homed in on the use of MI nanoparticles for the identification and eradication of senescent cells. MI serves as a marker to detect these cells, both within and outside the body. Multiple experiments, both in vitro and in vivo, have demonstrated no adverse reactions and a remarkable ability for MI nanoparticles to selectively attach to senescent cells [188]. Moreover, the authors have provided a compelling proof-of-concept for the in vitro utilization of MIPs to identify and eliminate senescent cells. They achieved this by incorporating specific drugs into the MIPs. Additionally, they presented a proof-of-concept for in vivo applications. In a related study, Yin and colleagues employed SA as a template to imprint onto gold nanorods. This innovative approach allowed for the detection of cancer cells in a unique and effective manner [189]. The researchers conducted an experiment using mice with tumors to assess the effectiveness of nanorods in detecting and eradicating senescent cells, which had been loaded with specific drugs. The outcomes demonstrated that these nanorods could specifically attach to cancer cells and inhibit their growth. When introduced into the bloodstream, the nanorods tended to accumulate near the tumor sites, simplifying their identification. This technology exhibited promise for applications in both laboratory and live animal scenarios. Subsequently, a near-infrared laser beam was precisely directed toward the tumor location, and the nanorods absorbed the photon energy and efficiently transformed it into heat. This resulted in the targeted thermal destruction of only those cells in direct contact with the laser light. Moreover, the possibility of integrating MIP sensors directly into scaffolds or employing them within an in vitro model emerged as a means to enhance control over the cellular microenvironment. In recent developments, researchers created protein sensors for the detection of cancer biomarkers through the electro-polymerization of peptides on a gold surface. This innovative approach holds promise for advancing cancer detection methods. As a further contribution, Lakshmi et al. [190] made a significant contribution by developing an electrochemical sensor based on MIP for detecting trimethylamine N-oxide (TMAO). TMAO is a metabolite produced by microbiota and is believed to be associated with various health conditions, including chronic kidney disease and cardiac disease. In their study, the authors leveraged the noncovalent interactions between the template polypyrene and the electroactive polymer TMAO to create the electroactive sensor. These biosensors exhibited broad linear detection ranges, impressive sensitivities, high selectivity, and rapid response times (approximately 20 min). Despite some initial successful examples, several challenges persist, limiting the translation of MI to TE. One such challenge is the difficulty in imprinting complex biomolecules like proteins, given their intricate nature. Additionally, there's a need to explore a wider range of noncytotoxic materials and cell-friendly processing conditions. Furthermore, optimizing fabrication techniques to achieve more consistent outcomes and facilitate industrial applications is essential. These challenges must be addressed in the near future to fully unlock the potential of MI in TE.

To conclude, MIPs have the potential to make a profound impact on the future of medicine through their advantages in drug delivery and TE. It is apparent that the precision, specificity, and tailored functionality that they offer hold great promise for enhancing therapeutic outcomes and tissue regeneration as a result. The journey ahead, however, will require a great deal of dedication, a lot of innovation, and a profound knowledge of both the science as well as the medical applications of the technology. There is no doubt that advances in the field of precision medicine may be forever transformed by the capabilities of MIPs if researchers and scientists manage to navigate the challenges involved.

15.6 Conclusion

In the field of drug delivery and TE, MIPs have emerged as a promising class of materials. As they mimic the specificity of natural antibodies while being more stable and cost effective, they have the potential to revolutionize these fields. This chapter examined the advantages and current challenges of using MIPs for drug delivery and TE, functioning as artificial antibodies. In drug delivery, MIPs offer several key advantages. Their highly customizable nature enables the precise design of binding sites, which allows selective recognition and controlled release of drugs. This specificity not only decreases off-target effects but also enhances the effectiveness of the therapy. A further advantage of MIPs is their remarkable stability across a wide range of environments, resulting in prolonged drug release over extended periods of time, thereby enhancing patient compliance. As a result of their tunable degradation kinetics, MIPs are particularly attractive as drug delivery agents. MIPs promote cell adhesion, growth, and differentiation and support tissue regeneration by imitating native ECM. They offer mechanical integrity and aid in tissue-specific function. In addition, MIPs can be engineered to sequester GFs, enabling spatiotemporal control over the development of tissues. The implementation of MIPs still presents challenges despite the potential for successful TE strategies that their combination of features enhances. It remains an intricate task to design MIPs with precise binding affinities and selectivities, which requires a profound understanding of molecular interactions. Developing MIPs with well-defined structures and optimal performance parameters can be a complex and time-consuming process. The issue of scaling up production while maintaining consistency also poses a challenge for widespread implementation. In conclusion, MIPs present a paradigm shift in drug delivery and TE. The ability of these agents to mimic natural antibodies with enhanced stability and cost-effectiveness holds great promise. In drug delivery, MIPs enable targeted and sustained release, minimizing side effects and improving therapeutic outcomes. MIPs mimic the ECM and facilitate tissue regeneration in TE. Although there are obstacles, ongoing research and development on MIP synthesis techniques and design principles are expected to overcome them. MIPs hold promise in revolutionizing medical science and healthcare by offering tailored and effective solutions for drug delivery and tissue regeneration.

References

1. Pan J, Chen W, Ma Y, Pan G (2018) Molecularly imprinted polymers as receptor mimics for selective cell recognition. Chem Soc Rev 47(15):5574–5587
2. BelBruno JJ (2018) Molecularly imprinted polymers. Chem Rev 119(1):94–119
3. Roy E, Patra S, Tiwari A, Madhuri R, Sharma PK (2017) Introduction of selectivity and specificity to graphene using an inimitable combination of molecular imprinting and nanotechnology. Biosens Bioelectron 89:234–248
4. Villa CC, Sánchez LT, Valencia GA, Ahmed S, Gutiérrez TJ (2021) Molecularly imprinted polymers for food applications: a review. Trends Food Sci Technol 111:642–669
5. Bayer CL, Peppas NA (2008) Advances in recognitive, conductive and responsive delivery systems. J Control Release 132(3):216–221
6. He S et al (2021) Advances of molecularly imprinted polymers (MIP) and the application in drug delivery. Eur Polym J 143:110179
7. Liu Z, Huang Y, Yang Y (2021) Molecularly imprinted polymers as advanced drug delivery systems: synthesis, character and application. Springer
8. Bărăian A-I, Iacob B-C, Bodoki AE, Bodoki E (2022) In vivo applications of molecularly imprinted polymers for drug delivery: a pharmaceutical perspective. Int J Mol Sci 23(22):14071
9. Zahedi P, Khatibi A, Fallah-Darrehchi M (2023) Antimicrobial electrospun membranes. Electrospun and nanofibrous membranes. Elsevier, pp 501–519
10. Dixit CK, Bhakta S, Reza KK, Kaushik A (2022) Exploring molecularly imprinted polymers as artificial antibodies for efficient diagnostics and commercialization: a critical overview. Hybrid Adv 1:100001
11. Fathi-Achachelouei M et al (2019) Use of nanoparticles in tissue engineering and regenerative medicine. Front Bioeng Biotechnol 7:113
12. Bonatti AF, De Maria C, Vozzi G (2021) Molecular imprinting strategies for tissue engineering applications: a review. Polymers 13(4):548
13. Zaidi SA (2020) Molecular imprinting: a useful approach for drug delivery. Mater Sci Energy Technol 3:72–77

14. Haupt K, Linares AV, Bompart M, Bui BTS (2011) Molecularly imprinted polymers. Chem Rev:1–28
15. Takeuchi T, Matsui J (1996) Molecular imprinting: an approach to "tailor-made" synthetic polymers with biomimetic functions. Acta Polym 47(11–12):471–480
16. Polyakov M (1931) Adsorption properties and structure of silica gel. Zhur Fiz Khim 2:799–805
17. Dickey FH (1949) The preparation of specific adsorbents. Proc Natl Acad Sci 35(5):227–229
18. Arshady R, Mosbach K (1981) Synthesis of substrate-selective polymers by host-guest polymerization. Die Makromolekulare Chem Macromol Chem Phys 182(2):687–692
19. Parisi OI, Ruffo M, Scrivano L, Malivindi R, Vassallo A, Puoci F (2018) Smart bandage based on molecularly imprinted polymers (MIPs) for diclofenac controlled release. Pharmaceuticals 11(4):92
20. Wulff G (1972) The use of polymers with enzyme-analogous structures for the resolution of racemates. Angrew Chem Int Ed 11:341–341
21. Tamayo F, Turiel E, Martín-Esteban A (2007) Molecularly imprinted polymers for solid-phase extraction and solid-phase microextraction: recent developments and future trends. J Chromatogr A 1152(1-2:32–40
22. Mickey B, Howard J (1995) Rigidity of microtubules is increased by stabilizing agents. J Cell Biol 130(4):909–917
23. Puoci F, Cirillo G, Curcio M, Parisi OI, Iemma F, Picci N (2011) Molecularly imprinted polymers in drug delivery: state of art and future perspectives. Expert Opin Drug Deliv 8(10):1379–1393
24. Chen L, Xu S, Li J (2011) Recent advances in molecular imprinting technology: current status, challenges and highlighted applications. Chem Soc Rev 40(5):2922–2942
25. Wang X, Chen G, Zhang P, Jia Q (2021) Advances in epitope molecularly imprinted polymers for protein detection: a review. Anal Methods 13(14):1660–1671
26. Ndunda EN (2020) Molecularly imprinted polymers—A closer look at the control polymer used in determining the imprinting effect: a mini review. J Mol Recognit 33(11):e2855
27. Hashim SN, Boysen RI, Schwarz LJ, Danylec B, Hearn MT (2014) A comparison of covalent and non-covalent imprinting strategies for the synthesis of stigmasterol imprinted polymers. J Chromatogr A 1359:35–43
28. Sullivan MV, Nanalal S, Dean BE, Turner NW (2023) Molecularly imprinted polymer hydrogel sheets with metalloporphyrin-incorporated molecular recognition sites for protein capture. Talanta 266:125083
29. Gao C et al (2023) Surface modification methods and mechanisms in carbon nanotubes dispersion. Carbon:118133
30. Mabrouk M, Hammad SF, Abdella AA, Mansour FR (2023) Tipps and tricks for successful preparation of molecularly imprinted polymers for analytical applications: a critical review. Microchem J:109152
31. Huang Y et al (2023) Strategy of choosing templates in molecular imprinting to expand the recognition width for family-selectivity. Anal Chem 95(29):11070–11077
32. Whitcombe MJ, Rodriguez ME, Villar P, Vulfson EN (1995) A new method for the introduction of recognition site functionality into polymers prepared by molecular imprinting: synthesis and characterization of polymeric receptors for cholesterol. J Am Chem Soc 117(27):7105–7111
33. Lusina A, Cegłowski M (2022) Molecularly imprinted polymers as state-of-the-art drug carriers in hydrogel transdermal drug delivery applications. Polymers 14(3):640
34. Zhang B, Fan X, Zhao D (2018) Computer-aided design of molecularly imprinted polymers for simultaneous detection of clenbuterol and its metabolites. Polymers 11(1):17
35. Pratama KF, Manik MER, Rahayu D, Hasanah AN (2020) Effect of the molecularly imprinted polymer component ratio on analytical performance. Chem Pharm Bull 68(11):1013–1024
36. Beltran A, Borrull F, Marcé R, Cormack P (2010) Molecularly-imprinted polymers: useful sorbents for selective extractions. TrAC Trends Anal Chem 29(11):1363–1375

37. Yan H, Row KH (2006) Characteristic and synthetic approach of molecularly imprinted polymer. Int J Mol Sci 7(5):155–178
38. Parisi OI et al (2022) The evolution of molecular recognition: from antibodies to molecularly imprinted polymers (MIPs) as artificial counterpart. J Funct Biomater 13(1):12
39. Beyazit S, Bui BTS, Haupt K, Gonzato C (2016) Molecularly imprinted polymer nanomaterials and nanocomposites by controlled/living radical polymerization. Prog Polym Sci 62:1–21
40. Hattori K et al (2004) Gate effect of theophylline-imprinted polymers grafted to the cellulose by living radical polymerization. J Membr Sci 233(1-2):169–173
41. Niu M, Pham-Huy C, He H (2016) Core-shell nanoparticles coated with molecularly imprinted polymers: a review. Microchim Acta 183:2677–2695
42. Zahedi P, Ziaee M, Abdouss M, Farazin A, Mizaikoff B (2016) Biomacromolecule template-based molecularly imprinted polymers with an emphasis on their synthesis strategies: a review. Polym Adv Technol 27(9):1124–1142
43. Ye L, Cormack PA, Mosbach K (1999) Molecularly imprinted monodisperse microspheres for competitive radioassay. Anal Commun 36(2):35–38
44. Jing T et al (2009) Determination of trace tetracycline antibiotics in foodstuffs by liquid chromatography–tandem mass spectrometry coupled with selective molecular-imprinted solid-phase extraction. Anal Bioanal Chem 393:2009–2018
45. Vasapollo G et al (2011) Molecularly imprinted polymers: present and future prospective. Int J Mol Sci 12(9):5908–5945
46. Chen L, Wang X, Lu W, Wu X, Li J (2016) Molecular imprinting: perspectives and applications. Chem Soc Rev 45(8:2137–2211
47. Hang H et al (2013) Selective separation of lambdacyhalothrin by porous/magnetic molecularly imprinted polymers prepared by pickering emulsion polymerization. J Sep Sci 36(19):3285–3294
48. Pan J et al (2014) Molecularly imprinted polymer foams with well-defined open-cell structure derived from Pickering HIPEs and their enhanced recognition of λ-cyhalothrin. Chem Eng J 253:138–147
49. Thangadurai TD, Nandhakumar M, Thomas S, Nzihou A (2022) Polymer nanocomposites for energy applications
50. Carter SR, Rimmer S (2004) Surface molecularly imprinted polymer core–shell particles. Adv Funct Mater 14(6):553–561
51. Belmont A-S, Jaeger S, Knopp D, Niessner R, Gauglitz G, Haupt K (2007) Molecularly imprinted polymer films for reflectometric interference spectroscopic sensors. Biosens Bioelectron 22(12):3267–3272
52. Tan CJ, Tong YW (2007) Molecularly imprinted beads by surface imprinting. Anal Bioanal Chem 389:369–376
53. Lu C-H, Zhou W-H, Han B, Yang H-H, Chen X, Wang X-R (2007) Surface-imprinted core–shell nanoparticles for sorbent assays. Anal Chem 79(14):5457–5461
54. Poma A, Turner AP, Piletsky SA (2010) Advances in the manufacture of MIP nanoparticles. Trends Biotechnol 28(12):629–637
55. Canfarotta F, Cecchini A, Piletsky S (2018) Nano-sized molecularly imprinted polymers as artificial antibodies. Mechanochemistry in materials: Royal Society of Chemistry. pp 1–27
56. Canfarotta F, Poma A, Guerreiro A, Piletsky S (2016) Solid-phase synthesis of molecularly imprinted nanoparticles. Nat Protoc 11(3):443–455
57. Biffis A, Graham NB, Siedlaczek G, Stalberg S, Wulff G (2001) The synthesis, characterization and molecular recognition properties of imprinted microgels. Macromol Chem Phys 202(1):163–171
58. Fuchs Y, Soppera O, Haupt K (2012) Photopolymerization and photostructuring of molecularly imprinted polymers for sensor applications—a review. Anal Chim Acta 717:7–20
59. Poma A, Guerreiro A, Whitcombe MJ, Piletska EV, Turner AP, Piletsky SA (2013) Solid–phase synthesis of molecularly imprinted polymer nanoparticles with a reusable template–"plastic antibodies". Adv Funct Mater 23(22):2821–2827

60. Medina Rangel PX et al (2019) Solid-phase synthesis of molecularly imprinted polymer nanolabels: affinity tools for cellular bioimaging of glycans. Sci Rep 9(1):3923
61. Song Z et al (2022) Molecularly imprinted polymers based materials and their applications in chromatographic and electrophoretic separations. TrAC Trends Anal Chem 146:116504
62. Bui BTS, Mier A, Haupt K (2023) Molecularly imprinted polymers as synthetic antibodies for protein recognition: the next generation. Small 19(13):2206453
63. Zhang N et al (2019) Molecularly imprinted materials for selective biological recognition. Macromol Rapid Commun 40(17):1900096
64. Laurichesse S, Avérous L (2014) Chemical modification of lignins: towards biobased polymers. Prog Polym Sci 39(7):1266–1290
65. Horikawa R, Sunayama H, Kitayama Y, Takano E, Takeuchi T (2016) A programmable signaling molecular recognition nanocavity prepared by molecular imprinting and post-imprinting modifications. Angew Chem 128(42):13217–13221
66. Jolly P, Tamboli V, Harniman RL, Estrela P, Allender CJ, Bowen JL (2016) Aptamer–MIP hybrid receptor for highly sensitive electrochemical detection of prostate specific antigen. Biosens Bioelectron 75:188–195
67. Saeki T et al (2020) Signalling molecular recognition nanocavities with multiple functional groups prepared by molecular imprinting and sequential post-imprinting modifications for prostate cancer biomarker glycoprotein detection. J Mater Chem B 8(35):7987–7993
68. Ye L (2015) Synthetic strategies in molecular imprinting. Molecularly imprinted polymers in biotechnology, pp 1–24
69. Ren Y, Fan Z (2023) Synthesis of fluorescent probe based on molecularly imprinted polymers on nitrogen-doped carbon dots for determination of tobramycin in milk. Food Chem 416:135792
70. Spivak DA (2005) Optimization, evaluation, and characterization of molecularly imprinted polymers. Adv Drug Deliv Rev 57(12):1779–1794
71. Boroznjak R, Reut J, Tretjakov A, Lomaka A, Öpik A, Syritski V (2017) A computational approach to study functional monomer-protein molecular interactions to optimize protein molecular imprinting. J Mol Recognit 30(10):e2635
72. Arabi M, Ostovan A, Ghaedi M, Purkait MK (2016) Novel strategy for synthesis of magnetic dummy molecularly imprinted nanoparticles based on functionalized silica as an efficient sorbent for the determination of acrylamide in potato chips: optimization by experimental design methodology. Talanta 154:526–532
73. Scorrano S, Mergola L, Del Sole R, Vasapollo G (2011) Synthesis of molecularly imprinted polymers for amino acid derivates by using different functional monomers. Int J Mol Sci 12(3):1735–1743
74. Wulff G (1995) Molecular imprinting in cross-linked materials with the aid of molecular templates—a way towards artificial antibodies. Angew Chem Int Ed Engl 34(17):1812–1832
75. Shen J et al (2021) Hydrogen bonding interactions between arsenious acid and dithiothreitol/dithioerythritol at different pH values: a computational study with an explicit solvent model. New J Chem 45(43):20181–20192
76. Dong C, Li X, Guo Z, Qi J (2009) Development of a model for the rational design of molecular imprinted polymer: computational approach for combined molecular dynamics/quantum mechanics calculations. Anal Chim Acta 647(1):117–124
77. Piletska EV, Guerreiro AR, Whitcombe MJ, Piletsky SA (2009) Influence of the polymerization conditions on the performance of molecularly imprinted polymers. Macromolecules 42(14):4921–4928
78. Puoci F, Cirillo G, Curcio M, Iemma F, Spizzirri U, Picci N (2007) Molecularly imprinted solid phase extraction for the selective HPLC determination of α-tocopherol in bay leaves. Anal Chim Acta 593(2):164–170
79. Cheong WJ, Yang SH, Ali F (2013) Molecular imprinted polymers for separation science: a review of reviews. J Sep Sci 36(3):609–628
80. Ertürk G, Mattiasson B (2017) Molecular imprinting techniques used for the preparation of biosensors. Sensors 17(2):288

81. Komiyama M, Mori T, Ariga K (2018) Molecular imprinting: materials nanoarchitectonics with molecular information. Bull Chem Soc Jpn 91(7):1075–1111
82. Beltran A, Marcé R, Cormack P, Borrull F (2010) Synthetic approaches to parabens molecularly imprinted polymers and their applications to the solid-phase extraction of river water samples. Anal Chim Acta 677(1):72–78
83. Kitayama Y, Yoshikawa K, Takeuchi T (2017) Post-cross-linked molecular imprinting with functional polymers as a universal building block for artificial polymeric receptors. Macromolecules 50(19):7526–7534
84. Manzari MT, Shamay Y, Kiguchi H, Rosen N, Scaltriti M, Heller DA (2021) Targeted drug delivery strategies for precision medicines. Nat Rev Mater 6(4:351–370
85. Samad AFA, Kamaroddin MF (2023) Innovative approaches in transforming microRNAs into therapeutic tools. Wiley Interdiscip Rev RNA 14(1):e1768
86. Reina G, González-Domínguez JM, Criado A, Vázquez E, Bianco A, Prato M (2017) Promises, facts and challenges for graphene in biomedical applications. Chem Soc Rev 46(15):4400–4416
87. Xu S, Wang L, Liu Z (2021) Molecularly imprinted polymer nanoparticles: an emerging versatile platform for cancer therapy. Angew Chem Int Ed 60(8):3858–3869
88. Liu R, Poma A (2021) Advances in molecularly imprinted polymers as drug delivery systems. Molecules 26(12):3589
89. Chahibi Y (2017) Molecular communication for drug delivery systems: a survey. Nano Commun Netw 11:90–102
90. Geraili A, Xing M, Mequanint K (2021) Design and fabrication of drug-delivery systems toward adjustable release profiles for personalized treatment. Viewpoints 2(5):20200126
91. Zaidi SA (2016) Molecular imprinted polymers as drug delivery vehicles. Drug Deliv 23(7):2262–2271
92. Maiti S, Sen KK (2017) Introductory chapter: drug delivery concepts. Adv Technol Deliv Ther:1–12
93. Xiong M-H, Bao Y, Yang X-Z, Zhu Y-H, Wang J (2014) Delivery of antibiotics with polymeric particles. Adv Drug Deliv Rev 78:63–76
94. Borges J, Rodrigues LC, Reis RL, Mano JF (2014) Layer-by-layer assembly of light-responsive polymeric multilayer systems. Adv Funct Mater 24(36):5624–5648
95. Farjadian F, Ghasemi A, Gohari O, Roointan A, Karimi M, Hamblin MR (2019) Nanopharmaceuticals and nanomedicines currently on the market: challenges and opportunities. Nanomedicine 14(1):93–126
96. Alvarez-Lorenzo C, Concheiro A (2004) Molecularly imprinted polymers for drug delivery. J Chromatogr B 804(1):231–245
97. Zheng X-F, Lian Q, Yang H, Wang X (2016) Surface molecularly imprinted polymer of chitosan grafted poly (methyl methacrylate) for 5-fluorouracil and controlled release. Sci Rep 6(1):21409
98. Norell MC, Andersson HS, Nicholls IA (1998) Theophylline molecularly imprinted polymer dissociation kinetics: a novel sustained release drug dosage mechanism. J Mol Recognit 11(1-6):98–102
99. Tuwahatu CA, Yeung CC, Lam YW, Roy VAL (2018) The molecularly imprinted polymer essentials: curation of anticancer, ophthalmic, and projected gene therapy drug delivery systems. J Control Release 287:24–34
100. Cunliffe D, Kirby A, Alexander C (2005) Molecularly imprinted drug delivery systems. Adv Drug Deliv Rev 57(12):1836–1853
101. Dargie W, Poellabauer C (2010) Fundamentals of wireless sensor networks: theory and practice. John Wiley & Sons
102. Liechty WB, Kryscio DR, Slaughter BV, Peppas NA (2010) Polymers for drug delivery systems. Ann Rev Chem Biomol Eng 1:149–173
103. Cristallini C, Gagliardi M, Barbani N, Giannessi D, Guerra GD (2012) Novel biodegradable, biomimetic and functionalised polymer scaffolds to prevent expansion of post-infarct left ventricular remodelling. J Mater Sci Mater Med 23:205–216

104. Martin-Esteban A (2001) Molecularly imprinted polymers: new molecular recognition materials for selective solid-phase extraction of organic compounds. Fresenius J Anal Chem 370:795–802
105. Cazares-Cortes E et al (2019) Recent insights in magnetic hyperthermia: from the "hot-spot" effect for local delivery to combined magneto-photo-thermia using magneto-plasmonic hybrids. Adv Drug Deliv Rev 138:233–246
106. Xie L, Xiao N, Li L, Xie X, Li Y (2020) Theoretical insight into the interaction between chloramphenicol and functional monomer (methacrylic acid) in molecularly imprinted polymers. Int J Mol Sci 21(11):4139
107. Bossi A, Bonini F, Turner A, Piletsky S (2007) Molecularly imprinted polymers for the recognition of proteins: the state of the art. Biosens Bioelectron 22(6):1131–1137
108. Sleisz Á, Raisz D (2017) Integrated mathematical model for uniform purchase prices on multi-zonal power exchanges. Electr Power Syst Res 147:10–21
109. Xu Y et al (2019) A novel controllable molecularly imprinted drug delivery system based on the photothermal effect of graphene oxide quantum dots. J Mater Sci 54(12:9124–9139
110. Dey A, Mukhopadhyay AK, Gangadharan S, Sinha MK, Basu D (2009) Weibull modulus of nano-hardness and elastic modulus of hydroxyapatite coating. J Mater Sci 44:4911–4918
111. Ahad A, Aqil M, Ali A (2014) Investigation of antihypertensive activity of carbopol valsartan transdermal gel containing 1, 8-cineole. Int J Biol Macromol 64:144–149
112. Ruela ALM, Figueiredo EC, Pereira GR (2014) Molecularly imprinted polymers as nicotine transdermal delivery systems. Chem Eng J 248:1–8
113. Bai J et al (2018) Synthesis and characterization of paclitaxel-imprinted microparticles for controlled release of an anticancer drug. Mater Sci Eng C 92:338–348
114. Pichayakorn W, Suksaeree J, Boonme P, Amnuaikit T, Taweepreda W, Ritthidej GC (2012) Nicotine transdermal patches using polymeric natural rubber as the matrix controlling system: effect of polymer and plasticizer blends. J Membr Sci 411:81–90
115. Pastore MN, Kalia YN, Horstmann M, Roberts MS (2015) Transdermal patches: history, development and pharmacology. Br J Pharmacol 172(9):2179–2209
116. Ruela ALM, de Figueiredo EC, Carvalho FC, de Araújo MB, Pereira GR (2018) Adsorption and release of nicotine from imprinted particles synthesised by precipitation polymerisation: optimising transdermal formulations. Eur Polym J 100:67–76
117. Olivier J-C, Rabouan S, Couet W (2003) In vitro comparative studies of two marketed transdermal nicotine delivery systems: Nicopatch® and Nicorette®. Int J Pharm 252(1-2):133–140
118. He W, Kapate N, Shields CW IV, Mitragotri S (2020) Drug delivery to macrophages: a review of targeting drugs and drug carriers to macrophages for inflammatory diseases. Adv Drug Deliv Rev 165:15–40
119. Yang Y et al (2011) Enzyme-responsive multifunctional magnetic nanoparticles for tumor intracellular drug delivery and imaging. Chem Asian J 6(6):1381–1389
120. Wang C et al (2021) Endogenous tumor microenvironment-responsive multifunctional nanoplatforms for precision cancer theranostics. Coord Chem Rev 426:213529
121. Zhang Y, Cui H, Zhang R, Zhang H, Huang W (2021) Nanoparticulation of prodrug into medicines for cancer therapy. Adv Sci 8(18):2101454
122. Cegłowski M, Jerca VV, Jerca FA, Hoogenboom R (2020) Reduction-responsive molecularly imprinted poly (2-isopropenyl-2-oxazoline) for controlled release of anticancer agents. Pharmaceutics 12(6):506
123. Cazares-Cortes E, Nerantzaki M, Fresnais J, Wilhelm C, Griffete N, Ménager C (2018) Magnetic nanoparticles create hot spots in polymer matrix for controlled drug release. Nano 8(10):850
124. Polo-Corrales L, Rinaldi C (2012) Monitoring iron oxide nanoparticle surface temperature in an alternating magnetic field using thermoresponsive fluorescent polymers. J Appl Phys 111(7)
125. Çakir P, Cutivet A, Resmini M, Bui BTS, Haupt K (2013) Protein-size molecularly imprinted polymer nanogels as synthetic antibodies, by localized polymerization with multi-initiators. Adv Mater 25(7):1048–1051

126. Hassanpour A, Irandoust M, Soleimani E, Zhaleh H (2019) Increasing the anticancer activity of azidothymidine toward the breast cancer via rational design of magnetic drug carrier based on molecular imprinting technology. Mater Sci Eng C 103:109771
127. Zaidi SA (2016) Latest trends in molecular imprinted polymer based drug delivery systems. RSC Adv 6(91):88807–88819
128. Luliński P (2017) Molecularly imprinted polymers based drug delivery devices: a way to application in modern pharmacotherapy. A review. Mater Sci Eng C 76:1344–1353
129. Xu J, Miao H, Wang J, Pan G (2020) Molecularly imprinted synthetic antibodies: from chemical design to biomedical applications. Small 16(27):1906644
130. Teixeira SP, Domingues RM, Shevchuk M, Gomes ME, Peppas NA, Reis RL (2020) Biomaterials for sequestration of growth factors and modulation of cell behavior. Adv Funct Mater 30(44):1909011
131. Shome A, Mugisho OO, Niederer RL, Rupenthal ID (2021) Blocking the inflammasome: a novel approach to treat uveitis. Drug Discov Today 26(12):2839–2857
132. Li S, Cao S, Whitcombe MJ, Piletsky SA (2014) Size matters: challenges in imprinting macromolecules. Prog Polym Sci 39(1):145–163
133. Kryscio DR, Peppas NA (2012) Critical review and perspective of macromolecularly imprinted polymers. Acta Biomater 8(2):461–473
134. Peltomaa R, Benito-Peña E, Moreno-Bondi MC (2018) Bioinspired recognition elements for mycotoxin sensors. Anal Bioanal Chem 410:747–771
135. Zidarič T, Finšgar M, Maver U, Maver T (2022) Artificial biomimetic electrochemical assemblies. Biosensors 12(1):44
136. Bergmann NM, Peppas NA (2008) Molecularly imprinted polymers with specific recognition for macromolecules and proteins. Prog Polym Sci 33(3):271–288
137. Mouw JK, Ou G, Weaver VM (2014) Extracellular matrix assembly: a multiscale deconstruction. Nat Rev Mol Cell Biol 15(12):771–785
138. Kim TG, Shin H, Lim DW (2012) Biomimetic scaffolds for tissue engineering. Adv Funct Mater 22(12):2446–2468
139. Foroutan Koudehi M, Zibaseresht R (2020) Synthesis of molecularly imprinted polymer nanoparticles containing gentamicin drug as wound dressing based polyvinyl alcohol/gelatin nanofiber. Mater Technol 35(1):21–30
140. Rosellini E et al (2020) Development of biomimetic alginate/gelatin/elastin sponges with recognition properties toward bioactive peptides for cardiac tissue engineering. Biomimetics 5(4):67
141. Jeon H, Kim G (2012) Effects of a cell-imprinted poly (dimethylsiloxane) surface on the cellular activities of MG63 osteoblast-like cells: preparation of a patterned surface, surface characterization, and bone mineralization. Langmuir 28(37):13423–13430
142. Criscenti G et al (2018) Soft-molecular imprinted electrospun scaffolds to mimic specific biological tissues. Biofabrication 10(4):045005
143. Rosellini E, Barbani N, Giusti P, Ciardelli G, Cristallini C (2010) Novel bioactive scaffolds with fibronectin recognition nanosites based on molecular imprinting technology. J Appl Polym Sci 118(6):3236–3244
144. Pan G, Guo Q, Ma Y, Yang H, Li B (2013) Thermo-responsive hydrogel layers imprinted with RGDS peptide: a system for harvesting cell sheets. Angew Chem Int Ed 27(52):6907–6911
145. Pan G et al (2017) An epitope-imprinted biointerface with dynamic bioactivity for modulating cell–biomaterial interactions. Angew Chem Int Ed 56(50):15959–15963
146. Piletsky S, Canfarotta F, Poma A, Bossi AM, Piletsky S (2020) Molecularly imprinted polymers for cell recognition. Trends Biotechnol 38(4):368–387
147. Moosazadeh Moghaddam M, Bonakdar S, Shokrgozar MA, Zaminy A, Vali H, Faghihi S (2019) Engineered substrates with imprinted cell-like topographies induce direct differentiation of adipose-derived mesenchymal stem cells into Schwann cells. Artif Cells Nanomed Biotechnol 47(1):1022–1035
148. DePorter SM, Lui I, McNaughton BR (2012) Programmed cell adhesion and growth on cell-imprinted polyacrylamide hydrogels. Soft Matter 8(40):10403–10408

149. Mahmoudi M et al (2013) Cell-imprinted substrates direct the fate of stem cells. ACS Nano 7(10):8379–8384
150. Mashinchian O et al (2014) Cell-imprinted substrates act as an artificial niche for skin regeneration. ACS Appl Mater Interfaces 6(15):13280–13292
151. Bonakdar S et al (2016) Cell-imprinted substrates modulate differentiation, redifferentiation, and transdifferentiation. ACS Appl Mater Interfaces 8(22):13777–13784
152. Saltzman WM, Olbricht WL (2002) Building drug delivery into tissue engineering design. Nat Rev Drug Discov 1(3):177–186
153. Mao C et al (2017) The controlled drug release by pH-sensitive molecularly imprinted nanospheres for enhanced antibacterial activity. Mater Sci Eng C 77:84–91
154. Naik SS, Anusha G, Leela KV, Ravi S (2023) Promising approaches in drug delivery against resistant bacteria. Advances in novel formulations for drug delivery, pp. 219–229
155. Gu Z, Dong Y, Xu S, Wang L, Liu Z (2021) Molecularly imprinted polymer-based smart prodrug delivery system for specific targeting, prolonged retention, and tumor microenvironment-triggered release. Angew Chem 133(5):2695–2699
156. Culver HR, Clegg JR, Peppas NA (2017) Analyte-responsive hydrogels: intelligent materials for biosensing and drug delivery. Acc Chem Res 50(2):170–178
157. Kurczewska J, Cegłowski M, Pecyna P, Ratajczak M, Gajęcka M, Schroeder G (2017) Molecularly imprinted polymer as drug delivery carrier in alginate dressing. Mater Lett 201:46–49
158. Aguilar-Colomer A et al (2021) Impact of the antibiotic-cargo from MSNs on gram-positive and gram-negative bacterial biofilms. Microporous Mesoporous Mater 311:110681
159. Cheng J, Jun Y, Qin J, Lee S-H (2017) Electrospinning versus microfluidic spinning of functional fibers for biomedical applications. Biomaterials 114:121–143
160. Li Q et al (2017) In situ sequestration of endogenous PDGF-BB with an ECM-mimetic sponge for accelerated wound healing. Biomaterials 148:54–68
161. Grier WK, Tiffany AS, Ramsey MD, Harley BA (2018) Incorporating β-cyclodextrin into collagen scaffolds to sequester growth factors and modulate mesenchymal stem cell activity. Acta Biomater 76:116–125
162. Lee SS et al (2015) Gel scaffolds of BMP-2-binding peptide amphiphile nanofibers for spinal arthrodesis. Adv Healthc Mater 4(1):131–141
163. Teixeira SP et al (2021) Epitope-imprinted nanoparticles as transforming growth factor-β3 sequestering ligands to modulate stem cell fate. Adv Funct Mater 31(4):2003934
164. Chen F-M, Zhang M, Wu Z-F (2010) Toward delivery of multiple growth factors in tissue engineering. Biomaterials 31(24):6279–6308
165. Lord MS, Berret JF, Singh S, Vinu A, Karakoti AS (2021) Redox active cerium oxide nanoparticles: current status and burning issues. Small 17(51):2102342
166. Ostovan A et al (2022) Greenificated molecularly imprinted materials for advanced applications. Adv Mater 34(42):2203154
167. Mamo SK, Elie M, Baron MG, Gonzalez-Rodriguez J (2020) Computationally designed perrhenate ion imprinted polymers for selective trapping of rhenium ions. ACS Appl Polym Mater 2(8):3135–3147
168. Kumar G, Sharma S, Shafiq N, Pandhi P, Khuller GK, Malhotra S (2011) Pharmacokinetics and tissue distribution studies of orally administered nanoparticles encapsulated ethionamide used as potential drug delivery system in management of multi-drug resistant tuberculosis. Drug Deliv 18(1):65–73
169. Bacakova L, Filova E, Parizek M, Ruml T, Svorcik V (2011) Modulation of cell adhesion, proliferation and differentiation on materials designed for body implants. Biotechnol Adv 29(6):739–767
170. Li J, Jiang X, Li H, Gelinsky M, Gu Z (2021) Tailoring materials for modulation of macrophage fate. Adv Mater 33(12):2004172
171. Przekora A (2019) The summary of the most important cell-biomaterial interactions that need to be considered during in vitro biocompatibility testing of bone scaffolds for tissue engineering applications. Mater Sci Eng C 97:1036–1051

172. Neves MI, Wechsler ME, Gomes ME, Reis RL, Granja PL, Peppas NA (2017) Molecularly imprinted intelligent scaffolds for tissue engineering applications. Tissue Eng Part B Rev 23(1):27–43
173. Kathuria N, Tripathi A, Kar KK, Kumar A (2009) Synthesis and characterization of elastic and macroporous chitosan–gelatin cryogels for tissue engineering. Acta Biomater 5(1):406–418
174. Das PS, Saha P (2017) Contact lenses: a development towards ocular drug delivery system. World J Pharm Res 6:207–216
175. Hilt JZ, Byrne ME (2004) Configurational biomimesis in drug delivery: molecular imprinting of biologically significant molecules. Adv Drug Deliv Rev 56(11):1599–1620
176. Allender C, Brain K, Heard C (1999) Progress in medicinal chemistry. Prog Med Chem 36:235–291
177. Konstas AG et al (2010) Twenty-four-hour intraocular pressure control with bimatoprost and the bimatoprost/timolol fixed combination administered in the morning, or evening in exfoliative glaucoma. Br J Ophthalmol 94(2):209–213
178. Kumar S, Jain S, Nehra M, Dilbaghi N, Marrazza G, Kim K-H (2020) Green synthesis of metal–organic frameworks: a state-of-the-art review of potential environmental and medical applications. Coord Chem Rev 420:213407
179. Yoo J, Park C, Yi G, Lee D, Koo H (2019) Active targeting strategies using biological ligands for nanoparticle drug delivery systems. Cancers 11(5):640
180. Liu M et al (2020) Combinatorial entropy behaviour leads to range selective binding in ligand-receptor interactions. Nat Commun 11(1):4836
181. Graves DB (2012) The emerging role of reactive oxygen and nitrogen species in redox biology and some implications for plasma applications to medicine and biology. J Phys D Appl Phys 45(26):263001
182. Ahmadi E, Harirchi P, Zahedi P, Bakhshandeh B, Zolfagharian S, Khatibi A (2023) Performance evaluation of poly (ε-caprolactone)/cellulose acetate nanofibrous scaffold containing graphene oxide used for neural tissue regeneration potentially. Fibers Polymers 24:2293–2303
183. Khatibi A, Zahedi P, Ghourchian H, Lari AS (2021) Development of microfluidic-based cellulose acetate phthalate nanoparticles containing omeprazole for antiulcer activity: in vitro and in vivo evaluations. Eur Polym J 147:110294
184. Carballido L, Karbowiak T, Cayot P, Gerometta M, Sok N, Bou-Maroun E (2022) Applications of molecularly imprinted polymers and perspectives for their use as food quality trackers. Chem 8:2330–2341
185. Kashani SY et al (2021) An integrated microfluidic device for stem cell differentiation based on cell-imprinted substrate designed for cartilage regeneration in a rabbit model. Mater Sci Eng C 121:111794
186. Demir B et al (2018) Tracking hyaluronan: molecularly imprinted polymer coated carbon dots for cancer cell targeting and imaging. ACS Appl Mater Interfaces 10(4):3305–3313
187. Panagiotopoulou M, Kunath S, Medina-Rangel PX, Haupt K, Bui BTS (2017) Fluorescent molecularly imprinted polymers as plastic antibodies for selective labeling and imaging of hyaluronan and sialic acid on fixed and living cells. Biosens Bioelectron 88:85–93
188. Sayes CM, Marchione AA, Reed KL, Warheit DB (2007) Comparative pulmonary toxicity assessments of C60 water suspensions in rats: few differences in fullerene toxicity in vivo in contrast to in vitro profiles. Nano Lett 7(8):2399–2406
189. Yin D, Li X, Ma Y, Liu Z (2017) Targeted cancer imaging and photothermal therapy via monosaccharide-imprinted gold nanorods. Chem Commun 53(50):6716–6719
190. Lakshmi G, Yadav AK, Mehlawat N, Jalandra R, Solanki PR, Kumar A (2021) Gut microbiota derived trimethylamine N-oxide (TMAO) detection through molecularly imprinted polymer based sensor. Sci Rep 11(1):1338

Importance of MIPs as Artificial Antibodies in Drug Delivery and Tissue Engineering

16

Sepehr Izadi, Hooman Pourbala,
Mohammad Mahdi Feizollahi, Kamran Banan,
Bahar Mostafiz, Mohammad-Ali Shahbazi,
and Fatemeh Ghorbani-Bidkorpeh

Abstract

Molecularly imprinted polymers (MIPs) are novel polymeric networks that exhibit valuable pharmaceutical properties, making them a promising tool for drug delivery systems (DDSs). These unique polymers offer an appropriate approach to increasing the bioavailability of drugs and protecting target molecules under harsh conditions. The double-imprinting technique is a convenient method to achieve favorable responses for targeted delivery and can reduce the undesired effects of drugs.

Numerous studies have utilized nanotechnology to design controlled-release systems. MIPs are effective reservoirs that optimize the release and bioavailability of medications. As stimuli-responsive DDSs, MIPs are widely used to achieve the precise release of target molecules caused by endo- or exogenous stimuli. Thus, the primary focus of this chapter is on implementing novel MIP-based DDSs and their main features. Moreover, MIPs have been utilized as theranostic agents to understand better how interventions impact the site of action and monitor the drug delivery process. The theranostic MIPs provide a valuable tool for

S. Izadi · H. Pourbala · M. M. Feizollahi · K. Banan · F. Ghorbani-Bidkorpeh (✉)
Department of Pharmaceutics and Pharmaceutical Nanotechnology, School of Pharmacy, Shahid Beheshti University of Medical Sciences, Tehran, Iran
e-mail: f.ghorbani@sbmu.ac.ir

B. Mostafiz
Department of Mechanical and Materials Engineering, University of Turku, Turku, Finland

M.-A. Shahbazi
Department of Biomedical Engineering, University Medical Center Groningen, University of Groningen, Groningen, The Netherlands

W.J. Kolff Institute for Biomedical Engineering and Materials Science, University of Groningen, Groningen, The Netherlands

S. Patra et al. (eds.), *Molecularly Imprinted Polymers: Path to Artificial Antibodies*, https://doi.org/10.1007/978-981-97-4379-7_16

achieving precise and real-time monitoring of drug release, enabling researchers to optimize drug delivery systems and improve patient outcomes.

In this chapter, we also aim to investigate the features of MIPs that aid tissue engineering by enabling the production of scaffolds with controlled release of growth factors or other active substances, cell binding ability, and the capability to contribute to cell migration and proliferation. MIPs are introduced as suitable alternatives to neutral antibodies due to their biocompatibility, lower toxicity, and ability to imprint a wide range of templates, including ions, small molecules, peptides, viruses, bacteria, and even cells. In addition, the simplicity of production, lower immunogenicity, ability to sustain release for an extended period, endurance of harsh conditions, and lower costs make these artificial antibodies suitable for enhancing drug delivery systems. Despite the exciting and unique features of MIPs mentioned above, further research is needed to optimize previous synthesis methods and propose new ones, especially for imprinting larger molecules like peptides, to determine long-term safety and compatibility with biological systems.

Keywords

Molecularly imprinted polymers (MIPs) · Artificial antibody · Drug delivery

Abbreviations

AMF	Alternative magnetic field
BAC	*N,N′*-bis(acryloyl) cystamine
BBB	Blood–brain barrier
CA 125	Cancer antigen 125
Ce6	Chlorin e6
CEA	Carcinoembryonic antigen
DDS	Drug delivery system
DFCR	Deoxy-5-fluorocytidine
DOX	Doxorubicin
DTT	Dithiothreitol
ECM	Extracellular matrix
EGFR	Epidermal growth factor receptor
Glu-FH	Fibroblast growth factor-inducible 14 modified with glucose
GSH	Glutathione
HER2	Human epidermal growth factor receptor-2
HKUST-1	Hong Kong University of Science and Technology
H. pylori	*Helicobacter pylori*
LC	Liquid crystalline
MIL-101	Matérial Institut Lavoisier
MIP	Molecularly imprinted polymer
MMP	Magnetic molecularly imprinted polymer

MOF	Metal–organic framework
MSN	Mesoporous silica nanoparticles
$NaBH_4$	Sodium borohydride
NIP	Nonimprinted polymer
PMIP	Photo-responsive molecularly imprinted polymer
RNase A	Ribonuclease A
ROS	Reactive oxygen species
SA	Sialic acid
S. aureus	*Staphylococcus aureus*
TF	Transferrin
UiO-66	*Universitetet i Oslo*
UV	Ultraviolet
ZIF-8	Zeolitic Imidazolate Framework
5-FU	5-fluorouracil

16.1 Introduction

Conventional therapeutics have made significant progress in treating various diseases over the years. However, with the constantly evolving needs of people today, there may be several limitations to what traditional treatments can provide. These challenges include limited efficacy, lack of specificity, poor solubility or stability, drug resistance, and potential side effects [1–5]. The human body possesses several physiological barriers that can impede the delivery of drugs, including the BBB and the gastrointestinal barrier [3, 4, 6–8]. Furthermore, certain drugs require sustained or controlled release over an extended period to maintain therapeutic levels. Also, each patient may have unique physiological characteristics or treatment requirements due to genetic variations, lifestyle factors, or co-existing conditions. Therefore, exploring innovative and personalized approaches to overcome these challenges is essential [9, 10].

MIPs are highly versatile and customizable synthetic polymers with various applications, attributed to their low toxicity, high biocompatibility, and cost-effective production techniques. These polymers have been extensively studied in various fields, such as sensors, chromatography, immunoassays, and catalysis [11–14].

In recent years, extensive research has been done to enhance the efficacy and safety profile of DDSs. By utilizing MIPs, researchers can develop specific binding sites to enhance loading capacities and adjust release profiles. This capacity of MIPs enables the development of novel DDSs and various methods of administration [14]. For instance, ocular MIPs have been employed in treating glaucoma by reducing the frequency of medication intake [3].

MIPs were developed to treat various diseases, such as cancers, infections, and inflammations. Numerous studies have revealed the potential of MIPs to enhance the recognition or delivery of specific molecules. This ability can lead to early detection, enhanced therapeutic outcomes, and reduced adverse events. Furthermore, MIPs have shown great promise in the diagnosis and treatment of various ailments, such as cancers, infections, and inflammations [3, 15–18].

Researchers have used MIPs as targeted DDS to transport drugs to specific areas of the body. These structures can be stimulant-responsive, dual-imprinted, enantioselective, and theranostic. Furthermore, there has been extensive research on enantioselective MIPs, with propranolol being one of the first templates in this category. This field aims to deliver specific enantiomers with enhanced therapeutic efficacy and reduced side effects [19].

Using MIPs as DDSs began with oral drugs in 1998 and contact lenses in 2002. Enantioselective systems were introduced in 2000 with MIP-based systems. The first in vivo experiments were conducted in 2005. In 2006, the first light-responsive systems used caffeine as the template. The evaluation of transdermal MIPs began in 2008, and injectable forms have been examined over time. Over the years, different systems have been studied, such as those that respond to electricity, magnetism, pH, and heat. Furthermore, multiresponsive systems were also investigated in 2016. Figure 16.1 shows the development and evolution of MIPs [15–17, 19–23].

After utilizing MIPs as a targeted DDS in 2015, a significant advancement occurred in 2019 with the introduction of multifunctional MIPs that integrate imaging, dual imprinting, MOF-based systems, and theranostic capabilities [18, 24].

Tissue engineering combines material engineering and biochemical agents to create new functional tissues to replace or repair damaged body parts. In recent years, this field has attracted the attention of many researchers. Tissue engineering uses a unique scaffold, a three-dimensional environment that provides a favorable setting for cell communications. This environment facilitates cell adhesion, migration, proliferation, and differentiation, which is crucial for regenerating new functional tissues successfully [25–27].

This chapter delves into the diverse applications of MIPs in DDSs and emphasizes their effectiveness as artificial antibodies compared to their natural counterparts. Furthermore, we also focused on the MIPs that have the potential to be utilized

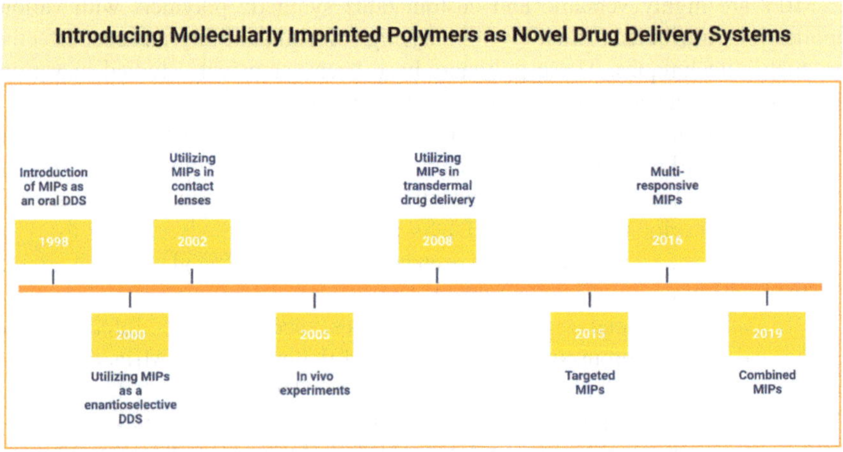

Fig. 16.1 An overview of progress in molecular imprinting technology

in tissue engineering through efficient drug delivery. We aim to fully evaluate how MIPs are used in these fields, meeting academic standards.

16.2 Responsive

Stimuli-responsive MIPs can react to stimuli like electric or magnetic interference, pH, photo irradiation, and redox environment, as shown in Fig. 16.2. The advantages of stimuli-responsive MIPs are enhancing targeted drug delivery and reducing side effects. Each stimuli-responsive MIP is explained in the following sections.

16.2.1 Magnetic-Responsive

Magnetic nanoparticles have garnered significant attention for their potential to be employed in magnetic targeting and magnetic hyperthermia applications. This capability arises from their ability to generate heat when subjected to an AMF without any limit to penetration depth. To use heat in this context, we need numerous nanoparticles capable of generating heat when exposed to a magnetic field. However, achieving this aim at a macroscopic level through systemic delivery is quite unrealistic. Consequently, new drug delivery strategies have recently been proposed [28].

MMIPs are designed to release drug molecules by utilizing an external magnetic field to a specific location. The MMIPs possess a super-paramagnetic iron oxide core, allowing them to navigate the body and gather in specific areas. Coating the magnetic core with drug molecules improves loading and release efficiency, preventing the nanoparticles from agglomeration [29, 30].

Conventional DDSs, especially those used in cancer chemotherapy, suffer from significant drawbacks. They lack selectivity, resulting in poor therapeutic outcomes, systemic toxicity, and adverse side effects. To address these challenges, researchers have been developing advanced systems that can accurately deliver medication at the right time and place [28–30].

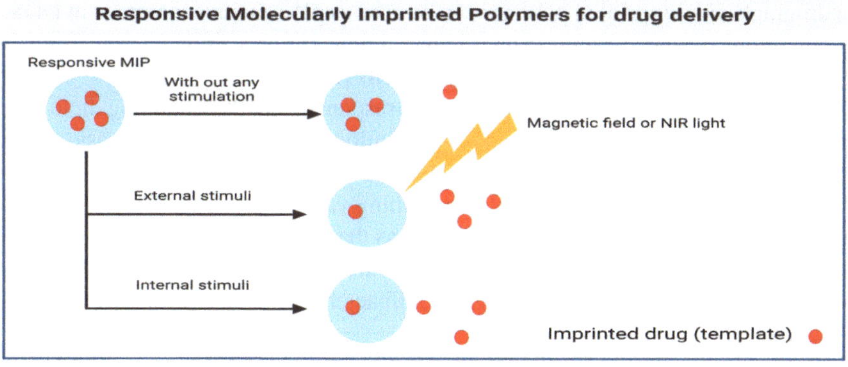

Fig. 16.2 Responsive MIPs

MMIPs have been developed using several templates, primarily in cancer treatment, including DOX, mitomycin, carbazole, 5-FU, and quercetin. Additionally, rivastigmine has been evaluated as a promising template target for Alzheimer's disease [28–32].

For instance, a novel MMIP was developed by utilizing Fe_2O_3 as a magnetic core and coating it with a MIP utilizing DOX as a template. This functional system resulted in the increased release of DOX in a magnetic field [28]. The magnetic carrier used in a study by Shadabfar et al. demonstrated a significant 61% cytotoxicity against cancer cells. This cytotoxicity was approximately 25 times greater than that observed with the untargeted approach. Additionally, it had no significant effect on normal cells [33].

In certain studies, magnetic nanoparticles have only been employed for separation rather than regulating drug release. MMIPs offer the advantage of easy and rapid separation during their preparation and synthesis, saving material and time that is usually lost in centrifugation or filtration processes [34].

Nonetheless, DDSs based on MMIPs can result in temperature elevation. While increasing the temperature, the AMF can disrupt the magnetic interactions between the drug molecules and the magnetic core, leading to drug release [28]. Overall, MMIPs represent a significant advancement in DDSs, offering precise and targeted drug release to improve treatment outcomes and minimize adverse effects [28].

16.2.2 pH-Responsive

Recently, pH-responsive MIPs have acquired attention for their specific drug release in certain pH conditions, like acidic tumor environments, colon drug delivery, and antimicrobial applications [35–37].

The tumor tissues are more acidic than normal human tissues, which have a pH of 7.4 [35]. Hence, using pH-responsive DDSs can reduce side effects on normal tissues and increase drug uptake by tumor cells, thus improving therapeutic effectiveness [38]. The swelling and solubility of MIPs are influenced by the pH level, ultimately impacting the release rate of drugs. The polymers used in magnetic andrographolide-imprinted polymer can be used as pH-sensitive materials in DDSs. Furthermore, the acidic tumor microenvironment triggers drug release from the polymer matrix and its leakage into cancer cells [39].

A study by Mohajeri et al. found that pH-responsive MIPs containing diclofenac exhibited a release rate of 14% in acidic solutions and up to 90% in basic conditions. These findings may impact the advancement of oral drug delivery [40]. Also, the synthesized pH-responsive MIPs containing vancomycin showed faster release and higher concentrations in acidic conditions than in neutral conditions, making it a promising option for suppressing bacterial infections. Additionally, the proposed MIP demonstrated significantly higher antibacterial efficacy (over 90%) against *S. aureus* [35].

Researchers have investigated oral DDSs to ensure that drugs are delivered to the colon with the appropriate concentration. Unfortunately, conventional methods of drug delivery often result in the drug being released in the gastric instead of reaching its desired target, leading to insufficient concentration in the colon. To tackle this problem, Men et al. created a pH-responsive MIP for delivering metronidazole to the colon. Therefore, with the help of these MIPs, the drug release rate can be effectively controlled, ensuring targeted release in the colon where it is needed [24].

Bacterial infections can lower the pH of wounds, making them more acidic. This acidic environment triggers the release of antibiotics from MIPs that specifically target bacterial growth [41].

16.2.3 Photo-Responsive

PMIPs have become popular lately due to their numerous advantages, including green synthesis, safety, ease of use, controllability, and high efficiency. The light as an actuator can control drug release. Moreover, the PMIPs have various applications, including catalysts, pollutant monitoring, phase extraction, and cancer drug delivery [42, 43].

Previous studies on photo-responsive MIPs utilized azobenzene functional monomers to convert *cis-* to *trans-optical* isomers under UV and visible light exposure. To overcome the tissue penetration of UV light, Liu et al. replaced UV with near-infrared light (NIR), providing suitable tissue permeability and reducing phototoxicity [20, 44].

The importance of targeted delivery, especially in cancer treatment, has increased because of the benefits described earlier. In this regard, special attention has been paid to photo-chemodynamic therapy. Photo-chemodynamics is a new treatment method in which fluorescent silica nanoparticles are used. Peng et al. used MIPs in the outer layer, along with fluorescent silica nanoparticles core and photosensitizer Ce6. When exposed to visible light, these materials generate ROS near cancer cells, ultimately destroying them. Additionally, using a dual treatment method, DOX and ROS resulted in better outcomes [43, 45].

16.2.4 Redox-Responsive

The redox-responsive materials are developed to release drugs in response to GSH, providing a targeted and controlled release mechanism that exploits the higher reductive stress levels in extracellular rather than intracellular environments [46]. Cross-linkers with disulfide bonds, such as BAC, are required for redox-responsive MIPs. In order to control drug release at the desired location with the correct dosage, chemical bonds must be broken or weakened in the presence of reducing agents, such as $NaBH_4$, DTT, GSH, and others [47].

In a reductive environment, disulfide bonds can be converted into thiol groups. This process results in the degradation of the carrier, and as a result, the encapsulated molecules would be released. This material degradation also aids in the removal of carriers after drug release. The reductive reactions of disulfide bonds rely on the presence of GSH, which is present in much higher concentrations within cells (1–10 mM) than in the extracellular environment (20–40 μM). Therefore, these reactions can only occur after the cells have internalized the materials [47].

Disulfide-based materials are more stable delivery systems outside the cell than pH-responsive or hydrolyzable polymers, which can quickly degrade in the blood circulation or ECM. Furthermore, disulfide-based materials can develop anticancer DDSs due to the higher GSH concentration in the tumor microenvironment [42, 47].

The study by Cegłowski et al. showed that a pH-redox-responsive MIP was a better option than a pH-responsive MIP for cancer treatment. Utilizing polymers that are cross-linkers with disulfide bonds increased the release of 5-FU in a redox environment. The release of 5-FU was observed to be 60% without the triggering agent (GSH). However, when GSH was added, the release increased by approximately 78% [46].

16.3 Targeted Drug Delivery

Over the past decade, many studies have investigated the potential advantages of utilizing MIPs as targeted DDSs. Scientific research facilitates the improvement of medication efficacy by focusing on precise delivery methods. This approach achieves more substantial therapeutic outcomes and reduces potential side effects.

As shown in Fig. 16.3, most MIPs designed for cancer therapy have focused on utilizing a tumor biomarker as a template. These MIPs can effectively load an anticancer drug, most commonly DOX. While certain MIPs were designed with a tumor biomarker as a template and loaded DOX without imprinting it [43, 48–55], others employed a dual-imprinted MIP with DOX as a secondary template [24, 45, 56, 57]. Some studies have also examined other antitumor medications like RNase A, DFCR, and bleomycin as templates [58–60].

The epitope imprinting method utilizes larger molecules, like protein tumor biomarker templates. Some examples include HER2, C125, P32, Glu-FH, CD59, CEA,

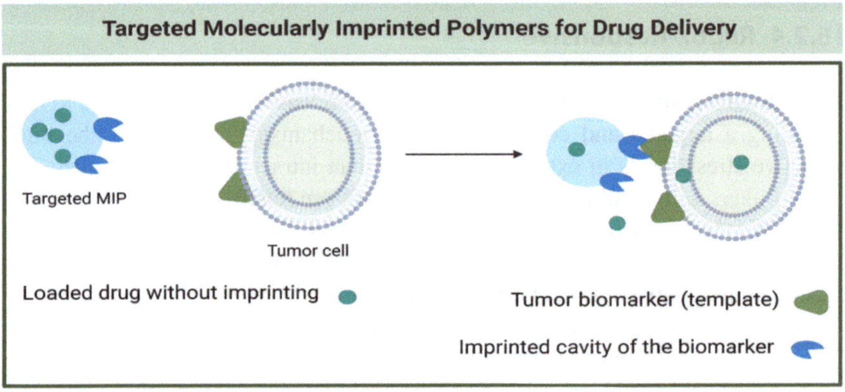

Fig. 16.3 Targeted MIPs

CD47, EGFR, and TF [24, 57]. This method enables direct printing from the exposed peptide sequence to reduce costs and improve efficiency [24].

Tumor biomarker imprinting began with the HER2 receptor, which is overexpressed in several malignancies and is often linked to a poor prognosis. Several anti-HER2 monoclonal antibodies, such as trastuzumab, are available [24, 57]. These antibodies may help compare the cost-effectiveness of the designed MIPs as synthetic antibodies with natural ones.

Two studies used dual imprinting MIPs using DOX and HER2 epitopes as the template. In vitro studies conducted by Wang et al. and Hashemi-Moghaddam et al. demonstrated a controlled release of DOX. Wang et al. proposed a pH-responsive release mechanism, allowing them for targeted drug release in acidic tumors. The in vivo study conducted by Hashemi-Moghaddam showed a significant decrease in tumor volume. The tumor tissue had higher levels of DOX than the control group, while the other organs had the lowest drug concentrations [24, 57].

SA is a glycoprotein side chain widely researched as a potential tumor biomarker for use as a template in targeted MIPs. It is overexpressed on cell membranes in certain pathological conditions, particularly cancer. This tumor biomarker is relatively small compared to protein biomarkers [49–52, 58, 59]. Additionally, Sarcosine is also a small-sized tumor biomarker that has been effectively used to deliver DOX to prostate tumors [48].

Many research studies have invested significant efforts in pursuing precise tumor treatment using SA. Most of these studies have loaded DOX as an anticancer drug. Liu et al. [49] and Han et al. [51] used a different method involving Ce6 to generate toxic ROS under laser irradiation for chemical/photodynamic tumor treatment. Liu et al. achieved a significant breakthrough in cancer treatment by utilizing TF and SA as templates for more accurate drug delivery. This study is the first to combine a dual-targeted MIP with chemical/photodynamic therapy.

Using MIPs for delivering therapeutic proteins in cancer treatment is a relatively new and promising method. For instance, Lu et al. employed an intracellular DDS to load RNase A, a therapeutic protein interrupting cell proliferation by cleaving cellular RNA. They successfully investigated the cellular uptake of the proposed MIP. The in vivo study provided further evidence, showing that the designed DDS is both safe and effective when compared to the control group and NIP [58].

Gu et al. suggested a dual imprinted MIP for targeted prodrug delivery to SA-dependent tumors. Their method for prodrug delivery, unlike conventional methods, was liver-independent. Instead, a two-step enzymatic reaction was done to convert the DFCR into 5-FU, an anticancer medication. Attempts like this, in combination with tumor-targeted MIPs, may result in more specific drug delivery toward tumor cells [59].

There is limited research on using MIPs for delivering antibiotics compared to anticancer drugs. A study on amoxicillin delivery through dual imprinting MIPs showed promising in vivo results for treating *H. pylori*. The details of this study can be found in the dual imprinting MIPs section [18].

To summarize, there is a growing demand for developing targeted MIPs, especially in cancer therapy. MIPs can impact various fields, from genetic disease treatment to drug delivery to vital organs.

16.4 Dual Imprinting

Dual imprinting MIPs use two templates in the printing process as shown in Fig. 16.4. These MIPs are commonly used in cancer therapy and antimicrobial treatments. They involve using a drug and a biomarker as a template to target specific tumor cells or bacteria [18, 24, 45, 49, 56, 59, 60].

The study by Topete et al. explored the effects of using an antibiotic and a non-steroidal anti-inflammatory drug (NSAID) together for post-cataract surgery medication. According to the study, the imprinting process using these templates did not affect the release profile compared to loading these drugs without imprinting them. However, it is speculated that this outcome may be attributed to an incorrect dosage of a cross-linker [61].

The study by Topete et al. is exceptional in dual imprinting MIPs because, in contrast to most other studies, it used multiple drugs as templates [61]. Other studies focused on single drugs, such as DOX [24, 45, 56], bleomycin [60], DFCR [59], and amoxicillin [18]. Furthermore, these studies have utilized tumor markers as another template for targeted drug delivery. These markers include HER2, the epitope of P32, Glu-FH, CD59, SA, and TF, or even a bacterium lipopolysaccharide [18, 24, 45, 49, 56, 59, 60].

Although most studies utilize one tumor biomarker and a drug as a template, Liu et al. used two tumor biomarkers as templates to deliver drugs specifically to tumor cells. Ce6 and DOX were loaded into MSN to inhibit tumor cell growth and ROS. The release of DOX was triggered by laser irradiation [61]. The core's

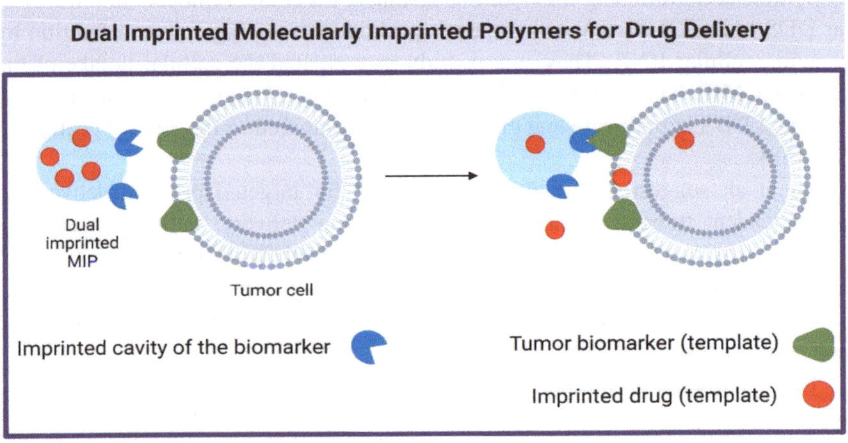

Fig. 16.4 Dual imprinted MIPs

porosity enables effective drug capture and release and offers a large surface area for interacting with their ligands. Therefore, it significantly reduces the risk of early drug leakage when loaded into the outer MIP layer.

To achieve GSH responsiveness, Liu et al. used functional monomers with a disulfide bond. This bond degrades in the tumor's redox environment [49]. Similarly, Peng et al. utilized Ce6 and DOX to produce oxygen radicals. However, they used DOX and the CD59 biomarker as the templates. In both in vivo experiments, the designed systems successfully reduced the tumor size and showed no toxic effects [45, 49].

Several studies have developed novel theragnostic MIPs for cancer therapy. They used fluorescent silicon nanoparticles as the emitting core, with an imprinting polymer cover. This approach enabled simultaneous treatment and diagnosis [24, 45, 56, 60]. Peng et al. utilized gadolinium-doped silicon quantum dots for tumor therapy and simultaneous fluorescent and MRI imaging [45].

Dual imprinted targeted MIPs have not been used against bacterial infections as much as anticancer therapy. However, it could be a promising approach because of the increased bacterial resistance to antibiotics [18]. Wu et al. used dual imprinting MIP for antibacterial therapy, using amoxicillin and *H. pylori* lipoprotein as templates. The study discovered that mice treated with the MIP exhibited no *H. pylori* colonies or urease, unlike those treated with amoxicillin [18].

Cui et al. took a unique approach to drug delivery using dual imprinting MIPs. They focused on patients with alpha-1 antitrypsin deficiency and aimed to reduce tissue damage by trypsin. A highly effective MIP was developed to inhibit trypsin. The designed MIP protected the basement membrane and extracellular matrix from damage. Besides trypsin, salicylic acid was also used as a template for its anti-inflammatory properties and trypsin inhibition. MIP's ability to inhibit enzymes and deliver drugs simultaneously could be a new treatment option for some hereditary diseases [62].

Dual imprinting MIPs are often used for targeted DDSs. Several studies have explored using them to release two drugs simultaneously to reduce the need for frequent medication consumption. Further studies are needed to explore using multiple templates and how their interactions affect release profiles.

16.5 Metal–Organic Frameworks

Metal–organic frameworks (MOFs) are porous materials made of metal clusters connected by organic ligands. Their adjustable structures, size, and loading capacities are ideal for controlled drug release [63]. Recently, remarkable attempts have been made to combine various MOFs, including ZIF-8, UIO-66, MIL-101, and HKUST-1, with MIPs to develop innovative DDSs [43, 48, 50, 53, 64–66]. Most MOF-based MIPs employed DOX as a drug template for cancer therapy. Figure 16.5 illustrates the interaction of MOF-based MIPs with a tumor biomarker.

Using MOFs as pH-responsive drug carriers for anticancer drugs has recently increased. However, this approach has some drawbacks, such as early drug leakage and toxicity to normal tissues. Researchers used the MOF core with a coated layer of MIP to address these issues to enhance adsorption capacity and prevent drug

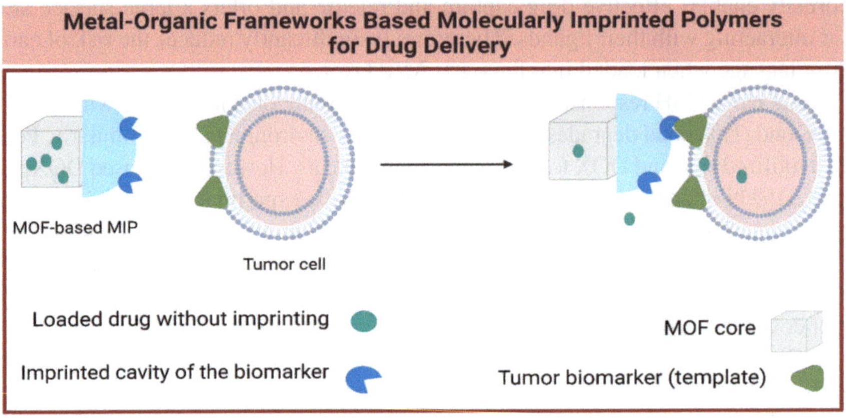

Fig. 16.5 MOF-based MIPs

leakage. This strategy also enhances specific binding site creation on the MIPs due to the large surface area of the MOF core [50].

MOF-based MIPs were used to create a DDS that is responsive to pH and GSH. The system utilizes the faster degradation of MOFs in mild acidic tumor environments and combines it with the GSH-responsive MIPs to achieve targeted drug release at the tumor site [48, 50, 53]. Qin et al. implemented a GSH-responsive DDS consisting of a fluorescent ZIF core for imaging purposes and controlled release of drugs to target cancer cells [53].

Zhang et al. developed a unique MOF-based MIP that improves the bioavailability of capecitabine compared to regular tablets. They used HKUST-1 as the MOF core and implemented LC monomers for MIPs to enhance floating properties. This study demonstrated a complete release of capecitabine from regular tablets within 7 h, while the LC monomer enhanced the controlled release characteristics of the drug [66].

Shi et al. created a unique MOF-based MIP to prove the necessity of the MIP coating layer. Their findings showed an impressive ability to release 28.03% and 46.19% of the loaded DOX at pH 7.4 and 5.5, respectively, within 72 h. This finding contrasts the uncoated MOF, MIL-101, which released 98% of the loaded DOX within 48 h at pH 7.4 [65].

The techniques used by Zhang et al. and Shi et al. may lead to premature drug leakage, hindering targeted drug delivery's effectiveness. To solve this problem, other researchers have taken a distinct approach compared to them, which focused on loading DOX into ZIF-8 or UiO-66 instead of MIP. They utilized a tumor biomarker as a template to create a MIP thin layer coat for their MOF core. This innovative approach enables precise drug delivery to target-specific tumor cells [43, 48, 50, 53]. For example, a study combined hyaluronic acid and ZIF-8 to increase the composite's hydrophilicity and circulation time. It also boosted targeted drug delivery through interaction with overexpressed CD44 receptors in tumor cells [48].

Overall, MOF-based MIPs can reduce early drug leakage and enhance surface area for the coating MIP layer, responsivity, and targeted drug delivery. Table 16.1

Table 16.1 The main features of responsive, targeted, dual imprinted, and MOF-based MIPs

Loaded drug	Monomers	Cross-linker	Template	Experiment	Indication	Route of administration	Response	Targeted	Dual imprinting	MOF core	Ther-anostics	Ref.
DOX	AMPS and VIM	BAC	Sarcosine	In vitro and in vivo	Anticancer	IV injection	pH, GSH, and hyaluronidase	Yes	No	ZIF-8	No	[48]
DOX and Ce6	Cys-DA and 4-MPBA	–	SA and TF	In vitro and in vivo	Anticancer	IV injection	pH, GSH, and light	Yes	Yes	No	No	[49]
Sorafenib	Allyl tryptophan and BAEDS	EGDMA	Sorafenib	In vitro	Anticancer	–	Magnet, pH, redox	No	No	No	No	[67]
Andrographolide	β-CD and AMAB	EGDMA and BAC	Andrographolide	In vitro	Anti-inflammatory, bacteriostatic, antiviral, immune regulation, and antitumor	–	pH, redox, magnet, and light	No	No	No	No	[42]
DOX	MMA	BIS	DOX	In vitro	Anticancer	–	Magnet and pH	No	No	No	No	[68]
4-aminopyridine	MAPDIA	EGDMA	4-aminopyridine	In vitro	Axonal demyelination	–	Light and temperature	No	No	No	No	[69]
Dox and Ce6	DA, 4-aminophenyl boronic acid, glycerol and lactide	–	SA	In vitro and in vivo	Anticancer	–	pH and light	Yes	No	No	No	[51]
DOX	MAM	BAC	SA	In vitro	Anticancer	–	pH and GSH	Yes	No	UiO-66	No	[50]
Formononetin	MAA	BAC	Formononetin	In vitro	Antioxidation, anti-atherosclerosis, osteoporosis, menopausal symptom and breast cancer	–	pH and redox	No	No	No	No	[70]
DOX	DA	–	EGFR epitope	In vitro	Anticancer	–	pH and light	Yes	No	ZIF-8	No	[43]
RNase A	APTES and BTEPDS	TEOS	SA	In vitro and in vivo	Anticancer	IV injection	GSH	Yes	No	No	Yes	[58]
CaO2	NIPAM, EG, and copper acrylate	BAC	CD47 epitope	In vitro	Anticancer	–	GSH, pH, and light	Yes	No	No	Yes	[71]
DOX	AAM, NIPAAM and NAPMAAm	BIS	SA	In vitro	Anticancer	–	pH	Yes	No	No	No	[52]

(continued)

Table 16.1 (continued)

Loaded drug	Monomers	Cross-linker	Template	Experiment	Indication	Route of administration	Response	Targeted	Dual imprinting	MOF core	Ther-anostics	Ref.
Amitriptyline hydrochloride	MAA	TRIM	Amitriptyline hydrochloride	In vitro	Depression and migraine		pH	No	No	No	No	[72]
DFCR	Boronic acids	–	SA and DFCR	In vitro and in vivo	Anticancer	IV injection	pH	Yes	Yes	No	No	[59]
PMX	DEAEMA and MAA	EDGMA	PMX	In vitro	Anticancer	–	Magnet and temperature	No	No	No	No	[73]
Moxifloxacin and diclofenac	AA or MAA or 4-VP or MAA + 4-VP	EGDMA	Moxifloxacin and diclofenac	In vitro and in vivo	Antibacterial and anti-inflammatory	Intraocular lens	–	No	Yes	No	No	[61]
Salicylic acid	NIPAAm	BIS	Trypsin, salicylic acid	In vitro	A1ATD	–	–	Yes	Yes	No	No	[62]
DOX	DA	–	CEA	In vitro	Anticancer	–	pH	Yes	No	No	No	[54]
5-FU	HEMA	EGDMA, TRIM	5-FU	In vitro	Anticancer	–	Redox	No	No	No	No	[46]
S-propranolol	MAA	DSDMA or EGDMA or BAC	S-propranolol	In vitro	–	–	Redox	No	No	No	No	[47]
DOX	HEMA	MBA	DOX	In vitro	Anticancer	–	pH	No	No	No	Yes	[74]
DOX and Ce6	NIPAm, AAM and TBAm	BIS	CD59 epitope and DOX	In vitro and in vivo	Anticancer	IP injection	Light	Yes	Yes	No	No	[45]
DOX	DMAEMA	BAC	CD59 epitope	In vitro and in vivo	Anticancer	IP injection	pH and GSH	Yes	No	ZIF-8	Yes	[53]
Aminoglutethimide	MAA	EGDMA	Aminoglutethimide	In vitro and in vivo	Anticancer	–	–	No	No	Metal organic gel	No	[64]
Paracetamol	MADPADSA	TEAMA	Paracetamol	In vitro	Analgesic	–	Light	No	No	No	No	[44]
DOX	APTES	TEOS	DOX	In vitro	Anticancer	–	pH	No	No	MIL-101	No	[65]
Bleomycin	ZnA, VPBA	EGDMA	Glu-FH epitope and bleomycin	In vitro and in vivo	Anticancer	IV injection	pH	Yes	Yes	No	Yes	[60]
DOX	NIPAAM, TBAm, TFMA	BIS	P32 epitope and DOX	In vitro and in vivo	Anticancer	SC injection	pH	Yes	Yes	No	Yes	[56]
Silybin	–	–	Silybin	In vitro	Hepatoprotective	–	pH and temperature	No	No	No	No	[75]
Ciprofloxacin	DA and CMC	EDC	Ciprofloxacin	In vitro	Anti-biotics	–	pH	No	No	No	No	[41]
5-FU	4-VP or AA	EGDMA	5-FU	In vitro	Anticancer	–	Magnet and pH	No	No	No	No	[76]

DOX	MMA, ZnA	EGDMA	HER2 epitope and DOX	In vitro	Anticancer	–	pH	Yes	Yes	No	Yes	[24]
Capecitabine	NIPAM or HEMA or 4-VP or GMA or AM	EGDMA	Capecitabine	In vitro and in vivo	Anticancer	Oral	Solvent-responsive	No	No	HKUST-1	No	[66]
DOX	DA	–	CA125	In vitro	Anticancer	–	pH	Yes	No	No	No	[55]
Dexamethasone sodium phosphate	AA and DPA	BIS	Dexamethasone sodium phosphate	In vitro	Anti-inflammatory	–	pH	No	No	No	No	[77]
DOX	AA, AAM	EGDMA	DOX	In vitro	Anticancer	–	Magnet	No	No	No	No	[78]
MTX	MAA	Divinylbenzene	MTX	In vitro	Anticancer	–	Magnet and temperature	No	No	No	No	[79]
Diclofenac	MAA	EGDMA	Diclofenac	In vitro	Anti-inflammatory	Topical	–	No	No	No	No	[80]
Metronidazole	Itaconic acid	EGDMA	Metronidazole	In vitro	Anti-biotics	Oral	pH	No	No	No	No	[81]
Acyclovir	ADDM	TEAMA	Acyclovir	In vitro	Antiviral	Transdermal	Light	No	No	No	No	[82]
Vancomycin	HEMA and DEAEMA	EGDMA	Vancomycin	In vitro	Anti-biotics	–	pH and light	No	No	No	No	[35]
DOX	DA	–	HER2 epitope and DOX	In vitro and in vivo	Anticancer	IV injection	–	Yes	Yes	No	No	[57]
DOX	HEMA	PEGDA	DOX	In vitro	Anticancer	–	pH	No	No	No	No	[38]
5-FU	Dopamine	TRIS	5-FU	In vitro and in vivo	Anticancer	–	Magnet	No	No	No	No	[29]
Quercetin	*N*-vinyl imidazole	Tragacanth gum	Quercetin	In vitro	Anticancer	–	Magnet	No	No	No	No	[32]
Rivastigmine	AMPS	EGDA	Rivastigmine	In vitro	Anti-alzheimer	Oral	Magnet	No	No	No	No	[31]
DOX	NIPAAM and MAA	EDGMA	DOX	In vitro	Anticancer		Magnet, pH, temperature	No	No	No	No	[83]
Letrozole	MAA	TRIM	Letrozole	In vitro	Anticancer	–	Magnet	No	No	No	No	[34]
DOX	DA	–	DOX	In vitro and in vivo	Anticancer		Magnet	No	No	No	No	[84]
5-FU	NIPAM	BIS	5-FU	In vitro	Anticancer	–	Magnet and temperature	No	No	No	No	[85]
DOX	MABHD	EGDMA	DOX	In vitro	Anticancer		pH and redox	No	No	No	No	[86]
Gatifloxacin	AA and NIPAAm	BIS	Gatifloxacin	In vitro and in vivo	Anti-biotics	IG injection (oral)	pH and temperature	No	No	No	No	[87]
Amoxicillin	AAM	BIS	*H. pylori* lipoprotein and amoxicillin	In vitro and in vivo	Antibacterial	Oral	–	Yes	Yes	No	No	[18]

(continued)

Table 16.1 (continued)

Loaded drug	Monomers	Cross-linker	Template	Experiment	Indication	Route of administration	Response	Targeted	Dual imprinting	MOF core	Ther-anostics	Ref.
DOX	AAM	EGDMA	DOX	In vitro	Anticancer	–	Magnet	No	No	No	No	[28]
1,4-dimethyl-6-hydroxy-9H--carbazole	MAA	EGDMA	1,4-dimethyl-6-hydroxy-9H--carbazole	In vitro	Anticancer	–	Magnet	No	No	No	No	[30]
DOX	4-VP and HEMA	Cupric ion and BIS	DOX	In vitro	Anticancer	–	pH	No	No	No	No	[88]
Pantoprazole	HEMA or MAA or MAM or 4-VP	EGDMA	Pantoprazole	In vitro	Gastric ulcers, gastroesophageal reflux, and *Helicobacter pylori*	Oral	pH	No	No	No	No	[39]
Vancomycin	HEMA	EGDMA	Propranolol	In vitro	–	–	Light	No	No	No	No	[89]
Mitomycin C	HEMA, MAH	EGDMA	Mitomycin C	In vitro	Anticancer	–	Magnet	No	No	No	No	[90]
Diclofenac	MAA, MAM or 4-VP	EGDMA	Diclofenac	In vitro	Analgesic, antipyretic, and anti-inflammatory	Oral	pH	No	No	No	No	[40]
Ibuprofen	DMAEMA	EGDMA	Ibuprofen	In vitro	Analgesic, antipyretic, and anti-inflammatory	Oral	pH	No	No	No	No	[91]
Caffeine	MPABA	TRIM	Caffeine	In vitro	–	–	Light	No	No	No	No	[20]

Abbreviations: *EGDMA* ethylene glycol dimethacrylate, *MAA* methacrylic acid, *TRIM* trimethylolpropane trimethacrylate, *SA* sialic acid, *HEMA* 2-hydroxyethyl methacrylate, *AA* acrylic acid, *NIPAAm* *N*-isopropyl acrylamide, *MAEMA* (methacrylamido)ethyl methacrylate amide, *4-VP* 4-vinylpyridin, *MAH* *N*-methacryloyl-(L)-histidine methyl ester, *EGFR* epidermal growth factor receptor, *β-CD* β-cyclodextrin, *DMAEMA* 2-(dimethylamino)ethyl methacrylate, *TEOS* tetramethyl orthosilicate, *DEAEMA* 2-(diethylamino) ethyl methacrylate, *BAEDS* 3,3-dithiodipropionic acid-(2-propionyloxy) ethyl ester, *APMS* 2-acrylamide-2-methylpropanesulfonic acid, *VIM* 1-vinylimidazole, *ZIF-8* zeolitic imidazolate framework-8, *BAC* *N,N′*-bis(acryloyl)cystamine, *DOX* doxorubicin, *Ce6* chlorin e6, *BIS* *N,N′*-methylenebisacrylamide, *APTES* (3-aminopropyl) triethoxysilane, *BTEPDS* bis[(3-triethoxysilyl propyl)]-disulfide, *DA* dopamine, *VPBA* 4-vinylphenylboronic acid, *MTX* methotrexate, *Cys-DA* 2-amino-*N*-(3,4-dihydroxyphenethyl)-3-mercaptopropanamide, *TF* transferrin, *AMAB* 4-amino-4′-methacrylamide azobenzene, *APMA* *N*-(3-aminopropyl) methacrylamide hydrochloride, *CMC* carboxymethyl cellulose, *4-MPBA* 4-mercaptophenylboronic acid, *TEAMA* triethanolamine trimethacrylate, *ADDM* 1,3-Adamantanediol dimethacrylate, *MABHD* 2-methacrylester hydroxyethyl disulfide, *MPABA* 4-[(4-methacryloyloxy) phenylazo]benzoic acid, *MADPADSA* 4-[(4-methacryloyloxy)-2,6-dimethyl phenylazo]-3,5-dimethyl benzene-sulfonic acid, *DFCR* deoxy-5-fluorocytidine, *EGDA* ethylene glycol diacrylate represents a detailed summary of related articles regarding applying MIPs as novel DDSs.

16.6 Enantioselective Drug Delivery

Enantiomers are different stereoisomers in which drugs can exist, and one might be more effective or have fewer side effects than the others. These beneficial enantiomers are known as eutomers, with the remaining ones referred to as distomers. For example, the S-enantiomer of propranolol is over a hundred times more potent than its R-enantiomer [92]. Also, the S-enantiomer of thalidomide leads to severe teratogenic side effects [93].

Scientists used MIPs as stereo-selective DDSs to deliver specific eutomers of interest molecules instead of racemic mixtures. This approach was done in response to the expensive production costs of specific enantiomers as medications.

Propranolol was extensively studied as a potential template for developing stereo-selective MIP for oral and transdermal drug delivery applications [19, 23, 92, 94, 95]. Also, eutomers of ketoprofen, ibuprofen, omeprazole, thalidomide, and sulpiride were used as templates for stereo-selective MIPs [93, 95–97].

Researchers attempt to create stereo-selective DDSs by imprinting the active eutomer of the target molecule. By washing the template, the remaining cavities can precisely detect and attach to the eutomer even in a racemic mixture, enhancing eutomer release and binding capacity. In the study conducted by Wang et al., the customized MIP showed a higher adsorption of 47.89 μmol/g for S-sulpiride compared to 4.10 μmol/g for R-sulpiride under similar conditions [97].

As shown in Fig. 16.6, enantioselective MIPs improve the release of the eutomer compared to the distomer. This phenomenon may be explained by the reversible binding of the eutomer to receptor sites on the MIP, allowing it to be transported through the polymer and eventually released [92, 94, 98].

Suedee et al. designed an S-ibuprofen MIP that prolongs the release of the S-enantiomer (eutomer) and enhances the release of the R-enantiomer (distomer). The slower release of the template molecule might be caused by the stronger affinity between S-ibuprofen MIP and its template, compared to the R-enantiomer [95]. However, other designed enantioselective MIPs for propranolol, omeprazole, and

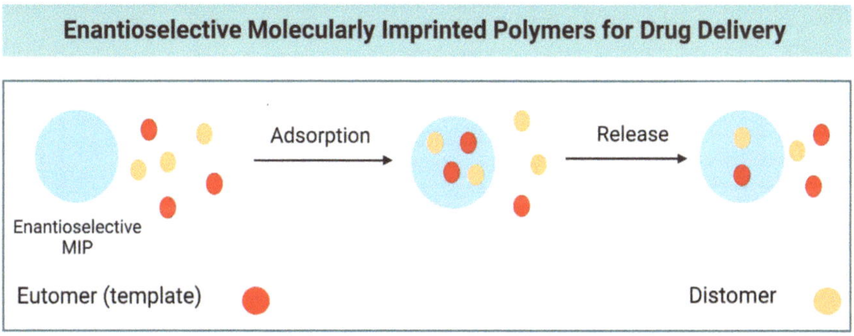

Fig. 16.6 Enantioselective MIPs

thalidomide had demonstrated an improved release rate for the template's isomers [11, 83, 86, 88].

Utilizing chiral functional monomers, such as *N*-acryloyl-tryptophan, can make the designed MIP more enantioselective [97]. Another option is to use cellulose derivative membranes, which have stereo-selectivity for propranolol through transdermal delivery [23, 92, 94].

During research on enantioselective drug release, it was found that the desired eutomer was not released exclusively, as some undesired distomer was also consistently observed to be released. Even a minor release of the distomer in medications with severe side effects can cause undesired adverse events and treatment failure. Therefore, the application of the proposed DDS may be reduced compared to the already available pure eutomer drugs.

Despite examining pharmacokinetics, previous research has yet to compare the in vivo effectiveness of pure eutomer and enantioselective MIPs [23, 98]. Instead, they suggest that their approach could be cost effective because researchers can use a racemic mixture in their DDS. However, using pure eutomer is still necessary when preparing MIPs for imprinting. Table 16.2 demonstrates a summary of enantioselective MIPs.

Table 16.2 The main features of enantioselective MIPs

Loaded drug	Monomers	Cross linker	Template	Experiment	Indication	Route of administration	Response	Targeted	Dual imprinting	MOF core	Theranostic	Ref.
S-Sulpiride	ATrp	EGDMA	S-Sulpiride	In vitro	Anti-psychotic	–	–	No	No	No	No	[97]
R-thalidomide	MAA	BIS	R-thalidomide	In vitro	Anti-cancer	–	Temperature	No	No	No	Yes	[93]
S-propranolol	MAA	EGDMA	S-propranolol	In vitro and in vivo	Hypertension and cardiovascular disease	Oral	–	No	No	No	No	[98]
S-omeprazole	MQN and MQD	EGDMA	S-omeprazole	In vitro	PPI anti-acid	Oral	pH	No	No	No	No	[96]
S-propranolol	MAA	EGDMA	S-propranolol	In vitro	Hypertension and cardiovascular disease	Transdermal	–	No	No	No	No	[94]
S-propranolol	MAA	EGDMA	S-propranolol	In vitro and in vivo	Hypertension and cardiovascular disease	Transdermal	–	No	No	No	No	[23]
S-propranolol	MAA	EGDMA	S-propranolol	In vitro	Hypertension and cardiovascular disease	Transdermal	–	No	No	No	No	[92]
S-Ketoprofen, R-propranolol, and S-ibuprofen	4-VPD or MAA	EGDMA	S-Ketoprofen, and R-propranolol, S-ibuprofen	In vitro	Hypertension and inflammation	Oral	–	No	No	No	No	[95]
R- or S-propranolol	MAA or NAA	EGDMA	R- or S-propranolol	In vitro	Hypertension and cardiovascular disease	Oral	–	No	No	No	No	[19]

Abbreviations: *ATrp* *N*-acryloyl-tryptophan, *EGDMA* ethyleneglycol dimethacrylate, *BIS* methacrylic acid (MAA), *N,N′*-methylenebisacrylamide, *MQN* methacryloyl quinine, *MQD* methacryloyl quinidine, *4-VPD* 4-vinylpyridine, *NAA* *N*-acryloyl-alanine polymer

16.7 Tissue Engineering

Tissue engineering is a field that combines various disciplines to repair and regenerate damaged tissues and organs. It involves using advanced materials, stem cells, and developmental biology. This field creates living tissues using synthetic biomaterials and cellular techniques. Regenerative engineering is essential for studying dynamic cellular mechanisms. It also presents new challenges and questions our understanding of tissue development and repair [25–27].

Scaffolds play a crucial role in tissue engineering, facilitating the growth and regeneration of tissues. Scaffold development utilizes different natural and synthetic polymers to enhance effectiveness. Regenerative engineering scaffolds encounter multiple challenges. These challenges require precise adjustment of porosity while maintaining the desired mechanical properties [25–27].

A scaffold must possess critical features like biocompatibility, optimized loading capacity, surface modification capabilities, and support for vital cellular activities, such as adhesion, migration, proliferation, and differentiation. The modulation of these processes is a complex and dynamic interaction influenced by a variety of small molecules and transcriptional factors [99]. Therefore, the scaffold should accurately imitate the ECM characteristics and regulate the degradation and absorption rates with specific kinetics. Additionally, it should easily integrate with target tissues and be flexible in altering shapes and sizes [25–27].

To achieve successful scaffold development, researchers use various manufacturing techniques, such as electrospinning, 3D printing, and self-assembly. Each method offers distinct advantages regarding scaffold architecture, porosity, and interconnectivity [25, 27].

Using both natural and synthetic polymers in scaffold fabrication can advance regenerative medicine. Researchers can create customized scaffolds using these biomaterials to improve tissue regeneration. This field might revolutionize healthcare and enhance the quality of life for many people [25–27]. Figure 16.7 illustrates the application of MIPs in tissue engineering and the various polymer options for creating proper scaffolds.

Natural polymers, derived from biological sources, have inherent biocompatibility and bioactivity. They possess various properties, including adjustable mechanical strength, degradation rates, and surface properties. Natural polymers are highly versatile and can be customized to meet specific tissue engineering requirements. Natural polymers such as alginate, gelatin, elastin, and apamin epitope were used to develop the scaffold in the mentioned studies [25–27].

Synthetic polymers allow precise control over scaffold design and properties through their chemical composition. These polymers can be tailored to meet specific mechanical, degradation, and surface properties. Moreover, synthetic polymers offer the advantage of batch-to-batch consistency, enabling reproducibility in scaffold fabrication [25–27]. Table 16.3 indicates the main characteristics of utilized MIPs for tissue engineering.

A wide range of synthetic polymers have been used to design scaffolds in the mentioned studies. These polymers include PLDLLA/MWCNTs, PLDLLA/

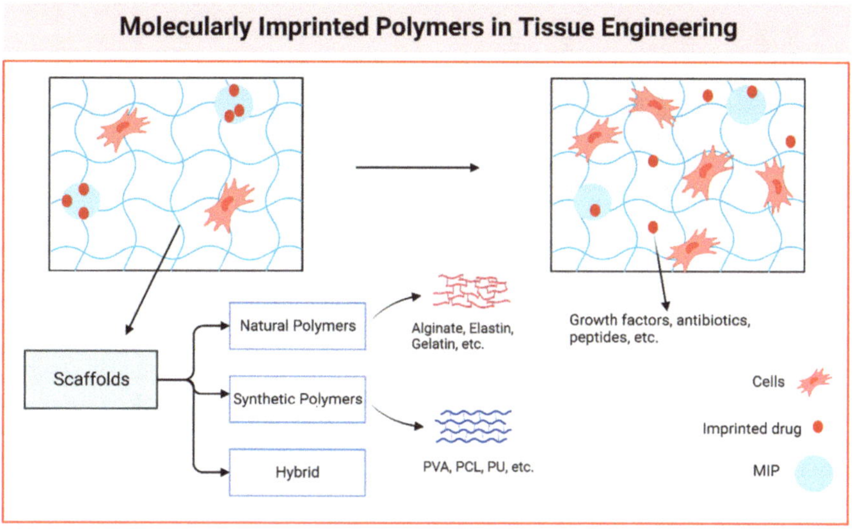

Fig. 16.7 The application of MIPs in tissue engineering

PAAc-10, PVL, PEG35k, VL/PEG with varying ratios, PVL-POE-PVL, PVA, PCL, PCL-POE-PCL, and also PU [100–108].

Combining natural and synthetic polymers in scaffolds offers benefits from both materials. The hybrid approach improves the scaffolds' strength, compatibility, and degradation profile. Adding bioactive molecules to these scaffolds facilitates cell adhesion, proliferation, and differentiation [25–27]. Foroutan-Koudehi et al. conducted an exceptional study in 2019 using this hybrid approach to design their scaffold. The study demonstrated that using PVA and gelatin in the hybrid approach for wound dressing significantly improved the healing process [104].

It is essential to consider other parameters alongside the scaffold for designing a MIP in tissue engineering. In this context, the key elements to consider are determining the appropriate template molecule and functional monomers. Choosing biocompatible and nontoxic monomers with a high affinity for the template is crucial. It is also crucial to consider template–monomer complexes' stability and thermodynamic favorability during reactions [27].

Most researchers have utilized MAA as a functional monomer for designing MIPs. Furthermore, AAM, AmEM, HEMA, VA, and PEG-EEMA were also among the monomers used. Cross-linking agents used in the mentioned studies were EGDMA, BIS, TRIM, PETRA, and GA [100–108].

Choosing an appropriate template is crucial when designing MIPs for differentiating stem cells. Stem cells can be grown in a lab and transformed into different cells for tissue regeneration purposes. Delivery and release profiles of growth factors are crucial for optimal differentiation of stem cells [26, 27]. The templates used in related studies included three general categories of drugs, peptides, and growth factors. Based on Table 16.3, vancomycin and gentamicin were employed to heal

Table 16.3 The application of MIPs for tissue engineering purposes

Monomer	Cross linker	Template	Scaffold	Synthesis method	Loading capacity	Binding capacity or IF	Release profile	Experiment	Regeneration target	Ref.
MAA	EGDMA	4-AP	PLDLLA/MWCNTs nanofibers	Suspension polymerization	~69%	1–1.78	~40% in first 12 h ~61% in 96 h	In vitro in—in vivo *(rat's sciatic)*	Peripheral nerve	[100]
AAm AmEM	BIS	TGF-β3	Epitope (Apamin)-imprinted nanoparticle	Inverse microemulsion polymerization	NM	TGF-β3: ~ 5 ng per mg NINP ~ 11.7 ng per MINP	NM	In vitro	Osteogenic + Adipogenic + Chondrogenic	[101]
HEMA	EGDMA	Dexa	PLDLLA/PAAc-10 nanofibers	Precipitation polymerization	26 ± 1.3–57 ± 1.2	~57% 1.02–1.78	~61% in 72 h	In vitro	Bone	[102]
MAA	PETRA	GRGDSP YIGSR	AGE sponges	Precipitation polymerization	NM	**MIPs:** **GRGDSP:** Q: 57.34 IF: 1.78 **YIGSR:** Q: 83.32 IF: 1.66 **Sponge + MIPs:** **GRGDSP:** Q: 72.7 IF: 1.99 **YIGSR:** Q: 203.7 IF: 2.09	45% after 7 days 82% after 90 days	In vitro In vivo *(cryoinjury rat model)*	Heart	[103]
VA	GA	Gentamycin	PVA/gelatin nanofibers	Precipitation polymerization	~88.6%	NM	~85% in the first 8 h ~ 98% in 110 h.	In vitro In vivo *(rat's skin)*	Skin	[104]

MAA	EGDMA	Vancomycin	Alginate matrix	Precipitation polymerization	~0.07 g drug per 1 g of MIP	NM	After using the double barrier system: ~87% to 47% in the first 24 h MIP: ~80% in 12 h MIP/alginate: ~80% in 5 days	In vitro	Skin	[105]
MAA PEG-EEMA	TRIM	MMP-9	PVL PEG35k PEG20k VL/PEG (90:10/85:15/80:20) PVL-POE-PVL films PVL-POE-PVL/gelatin blend (90:10)	Precipitation polymerization	56.95%	1.15 Specific rebinding: 18.67	After 40 days: For pure copolymers: 10% For blend: 37%	In vitro	Heart	[106]
MAA	TRIM	YIGSR GRGDS	PCL PCL-FOE-PCL PU 2% w/v films	Precipitation polymerization	YIGSR: ~80.8%	**MIPs:** **YIGSR:** Q: 7.32 IF: 1.2 **GRGDS:** Q: 1.76 IF: 2.33 **Films + MIPs:** **YIGSR:** PCL: Q: 37.94 IF: 1.73 PCL-POE-PCL: Q: 61.16 IF: 2.77 PU 2%: Q: 50.31 IF: 2.08	~14% in 20 days	In vitro	Myoblasts	[107]

(continued)

Table 16.1 (continued)

Monomer	Cross linker	Template	Scaffold	Synthesis method	Loading capacity	Binding capacity or IF	Release profile	Experiment	Regeneration target	Ref.
MAA	PETRA	RGDSPASSKP GRGDS	PCL PCL-POE-PCL PU 2% w/v films	Precipitation polymerization	GRGDS: ~86.1% (22.4 μmol/g particles)	**MIPs:** **RGDSPASSKP:** Q: 0.096 IF: 1.39 **GRGDS:** Q: 7.69 IF: 1.34 **Films + MIPs:** **GRGDS:** PCL: Q: 28.88 IF: 1.77 PCL-POE-PCL: Q: 34.04 IF: 1.67 PU 2%: Q: 36.09 IF: 1.82	NM	In vitro	Myoblasts	[108]

Abbreviations: *NM* not mentioned, *Q* Quantitative binding capacity, *IF* Imprinting factor, *PLDLLA* poly (L-lactide-co-D, L-lactide), *MWCNTs* multiwalled carbon nanotubes, *4-AP* 4-aminopyridine, *EGDMA* ethyleneglycol dimethacrylate, *MAA* methacrylic acid, *BIS N,N′*-methylenebisacrylamide, *TGF-β3* transforming growth factor-β3, *TRIM* trimethylolpropane trimethacrylate, *AGE* alginate-gelatin-elastin, *PAAc-10* poly (acrylic acid) 10 wt%, *HEMA* poly (2-hydroxyethyl methacrylate), *AAm* acrylamide, *AmEM* 2-aminoethyl methacrylate, *Dexa* dexamethasone, *MINP* imprinted polymer nanoparticle, *NINP* non-imprinted polymer nanoparticle, *PETRA* pentaerythritol triacrylate, *PVA* poly (vinyl alcohol), *VA* vinyl alcohol, *GA* glutaraldehyde, *GRGDS, GRGDSP, RGDSPASSKP* peptide sequences derived from fibronectin, *YIGSR* a peptide sequence derived from laminin, *MMP-9* matrix metalloproteinase-9, *PCL* poly (caprolactone), *PCL–POE–PCL* poly (caprolactone)-poly (ethylene oxide)-poly (caprolactone), *PU* poly(ether ester urethane), *PEG35k* poly(ethylene glycol) 35 kilo Daltons, *PEG20k* poly(ethylene glycol) 20 kilo Daltons, *PVL* poly (δ-valerolactone), *VL δ-valerolactone*, *PVL-POE-PVL* poly (δ-valerolactone)-poly (ethylene oxide)-poly (δ-valerolactone), *PEG-EEMA* poly(ethylene glycol)-ethyl ether methacrylate skin wounds [104, 105]. Dexamethasone was also used to facilitate the regeneration of bone tissue [102].

Moreover, 4-aminopyridine was employed to promote the regeneration of peripheral nervous tissue [100]. Several peptides have been utilized in various studies for heart tissue regeneration. YIGSR, GRGDS, RGDSPASSKP, and GRGDSP are among the peptides employed for this purpose [103, 107, 108]. Also, the template used in Teixeira's study was TGF-β3 to elicit osteogenic, chondrogenic, and adipogenic properties [101].

Regenerative medicine studies have concentrated on regenerating cardiac myoblasts and healing skin wounds [103–108]. Another primary purpose of the studies was to differentiate various stem cells [101, 102]. However, Fallah-darrehchi's study in 2021 shifted its focus to regenerating peripheral nerve tissue [100].

Cristallini et al.'s study in 2012 had differences compared to other studies mentioned. Their template was the enzyme MMP-9, which is involved in cardiac inflammatory reactions. Then, they used MAA and PEG-EEMA as functional monomers and TRIM as a cross-linker. They aimed to trap MMP-9 to prevent left ventricular remodeling and, thus, the formation of new cardiac tissues. For this purpose, they designed and tested several types of scaffolds to synthesize multilayered 3D scaffolds. The designed scaffolds included PVL, PEG20k, PEG35k, VL/PEG with varying ratios, and PVL-POE-PVL. Finally, after multiple tests and adding gelatin polymer, they successfully obtained the PVL-POE-PVL (85:15)/GEL (90:10% w/w) blend. The pure copolymer samples lost 10% of their mass after 40 days, while the micro-fabricated system lost 37%. The higher degradation rate of the designed structure is because of the gradual release of the GEL component during degradation time. No significant decrease in the weight of the samples of both synthetic and biosynthetic groups was detected after sterilization, as it was less than 2%. This study revealed that MMP-9 had a specific rebinding value of 18.67 with the MIP, showing that the enzyme remained immobile on the scaffold [106].

The selectivity of MIPs depends on the polymer matrix used, as proteins have various binding sites and regions. The polymer matrix can either facilitate or impede cross-reactivity with proteins similar to the template molecule [26, 27].

In summary, several essential factors affect the effectiveness of MIPs in drug delivery for tissue engineering. These factors include loading capacity, release profiles, imprinting factor, and experiment type (in vitro or in vivo). Table 16.3 demonstrates the detailed information for the included articles.

16.8 Conclusion and Future Perspective

This chapter reviews the history and applications of MIPs for drug delivery and tissue engineering and highlights notable studies. MIPs, as artificial antibodies, could replace their natural counterparts because of their cost-effectiveness, easy synthesis, and ability to resist harsh physiological conditions. This advancement would overcome the limitations of natural antibodies, such as poor purification, immunogenicity, high production costs, and instability.

Researchers can explore new opportunities for therapeutic applications and the intricate mechanisms of diseases by broadening the scope of potential targets of multifunctional MIPs or using different templates for drug delivery purposes. For instance, imprinting larger templates, such as protein biomarkers, remains challenging, requiring further investigation into the pharmacokinetics, immunogenicity, and toxicity of new DDSs.

While several studies have indicated promising results of utilizing MIPs in tissue engineering and regenerative medicine, further research is needed to explore their potential applications and involve a broader range of targets in this field. Specifically, further research is needed to elucidate the potential of MIPs in DDSs for activating stem cells. Furthermore, researchers can enhance the effectiveness of DDSs by designing more sophisticated scaffolds to represent the body's physiological conditions with greater accuracy.

In conclusion, the future perspectives for employing MIPs in various applications are promising, such as enhanced drug delivery for advanced cancers, combating antibiotic resistance, and facilitating the healing of skin injuries. Moreover, the potential of MIPs can extend to the treatment of chronic diseases, including neurodegenerative and cardiovascular diseases. However, further investigations are necessary to fully understand the nuances of these applications and optimize their efficacy. Through continued research and innovation, MIPs have the potential to impact the field of medicine and significantly improve therapeutic outcomes.

References

1. Parodi A, Buzaeva P, Nigovora D, Baldin A, Kostyushev D, Chulanov V et al (2021) Nanomedicine for increasing the oral bioavailability of cancer treatments. J Nanobiotechnol 19(1):1–19
2. Li X, Peng X, Zoulikha M, Boafo GF, Magar KT, Ju Y, He W (2024) Multifunctional nanoparticle-mediated combining therapy for human diseases. Signal Transduct Target Ther 9(1):1
3. Ren Y, Nie L, Zhu S, Zhang X (2022) Nanovesicles-mediated drug delivery for oral bioavailability enhancement. Int J Nanomedicine 17:4861
4. Rizvi SA, Saleh AM (2018) Applications of nanoparticle systems in drug delivery technology. Saudi Pharm J 26(1):64–70
5. Tangsiri M, Hheidari A, Liaghat M, Razlansari M, Ebrahimi N, Akbari A et al (2024) Promising applications of nanotechnology in inhibiting chemo-resistance in solid tumors by targeting epithelial-mesenchymal transition (EMT). Biomed Pharmacother 170:115973
6. Omeh RC, Ugwueze ME, Offiah RO, Mbah CC, Momoh AM, Onyishi IV et al (2022) Oral drug delivery: gastrointestinal tract adaptations, barriers and strategies for delivery enhancement-a review. Bio-Res 20(3):1685–1698
7. Hersh DS, Wadajkar AS, Roberts N, Perez JG, Connolly NP, Frenkel V et al (2016) Evolving drug delivery strategies to overcome the blood brain barrier. Curr Pharm Des 22(9):1177–1193
8. Wu D, Chen Q, Chen X, Han F, Chen Z, Wang Y (2023) The blood–brain barrier: structure, regulation, and drug delivery. Signal Transduct Target Ther 8(1):217
9. Murar M, Albertazzi L, Pujals S (2022) Advanced optical imaging-guided nanotheranostics towards personalized cancer drug delivery. Nano 12(3):399
10. Zhou K, Li Z-Z, Cai Z-M, Zhong N-N, Cao L-M, Huo F-Y et al (2023) Nanotheranostics in cancer lymph node metastasis: the long road ahead. Pharmacol Res 198:106989

11. Bodoki AE, Iacob B-C, Dinte E, Vostinaru O, Samoila O, Bodoki E (2021) Perspectives of molecularly imprinted polymer-based drug delivery systems in ocular therapy. Polymers 13(21):3649
12. Byrne ME, Park K, Peppas NA (2002) Molecular imprinting within hydrogels. Adv Drug Deliv Rev 54(1):149–161
13. Schirhagl R (2014) Bioapplications for molecularly imprinted polymers. Anal Chem 86(1):250–261
14. Yilmaz E, Haupt K, Mosbach K (2000) The use of immobilized templates—a new approach in molecular imprinting. Angew Chem Int Ed 39(12):2115–2118
15. Alvarez-Lorenzo C, Hiratani H, Gomez-Amoza JL, Martínez-Pacheco R, Souto C, Concheiro A (2002) Soft contact lenses capable of sustained delivery of timolol. J Pharm Sci 91(10):2182–2192
16. Alvarez-Lorenzo C, Yanez F, Barreiro-Iglesias R, Concheiro A (2006) Imprinted soft contact lenses as norfloxacin delivery systems. J Control Release 113(3):236–244
17. Oliveira D, Gomes CP, Dias RC, Costa MRP (2016) Molecular imprinting of 5-fluorouracil in particles with surface RAFT grafted functional brushes. React Funct Polym 107:35–45
18. Wu Z, Hou J, Wang Y, Chai M, Xiong Y, Lu W, Pan J (2015) Preparation and evaluation of amoxicillin loaded dual molecularly imprinted nanoparticles for anti-helicobacter pylori therapy. Int J Pharm 496(2):1006–1014
19. Suedee R, Srichana T, Martin GP (2000) Evaluation of matrices containing molecularly imprinted polymers in the enantioselective-controlled delivery of β-blockers. J Control Release 66(2–3):135–147
20. Gong C, Lam MW, Yu H (2006) The fabrication of a photoresponsive molecularly imprinted polymer for the photoregulated uptake and release of caffeine. Adv Funct Mater 16(13):1759–1767
21. Hiratani H, Fujiwara A, Tamiya Y, Mizutani Y, Alvarez-Lorenzo C (2005) Ocular release of timolol from molecularly imprinted soft contact lenses. Biomaterials 26(11):1293–1298
22. Norell MC, Andersson HS, Nicholls IA (1998) Theophylline molecularly imprinted polymer dissociation kinetics: a novel sustained release drug dosage mechanism. J Mol Recognit 11(1–6):98–102
23. Suedee R, Bodhibukkana C, Tangthong N, Amnuaikit C, Kaewnopparat S, Srichana T (2008) Development of a reservoir-type transdermal enantioselective-controlled delivery system for racemic propranolol using a molecularly imprinted polymer composite membrane. J Control Release 129(3):170–178
24. Wang H-Y, Cao P-P, He Z-Y, He X-W, Li W-Y, Li Y-H, Zhang Y-K (2019) Targeted imaging and targeted therapy of breast cancer cells via fluorescent double template-imprinted polymer coated silicon nanoparticles by an epitope approach. Nanoscale 11(36):17018–17030
25. Bonatti AF, De Maria C, Vozzi G (2021) Molecular imprinting strategies for tissue engineering applications: a review. Polymers 13(4):548
26. Neves MI, Wechsler ME, Gomes ME, Reis RL, Granja PL, Peppas NA (2017) Molecularly imprinted intelligent scaffolds for tissue engineering applications. Tissue Eng Part B Rev 23(1):27–43
27. Clegg JR, Wechsler ME, Peppas NA (2017) Vision for functionally decorated and molecularly imprinted polymers in regenerative engineering. Regen Eng Transl Med 3:166–175
28. Griffete N, Fresnais J, Espinosa A, Wilhelm C, Bée A, Ménager C (2015) Design of magnetic molecularly imprinted polymer nanoparticles for controlled release of doxorubicin under an alternative magnetic field in athermal conditions. Nanoscale 7(45):18891–18896
29. Hashemi-Moghaddam H, Kazemi-Bagsangani S, Jamili M, Zavareh S (2016) Evaluation of magnetic nanoparticles coated by 5-fluorouracil imprinted polymer for controlled drug delivery in mouse breast cancer model. Int J Pharm 497(1–2):228–238
30. Parisi OI, Morelli C, Puoci F, Saturnino C, Caruso A, Sisci D et al (2014) Magnetic molecularly imprinted polymers (MMIPs) for carbazole derivative release in targeted cancer therapy. J Mater Chem B 2(38):6619–6625

31. Hemmati K, Sahraei R, Ghaemy M (2016) Synthesis and characterization of a novel magnetic molecularly imprinted polymer with incorporated graphene oxide for drug delivery. Polymer 101:257–268
32. Hemmati K, Masoumi A, Ghaemy M (2016) Tragacanth gum-based nanogel as a superparamagnetic molecularly imprinted polymer for quercetin recognition and controlled release. Carbohydr Polym 136:630–640
33. Shadabfar M, Abdouss M, Khonakdar HA (2020) Synthesis, characterization, and evaluation of a magnetic molecular imprinted polymer for 5-fluorouracil as an intelligent drug delivery system for breast cancer treatment. J Mater Sci 55:12287–12304
34. Kazemi S, Sarabi AA, Abdouss M (2016) Synthesis and characterization of magnetic molecularly imprinted polymer nanoparticles for controlled release of letrozole. Korean J Chem Eng 33:3289–3297
35. Mao C, Xie X, Liu X, Cui Z, Yang X, Yeung K et al (2017) The controlled drug release by pH-sensitive molecularly imprinted nanospheres for enhanced antibacterial activity. Mater Sci Eng C 77:84–91
36. Chen W, Ma Y, Pan J, Meng Z, Pan G, Sellergren B (2015) Molecularly imprinted polymers with stimuli-responsive affinity: Progress and perspectives. Polymers 7(9):1689–1715
37. Men J, Gao B, Yao L, Zhang Y (2014) Preparation and characterization of metronidazole-surface imprinted microspheres MIP-PSSS/CPVA for colon-specific drug delivery system. J Macromol Sci A 51(11):914–923
38. Çetin K, Alkan H, Bereli N, Denizli A (2017) Molecularly imprinted cryogel as a pH-responsive delivery system for doxorubicin. J Macromol Sci A 54(8):502–508
39. Mohajeri SA, Sajadi Tabassi SA, Hassanpour Moghadam M, Mohajeri SA (2014) Preparation of a pH-sensitive pantoprazole-imprinted polymer and evaluation of its drug-binding and-releasing properties. SCIENCE CHINA Chem 57:857–865
40. Mohajeri SA, Malaekeh-Nikouei B, Sadegh H (2012) Development of a pH-responsive imprinted polymer for diclofenac and study of its binding properties in organic and aqueous media. Drug Dev Ind Pharm 38(5):616–622
41. Yan Q, Liu L, Wang T, Wang H (2019) A pH-responsive hydrogel system based on cellulose and dopamine with controlled hydrophobic drug delivery ability and long-term bacteriostatic property. Colloid Polym Sci 297:705–717
42. Liu L, Tan Y, Bao Y, Liao S, Wang D, Ye Q et al (2023) A new pH/redox/light stimuli-responsive magnetic molecularly imprinted polymer based on β-cyclodextrin for recognition and controlled release of andrographolide. React Funct Polym 182:105485
43. Liu H, Deng Z, Bu J, Zhang Y, Zhang Z, He Y et al (2021) Capsule-like molecular imprinted polymer nanoparticles for targeted and chemophotothermal synergistic cancer therapy. Colloids Surf B: Biointerfaces 208:112126
44. Liu L-t, Chen M-j, Yang H-l, Huang Z-j, Tang Q, Chow C-f et al (2020) An NIR-light-responsive surface molecularly imprinted polymer for photoregulated drug release in aqueous solution through porcine tissue. Mater Sci Eng C 106:110253
45. Peng H, Qin Y-T, He X-W, Li W-Y, Zhang Y-K (2020) Epitope molecularly imprinted polymer nanoparticles for chemo‑/photodynamic synergistic cancer therapy guided by targeted fluorescence imaging. ACS Appl Mater Interfaces 12(11):13360–13370
46. Cegłowski M, Jerca VV, Jerca FA, Hoogenboom R (2020) Reduction-responsive molecularly imprinted poly (2-isopropenyl-2-oxazoline) for controlled release of anticancer agents. Pharmaceutics 12(6):506
47. Zhao Y, Simon C, Attieh MD, Haupt K, Falcimaigne-Cordin A (2020) Reduction-responsive molecularly imprinted nanogels for drug delivery applications. RSC Adv 10(10):5978–5987
48. Song Y, Han S, Liu S, Sun R, Zhao L, Yan C (2023) Biodegradable imprinted polymer based on ZIF-8/DOX-HA for synergistically targeting prostate cancer cells and controlled drug release with multiple responses. ACS Appl Mater Interfaces 15:25339
49. Liu S, Han S, Song Y, Sun R, Zhao L, Yan C (2023) Disulfide-bridged dendritic organosilicas-based biodegradable molecularly imprinted polymers for multiple targeting and pH/redox-

responsive drug release toward chemical/photodynamic synergistic tumor therapy. Adv Healthc Mater 12:2300184
50. Han S, Song Y, Liu S, Zhao L, Sun R (2022) Dual responsive molecularly imprinted polymers based on UiO-66-DOX for selective targeting tumor cells and controlled drug release. Eur Polym J 171:111219
51. Han S, Liu S, Song Y, Jiang H (2022) Biodegradable magnesium ion-doped silica-based molecularly imprinted nanoparticles for targeting tumor cells to drugs controlled release and recognition mechanism research. Colloids Surf B: Biointerfaces 217:112665
52. Yin Y, Guan L, Wang Y, Ma Y, Pan J, Peng Y, Pan G (2021) Sialic acid-imprinted mesoporous nanocarriers for tumor cell targeted drug delivery. Coll Interf Sci Commun 42:100421
53. Qin Y-T, Feng Y-S, Ma Y-J, He X-W, Li W-Y, Zhang Y-K (2020) Tumor-sensitive biodegradable nanoparticles of molecularly imprinted polymer-stabilized fluorescent zeolitic imidazolate framework-8 for targeted imaging and drug delivery. ACS Appl Mater Interfaces 12(22):24585–24598
54. Han S, Teng F, Wang Y, Su L, Leng Q, Jiang H (2020) Drug-loaded dual targeting graphene oxide-based molecularly imprinted composite and recognition of carcino-embryonic antigen. RSC Adv 10(19):10980–10988
55. Han S, Su L, Zhai M, Ma L, Liu S, Teng Y (2019) A molecularly imprinted composite based on graphene oxide for targeted drug delivery to tumor cells. J Mater Sci 54(4):3331–3341
56. Qin Y-T, Peng H, He X-W, Li W-Y, Zhang Y-K (2019) Highly effective drug delivery and cell imaging using fluorescent double-imprinted nanoparticles by targeting recognition of the epitope of membrane protein. Anal Chem 91(20):12696–12703
57. Hashemi-Moghaddam H, Zavareh S, Karimpour S, Madanchi H (2017) Evaluation of molecularly imprinted polymer based on HER2 epitope for targeted drug delivery in ovarian cancer mouse model. React Funct Polym 121:82–90
58. Lu H, Xu S, Guo Z, Zhao M, Liu Z (2021) Redox-responsive molecularly imprinted nanoparticles for targeted intracellular delivery of protein toward cancer therapy. ACS Nano 15(11):18214–18225
59. Gu Z, Dong Y, Xu S, Wang L, Liu Z (2021) Molecularly imprinted polymer-based smart prodrug delivery system for specific targeting, prolonged retention, and tumor microenvironment-triggered release. Angew Chem 133(5):2695–2699
60. Jia C, Zhang M, Zhang Y, Ma Z-B, Xiao N-N, He X-W et al (2019) Preparation of dual-template epitope imprinted polymers for targeted fluorescence imaging and targeted drug delivery to pancreatic cancer BxPC-3 cells. ACS Appl Mater Interfaces 11(35):32431–32440
61. Topete A, Barahona I, Santos LF, Pinto CA, Saraiva JA, Serro AP, Saramago B (2021) The effects of addition of functional monomers and molecular imprinting on dual drug release from intraocular lens material. Int J Pharm 600:120513
62. Cui Y, Li M, Hong X, Du D, Ma Y (2020) Solid-phase interfacial synthesis of dual-imprinted colloid particles for multifunctional nanomedicine development. Coll Interf Sci Commun 36:100267
63. Wu MX, Yang YW (2017) Metal–organic framework (MOF)-based drug/cargo delivery and cancer therapy. Adv Mater 29(23):1606134
64. Zhao L, Chai M-H, Yao H-F, Huang Y-P, Liu Z-S (2020) Molecularly imprinted polymers doped with carbon nanotube with aid of metal-organic gel for drug delivery systems. Pharm Res 37:1–14
65. Shi Y, Wang Y, Zhu J, Liu W, Khan MZH, Liu X (2020) Molecularly imprinting polymers (MIP) based on nitrogen doped carbon dots and mIL-101 (Fe) for doxorubicin hydrochloride delivery. Nano 10(9):1655
66. Zhang LP, Mo CE, Huang YP, Liu ZS (2019) Preparation of liquid crystalline molecularly imprinted polymer coated metal organic framework for capecitabine delivery. Part Part Syst Charact 36(1):1800355
67. Wang D, Bao Y, Tan Y, Liu L, Ye Q, Zeng C, Tan N (2023) A novel smart stealth sorafenib delivery system based on the magnetic imprinting material modified by polyethylene glycol. Polym Adv Technol 34(5):1515–1528

68. Saboury A, Mohammadi R, Javanbakht S, Ghorbani M (2023) Doxorubicin imprinted magnetic polymethacrylamide as a pH-sensitive anticancer nanocarrier. J Drug Deliv Sci Technol 79:103998
69. Rahimi Haji Abadi F, Tadayon F, Saber Tehrani M, Ahmad PH (2023) Synthesis and characterization of the photoresponsive and thermoresponsive molecularly imprinted polymer with a novel functional monomer for controlled release of 4-Aminopyridine. Int J Polymer Mater Polymer Biomater 72(6):425–432
70. Tan N, Ding Z, Chen C, Yang Y, He L, Liao S et al (2022) A new pH/redox dual stimulus-responsive formononetin imprinting material. React Funct Polym 170:105141
71. Wang H-Y, Su Z-C, He X-W, Li W-Y, Zhang Y-K (2021) H 2 O 2 self-supplying degradable epitope imprinted polymers for targeted fluorescence imaging and chemodynamic therapy. Nanoscale 13(29):12553–12564
72. Mohebali A, Abdouss M, Kazemi Y, Daneshnia S (2021) Fabrication and characterization of pH-responsive poly (methacrylic acid)-based molecularly imprinted polymers nanosphere for controlled release of amitriptyline hydrochloride: fabrication of MIP for amitriptyline hydrochloride. Polym Adv Technol 32(11):4386–4396
73. Kubo T, Shimonaka M, Watabe Y, Akiyoshi K, Balachandran J, Otsuka K (2021) Study on magnetic thermal seeds coated with thermal-responsive molecularly imprinted polymers. Nano 7(1):215–225
74. Javanbakht S, Saboury A, Shaabani A, Mohammadi R, Ghorbani M (2020) Doxorubicin imprinted photoluminescent polymer as a pH-responsive nanocarrier. ACS Appl Bio Mater 3(7):4168–4178
75. Tan N, Chen C, Ji K, Liao S, Liu Y, Hu L et al (2021) Preparation and properties of hollow magnetic liquid crystal molecularly imprinted polymers as Silybin sustained-release carriers. ChemistrySelect 6(34):9024–9031
76. Talavat L, Güner A (2019) Thermodynamic computational calculations for preparation 5-fluorouracil magnetic moleculary imprinted polymers and their application in controlled drug release. Inorg Chem Commun 103:119–127
77. Wei P, Song R, Chen C, Li Z, Zhu Z, Li S (2019) A pH-responsive molecularly imprinted hydrogel for dexamethasone release. J Inorg Organomet Polym Mater 29:659–666
78. Cazares-Cortes E, Nerantzaki M, Fresnais J, Wilhelm C, Griffete N, Ménager C (2018) Magnetic nanoparticles create hot spots in polymer matrix for controlled drug release. Nano 8(10):850
79. Kubo T, Tachibana K, Naito T, Mukai S, Akiyoshi K, Balachandran J, Otsuka K (2018) Magnetic field stimuli-sensitive drug release using a magnetic thermal seed coated with thermal-responsive molecularly imprinted polymer. ACS Biomater Sci Eng 5(2):759–767
80. Parisi OI, Ruffo M, Scrivano L, Malivindi R, Vassallo A, Puoci F (2018) Smart bandage based on molecularly imprinted polymers (MIPs) for diclofenac controlled release. Pharmaceuticals 11(4):92
81. Marcelo G, Ferreira IC, Viveiros R, Casimiro T (2018) Development of itaconic acid-based molecular imprinted polymers using supercritical fluid technology for pH-triggered drug delivery. Int J Pharm 542(1–2):125–131
82. Liu L, Li N, Chen M, Yang H, Tang Q, Gong C (2018) Visible-light-responsive surface molecularly imprinted polymer for acyclovir through chicken skin tissue. ACS Appl Bio Mater 1(3):845–852
83. Kaamyabi S, Habibi D, Amini MM (2016) Preparation and characterization of the pH and thermosensitive magnetic molecular imprinted nanoparticle polymer for the cancer drug delivery. Bioorg Med Chem Lett 26(9):2349–2354
84. Zavareh S, Mahdi M, Erfanian S, Hashemi-Moghaddam H (2016) Synthesis of polydopamine as a new and biocompatible coating of magnetic nanoparticles for delivery of doxorubicin in mouse breast adenocarcinoma. Cancer Chemother Pharmacol 78:1073–1084
85. Li L, Chen L, Zhang H, Yang Y, Liu X, Chen Y (2016) Temperature and magnetism bi-responsive molecularly imprinted polymers: preparation, adsorption mechanism and prop-

erties as drug delivery system for sustained release of 5-fluorouracil. Mater Sci Eng C 61:158–168
86. Zhang K, Guan X, Qiu Y, Wang D, Zhang X, Zhang H (2016) A pH/glutathione double responsive drug delivery system using molecular imprint technique for drug loading. Appl Surf Sci 389:1208–1213
87. Wang X-L, Yao H-F, Li X-Y, Wang X, Huang Y-P, Liu Z-S (2016) pH/temperature-sensitive hydrogel-based molecularly imprinted polymers (hydroMIPs) for drug delivery by frontal polymerization. RSC Adv 6(96):94038–94047
88. Zhang Q, Zhang L, Wang P, Du S (2014) Coordinate bonding strategy for molecularly imprinted hydrogels: toward pH-responsive doxorubicin delivery. J Pharm Sci 103(2):643–651
89. Ma Y, Li H, He S, Zhang H, Zhang H (2014) Efficient one-pot synthesis of water-compatible and photoresponsive molecularly imprinted polymer nanoparticles by facile RAFT precipitation polymerization. J Polym Sci A Polym Chem 52(14):1941–1952
90. Türkmen D, Bereli N, Çorman ME, Shaikh H, Akgöl S, Denizli A (2014) Molecular imprinted magnetic nanoparticles for controlled delivery of mitomycin C. Artific Cells Nanomed Biotechnol 42(5):316–322
91. da Silva MS, Viveiros R, Morgado PI, Aguiar-Ricardo A, Correia IJ, Casimiro T (2011) Development of 2-(dimethylamino) ethyl methacrylate-based molecular recognition devices for controlled drug delivery using supercritical fluid technology. Int J Pharm 416(1):61–68
92. Bodhibukkana C, Srichana T, Kaewnopparat S, Tangthong N, Bouking P, Martin GP, Suedee R (2006) Composite membrane of bacterially-derived cellulose and molecularly imprinted polymer for use as a transdermal enantioselective controlled-release system of racemic propranolol. J Control Release 113(1):43–56
93. Suksuwan A, Lomlim L, Rungrotmongkol T, Nakpheng T, Dickert FL, Suedee R (2015) The composite nanomaterials containing (R)-thalidomide-molecularly imprinted polymers as a recognition system for enantioselective-controlled release and targeted drug delivery. J Appl Polym Sci 132(18)
94. Jantarat C, Tangthong N, Songkro S, Martin GP, Suedee R (2008) S-propranolol imprinted polymer nanoparticle-on-microsphere composite porous cellulose membrane for the enantioselectively controlled delivery of racemic propranolol. Int J Pharm 349(1–2):212–225
95. Suedee R, Srichana T, Rattananont T (2002) Enantioselective release of controlled delivery granules based on molecularly imprinted polymers. Drug Deliv 9(1):19–30
96. Suedee R, Jantarat C, Lindner W, Viernstein H, Songkro S, Srichana T (2010) Development of a pH-responsive drug delivery system for enantioselective-controlled delivery of racemic drugs. J Control Release 142(1):122–131
97. Wang L, She X, Chen Z, Quan S, Liu Y, Mai X et al (2021) Preparation and characterization of a chiral molecularly imprinted polymer with a novel functional monomer for controlled release of S-sulpiride. Int J Pharm 601:120526
98. Yin J, Cui Y, Yang G, Wang H (2010) Molecularly imprinted nanotubes for enantioselective drug delivery and controlled release. Chem Commun 46(41):7688–7690
99. Biglari N, Mehdizadeh A, Mastanabad MV, Gharaeikhezri MH, Afrakoti LGMP, Pourbala H et al (2023) Application of mesenchymal stem cells (MSCs) in neurodegenerative disorders: history, findings, and prospective challenges. Pathol Res Pract 247:154541
100. Fallah-Darrehchi M, Zahedi P, Safarian S, Ghaffari-Bohlouli P, Aeinehvand R (2021) Conductive conduit based on electrospun poly (L-lactide-co-D, L-lactide) nanofibers containing 4-aminopyridine-loaded molecularly imprinted poly (methacrylic acid) nanoparticles used for peripheral nerve regeneration. Int J Biol Macromol 190:499–507
101. Teixeira SP, Domingues RM, Babo PS, Berdecka D, Miranda MS, Gomes ME et al (2021) Epitope-imprinted nanoparticles as transforming growth factor-β3 sequestering ligands to modulate stem cell fate. Adv Funct Mater 31(4):2003934
102. Ghaffari-Bohlouli P, Zahedi P, Shahrousvand M (2020) Enhanced osteogenesis using poly (l-lactide-co-d, l-lactide)/poly (acrylic acid) nanofibrous scaffolds in presence of dexamethasone-loaded molecularly imprinted polymer nanoparticles. Int J Biol Macromol 165:2363–2377

103. Rosellini E, Madeddu D, Barbani N, Frati C, Graiani G, Falco A et al (2020) Development of biomimetic alginate/gelatin/elastin sponges with recognition properties toward bioactive peptides for cardiac tissue engineering. Biomimetics 5(4):67
104. Foroutan Koudehi M, Zibaseresht R (2020) Synthesis of molecularly imprinted polymer nanoparticles containing gentamicin drug as wound dressing based polyvinyl alcohol/gelatin nanofiber. Mater Technol 35(1):21–30
105. Kurczewska J, Cegłowski M, Pecyna P, Ratajczak M, Gajęcka M, Schroeder G (2017) Molecularly imprinted polymer as drug delivery carrier in alginate dressing. Mater Lett 201:46–49
106. Cristallini C, Gagliardi M, Barbani N, Giannessi D, Guerra GD (2012) Novel biodegradable, biomimetic and functionalised polymer scaffolds to prevent expansion of post-infarct left ventricular remodelling. J Mater Sci Mater Med 23:205–216
107. Rosellini E, Barbani N, Giusti P, Ciardelli G, Cristallini C (2010) Novel bioactive scaffolds with fibronectin recognition nanosites based on molecular imprinting technology. J Appl Polym Sci 118(6):3236–3244
108. Rosellini E, Barbani N, Giusti P, Ciardelli G, Cristallini C (2010) Molecularly imprinted nanoparticles with recognition properties towards a laminin H–Tyr–Ile–Gly–Ser–Arg–OH sequence for tissue engineering applications. Biomed Mater 5(6):065007

MIX
Papier aus verantwortungsvollen Quellen
Paper from responsible sources
FSC® C105338
FSC
www.fsc.org

If you have any concerns about our products,
you can contact us on
ProductSafety@springernature.com

In case Publisher is established outside the EU,
the EU authorized representative is:
Springer Nature Customer Service Center GmbH
Europaplatz 3, 69115 Heidelberg, Germany

Printed by Libri Plureos GmbH
in Hamburg, Germany